D0218446

Prefixes for Powers of 10*

Multiple	Prefix	Abbreviation
10^{24}	yotta	Y
10^{21}	zetta	Z
10^{18}	exa	E
10^{15}	peta	P
10^{12}	tera	T
10^{9}	giga	G
10^{6}	mega	M
10^{3}	kilo	k
10^{2}	hecto	h
10^{1}	deka	da
10^{-1}	deci	d
10^{-2}	centi	c
10^{-3}	milli	m
10^{-6}	micro	μ
10^{-9}	nano	n
10^{-12}	pico	p
10^{-15}	femto	f
10^{-18}	atto	a
10^{-21}	zepto	z
10^{-24}	yocto	y

* Commonly used prefixes are in red. All prefixes are pronounced with the accent on the first syllable.

The Greek Alphabet

Alpha	A	α	Nu	N	ν
Beta	B	β	Xi	Ξ	ξ
Gamma	Γ	γ	Omicron	O	o
Delta	Δ	δ	Pi	Π	π
Epsilon	E	ϵ, ε	Rho	P	ρ
Zeta	Z	ζ	Sigma	Σ	σ
Eta	H	η	Tau	T	τ
Theta	Θ	θ	Upsilon	Υ	υ
Iota	I	ι	Phi	Φ	φ
Kappa	K	κ	Chi	X	χ
Lambda	Λ	λ	Psi	Ψ	ψ
Mu	M	μ	Omega	Ω	ω

Terrestrial and Astronomical Data*

Acceleration of gravity at Earth's surface	g	9.81 m/s^2
Radius of Earth R_E	R_E	$6.38 \times 10^6 \text{ m}$
Mass of Earth	M_E	$5.98 \times 10^{24} \text{ kg}$
Mass of the Sun		$1.99 \times 10^{30} \text{ kg}$
Mass of the Moon		$7.35 \times 10^{22} \text{ kg}$
Escape speed at Earth's surface		$11.2 \text{ km/s} = 6.95 \text{ mi/s}$
Standard temperature and pressure (STP)		$0°C = 273.15 \text{ K}$ $1 \text{ atm} = 101.3 \text{ kPa}$
Earth–Moon distance†		$3.84 \times 10^8 \text{ m} = 2.39 \times 10^5 \text{ mi}$
Earth–Sun distance (mean)†		$1.50 \times 10^{11} \text{ m} = 9.30 \times 10^7 \text{ mi}$
Speed of sound in dry air (20°C, 1 atm)		344 m/s
Density of dry air (STP)		1.29 kg/m^3
Density of dry air (20°C, 1 atm)		1.20 kg/m^3
Density of water (4°C, 1 atm)		1000 kg/m^3
Latent heat of fusion of water (0°C, 1 atm)	L_F	334 kJ/kg
Latent heat of vaporization of water (100°C, 1 atm)	L_V	2260 kJ/kg

* Additional data on the solar system can be found in Appendix B and at http://nssdc.gsfc.nasa.gov/planetary/planetfact.html.
† Center to center.

Mathematical Symbols

$=$	is equal to
$\equiv$	is defined by
$\neq$	is not equal to
$\approx$	is approximately equal to
$\sim$	is of the order of
$\propto$	is proportional to
$>$	is greater than
$\geq$	is greater than or equal to
$>>$	is much greater than
$<$	is less than
$\leq$	is less than or equal to
$<<$	is much less than
Δx	change in x
dx	differential change in x
$\lvert x \rvert$	absolute value of x
$n!$	$n(n-1)(n-2)\ldots1$
Σ	sum
$\lim$	limit
$\Delta t \rightarrow 0$	Δt approaches zero
dx/dt	derivative of x with respect to t
$\partial x/\partial t$	partial derivative of x with respect to t
$\int_{x_1}^{x_2} f(x)\, dx$	definite integral $= F(x) \big\rvert_{x_1}^{x_2} = F(x_2) - F(x_1)$

Abbreviations for Units

A	ampere	h	hour	N	newton
Å	angstrom (10^{-10} m)	Hz	hertz	nm	nanometer (10^{-9} m)
atm	atmosphere	in.	inch	Pa	pascal
BTU	British thermal unit	J	joule	rad	radians
Bq	becquerel	K	kelvin	rev	revolution
C	coulomb	kg	kilogram	R	roentgen
°C	degree Celsius	km	kilometer	Sv	sievert
cal	calorie	keV	kilo-electron volt	s	second
Ci	curie	lb	pound	T	tesla
cm	centimeter	L	liter	u	unified mass unit
eV	electron volt	m	meter	V	volt
°F	degree Fahrenheit	MeV	mega-electron volt	W	watt
fm	femtometer, fermi (10^{-15} m)	mi	mile	Wb	weber
ft	foot	min	minute	y	year
G	gauss	mm	millimeter	μm	micrometer (10^{-6} m)
Gy	gray	mmHg	millimeters of mercury	μs	microsecond
g	gram	mol	mole	μC	microcoulomb
H	henry	ms	millisecond	Ω	ohm

Some Conversion Factors

Length

1 m = 39.37 in. = 3.281 ft = 1.094 yard

1 m = 10^{15} fm = 10^{10} Å = 10^9 nm

1 km = 0.6214 mi

1 mi = 5280 ft = 1.609 km

1 light-year = 1 $c \cdot$ y = 9.461 $\times$ 10^{15} m

1 in. = 2.540 cm

Volume

1 L = 10^3 cm^3 = 10^{-3} m^3 = 1.057 qt

Time

1 h = 3600 s = 3.6 ks

1 y = 365.24 day = 3.156 $\times$ 10^7 s

Speed

1 km/h = 0.278 m/s = 0.6214 mi/h

1 ft/s = 0.3048 m/s = 0.6818 mi/h

Angle–angular speed

1 rev = 2π rad = 360°

1 rad = 57.30°

1 rev/min (rpm) = 0.1047 rad/s

Force–pressure

1 N = 10^5 dyn = 0.2248 lb

1 lb = 4.448 N

1 atm = 101.3 kPa = 1.013 bar = 760 mmHg = 14.70 lb/in.^{2}

Mass

1 u = [(10^{-3} mol^{-1})/N_A] kg = 1.661 $\times$ 10^{-27} kg

1 tonne = 10^3 kg = 1 Mg

1 kg = 2.205 lb

Energy–power

1 J = 10^7 erg = 0.7376 ft $\cdot$ lb = 9.869 $\times$ 10^{-3} L $\cdot$ atm

1 kW $\cdot$ h = 3.6 MJ

1 cal = 4.186 J

1 L $\cdot$ atm = 101.325 J = 24.22 cal

1 eV = 1.602 $\times$ 10^{-19} J

1 BTU = 778 ft $\cdot$ lb = 252 cal = 1054 J

1 horsepower = 550 ft $\cdot$ lb/s = 746 W

Thermal conductivity

1 W/(m $\cdot$ K) = 6.938 BTU $\cdot$ in./(h $\cdot$ ft^2 $\cdot$ °F)

Magnetic field

1 T = 10^4 G

Viscosity

1 Pa $\cdot$ s = 10 poise

PRELIMINARY EDITION

UNIVERSITY PHYSICS

for the Physical and Life Sciences

Volume II

Philip R. Kesten   David L. Tauck

W. H. FREEMAN AND COMPANY
New York

THIS PRELIMINARY EDITION DOES NOT INCLUDE COMPLETE END-OF-CHAPTER PROBLEM SETS OR ANSWERS

Executive Editor: Jessica Fiorillo
Marketing Manager: Alicia Brady
Market Development Manager: Kirsten Watrud
Associate Market Development Manager: Kerri Russini
Developmental Editors: Kharissia Pettus, Blythe Robbins
Media and Supplements Editor: Dave Quinn
Senior Media Producer: Keri Fowler
Editorial Assistants: Heidi Bamatter, Nicholas Ciani
Marketing Assistant: Joanie Rothschild
Senior Project Editor: Georgia Lee Hadler
Copy Editor: Connie Parks
Photo Editor: Ted Szczepanski
Photo Researcher: Christina Micek
Text and Cover Designer: Blake Logan
Senior Illustration Coordinator: Bill Page
Illustration Coordinator: Janice Donnola
Illustrations: Precision Graphics
Production Coordinator: Paul Rohloff
Composition: Preparé Inc.
Printing and Binding: Quad Graphics

Cover photo by Tim Fitzharris

Library of Congress Control Number: 2011928382

Preliminary Edition Vol. I: 1-4641-1524-9
Preliminary Edition Vol. II: 1-4641-1525-7

©2012 by W.H. Freeman and Company
All rights reserved

Printed in the United States of America
First printing

MCAT® is a registered trademark of the Association of American Medical Colleges. MCAT material included is printed with permission of the AAMC. Additional duplication is prohibited without written permission of the AAMC. The appearance of MCAT material in this book does not constitute sponsorship or endorsement by the AAMC.

W. H. Freeman and Company
41 Madison Avenue
New York, NY 10010
Houndmills, Basingstoke RG21 6XS, England
www.whfreeman.com

Contents in Brief

Contents

Biological Applications

These unique and fully integrated physiological and biological applications are written and carefully explained by a biologist so that they are easy for a physicist to teach.

Chapter 1
Lung volumes (unit conversion)
Hair growth (unit conversion)

Chapter 2
Muscle contraction (velocity)
Cheetah running (acceleration)

Chapter 3
Animal navigation (vectors)
Electrocardiograms (vector addition)
Vestibular system (uniform circular motion)

Chapter 4
Microtubules (Newton's second law)
Fish propulsion (Newton's third law)

Chapter 5
Gecko feet (static friction)
Cornea and eyelids (static friction)
Wrist bones (static friction)
Fainting (forces and uniform circular motion)

Chapter 6
Kangaroo hopping (energy transfer)
Rigor mortis (energy transfer)
Actin in horseshoe crab sperm (work–kinetic energy theorem)
ATPsynthesis (work energy)
Weight lifting (potential energy)
Arteries (conservation of energy)
Photosynthesis (energy conversion)

Chapter 7
Squid propulsion (momentum)
Leaping ballet dancer (center of mass)

Chapter 8
Muscle and bone geometry (lever arms)
Alligator death roll (angular momentum)
Falling cat (angular momentum)

Chapter 9
Arterial blood flow (elasticity)
Anterior cruciate ligament (Young's modulus)
Decompression sickness (volume stress and strain)

Properties of lung tissue (bulk modulus)
Endothelial cells in arteries (shear stress and strain)
Connective tissue (elasticity)
Anterior cruciate ligament (tensile stress, failure, yield strength)

Chapter 11
Bones, muscle, and feathers (density)
Cerebrospinal fluid (pressure and depth)
Blood pressure (pressure and depth)
Breathing (pressure differences)
Body fat assessment (Archimedes' principle)
Blood flow (equation of continuity)
Chronic mountain sickness (viscosity)
Blood flow during exercise (laminar flow)
Ischemic heart disease (Poiseuille's law)

Chapter 12
Intracellular Ca^{2+} concentration (oscillations)
Lung volumes during breathing (phase angle)
Eardrum movement (phase angle)
Heart rate (frequency and period)
Swimming dolphin (potential energy to kinetic energy)
Walking (simple pendulum)
Knee-jerk reflex (physical pendulum)
Hip joint (moment of inertia)
Sensory cells and maintaining balance (damped oscillator)
Beating insect wings (resonance)
Buzzing mosquitoes (natural frequency)

Chapter 13
Horned desert viper (surface waves)
Hearing (frequency and period)
Flagellum (wave speed)
Elephant vocalizations (interference)
Bornean tree-hole frog (longitudinal standing waves)
Eardrum (natural frequency and sound level)
Speed of blood flow (Doppler shift)
Bats and dolphins hunting prey (Doppler shift)
Ultrasonic imaging (Doppler shift)

Chapter 14
Fire beetles (infrared receptors)
Body temperature regulation (temperature)
Hypoxia at high altitude (partial pressure)
Water vapor and respiration (partial pressure)
Sweating (heat flow)
Surviving in harsh environments (radiation, convection, conduction)

Hummingbird torpor (heat exchange)
Dolphin flippers (countercurrent heat exchange)
Polar bear fur (thermal conductivity)

Chapter 15
Bombardier beetle (pressure, temperature, work)
Pressure and volume in the heart (first law of thermodynamics)
Metabolic efficiency (first law of thermodynamics)
Staying warm with fur or feathers (adiabatic processes)
Energy transfer in mitochondria (second law of thermodynamics, heat engine)
Cellular respiration (entropy)
Organisms (entropy)

Chapter 16
DNA and protein structure (electrostatic forces)
Urine formation (electrostatic forces)
Immune system (electrostatic forces)
Pollination of flowers (electric charge)
Red blood cells (Coulomb's law)
Membrane transport mechanisms (conductors and insulators)
Sea urchin spines (electric field lines)
Electrophoresis (electric fields)

Chapter 17
Electrostatic cell sorters
Phospholipid bilayers (capacitance)
Action potentials (potential difference)
Myelin (capacitance)
Insulin release (capacitance)
Defibrillators (energy storage on capacitors)
Membrane lipids (dielectric constant)

Chapter 18
Ion channels (resistance)
Squid giant axons (resistance)
Bioelectricity

Chapter 19
Bacterial navigation (magnetic field)

Magnetic resonance imaging (magnetic field)
Magnetocardiograms and magnetoencephalograms (magnetic field)
Navigation by pigeons, honeybees, and sea turtles (magnetic field)

Chapter 20
Navigation by sharks (magnetic induction)
Measuring eye movements (magnetic field generation)
Electromagnetic blood flow meters (Faraday's law of induction)
Magnetic induction tomography (Faraday's law of induction)
Hearing loops (Faraday's law of induction)
Transcranial magnetic stimulation (inductance)

Chapter 21
Magnetic resonance imaging (*LRC* circuit)

Chapter 22
Damselfish coloration (ultraviolet waves)
DNA damage and cancer (ionizing radiation)

Chapter 23
Laser lithotripsy and kidney stones (refraction)
Honeybee navigation (polarization)
Eye-shine and the tapetum lucidum (thin film interference)
Iridescent wings of the blue morpho butterfly (thin film interference)
Glare reduction in eyeglasses (destructive interference)
Pupils (circular apertures)

Chapter 24
Tapetum lucidum (retroreflector)
Presbyopia (lens)
Lens and cornea (lens)
Eye (spherical mirror)

Chapter 25
Caenorhabditis elegans and imaginary spaceship (time dilation)

About the Authors

DR. PHILIP KESTEN

Dr. Philip Kesten, Associate Professor of Physics and Associate Vice Provost for Undergraduate Studies at Santa Clara University, Santa Clara, CA, received a B.S. in physics from the Massachusetts Institute of Technology and received his Ph.D. in high energy particle physics from the University of Michigan. Since joining the Santa Clara faculty in 1990, Dr. Kesten has also served as Chair of Physics, Faculty Director of the ATOM and da Vinci Residential Learning Communities, and Director of the Ricard Memorial Observatory. He has received awards for teaching excellence and curriculum innovation, was Santa Clara's Faculty Development Professor for 2004–2005, and was named the California Professor of the Year in 2005 by the Carnegie Foundation for the Advancement of Education. Dr. Kesten has also served as the Senior Editor for *Modern Dad*, a newsstand magazine, and was co-founder of the Internet software company Docutek, a SirsiDynix Company.

DR. DAVID TAUCK

Dr. David Tauck, Associate Professor of Biology at Santa Clara University, Santa Clara, CA, holds both a B.A. in biology and an M.A. in Spanish from Middlebury College. He earned his Ph.D. in physiology at Duke University and completed post-doctoral fellowships at Stanford University and Harvard University in anesthesia and neuroscience, respectively. Since joining the Santa Clara University faculty in 1987, he has served as Chair of the Biology Department, and as President of the local chapter of Phi Beta Kappa. Dr. Tauck currently serves as the Faculty Director in Residence of the da Vinci Residential Learning Community.

A Letter from the Authors

As faculty members we understand the frustration and hassles caused by errors in a textbook. We also know that no matter how carefully reviewers as well as authors and their editors hunt for mistakes, the first edition of any book is likely to have at least a few typographical errors or other small problems. That's why we convinced our publisher to print a preliminary edition of our book. Distributing a pre-first edition to potential adopters buys us time to send the entire book through another round of reviews by physicists, editors, proof readers, and us. As we approach the publication date of the first edition, it's especially important to us to continue doing every last thing possible to ensure that the book our students use is as close to perfect as possible. So if you come across any problems as you peruse chapters, please alert one of us directly at kestenandtauck@whfreeman.com, or your sales rep if that's easier for you. We would be grateful!

We hope that, by scrutinizing every detail of the preliminary edition one more time, when the full first edition is published it will read much more like a typical second edition. Thank you for considering the preliminary edition of our book when you evaluate textbooks for the upcoming year.

Sincerely,

Philip R. Kesten, Ph.D.

David L. Tauck, Ph.D.

Preface

University Physics for the Physical and Life Sciences

Nearly half of all undergraduate students enrolled in introductory physics courses major in biology and biomedical sciences. Although there are several excellent calculus-based books aimed at physics and engineering students, the same is not true for those in the life sciences. Books for such students are typically either too basic to prepare students for the Medical College Admissions Test (MCAT) or only differ from books geared for physics and engineering students by having a few biomedical examples layered over a traditional university physics text. This book fills that void by being the first calculus-based physics text written specifically for life sciences students, and the first written by both a physicist and a physiologist.

A Seamless Blend of Physics and Biology

Like traditional physics textbooks, this one offers a rigorous presentation of the fundamentals of introductory physics, with reliance on calculus where appropriate. This book integrates biological examples and clinical correlations into the physics discussions; physiological examples are not merely sprinkled throughout as in a standard physics text. Physical principles are often introduced with a brief, qualitative discussion of a biological application. In this way, the book not only encourages students to learn the fundamentals of physics, it also offers them reasons why learning physics should be important to them. The book strives to instill in students an appreciation of physics by showing them how it determines many characteristics of living systems. This is consistent with the recommendation of the American Association for the Advancement of Science (AAAS) to provide students with an education that enables them to integrate knowledge from different fields of science. In addition, the text entices students to see more deeply into the material by using examples and problems from the physics in students' daily lives, including the modern technology they use every day, the natural phenomena they have experienced throughout their lives, and the sports they commonly watch and play.

Just-in-time Presentation of Calculus

In many instances, using calculus and other slightly more advanced mathematical tools ultimately makes physics easier to grasp. For those students with a weaker background in calculus, the text adapts a just-in-time teaching style to explain calculus as needed to meet the Association of American Medical College's (AAMC) recommendation of developing interdisciplinary science curricula and increased scientific rigor at the undergraduate level. While not all medical schools require calculus, many of the best ones do. With this approach, everyone should be able to understand the ideas behind derivations as well as the concepts expressed in the resulting equations, regardless of their mathematical acumen.

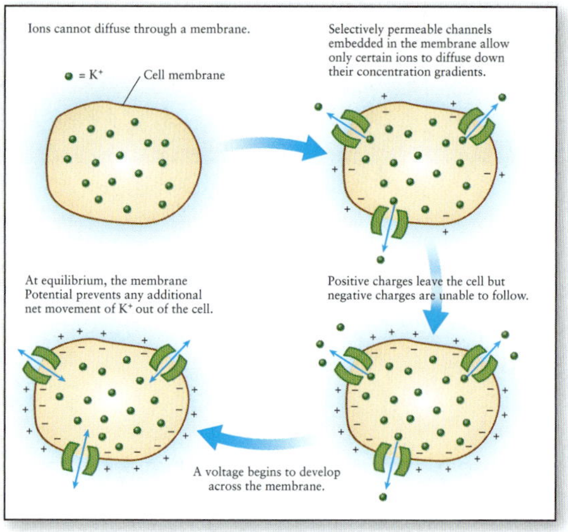

Ions cannot diffuse through a membrane.

= K⁺   Cell membrane

Selectively permeable channels embedded in the membrane allow only certain ions to diffuse down their concentration gradients.

At equilibrium, the membrane Potential prevents any additional net movement of K⁺ out of the cell.

Positive charges leave the cell but negative charges are unable to follow.

A voltage begins to develop across the membrane.

Physicist's Toolbox 4-1 Free-Body Diagrams

SET UP

1. Start a free-body diagram by drawing the object or objects on which forces act. Objects can be represented by simple shapes (for example, rectangles or circles) but it is important to get the orientations of the objects approximately correct (Figure 4-9).

2. Identify the origin or agent of any force before adding it to a free-body diagram. In most of the problems that we will encounter, two objects must be touching for one to exert a force on the other. The one exception for now is the force due to gravity. We'll come across some other exceptions later, such as the magnetic force and the force between electric charges.

SOLVE

3. Draw each force acting on an object as an arrow, where the tail end of the arrow starts at the center of the object *on which the force acts.*

4. Point the arrowhead in the direction in which the force acts.

5. Draw the arrow longer or shorter in order to represent the magnitude of a force relative to the other forces.

6. Label each force.

7. Draw and label the coordinate axes.

REFLECT

8. If there is more than one object to be studied in a system, draw all objects and forces in the same free-body diagram. Although some physicists draw a separate diagram f... the relationship show the com... lines for arrow... forces with the...

9. One way to ch... ticular object t... sure that all of... make sure that... is at least one... the magnitude...

Math Box 4-1 Sigma

Physicists often use mathematical notation in order to represent operations that are either long or of a length not well defined. The capital Greek letter Σ (sigma), for example, is mathematical shorthand for adding a series of terms. If N is to be found by adding four different values of n or

$$N = n_1 + n_2 + n_3 + n_4$$

the addition can be equivalently written using the summation sign

$$N = \sum_{i=1}^{4} n_i$$

Notice the role of the notation above and below the summation sign; we show the starting value of the variable that counts the terms below it and the final value above. We might not know how many terms will be in the sum. In that case, we write

$$N = $$

We can find that the acceleration vector will point at an angle

$$\phi = \tan^{-1}\left(\frac{d_y}{d_x}\right)$$

counterclockwise from the positive x axis by using Equation 3-4.

√x̄ See the Math Tutorial for more information on Trigonometry

Detailed Artwork Designed to Promote Visual Learning

In many textbooks figures contribute little to helping students learn the material. This text uses artwork as a teaching tool. Rather than confining descriptive information in figure legends, as much information as is practical often appears directly on the figures. The result is an annotated figure that reinforces the physics presented in the flow of the text. Moreover, the figures themselves are simple, colorful, and approachable, inviting students to explore them rather than intimidating students into ignoring them.

Text Features

This text includes a problem-solving approach that provides students with all the support they will need, including tools to overcome misconceptions and tackle problems and assessment so students can check for understanding.

Physicist's Toolbox

It's essential for students to have as many tools as possible in their arsenal for tackling the problems found in worked examples and end of chapter problems. **Physicist's Toolboxes** teach the important steps that students can use to successfully set up and solve problems.

Math Box and Math Tutorial

Students may come across a mathematical concept that needs elaboration to allow for deeper understanding of the chapter concepts. **Math Boxes** elaborate on these mathematical concepts by discussing them in the text. Examples of Math Box topics include the dot product and computing integrals by substitution.

Sometimes it's been a while since students have studied a particular math concept and they need a reminder. Margin notes link concepts in the text to the **Math Tutorial** found in the back of the book. The Math Tutorial serves as a refresher and includes worked examples as well as practice problems. Topics covered by the tutorials include trigonometry, differential calculus, and integrals.

Got the Concept?

Most life science students are accustomed to a conceptual approach to learning. Conceptual questions appear as part of the text itself so that students can confirm their readiness to move on to the next topic.

? **Got the Concept 6-4**
  How High?

A block is released from rest at vertical height H on one side of an asymmetric skateboard ramp, as shown in **Figure 6-16.** The angle from the horizontal of the left ramp is twice the angle of the right ramp. If the ramp is frictionless, how high (vertically) does the block rise on the right side before stopping and sliding back down?

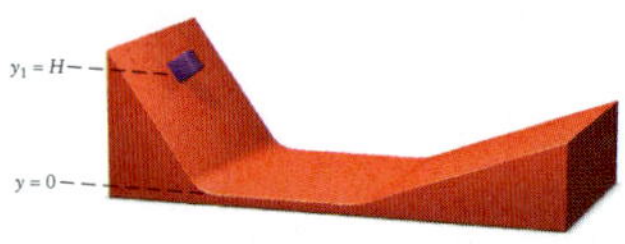

Figure 6-16 A block is released from rest at vertical height H on one side of an asymmetric skateboard ramp.

$y_1 = H$
$y = 0$

Watch Out!

Having taught physics for many years, the authors know which topics are often difficult to learn. Students are often puzzled, for example, by a perception that mirrors reverse images horizontally but not vertically. By tackling such misconceptions directly, the **Watch Out!** feature helps students avoid common pitfalls and gain a deeper understanding of the physics.

! **Watch Out**
  Numeric values can be deceiving.

You might have wondered if there was an error in the two examples of volume in Table 1-2. Could it be that a giant party balloon (**Figure 1-2**) has a volume of 0.10 m³ while a typical human head is only 0.005 m³? The volume of a sphere is given by $\frac{4}{3}\pi r^3$, so the radius of the party balloon is

$$r = \left(\frac{3(0.10\ \text{m}^3)}{4\pi}\right)^{1/3}$$

or about one-third of a meter. That value seems reasonable. How about a head? Try measuring your own—typical values for an adult human head are 17 cm high, 17 cm across, and 21 cm front to back. Taking the shape as box-like, a typical volume is 0.17 m × 0.17 m × 0.21 m = 0.006 m³, so the round number in Table 1-2 is also reasonable.

Problem-Solving Strategy: Set up, Solve, Reflect in Worked Examples

In addition to teaching physics, this text promotes the development of students' problem-solving skills. Too many students would rather memorize equations than comprehend the underlying physics, a mistake that some textbooks encourage in the way they present material. This book models problem-solving skills for students by applying several common steps to all worked example problems. This procedure, summarized by the key phrases "**Set Up**," "**Solve**," and "**Reflect**," mirrors the approach scientists take in attacking problems.

SET UP. The first step in each problem is to determine an overall approach and to gather together the necessary pieces. These might include sketches, equations related to the physics, and concepts. This step relies heavily on diagrams; many sample problems are accompanied by one or more visual aids.

SOLVE. Rather than simply summarizing the mathematical manipulations required to move from first principles to the final answer, the "Solve" section shows many intermediate steps in working out solutions to the sample problems. Other authors too often omit these intermediate steps, either as "an exercise for the student" or perhaps because they appear obvious. Students, however, often do not find the missing steps obvious, and, as a result, simply pass over what might otherwise be a valuable learning experience. In addition, as the authors do during their classroom lectures, the text presents the reasoning and thought process as well as the mathematical steps involved in solving problems. In this way, the text leads students to develop their own problem-solving skills and habits.

REFLECT. An important part of the process of solving a problem is to reflect on the meaning, implications, and validity of the answer. Is it physically reasonable? Do the units make sense? Is there a deeper or wider understanding that can be drawn from the result? This step addresses these and related questions when appropriate. Sometimes the reflection suggests that a second, related problem should be considered; in those cases, a second problem might be tackled or some conceptual questions suggested that allow students to gain greater understanding into the answer.

Associated Practice Problems

Associated **Practice Problems** follow worked examples in the text and reinforce problem-solving skills by asking students to apply the strategies used in the example to a new problem.

Estimate It!

Rough estimations can be a powerful tool in doing science, especially when just starting a new problem. Students will find that it helps them grasp the science at a more intuitive level. The **Estimate It!** feature leads students to develop their estimation skills by modeling the process for them.

What's Important

At the end of each chapter section, the **What's Important** feature reminds students about the main points of each section.

Example 6-2 Pushing a Car

Three people each apply a constant force of 200 N horizontally to a car that has run out of gas. To understand the magnitude of a joule, determine the work done on the car by the three people to move it 5 m.

SET UP
The applied force is constant and in the direction in which the car moves, as shown in Figure 6-2, so Equation 6-1 applies. Note that this figure is not a free-body diagram, but instead shows the relevant force and the direction of motion.

SOLVE
We substitute the given values of force and distance into Equation 6-1 to determine the work

$$W = (3)(200\,\text{N})(5\,\text{m}) = 3000\,\text{J}$$

Together the three people do 3000 J of work on the car.

Figure 6-2 Equation 6-1 applies because the applied force is constant and in the direction in which the car moves.

REFLECT
Even if you have never had to push a car that has run out of gas, try to imagine the amount of effort required. Each of the three people in this problem did 1000 J of work on the car.

Practice Problem 6-2 Each of three people apply a constant force of 200 N horizontally to a car. How far will they move it if each does 1 J of work on the car?

Estimate It! 7-1 Enormous Force!

When major league baseball player Derek Jeter hits a home run, the baseball is in contact with his bat for about 0.0007 s. Assuming that the ball leaves Jeter's bat at about the same speed as the pitch (about 95 mph) estimate the average force Jeter's bat exerts on the ball. Major league baseballs weigh between 5.00 and 5.25 oz.

SET UP
We use Equations 7-23 and 7-24 to write the average force as a vector

$$\vec{F}_{avg} = \frac{\Delta\vec{p}}{\Delta t} = \frac{\vec{p}_f - \vec{p}_i}{\Delta t}$$

So

$$\vec{F}_{avg} = \frac{m\vec{v}_f - m\vec{v}_i}{\Delta t}$$

The initial and final velocity vectors are in opposite directions, so the velocities have opposite signs. We choose to let v_f be positive so v_i is negative, and then

$$F_{avg} = \frac{mv_f + mv_i}{\Delta t}$$

SOLVE
We have assumed that the ball strikes and leaves the bat at the same speed, so $v_f = v_i \approx 95$ mph or 42 m/s. Let m equal 5.125 oz or 0.15 kg and $\Delta t = 0.0007$ s. So

$$F_{avg} = \frac{(0.15\,\text{kg})(42\,\text{m/s}) + (0.15\,\text{kg})(42\,\text{m/s})}{0.0007\,\text{s}}$$

$$= 1.8 \times 10^4\,\text{N}$$

REFLECT
The average force exerted on the ball is 1.8×10^4 N—over 2 tons. That's an enormous force to be generated by a human (even a professional baseball player!), but it only lasts for a short time.

✸ What's Important 7-4

The average force that one object exerts on another during a collision is proportional to the change in momentum and inversely proportional to the contact time, which is the length of time objects touch each other during a collision. The difference between the final and initial values of momentum is called impulse.

SUMMARY

Topic	Summary	Equation or Symbol	
acceleration	Acceleration is the rate of change of velocity versus time. In this chapter, we have considered only cases in which acceleration is constant. The SI unit of acceleration is meters per second per second, or meters per second squared (m/s^2).	$a = \dfrac{dv}{dt}$	
average acceleration	The average acceleration of an object over time interval $t - t_0$ is the change in velocity $v - v_0$ divided by that time interval. Average acceleration represents the average rate of change in velocity over a time interval.	$a = \dfrac{v - v_0}{t - t_0}$	(2-22)
average speed	The average speed of an object over time interval $t - t_0$ is the net distance the object has traveled divided by that time interval.	$s_{avg} = \dfrac{\text{total distance}}{\Delta t}$	(2-14)
average velocity	The average velocity of an object over time interval $t - t_0$ is the net displacement of the object Δx divided by that time interval. Average velocity equals instantaneous velocity when the velocity of an object is constant. Average velocity represents the average rate of change in position over a time interval.	$v_{avg} = \dfrac{\Delta x}{\Delta t}$	(2-13)

End of Chapter Summary

The end of chapter **Summary** incorporates the key concepts from each chapter as well as any associated equation or symbol into an easy-to-follow table.

End of Chapter Questions and Problems

To reinforce the problem-solving skills taught in the chapters, the end of chapter **Questions and Problems** incorporate conceptual questions and physics problems in the following format:

> Conceptual Problems
> Multiple-Choice Problems
> Estimation/Numerical Analysis
> Problems by Chapter Section
> General Problems

Problems include three levels of difficulty: basic, single-concept problems (•), intermediate problems that may require synthesis of concepts and multiple steps (••), and challenging problems (•••).

MCAT® Section

The section that follows includes material from previously administered MCAT® items and is reprinted with permission of the Association of American Medical Colleges (AAMC).

Passage 13 (81–85)
Tennis balls must pass a rebound test before they can be certified for tournament play. To qualify, balls dropped from a given height must rebound within a specified range of heights. Measuring rebound height can be difficult because the ball is at its maximum height for only a brief time. It is possible to perform a simpler indirect measurement to calculate the height of rebound by measuring how long it takes the ball to rebound and hit the floor again. The diagram below illustrates the experimental setup used to make the measurement.

The ball is dropped from a height of 2.0 m, and it hits the floor and then rebounds to a height h. A microphone detects the sound of the ball each time it hits the floor, and a timer connected to the microphone measures the time (t) between the two impacts. The height of the rebound is $h = gt^2/8$ where $g = 9.8 \text{ m/s}^2$ is the acceleration due to gravity. Care must be taken so that the measured times do not contain systematic error. Both the speed of sound and the time of impact of the ball with the floor must be considered.

Four balls were tested using the method. The results are listed in the table below.

Ball	Time (s)
A	1.01
B	1.05
C	0.97
D	1.09

81. If NO air resistance is present, which of the following quantities remains constant while the ball is in the air between the first and second impacts?
A. Kinetic energy of the ball
B. Potential energy of the ball
C. Momentum of the ball
D. Horizontal speed of the ball

To help pre-med students prepare for the MCAT® exam, some chapters include problems from actual former MCAT® exams.

Student Ancillary Support

Supplemental learning materials allow students to interact with concepts in a variety of scenarios. By analyzing figures, reinforcing problem-solving methods, and reviewing chapter objectives, students obtain a practical understanding of the core concepts. With that in mind, W. H. Freeman has developed the most comprehensive student learning package available.

Printed Resources

• *Problem-Solving Guide with Solutions* by Timothy A. French, DePaul University

Volume I (Chapters 1–15) ISBN: 1-4641-0096-9

Volume II (Chapters 16–28) ISBN: 1-4641-0097-7

The *Problem-Solving Guide with Solutions* takes a unique approach to promoting students' problem-solving skills by providing detailed and annotated solutions to selected problems. This guide follows the "Set-up," "Solve," and "Reflect" format outlined in the worked examples for solutions to selected odd-numbered End-of-Chapter problems in the textbook. It also includes integrated media icons which point to selected problem-solving tools that can be accessed.

Free Media Resources

• Book Companion Website

The *University Physics for the Physical and Life Sciences* Companion Website, www.whfreeman.com/universityphysics1e, provides a range of tools for problem-solving and conceptual support:

• Student self-quizzes

• Flashcards

• Recommended media assets for self-study

• Access to the Premium Multimedia Resources, which can be purchased for a nominal fee.

Premium Media Resources

The Premium Media Resources can be purchased directly from the Book Companion Website for a small fee, and are also embedded in the multimedia-enhanced eBook and the WebAssign online homework system.

- **P'Casts** are videos that emulate the face-to-face experience of watching an instructor work a problem. Using a virtual whiteboard, the P'Cast tutors demonstrate the steps involved in solving key worked examples, while explaining concepts along the way. The worked examples were chosen with the input of physics students and instructors across the U.S. and Canada. P'Casts can be viewed online or downloaded to portable media devices.

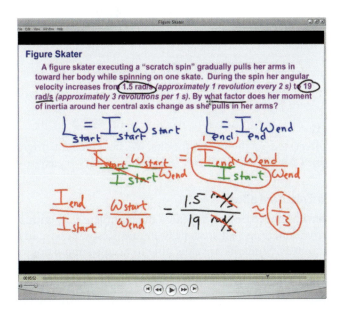

- **Interactive Exercises** are active learning, problem-solving activities. Each Interactive Exercise consists of a parent problem accompanied by a Socratic-dialog "help" sequence designed to encourage critical thinking as users do a guided conceptual analysis before attempting the mathematics. Immediate feedback for both correct and incorrect responses is provided through each problem-solving step.

Go to Interactive Exercise X.Y for more practice

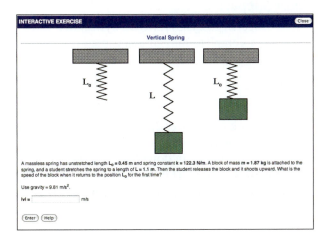

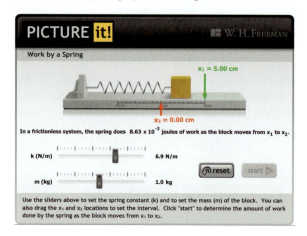

- **Picture Its** help bring static figures from the text to life. By manipulating variables within each animated figure, students will be able to visualize and better understand a variety of physics concepts.

Go to Picture It X.Y
for more practice

eBook Options

For students interested in digital textbooks, W. H. Freeman offers the complete *University Physics for the Physical and Life Sciences* in two easy-to-use formats.

The Multimedia-Enhanced eBook

The **Multimedia-Enhanced eBook** contains the complete text with a wealth of helpful interactive functions. All student Premium Resources are linked directly from the eBook pages. Students are thus able to access supporting resources when they need them, taking advantage of the "teachable moment" as they read. Customization functions include instructor and student notes, highlighting, document linking, and editing capabilities. Access to the Multimedia-Enhanced eBook can be purchased from the book companion website.

The CourseSmart eTextbook

The **CourseSmart eTextbook** provides the full digital text, along with tools to take notes, search, and highlight passages. A free app allows access to CourseSmart eTextbooks on Android and Apple devices, such as the iPad. They can also be downloaded to your computer and accessed without an Internet connection, removing any limitations for students when it comes to reading digital text. The CourseSmart eTextbook can be purchased at www.coursesmart.com.

Instructor Ancillary Support

For instructors using *University Physics for the Physical and Life Sciences*, W. H. Freeman provides a complete suite of assessment tools and course materials for the taking.

Computerized Test Bank

ISBN: 1-4641-0098-5

The **Computerized Test Bank** offers over 2,000 multiple-choice questions, tackling both core physics concepts and various life science applications. While the Test Bank is also available in downloadable Word files on the book companion website, the easy-to-use CD includes Windows and Macintosh versions of the widely used Diploma test generation software, allowing instructors to add, edit, and sequence questions to suit their testing needs.

Electronic Instructor Resources

Instructors can access valuable teaching tools through www.whfreeman.com/universityphysics1e. These password-protected resources are designed to enhance lecture presentations, and include Textbook Images (available in .JPEG and PowerPoint format), Clicker Questions, Lecture PowerPoints, Instructor Solutions, and a Biology Appendix to provide deeper insight into biological processes.

Course Management System Cartridges

W. H. Freeman provides seamless integration of resources in your Course Management Systems. Four cartridges are available (Blackboard, WebCT, Desire2Learn, and Angel), and other select cartridges (Moodle, Sakai, etc.) can be produced upon request.

Online Learning Environments

W. H. Freeman offers the widest variety of online homework options on the market.

WebAssign Premium

For instructors interested in online homework management, **WebAssign Premium** features a time-tested secure online environment already used by millions of students worldwide. Featuring algorithmic problem generation and supported by a wealth of physics-specific learning tools, WebAssign Premium for *University Physics for the Physical and Life Sciences* presents instructors with a powerful assignment manager and student environment. WebAssign Premium provides the following resources:

- Algorithmically generated problems: Students receive homework problems containing unique values for computation, encouraging them to work out the problems on their own.

- Complete access to the interactive eBook is available from a live table of contents, as well as from relevant problem statements.

- Links to the Premium Multimedia Resources are provided as hints and feedback to ensure a clearer understanding of the problems and the concepts they reinforce.

Sapling Learning

Sapling Learning provides highly effective interactive homework and instruction that improve student learning outcomes for the problem-solving disciplines. They offer an enjoyable teaching and effective learning experience that is distinctive in three important ways:

- Ease of Use: Sapling Learning's easy to use interface keeps students engaged in problem-solving, not struggling with the software.

- Targeted Instructional Content: Sapling Learning increases student engagement and comprehension by delivering immediate feedback and targeted instructional content.

- Unsurpassed Service and Support: Sapling Learning makes teaching more enjoyable by providing a dedicated Masters or Ph.D. level colleague to service instructors' unique needs throughout the course, including content customization.

Acknowledgments

Creating a first-edition textbook requires the coordinated effort of an enormous number of talented professionals. We are grateful for the dedicated support of our in-house team at W. H. Freeman; thank you for transforming our concept into a beautiful book.

We especially want to thank our highly-talented developmental editor, Kharissia Pettus, for guiding us through the process of creating a textbook and for sharing her understanding of publishing, editorial prowess, and expertise in thermodynamics. We would also like to thank executive editor Jessica Fiorillo for encouraging us and leading our editorial team, developmental editor Blythe Robbins for coordinating the work of many people and organizing a complicated schedule, media and supplements editor Dave Quinn for producing gorgeous ancillaries and contributing significantly to the design of the book and cover, and of course, we thank editorial assistants Heidi Bamatter and Nick Ciani for their enthusiasm, and hard work on the project. Special thanks also go to our skilled in-house production team, Georgia Hadler, Paul Rohloff, Ted Szczepanski, Bill Page, Janice Donnola, and Blake Logan, for their patience, dedication, and attention to detail.

We are particularly grateful for the substantial contributions of Todd Ruskell, as well as copy editor Connie Parks, accuracy checker Valerie Walters, problems editor Mark Hollabaugh, solutions manual author Tim French, and test bank author Don Franklin. Francesca Monaco of Preparé Inc. also deserves a special thank you for her remarkable attention to details.

Finally, we are grateful to Clancy Marshall both for finding us before we signed with another publisher and for convincing us that we belong on the W. H. Freeman team.

Friends and Family

One of us (PRK) would like to acknowledge valuable and insightful conversations on physics and physics teaching with Richard Barber, John Birmingham, J. Patrick Dishaw, Guy Ramon, and Chris Weber of Santa Clara University, and to offer these colleagues my gratitude. Finally, I offer my gratitude to my wife Kathy and my children Sam and Chloe, for their unflagging support during the arduous process that led to the book you hold in your hands.

One of us (DLT) thanks his family and friends for accommodating my tight schedule during the years that writing this book consumed. I especially want to thank my parents, Bill and Jean, for their boundless encouragement and support, and for teaching me everything I've ever really needed to know; they've shown me how to live a good life, be happy and age gently. I greatly appreciate my sister for encouraging me not to abandon a healthy lifestyle just to write a book. I also want to thank my non-biological family, Holly and Geoff, for leading me to Sonoma County and for making Sebastopol feel like home.

Advisory Board

We sincerely appreciate the following physicists who graciously provided their creative input to the development of the text by serving as members of our Advisory Board:

Timothy A. French, *Harvard University*
Andrew Pelling, *University of Ottawa*
Ryan Snow, *University of California, Davis*
Raluca Teodorescu, *Massachusetts Institute of Technology*
Brian Woodahl, *Indiana University/Purdue University*

Accuracy Review

We know that instructors have particular concerns about using first edition texts because they are prone to errors. We have done everything in our power to alleviate this concern in using our text by submitting chapters and End of Chapter problem sets to several rounds of accuracy reviews and detailed error-checking, by the following:

Wayne R. Anderson
Sacramento City College

Linghong Li
The University of Tennessee at Martin

Marisa Bauza Roman
Drexel University

Alan Meert
University of Pennsylvania

Kevin W. Cooper
Ohio University

Todd G. Ruskell
Colorado School of Mines

Mark Hollabaugh
Normandale Community College

Luke Somers
University of Pennsylvania

Guy Letteer
*Sacred Heart Preparatory and
Santa Clara University*

Valerie A. Walters
V Walters Consulting, LLC

Class Testers

We thank the faculty below and their 2000+ students for class testing the text.

Neil Alberding, *Simon Fraser University*
George Alexandrakis, *University of Miami*
Philip Backman, *University of New Brunswick*
Philip Blanco, *Grossmont College*
Ethan Dolle, *Northern Arizona University*
Diana Driscoll, *Case Western Reserve University*
John Evans, *Georgia State University*
Melissa Franklin, *Harvard University*
Timothy A. French, *Harvard University*
Logan McCarty, *Harvard University*
Andrew Pelling, *University of Ottawa*
Greg Thompson, *Adrian College*
Carolyne Van Vliet, *University of Miami*
Brian Woodahl, *Indiana University-Purdue University*

Reviewers

We would also like to thank the many colleagues who carefully reviewed chapters for us. Their insightful comments significantly improved our book.

Victor O. Aimiuwu, *Central State University*
Bijaya Aryal, *University of Minnesota*
Quentin G. Bailey, *Embry-Riddle Aeronautical University*
Pradip Bandyopadhyay, *Pennsylvania State University, Berks Campus*
Arun Bansil, *Northeastern University*
David Baxter, *Indiana University*
Kalayu Belay, *Florida A&M University*
Bereket Berhane, *Embry-Riddle Aeronautical University*
Luca Bertello, *University of California, Los Angeles*
Nancy Beverly, *Mercy College*
Angela Biselli, *Fairfield University*
Arie Bodek, *University of Rochester*
Robert Boivin, *Auburn University*
Suzanne White Brahmia, *Rutgers University–Busch Campus*
John Broadhurst, *University of Minnesota*
Yorke Brown, *Dartmouth College*

Terry Buehler, *University of California, Berkeley*
James J. Butler, *Pacific University*
Frank Capozzi, *Bradley University, Bridgewater State University*
Duncan Carlsmith, *University of Wisconsin–Madison*
B. Ross Carroll, *Arkansas State University*
Tom Carter, *College of DuPage*
Ryan Case, *Elizabethtown Community and Technical College*
Paola M. Cereghetti, *Lehigh University*
John Cerne, *State University of New York–Buffalo*
Jan L. Chaloupka, *University of Northern Colorado*
Prem Chapagain, *Florida International University*
Saumitra Kumar Chowdhury, *University of Connecticut*
Michael Colaneri, *SUNY at Old Westbury*
John S. Conway, *University of California, Davis*
Susan Coppersmith, *University of Wisconsin–Madison*
Jay Demas, *St. Olaf College*
Nasser Demir, *Duke University*
Richard Di Dio, *LaSalle University*
Margaret Dobrowolska, *University of Notre Dame*
Rodney Dunning, *Longwood University*
Michael Eads, *University of St. Francis*
Nayer Eradat, *San Jose State University*
Milton W. Ferguson, *Norfolk State University*
Jane Flood, *Muhlenberg College*
Lewis Ford, *Texas A&M University*
Ted Forringer, *LeTourneau University*
Tim French, *Harvard University*
Lev Gasparov, *University of North Florida*
Vadas Gintautas, *Chatham University*
Yvonne Glanville, *Penn State Worthington*
Michael J. Graf, *Boston College*
Benjamin Grinstein, *University of California, San Diego*
Puru Gujrati, *University of Akron*
Ajawad (A.J.) Haija, *Indiana University of Pennsylvania*
Katrina Hay, *Pacific Lutheran University*
Ken Henisey, *Pepperdine University*
Jim Henriques, *Cerritos College*
Andrew S. Hirsch, *Purdue University*
John T. Ho, *University at Buffalo*
Mark Hollabaugh, *Normandale Community College*
John D. Hopkins, *Penn State University*
David Hough, *Trinity University*
Bill Ingham, *James Madison University*
David C. Ingram, *Ohio University*
Sai Iyer, *Washington University*
Bob Jacobsen, *University of California, Berkeley*
Mark C. James, *Northern Arizona University*
Pu Chun Ke, *Clemson University*
Ed Kearns, *Boston University*
Peter Kernan, *Case Western Reserve University*
Kevin R. Kimberlin, *Bradley University*
Jeremy King, *Clemson University*
Yury Kolomensky, *University of California, Berkeley*
J.K. Krebs, *Franklin and Marshall College*
Andrew Kunz, *Marquette University*
David Lamp, *Texas Tech University*

Mark Lattery, *University of Wisconsin–Oshkosh*
Shelly R. Lesher, *University of Wisconsin–La Crosse*
Judah Levine, *University of Colorado*
Hong Lin, *Bates College*
Ramon E. Lopez, *University of Texas at Arlington*
Enrico Lunghi, *Indiana University*
Robert C. Mania Jr., *Kentucky State University*
Muhammed Maqbool, *Mount Olive College*
Eric C. Martell, *Millikin University*
Dario Martinez, *Rice University*
Donald Mathewson, *Kwantlen University College*
Mark Matlin, *Bryn Mawr College*
Dan Mazilu, *Washington and Lee University*
Jeffrey McGuirk, *Simon Fraser University*
James G. McLean, *SUNY Geneseo*
Edwin F Meyer, *Baldwin Wallace College*
Mark Morgan-Tracy, *Central New Mexico Community College*
Jeffrey S. Olafsen, *Baylor University*
John Parsons, *Columbia University*
Andrew E. Pelling, *University of Ottawa*
Chris Petrie, *Brevard Community College*
Jason Pinkney, *Ohio Northern University*
Amy Pope, *Clemson University*
B. E. Powell, *University of West Georgia*
Gurcharan S. Rahi, *Fayetteville State University*
Roberto Ramos, *Indiana Wesleyan University*
Tilo Reinert, *University of North Texas*
Paul Rider, *Grand View University*
Stephen Robinson, *Belmont University*
S. Clark Rowland, *Andrews University*
Mark Rupright, *Birmingham-Southern College*
Mehmet Alper Sahiner, *Seton Hall University*
Stiliana Savin, *Barnard College*
Morton P. Seitelman, *Farmingdale State College*
George T. Shubeita, *University of Texas, Austin*
Kanwal Singh, *Sarah Lawrence College*
Earl F. Skelton, *Georgetown University*
Ryan Snow, *University of California, Davis*
Michael Sobel, *Brooklyn College*
C.E. Sosolik, *Clemson University*
Achilles Speliotopoulos, *University of California, Berkeley*
Zbigniew M. Stadnik, *University of Ottawa*
J. Scott Steckenrider, *Illinois College*
Jason Stevens, *Deerfield Academy*
Oleg Tchernyshyov, *Johns Hopkins University*
Alem A. Teklu, *College of Charleston*
Raluca E. Teodorescu, *Massachusetts Institute of Technology*
Beth A. Thacker, *Texas Tech University*
Gregory B. Thompson, *Adrian College*
Christos Velissaris, *University of Central Florida*
E. Prasad Venugopal, *University of Detroit, Mercy*
Chuji Wang, *Mississippi State University*
David A. Ward, *Union University*
Luc T. Wille, *Florida Atlantic University*
Gary A. Williams, *University of California, Los Angeles*
Shannon Willoughby, *Montana State University*

Jeff Allen Winger, *Mississippi State University*
Scott W. Wissink, *Indiana University–Bloomington*
Carey Witkov, *Broward College*
Gregory G. Wood, *CSU Channel Islands*
Brian Woodahl, *Indiana University–Purdue University Indianapolis*
Ruqian Wu, *University of California, Irvine*
Alexander Wurm, *Western New England University*
De-Ping Yang, *College of the Holy Cross*
Chadwick Young, *Nicholls State University*
J. Yu, *Fitchburg State University*
Tanya Zelevinsky, *Columbia University*
Ulrich Zurcher, *Cleveland State University*

16 Electrostatics I

(Research Collaboratory for Structural Bioinformatics. Shinoda, T., Ogawa, H., Cornelius, F., Toyoshima, C. (2009) Crystal structure of the sodium-potassium pump at 2.4 A resolution. Nature 459: 446–50.)

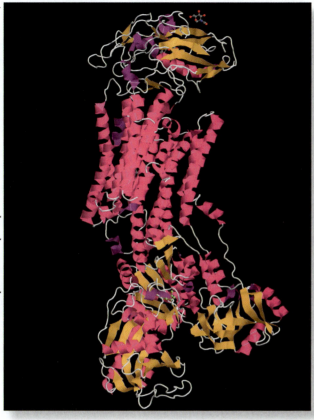

Proteins are essential for life—among their many functions they enable you to move, determine what gets in and out of your cells, help you fight infections, transport oxygen and cholesterol in your blood, and catalyze the chemical reactions that make life possible. All proteins are synthesized as a linear chain of amino acids, but every protein has a unique shape that determines its function, including the sodium-potassium pump depicted here. The forces that cause the protein to twist, coil, and fold on itself—and that ultimately determine protein structure—arise to a large extent from interactions between positive and negative charges on the protein's amino acids. These *electrostatic forces* result in interactions both between charges on the proteins themselves and also between proteins and the aqueous environment in which most proteins exist. In this particular example, the bright pink helices at the center of the image represent non-charged regions of the protein that extend through the oily, non-charged interior of a membrane. The structures above and below the helices in this drawing represent the regions of the protein that project into the aqueous environment on either side of the membrane.

In addition to its critical role in determining the structure and therefore the function of proteins, *electric charge* influences many other important biological and biochemical phenomena. For example, the structures of many molecules, including DNA and proteins, are stabilized by the hydrogen bonds that form when two electronegative atoms such as N and O weakly bind to the same hydrogen atom. *Electrostatic forces* also mediate the interactions between antibodies and their antigens. Cells of the immune system called *leukocytes* ordinarily repel one another because they carry net negative charges on their surfaces. However, the removal of the charges allows activated leukocytes to expose receptors on their surfaces and mount an immune response. Electrostatic interactions are important in the kidney, too; the presence of a net negative charge between the glomerulus and Bowman's capsule helps prevent the filtration of proteins from the blood into the fluid that will become urine. Finally, the reason that ATP (adenosine triphosphate) is the energy-storing molecule in living things has everything to do with electric charge. Forming a chemical bond between phosphate and ADP (adenosine diphosphate) to make ATP requires a lot of energy because the negative charge on phosphate strongly repels the two negatively charged phosphate groups on ADP. As we saw in the previous chapter, when cells break such high-energy chemical bonds, they harness a lot of the released energy to do work. In this chapter, we explore the nature of electric charge and electrostatic interactions.

Figure 16-1 A spark jumps between someone's finger and a doorknob on a dry winter's day. *(Richard Megna/ Fundamental Photographs)*

16-1 Electric Charge

At one time or another you've probably walked across a carpet on a dry day and seen (and felt!) a spark jump between your finger and a doorknob (**Figure 16-1**). Or like the girl in **Figure 16-2**, you've probably rubbed a balloon on your sleeve and then observed how the balloon attracts your hair or sticks to a wall. A similar effect causes pollen to be attracted to the stigmas of flowers, which aids in the process of pollination. These phenomena result from the displacement of *electric charges* present in every object, including living things.

Electric charge comes in two forms, negative and positive. All objects are composed of atoms, which are in turn composed of electrons, protons, and (except for most hydrogen atoms) neutrons. Electrons are negatively charged, protons carry net positive charges, and neutrons, as their name suggests, are charge neutral. Both protons and neutrons are composed of constituent particles, called *quarks,* which do carry either negative or positive charge. However, while the overall charge of a proton is positive, the charges of the quarks that comprise a neutron cancel.

Rubbing a balloon on your clothes transfers charges from your clothes to the balloon, causing it to acquire an overall negative charge. An attractive force arises between objects that carry different charges, that is, between a negative charge and a positive charge. Because a neutral object is nevertheless made up of both negatively and positively charged constituents, it is also possible for an attractive force to arise between a neutral object and a charged one, when charges become separated in the neutral object. For example, hair tends to have no net charge; the difference in net charge between the charged balloon and a neutral strand of hair results in the two being attracted to each other.

The magnitude of the charge of both electrons and protons is the same. This **fundamental charge** *e* is

$$e = 1.602 \times 10^{-19} \text{ coulombs}$$

Figure 16-2 Rubbing a balloon on your sleeve can cause it to make your hair stand up. *(Image Source/Getty Images)*

The SI unit of charge is the coulomb, named after the 18th-century French physicist Charles-Augustin de Coulomb who uncovered the fundamental law that governs the interaction of charges. Using q as the variable to represent charge, the charge of an electron is $q_e = -e = -1.602 \times 10^{-19}$ C, and the charge of the proton is $q_p = +e = +1.602 \times 10^{-19}$ C. Although the charges that make up protons and neutrons are themselves smaller in magnitude than *e*, quarks have never been observed experimentally except when grouped together into larger particles. For our purposes then, the smallest—fundamental—unit of charge is *e*.

Any object likely contains an enormous number of charges, but it is unlikely to have a *net electrical charge* because under normal circumstances, the number of negative charges equals the number of positive ones. An excess of either negative charge or positive charge in an object results in a net charge. In our example above, the balloon picks up electrons as it is rubbed on cloth and in that way acquires a net negative charge. The net charge on an object is the sum of all of the negative charges and all of the positive charges. As we'll discover, however, two objects of the same charge repel each other. If we attempt to put a large number of charges on an object, each will be repelled by all of the others; this large repulsive force makes it relatively difficult to amass a large net charge on an object. In addition, two objects of opposite charge attract each other. So an object that is carrying a net positive charge, for example, will tend to collect negative charges from whatever is nearby, thereby bringing the object (eventually) to a charge neutral state.

Charge is conserved. That's why rubbing a balloon on your sleeve leaves a net positive charge—the transfer of negative charges to the balloon leaves your shirt and you with a deficit of negative charges, which is equivalent to a net positive charge. You probably won't notice this slight positive charge, however—think of how many electrons were available for you to give! You also won't stay positively charged for long, because electrons in the air and dust around you get pulled in and cancel out the imbalance. In the end, however, although the distribution of charges may change in a process, the net amount of charge will remain the same.

Estimate It! 16-1   Electrons in Water

The molar density of water is approximately 20 g/mol. (Its precise value is slightly greater than 18 g/mol.) Use this to estimate the number of electrons in 1-L bottle of water.

SET UP

A water molecule is composed of two hydrogen atoms, each of which has one electron, and an oxygen atom, which has eight electrons. To estimate the number of electrons in the bottle of water, we only need to estimate the number of H_2O molecules in the bottle, and then multiply by ten.

A 1-L bottle of water has a mass of 1000 g. For this estimate, we'll say it's pure water and nothing else.

SOLVE

The number of moles of H_2O molecules in 1000 g of water is

$$n \approx \frac{1000 \text{ g}}{20 \text{ g/mol}} = 50 \text{ mol}$$

According to Avogadro's number (N_A), 1 mol contains about 6×10^{23} molecules, so the number of H_2O molecules in 1 L of water is approximately

$$N \approx (50 \text{ mol}) N_A = (50 \text{ mol})(6 \times 10^{23} \text{ molecules/mol}) = 3 \times 10^{25} \text{ molecules}$$

There are therefore approximately 3×10^{26} electrons in 1 L of water.

REFLECT

How large a number is 3×10^{26}? The human body contains about 10^{14} cells. Even if you include all of the cells in all of the approximately 7×10^9 people in the world, that's only about 10^{24} cells—still far fewer than the number of electrons in our 1-L bottle of water. How about the number of liters of water in all of Earth's oceans?

The U.S. National Oceanic and Atmospheric Administration estimates there are more than 10^{21} L of water in the oceans, still far fewer than the number of electrons in just one of those liters. There are a lot of charges in things!

> ### ✱ What's Important 16-1
> Electric charge comes in two forms, negative and positive. Electrons are negatively charged, protons carry positive charge. The magnitude of the charge of both electrons and protons is the fundamental charge e, equal to 1.602×10^{-19} C. Net charge is conserved.

16-2 Coulomb's Law

Opposites attract. Philosophers and casual observers have debated the truth of this statement in human relationships for centuries. There is no doubt, however, that opposites attract when charges interact; the force between a negatively charged object and a positively charged one draws the two objects closer together. Conversely, the force between two charges of the same sign pushes them apart. The net negative charge on the surface of red blood cells, for example, causes them to repel each other and in so doing helps prevent them from clumping together unless they are stimulated to form a blood clot.

The force between two point charges is proportional to the amount or magnitude of charge present. The greater the point charge, the larger the force. The force also depends on the distance between the point charges; the closer the point charges, the larger the force. The force on a point charge q_2 due to q_1, separated by a distance r is given by **Coulomb's law**:

$$\vec{F} = \frac{kq_1q_2}{r^2}\hat{r} \tag{16-1}$$

where the *Coulomb constant k* is equal to $8.99 \times 10^9 \, \text{N} \cdot \text{m}^2/\text{C}^2$. The radial unit vector $\hat{r}$ points along the line that connects q_1 to q_2. The direction of $\hat{r}$ is away from q_1 and toward q_2.

Notice the similarity between Coulomb's law and Newton's law of universal gravitation (Equation 10-1). In both, the force that one object exerts on the other is inversely proportional to the square of the distance between them. In addition, the gravitational force is proportional to the product of the masses of the two particles, while the electric force is proportional to the product of the charges of the two particles. Also, notice that the presence of $\hat{r}$ in both Newton's law of universal gravitation and Coulomb's law reminds us that the force is directed along the line that connects the two particles.

In what other ways are the gravitational and electric forces similar? Like any force, the net electric force on an object is the vector sum of every separate electric force that acts on it. We can therefore treat a solid, charged object as an assemblage of infinitesimal point charges, which leads directly to our using the shell theorem (Section 10-2). For spherically symmetric charged objects, the electric force can be found by treating each object as if all of its charges were concentrated at its center. In addition, according to the shell theorem, a hollow charged sphere exerts no electric force on a charged object inside it. Furthermore, Coulomb's law, like Newton's law of universal gravitation, is symmetric. When point charges q_1 and q_2 interact, the force that q_1 exerts on q_2 is equal in magnitude to the force that q_2 exerts on q_1. The two forces will, of course, be in opposite directions.

Are there any ways in which Newton's universal law of gravitation (Equation 10-1) and Coulomb's law (Equation 16-1) are not similar? Compare the forms of these laws again. Newton's law of universal gravitation includes a minus sign while Coulomb's law does not. A negative force is an attractive one. The Newton's law of universal gravitation equation uses the $-\hat{r}$ direction; $-\hat{r}$ points inward (that is, from one object toward the other) because $+\hat{r}$ is defined as radially outward. We have not written a minus sign in Coulomb's law, but the equation's lack of a minus sign does not mean that the force between point charges is always repulsive. Rather, the sign (and therefore the direction of the force) is determined by the product of the signs of the two charges. When q_1 and q_2 have opposite signs, their product is negative. The overall sign of the force is also negative, indicating an attractive force. When both carry the same sign charge, either both negative or both positive, the product of q_1 and q_2 is positive, and the force is repulsive. This possibility does not occur for the gravitational force because mass, unlike charge, is described only by nonnegative values.

❓ Got the Concept 16-1
Forces on Charges

Figure 16-3 shows a charged object q placed near a second, positively charged object $+Q$ that is fixed in place. Make your own drawing of this situation, and add a vector which shows the force on q if q is negatively charged. Make a second drawing showing the force on q if q is positively charged.

+Q   q

Figure 16-3 Use this figure to draw the force vector on charge q, first if q is positively charged and then if q is negatively charged.

❓ Got the Concept 16-2
Red Blood Cells

Red blood cells carry oxygen from the lungs to the cells of your body. The relatively large, net negative charge that resides on the surfaces of red blood cells helps prevent them from clumping together. The image on the left in Figure 16-4 shows normal red blood cells. The cells on the right clump together after being treated with an enzyme that reduces the excess charge. Why would excess charge on the surface of red blood cells have this effect?

10 μm

Figure 16-4 Red blood cells normally carry a net negative surface charge. (*Role of Surface Electric Charge in Red Blood Cell Interactions, Kung-Ming Jan and Shu Chien, The Journal of General Physiology, Volume 61, 1973 pages 638–654*)

! Watch Out

Don't assume the electric force between two charges is repulsive, even if no minus sign is evident.

The sign (and therefore the direction of the electric force between two point charges) is determined by the sign of the product of the charges. When the sign of a point charge is not immediately known (for example, if the charge is given as q_1), you will need to look for other clues to determine whether the force it exerts, or the force exerted on it, is attractive or repulsive. See Got the Concept 16-2 for an example.

? Got the Concept 16-3
Zero Net Force

When point charge q is placed halfway between two identical particles carrying positive charge $+Q$ (**Figure 16-5**), each of the three charges experiences a net force of zero. What is the sign of point charge q?

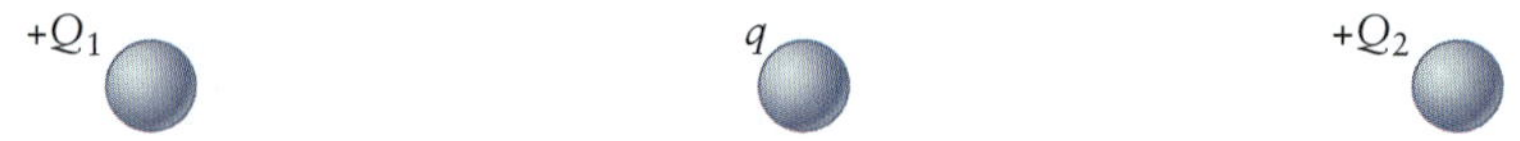

Figure 16-5 A point charge q is placed halfway between two identical particles carrying positive charge $+Q$. What is the sign of q so that the net force on each of the three separately is zero?

Go to Interactive Exercise 16-1 for more practice dealing with Coulomb's law.

Example 16-1 Three Charges, Again

In Got the Concept 16-3, we considered a negative point charge q placed halfway between two identical particles carrying positive charge $+Q$ (Figure 16-5) and required that each point charge experiences a net force of zero. What is the magnitude of charge q in terms of Q? In Figure 16-5, the positive point charges are labeled $+Q_1$ and $+Q_2$ to distinguish one from the other, but the magnitudes of Q_1 and Q_2 are the same (that is, $Q_1 = Q_2 = Q$).

SET UP
The net force on each of the three point charges is zero; in other words, the sum of the forces on any one point charge due to the other two equals zero. For example, the sum of the force of Q_1 on Q_2, $\vec{F}_{Q_1 \to Q_2}$, and the force of q on Q_2, $\vec{F}_{q \to Q_2}$, equals zero:

$$\vec{F}_{Q_1 \to Q_2} + \vec{F}_{q \to Q_2} = 0$$

This equation requires that the magnitude of the two forces be the same:

$$F_{Q_1 \to Q_2} = F_{q \to Q_2}$$

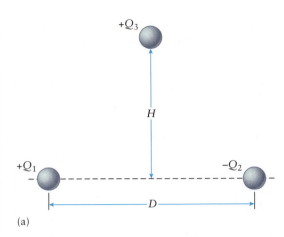

(a)

Figure 16-6 **Figure 16-6** (a) Three point charges are fixed in space. What is the magnitude and direction of the force on Q_3 due to the other two charges? (b) The force on Q_3 due to Q_1 is repulsive. The force on Q_3 due to Q_2 is attractive. (c) The forces on Q_3 are added by component; the vertical components of the two forces are equal and opposite, so they cancel.

Because Q_1 and Q_3 are the same sign, the force of charge Q_1 on Q_3 is repulsive.

Because Q_2 and Q_3 are of opposite sign, the force of charge Q_2 on Q_3 is attractive.

(b)

Because the magnitudes of $\vec{F}_{1\rightarrow3}$ and $\vec{F}_{2\rightarrow3}$ are the same, by the symmetry inherent in the positions of the charges, the y components of the two forces are equal and in the opposite direction.

Because the magnitudes of $\vec{F}_{1\rightarrow3}$ and $\vec{F}_{2\rightarrow3}$ are the same, by the symmetry inherent in the positions of the charges, the x components of the two forces are equal and in the same direction.

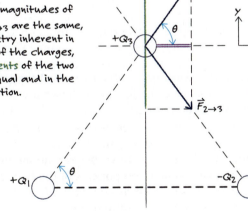

(c)

Using Coulomb's law (Equation 16-1),

$$F_{\text{net}} = \left(\frac{kQ_1Q_3}{\left(\dfrac{D}{2}\right)^2 + H^2} \right)\cos\theta + \left(\frac{kQ_2Q_3}{\left(\dfrac{D}{2}\right)^2 + H^2} \right)\cos\theta$$

Note that the distance between each pair of charges is $\sqrt{(D/2)^2 + H^2}$. Also, because all three charges have magnitude Q_0,

$$F_{\text{net}} = \left(\frac{2kQ_0^2}{\left(\dfrac{D}{2}\right)^2 + H^2} \right) \cos \theta$$

Finally, we can use the distances D and H to determine $\cos \theta$:

$$\cos \theta = \frac{D/2}{\sqrt{\left(\dfrac{D}{2}\right)^2 + H^2}}$$

So

$$F_{\text{net}} = \left(\frac{2kQ_0^2}{\left(\dfrac{D}{2}\right)^2 + H^2} \right)\left(\frac{D/2}{\sqrt{\left(\dfrac{D}{2}\right)^2 + H^2}} \right) = \frac{kDQ_0^2}{\left(\left(\dfrac{D}{2}\right)^2 + H^2\right)^{3/2}}$$

Here

$$Q_0^2 = \left(\frac{5\sqrt{5}}{3} C \right)^2 = \frac{125}{9} C^2$$

and

$$\left(\left(\frac{D}{2}\right)^2 + H^2\right)^{3/2} = \left(\sqrt{(3 \text{ m})^2 + (4 \text{ m})^2} \right)^3 = (5 \text{ m})^3 = 125 \text{ m}^3$$

So

$$F_{\text{net}} = \frac{kDQ_0^2}{\left(\left(\dfrac{D}{2}\right)^2 + H^2\right)^{3/2}} = \frac{(8.99 \times 10^9 \text{ N} \cdot \text{m}^2/\text{C}^2)(6 \text{ m})(125/9 \text{ C}^2)}{125 \text{ m}^3}$$

or

$$F_{\text{net}} = 5.99 \times 10^9 \text{ N}$$

REFLECT

The net force is the vector sum of the force on Q_3 due to each of the other charges separately. We simplified the vector sum of the forces by drawing a force diagram and noting that the symmetry causes the y components to cancel and the x components to add.

Practice Problem 16-3 Three point charges are fixed in space, as shown in Figure 16-6a. All three carry charge of magnitude $Q_0 = (5\sqrt{5})/3$ C; all three charges are positive. Q_1 and Q_2 are 6 m apart (D) and Q_3 is located 4 m (H) above the middle of the line that connects Q_1 and Q_2. Find the magnitude and direction of the force on Q_3 due to the other two charges.

The gravitational force and the electric force differ in that while the electric force can either attract charges to one another or repel them, the gravitational force is only attractive. The two forces differ in another way, too, although it is more subtle. Because we easily experience gravitational force every moment of our lives, we tend to think of gravity as the strongest force in nature. Not so. As we find in the next Estimate It problem, for subatomic particles on which the two forces can be compared directly, the gravitational force is many orders of magnitude weaker than the Coulomb force.

Estimate It! 16-2 Ratio of the Coulomb Force to the Gravitational Force

Electrons and protons have both mass and charge. Estimate the ratio of the Coulomb force between an electron and a proton to the gravitational force between them. For this quick estimation, we'll take all constants just to their order of magnitude, so $m_e \approx 10^{-30}$ kg, $m_p \approx 10^{-27}$ kg, and $e \approx 10^{-19}$ C. In the same way, G is approximately 10^{-10} N·m^2/kg^2 and the Coulomb constant k is approximately 10^{10} N·m^2/C^2.

SET UP

Newton's universal law of gravitation (Equation 10-1) describes the gravitational force that one object exerts on another; the magnitude of the gravitational force between an electron and a proton separated by distance r is therefore

$$F_{\text{grav}} = \frac{Gm_e m_p}{r^2}$$

The magnitude of the Coulomb force that one exerts on the other is, according to Equation 16-1,

$$F_{\text{Coul}} = \frac{kq_e q_p}{r^2}$$

The ratio is then

$$\frac{F_{\text{Coul}}}{F_{\text{grav}}} = \frac{kq_e q_p}{r^2} \bigg/ \frac{Gm_e m_p}{r^2} = \frac{kq_e q_p}{Gm_e m_p}$$

SOLVE

So the ratio of the Coulomb force to the gravitational force is approximately

$$F_{\text{Coul}}/F_{\text{grav}} \approx \frac{(10^{10}\,\text{N·m}^2/\text{C}^2)(10^{-19}\,\text{C})^2}{(10^{-10}\,\text{N·m}^2/\text{kg}^2)(10^{-30}\,\text{kg})(10^{-27}\,\text{kg})}$$

or

$$F_{\text{Coul}}/F_{\text{grav}} \approx 10^{39}$$

REFLECT

The Coulomb force is about 10^{40} times more powerful than the gravitational force. Not 40 times more powerful, but 10 to the 40th power times as powerful. On the scale of particles such as electrons and protons, the gravitational force is insignificant.

★ What's Important 16-2

The force between a negatively charged object and a positively charged one draws the two objects closer together; opposite charges attract. Conversely, the force between two charges of the same sign pushes them apart; like charges repel. Coulomb's law describes the force that one point charge exerts on another. The force is inversely proportional to the square of the distance between them.

16-3 Conductors and Insulators

From your experiences you probably know that some materials, such as metals, allow charge to move from one place to another. We are not yet ready to consider the physics of moving charges, but certain charges move more easily in some materials than others. The charged phosphate groups on the inner and outer surfaces of cell membranes, for example, readily interact with other charges; the oily interior of the membrane, however, forms a barrier through which charged particles cannot cross, at least not without the help of some specialized protein transport mechanism such as the one mediated by the sodium-potassium pump shown on the opening page of this chapter. This protein uses energy from ATP to pump Na^+ out of the cell and K^+ into the cell, both against a concentration gradient.

Whether charges move or stay fixed in place depends on how easily they move on and in an object. A **conductor** is a material in which charge can move easily. In nearly all biological systems, charge carriers are ions suspended in aqueous solutions. Because water (H_2O) has slightly more negative charge near the oxygen atom of the molecule and slightly more positive charge near the hydrogen atoms, water tends to dissolve ionic compounds such as sodium chloride (NaCl) into separate ions, Na^+ and Cl^- in this case, each surrounded by water molecules. The ions are free to move in solution, so the aqueous solutions which make up living things are conductors. In nearly all other cases, the moving charge carriers are electrons; any material that has many electrons which are free to move is a good conductor. The physics is the same whether charge is carried by an electron or by an ion. To make things easier to understand, we'll focus on the simpler, nonbiological systems and discuss current from the perspective of the flow of electrons rather than the flow of ions in aqueous solution.

In the atoms of most materials, negatively charged electrons are bound to nuclei, which are positively charged because they are made of protons. Electrons bind tightly to a nucleus because of the attractive force between them and the protons. In some materials, however, some electrons are loosely bound and therefore are free to move around inside the material. In a metal such as copper, for example, the electric force on the outermost, or valence, electron is weak not only because of its distance from the positive protons but also because the many electrons closer to the nucleus tend to shield the valence electron from the positive charge of the nucleus. The valence electrons are therefore weakly bound to the atomic nuclei and so can move between copper atoms relatively freely.

Electrons cannot move freely in an **insulator**, a material which is nonconducting. In insulating materials such as wood or rubber, all of the electrons are bound tightly to the nuclei. Any excess charge added to an insulator tends to stay wherever it is placed; in contrast, excess charge added to a conductor redistributes itself as a result of the repulsive Coulomb forces. We can easily arrange for a chunk of rubber to have a nonuniform distribution of charge, but not so for a block of copper. No matter where we add excess electrons to a block of copper, they will end up evenly distributed on the surface of the copper in order to be as far from the other electrons as possible.

Materials that are good electrical conductors are also good thermal conductors, and materials that are good electrical insulators are good thermal insulators. Both rubber and fat, for example, are good thermal insulators for the same reason that they don't conduct electricity. Conduction (Section 14-7) requires charged particles that can move. In Chapter 14, we saw the relationship between the kinetic energy of particles in a substance and its temperature. For heat to be transferred, the faster moving particles in the hotter part of an object must transfer kinetic energy through collisions with other particles so that the kinetic energy of cooler, slower moving particles in the object increases. The transport of both heat and electrical current therefore requires particles that are free to move; free electrons make metals good thermal conductors for the same reason that they are good conductors of electricity.

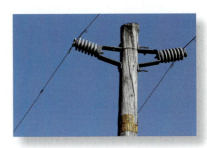

> **? Got the Concept 16-4**
> **Insulators and Conductors**
>
> Porcelain is often used as an electrical insulator to isolate electrical wires from their supporting poles (**Figure 16-7**). Would you expect the thermal properties of porcelain to be more similar to copper or wood? *Hint:* Would you rather live in a house built from copper or wood? Yes, copper might look nicer, but that's not what we mean!

Figure 16-7 Porcelain is often used as an electrical insulator to isolate high voltage electrical wires from their supporting poles. *(Courtesy of David Tauck.)*

> **★ What's Important 16-3**
>
> A conductor is a material in which charge can move easily. Electrons cannot move freely in an insulator.

16-4 Electric Field

Most forces in our daily experiences arise only when one object contacts another object directly. Nevertheless, some forces, such as the gravitational force and the Coulomb force, appear to act even between two objects separated by a distance. The fundamental theory of interactions between particles suggests that objects exert forces on each other by exchanging small units of energy or matter. For our purposes, however, a more useful approach employs the concept of the *field*. We will consider every charged object to be surrounded by an **electric field**, which is strongest closest to the object but extends infinitely far away. In contrast, an electric force is an interaction between a given charge and the electric field surrounding one or more other charges.

A charge q is placed in an electric field that has a magnitude E. The magnitude of the electric force is proportional to q and the magnitude of the field:

$$F = qE$$

The SI units of electric field are newtons per coulomb (N/C). Because force and electric field are vectors, we need to write the relationship between electric force and electric field as a vector relationship:

$$\vec{F} = q\vec{E} \tag{16-2}$$

We have seen that the force on a charge can be either attractive or repulsive depending on what causes it. Using Equation 16-2, the direction of $\vec{E}$ directly determines the direction of the force.

In the previous section, we noted a number of parallels between Newton's universal law of gravitation and Coulomb's law. Let's consider another parallel now. Although we didn't describe it in this way earlier, we encountered the field description in the study of gravitational force. Recall that weight is the force with which the gravitational force acts on an object, and that

$$W = mg \tag{4-3}$$

Writing weight as force and noting that both weight and acceleration are vectors,

$$\vec{F} = m\vec{g}$$

Notice the similarity between this relationship and the one between electric force and electric field. The vector $\vec{g}$ represents the gravitational field in a similar way that $\vec{E}$ represents the electric field.

We don't have to specify what gives rise to the electric field in order to understand the force on a charge; we simply observe that the charge experiences a force

Figure 16-8 The electric field surrounding a point charge $+Q$ can be mapped out by imagining the force on positive test charges placed around it.

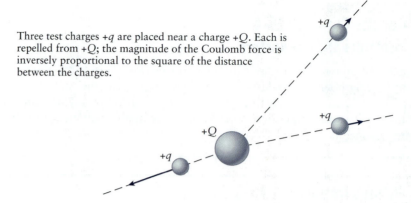

Three test charges $+q$ are placed near a charge $+Q$. Each is repelled from $+Q$; the magnitude of the Coulomb force is inversely proportional to the square of the distance between the charges.

in the presence of a field. Rearranging Equation 16-2 makes this relationship clearer:

$$\vec{E} = \frac{\vec{F}}{q} \tag{16-3}$$

Through this relationship the electric field $\vec{E}$ is seen to be a field of vectors defined by the force exerted on a charge q placed in it.

Because an electric field surrounds every charged object, representing the field pictorially helps us understand the interaction between fields and charges, much in the same way that free-body diagrams help us understand forces. Imagine a number of test charges $+q$ placed near a point charge $+Q$ as in **Figure 16-8**. Each charge is infinitesimally small, although we have exaggerated their sizes in order to make the figure clearer. In this case, the Coulomb force exerted by $+Q$ repels each test charge, so the direction of each force, shown in blue, points radially away from $+Q$. The magnitudes of the forces decrease as the distance between $+Q$ and $+q$ increases. (The magnitudes are proportional to the distance squared, so because the farthest charge is three times farther than the closest one, the force on it is nine times weaker.)

We could expand the view presented in Figure 16-8 by adding many more test charges and drawing the force vector on each, as shown in **Figure 16-9a**. This

We could represent the field in Figure 16-8 by drawing a large number of vectors, each representing a force.

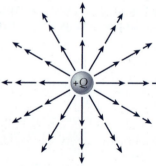

(a)

Figure 16-9 (a) The electric field surrounding a positive charge is mapped for many points. Each vector arrow gives the direction of the field at the point marked by the tail of the vector, and the length of the arrow gives the magnitude of the field at that point. (b) In this simpler and more standard mapping of the electric field due to a positive charge, the density of arrows (the number of arrows passing through a given area) gives the magnitude of the electric field—the closer the lines are in some region, the greater the magnitude of the electric field. The direction of the field is given by the direction of the vector arrows.

In the standard representation of a field, the direction of the arrows show the direction of $\vec{E}$, and the density of lines (how close together they are) gives the magnitude.

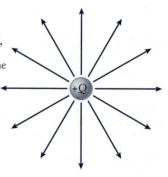

(b)

Figure 16-10 The electric field
lines emanate radially outward
from a positive point charge,
which means that the field is
three-dimensional. The lines
of field look much like the
spines on a sea urchin. *(NOAA.)*

representation of the electric field provides both the magnitude and direction of
the field at every point in space. It's somewhat cumbersome to draw, however;
Figure 16-9b shows a simpler approach. Here the direction of the arrows represents the direction of the field again, but in this case the density of the arrows gives
its magnitude. The density of arrows is related to the magnitude of the electric
field—the closer the lines are in some region, the greater the magnitude of the electric field.

Figure 16-19b emphasizes that the lines of electric field emanate radially
outward from the positive charge $+Q$. This figure is somewhat misleading, however, because it implies that the field lines lie only in a plane. Not so! The electric
field completely surrounds the charge $+Q$; it is three-dimensional. The arrows of
the field surrounding the positive charge look much like the spines on the sea urchin
in **Figure 16-10**, pointing out radially in all directions.

The arrows of an electric field point in the direction of force, as we see from
Figures 16-8, 16-9a, and 16-9b. We started the pro-
cess of making these figures by placing positive point
charges $+q$ near $+Q$, so the force on each, and therefore the field, points radially outward. You may wonder, however, why we couldn't have tested the field by
placing negative point charges around $+Q$. The force
on the negative test charge would point radially
inward. Would that imply that the field lines would
also point inward? No! By convention, the field lines
always point in the direction that a positive test charge
would experience an electric force. The direction of
the field due to charge $+Q$ does not depend on what
other charges are actually present.

Figure 16-11 shows the two-dimensional representation of the electric field due to positive charge $+Q$
side by side with a two-dimensional representation of
the electric field due to a negative charge $-Q$. Because
a positive test charge would be attracted to $-Q$, the

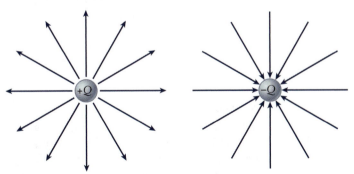

Figure 16-11 A two-dimensional representation of the
electric field due to positive charge $+Q$ (on the left) and a
two-dimensional representation of the electric field due to a
negative charge $-Q$ (on the right) show that the electric
field points away from a positive charge and toward a
negative charge.

field lines point inward for the negative charge. Remember that in both cases the actual field is three-dimensional, with the arrows extending like the spines of a sea urchin.

? Got the Concept 16-5
Electric Field Shape

A positive charge is placed directly in front of you. Describe the electric field vector at a point directly between you and the charge. How would your answer change if the charge were negative?

When there are two or more point charges present, the electric field at any point in space is the vector sum of the fields due to each charge separately, as in **Figure 16-12**. So, the force on a test charge is the vector sum of the forces exerted on it by each of the individual point charges separately. **Figure 16-13** shows the electric field due to two point charges of the same magnitude but opposite signs, a configuration known as an **electric dipole**.

Because the field due to a complex configuration of charge is the (vector) sum of the fields due to each point charge (or each infinitesimally small element of charge in the system), we need to know the field due to a single point charge. If we place a charge q_{test} a distance r from charge q, the Coulomb force on it is $\vec{F} = (kqq_{test}/r^2)\hat{r}$, from Equation 16-1. The electric field, from Equation 16-3, is then

$$\vec{E} = \frac{\vec{F}}{q_{test}} = \frac{\dfrac{kqq_{test}}{r^2}\hat{r}}{q_{test}}$$

or

$$\vec{E} = \frac{kq}{r^2}\hat{r} \qquad (16\text{-}4)$$

This is the **electric field at distance r from a point charge q**. The unit vector $\hat{r}$ reminds us that the electric field points radially toward or away from a point charge.

Let's now find the electric field at an arbitrary point along the midline of a dipole formed by charges $+q$ and $-q$ separated by a distance $2d$. The midline is

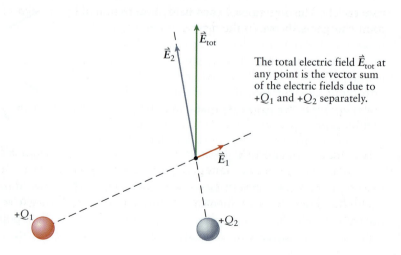

The total electric field $\vec{E}_{tot}$ at any point is the vector sum of the electric fields due to $+Q_1$ and $+Q_2$ separately.

Figure 16-12 The total electric field at any point is the vector sum of the electric fields due to each charge separately.

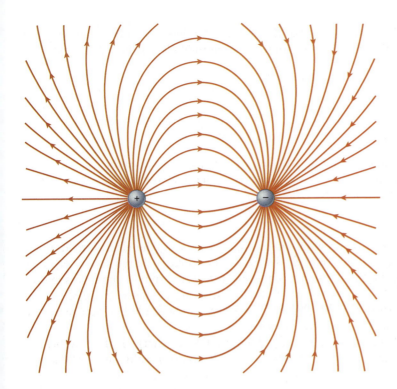

Figure 16-13 The lines of electric field of an electric dipole, a configuration of two point charges of the same magnitude but opposite signs, point away from the positive charge and toward the negative charge.

perpendicular to the line connecting the charges and halfway between them, as shown in **Figure 16-14**. We will find the electric field a distance y from the line connecting the charges along the midline.

The total electric field $\vec{E}$ at the point of interest is the vector sum of $\vec{E}_+$ and $\vec{E}_-$, the electric fields due to the positive and negative charges, respectively. Both charges are the same distance r from the point, so from Equation 16-4, the fields due to $+q$ and $-q$ are

$$\vec{E}_+ = +\frac{kq}{r^2}\hat{r}$$

and

$$\vec{E}_- = -\frac{kq}{r^2}\hat{r}$$

respectively. The direction of each field, as determined by the sign of the point charge, is shown in the figure.

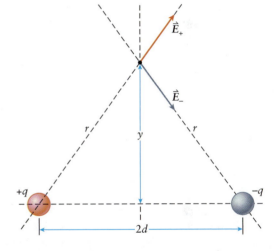

Figure 16-14 At any point in space the electric field of a dipole is the vector sum of the fields due to each of the point charges. In this sketch we find the field at a point along the midline of a dipole.

! Watch Out

Adding the magnitudes of vectors does not give the vector sum.

You might want to refer back to Section 3-3 to remind yourself that vectors are added component by component. For example, $\vec{E}_+$ and $\vec{E}_-$ have the same magnitude (kq/r^2) and opposite sign, but their sum is not zero. Because $\vec{E}_+$ and $\vec{E}_-$ don't point in directly opposite directions, the two vectors do not cancel.

To determine the vector sum of $\vec{E}_+$ and $\vec{E}_-$, we must add their x and y components separately. The x and y components of the total electric field $\vec{E}$ are

$$E_x = E_{+,x} + E_{-,x}$$

and

$$E_y = E_{+,y} + E_{-,y}$$

As you can see in **Figure 16-15a**, the vertical components cancel and the horizontal components add. Thus,

$$E_x = \frac{kq}{r^2}\cos\theta + \frac{kq}{r^2}\cos\theta = 2\frac{kq}{r^2}\cos\theta$$

and

$$E_y = 0$$

So the total electric field at the point of interest is

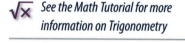
See the Math Tutorial for more information on Trigonometry

$$E = \sqrt{E_x^2 + E_y^2} = \sqrt{\left(2\frac{kq}{r^2}\cos\theta\right)^2 + 0} = 2\frac{kq}{r^2}\cos\theta$$

Go to Picture It 16-1 for more practice dealing with electric fields.

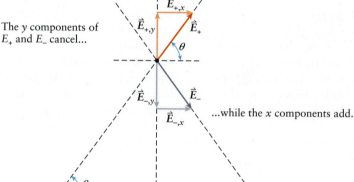

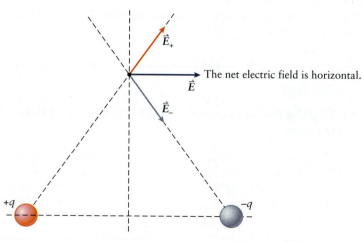

Figure 16-15 (a) The vertical components of the electric fields due to the two charges cancel. The horizontal components of the electric fields due to the two charges add. (b) The net electric field is horizontal.

Because E has no y component, the magnitude of E is just twice the magnitude of the x component of either of the fields separately. To put the result into a form that depends only on the parameters originally given—q, d, and y—note that

$$\cos \theta = \frac{d}{r}$$

and

$$r = \sqrt{d^2 + y^2}$$

So the electric field due to a dipole of charge magnitude q and separation $2d$, at a distance y along the midline is

$$E = 2\frac{kq}{r^2}\frac{d}{r} = 2\frac{kqd}{r^3} = 2\frac{kqd}{(\sqrt{d^2 + y^2})^3}$$

or

$$E = \frac{2kqd}{(d^2 + y^2)^{3/2}} \tag{16-5}$$

The direction of the net electric field vector, shown in **Figure 16-15b**, is perpendicular to the midline, or horizontal using the orientation presented in the figure.

Notice that when $y = 0$, that is, when the point of interest lies on the line that connects the two charges, Equation 16-5 gives $E = (2kqd)/(d^2 + 0)^{3/2} = (2kqd/d^3) = (2kq/d^2)$ as the magnitude of the electric field. As we noted above, this quantity is just twice the value of the field due to each charge separately. Also, when the point of interest is far from the dipole (that is, when $y \gg d$), we can neglect d when compared to y and find that

$$\lim_{y \gg d} E = \frac{2kqd}{(y^2)^{3/2}} = \frac{2kqd}{y^3} \tag{16-6}$$

Far from the dipole, the field decreases as the cube of the distance.

❓ Got the Concept 16-6
No Electric Field

Two positive point charges of different magnitudes are placed in fixed positions. No other charges are present. Identify all possible locations where the net electric field could be zero.

Example 16-4 Electric Field Is Zero

A point charge $Q_1 = +4$ C is placed 0.5 m from point charge $Q_2 = +9$ C (**Figure 16-16**). Find the point, relative to Q_1, at which the net electric field is zero.

The total distance between the charges is D, so if we label the distance from Q_1 to P as x, the distance from Q_2 to P is $D-x$.

 Go to Interactive Exercise 16-2 for more practice dealing with electric fields.

Figure 16-16 At what point P between two charges is the electric field equal to zero?

SET UP

The point at which the net field is zero must lie between the charges (see Got the Concept 16-6). In Figure 16-16, this point (P) has been placed at an arbitrary location between Q_1 and Q_2. We label the distance between Q_1 and point P as x; the total distance between the charges is D. To solve for x, we set the fields at P due to each charge equal to each other. We know the fields point in opposite directions, because the two charges are in opposite directions from P. Thus, the net field is zero at the point where the magnitudes of the two fields are the same.

SOLVE

Using Equation 16-4, the magnitude of the electric field due to each charge is

$$E_1 = \frac{kQ_1}{x^2}$$

and

$$E_2 = \frac{kQ_2}{(D-x)^2}$$

Setting the expressions equal gives

$$\frac{kQ_1}{x^2} = \frac{kQ_2}{(D-x)^2}$$

or

$$\frac{(D-x)^2}{x^2} = \frac{Q_2}{Q_1}$$

Taking the square root of both sides of the equation gives

$$\frac{D-x}{x} = \pm\sqrt{\frac{Q_2}{Q_1}} \tag{16-7}$$

We will work with only the positive root here, and then consider the negative root in the Reflect part of this example. So

$$D - x = \sqrt{\frac{Q_2}{Q_1}}x$$

or

$$D = \left(1 + \sqrt{\frac{Q_2}{Q_1}}\right)x = \frac{\sqrt{Q_1} + \sqrt{Q_2}}{\sqrt{Q_1}}x \tag{16-8}$$

To determine x, the preceding equation can be rearranged and the values from the problem statement included in the resulting equation:

$$x = \frac{\sqrt{Q_1}}{\sqrt{Q_1} + \sqrt{Q_2}}D = \frac{\sqrt{4\,\text{C}}}{\sqrt{4\,\text{C}} + \sqrt{9\,\text{C}}}(0.5\text{ m})$$

or

$$x = 0.2\text{ m}$$

REFLECT

Because charge Q_1 is smaller than Q_2 its field decreases more quickly with distance. The location of the point at which the net electric field is zero must therefore be closer to Q_1 than Q_2. Our result for x is 40% of the total distance, so the point of zero field is indeed closer to Q_1 than Q_2.

Notice that if we had used the negative root in Equation 16-7, Equation 16-8 would become

$$D = \left(1 - \sqrt{\frac{Q_2}{Q_1}}\right)x = \frac{\sqrt{Q_1} - \sqrt{Q_2}}{\sqrt{Q_1}}x$$

or

$$x = \frac{\sqrt{Q_1}}{\sqrt{Q_1} - \sqrt{Q_2}}D = \frac{\sqrt{4\,C}}{\sqrt{4\,C} - \sqrt{9\,C}}(0.5\text{ m})$$

Because $\sqrt{9}$ is greater than $\sqrt{4}$, the result is negative. This result would then suggest an answer on the other side of Q_1 from Q_2. However, we know that the point at which the net field is zero must lie *in between* the two charges, so we can safely ignore this result.

Practice Problem 16-4 A point charge $Q_1 = +4$ C is placed 0.5 m from point charge $Q_2 = -9$ C (Figure 16-16). Find the point, relative to Q_1, at which the net electric field is zero.

When a charged particle encounters a nonzero electric field, it experiences a force and therefore accelerates. The force on a particle of charge q and mass m immersed in an electric field $\vec{E}$ is, from Equation 16-2,

$$\vec{F} = q\vec{E}$$

If this force is the only one acting on the particle, Newton's second law requires that

$$\vec{F} = m\vec{a}$$

so

$$m\vec{a} = q\vec{E}$$

or

$$\vec{a} = \frac{q}{m}\vec{E} \tag{16-9}$$

Any charged particle has a specific charge q and a specific mass m, so the charge-to-mass ratio q/m will be a fixed, constant number for a given particle. For example, the magnitude of the charge-to-mass ratio of an electron is

$$q_e/m_e = 1.7588 \times 10^{11}\text{ C/kg}$$

and has been measured to many digits of significance.

? Got the Concept 16-7
Sorting DNA

Electrophoresis is a process used to separate large molecules such as proteins or DNA fragments by length. When DNA fragments are placed in an agarose–acrylamide gel and then accelerated by an electric field, shorter molecules move more quickly and farther than longer ones (**Figure 16-17**). Although each separate fragment has a different mass and a different charge, these differences do not affect the sensitivity of electrophoresis to separate DNA fragments. What explanation can you give for this process to be independent of the mass and charge separately?

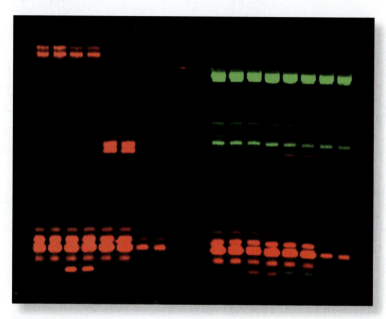

Figure 16-17 An electric field accelerates DNA fragments through a viscous gel. Shorter molecules move more quickly and farther than longer ones, from the top to the bottom of the gel in this photograph. *(Courtesy of Dr. Ángel Islas.)*

Example 16-5   A Curious Particle

When released from rest in a uniform electric field of strength 10^4 N/C, a certain charged particle of mass 6.64×10^{-27} kg travels the 2.00 cm between two charged plates in 2.84×10^{-7} s. Find the charge of the particle.

SET UP

The distance Δx that an object travels in time t while experiencing an acceleration a is (from Equation 2-26)

$$\Delta x = \frac{1}{2}at^2$$

From Equation 16-9, the magnitude of the acceleration is given by the charge-to-mass ratio and the electric field

$$a = \frac{q}{m}E$$

thus

$$\Delta x = \frac{1}{2}\frac{q}{m}Et^2$$

SOLVE

Rearranging the equation for the charge q gives

$$q = \frac{2m\Delta x}{Et^2}$$

Substituting the given values into the equation yields

$$q = \frac{2(6.64 \times 10^{-27}\,\text{kg})(0.020\,\text{m})}{(10^4\,\text{N/C})(2.84 \times 10^{-7}\,\text{s})^2} = 3.29 \times 10^{-19}\,\text{C}$$

Because all values are in SI units, we know that the calculation results in a value in coulombs.

REFLECT

The fundamental charge e is 1.602×10^{-19} C. The particle in this example, a helium nucleus, is composed of two protons and two neutrons; the two protons give it an overall charge of $+2e$. A helium nucleus, that is, a helium atom without associated electrons, is called an *alpha particle*.

Practice Problem 16-5 A certain charged subatomic particle of mass 2.20×10^{-27} kg and charge $+2e$ is released from rest in a uniform electric field of strength 10^4 N/C. How long does it take the particle to move 0.500 cm?

★ ## What's Important 16-4

An electric field surrounds every charged object; it is strongest closest to the object but extends infinitely far away in three dimensions. The shape of the electric field depends on the distribution of charge; for example, the electric field due to two point charges of the same magnitude but opposite signs, a configuration known as an electric dipole, decreases as the cube of the distance.

16-5 Electric Field for Some Objects

In Chapter 8, we uncovered a relationship for the moment of inertia of a rotating object. Determining the relationship required us to imagine breaking the object into an infinite number of infinitesimally small pieces, and then adding the contribution each piece makes to the total moment of inertia. Similarly, we noted in Chapter 10 that although Newton's universal law of gravitation describes the force between two pointlike objects, we can apply it to large objects such as a star or planet if we consider them as being composed of an infinite number of infinitesimally small pieces and then finding the vector sum of the forces on another object due to each small piece. We are faced with the same issue for charges and fields, because while Equation 16-4 gives the electric field due to a point charge, the charged objects we encounter in the real world will be extended and possibly complex. You have likely guessed the solution to this dilemma. Consider a charged object as an infinite number of point charges and find the total electric field by calculating the vector sum of the electric fields of each point charge.

Electric Field along the Axis of a Charged Rod

We start by finding the electric field of a rod of length L carrying a uniform charge $+Q$, a distance d away from one end along the axis of the rod. The rod and the parameters are shown in **Figure 16-18a**. We imagine that the rod is divided into an infinite number of point charges dq, each of which results in a contribution $d\vec{E}$ to the total field at any point in space. One representative dq, along with its associated $d\vec{E}$, is shown in **Figure 16-18b**. We exaggerated the size of dq in the figure for the sake of clarity; the length of dq is infinitesimally small.

Notice that we arbitrarily selected the representative point charge. We didn't select one at either end, or exactly in the middle; the choice of one of those could lead to a conclusion not true for any arbitrarily selected charge dq. Also, note that we labeled the distance x from the point of interest to dq in order to determine $d\vec{E}$ using Equation 16-4. To find the total electric field using that equation, we determine the vector sum of the contribution of each point charge dq. The electric field due to each point charge can be represented by

$$d\vec{E} = \frac{kdq}{x^2}\hat{x} \tag{16-10}$$

Because we are doing this problem along the x axis we changed the notation of Equation 16-4 from r and $\hat{r}$ to x and $\hat{x}$. This helps us remember that the field lines point along the x axis. To add the contributions of each point charge to determine the total electric field, we need to consider first the vector components of each $d\vec{E}$. Using Figure 16-18b, you can see that every $d\vec{E}$ at the point of interest is directed along the axis of the rod. Each $d\vec{E}$ is in the same direction, each has a vector component only in that direction, and each component is the full magnitude of each vector. So in this case, we find the total field by directly adding the magnitudes of the $d\vec{E}$ vectors, secure in the knowledge that the total electric field points along the axis of the rod. We will therefore carry out the integration of Equation 16-10 using vector magnitudes.

The total electric field is, by using Equation 16-10,

$$E = \int dE = \int \frac{k\,dq}{x^2} \tag{16-11}$$

Recall that the process of integration results in the summation of each individual term. We will carry out the integration over the length of the rod in order to include the contribution of every possible charge element dq. What bounds should be applied to the integral in Equation 16-11 in order to accomplish this?

Figure 16-18 (a) What is the electric field a distance d from one end of a charged rod? (b) The electric field is the vector sum of the contributions $d\vec{E}$ to the total electric field made by each infinitesimal charge element dq of the rod. The length of each charge element is dx.

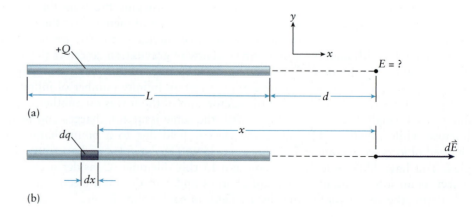

> ## ❗ Watch Out
> ### The limits of an integral must match the variable used as the differential.
>
> To integrate from the right side of the rod, where $x = d$, to the left side, where $x = d + L$, you might also be tempted to insert d to $d + L$ as the limits. Don't use those limits. The limits on an integral must match the differential variable! The variable of integration is dq in Equation 16-11, so the limits would need to be the smallest and largest value that q can be. The values d and $d + L$ are positions, not charges, and are therefore not valid as bounds on an integral over dq.

To write the integral in Equation 16-11 in a way that can be directly evaluated, we look to rewrite the variable of integration dq in terms of other variables. Because the rod is uniformly charged, every small piece of a given length carries the same fraction of the total charge Q, that is,

$$dq = \frac{dx}{L} Q \tag{16-12}$$

where dx/L is the fraction of the total length L of the rod occupied by a slice of the rod of length dx. You might prefer to think of the relationship expressed in Equation 16-12 as a proportion, that is, the fraction of charge in the small piece in proportion to the total charge is the same as the fraction of the length of the piece relative to the total length of the rod:

$$\frac{dq}{Q} = \frac{dx}{L}$$

This is equivalent to Equation 16-12.

Notice that dx, the length of the piece of the rod, is an infinitesimally small, or differential, element of the distance x between the point of interest and dq. We can therefore rewrite the electric field integral (Equation 16-11):

$$E = \int_d^{d+L} \frac{k}{x^2} \frac{dx}{L} Q \tag{16-13}$$

 See the Math Tutorial for more information on Integrals

Here, d to $d + L$ is the appropriate set of limits required to add up all of the slices of the rod from $x = d$ to $x = d + L$, so

$$E = \frac{kQ}{L} \int_d^{d+L} \frac{dx}{x^2} = \frac{kQ}{L} \left(-\frac{1}{x} \Big|_d^{d+L} \right) = \frac{kQ}{L} \left(-\frac{1}{d + L} + \frac{1}{d} \right)$$

This equation simplifies to

$$E = \frac{kQ}{d(d + L)} \tag{16-14}$$

 Go to Interactive Exercise 16-3 for more practice dealing with electric fields.

The total electric field at our point of interest is then $\vec{E} = kQ/[d(d + L)]\hat{x}$, where $\hat{x}$ points along the axis of the rod. The direction of $\vec{E}$ is away from the rod because we set the charge of the rod to be positive. Were the rod negatively charged, $\vec{E}$ would point toward it.

Of more interest, perhaps, is the field at a distance d far from the end of the rod, where $d \gg L$. Using Equation 16-14, we can ignore L whenever it appears combined with d, so

$$E_{\text{far}} = \lim_{d \gg L} \frac{kQ}{d(d + L)} = \frac{kQ}{d(d)} = \frac{kQ}{d^2}$$

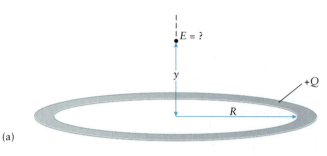

(a)

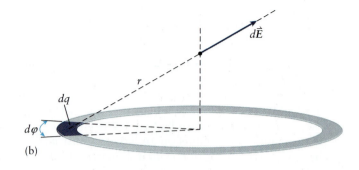

(b)

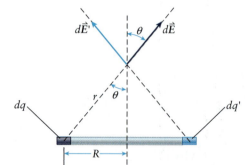

(c)

Figure 16-19 (a) What is the electric field at a distance y above a thin, charged ring, along the central axis of the ring? (b) The electric field is the vector sum of the contributions $d\vec{E}$ to the total electric field made by each infinitesimal charge element dq of the ring. (c) The horizontal components of the electric field contributions from opposite sides of the ring cancel, so that the total electric field at the point of interest points vertically.

You recognize this expression as the electric field a distance d from a point charge Q. In other words, from far away and along its axis, a charged rod appears to be a point charge.

Electric Field along the Central Axis of a Charged Ring

We now consider the electric field a distance y along the central axis of a thin ring of radius R and uniformly distributed charge $+Q$, as shown in **Figure 16-19a**. Again, we imagine the ring divided into an infinite number of point charges dq, each of which contributes $d\vec{E}$ to the total field at any point in space. **Figure 16-19b** shows a representative and arbitrarily selected dq along with its associated field $d\vec{E}$. We exaggerated the thickness of the ring and the size of dq in order to make the figure clearer.

As we did to find the electric field of the rod at a point along its axis, we add up the field $d\vec{E}$ that each slice of the ring contributes to the total. Using Equation 16-4, the electric field due to any dq is

$$d\vec{E} = \frac{k\,dq}{r^2}\hat{r} \tag{16-15}$$

where r is the distance from the point of interest to the charge element dq, as shown in Figure 16-19b. [Notice that r and R (the radius of the ring) are not the same in this problem.] To find the total electric field, we must determine the vector sum of the contributions of each slice of the ring.

Consider the $d\vec{E}$ contributions. Unlike in the previous example of the field at a point on the axis of the rod, here the electric field vectors do not all point in the same direction. To see how the vector addition works, we added a second electric field vector $d\vec{E}'$ due to dq' in **Figure 16-19c**. We carefully selected dq' to be on the opposite side of the central axis of the ring, at the same distance as dq from the axis. In this way the horizontal components in the sum $d\vec{E} + d\vec{E}'$ cancel, as you can verify by inspection of Figure 16-19c. Because a charge element dq' on one side of the ring exists for every dq selected on the other side, the horizontal component of the total electric field at the point of interest is zero. We draw two conclusions from this statement. First, the electric field $\vec{E}$ at the point of interest points vertically. Second, we need only add up the vertical components of each $d\vec{E}$ to find the total field. Using the angle θ defined in Figure 16-19c, the vertical component $d\vec{E}_y$ is

$$dE_y = \frac{k\,dq}{r^2}\cos\theta$$

Using Equation 16-15, the total electric field is then

$$E = E_y = \int dE_y = \int \frac{k\,dq}{r^2}\cos\theta \qquad (16\text{-}16)$$

Like the rod, because the charge on the ring is uniformly distributed, dq can be expressed as a fraction of the total charge Q. As shown in Figure 16-19c, each dq is of angular extent $d\phi$ and so accounts for the fraction $d\phi/2\pi$ of the total charge:

$$dq = \frac{d\phi}{2\pi}Q \qquad (16\text{-}17)$$

As in our treatment of the charged rod, we could also express this relationship by noting the charge of the piece is in the same proportion to the total charge as the angular extent of the piece is to 2π, the total angle of the ring:

$$\frac{dq}{Q} = \frac{d\phi}{2\pi}$$

By substituting Equation 16-17 into the electric field integral (Equation 16-16), we arrive at the equation

$$E = \int_0^{2\pi} \frac{k}{r^2}\frac{d\phi}{2\pi}Q\cos\theta \qquad (16\text{-}18)$$

To take into account every dq around the ring, the variable ϕ must cover a full circle or $\phi = 0$ to $\phi = 2\pi$. Also, note from Figure 16-19c that the angles ϕ and θ are not the same. In particular, θ does not change over the limits of integration while ϕ clearly does.

The magnitude of the electric field is then

$$E = \frac{kQ}{2\pi r^2}\cos\theta\int_0^{2\pi} d\phi = \frac{kQ}{2\pi r^2}\cos\theta\,\phi\,\Big|_0^{2\pi}$$

or

$$E = \frac{kQ}{r^2}\cos\theta$$

We must express both r and θ in terms of the given parameters R and y. Using the triangle evident in Figure 16-19c, $\cos\theta = y/r$ and $r = (R^2 + y^2)^{1/2}$, thus

$$E = \frac{kQ}{(R^2 + y^2)}\left(\frac{y}{(R^2 + y^2)^{1/2}}\right) = \frac{kQy}{(R^2 + y^2)^{3/2}}$$

The total electric field is then

$$\bar{E} = \frac{kQy}{(R^2 + y^2)^{3/2}}\,\hat{y} \qquad (16\text{-}19)$$

where $\hat{y}$ points along the central axis of the ring. The direction of $\bar{E}$ is away from the ring because we set the charge of the ring to be positive. If the ring were negatively charged, the electric field $\bar{E}$ would point toward the center of the ring at points along the central axis. This conclusion would be true both above and below the ring.

The electric field at large distances from the ring, or when $y \gg R$, provides a good check on the validity of our result. In this limit, we treat R as negligible whenever it appears combined with y in Equation 16-19, which gives

$$E_{\text{far}} = \lim_{y \gg R} \frac{kQy}{(R^2 + y^2)^{3/2}} = \frac{kQy}{(y^2)^{3/2}} = \frac{kQ}{y^2}$$

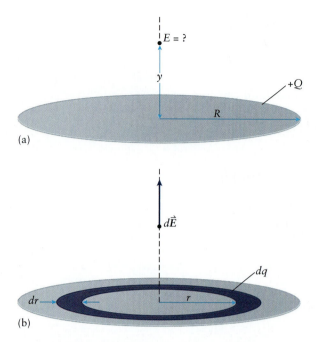

(a)

(b)

Figure 16-20 (a) What is the electric field at a point y above a thin, uniformly charged disk, along the central axis of the disk? (b) The electric field is the vector sum of the contributions $d\vec{E}$ to the total electric field made by each infinitesimally small annulus of charge dq that makes up the disk.

This expression is just the electric field at a distance y from a point charge Q. From far away and along the central axis, a charged ring appears to be a point charge, which is certainly reasonable.

Electric Field along the Central Axis of a Charged Disk

We now consider the electric field a distance y along the central axis of a thin, uniformly charged disk of radius R and charge $+Q$, as shown in **Figure 16-20a**. We'll consider the disk to be so thin that it can be treated as a two-dimensional object (the object has area but no thickness). As we did for the rod and ring, we find the electric field of the disk by dividing it into an infinite number of infinitesimally small charges dq, each of which makes a contribution $d\vec{E}$ to the total field at any point in space.

How should dq be selected? We require that dq be infinitesimally small, so that the distance between the point of interest and any point on (or in) dq is the same. Ideally, we would also like to define dq so that the electric field due to each dq will point in a direction that is easy to use. To guarantee that dq is small enough to satisfy these criteria, we generally consider charge elements that are infinitesimally small in every direction. A charge element that is infinitesimally small in the direction in which we will integrate is satisfactory even if it is not infinitesimally small in other directions. This is the case here, in which we have chosen dq to be an annulus of radius r and width dr (**Figure 16-20b**). Thus the annulus is infinitesimally small in the radial (integration) direction. In addition, although the circumference of the annulus is not infinitesimally small, dr is. (The thickness of the annulus has been exaggerated to make the figure clearer.) Therefore the area of the annulus, which must be proportional to the product of r and dr, is also infinitesimal.

The thin charged annulus we selected as dq is effectively a charged ring. The electric field due to each annulus therefore points vertically according to Equation 16-19. The contribution $d\vec{E}$ that the representative dq makes to the total field is shown in Figure 16-20b. To find the total electric field, we add up $d\vec{E}$ due to each ring-shaped element. Using Equation 16-19, the magnitude of the electric field due to each dq is

$$dE = \frac{k \, dq \, y}{(r^2 + y^2)^{3/2}}$$

Notice that Q in Equation 16-19 has been replaced by dq, because here Q represents the total charge of the disk. In addition, R has been replaced by r. This replacement is an important distinction, because while R is the radius of the entire disk, a constant, r is the variable distance from the center of the disk out to each annulus. Adding the contributions each dq makes to the total field gives

$$E = \int dE = \int \frac{k \, dq \, y}{(r^2 + y^2)^{3/2}} \tag{16-20}$$

We don't need to account for the fact that the contributions are vectors, because the electric field due to each dq points only in the vertical direction. So, the sum will also point only in the vertical direction.

The charge is uniformly distributed on the disk, so each tiny dq carries the same fraction of the total charge. That fraction is the ratio of the area of dq divided by the total area of the disk:

$$dq = \frac{dA_{dq}}{A_{disk}} Q \tag{16-21}$$

The annulus that defines dq is the area between a circle of radius $r + dr$ and one of radius r, so

$$dA_{dq} = \pi(r + dr)^2 - \pi r^2 = \pi r^2 + 2\pi r\, dr + \pi\, dr^2 - \pi r^2$$

or

$$dA_{dq} = 2\pi r\, dr + \pi\, dr^2$$

However, because dr is really small, dr^2 is really, *really* small and can be treated as negligible. Thus

$$dA_{dq} = 2\pi r\, dr \tag{16-22}$$

Rather than memorize this result, which we'll use in other contexts, you can think of dA_{dq} as the area of a rectangle. Because dr is infinitesimally small, if we cut the annulus open and were able to unroll it, the annulus would approximate a rectangle of height dr and width $2\pi r$ (the circumference of the annulus). The area of that rectangle is $2\pi r\, dr$.

Using Equation 16-22, Equation 16-21 becomes

$$dq = \frac{2\pi r\, dr}{\pi R^2} Q$$

so the integral by which we add the electric field contributions (Equation 16-20) is

$$E = \int dE = \int_0^R \frac{ky}{(r^2 + y^2)^{3/2}} \frac{2\pi r\, dr}{\pi R^2} Q$$

Notice that the integral now has the variable of integration dr; to include every possible annular dq, we integrate using limits from $r = 0$, the smallest possible annulus, to $r = R$, the largest one. Again, we have been careful to distinguish between r and R.

The magnitude of the electric field is then

$$E = \frac{2\pi kyQ}{\pi R^2} \int_0^R \frac{r\, dr}{(r^2 + y^2)^{3/2}} \tag{16-23}$$

This integral is easier than it looks! If we let $u = (r^2 + y^2)^{-1/2}$ then

$$du = -\frac{1}{2}(r^2 + y^2)^{-3/2}(2r\, dr) = -\frac{r\, dr}{(r^2 + y^2)^{3/2}}$$

In other words, the integral of Equation 16-23 is of the form $\int du$; employing $\int du = u$, the integral becomes

$$E = \frac{2kyQ}{R^2}\left(- (r^2 + y^2)^{-1/2}\Big|_0^R\right) = \frac{2kyQ}{R^2}\left(-\frac{1}{(R^2 + y^2)^{1/2}} + \frac{1}{(y^2)^{1/2}}\right)$$

or

$$E = \frac{2kyQ}{R^2}\left(\frac{1}{y} - \frac{1}{(R^2 + y^2)^{1/2}}\right) = \frac{2kQ}{R^2}\left(1 - \frac{y}{(R^2 + y^2)^{1/2}}\right)$$

The total electric field of a uniformly charged disk at a point along its central axis is then

$$\vec{E} = \frac{2kQ}{R^2}\left(1 - \frac{y}{(R^2 + y^2)^{1/2}}\right)\hat{y} \tag{16-24}$$

where $\hat{y}$ points along the central axis of the ring. As with the charged ring, $\vec{E}$ points away because the disk is positively charged. Had we made the disk negatively charged, $\vec{E}$ would point toward the disk.

The result just obtained hides a surprise. First note that for points close to the disk or when the disk is large, that is, where $y \ll R$,

$$E_{\text{close}} = \lim_{y \ll R} \frac{2kQ}{R^2}\left(1 - \frac{y}{(R^2 + y^2)^{1/2}}\right) = \frac{2kQ}{R^2} \qquad (16\text{-}25)$$

In the limit where y is very small compared to R, the second term in the parentheses can be neglected. Because every term in E_{close} is a constant, E_{close} is itself constant—close to the charged disk, or when the disk gets large, the electric field does not depend on distance from the disk. In addition, our study of the electric field due to the disk has been along the central axis, but for points close to the disk we can generalize our description of the field to points off the central axis.

For points far from the disk, we can tell if a given point above the disk is on the central axis by asking whether it is the same distance from the edges of the disk in opposite directions. In terms of the electric field, we ask whether the charge at those opposite edges contributes the same field to the total. However, when the point of interest is close to the disk, the edges are far away relative to the distance above the disk. The resulting contributions to the total electric field from those far-flung charge elements are so small, compared to the field due to nearby charge elements, that it's hard to distinguish a point *on* the central axis from one *near* the central axis. As you can imagine, the closer the point of interest gets to the disk, the harder it becomes to tell whether the point is on the axis, which means that E_{close} applies to any point in the region far from the edge of the disk. Therefore (this is the surprise), the electric field close to a disk of uniform charge and far from the edge is constant, regardless of whether or not the point of interest is on the central axis.

The field close to a charged disk (Equation 16-25), which also applies to points near a large charged disk, is often rewritten in a simpler form. We described the disk as having total charge Q spread out uniformly over the surface area πR^2. The (surface) charge density of the disk is therefore

$$\sigma = \frac{Q}{\pi R^2}$$

Surface charge density describes the amount of charge that resides on a surface per unit area. Using this expression, Equation 16-25 becomes

$$E_{\text{close}} = 2\pi k\sigma \qquad (16\text{-}26)$$

Because both k and σ are constants, this form emphasizes our conclusion that the field close to a uniformly charged disk is constant.

★ **What's Important 16-5**
From far away and along its axis, a charged rod appears to be a point charge. From far away and along the central axis, a charged ring appears to be a point charge. The field close to a uniformly charged disk is constant.

16-6 Gauss' Law

In Chapter 11, we found that the product of the velocity of fluid flow and the cross-sectional area of the tube through which the fluid flows remains constant. This *equation of continuity* (Equation 11-17) explains, for example, why the velocity of water flow increases when you put your thumb over the end of a garden hose and why blood flows faster as it gets closer to the heart, where the combined cross-sectional area of the veins approaches a minimum. We can generalize *flux*, the quantity that describes flow across an area, so that it has applications beyond fluids, for example, to the study of charges and fields. Imagine that the flow of the water in a hose or blood in a vein is represented by a field of vectors, each giving the velocity of the fluid at a specific point in the flow. The product of the magnitude or a component of each vector and a small bit of the cross-sectional area of the flow leads to the flux. However, we can carry out the same mathematical operation for a vector field that does not represent the motion of fluid, but, like $\vec{E}$, the strength of a field at a particular point in space. In particular, the relationship between the electric field and the charge from which it arises allows the determination of electric field in a way that complements the Coulomb approach. This relationship, referred to as *Gauss' law* after the 19th-century German mathematician and scientist Carl Gauss, is the subject of this section.

Figure 16-21a shows a uniform electric field E passing through an imaginary rectangular frame of area A. With the frame shown from the side, we see more clearly that the field is oriented perpendicular to the plane of the frame (Figure 16-21b). We define the quantity **electric flux**, Φ, to represent the strength of the electric field as it crosses the area A:

$$\Phi = EA$$

Electric flux describes how much of the electric field—how many field lines—pass through an area A, in this case the area of the frame. Suppose we now tilt the imaginary frame with respect to the direction of the field $\vec{E}$, as in Figure 16-21c. Fewer lines of field pass through the frame, indicating a corresponding decrease in the electric flux, in spite of the fact that neither the strength of the field nor the area of the frame changed. To accommodate the decrease in flux, we modify the definition of flux guided by Figure 16-22. The height of the imaginary frame is h, but the effective height that determines how much of the field passes through the opening is $h \cos \theta$. Angle θ describes the tilt of the frame with respect to the direction of $\vec{E}$. This angle is measured between $\vec{E}$ and $\vec{A}$, a vector that points in the direction perpendicular to the plane of the frame and has magnitude equal to its area.

> The **area vector** of a surface $\vec{A}$ with a vector on top points in a direction perpendicular to the surface and has magnitude equal to the area of the surface. If the surface is closed, that is, if it divides the universe into "inside" and "outside," by convention the area vector points outward.

The area of the opening of the frame is always hw, where w is the width of the opening (not shown in the figure). When the frame is tilted, however, the effective area A_{eff} of the opening is $h(\cos \theta)w$, or

$$A_{\text{eff}} = h(\cos \theta)w = A \cos \theta$$

The strength of the electric field as it crosses the area of the opening, the flux Φ, becomes

$$\Phi = EA_{\text{eff}} = EA \cos \theta$$

The field is a vector, and we have now defined area as represented by a vector, so this expression can be written as a dot product (Equation 6-4):

$$\Phi = \vec{E} \cdot \vec{A} \qquad (16\text{-}27)$$

According to Equation 16-27, the electric flux is greatest when a field is perpendicular to a surface through which it passes and zero when

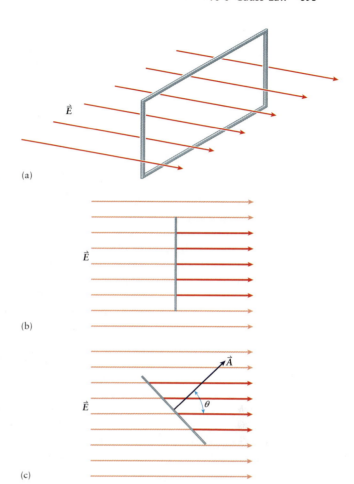

(a)

(b)

(c)

Figure 16-21 (a) A uniform electric field passes through an imaginary rectangular frame. (b) A side view makes it clear that the electric field is perpendicular to the plane of the frame. (c) When the frame is tilted so that the direction of $\vec{E}$ is not perpendicular to the plane of the frame, fewer lines of field pass through it.

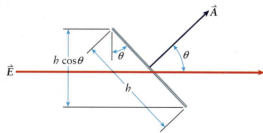

Figure 16-22 When the imaginary rectangular frame through which an electric field passes is tilted at angle θ with respect to the direction of $\vec{E}$, the effective height of the frame is reduced from h to $(h \cos \theta)$.

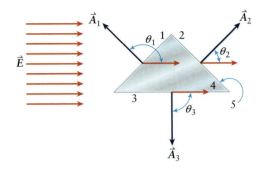

Figure 16-23 Electric flux passes through a triangular prism, viewed looking directly at one of the triangular sides. The net electric flux is the sum of the electric flux through each of the five sides separately.

the field and surface are parallel. In the first case, θ is 0 and $\cos \theta$ equals 1, because $\vec{A}$ is itself perpendicular to the surface; for the same reason, θ is 90° and $\cos \theta$ equals 0 in the second case.

Electric flux is additive. Consider an electric field passing through a triangular prism, a solid object formed by connecting a triangular top and bottom with three rectangular side surfaces. Figure 16-23 shows one surface of a triangular prism as seen by looking directly at one of the triangular ends. We labeled the rectangular side surfaces 1, 2, and 3; the triangular end surfaces are labeled 4 and 5. Surface 4 is closest to you as viewed in the figure and surface 5 is directly beneath it. Side surfaces 1 and 2 are equal in area. The total electric flux is then

$$\Phi = \Phi_1 + \Phi_2 + \Phi_3 + \Phi_4 + \Phi_5$$

Three of these fluxes are immediately seen to be zero. θ_3 is 90°. Although $\vec{A}_4$ and $\vec{A}_5$ are not shown in the figure, they are perpendicular to the field, so θ_4 and θ_5 are also 90°. Φ_3, Φ_4, and Φ_5 are therefore zero because $\cos 90° = 0$. Each of the remaining flux terms, Φ_1 and Φ_2, can be expressed, using Equation 16-27, as $\Phi_i = \vec{E} \cdot \vec{A}_i = EA_i \cos \theta_i$. The net flux is then

$$\Phi = \Phi_1 + \Phi_2 = EA_1 \cos \theta_1 + EA_2 \cos \theta_2 \qquad (16\text{-}28)$$

Notice that θ_1 and θ_2 are related; $\theta_1 + \theta_2 = 180°$, so $\cos \theta_1 = -\cos \theta_2$. (You should verify this relationship.) Because the areas A_1 and A_2 are equal, the two terms in Equation 16-28 are equal in magnitude but of opposite sign. The two terms cancel, and *the net electric flux through the prism is zero*. Think of the net electric flux in this way: Whatever electric field enters a closed surface must also leave it, so the net flux is zero.

> **Watch Out**
>
> **Electric field and electric flux are not the same!**
>
> Although electric charge gives rise to electric field, we don't need to know what charge gave rise to a particular field. There can be a region of space in which there is an electric field and nothing else. The definition of flux, however, requires two things to be present—an electric field and a surface through which it passes. If that surface is closed (that is, it divides the universe into "inside" and "outside"), then as long as it encloses no net charge the net flux *through the surface* is zero. Whatever electric flux passes into the region also passes out. In this way, we can have zero flux but a nonzero field.

Electric flux is not a vector, but it carries sign. Because flux has a sign and because flux is additive, the flux entering the prism from the left in Figure 16-23 cancels the flux leaving the prism on the right. As a general rule, whatever electric flux enters a region bounded by a surface also leaves through that surface, so

When there is no source of electric field inside a closed surface, the net electric flux through the surface is zero.

To find the total electric flux through the triangular prism, we added the flux through each of the five sides. More generally, when a surface can be broken into N separate sections, the total flux is given by

$$\Phi = \sum_{i=1}^{N} \vec{E} \cdot \vec{A}_i$$

Even more generally, to calculate the flux when either the field or the surface shape is not constant, we must break the flux into infinitesimally small pieces in order for E, A, and the angle between them to be constant over each region of the surface:

$$\Phi = \int \vec{E} \cdot d\vec{A} \qquad (16\text{-}29)$$

where $d\vec{A}$ is an infinitesimal piece of the area of the surface.

What if there *is* a source of electric field inside a closed surface? The simplest case we can consider is a point charge q_{encl} enclosed by a spherical surface of radius R as in **Figure 16-24**. For simplicity we have placed the charge at the center of the sphere. The electric field lines emanate radially from the charge, like the spines on a sea urchin, although we only show one representative line of electric field in the figure. The area vector $\vec{A}$ is also radial; it points outward and is perpendicular to the surface of the sphere. So, at the point on the surface where the electric field $\vec{E}$ passes through it, $\vec{A}$ is aligned with $\vec{E}$, as shown in the figure. Thus, the angle between $\vec{E}$ and $\vec{A}$ is zero, and the electric flux is

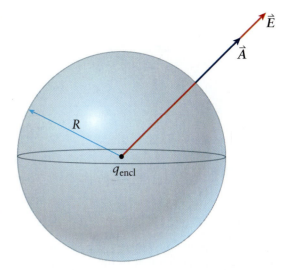

Figure 16-24 What is the electric flux through a closed spherical boundary that encloses a point charge at its center?

$$\Phi = \vec{E} \cdot \vec{A} = EA \cos \theta = EA$$

The surface area of a sphere is $4\pi R^2$, and the electric field is described by Equation 16-4 or $E = kq_{encl}/R^2$, thus

$$\Phi = \frac{kq_{encl}}{R^2} 4\pi R^2 = 4\pi k q_{encl} \qquad (16\text{-}30)$$

Notice that the radius of the sphere does not appear in the equation. Because the magnitude of the electric field decreases as the square of the sphere's radius and the surface area increases at the same rate, their product (the flux) is constant and does not depend on the radius. Regardless of the size of the sphere, however, the electric flux depends on the charge enclosed.

The Coulomb constant k is often defined in terms of ϵ_0, the **permittivity of free space**. Permittivity is related to how an electric field affects a region or material; ϵ_0 has the value $8.85 \times 10^{-12}\,\mathrm{C^2/(N \cdot m^2)}$ and is related to the Coulomb constant by

$$\epsilon_0 = \frac{1}{4\pi k}$$

The electric flux through the spherical surface that encloses charge q_{encl} (Equation 16-30) becomes

$$\Phi = 4\pi \frac{1}{4\pi \epsilon_0} q_{encl} = \frac{q_{encl}}{\epsilon_0}$$

Although this result could apply only to this point charge and this spherical surface, what we have uncovered is far more general. Named after the 19th-century German mathematician and scientist, **Gauss' law** states that

The net electric flux through a closed surface equals the charge enclosed divided by the permittivity ϵ_0.

This equation summarizes Gauss' law:

$$\oint \vec{E} \cdot d\vec{A} = \frac{q_{encl}}{\epsilon_0} \qquad (16\text{-}31)$$

Notice that the left-hand side of the equation is slightly different from the general definition we established for electric flux (Equation 16-29). The integral sign with

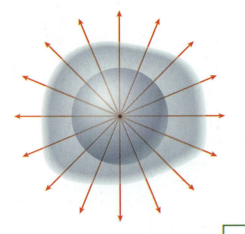

Figure 16-25 Gauss' law gives the same answer independent of the shape or size of the Gaussian boundary.

a circle through it indicates that the sum of flux contributions is taken over a closed surface, that is, a boundary that has an inside and an outside. A surface used to enclose charge in order to apply Gauss' law is referred to as a **Gaussian surface** or **Gaussian boundary**.

Although we found Gauss' law by considering a point charge at the center of a sphere, it applies to a boundary surface of any shape or size. **Figure 16-25** shows a point charge enclosed in both a spherical boundary and also one of odd shape. The same number of electric field lines pass through both boundaries, so we conclude that the flux through each is the same. In addition, Gauss' law does not depend on the location of the charge or charges within the boundary.

> **? Got the Concept 16-8**
> **Electric Flux I**
>
> When a certain point charge is placed at the center of a sphere, the net electric flux through the surface is Φ_0. If the radius of the sphere were to be doubled, would the new flux be one-quarter of the original ($\Phi_0/4$), one-half of the original ($\Phi_0/2$), the same as the original (Φ_0), twice the original ($2\Phi_0$), or 2^2 times the original ($4\Phi_0$)?

> **? Got the Concept 16-9**
> **Electric Flux II**
>
> When a certain point charge is placed at the center of a sphere, the net flux through the surface is Φ_0. If the sphere were elongated into an oblate spheroid (such as a rugby ball or an American football), would you expect the electric flux through the surface to be lower than Φ_0, the same as Φ_0, or higher than Φ_0?

> **? Got the Concept 16-10**
> **Electric Flux III**
>
> When a certain point charge is placed at the center of a sphere, the net flux through the surface is Φ_0. Imagine that the charge were moved so that it was nearly touching the inside surface of the sphere; in this case would you expect the flux through the surface to be lower than Φ_0, the same as Φ_0, or higher than Φ_0?

Both Coulomb's law and Gauss' law allow us to determine the electric field that arises due to charge. Problems that involve symmetry are often more straightforward to solve using Gauss' law. Physicists also consider Gauss' law to be the more elegant and the more fundamental law. In addition, although we won't have the occasion to see it now, Coulomb's law applies only to stationary charges, while Gauss' law holds for moving as well as stationary charges.

⁕ **What's Important 16-6**

Electric flux, Φ, represents the strength of the electric field across an area. It is greatest when the field is perpendicular to the surface through which it passes and zero when the field and surface are parallel. The area vector of a surface used to determine flux points in a direction perpendicular to the surface and has magnitude equal to the area of the surface. According to Gauss' law the electric flux through a closed boundary is proportional to the charge enclosed by the boundary.

16-7 Applications of Gauss' Law

Physicists like simplicity! Why do a problem a hard way when an easy way is available? This issue often confronts us when trying to determine the electric field due to a charge configuration. For many problems, Gauss' law is a better choice than Coulomb's law. We'll see that while it might be easier to set up a problem using Coulomb's law by defining a small charge element and then setting up an integral to add up the contribution each element makes to the electric field, the integration can be challenging. Conversely, it takes a bit of analysis to set up properly a solution using Gauss' law, but once the solution is set up, it tends to be relatively straightforward.

In Section 16-5, we determined the electric field near a large, uniformly charged disk. Glance back at the work required. First, we determined the electric field due to a charged ring, and then used that result to find the field due to the disk. The integrals were easy to set up but challenging to carry out. In order to compare the two methods we'll do the same problem again using Gauss' law.

Figure 16-26 shows a large, thin, uniformly charged disk of radius R viewed from the side. The disk carries an excess charge $+Q$. We want to determine the electric field it generates. In order to apply Gauss' law (Equation 16-31) we need to enclose the charge, or a representative element of the charge, in a Gaussian surface and evaluate the flux by the surface integral $\oint \vec{E} \cdot d\vec{A}$. The integral depends on $\vec{E}$, $\vec{A}$, and the angle between them; to simplify the integral, we look for a boundary that offers some or all of these characteristics:

- E is constant on some or all regions of the boundary.
- The angle between $\vec{E}$ and $\vec{A}$ is either 0° or 90°, on some or all regions of the boundary, so that the cosine is either 1 or 0, respectively, for those regions.
- Some or all regions of the boundary have a shape for which we know how to determine the area. Examples include the surface of a sphere, a cylinder, and a cube.

In addition, keep in mind that any point at which the field is to be determined must lie on the surface of the boundary.

The first two criteria for selecting a Gaussian boundary present a conundrum. How can we determine whether E is constant on a boundary region, and how can we decide whether the angle between $\vec{E}$ and $\vec{A}$ is convenient without first knowing the electric field, which is what we're trying to find? The answer lies in seeking out symmetry in the charge configuration. Symmetry is a pattern, a similarity of one region of the field to another, from which we can guess general characteristics of the field.

Consider the electric field at a point close to the disk in Figure 16-26 and far from its edges. The symmetry of the distribution of charge allows us to choose pairs of charge elements on the surface, such as the green and blue charges and the

Figure 16-26 The electric field close to a large, thin, uniformly charged disk is perpendicular to the plane of the disk.

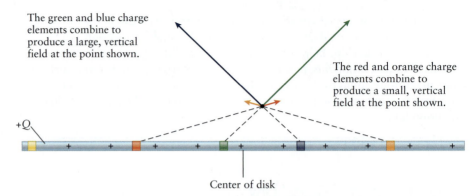

A disk carries an excess charge +Q.

The green and blue charge elements combine to produce a large, vertical field at the point shown.

The red and orange charge elements combine to produce a small, vertical field at the point shown.

+Q

Center of disk

The yellow charge element results in a horizontal contribution to the total field, but it is so far from the point of interest that its field contribution can be neglected.

The net field far from the edge of the disk is vertical. A vertical field has constant magnitude.

red and orange charges, for which the horizontal components cancel when the fields due to each element of the pair are added. It is true that unless the point of interest is on the central axis there will be charge elements, like the one shown in yellow, for which there isn't a corresponding charge that cancels its horizontal field component. However, for a large enough disk, or for points close to the surface, the fields due to such charge elements are so small that they can be neglected. Symmetry, then, convinces us that far from the edges, the field is vertical. Vertical lines are by definition parallel to each other, which means that the density of electric field lines remains the same. The electric field is therefore constant far from the edges of the disk—it doesn't depend on the distance from the disk or the exact point of interest.

To take advantage of our conclusion that the field is vertical and constant far from the edges of the disk, we choose a cylindrical surface of radius r as the Gaussian boundary that passes vertically through the charged disk, as shown in **Figure 16-27a**. The field $\vec{E}$ is the same everywhere on the top and bottom of the dark grey cylinder, and because both $\vec{E}$ and $\vec{A}$ are perpendicular to the top and bottom surfaces of the cylinder (as you can see in **Figure 16-27b**), the angle between $\vec{E}$ and $\vec{A}$ is 0°. (Only one representative field vector in each direction is shown in the figure.) Around the side of the cylinder, $\vec{E}$ and $\vec{A}_{\text{side}}$ are perpendicular to each other. We have therefore satisfied the criteria for selecting a Gaussian boundary: $\vec{E}$ is constant on some part of the boundary, the angle between $\vec{E}$ and $\vec{A}$ is either 0° or 90° everywhere, and we know how to determine the area of the various parts of the surface.

The total electric flux is

$$\oint \vec{E} \cdot d\vec{A} = \int_{\text{top}} \vec{E} \cdot d\vec{A} + \int_{\text{side}} \vec{E} \cdot d\vec{A} + \int_{\text{bottom}} \vec{E} \cdot d\vec{A} \qquad (16\text{-}32)$$

Because the angle between $\vec{E}$ and $\vec{A}_{\text{top}}$ is 0°,

$$\int_{\text{top}} \vec{E} \cdot d\vec{A} = \int \vec{E} \cdot d\vec{A}_{\text{top}} = E A_{\text{top}} \cos 0° = E A_{\text{top}}$$

The same is true for the bottom of the cylinder, and because $A_{\text{top}} = A_{\text{bottom}} = \pi r^2$,

$$\int_{\text{top}} \vec{E} \cdot d\vec{A} = \int_{\text{bottom}} \vec{E} \cdot d\vec{A} = E \pi r^2$$

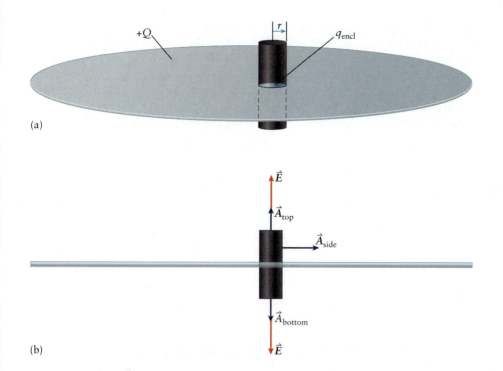

(a)

(b)

Figure 16-27 (a) The electric field close to a large, thin, uniformly charged disk is perpendicular to the plane of the disk, so a cylindrical surface is a good choice for a boundary over which to apply Gauss' law. (b) The electric field is parallel to the area vectors of the top and bottom surfaces of the cylinder, and perpendicular to the area vector associated with the curved side of the cylinder.

Finally, because the angle between $\vec{E}$ and $\vec{A}_{side}$ is 90°,

$$\int_{side} \vec{E} \cdot d\vec{A} = \int \vec{E} \cdot d\vec{A}_{side} = E A_{side} \cos 90° = 0$$

The integral on the left side of Gauss' law, by using Equation 16-32, is

$$\oint \vec{E} \cdot d\vec{A} = E\pi r^2 + 0 + E\pi r^2 = 2E\pi r^2$$

$\sqrt{x}$ *See the Math Tutorial for more information on Integrals*

To determine the right side of Gauss' law we need to know the charge enclosed in the cylindrical surface. That charge, q_{encl} in Figure 16-27a, is the same fraction of the total charge Q as the area of the surface inside the cylinder is to the total area of the disk. That is,

$$q_{encl} = \frac{\pi r^2}{\pi R^2} Q$$

Setting the two sides of Gauss' law equation equal gives

$$2E\pi r^2 = \frac{1}{\epsilon_0} \frac{\pi r^2}{\pi R^2} Q$$

or

$$E = \frac{1}{2\epsilon_0} \frac{Q}{\pi R^2}$$

The ratio of the total charge to the total area is just the area charge density σ, so

$$E = \frac{\sigma}{2\epsilon_0}$$

This is, not surprisingly, the same result that we got using Coulomb's law to determine the field close to a large disk (Equation 16-26).

Let's compare the work required to find the electric field near the disk using the approaches of Coulomb's law and Gauss' law. Setting up the problem using

Coulomb's law, by breaking the disk into an infinite number of infinitesimal charge elements and then adding their contributions to the electric field, was relatively straightforward. However, dealing with the vector nature of the sum and carrying out the integrations were perhaps a bit challenging. Using Gauss' law, we had to think carefully about how to set up the Gaussian boundary by taking symmetry into account. But once we determined the Gaussian boundary, perhaps the hardest part of doing the integration was remembering the area of a circle!

> ## ? Got the Concept 16-11
> ### Size of a Gaussian Surface
>
> In our determination of the electric field far from the edge of the charged disk, suppose the radius of the Gaussian cylinder were doubled. Would the electric field decrease by one-half, stay the same, or double in magnitude?

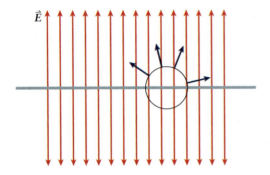

Figure 16-28 A spherical boundary is not the best choice over which to evaluate Gauss' law, because the angle between the electric field $\vec{E}$ and the area vector $\vec{A}$ varies, as shown at four points on the sphere as dark blue arrows.

Could we have used a boundary other than a cylinder to find the electric field of the disk? Certainly. A boundary of any shape and size must give the same result. Integrating could be far more difficult, however, using a boundary that did not satisfy the criteria we established. For example, we might have selected a sphere as the Gaussian boundary, as in **Figure 16-28**. (The disk and sphere are shown from the side.) Although the direction and magnitude of the electric field vectors are the same everywhere on the sphere, the angle varies between the electric field $\vec{E}$ and the area vector $\vec{A}$, as shown at various points on the sphere. The integral could still be done, but would be more complicated.

> ## ? Got the Concept 16-12
> ### Spherical Symmetry
>
> We would like to determine the electric field inside a sphere carrying a uniform charge density. What would be a good choice for the boundary with which to evaluate Gauss' law?

Example 16-6 A Long Charged Wire

A long, straight wire carries a uniform, positive linear charge density λ. Use Gauss' law to find the electric field at a point a distance r from the wire, close to it and far from its ends.

SET UP

To apply Gauss' law, we need to set up a Gaussian boundary surface on which to evaluate $\oint \vec{E} \cdot d\vec{A}$ and with which to enclose charge. Any choice will result in the correct determination of the field, but the integration becomes more straightforward to carry out when $\vec{E}$ is constant or easily described on the boundary, and when the angle between $\vec{E}$ and $\vec{A}$ is 0° or 90°. To assess a potential choice for the boundary surface, we appeal to symmetry to gain an understanding of the shape of the field. As you follow this analysis, make your own sketch!

We start by selecting an arbitrary point near the wire at which to consider the field. The wire is a long, straight line; pick a point far from the ends and relatively close to the wire. The quantity r is the distance to the point along the shortest line from it to the wire. Much like we did in Figure 16-26, pick some representative charge elements on the wire, and sketch the field due to each at the point of interest. Just like in that figure, the arrows of $\vec{E}$ start at the point of interest and point away from the charge elements that you selected.

For most charge elements you might choose, there will be another that is its reflection about the line that connects the point of interest and the wire. The electric field due to the second, reflected charge element has the same component in the direction away from the wire but its component parallel to the wire cancels the parallel component due to the first charge element. This symmetry results in a field that points away from the wire when we find the vector sum of the two field contributions. Yes, because the point of interest is not necessarily in the exact middle of the wire, there will be charge elements that do not have a pair that cancels the parallel component. Because the wire is long, however, the field due to those charge elements, which must be out near an end of the wire, is so small that we can neglect it. The field due to the wire points away from it in all directions. The field looks like the bristles on a round bottle brush (**Figure 16-29a**) or the spines on the flower of the bottlebrush tree (**Figure 16-29b**).

SOLVE

The best choice of a boundary with which to apply Gauss' law for the long straight wire is a cylinder, with the wire running along the cylinder's central axis. This setup, along with two representative field lines and the corresponding area vectors on the boundary surface, is shown in **Figure 16-30**. On the two ends of the cylinder $\vec{E}$ and $\vec{A}_{end}$ are perpendicular, so $\vec{E} \cdot \vec{A}_{end} = EA_{end} \cos 90° = 0$. On the curved side of the cylinder $\vec{E}$ and $\vec{A}_{end}$ are parallel, so $\vec{E} \cdot \vec{A}_{side} = EA_{side} \cos 0° = EA_{side}$; thus,

$$\oint \vec{E} \cdot d\vec{A} = \int_{end} \vec{E} \cdot d\vec{A} + \int_{side} \vec{E} \cdot d\vec{A} + \int_{end} \vec{E} \cdot d\vec{A}$$

$$= 0 + EA_{side} + 0$$

The area of the curved side of the cylinder is the product of its circumference, $2\pi r$, and its length L, so

$$\oint \vec{E} \cdot d\vec{A} = E\, 2\pi r L$$

(a)

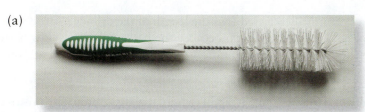

(b)

Figure 16-29 The electric field due to a long, uniformly charged wire resembles (a) the bristles on a round bottle brush and (b) the spines on the flower of the bottlebrush tree. *(Courtesy of David Tauck.)*

Figure 16-30 A cylindrical surface is an ideal boundary over which to apply Gauss' law to a long, straight wire carrying a uniform, linear charge density. When the central axis of the cylinder is aligned with the wire, the electric field is perpendicular to the area vector of the ends of the cylinder and parallel to the area vector associated with the curved side of the cylinder.

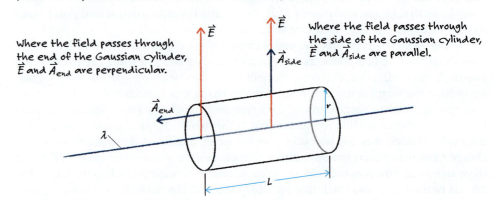

The electric field due to a long straight wire carrying uniform linear charge density λ points radially from the wire in all directions. Two representative lines of field are shown.

Where the field passes through the end of the Gaussian cylinder, $\vec{E}$ and $\vec{A}_{end}$ are perpendicular.

Where the field passes through the side of the Gaussian cylinder, $\vec{E}$ and $\vec{A}_{side}$ are parallel.

The charge enclosed in the cylinder is that charge on the part of the wire within the cylinder. The charge per unit length of the wire is λ, so the right side of Gauss' law equation (Equation 16-31) is

$$\frac{q_{encl}}{\epsilon_0} = \frac{\lambda L}{\epsilon_0}$$

Setting the expressions equal gives

$$E\,2\pi r L = \frac{\lambda L}{\epsilon_0}$$

or

$$E = \frac{\lambda}{2\pi \epsilon_0 r}$$

The electric field decreases in proportion to the distance from the wire.

REFLECT
Notice that we arbitrarily selected L as the length of the cylinder. Any choice would have been acceptable for the length, because the nature of Gauss' law is such that the electric field does not depend on the size of the boundary used to evaluate the integral.

Practice Problem 16-6 A long, straight wire carrying a uniform, positive linear charge density λ runs down the axis of a narrow conducting cylinder that carries a uniform, negative charge density $-\lambda$. Use Gauss' law to find the electric field at a point a distance r from the wire, outside but close to the cylinder, and far from its ends.

Gauss' law leads to a curious conclusion about conducting materials to which excess charge is added. Because copper is a conductor, for example, any piece of copper contains many electrons which can move about the material relatively freely. Each electron repels all the others, so the free electrons move. This motion continues until equilibrium is reached, that is, until the net force on each electron is zero.

Imagine that we add *extra* electrons to a block of copper so that the block acquires a net charge. The electrons are repelled by the other electrons as well, so they also move until the net force on each one is zero. Because electric force and electric field are directly related (Equation 16-2), when the electrons experience no net force, the net electric field is zero. We must conclude, then, that the electric field inside the block of copper is zero—otherwise excess electrons would experience a force and would accelerate to some new location. Note that this statement only applies inside the block. Outside the charged block of copper the field is not zero.

Where does Gauss' law come in? Imagine we set up a Gaussian boundary inside the block and right beneath the surface, along all of its sides. Gauss' law demands that

$$\oint \vec{E} \cdot d\vec{A} = \frac{q_{encl}}{\epsilon_0} \qquad (16\text{-}31)$$

When the electric flux through a boundary is zero, as it must be here because the electric field is zero, the charge enclosed by the boundary is zero as well. Where did all of the excess charge go if it's not enclosed by the Gaussian boundary we established? Because the boundary is beneath the surface of the block, the only region in the block not within the Gaussian boundary is the surface itself. Gauss' law tells us that excess charge added to a conductor resides only on the surface! (This may not be true if forces not included in the current discussion push charges away from the surface.)

Example 16-7 Surface Charge

Figure 16-31a shows a sphere carrying excess charge +7 C enclosed within a spherical conducting shell that carries a net charge +2 C. Find the charge on the inside and outside surfaces of the conducting shell.

SET UP
The electric field inside the material of the conducting shell must be zero. So if we imagine a spherical Gaussian boundary that sits inside the shell, as shown in Figure 16-31b, it must enclose a net charge of zero. The charge on each surface of the shell will arrange itself in order to make that happen.

SOLVE
The charge enclosed inside the Gaussian boundary we selected includes the charge of the central sphere and whatever charge is induced on the inner surface of the shell. These charges add to zero:

$$q_{encl} = q_{sphere} + q_{inner}$$

For $q_{encl} = 0$,

$$q_{inner} = -q_{sphere} = -7\,C$$

The total charge on the shell is +2 C. Because all of the excess charge resides on the surface of a conductor, this +2 C is the sum of the charge on the inner and outer surfaces:

$$q_{shell} = q_{inner} + q_{outer}$$

So

$$q_{outer} = q_{shell} - q_{inner} = +2\,C - (-7\,C) = +9\,C$$

The inner and outer surfaces of the shell carry charge −7 C and +9 C respectively, for a total of +2 C.

REFLECT
It may look odd to you that the charge on the outer surface of the shell is greater than the total charge on the inner surface. Recall, however, that there is an enormous number of charges in even a small amount of matter; some of the charge has

Figure 16-31 (a) A charged sphere is enclosed within a charged, spherical conducting shell. What is the charge on the inside and outside surfaces of the conducting shell? (b) Because the electric field inside the material of the conducting shell must be zero, a spherical Gaussian boundary that sits inside the shell, shown here as a dashed red line within the wall of the spherical shell, must enclose a net charge of zero.

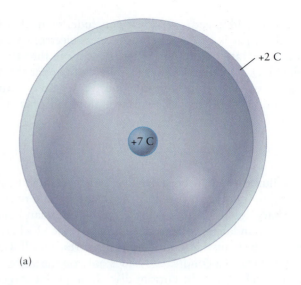

(a)

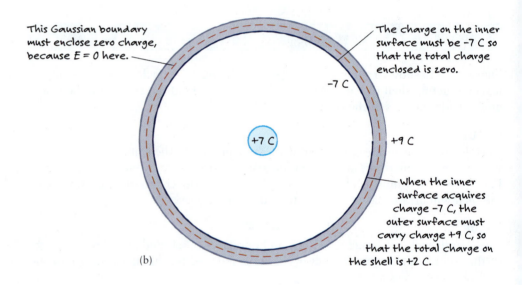

This Gaussian boundary must enclose zero charge, because E = 0 here.

The charge on the inner surface must be −7 C so that the total charge enclosed is zero.

−7 C

+7 C

+9 C

When the inner surface acquires charge −7 C, the outer surface must carry charge +9 C, so that the total charge on the shell is +2 C.

(b)

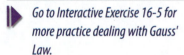

Go to Interactive Exercise 16-4 for more practice dealing with Gauss' Law.

Go to Interactive Exercise 16-5 for more practice dealing with Gauss' Law.

rearranged itself under the influence of the Coulomb force. When the positively charged sphere is placed inside the shell, electrons are drawn to the shell's inner surface. This leaves a deficit of electrons—a positive charge—on the outside surface. Regardless of the arrangement of the charges, however, the net charge on the conducting shell must remain the same, which results in the conclusion that the sum of the charges on the inner and outer surfaces equals the total charge on the shell.

Practice Problem 16-7 Suppose that the spherical conducting shell in Figure 16-31a carries a net charge −9 C. (Let the conducting sphere it encloses carry excess charge +7 C.) Find the charge on the inside and outside surfaces of the conducting shell.

✳ What's Important 16-7

The net electric flux through a boundary is zero when it encloses no net charge. Gauss' law tells us that in the absence of external forces, excess charge added to a conductor resides only on the surface.

Answers to Practice Problems

16-1 $Q/2$

16-2 1.4×10^{-2} N, 100 times larger

16-3 7.99×10^9 N, in the $+y$ direction

16-4   1 m to the left of Q_1

16-5 8.29×10^{-8} s

16-6   0 N/C

16-7 $q_{inner} = -7$ C, $q_{outer} = -2$ C

Answers to Got the Concept Questions

16-1   When Q is positive and q is negative the product of the two is negative. So the force, which according to Coulomb's law is proportional to the product of the charges, has an overall minus sign. A negative force is an attractive one, so $-q$ is attracted to $+Q$. When both Q and q are positive the product of the two is positive. In this case, the force has an overall plus sign. A positive force is repulsive, so $+q$ is repelled from $+Q$. The force vector in each case is shown in **Figure 16-32**.

Figure 16-32

16-2   Because all red blood cells carry a net negative charge, they repel each other. (Opposite sign charges attract, like sign charges repel.)

16-3   Because the point charges labeled $+Q_1$ and $+Q_2$ ($Q_1 = Q_2 = Q$) carry charge of the same sign, the force between them is positive and therefore repulsive. In the absence of point charge q, the two particles would move away from each other. Consider $+Q_2$, for example. This point charge is repelled by $+Q_1$, so the force of $+Q_1$ on $+Q_2$ points to the right. For the net force on $+Q_2$ to be zero, point charge q must exert a force in the opposite direction or to the left. Thus, the force of q on $+Q_2$ must be attractive. We conclude that the sign of q must be opposite to the sign of $+Q_2$; that is, q must be negative. The same reasoning can be applied to $+Q_1$. Finally, notice that when q is negatively charged, it is attracted to both $+Q_1$ and $+Q_2$. Because the magnitudes of $+Q_1$ and $+Q_2$ are the same, the forces on q due to $+Q_1$ and $+Q_2$ are equal

and cancel out. All three point charges therefore experience a net force of zero when q is negatively charged.

16-4   Porcelain is a good thermal insulator for the same reason that it is a good electrical insulator. Its thermal properties are similar to those of wood, also a good electrical and thermal insulator. Copper is a good electrical conductor, but compared to porcelain it is a poorer thermal insulator by a factor of hundreds. A house built from copper would be cold in the winter and hot in the summer!

16-5   Arrows that represent the electric field vectors around a positive point charge point radially away from it, much like the spines of a sea urchin. So the field vector at a point directly between you and the charge points directly toward you. Because the electric field points toward a negative charge, in the second case the field vector at a point directly in front of you would point away from you and toward the negative point charge.

16-6   First, any point in space at which the net electric field is zero must lie on the line that connects the two charges. At any point not on the line, for example, the point shown in **Figure 16-33a**, the fields due to the two charges separately could not possibly point in the same direction. Although it could happen that the components of the fields in one or two directions cancel, it would be impossible for the components in all three directions to cancel. So, the net field, which is found by determining the vector sum of the two fields, cannot be zero for points not on the line that connects the two charges. Points of zero net field also cannot lie outside of the region between the two charges, as you can see in **Figure 16-33b**. At any such point, the fields due to each charge point in the same direction, so again, no cancellation is possible. Only at a point between the charges (**Figure 16-33c**), where the fields due to each charge point in opposite directions, can there be complete cancellation when the vector sum of the fields is calculated.

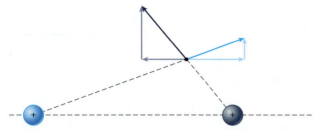

(a) The horizontal components of the two fields cancel, but for points not on the line connecting the charges, it is not possible to cancel the net field in every direction.

(b) For points outside of the region between the charges but on the line that connects them, the field due to each one points in the same direction, so there is no way for the fields to cancel when added as vectors.

(c) There is a point between the charges where the separate fields point in opposite directions, with equal magnitude. At this point the net field is zero.

Figure 16-33

16-7 The acceleration that a charged object undergoes in an electric field depends not on the charge or the mass separately, but on the ratio of the charge to the mass. DNA is a polymer with a repeating structure, so fragments of any size and composition tend to have a similar charge-to-mass ratio. For that reason, the contribution to a DNA fragment's acceleration due to the electric field is the same for all fragments. The rate at which fragments move through a particular gel is therefore determined primarily by the drag force each fragment experiences, which depends on the viscosity of the gel and the fragment size.

16-8 The flux remains the same, Φ_0. According to Gauss' law, the electric flux due to an enclosed charge does not depend on the size of the boundary surface.

16-9 The flux remains the same, Φ_0. According to Gauss' law, the electric flux due to an enclosed charge does not depend on the shape of the boundary surface.

16-10 The flux remains the same, Φ_0. According to Gauss' law, the electric flux due to an enclosed charge does not depend on the location of the charge within the region bounded by the surface.

16-11 The electric field stays the same. Notice that the result does not depend on the radius of the cylinder we chose. Gauss' law assures us that the electric field does not depend on the size—or shape—of the boundary used to evaluate the integral.

16-12 The shell theorem tells us that a charged sphere is treated as if all of the charge were concentrated at the center of the sphere. Even if we didn't already know that the electric field due to a point charge is radial and is the same in all directions, symmetry would lead us to the same conclusion. How could the field not be the same in every direction? Symmetry also suggests that the magnitude of the field is the same at all points which are the same distance from the center. A spherical surface, which is of course the surface defined as all points the same distance from a common center, is therefore the best choice. We satisfy the first criterion for simplifying the Gauss' law integral because the magnitude of the field is constant on the sphere. The surface of a sphere is perpendicular to the radial direction everywhere, so the second criterion is satisfied. Finally, we know how to determine the surface area of a sphere. Therefore, a spherical surface meets all three criteria for the Gaussian boundary in this example.

SUMMARY

Topic	Summary	Equation or Symbol
area vector	The area vector of a surface points in a direction perpendicular to the surface and has magnitude equal to the area of the surface.	$\vec{A}$
conductor	A conductor is a material in which charge can move easily.	
Coulomb's law	The electric force between two point charges q_1 and q_2, separated by a distance r, is given by Coulomb's law. The electric force points along the line that connects the charges.	$\vec{F} = \dfrac{kq_1q_2}{r^2}\hat{r}$ (16-1)
electric charge	Electric charge is a property of fundamental particles. Electrons carry negative charge, protons carry positive charge.	

Topic	Summary	Equation or Symbol	
electric dipole	An electric dipole is formed by separating charges of equal magnitude and opposite sign. The simplest electric dipole is two point charges $-q$ and $+q$ separated by a finite distance.		
electric field	An electric field surrounds every charged object; it is strongest closest to the object but extends infinitely far away. The electric field represents the force that the object would exert on another charged object at every point in space.	$\vec{E}$	
electric field, point charge	The electric field of a point charge q falls off as the inverse of the square of the distance r from the charge. The field vector points radially away from a positive charge and radially toward a negative charge.	$\vec{E} = \dfrac{kq}{r^2}\hat{r}$	(16-4)
electric flux	Electric flux Φ represents the strength of the electric field across an area.	$\Phi = \displaystyle\int \vec{E} \cdot d\vec{A}$	(16-29)
fundamental charge	The fundamental charge e is the magnitude of the charge of both an electron and a proton.	$e = 1.602 \times 10^{-19}\ \text{C}$	
Gauss' law	Gauss' law is a relationship between the electric flux across a closed boundary and the charge enclosed by that boundary. The integral sign with a superimposed circle in Gauss' law is a reminder that the integral is carried out over a closed boundary.	$\displaystyle\oint \vec{E} \cdot d\vec{A} = \dfrac{q_{\text{encl}}}{\epsilon_0}$	(16-31)
Gaussian boundary (or Gaussian surface)	A surface used to enclose charge in order to apply Gauss' law is referred to as a Gaussian boundary or Gaussian surface.		
insulator	In an insulator, a material which is nonconducting, electrons cannot move freely.		
permittivity of free space	Permittivity is related to how an electric field affects a region or material. The permittivity of free space is given by ϵ_0.	$\epsilon_0 = 8.85 \times 10^{-12}\ \text{C}^2/(\text{N} \cdot \text{m}^2)$	

QUESTIONS AND PROBLEMS

In a few problems, you are given more data than you actually need; in a few other problems, you are required to supply data from your general knowledge, outside sources, or informed estimate.

Interpret as significant all digits in numerical values that have trailing zeros and no decimal points.

For all problems, use $g = 9.8\ \text{m/s}^2$ for the free-fall acceleration due to gravity. Neglect friction and air resistance unless instructed to do otherwise.

• Basic, single-concept problem

•• Intermediate-level problem, may require synthesis of concepts and multiple steps

••• Challenging problem

SSM *Solution is in Student Solutions Manual*

Conceptual Questions

1. •Discuss the similarities and differences between the gravitational and electric forces.

2. •Why is the gravitational force ignored in problems on the scale of particles such as electrons and protons?

3. •How, if at all, would the physical universe be different if the proton were negatively charged and the electron were positively charged?

4. •How, if at all, would the physical universe be different if the proton's charge was very slightly larger in magnitude than the electron's charge?

5. •When an initially electrically neutral object acquires a net positive charge, does its mass increase or decrease? SSM

6. •When you remove socks from a hot dryer, they tend to cling to everything. Two identical socks, however, usually repel. Why?

7. •Describe a set of experiments that might be used to determine if you have discovered a third type of charge other than positive and negative.

8. •How does a person become "charged" as they shuffle across a carpet, wearing cloth slippers, on a dry winter day?

9. •After combing your hair with a plastic comb, you find that when you bring the comb near a small bit of paper, the bit of paper moves toward the comb. Then, shortly after the paper touches the comb, it moves away from the comb. Explain these observations. SSM

10. •After combing your hair with a plastic comb you find that when you bring the comb near an empty aluminum soft-drink can that is lying on its side on a nonconducting tabletop, the can rolls toward the comb. After being touched by the comb the can is still attracted by the comb. Explain these observations.

11. •(a) A positively charged glass rod attracts a lighter object suspended by a thread. Does it follow that the object is negatively charged? (b) If, instead, the rod repels it, does it follow that the suspended object is positively charged?

12. •Some days it can be frustrating to attempt to demonstrate electrostatic phenomena for a physics class. An experiment that works beautifully one day may fail the next day if the weather has changed. Air-conditioning helps a lot while demonstrating the phenomena during the summer. Why?

13. •(a) What are the advantages of thinking of the force on a charge at a point P as being exerted by an electric field at P, rather than by other charges at other locations? (b) Is the convenience of the field as a calculation device worth inventing a new physical quantity? Or is there more to the field concept than that? SSM

14. •Do electric field lines point along the trajectory of positively charged particles? Why or why not?

15. •An electron and a proton are released in a region of space where the electric field is vertically downward. How do the electric forces acting on the electron and proton compare?

16. •Explain why integrating over a charge distribution works to find the net electric field at some point P due to that charge distribution.

17. •Is the electric field $\vec{E}$ in Gauss' law only the electric field due to the charge inside the Gaussian surface, or is it the total electric field due to all charges both inside and outside the surface? Explain your answer. SSM

18. •Inside a spherical charge distribution of constant volume charge density, why is it that as one moves out from the center, the electric field increases as r rather than decreases as $1/r^2$?

19. •If the net electric flux out of a closed surface is zero, does that mean the charge density must be zero everywhere inside the surface? Explain your answer.

20. ••Why is the expression $E = 2\pi k\sigma \left[1 - |x|/\sqrt{x^2 + R^2}\right]$ for the electric field along the axis of a uniformly charged thin disk, which we derived using Coulomb's law, different from the expression $E = 2\pi k\sigma$ for the electric field of a uniformly charged infinite plane, when Gauss' law seemingly gives the same result for both the charged disk and infinite plane?

Multiple-Choice Questions

21. •Electric charges of the opposite sign
   A. exert no force on each other.
   B. attract each other.
   C. repel each other.
   D. repel and attract each other.
   E. repel and attract each other depending on the magnitude of the charges.

22. •If two uncharged objects are rubbed together and one of them acquires a negative charge, then the other one
   A. remains uncharged.
   B. also acquires a negative charge.
   C. acquires a positive charge.
   D. acquires a positive charge equal to twice the negative charge.
   E. acquires a positive charge equal to half the negative charge.

23. •Metal sphere A has a charge of $-Q$. An identical metal sphere B has a charge of $+2Q$. The magnitude of the electric force on B due to A is F. The magnitude of the electric force on A due to B is
   A. $F/4$
   B. $F/2$
   C. F
   D. $2F$
   E. $4F$ SSM

24. •A balloon can be charged by rubbing it with your sleeve while holding it in your hand. You can conclude from this that balloon is a(n)
   A. conductor.

This page is intentionally left blank.

For complete end of chapter problem sets, please go to
www.whfreeman.com/kestentauck

Use a graphing calculator or spreadsheet.

r (m)	F (N)
0.003	5500
0.004	3000
0.005	2000
0.010	600
0.020	175
0.040	30
0.050	10
0.080	6
0.100	5
0.200	1
0.300	0.5
0.400	0.35
0.500	0.25
0.600	0.15

Problems

16-1: Electric Charge

38. •Five electrons are added to 1.0 C of positive charge. What is the net charge of the system?

39. •The nucleus of a copper atom has 29 protons and 35 neutrons. What is the total charge of the nucleus?

40. •An ion has 17 protons, 18 neutrons, and 18 electrons. What is the net charge of the ion?

41. ••How many coulombs of negative charge are there in 0.5 kg of water? SSM

42. •How many electrons must be transferred from an object to produce a charge of 1.6 C?

43. ••The charge per unit length on a glass rod is 0.0050 C/m. If the rod is 1 mm long, how many electrons have been removed from the glass rod?

16-2: Coulomb's Law

44. •Two point charges are separated by a distance of 20 cm. The numerical value of one charge is twice that of the other. If each charge exerts a force of magnitude 45 N on the other, find the magnitude of the charges.

45. •The mass of an electron is 9.11×10^{-31} kg. How far apart would two electrons have to be in order for the electric force exerted by each on the other to be equal to the weight of an electron? SSM

46. •Charge A, $+5$ μC, is positioned at the origin of a coordinate system (**Figure 16-36**). Charge B, -3 μC, is fixed at $x = 3$ m. (a) Determine the magnitude and direction of the force that charge B exerts on charge A. (b) What is the force that charge A exerts on charge B?

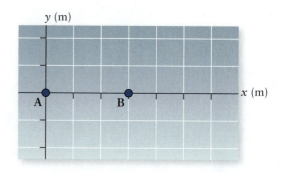

Figure 16-36 Problem 46

47. ••Determine the net force on charge A, located at the origin, due to charges B, C, and D (**Figure 16-37**).

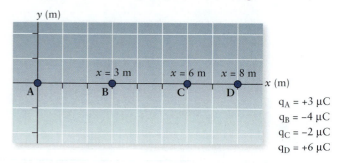

Figure 16-37 Problem 47

48. ••A charge of $+3.00$ μC is located at the origin and a second charge of -2.00 μC is located on the x–y plane at the point (30.0 cm, 20.0 cm). Determine the electric force exerted by the -2.00 μC charge on the 3.00 μC charge.

49. ••At what point along the x axis will the net force on a $+2$ nC charge be exactly zero due to the two fixed charges shown in **Figure 16-38**? SSM

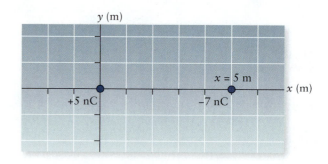

Figure 16-38 Problem 49

50. ••Two charges lie on the x axis, a -2.00 μC charge at the origin and a $+3.00$ μC charge at $x = 0.100$ m. At what position x, if any, is the force due to the charges on a $+4.00$ μC charge equal to zero?

51. ••A charge q_1 equal to 0.6 μC is at the origin, and a second charge q_2 equal to 0.8 μC is on the x axis at 5.0 cm. (a) Find the force (magnitude and direction) that each charge exerts on the other. (b) How would your answer change if q_2 were $-0.8 \ \mu C$?

52. ••In **Figure 16-39**, $\vec{F}_1$ is the force on q due to q_1, and $\vec{F}_2$ is the force on q due to another charge q_2. The total force on charge q is $\vec{F} = \vec{F}_1 + \vec{F}_2$. (a) If q is a positive charge, what is the sign of q_1? (b) Which of q_1 or q_2 is larger in magnitude?

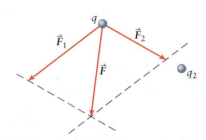

Figure 16-39 Problem 52

53. ••Suppose three charges are placed in the following charge distribution (**Figure 16-40**). Find the net force on each charge due to the others. SSM

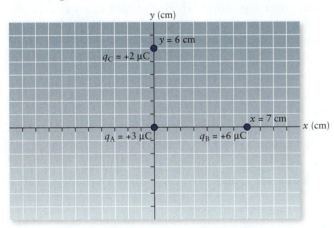

Figure 16-40 Problem 53

16-3: Conductors and Insulators

54. •A thin, flat copper plate has dimensions of 4 cm × 4 cm. If $-10 \ \mu C$ of charge is added to the metal conductor, find the average distance between the electrons on one side of the plate.

55. •Suppose 2 C of positive charge is distributed evenly throughout a sphere of 1.27 cm radius. (a) What is the charge per unit volume, ρ, for this situation? (b) Is the sphere insulating or conducting? How do you know?

56. •The maximum amount of charge that can be collected on a Van de Graaff generator's conducting sphere (30-cm diameter) is about 30 μC. Calculate the surface charge density, σ, of the sphere in C/m².

57. •**Biology** Most workers in nanotechnology are actively monitored for excess static charge buildup. The human body acts like an insulator as one walks across a carpet, collecting -50 nC per step. (a) What charge buildup will a worker in a manufacturing plant accumulate if she walks 25 steps? (b) How many electrons are present in that amount of charge? (c) If a delicate manufacturing process can be damaged by an electrical discharge greater than 10^{12} electrons, what is the maximum number of steps that any worker should be allowed to take before touching the components? SSM

16-4: Electric Field

58. ••Draw the electric field lines for the five charge distributions in **Figure 16-41**.

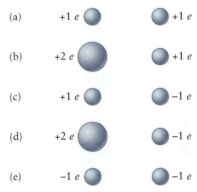

Figure 16-41 Problem 58

59. •At point P in **Figure 16-42**, the electric field is zero. (a) What are the signs of q_1 and q_2? (b) Describe their magnitudes.

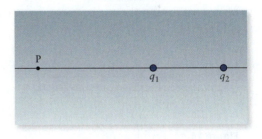

Figure 16-42 Problem 59

60. •Near the surface of Earth an electric field points radially downward and has a magnitude of approximately 100 N/C. What charge (magnitude and sign) would have to be placed on a penny that has a mass of 3 g to cause it to rise into the air with an upward acceleration of 0.19 m/s²?

61. ••Two charges are placed on the x axis, $+5 \ \mu C$ at the origin and $-10 \ \mu C$ at $x = 10$ cm. (a) Find the electric field on the x axis at $x = 6$ cm. (b) At what point(s) on the x axis is the electric field zero?

62. ••In **Figure 16-43**, the electric field at the origin is zero. If q_1 is 10^{-7} C, what is q_2?

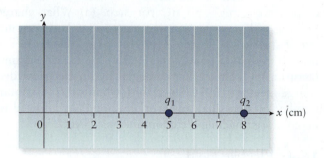

Figure 16-43 Problems 62, 63, and 64

63. •In **Figure 16-43**, if $q_1 = 10^{-7}$ C and $q_2 = 2 \times 10^{-7}$ C, (a) what is the electric field $\vec{E}$ at the point $(x, y) = (0$ m, 3 m)? (b) What is the force $\vec{F}$ acting on an electron at that position?

64. •In Figure 16-43, if $q_1 = 10^{-7}$ C and $q_2 = 2 \times 10^{-7}$ C, (a) what is the electric field $\vec{E}$ at the point $(x, y) = (6$ m, 3 m)? (b) What is the force $\vec{F}$ acting on a proton at that position?

65. ••In the Bohr model, the hydrogen atom consists of an electron in a circular orbit of radius $a_0 = 5.29 \times 10^{-11}$ m around the nucleus. Using the model, what is the speed of the electron?

16-5: Electric Field for Some Objects

66. ••**Calc** (a) What is the electric field at the point P beside the charged rod (length L, charge $+Q$) (**Figure 16-44**)? (b) If L equals 1.25 m, Q equals $+20$ mC, and D equals 3.25 m, determine the magnitude of the electric field at point P.

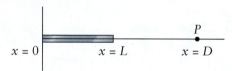

Figure 16-44 Problem 66

67. ••A rod has a length of 1.0 m and a positive charge of 1.0 C evenly distributed throughout (**Figure 16-45**). What is the electric field strength at point P, which is 0.50 m from the middle of the rod along a line bisecting the rod? SSM

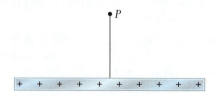

Figure 16-45 Problem 67

68. ••What is the electric field at the center of a circular ring that has a radius of 16 cm and a uniformly distributed charge of 0.5 mC?

69. ••A circular ring has a radius of 20 cm and a uniformly distributed charge of 0.5 μC. What is the magnitude of the electric field on the axis of the ring that is 10 cm from the center?

70. •••**Calc** Find the electric field at point P on the y axis, 3.00 m from the origin, due to the charge on a wire that runs along the x axis between 1.00 m and 3.00 m and which carries a uniform charge per unit length of 4.00 μC/m.

71. ••**Calc** The electric field at a point along the central axis of a ring of charge Q is

$$E_y = \frac{kQy}{(R^2 + y^2)^{3/2}}$$

(a) Determine the location y for the maximum value of E_y and (b) determine the value of the maximum electric field. SSM

16-6: Gauss' Law

72. •A rectangular area is rotated in a uniform electric field, from a position where the maximum electric flux goes through it, to an orientation where only half the flux goes through it. What is the angle of rotation?

73. •A point charge of 4.0×10^{-12} C is located at the center of a cubical Gaussian surface. What is the electric flux through each face of the cube?

74. •The net electric flux from a box that is a cube and has sides that are 20.0 cm long is 4.80×10^3 N·m²/C. What charge is enclosed by the box?

75. ••A 10.0-cm-long uniformly charged plastic rod is sealed inside a plastic bag. The net electric flux through the bag is 7.5×10^5 N·m²/C. What is the linear charge density on the rod? SSM

76. ••**Figure 16-46** shows a prism-shaped object that is 40.0 cm high, 30.0 cm deep, and 80.0 cm long. The prism is immersed in a uniform electric field of $(500$ N/C$)\hat{x}$. (a) Calculate the electric flux out of each of its five faces and (b) the net electric flux out of the entire closed surface. (c) If in addition to the given electric field the prism also enclosed a point charge of -2.00 μC, qualitatively how would your answers above change, if at all?

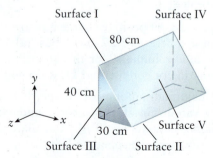

Figure 16-46 Problems 76 and 77

77. ••The prism in Figure 16-46 is now immersed in a uniform electric field of $\vec{E} = (500 \text{ N/C})\hat{x} + (400 \text{ N/C})\hat{y}$. Calculate (a) the electric flux out of each of its five faces and (b) the net electric flux out of the entire closed surface.

16-7: Applications of Gauss' Law

78. •Determine the charge density for each of the following cases (assume that all densities are uniform): (a) a solid cylinder that has a length L, has a radius R, and carries a charge Q throughout its volume, (b) a flat plate (very thin) that has a width W, has a length L, and carries a charge Q on its surface area, (c) a solid sphere of radius R carrying a charge Q throughout its volume, and (d) a hollow sphere of radius R carrying a charge Q over its surface area.

79. ••Use Gauss' law to find the electric field just outside the surface of a sphere carrying a uniform surface charge density σ (charge per unit area). SSM

80. ••Consider an infinite plane that carries a uniformly distributed charge Q. (a) Use Gauss' law to find the electric field due to the plane. (b) What field would be created by two equal but oppositely charged parallel planes? Consider the region between the planes as well as the two regions outside the planes. The simplest way to express your answers is in terms of the surface charge density σ (the charge per unit area) on the plane.

81. ••A very long, hollow, charged cylinder has an inner radius a equal to 3.00 cm, an outer radius b equal to 5.00 cm, and a uniform charge density ρ equal to $+42.0 \ \mu\text{C/m}^3$. Determine the electric field for all radii $r \leq a$, $a \leq r \leq b$, and $r \geq b$. SSM

82. •••Calc A sphere of radius R carries a charge distribution that varies with radius according to $\sigma = (-23.0 \ \mu\text{C/m}^4)r$. Derive an expression for the electric field at all radii from the center of the sphere.

83. ••An electric field of magnitude 400 N/C exists at all points just outside the surface of a 2.00-cm-diameter steel ball bearing. Assuming the ball bearing is in electrostatic equilibrium, (a) what is the total charge on the ball? (b) What is the surface charge density on the ball?

84. ••A $-3.20 \ \mu\text{C}$ charge sits in static equilibrium in the center of a conducting spherical shell that has an inner radius 2.50 cm and an outer radius 3.50 cm. The shell has a net charge of $-5.80 \ \mu\text{C}$. Determine the charge on each surface of the shell and the electric field just outside the shell.

85. ••Calc Using Gauss' law, calculate the electric field vector at the point P due to a long, straight rod that possesses a linear charge density of $+12 \ \mu\text{C/m}$ (**Figure 16-47**). The point P is 10 cm radially out from the central axis of the rod. SSM

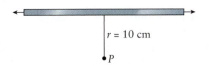

Figure 16-47 Problem 85

General Problems

86. ••A spherical birthday-party balloon that is 25 cm in diameter contains helium at room temperature (20°C) and at a pressure of 1.3 atm. If one electron could be stripped from every helium atom in the balloon and removed to a satellite orbiting Earth 22,000 mi (32,187 km) above the planet, with what force would the balloon and the satellite attract each other?

87. ••A plutonium-242 atom has a nucleus of 94 protons and 148 neutrons and has 94 electrons. The diameter of its nucleus is approximately 15×10^{-15} m. (a) Make a reasonable physical argument why we can treat the nucleus as a point charge for points outside of it. (b) Plutonium decays radioactively by emitting an *alpha particle* from its nucleus. The mass of the alpha particle is 6.6×10^{-27} kg and the particle has two protons and two neutrons. If the alpha particle comes from the surface of the ^{242}Pu nucleus, what is its greatest acceleration?

88. ••When a test charge of $+5$ nC is placed at a certain point, the force that acts on it has a magnitude of 0.08 N and is directed northeast. (a) If the test charge were -2 nC instead, what force would act on it? (b) What is the electric field at the point in question?

89. ••Biology As indicated in Example 16-2, a red blood cell typically carries an excess charge of about -2.5×10^{-12} C distributed uniformly over its surface. The cells, modeled as spheres, are approximately 7.5 μm in diameter and have a mass of 9.0×10^{-14} kg. (a) How many excess electrons does a typical red blood cell carry? (b) Does the mass of the extra electrons appreciably affect the mass of the cell? To find out, calculate the ratio of the mass of the extra electrons to the mass of the cell without the excess charge. (c) What is the surface charge density σ on the red blood cell? Express your answer in C/m^2 and in electrons/m^2. SSM

90. ••Biology Find the magnitude and direction of the electric field produced by a red blood cell at the following locations. Use the necessary quantities from Example 16-2 and the previous problem. (a) Inside the cell at a distance of 3.0 μm from the center. (b) Just inside the surface of the cell. (c) Just outside the surface of the cell. (d) At a point outside the cell 3.0 μm from its surface.

91. ••Two small spheres each have a mass m of 0.10 g and are suspended as pendulums by light insulating strings from a common point, as shown in **Figure 16-48**. The spheres are given the same electric charge, and the two come to equilibrium when each string is at an angle of $\theta = 3.0°$ with vertical. If each string is 1.0 m long, what is the magnitude of the charge on each sphere?

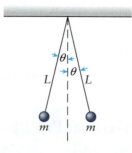

Figure 16-48 Problem 91

92. ••Three point charges are placed on the x–y plane: a +50.0 nC charge at the origin, a −50.0 nC charge on the x axis at 10.0 cm, and a +150 nC charge at the point (10.0 cm, 8.00 cm). (a) Find the total electric force on the +150 nC charge due to the other two. (b) What is the electric field at the location of the +150 nC charge due to the presence of the other two charges?

93. ••**Biology** The 9-inch-long elephant nose fish in the Congo River generates a weak electric field around its body using an organ in its tail. When small prey (or even potential mates) swim within a few feet of the fish, they perturb the electric field. The change in the field is picked up by electric sensor cells in the skin of the elephant nose. These remarkable fish can detect changes in the electric field as small as 3.0 μN/C. (a) How much charge (modeled as a point charge) in the fish would be needed to produce such a change in the electric field at a distance of 75 cm? (b) How many electrons would be required to create the charge? **SSM**

94. ••A small 1.0-g plastic ball that has a charge q of 1 C is suspended by a long string that has a length L of 1.0 m in a uniform electric field, as shown **Figure 16-49**. If the ball is in equilibrium when the string makes a 9.8° angle with the vertical as indicated by θ, what is the electric field strength?

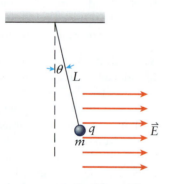

Figure 16-49 Problem 94

95. •••**Calc** A small 2.50-kg ball is attached to a thin flexible wire to make a simple pendulum. The pendulum is now taken for use in the orbiting space station. (a) What will be its period on the space station? (b) You want the pendulum to have the same period in the space station as on Earth, so you put a charge of −8.50 μC uniformly over its surface and place it in a uniform external electric field. How should the field be oriented relative to the pendulum—which way should it point—and what should be its magnitude? (c) In what significant way does your space-based pendulum differ from its behavior on Earth?

96. •••Three charges (q_A, q_B, and q_C) are placed at the vertices of the equilateral triangle that has sides of length s in **Figure 16-50**. Derive expressions for the electric field at (a) X (at the center of the triangle), (b) Y (at the midpoint of the side between q_B and q_C), and (c) Z (at the midpoint of the side between q_A and q_C). (d) Now use the following numerical values and calculate the electric field at those same points: $s = 10$ cm, $q_A = +20$ nC, $q_B = −8$ nC, and $q_C = −10$ nC.

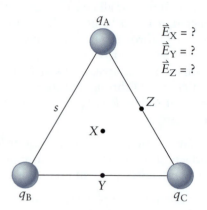

Figure 16-50 Problem 96

97. ••Calculate the electric field at the center of the hexagon shown in **Figure 16-51**. Assume the sides of the hexagon are all 5 cm long. **SSM**

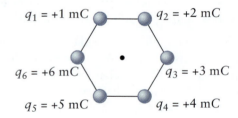

Figure 16-51 Problem 97

98. •••A nonconducting flexible ring is 4.50 cm in diameter and carries a charge distributed uniformly along it. At a point on the axis of the ring that is 1.50 cm from its center, the electric field is measured to be 6850 N/C pointing toward the ring. The ring is now carefully cut and straightened into a thin rod without disturbing the charge. Find the magnitude and direction of the electric field produced by the rod (a) 1.50 cm from one end of the rod along its axis and (b) 1.50 cm from the middle of the rod measured along a line perpendicular to the rod at its midpoint.

99. •••**Calc** Derive an expression for the electric field at the point C at the center of the circle that defines the circular arc of charge with a density of λ (**Figure 16-52**). The angle $ACB = \theta$ and $AC = BC = r$.

Figure 16-52 Problem 99

100. •••Calc Calculate the electric field at the point P, 10 cm above a ring of charge with a radius of 15 cm (**Figure 16-53**). Assume the point P is centered on the axis of the ring and there is a total of $-50\ \mu C$ spread uniformly on the ring.

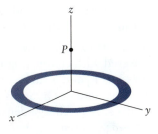

Figure 16-53 Problem 100

101. ••During electrical storms, lightning can strike from one part of a cloud to another. Measurements made within clouds during storms indicate that the maximum electric field strength is about $2.0 \times 10^5\ N/C$. (a) In one model, the cloud is viewed as two round, parallel, oppositely charged sheets that are 5.0 km in diameter and a distance of 6.0 km apart. If the field midway between the two sheets is equal to the maximum electric field, how much charge does each sheet carry, if they have equal magnitude charges? (b) In another model, the charges are viewed as equal but opposite point charges instead of sheets, separated by 6.0 km. If the field midway between the two charges is equal to the maximum electric field, what is the magnitude of the charges? **SSM**

102. ••Electric fields up to $2.00 \times 10^5\ N/C$ have been measured inside of clouds during electrical storms. Neglect the drag force due to the air in the cloud. (a) What acceleration does the maximum electric field produce for protons in the cloud? Express your answer in SI units and as a fraction of g. (b) If the electric field remains constant, how far will the proton have to travel to reach 10% of the speed of light if it started with negligible speed? (c) Can you neglect the effects of gravity? Explain your answer.

103. ••An electron with an initial speed of $5.0 \times 10^5\ m/s$ enters a region in which there is an electric field directed along its direction of motion. If the electron travels 5.0 cm in the field before being stopped, what are the magnitude and direction of the electric field?

104. ••An electron, released in a region where the electric field is uniform, is observed to have an acceleration of $3.00 \times 10^{14}\ m/s^2$ in the positive x direction. (a) Determine the electric field producing the acceleration. (b) Assuming the electron is released from rest, determine the time required for it to reach a speed of 11,200 m/s, the escape speed from Earth's surface.

105. ••Chemistry The iron atom (Fe) has 26 protons, 30 neutrons, and 26 electrons. The diameter of the atom is approximately 1.0×10^{-10} m, while the diameter of its nucleus is about 9.2×10^{-15} m. (You can reasonably model the nucleus as a uniform sphere of charge.) What are the magnitude and direction of the electric field that the nucleus produces (a) just outside the surface of the nucleus and (b) at the distance of the outermost electron? (c) What would

be the magnitude and direction of the acceleration of the outermost electron due only to the nucleus, neglecting any force due to the other electrons? **SSM**

106. •••An electron with kinetic energy K is traveling to the right along the $+x$ axis, which is along the axis of a cathode-ray tube as shown in **Figure 16-54**. There is an electric field $\vec{E} = (2.00 \times 10^4\ N/C)\hat{y}$ between the deflection plates, which are 0.06 m long and are separated by 0.02 m. Determine the minimum initial kinetic energy the electron can have and still avoid colliding with one of the plates.

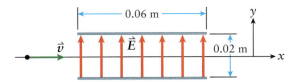

Figure 16-54 Problem 106

107. ••In the famous Millikan oil-drop experiment, tiny spherical droplets of oil are sprayed into a uniform vertical electric field. The drops get a very small charge (just a few electrons) due to friction with the atomizer as they are sprayed. The field is adjusted until the drop (which is viewed through a small telescope) is just balanced against gravity and therefore remains stationary. Using the measured value of the electric field, we can calculate the charge on the drop and from this calculate the charge e of the electron. In one apparatus, the drops are 1.10 μm in diameter and the oil has a density of $0.850\ g/cm^3$. (a) If the drops are negatively charged, which way should the electric field point to hold them stationary (up or down)? (b) Why? (c) If a certain drop contains four excess electrons, what magnitude electric field is needed to hold it stationary? (d) For another drop, you measure that the field needed to balance it is 5183 N/C. How many excess electrons are on the drop?

108. •••The electric field is zero everywhere except in the region $0 \le x \le 3.00$ cm, where there is a uniform electric field of 100 N/C in the $+y$ direction. A proton is moving along the $-x$ axis with a velocity $\vec{v} = (1.00 \times 10^6\ m/s)\hat{x}$. When the proton passes through the region $0 \le x \le 3.00$ cm, the electric field exerts a force on it. (a) When the x coordinate of the proton's position is 3.00 cm, what is its velocity and what is the y coordinate of its position? (b) When the x coordinate of its position equals 10.0 cm, what is its velocity and what is the y coordinate of its position?

109. ••A sphere of radius R has uniform volume charge density ρ. Consider the Gaussian surface shown in **Figure 16-55**, consisting of a circular disk and a hemisphere; the disk and the hemisphere have a radius r, are concentric with the charged sphere, and together form a closed surface. (a) What is the flux out of each segment of the

Gaussian surface? That is, what is the flux out of the disk and what is the flux out of the hemisphere? (b) Show that the net flux out of the Gaussian surface equals the charge inside it divided by ϵ_0. SSM

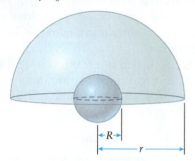

Figure 16-55 Problem 109

110. •••A line of charge with uniform linear charge density $\lambda = 4.00$ nC/m extends from $x = -2.00$ m to $x = +2.00$ m on the x axis. Evaluate your results to five significant figures. (a) What is the total charge Q on the line? (b) Estimate the electric field at point P on the y axis at 120 m by assuming the entire charge Q is located at the origin. (c) Compare your estimate with the actual electric field at P due to the line charge. (d) Estimate the electric field at point P' on the y axis at 2.00 cm by assuming the line charge is infinitely long with the same linear charge density. (e) Compare your estimate with the actual electric field at P' due to the line charge.

111. •••A semicircular ring of charge that has a radius R equal to 3 cm and is centered on the origin is situated as shown in (**Figure 16-56**). The charge is not uniform, but has a linear charge density that depends on the angle θ according to $\lambda = -(3\,\mu\text{C/m})\cos\theta$,

Figure 16-56 Problem 111

where θ is measured from the $+y$ axis as indicated. (a) Find the electric field due to the charge distribution at point P located at $y = -4$ cm. (b) If the semicircle were extended to be a complete ring, with the same function for the linear charge distribution, describe qualitatively how you would expect the electric field at point P to change, if at all.

112. •••Calc Determine the electric field at the points P, Q, and R. Point P is located inside the inner sphere of radius R_i, point Q is located in between the inner sphere and the outer spherical shell

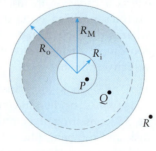

Figure 16-57 Problem 112

of inner radius R_M and outer radius R_o, and point R is located outside the spherical shell (**Figure 16-57**).

113. •••Two hollow, concentric, spherical shells are covered with charge (**Figure 16-58**). The inner sphere has a radius R_i and a surface charge density of $+\sigma_i$, while the outer sphere has a radius R_o and a surface charge density of $-\sigma_o$. Derive an expression for the electric field in the following three radial regions: (a) $r < R_i$, (b) $R_i < r < R_o$, and (c) $r > R_o$. (d) If $\sigma_i = +20\,\mu\text{C/cm}^2$, $\sigma_o = -14\,\mu\text{C/cm}^2$, $R_i = 5$ cm, and $R_o = 8$ cm, calculate the magnitude of the electric field in the three regions. SSM

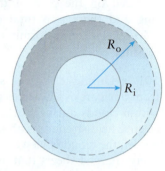

Figure 16-58 Problem 113

114. ••A charge $q_1 = +2q$ is at the origin and a charge $q_2 = -q$ is on the x axis at $x = a$. Find expressions for the total electric field on the x axis in each of the regions (a) $x < 0$; (b) $0 < x < a$; and (c) $a < x$. (d) Determine all points on the x axis where the electric field is zero. (e) Use a graphing calculator or spreadsheet to make a plot of E_x versus x for all points on the x axis, and (f) qualitatively discuss what happens for $-\infty < x < \infty$.

115. •••Calc A nonconducting spherical shell has an inner radius of R_i and an outer radius of R_o (**Figure 16-59**). There is no charge inside the inner cavity ($r < R_i$) and there is a nonuniform volume charge density between R_i and R_o given by $\rho(r) = \rho_0(R_i/r)$ where ρ_0 is a positive constant. Determine the electric field at the point P, for $r > R_o$.

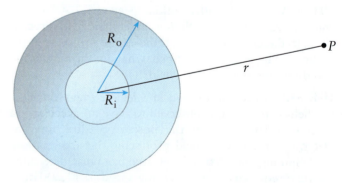

Figure 16-59 Problem 115

116. ••Calc A very long, solid cylinder possesses a volume charge density of ρ. The radius of the cylinder is R. Derive an expression for the electric field (a) at points located a distance $r < R$ from the axis of the cylinder and (b) at points located a distance $r > R$ from the axis of the cylinder. Assume that the length of the cylinder is much greater than r in both cases.

17 Electrostatics II

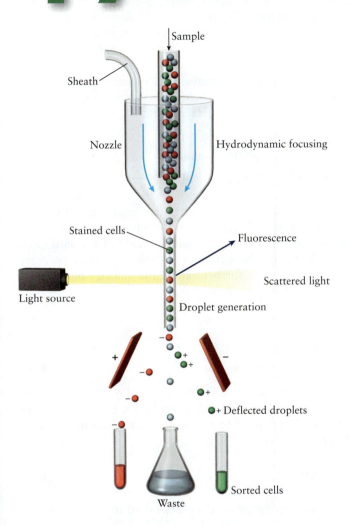

Sample

Sheath

Nozzle

Hydrodynamic focusing

Stained cells

Fluorescence

Scattered light

Light source

Droplet generation

+ −

−

+ +

+

+ Deflected droplets

Waste

Sorted cells

Fluorescence activated cell sorters rapidly separate cells of different types. Antibodies attached to fluorescent dyes only bind to specific cell types. As the cells that are suspended in a saline solution pass through the machine, it forms tiny drops of saline, each containing a single cell. The droplets then pass through a beam of light; those that fluoresce contain cells bound to the antibody. The cell sorter applies an electric charge to droplets containing a fluorescent cell. As the stream of droplets flows between two oppositely charged, high voltage plates, a strong electric field created by the plates deflects the charged droplets, containing the cells of interest, away from the path followed by uncharged droplets which contain the unlabeled cells. Applications for the resulting purified populations of cells vary from diagnosing and treating disease to engineering new proteins in animal and microbial cells.

Cardiologists assess cardiac health by studying electrical signals that result from charges spreading across the heart, causing it to contract. Similar processes occur in neurons and muscle cells. The energy driving such electrical messages is partially stored in the electric field created by charge accumulated across membranes. The electric potential energy is analogous to gravitational potential energy and spring potential energy that we have already discussed. We will explore electric potential energy, and related quantities, in this chapter.

17-1 Electric Potential

The change in an object's potential energy ΔU is directly related to the work done on the object. Using Equation 6-23,

$$\Delta U = -\int_a^b \vec{F} \cdot d\vec{s}$$

$\sqrt{x}$ *See the Math Tutorial for more information on Integrals*

where the integral, from Equation 6-17, is work:

$$W = \int_a^b \vec{F} \cdot d\vec{s} \tag{17-1}$$

We wrote these last two equations in a form slightly more general than in Chapter 6, by using the variable s as a general variable with which to describe a path. The magnitude of the $\vec{s}$ vector is distance and the direction of the vector describes the direction of travel along the path at any point. The $d\vec{s}$ vector is then an infinitesimally small piece of the path from a to b.

In Chapter 6, we saw that an object's potential energy increases when work is done to lift the object against a gravitational pull. Setting the force in Equation 6-23 equal to mg, the force of gravity near Earth's surface, the change in gravitational potential energy near the surface is

$$\Delta U_{\text{gravity}} = mgy_b - mgy_a$$

where y_a and y_b are two different heights along the object's path.

In the last chapter, we noticed that the magnitude of the electric force F on a charge q in the presence of an electric field E, $F = qE$, is analogous to the gravitational force on an object on the surface of Earth, $F = mg$. Both forces are conservative; in other words, the work done by either one is independent of the path along which the force is applied. Therefore, the definition of the change in potential energy applies (Equation 6-23), so we can further extend the analogy between the two forces by defining the change in **electric potential energy**:

$$\Delta U_{\text{electric}} = q_0 E x_b - q_0 E x_a$$

or

$$\Delta U_{\text{electric}} = q_0 E \Delta x \tag{17-2}$$

▶   *Go to Picture It 17-1 for more practice dealing with electric potential energy*

where a charge q_0 moves a distance Δx under the influence of a constant electric field E. Although this definition is useful, a related quantity, the change in electric potential energy divided by the charge used to test the field, is often more useful. The **electric potential difference** ΔV is

$$\Delta V = \frac{\Delta U_{\text{electric}}}{q_0} = E \Delta x \tag{17-3}$$

Electric potential difference does not depend on the charge used to test the strength of the field.

It is common to label electric potential as V rather than ΔV. However, electric potential is nevertheless the difference between a quantity evaluated at two different positions. You might recall that in Chapter 6 we noted that although the change in potential energy is physically more meaningful than potential energy itself, we often refer to "potential energy" as a shorthand, always remembering that it must be defined relative to a known reference. The same is true for electric potential difference. For this reason, we will also refer to this quantity as *electric potential*. Moreover, when the context makes it clear that "electric" is implied, we will sometimes refer to the quantity V as potential. It is also common and therefore convenient to refer to electric potential difference as **voltage**.

The SI unit of electric potential is the *volt* (V), named after the Italian scientist Alessandro Volta. Yes, the units are the same ones that you've learned to match with the batteries that you put into a flashlight, a wireless keyboard, or the controller for your gaming system. Volta invented the battery in 1800.

Notice that for a constant electric field the electric potential V depends only on the strength of the field E and the distance Δx over which it extends. That's why electric potential is a powerful concept: Knowing how the electric potential varies over a region of space enables us to determine how any charged object will move in that region, regardless of the magnitude of its charge.

 Watch Out

Electric potential and electric potential energy are not the same.

The quantities electric potential V and electric potential energy U are related but different. Both address the question of how much work is required to move an object from one place to another within a field. However, while the change in electric potential energy depends directly on the charge of the object placed in the electric field, this charge (labeled q_0 in Equation 17-2) does not appear in the final relationship for electric potential (the term on the right side of Equation 17-3). Consider the equation for gravitational potential energy. When an object of mass m is raised to height h above (and close to) Earth's surface, the potential energy increases by $\Delta U = mgh$. This change in potential energy depends on the mass of the object. We know, however, that a description of Earth's gravitational field does not depend on the mass of an object with which it interacts. For that reason, we could use a relationship similar to Equation 17-3 to define gravitational potential energy, that is,

$$V = \frac{\Delta U}{m} = gh$$

This description does not depend on the object placed in the field; it describes the magnitude of the gravitational field itself. Similarly, the electric potential V describes the strength of the electric field itself, independent of the magnitude of any test charge with which it interacts. Perhaps this isn't surprising, because the electric potential depends on the electric field, which itself is independent of the test charge used to measure it.

Remember that potential energy in general can be described in terms of work. During our discussion of gravitational forces, for example, we found that the change in gravitational potential energy of an object is equal in magnitude to the work done by the force acting on it:

$$\Delta U_{gravity} = -W_{gravity} \tag{6-22}$$

We should therefore be able to express electric potential energy in terms of work too, that is,

$$\Delta U_{electric} = -W_{electric}$$

Furthermore, we can also express the electric potential in terms of the work done by an electric field on a charged object:

$$V_a - V_b = \frac{-W_{a \to b}}{q_0}$$

Here $W_{a \to b}$ is the work done by the electric field as a charge moves from one position to another. Expressed as the difference in electric potential at point a and the electric potential at point b, this relationship reinforces the notion that electric potential is the difference between a quantity evaluated at two points. In the same way as for potential energy, however, we will write electric potential $V_a - V_b$ as V, so

$$V = \frac{-W_{a \to b}}{q_0} \tag{17-4}$$

Figure 17-1 Electric field is a vector. The field lines point in the direction that a positive test charge would feel an electric force. (a) We must do work on a positive test charge in order to move it closer to $+Q$; the electric potential difference between points a and b is positive. (b) The field does work on a positive test charge in order to move it closer to $-Q$; the electric potential difference between points a and b is negative.

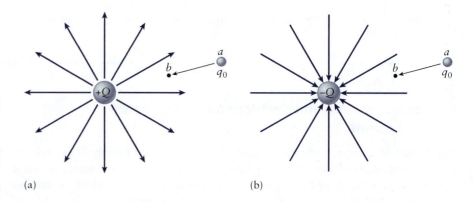

(a)   (b)

Electric potential is not a vector, although it can have both positive and negative values. The sign of the potential indicates whether the field will do positive or negative work on a charge. By convention, we describe an electric field according to how it affects a positive test charge. Imagine, then, that we try to move a positive test charge q_0 from position a to position b in the field generated by $+Q$, as shown in **Figure 17-1a**. As the lines of field suggest, a force pushes q_0 away from $+Q$. To move q_0 closer to $+Q$ (from a to b) requires us to do work on the charge, a positive quantity, which means that the work done by the field on q_0, $W_{a \to b}$, is a negative quantity. The resulting potential between a and b, from Equation 17-4, is therefore positive. Conversely, the force due to $-Q$ pulls a test charge q_0 toward it, as in **Figure 17-1b**. In this case, the work that the field does on q_0 is positive ($W_{a \to b}$ is positive) and the potential difference between a and b is negative.

We have previously defined the SI units of electric field, from $F = qE$, to be newtons per meter (N/m). As suggested by rearranging Equation 17-3 for E ($E = V/\Delta x$), an equally good set of units for electric field is volts per meter (V/m).

❗ Watch Out

● **Electric potential is not a force.**

Electric potential is not a force. It doesn't have units of force and doesn't obey Newton's laws. However, the potential difference between two points in space is a useful way to characterize the force that causes a charge to accelerate in the presence of an electric field. In the next chapter, for example, we'll see that a source of voltage, such as a battery, drives electrons through an electronic circuit.

❗ Watch Out

● **Electric potential is not a vector quantity.**

Although electric potential can be either negative or positive, there is no direction associated with it. Electric potential is not, therefore, a vector.

Equation 17-3 only applies when the electric field is constant. We can generalize Equation 17-3 for cases in which the field is not constant by recalling the general relationship between work, force, and displacement. Note that the net work done does not depend on the specific path chosen from a to b. (Look back at

Section 6-7.) Using the definition of work presented in Equation 17-1 and the relationship between potential energy and work, Equation 17-4 becomes

$$V = \frac{-\int_a^b \vec{F} \cdot d\vec{s}}{q_0}$$

or

$$V = \frac{-\int_a^b q_0 \vec{E} \cdot d\vec{s}}{q_0}$$

A more general form of Equation 17-3 which applies to a nonuniform $\vec{E}$ or a path over which the angle between $\vec{E}$ and the direction of the path changes from point to point is

$$V = -\int_a^b \vec{E} \cdot d\vec{s} \qquad (17\text{-}5)$$

Notice that although the integral involves vector quantities, the result of a dot product is a scalar not a vector. This equation is consistent with our earlier claim that potential is not a vector.

The dot product in Equation 17-5 can be expanded to show explicitly the dependence on the angle θ between $\vec{E}$ and $d\vec{s}$ in determining V. At any point along the path,

$$V = -\int_a^b E ds \cos \theta \qquad (17\text{-}6)$$

$\sqrt{x}$ See the Math Tutorial for more information on Trigonometry

Figure 17-2 shows an arbitrary path between two points in a nonuniform electric field. Notice that both $\vec{E}$ and $d\vec{s}$, as well as the angle between them, all change along the path. To determine V in this case requires the integral form of the definition of electric potential.

The electric field vectors and the path differential are shown at two points along a path from a to b in an electric field.

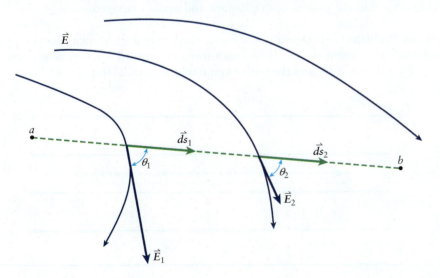

Figure 17-2 Both $\vec{E}$ and $d\vec{s}$, as well as the angle between them, change along an arbitrary path in a nonuniform electric field.

Example 17-1 Potential in a Uniform Field

Determine the electric potential difference between a and b in the uniform electric field $\vec{E}$ as shown in **Figure 17-3**. The points are separated by a distance D, and the path from a to b is antiparallel to the field lines.

SET UP

We will apply Equation 17-6:

$$V = -\int_a^b E ds \cos \theta$$

Go to Picture It 17-2 for more practice dealing with electric potential

to the path shown in the figure. To determine the potential, notice that the path from a to b is antiparallel to the field lines; that is, $\vec{E}$ points from left to right while the path runs from right to left. In other words, the angle θ between $\vec{E}$ and $d\vec{s}$ is 180°. In addition, because the electric field is constant everywhere, it is certainly constant along the path of integration; E can therefore be taken out of the integral in Equation 17-6.

SOLVE

Taking E out of the integral and setting $\cos \theta = \cos 180° = -1$ gives

$$V = E \int_a^b ds$$

The integral $\int ds$ is equal to s, the path length D, so,

$$V = Es \Big|_a^b = ED$$

The length of the path from a to b is the distance D.

REFLECT

V depends both on the electric field and also on the region over which the field extends. For an electric field of a given strength, the electric potential associated with it is larger the longer the distance over which the field can be maintained.

It's also worth noting that although we could have selected any path and obtained the same result for V, the path from a to b as shown is perhaps the best choice, because it is the most straightforward. The electric field does not vary along the path, so E can be taken out of the integral, and the angle between the electric field vector and the path is both constant and easily managed.

Practice Problem 17-1 Determine the electric potential difference between points b and a in the uniform electric field $\vec{E}$ as shown in Figure 17-3. The points are separated by a distance D, and the path from b to a is parallel to the field lines.

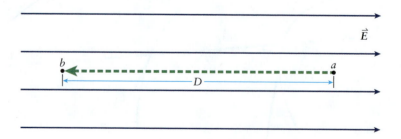

Figure 17-3 What is the electric potential difference between points a and b?

Example 17-2 Potential in a Uniform Field, Again

Determine the electric potential difference between a and c in the uniform electric field $\vec{E}$ as shown in **Figure 17-4**. The points are separated by a distance D along the direction of the field (b to a), and the path from a to c makes angle φ to the field lines as measured from the direction in which the field points.

SET UP

By Equation 17-6,

$$V = -\int_a^c Eds \cos\varphi$$

The electric field is uniform, so E can be taken out of the integral as in the previous example. The angle φ between $\vec{E}$ and $d\vec{s}$ is constant in this problem, so $\cos\varphi$ is also constant and can be taken out of the integral. The electric potential is therefore

$$V = -E\cos\varphi \int_a^c ds \qquad (17\text{-}7)$$

 See the Math Tutorial for more information on Trigonometry

SOLVE

The integral $\int_a^c ds$ is equal to the length of the path from a to c, which we can find from the triangle formed by points a, b, and c. In the triangle, $\cos(180° - \varphi)$ is equal to $D/$(path length) and $\cos(180° - \varphi)$ is equal to $-\cos\varphi$, so

$$-\cos\varphi = \frac{D}{\text{path length}}$$

and

$$\int_a^c ds = \text{path length} = \frac{-D}{\cos\varphi}$$

Equation 17-7 becomes

$$V = -E\cos\varphi \frac{-D}{\cos\varphi} = ED$$

REFLECT

Notice that the potential difference between a and c is the same as the potential difference between a and b (from the previous example). The dot product in Equation 17-5 ensures that potential difference depends only on displacement along the direction of the electric field. This result is similar to the conclusion we drew for gravitational potential energy near the surface of Earth. In Section 6-7, for example, we saw that the change in gravitational potential energy of two mountain climbers depended only on the difference between their initial and final elevations.

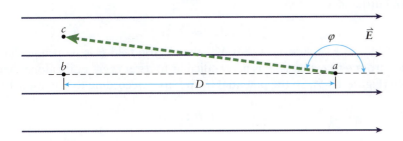

Figure 17-4 What is the electric potential difference between points a and c?

We could represent the two climbers coming down the mountain by rotating Figure 17-4 counterclockwise by 90°; the lines of $\vec{E}$ could easily be lines of gravitational field, and the paths a to b and a to c represent two paths down the mountain. The change in potential energy, and the difference in potential, depends only on the distance in the direction of the field. Said another way, points b and c are at the same potential in this field.

Practice Problem 17-2　Draw a copy of Figure 17-4, but place point c directly above point a. Determine the electric potential difference between b and c in the uniform electric field $\vec{E}$. Points b and a are separated by a distance D along the direction of the field, and the path from b to c makes angle $(180° - \varphi)$ to the field lines as measured from the direction in which the field points.

✳ What's Important 17-1

The electric potential depends only on the constant strength of the electric field and the distance over which it extends. Electric potential difference does not depend on the charge used to test the strength of the field. We describe an electric field according to how it would affect a positive test charge.

17-2 Equipotential Surfaces

Look back at the two examples in the previous section. Although the electric field is the same in both, the distance from a to c in the second example is clearly longer than the distance from a to b in the first. The potential difference depends in part on the distance over which an electric field extends, yet the potential difference between points a and b in the first example and a and c in the second is the same. Should we have expected that result?

The definition of potential difference,

$$V = -\int_a^b Eds \cos \theta \tag{17-6}$$

depends on an integration, the summing up of the contribution at each point along the path from a to b of the product of the electric field at that point with the small piece of the length of the path. We are free to choose any path, and to break up the path in any convenient way. If the path can be divided into sections for which the magnitude of E and the angle θ between $\vec{E}$ and $d\vec{s}$ are both constant, then

$$V = V_1 + V_2 + V_3 + \cdots$$

Here V_1, V_2, and so on represent the potential difference between the end points of each infinitesimal section of the path. (Although each section of the path is infinitesimally small, just as each one has a length, which is ds, it also has two ends!) As a result, the potential between a and c in Figure 17-4 can be determined by following a path that goes first from a to b and then from b to c, as in **Figure 17-5**. The potential difference is then

$$V_{ac} = V_{ab} + V_{bc}$$

We found V_{ab} in Example 17-1. For V_{bc}, the electric field $\vec{E}$ is perpendicular to the direction from b to c; that is, $\vec{E}$ is perpendicular to $d\vec{s}$ at every point along that leg of the path (Figure 17-5). So the potential from b to c, according to Equation 17-6, is

$$\dot{V}_{bc} = -\int_b^c Eds \cos 90° = 0$$

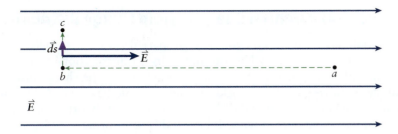

Figure 17-5 Electric potential difference is independent of the path taken from one point to another. The electric potential difference between points a and c can be determined by following a path directly from a to c, or by following a path from a to b and then from b to c. Because $\vec{E}$ is perpendicular to $d\vec{s}$ along the path from b to c, the contribution to the electric potential difference along that path is zero. The blue arrows represent equipotential lines.

The potential difference between a and c is then

$$V_{ac} = V_{ab} + 0$$

The potential between a and c is indeed the same as the potential between a and b.

? Got the Concept 17-1
Equal Potential Differences

In Chapter 6, we compared the work done by two mountain climbers taking different paths to the top of a mountain and found the work done on and by each is the same. Can you relate that result with the conclusion above that $V_{ac} = V_{ab}$ in Figure 17-5?

Because the path from b to c in Figure 17-5 is perpendicular to the electric field, moving a test charge from b to c requires zero work. The potential difference between b and c is therefore zero as well, that is, the potential is the same everywhere along the path from b to c. The path is an *equipotential line, equipotential curve,* or simply, an **equipotential**. Any path that is perpendicular to the lines of field is an equipotential. A surface that is perpendicular to a field everywhere on it is an **equipotential surface**. There is no cost (in energy—the currency of physics) to move a charge from one end of an equipotential path to the other.

? Got the Concept 17-2
Shape of Equipotential Surfaces

The lines of electric field due to a point charge are radial, much like the spines on a sea urchin (Figure 16-10). What is the shape of the equipotential surfaces around a point charge?

✴ What's Important 17-2
The potential is the same everywhere along an equipotential line, sometimes referred to as an equipotential curve or, simply, an equipotential. Any path that is perpendicular to the lines of field is an equipotential. A surface that is perpendicular to a field everywhere on it is an equipotential surface.

17-3 Electrical Potential Due to Certain Charge Distributions

The potential difference between two points in space depends on the electric field in that region (Equation 17-5) which is determined by the distribution of charge, so the electric potential also depends on the shape and extent of the charge distribution. For example, because the electric field around a dipole has a markedly different shape than the field near a charged disk, we should expect that the potential difference between two points near each will have a different form. We start this section by finding the potential difference in the region around a point charge, and then examine the electric potential that results from more complex charge distributions.

In **Figure 17-6**, points a and b are a distance r_a and r_b, respectively, from a point charge $+Q$. To determine the electric potential, we apply

$$V = -\int_a^b \vec{E} \cdot d\vec{s} \qquad (17\text{-}5)$$

along any path from a to b. As shown in the figure, we choose a path which first follows a circular arc from a to an intermediate point c, and then a radial path from c to b. Because $\vec{E}$ is radial, too, $d\vec{s}$ is perpendicular to the field everywhere along the arc from a to c, and antiparallel to the field along the path from c to b. We can add up the contributions to the potential difference in any grouping we like, so the potential difference between a and b is

$$V_{ab} = -\left[\int_a^c \vec{E} \cdot d\vec{s} + \int_c^b \vec{E} \cdot d\vec{s} \right]$$

or

$$V_{ab} = -\left[\int_{r_a}^{r_c} Eds \cos 90° + \int_{r_c}^{r_b} Eds \cos 180° \right]$$

where r_c is equal to r_a, because both a and c are the same distance from the point charge. We replace ds with $-dr$ because the direction of the path from c to b, which is the direction of $d\vec{s}$, is radially inward, so $ds \cos 180°$ equals dr. Then

$$V_{ab} = -\int_{r_c}^{r_b} Edr \qquad (17\text{-}8)$$

We know from Equation 16-4 that the magnitude of the electric field due to the point charge is

$$E = \frac{kQ}{r^2}$$

Figure 17-6 The electric potential difference between points a and b near a point charge $+Q$ is found by following a circular path from a to c and then a radial path from c to b. Because the path from a to c lies on an equipotential surface, the contribution to the electric potential difference along that path is zero.

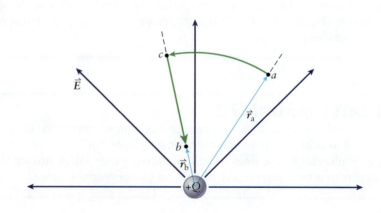

Therefore, Equation 17-8 becomes

$$V_{ab} = -\int_{r_a}^{r_b} \frac{kQ}{r^2}dr = -kQ\left(-\frac{1}{r}\Big|_{r_a}^{r_b}\right)$$

or

$$V_{ab} = kQ\left(\frac{1}{r_b} - \frac{1}{r_a}\right) \tag{17-9}$$

As we expect, the potential difference between a and b depends on the distance of a and b from the charge.

Recall that when considering potential energy, the physically meaningful quantity is the difference in potential energy from one point in space to another. Electric potential is related to electric potential energy, so it is the difference in potential that determines how charges behave in an electric field. We are free, however, to select any reference point that is convenient for a given problem. For example, we found that when considering gravitational potential energy as it applies to objects that have large masses ($U = -Gm_1m_2/r$, Equation 10-8), it is both convenient and conventional to select r equal to ∞ as the reference, because U equals zero there. The same approach is the standard for electrical potential, too. Using the choice $r_a = \infty$, Equation 17-9 becomes

$$V_{\infty b} = kQ\left(\frac{1}{r_b} - \frac{1}{\infty}\right)$$

The second term in the parentheses is equal to zero, so we write, in general,

$$V(r) = \frac{kQ}{r} \tag{17-10}$$

This equation is the electric potential at a distance r from a point charge $+Q$. Again, potential *difference* is the physically meaningful quantity, so it is important to keep in mind that implicit in Equation 17-10 is the choice of infinity as the reference point. So even though the value of distance equal to infinity does not appear in Equation 17-10, this equation should be read as the difference between the potential at an infinite distance from a point charge and a distance r away. The electric potential due to a point charge decreases as distance from the charge increases.

To determine the electric potential due to a charged object that is larger or more complex in shape than a point charge, we imagine breaking it into an infinite number of infinitesimally small pieces—point charges—and then adding the contribution each piece makes to the potential. We used this approach in Chapter 8 to find the moment of inertia for a rotating object, once we discovered the moment of inertia of a rotating particle. In the last chapter, we found the electric field generated by a charged object that is not a point charge by treating it as an assembly of an infinite number of point charges and adding the field due to each point charge. We will do the same here to determine the electric potential due to a charged ring and a charged disk.

Electrical potential is a scalar not a vector. For that reason, no direction is associated with potential and no vector considerations are required when adding contributions to the potential from elements of a charge distribution. This is not the case when finding electric field! Electric field *is* a vector, so when finding the field due to, for example, a charged ring or a charged disk, the direction of the field due to each infinitesimal point charge element must be taken into account.

Electric Potential along the Central Axis of a Charged Ring

Let's find the electric potential at a distance y along the central axis of a thin ring of radius R and uniformly distributed charge $+Q$, as shown in **Figure 17-7a**. We

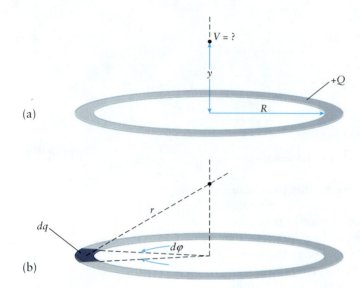

(a)

(b)

Figure 17-7 (a) What is the electric potential at a point a distance y above a thin ring of radius R and carrying charge $+Q$, along the central axis of the ring? (b) The electric potential is the sum of the contribution dV to the total electric potential made by each infinitesimal charge element dq of the ring.

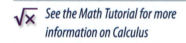 *See the Math Tutorial for more information on Calculus*

imagine the ring divided into an infinite number of point charges dq, such as the one shown in **Figure 17-7b**, each of which contributes to the potential at any point in space. We exaggerated the thickness of the ring and the size of dq to make the figure clearer.

Each charge element of the ring results in a contribution dV to the potential at the point of interest. The total potential is then

$$V = \int_{\text{ring}} dV$$

where "ring" indicates that we will integrate around the ring to include the contribution from every possible charge element dq. The potential due to one slice of the ring is, from Equation 17-10,

$$dV = \frac{k\,dq}{r} \tag{17-11}$$

so

$$V = \int_{\text{ring}} \frac{k\,dq}{r}$$

Because the charge on the ring is uniformly distributed, dq can be expressed as a fraction of the total charge Q. As shown in Figure 17-7b, each dq is of angular extent $d\varphi$ and so accounts for the fraction $d\varphi/2\pi$ of the total charge. We can therefore rewrite dq in terms of $d\varphi$:

$$dq = \frac{d\varphi}{2\pi}Q$$

We could also express this relationship by noting the charge of the slice of the ring is in the same proportion to the total charge as the angular extent of the piece is to 2π:

$$\frac{dq}{Q} = \frac{d\varphi}{2\pi}$$

To apply bounds to the integral, we substitute for dq and then sum the contributions of each charge element around the ring

$$V = \int_0^{2\pi} \frac{k\,d\varphi}{r\,2\pi}Q \tag{17-12}$$

For a fixed vertical height above the plane of the ring, r does not change as we integrate around the ring. (That is, r, which equals $\sqrt{R^2 + y^2}$, is constant.) We are therefore allowed to pull r, as well as the other constants k, Q, and 2π, from the integral:

$$V = \frac{kQ}{2\pi r}\int_0^{2\pi} d\varphi$$

Finally, $\int_0^{2\pi} d\varphi = \varphi\Big|_0^{2\pi} = 2\pi$, so

$$V = \frac{kQ}{2\pi r}2\pi = \frac{kQ}{r} = \frac{kQ}{\sqrt{R^2 + y^2}} \tag{17-13}$$

Compare the derivation of the electric potential due to a charged ring (along the central axis) with the derivation of the electric field of a ring in the previous chapter. Note the similarities, for example, the process of breaking the ring into small pieces and then adding the contribution each piece makes to the field and potential. The notable difference between our treatment of the field and the potential is that because electric field is a vector quantity but electric potential is not, no aspects of vectors or vector components appear in the derivation of the electric potential.

 Watch Out

The electric potential at the center of a charged ring is not zero.

From Chapter 16, you know that the electric field due to a charged ring is zero at the center of the ring; you might guess that the electric potential is zero there as well, but electric potential is not zero. The *field* is zero at the center because the electric field due to every small charge element is canceled by the field due to the charge element that is directly across the ring. This cancellation occurs because electric field is a vector; the field vectors due to each pair of charge elements point in opposite directions and therefore cancel. However, electric potential is not a vector, so no such cancellation occurs. As you can verify from Equation 17-13, the potential at the center of the charged ring, where $y = 0$, is

$$V = \frac{kQ}{\sqrt{R^2 + 0^2}} = \frac{kQ}{R}$$

Electric Potential along the Central Axis of a Charged Disk

We now consider the electric potential a distance y along the central axis of a thin disk of radius R and charge $+Q$ (**Figure 17-8**). We consider the disk to be so thin that it can be treated as two-dimensional, that is, as having surface area but no discernible thickness. As we did for the ring, we find the potential of the disk by dividing it into an infinite number of infinitesimally small pieces and adding up the contribution each makes to the total potential at any point in space. The contributions dV are described by

$$dV = \frac{kdq}{r} \tag{17-11}$$

where dq is the charge of any small piece of the disk. We require that dq be so small that the distance between the point of interest and any point on (or in) dq is the same. As we found when considering the electric field of a charged disk in Chapter 16, it is convenient to select an annulus of radius r and width dr to be dq. This selection not only satisfies the distance requirement, but has the advantage that we already know the contribution each annulus makes to the total potential because each one is effectively a ring.

To find the total electric potential, we add up dV due to each ring-shaped charged annulus. Using Equation 17-13, the magnitude of the electric potential due to each dq is

$$dV = \frac{kdq}{\sqrt{r^2 + y^2}}$$

We replaced Q in Equation 17-13 by dq, the charge on one infinitesimally small annulus. We also replaced R, which in the first equation is the radius of the ring, with the variable r, the (changing)

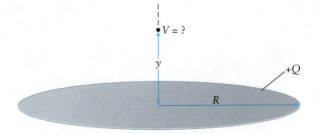

Figure 17-8 What is the electric potential at a point a distance y above a thin, uniformly charged disk of radius R and carrying charge $+Q$, along the central axis of the disk?

radius of each of the infinitesimal rings that make up the disk. Adding the contributions each dq makes to the total potential therefore takes the form

$$V = \int dV = \int \frac{k\,dq}{\sqrt{r^2 + y^2}} \tag{17-14}$$

The charge is uniformly distributed on the disk, so each dq carries the same fraction of the total charge. That fraction is the ratio of the area of dq divided by the total area of the disk:

$$dq = \frac{dA_{dq}}{A_{\text{disk}}} Q \tag{16-21}$$

The area of the annulus is $dA_{dq} = 2\pi r\,dr$ (Equation 16-22) and the area of the disk is $A_{\text{disk}} = \pi R^2$, so dq becomes

$$dq = \frac{2\pi r\,dr}{\pi R^2} Q$$

Equation 17-14 is then

$$V = \int dV = \int_0^R \frac{k}{\sqrt{r^2 + y^2}} \frac{2\pi r\,dr}{\pi R^2} Q \tag{17-15}$$

The integral is over dr; to include the contribution from every annulus that makes up the disk, we integrate from $r = 0$ to $r = R$. Thus (see Math Box 17-1 for more details),

$$V = \frac{2kQ}{R^2} \int_0^R \frac{r\,dr}{\sqrt{r^2 + y^2}} = \frac{2kQ}{R^2} \sqrt{r^2 + y^2}\,\Big|_0^R$$

or

$$V = \frac{2kQ}{R^2}\left(\sqrt{R^2 + y^2} - y\right) \tag{17-16}$$

Math Box 17-1 Computing Integrals by Substitution

Equation 17-15 becomes

$$V = \frac{2kQ}{R^2} \int_0^R \frac{r\,dr}{\sqrt{r^2 + y^2}} \tag{17-17}$$

The integral can be transformed into a straightforward form by defining u such that

$$u = \sqrt{r^2 + y^2} = (r^2 + y^2)^{1/2}$$

As you can verify from the Math Tutorial, when $u = f(x)^n$ the derivative with respect to x is

$$\frac{du}{dx} = (n)f(x)^{n-1}\frac{df}{dx}$$

So in our example

$$\frac{du}{dr} = \frac{1}{2}(r^2 + y^2)^{-1/2}(2r)$$

Thus

$$du = \frac{r\,dr}{\sqrt{r^2 + y^2}}$$

The integral in Equation 17-17 is seen to be of the form $\int du$, and $\int du = u$. So with $u = \sqrt{r^2 + y^2}$, Equation 17-17 becomes

$$V = \frac{2kQ}{R^2}\sqrt{r^2 + y^2}\bigg|_0^R$$

or

$$V = \frac{2kQ}{R^2}\left(\sqrt{R^2 + y^2} - y\right)$$

Got the Concept 17-3
Electric Potential

Does the result for the electric potential due to a charged disk (Equation 17-16) make sense in the limit where the distance from the plane of the disk is large, that is, when $y \gg R$?

What's Important 17-3

Electric potential is a scalar, not a vector, quantity. The electric potential due to a point charge decreases as distance from the charge increases.

17-4 Capacitance

The surface of every cell is a *membrane* composed of a phospholipid bilayer that separates the intracellular and extracellular fluids. Because the membrane is thin, negative charge that accumulates on its intracellular surface attracts positive charge on the extracellular surface. Excess charge stored on the cell's surfaces results in a potential difference across the membrane; the membrane potential is the source of the electrical signal neurons use to code, process, and transmit information.

The amount of charge that can be stored on the surfaces of a cell membrane depends on the thickness and surface area of the membrane. Axons, thin extensions of neurons, carry electrical signals called *action potentials* from one cell to another. Specialized support cells tightly wrap many additional layers of membrane (myelin) around some axons, increasing the distance between the intracellular surface of the membrane and the extracellular fluid. Myelin decreases the ability of the axons to store charge and is a primary reason that action potentials propagate faster in axons with myelin than in those without it.

A system that consists of two layers, or plates, that can store charge is called a **capacitor**. Stretching a spring increases its potential energy, and that stored energy can later be converted to motion—kinetic energy. In much the same way, the charge stored on the plates of a capacitor gives rise to a potential difference between the plates, and the associated potential energy can later be converted to the movement of charges. Cell membranes, for example, behave like a capacitor. During an action potential, ions move across the membrane partially as a result of the potential that results from stored charge on the inner and outer membrane surfaces. (The motion of ions is also a result of a difference in the ion concentrations on the two sides of the membrane.)

Figure 17-9a shows two large, parallel, and charged plates; one plate carries excess charge $+q$, the other plate carries excess charge $-q$. Close to a large charged

Two large plates (viewed from the side) carry charge $+q$ and $-q$. Close to the plates the electric fields are nearly constant, so we represent them by straight, parallel field lines.

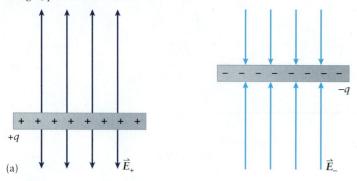

(a)

Because the fields are nearly constant, the magnitude of the field due to each plate is the same at any point near the plates...

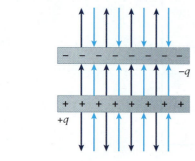

(b)

... so when the plates are placed close together, the fields cancel in the region outside the plates. The fields add in between the plates.

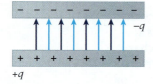

(c)

Figure 17-9 (a) Far from the edges, the electric field due to a large, charged plate is uniform, pointing away from a positively charged plate (on the left) and toward a negatively charged plate (on the right). (b) The plates are parallel and have been brought close together. (c) The electric fields outside the plates cancel.

disk and far from its edges the electric field is constant (Equation 16-26), so we have drawn the fields $\vec{E}_+$ and $\vec{E}_-$ as straight, parallel arrows. The plates have been brought close together in **Figure 17-9b**; notice that because the magnitude of the fields does not depend on distance from a plate, $\vec{E}_+$ cancels $\vec{E}_-$ in the region outside the plates (**Figure 17-9c**). In addition, the field between the plates is the sum of $\vec{E}_+$ and $\vec{E}_-$:

$$\vec{E} = \vec{E}_+ + \vec{E}_-$$

$\vec{E}_+$ and $\vec{E}_-$ point in the same direction, so from Equation 16-26, the magnitude of the field between the plates of the capacitor is

$$E = 2\pi k \frac{q}{A} + 2\pi k \frac{q}{A} = 4\pi k \frac{q}{A}$$

where A is the area of each plate. We note that both this statement, and the claim that $\vec{E}_+$ and $\vec{E}_-$ cancel outside the capacitor, are true only in the ideal case, in which the plates are large and close together, and when we consider points in space far from the edges of the capacitor. This idealization is physically reasonable for the kinds of problems we will encounter.

Recall that the Coulomb constant k is related to another physical constant ϵ_0, the permittivity of free space. Because the relationship between k and ϵ_0 involves a factor of 4π ($k = 1/4\pi\epsilon_0$), the field between the plates takes on a simpler form when expressed in terms of ϵ_0:

$$E = \frac{q}{\epsilon_0 A} \qquad (17\text{-}18)$$

We will use this relationship as we determine the potential difference between the plates of a capacitor. We begin with the equation for calculating the potential between any two points in an electric field $V = -\int_a^b \vec{E} \cdot d\vec{s}$ (Equation 17-5), where E is the electric field between the plates. The quantity $d\vec{s}$ is an infinitesimally small piece of the path from one plate to the other, so setting one plate a distance d away from the other, the limits of the integral become from 0 to d:

$$V = -\int_0^d \vec{E} \cdot d\vec{s}$$

Because the plates are symmetric, we can follow the field lines in either direction. From the negative plate to the positive plate, for example, $\vec{E}$ and $d\vec{s}$ are antiparallel (parallel but in opposite directions), so $\vec{E} \cdot d\vec{s} = Eds \cos 180° = -Eds$. The potential difference is then

$$V = \int_0^d \frac{q}{\epsilon_0 A} ds = \frac{q}{\epsilon_0 A} d$$

For a given capacitor of fixed plate area A and plate separation d, the potential difference between the plates V is proportional to the magnitude of the charge q on each plate. Perhaps more importantly, when a voltage difference V is applied to the initially uncharged capacitor, as shown in **Figure 17-10**, a net charge of $-q$ is driven to one of the plates. The standard symbol for a capacitor in the diagram of a circuit, as shown in the figure, is two parallel lines, meant to represent two parallel plates, connected to the circuit by straight lines, representing wires. A capacitor in a circuit diagram therefore looks like ⊣⊢. (The standard symbol for a voltage source in the diagram of a circuit, looks like ⊣⊢; the shorter of the two lines represents the negative side of the potential difference. We introduce this symbol starting in Figure 17-16.) The net charge on one plate in turn induces a net $+q$ charge on the other plate; the net charge on the capacitor must remain zero. The magnitude of the charge each plate acquires is proportional to the applied voltage:

$$q = \frac{\epsilon_0 A}{d} V$$

where the proportionality constant $\epsilon_0 A/d$, called the **capacitance**, depends only on the size of plates and the distance between them. Capacitance is therefore a compact way to summarize the properties of a capacitor, in particular the amount of charge that can be stored on a capacitor held at a given voltage. So,

$$q = CV \qquad (17\text{-}19)$$

where the capacitance C is

$$C = \frac{\epsilon_0 A}{d} \qquad (17\text{-}20)$$

This particular relationship applies specifically to two parallel plates of area A held a distance d apart.

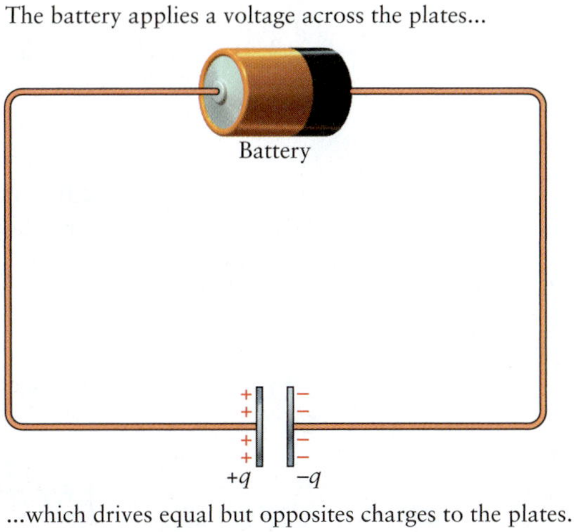

The battery applies a voltage across the plates...

Battery

+q   −q

...which drives equal but opposites charges to the plates.

Figure 17-10 When a voltage difference is applied across the plates of a capacitor, the excess charge that appears on the plates is the same in magnitude and opposite in sign.

An insulin-producing beta cell from the pancreas

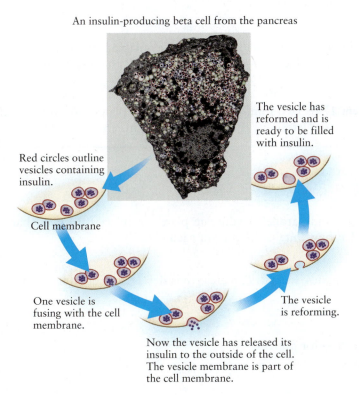

Red circles outline vesicles containing insulin.

Cell membrane

One vesicle is fusing with the cell membrane.

Now the vesicle has released its insulin to the outside of the cell. The vesicle membrane is part of the cell membrane.

The vesicle is reforming.

The vesicle has reformed and is ready to be filled with insulin.

Figure 17-11 Pancreatic β-cells synthesize insulin and store it in bubble-like vesicles made of membrane. The capacitance of a β-cell membrane increases when the vesicles fuse with it in order to release the insulin. (Modified from Sven Göpel et al. (2004) Capacitance measurements of exocytosis in mouse pancreatic α-, β- and ō-cells within intact islets of Langerhans. *J. Physiol.* 556(3):711-726.)

The SI unit of capacitance is the farad (F), after the nineteenth-century English physicist Michael Faraday. The capacitors you'll find in your neighborhood electronics supply store are typically in the range of 10 pF (10×10^{-12} F) to 1000 μF (1000×10^{-6} F); 1 F is an extremely large capacitance.

The value of ϵ_0 is 8.85×10^{-12} $C^2/(N \cdot m^2)$. (Here C is Coulombs, not capacitance.) The units are useful when addressing electric force and electric field problems. But notice that we can express ϵ_0 in different units by rearranging Equation 17-20:

$$\epsilon_0 = \frac{d}{A} C$$

We see that ϵ_0 can also be expressed in units of farads per meter (F/m).

Example 17-3 Parallel Plates

The dimensions of two thin, parallel conducting plates separated by 0.1 mm are 5 cm by 5 cm. Determine the capacitance of this configuration.

SET UP
We use the relationship between capacitance and the geometry of the plates:

$$C = \frac{\epsilon_0 A}{d} \tag{17-20}$$

Here, the area of each plate is A is 0.05 m $\times$ 0.05 m $= 2.5 \times 10^{-3}$ m^2 and d is 1.0×10^{-4} m.

SOLVE
So, substituting in known values allows us to calculate an answer

$$C = \frac{(8.85 \times 10^{-12} \text{ F/m})(2.5 \times 10^{-3} \text{ m}^2)}{1.0 \times 10^{-4} \text{ m}} = 2.2 \times 10^{-10} \text{ F}$$

Recall that the prefix "p" or pico is 10^{-12}, so $C = 220$ pF.

REFLECT
Capacitors of this size are well within the range available from your local electronics supply store, and are likely found in calculators, cell phones, and portable music players.

Practice Problem 17-3 Two thin, parallel conducting plates, 10 cm by 10 cm, are placed 0.4 mm apart. Determine the capacitance of this configuration.

Example 17-4 Parallel Plates, Again

Two thin, square, and parallel conducting plates are placed 0.1 mm apart. If the capacitance is 1 F, how long is the edge of each plate?

SET UP
If each side of the plates is length L, plate area is $A = L^2$, so Equation 17-20 becomes

$$C = \frac{\epsilon_0 L^2}{d}$$

Solving the equation for L gives

$$L = \sqrt{\frac{Cd}{\epsilon_0}}$$

SOLVE

To obtain a capacitance of 1 F with plates separated by 1.0×10^{-4} m, the sides of the plates must then be

$$L = \sqrt{\frac{(1 \text{ F})(1.0 \times 10^{-4} \text{ m})}{8.85 \times 10^{-12} \text{ F/m}}} = \sqrt{1.13 \times 10^{7} \text{ m}^2}$$

or

$$L = 3.4 \times 10^{3} \text{ m}$$

REFLECT

The plates are about 3.4 km, or about 2 mi, on a side. You'd have to place about 2200 football fields or soccer pitches together to form a square that big, and such large plates would have the capacity to hold lots of charge. One farad (1 F) is a large capacitance.

Practice Problem 17-4 Two thin, square, and parallel conducting plates are placed 0.200 mm apart. If the capacitance is 550 pF, how long is the edge of each plate?

Estimate It! 17-1 Insulin Release

The hormone *insulin* minimizes variations in blood glucose levels. Pancreatic β-cells synthesize insulin and store it in vesicles, bubble-like organelles approximately 150 nm in radius within the cytoplasm of the cells. To release insulin, vesicles fuse with the cell membrane. This increases the surface area of the β-cell by the surface area of the vesicle. The thickness of the cell membrane does not change, so the increase in surface area increases the capacitance of the β-cell membrane by the capacitance of the vesicle. This process is shown in **Figure 17-11**. **Figure 17-12** shows the change in capacitance of a β-cell as a function of time as insulin is released. If the capacitance of a biological membrane is approximately 1 μF/cm², estimate the number of vesicles that fuse with the membrane during 1 s of insulin release.

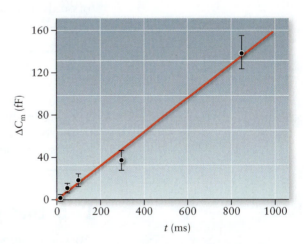

SET UP

Each vesicle that fuses with the cell membrane increases the capacitance of the membrane by the capacitance of the vesicle, which is the product of the surface area of the vesicle $SA_{vesicle}$ and its capacitance per unit surface area:

$$C_{vesicle} = \frac{C_{cell \text{ membrane}}}{\text{surface area}} SA_{vesicle} = \frac{C_{cell \text{ membrane}}}{\text{surface area}} 4\pi r_{vesicle}^{2}$$

Approximate values of $C_{cell \text{ membrane}}/(\text{surface area})$ and $r_{vesicle}$ are given in the problem statement.

We can estimate the increase in the cell's capacitance in a given amount of time from Figure 17-12. The change in β-cell membrane capacitance in 1 s divided by the change in capacitance due to one vesicle gives the number of vesicles that fuse with the membrane per second:

$$\frac{N_{vesicles}}{s} = \frac{\Delta C_{cell \text{ membrane}}}{1 \text{ s}} \bigg/ C_{vesicle}$$

Figure 17-12 The number of vesicles that fuse with the cell membrane as a function of time can be estimated from the plot of the change in capacitance of a β-cell membrane versus time as insulin is released. (Modified from Sven Göpel et al. (2004) Capacitance measurements of exocytosis in mouse pancreatic α-, β- and ō-cells within intact islets of Langerhans. *J. Physiol.* *556(3)*:711-726.)

SOLVE

The vesicle capacitance per unit surface area is approximately $1\ \mu F/cm^2$ or $10^{-2}\ F/m^2$. The average radius of a vesicle is approximately 150 nm or 1.5×10^{-7} m. The vesicle capacitance is then

$$C_{vesicle} = (10^{-2}\ F/m^2)4\pi(1.5 \times 10^{-7}\ m)^2 = 2.83 \times 10^{-15}\ F$$

So each vesicle changes the membrane capacitance by about 2.83×10^{-15} F. We'll keep one extra significant digit until we compute the final answer.

Now look at Figure 17-12. Because the straight line passes through (0, 0), we can read the change in capacitance in any time interval simply by reading the time and change in capacitance for a point on the line. Any big, round number will do, so we pick t equal to 300 ms, for which the change in capacitance is about 50 fF. A femtofarad (fF) is 10^{-15} F, so in 1 s the change in capacitance of the β-cell membrane is

$$\Delta C_{\beta\text{-cell membrane}} = \frac{50\ fF}{300\ ms} = \frac{50\ fF}{0.3\ s} = 167\ fF/s$$

Each vesicle contributes 2.8×10^{-15} F or 2.8 fF to the capacitance of the β-cell membrane. For the capacitance to change by 167 fF every 1 s requires

$$\frac{N_{vesicles}}{s} = \frac{167\ fF/s}{2.83\ fF} = 58.9$$

We're using big round numbers for the estimate, so to one significant digit, we estimate that 60 vesicles fuse with the cell membrane every second during insulin release.

REFLECT

The ability to detect changes in capacitance with this resolution allows scientists the opportunity to study the molecular mechanisms of hormone release from single cells and to verify previous biochemical measurements of insulin release from β-cells. After insulin release, new vesicles form by pinching inward from the cell membrane and are refilled with insulin. By recycling the vesicle membrane, secretory cells such as the β-cell maintain their size over the long term.

✱ What's Important 17-4

A capacitor stores energy in the form of an electric field. The charge stored on the plates of a capacitor gives rise to a potential difference between the plates; the associated potential energy can later be converted to the movement of charges. For a given capacitor, the potential difference between the plates is proportional to the magnitude of the charge on each plate.

17-5   Energy Stored in a Capacitor

Capacitors store energy. An applied voltage V such as the potential difference supplied by the battery in Figure 17-10 charges a capacitor by effectively pulling negative charges from one of the plates and depositing them on the other. To move the charges requires work, and it is this work that results in potential energy being stored in the capacitor. At a later time, the potential energy can be converted to the kinetic energy of charge leaving the capacitor.

Applying a voltage V to a capacitor causes charges to move from one plate to the other; from Equation 17-3,

$$V = \frac{dU}{dq}$$

or

$$dU = Vdq$$

Each small bit of charge dq pulled from one plate of the capacitor and deposited on the other results in an increase in the potential energy dU.

To find the total change in potential energy ΔU when charge Q is added to an initially uncharged capacitor, move the charge one small dq at a time and add together the contribution each makes to the total change in energy:

$$\Delta U = \int dU = \int_0^Q Vdq$$

We can express potential energy in terms of the capacitance using Equation 17-19:

$$\Delta U = \int dU = \int_0^Q \frac{q}{C}dq$$

which gives us

$$\Delta U = \frac{1}{C}\frac{q^2}{2}\Big|_0^Q = \frac{1}{2}\frac{Q^2}{C} \qquad (17\text{-}21)$$

For any given capacitor, then, the energy stored is proportional to the square of the magnitude of the charge stored.

Example 17-5 Defibrillator

Defibrillators (**Figure 17-13**) save lives during cardiac emergencies, such as heart attacks and arrhythmias, by delivering an electric shock to the heart to jump-start it back into its normal rhythm. A defibrillator is essentially a capacitor that can be charged by a high voltage source and then deliver the stored energy to the heart, often through paddles pressed against the patient's chest. (a) How much charge does the 80 μF capacitor in a certain defibrillator store when it is fully charged by applying 2500 V? (b) How much energy can the defibrillator deliver?

Go to Picture It 17-3 for more practice dealing with capacitance

Go to Interactive Exercise 17-1 for more practice dealing with capacitors

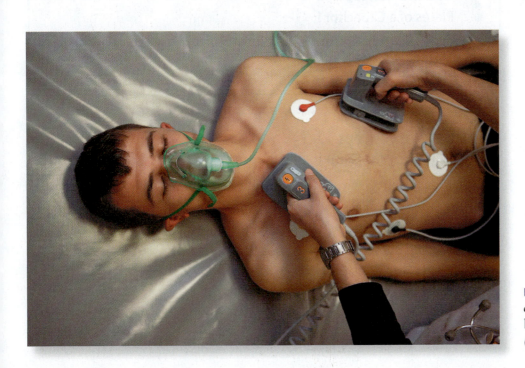

Figure 17-13 A defibrillator can jump-start a patient's heart back into its normal rhythm. *(Murat Seyit/ iStockphoto.)*

SET UP

The energy the defibrillator can deliver, ignoring any losses in the delivery mechanism, is equal to the energy stored in the capacitor. So from Equation 17-21,

$$\Delta U = \frac{1}{2}\frac{Q^2}{C}$$

The energy, in turn, depends on the maximum charge that can be stored on the capacitor, given by Equation 17-19:

$$Q = CV$$

So the energy that can be stored on the capacitor is

$$\Delta U = \frac{1}{2}\frac{(CV)^2}{C} = \frac{1}{2}CV^2$$

SOLVE

(a) Using Equation 17-19, when 2500 V is applied across an 80 μF capacitor, the stored charge is

$$Q = (80\ \mu F)(2500\ V) = 0.20\ C$$

(b) The increase in the electric potential energy stored in the capacitor is

$$\Delta U = \frac{1}{2}(80\ \mu F)(2500\ V)^2 = 250\ J$$

REFLECT

The American Heart Association recommends that a defibrillator shock should deliver between 40 and 360 J in order to be effective.

Practice Problem 17-5 A 125 μF capacitor is fully charged by applying 2750 V across it. (a) How much charge is stored when the capacitor is fully charged? (b) How much energy is stored when the capacitor is fully charged?

❓ Got the Concept 17-4
Plates of a Capacitor I

If you increase the distance between the plates of a charged capacitor without changing the amount of stored charge, will the energy stored in the capacitor decrease, remain the same, or increase?

❓ Got the Concept 17-5
Plates of a Capacitor II

If you increase the distance between the plates of a capacitor while holding the potential applied across them constant, will the maximum amount of energy the capacitor can store decrease, remain the same, or increase relative to the amount of energy it could store initially?

✳ What's Important 17-5

The energy stored in a capacitor is proportional to the square of the magnitude of the charge stored.

17-6 Capacitors in Series and Parallel

In both biological systems and electric circuits, it is not uncommon for more than one capacitor to be connected together in some way. In this section, we examine how capacitors act when combined.

Whenever two or more capacitors are connected, our goal will be to determine the **equivalent capacitance**, the capacitance of a single capacitor that is the equivalent replacement of all of capacitors actually present. The equivalent capacitance C_{equiv} must satisfy the known relationship between applied voltage and stored charge:

$$q = C_{\text{equiv}} V \qquad (17\text{-}22)$$

(from Equation 17-19), where q is the total charge that would appear on the equivalent capacitor and V is the voltage difference that would be applied across it.

Consider the three, initially uncharged capacitors C_1, C_2, and C_3 connected together as in **Figure 17-14**. The battery applies a voltage V across all three together after a switch is closed. Capacitors connected in this way are said to be in **series**.

What can we say about the charge on and the voltage difference across each capacitor? First, each capacitor separately obeys Equation 17-19:

$$q_1 = C_1 V_1$$
$$q_2 = C_2 V_2 \qquad (17\text{-}23)$$
$$q_3 = C_3 V_3$$

Three capacitors are connected to a voltage source in series.

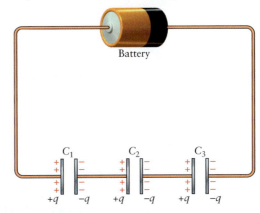

Battery

- The magnitude of the charge on each capacitor must be the same.
- The voltage across each capacitor is NOT the same.
- The sum of the voltage drops across each capacitor separately equals the voltage of the battery.

Figure 17-14 The charge on each of three capacitors connected in series is the same, but the voltage across each is not. The sum of the voltage differences across each of the three capacitors equals the total voltage supplied by the battery.

The magnitudes of the charge on the two plates of any capacitor must be equal. So, as the battery draws negative charge from the left plate of C_1, whatever positive charge $+q$ that plate acquires must be balanced by charge $-q$ on the right plate. The charging of the right plate of C_1 occurs as negative charge is drawn from the left plate of C_2, and because this whole section is initially uncharged, the left plate of C_2 acquires charge $+q$. The same reasoning is applied to C_3; the net result is that all three capacitors acquire the same charge. Thus,

$$q_1 = q_2 = q_3 = q \qquad (17\text{-}24)$$

Using Equation 17-24, the descriptions of the capacitors separately (Equation 17-23) become

$$q = C_1 V_1$$
$$q = C_2 V_2 \qquad (17\text{-}25)$$
$$q = C_3 V_3$$

We consider the voltage drops across any connecting wires in the circuit to be negligible, so the total voltage is the sum of the voltage across each capacitor separately:

$$V = V_1 + V_2 + V_3 \qquad (17\text{-}26)$$

We can sum the voltage differences across each capacitor by rearranging the three relationships in Equation 17-25 for voltage:

$$V_1 = \frac{q}{C_1}$$

$$V_2 = \frac{q}{C_2}$$

$$V_3 = \frac{q}{C_3}$$

Equation 17-26 is then

$$V = \frac{q}{C_1} + \frac{q}{C_2} + \frac{q}{C_3}$$

or

$$V = q\left(\frac{1}{C_1} + \frac{1}{C_2} + \frac{1}{C_3}\right)$$

To see more clearly the equivalent capacitance, we rearrange this equation to match the form $q = C_{\text{equiv}}V$ (Equation 17-22):

$$q = \frac{1}{\dfrac{1}{C_1} + \dfrac{1}{C_2} + \dfrac{1}{C_3}}V$$

To replace three capacitors in series with a single, equivalent capacitor, the equivalent capacitor must have a capacitance of

$$C_{\text{equiv}} = \frac{1}{\dfrac{1}{C_1} + \dfrac{1}{C_2} + \dfrac{1}{C_3}}$$

More generally, for N capacitors connected in series,

$$C_{\text{equiv}} = \frac{1}{\displaystyle\sum_{i=1}^{N} \dfrac{1}{C_i}} \tag{17-27}$$

❗ Watch Out

When adding capacitors in series, you must do the two inverse operations separately.

Sometimes students lump together the 1's in the two separate numerators in Equation 17-27, turning the sum in the denominator into $C_{\text{equiv}} = \sum_{i=1}^{N} C_i$. Of course, such a manipulation is not mathematically correct! To determine C_{equiv} correctly you must first add the inverse of each of the capacitances separately and then take the inverse of the sum.

 Got the Concept 17-6

Capacitors in Series

When N identical capacitors are combined in series, is the equivalent capacitance smaller than, the same as, or larger than the capacitance of any one single capacitor? Explain your answer in the context of the charge stored on and the voltage across the capacitors compared to that of the equivalent capacitor.

Now let's consider the three, initially uncharged capacitors C_1, C_2, and C_3 as connected together in **Figure 17-15**. The battery applies a voltage V across all three together after a switch is closed. Capacitors connected in this way are said to be in **parallel**. Notice the difference in the arrangement of the capacitors in parallel compared to capacitors in series by comparing Figures 17-15 and 17-14. In a series arrangement, the right plate of one capacitor is connected to the left plate of the capacitor to the right. When the capacitors are connected in parallel, all of the right plates are connected, and all of the left plates are connected separately.

As in the case of capacitors connected in series, each capacitor in the parallel arrangement separately is described by Equation 17-19:

$$q_1 = C_1 V_1$$

$$q_2 = C_2 V_2 \qquad (17\text{-}23)$$

$$q_3 = C_3 V_3$$

The total charge acquired by all three capacitors is

$$q = q_1 + q_2 + q_3$$

or

$$q = C_1 V_1 + C_2 V_2 + C_3 V_3 \qquad (17\text{-}28)$$

As in the case of capacitors connected in series, we consider the voltage drops across any connecting wires in the circuit to be negligible. For that reason, and because all of the right plates are connected to one terminal of the battery and all of the left plates are connected to the other terminal, the potential difference across each capacitor must be the same. That voltage difference is V, the voltage supplied by the battery. Thus, the voltage drop across each capacitor must be the same:

$$V_1 = V_2 = V_3 = V \qquad (17\text{-}29)$$

and by substitution, Equation 17-28 becomes

$$q = C_1 V + C_2 V + C_3 V = (C_1 + C_2 + C_3)V$$

Comparing this equation with Equation 17-22 for equivalent capacitance ($q = C_{\text{equiv}} V$), we see that in order to replace three capacitors in parallel with a single, equivalent capacitor,

$$C_{\text{equiv}} = C_1 + C_2 + C_3$$

More generally, for N capacitors connected in parallel,

$$C_{\text{equiv}} = \sum_{i=1}^{N} C_i \qquad (17\text{-}30)$$

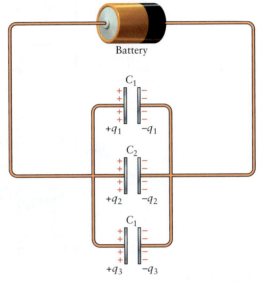

Three capacitors are connected to a voltage source in parallel.

- The voltage drop across each capacitor must be the same.
- The charge on each capacitor may not be the same.
- The sum of the charges on each capacitor separately equals the total charge on the equivalent capacitor.

Figure 17-15 The voltage difference across each of three capacitors connected in parallel is the same, and equal to the voltage supplied by the battery. The charge on each capacitor may be different.

 Got the Concept 17-7

Capacitors in Parallel

When N identical capacitors are combined in parallel, is the equivalent capacitance smaller than, the same as, or larger than the capacitance of any one single capacitor?

Table 17-1   Expressions for the Equivalent Capacitance of Two Capacitors Connected in Series and in Parallel

	Two Capacitors in Series	Two Capacitors in Parallel
Different ($C_1 \neq C_2$)		
Same ($C_1 = C_2$)		

Example 17-6 Two Capacitors

Two capacitors can be connected either in series or in parallel. Fill in Table 17-1 with relationships based on Equations 17-27 and 17-30 for the cases when the capacitors C_1 and C_2 have different values and when they have the same value ($C_1 = C_2 = C$).

SET UP
We want to apply Equations 17-27 and 17-30 to these special cases.

SOLVE
Using Equation 17-27, the equivalent capacitance of two capacitors connected in series is

$$C_{equiv} = \frac{1}{\frac{1}{C_1} + \frac{1}{C_2}}$$

So

$$C_{equiv} = \frac{1}{\frac{C_1 + C_2}{C_1 C_2}} = \frac{C_1 C_2}{C_1 + C_2}$$

When both capacitances are equal to C,

$$C_{equiv} = \frac{CC}{C + C} = \frac{C^2}{2C} = \frac{C}{2}$$

Using Equation 17-30, two capacitors connected in parallel have equivalent capacitance

$$C_{equiv} = C_1 + C_2$$

When both capacitances are equal to C,

$$C_{equiv} = C + C = 2C$$

REFLECT
Our results for two equal capacitors in series and in parallel support the conclusions we drew in Got the Concept 17-6 and 17-7. Connecting capacitors in series reduces the capacitance and connecting them in parallel increases the capacitance.

Any real network of capacitors is likely to have a mixture of capacitors connected in series and in parallel. To find the equivalent capacitance of such a network, we first identify any small grouping of capacitors that are either entirely in series or entirely in parallel, then find the equivalent capacitance of each group, and finally, using the series and parallel rules, we combine them in larger and larger groupings, until we have accounted for all of the capacitors in the network.

❓ Got the Concept 17-8

Series or Parallel?

Identify the groupings of capacitors you would use to determine the equivalent capacitance in **Figure 17-16**. For each, note whether the capacitors are connected in series or in parallel.

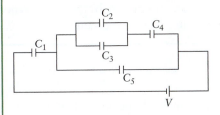

Figure 17-16 In what sequence can the capacitors be grouped in order to determine the equivalent capacitance?

Example 17-7   Multiple Capacitors

Find an expression for the equivalent capacitance of the three capacitors shown in **Figure 17-17a**, and evaluate the equation for $C_1 = 1\ \mu F$, $C_2 = 2\ \mu F$, and $C_3 = 6\ \mu F$.

SET UP

Whenever capacitors are combined in ways other than purely in series or purely in parallel, we look for groupings of capacitors that *are* either in parallel or in series, and then combine the groups one at a time. Notice in Figure 17-17a that C_1 and C_2 are in parallel because their two right plates are connected, as are their two left plates. We can therefore find C_{12} (the equivalent capacitance of C_1 and C_2) by using our relationship for capacitors in parallel (Equation 17-30). We draw the capacitors as in **Figure 17-17b**, in which C_1 and C_2 have been replaced by C_{12}. This diagram makes it easier to see that C_{12} and C_3 are connected in series. To find C_{123}, the equivalent capacitance of all three capacitors, we therefore combine C_{12} and C_3 using Equation 17-27.

  Go to Interactive Exercise 17-2 for more practice dealing with multiple capacitors

SOLVE

Using Equation 17-30, C_1 and C_2 in parallel gives

$$C_{12} = C_1 + C_2$$

Using Equation 17-27, the equivalent capacitor C_{12} in series with C_3 results in

$$C_{123} = \frac{1}{\dfrac{1}{C_{12}} + \dfrac{1}{C_3}} = \frac{C_{12}C_3}{C_{12} + C_3} = \frac{(C_1 + C_2)C_3}{C_1 + C_2 + C_3}$$

Using the values of capacitance provided in the problem statement,

$$C_{123} = \frac{(1\ \mu F + 2\ \mu F)(6\ \mu F)}{1\ \mu F + 2\ \mu F + 6\ \mu F} = \frac{(3\ \mu F)(6\ \mu F)}{9\ \mu F} = 2\ \mu F$$

Notice that the units are $\mu F^2/\mu F = \mu F$.

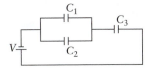

(a)

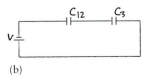

(b)

Figure 17-17 (a) What is the equivalent capacitance of the three capacitors? (b) Capacitors C_1 and C_2 are combined in parallel.

REFLECT

Don't be surprised because the result is smaller than the 6 μF capacitor, or smaller than the 1 μF and 2 μF capacitors combined. (C_1 and C_2 in parallel, called C_{12} in our solution, equals 3 μF.) As we uncovered in Got the Concept 17-6, when capacitors are connected in series, the equivalent capacitance is always smaller than the smallest individual capacitor.

Practice Problem 17-7 Find the equivalent capacitance of the three capacitors shown in Figure 17-17a, (a) for $C_1 = 1.00\ \mu$F, $C_2 = 200\ \mu$F, and $C_3 = 6.00\ \mu$F and (b) for $C_1 = 1.00\ \mu$F, $C_2 = 2.00\ \mu$F, and $C_3 = 600\ \mu$F. Compare your results to the answer to Example 17-7, noting that in part a) the capacitance of C_2 is 100 times that in Example 17-7 while C_1 and C_3 are the same, and in part b) the capacitance of C_3 is 100 times that in Example 17-7 while C_1 and C_2 are the same.

✳ What's Important 17-6

Whenever two or more capacitors are connected, the equivalent capacitance is the capacitance of a single capacitor that is the equivalent replacement of the individual capacitors. When capacitors are connected in series, the right plate of one capacitor is connected to the left plate of the capacitor to its right. When the capacitors are connected in parallel, all of the right plates are connected and (separately) all of the left plates are connected. The equivalent capacitance of capacitors connected in series is always smaller than the capacitance of the smallest individual capacitor. Connecting capacitors in parallel results in a larger capacitance.

17-7 Dielectrics

In all cells, the internal concentration of potassium (K^+) is greater than its external concentration; because the membrane is selectively permeable to K^+, negative charges are unable to follow K^+ ions as they diffuse down their concentration gradient, out of the cell. Therefore, the inner and outer surfaces of the membrane acquire opposite charge—negative charge on the inside and positive on the outside. In other words, a cell membrane can be modeled as a parallel plate capacitor! However, even though good measurements of the thickness and surface area of cell membranes exist, a calculation using $C = \epsilon_0 A/d$ (Equation 17-20) predicts a capacitance seven or more times lower than the measured value. To calculate the correct capacitance, we need to account for the electrical properties of the material that comprises the membrane.

The capacitance of an arrangement of charged plates or layers, such as those of a cell membrane, depends on the net electric field between the plates. For a parallel plate capacitor, capacitance is related to the potential difference between the plates:

$$q = CV \tag{17-19}$$

Using Equation 17-5, the potential difference across the capacitor gap (distance d) is related to the electric field:

$$V = -\int_0^d \vec{E} \cdot d\vec{s} = Ed$$

By rearranging Equation 17-19 and substituting the result for V we see that

$$C = \frac{q}{V} = \frac{q}{Ed} \tag{17-31}$$

Capacitance is therefore inversely proportional to the electric field; for example, capacitance increases as the field decreases.

The plates in both biological capacitors (membranes) and the kind found in electronic circuits are typically separated by nonconducting material, referred to as a **dielectric.** Although charge does not flow across the dielectric, the electric field $\vec{E}_0$ that arises due to the charge on the plates can cause a noticeable effect on the material that fills the gap between the plates. In some dielectric materials, such as the one shown in **Figure 17-18a**, the molecules that make up the materials are *polar*, that is, more negative charge is concentrated on one end of each molecule and more positive charge on the other end. The negative ends of the dipoles are attracted to the positive plate of the capacitor, and the positive ends to the negative plate, as in **Figure 17-18b**. The rearrangement of charge gives rise to an induced electric field $\vec{E}_{ind}$ pointing in the opposite direction of $\vec{E}_0$. The net field $\vec{E}$ inside the capacitor, the vector sum of $\vec{E}_0$ and $\vec{E}_{ind}$, is therefore lower due to the presence of the dielectric material. The lowered field results in a corresponding increase in capacitance.

The electric field within a capacitor can be lowered by the presence of a dielectric even if the material is not composed of polar molecules (**Figure 17-18c**). In these materials, the field due to the plates distorts the electron distribution around each atom. As suggested in **Figure 17-18d**, the induced separation of positive and negative charge generates an atomic electric dipole. So in this case, too, the rearrangement of charge gives rise to an induced electric field that reduces the net field in the capacitor.

Quantitatively, how does the capacitance change due to the presence of a dielectric? Consider a set of plates separated by a vacuum. The capacitance of the plates is C_0, and when the plates are charged the field between them is $\vec{E}_0$. By using Equation 17-31,

$$C_0 = \frac{q}{E_0 d}$$

Including a dielectric reduces the field to a fraction of E_0:

$$E = f E_0$$

where f is less than 1. If we consider the case in which the net charge on the capacitor plates remains unchanged, the capacitance becomes

$$C = \frac{q}{Ed} = \frac{q}{f E_0 d} = \frac{1}{f} C_0$$

Because f is less than 1, $1/f$ is greater than 1, so

$$C = \kappa C_0 \qquad (17\text{-}32)$$

where the **dielectric constant** κ (the Greek letter kappa) is greater than 1 and depends on the specific dielectric material. The dielectric constant of a vacuum is exactly 1.0, consistent with the assumption that C_0 is the capacitance when the plates are separated by a vacuum. The dielectric constant of air, $\kappa_{air} = 1.00058$ at room temperature and a pressure of 1 atm, is only slightly different from that of a vacuum. Lipids, the primary constituent of the interior of cell membranes, have a dielectric constant of 2.2. Water, which is of course found closely associated with the surfaces of cell membranes, has a dielectric constant of 80. Typical values of the dielectric constants of a variety of materials are listed in **Table 17-2**.

We now understand why the capacitance of a cell membrane is many times higher than what we would predict using $C = \epsilon_0 A/d$. Membranes are composed of

The orientations of the polar molecules in a dielectric material filling a capacitor are random when the capacitor is uncharged.

(a)

When the capacitor is charged, an induced electric field arises as the charged end of these dipoles are attracted to the oppositely charged plates.

(b)

The distribution of charge in the atoms of a material filling a capacitor is uniform when the capacitor is uncharged.

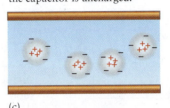

(c)

When the capacitor is charged, an induced electric field arises as the atomic charge separates slightly.

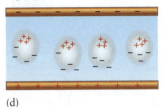

(d)

Figure 17-18 (a, b) When a capacitor filled with a dielectric material is charged, the polar molecules that make up the dielectric tend to align with the electric field that arises. This reduces the net field in the capacitor and thereby increases the capacitance. (c, d) A similar effect can occur in materials that do not contain polar molecules when the positive and negative charges in atoms are slightly separated, inducing an atomic electric dipole.

Table 17-2

Dielectric Constants (at 20°C and 1 atm)

Material	κ
vacuum	1
air	1.00058
lipid	2.2
paraffin	2.2
paper	2.7
ceramic (porcelain)	5.8
water	80

lipids sandwiched between charged phosphate head groups. There is also water on the membrane surfaces; the dielectric properties of these materials substantially increase membrane capacitance.

Example 17-8 Multiple Dielectrics

A capacitor filled with layers of three different dialectrics...

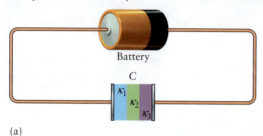

...is equivalent to three separate capacitors in series.

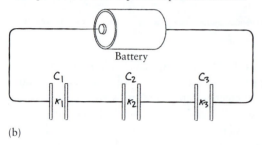

Figure 17-19 (a) What is the effective dielectric constant of a capacitor filled with three dielectric layers? (b) The capacitor can be treated as three separate capacitors in series.

A parallel plate capacitor of plate area A and plate separation d is filled with three dielectric materials layered as shown in **Figure 17-19a**. The dielectric constant and thickness of the layers are κ_1, κ_2, and κ_3, and d_1, d_2, and d_3, respectively. Show that the effective dielectric constant is

$$\kappa = \frac{d}{\dfrac{d_1}{\kappa_1} + \dfrac{d_2}{\kappa_2} + \dfrac{d_3}{\kappa_3}}$$

SET UP

Because we can consider a layer of charge to be induced on the surfaces of each dielectric layer, we can treat the capacitor as three separate capacitors in series as shown in **Figure 17-19b**. Using Equation 17-27, the equivalent capacitance is then

$$C = \frac{1}{\dfrac{1}{C_1} + \dfrac{1}{C_2} + \dfrac{1}{C_3}}$$

For each separate layer, we apply $C = \kappa C_0$ (Equation 17-32). C_0, the capacitance of each separate capacitor with the dielectric removed, depends on the plate area, which is the same for all three, and the plate separation, which is different for each. That is, for the three capacitor layers,

$$C_1 = \frac{\epsilon_0 A}{d_1} \kappa_1, \quad C_2 = \frac{\epsilon_0 A}{d_2} \kappa_2, \quad \text{and} \quad C_3 = \frac{\epsilon_0 A}{d_3} \kappa_3$$

SOLVE

So, the equivalent capacitance is

$$C = \frac{1}{\dfrac{d_1}{\epsilon_0 A \kappa_1} + \dfrac{d_2}{\epsilon_0 A \kappa_2} + \dfrac{d_3}{\epsilon_0 A \kappa_3}}$$

or

$$C = \frac{1}{\dfrac{1}{\epsilon_0 A} \left[\dfrac{d_1}{\kappa_1} + \dfrac{d_2}{\kappa_2} + \dfrac{d_3}{\kappa_3} \right]}$$

We need to compare this result with $C = \kappa C_0$ to see which terms form the effective dielectric constant. When not filled with a dielectric, the capacitance C_0 of plates that have an area A and a separation d is $\epsilon_0 A / d$. So, we multiply the numerator and denominator by d to bring out C_0:

$$C = \frac{d}{\dfrac{d}{\epsilon_0 A} \left[\dfrac{d_1}{\kappa_1} + \dfrac{d_2}{\kappa_2} + \dfrac{d_3}{\kappa_3} \right]}$$

▶ Go to Interactive Exercise 17-3 for more practice dealing with dialectrics

or

$$C = \frac{d}{\dfrac{d_1}{\kappa_1} + \dfrac{d_2}{\kappa_2} + \dfrac{d_3}{\kappa_3}} \left(\frac{\epsilon_0 A}{d}\right)$$

The term in parentheses is equal to C_0, so the effective dielectric constant by using $C = \kappa C_0$ is

$$\kappa = \frac{d}{\dfrac{d_1}{\kappa_1} + \dfrac{d_2}{\kappa_2} + \dfrac{d_3}{\kappa_3}} \qquad (17\text{-}33)$$

REFLECT

The effective dielectric is a function of the thickness and dielectric constant of each layer. When there is more than one layer of a given dielectric between the plates of a capacitor, each term in the denominator can be generalized to represent the combined thickness of each dielectric material. If, for example, κ_3 were equal to κ_2 in this problem, then Equation 17-33 would be

$$\kappa = \frac{d}{\dfrac{d_1}{\kappa_1} + \dfrac{d_2}{\kappa_2} + \dfrac{d_3}{\kappa_2}} = \frac{d}{\dfrac{d_1}{\kappa_1} + \dfrac{d_2 + d_3}{\kappa_2}}$$

Here, $d_2 + d_3$ is the combined thickness of layers of dielectric constant κ_2. You can interpret the thicknesses d_1, d_2, and d_3 in Equation 17-33 as the combined thickness of each dielectric material in between the plates of a capacitor.

Practice Problem 17-8 A parallel plate capacitor of plate area A and plate separation d is filled with three, layered dielectric materials as shown in Figure 17-19a. The dielectric constant and thickness of the layers are κ_1, κ_2, and κ_3, and d_1, d_2, and d_3, respectively. If κ_1, κ_2, and κ_3 are all equal to κ, show that capacitance of the capacitor is given by Equation 17-31.

Example 17-9 Capacitance of Membranes

Find the effective dielectric constant of the membrane shown in **Figure 17-20**, which consists of a layer of lipid surrounded by layers of a polarized phosphate molecules and water, each of which acts as a capacitor filled with a dielectric.

SET UP

In the previous example, we found the effective dielectric constant for a capacitor containing layers of material with different dielectric constants:

$$\kappa = \frac{d}{\dfrac{d_1}{\kappa_1} + \dfrac{d_2}{\kappa_2} + \dfrac{d_3}{\kappa_3}} \qquad (17\text{-}33)$$

We concluded in the **Reflect** section of Example 17-8 that the thicknesses d_1, d_2, and d_3 in Equation 17-33 represent the combined thickness of each material. So for the membrane shown in Figure 17-20, we will find the combined thickness of water layers, polar layers, and lipid layer in order to determine the effective dielectric constant.

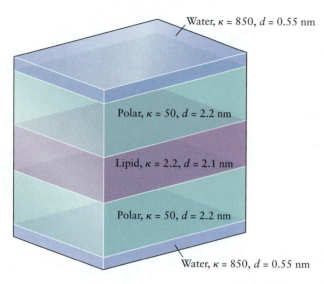
Water, $\kappa = 850$, $d = 0.55$ nm
Polar, $\kappa = 50$, $d = 2.2$ nm
Lipid, $\kappa = 2.2$, $d = 2.1$ nm
Polar, $\kappa = 50$, $d = 2.2$ nm
Water, $\kappa = 850$, $d = 0.55$ nm

Figure 17-20 Membranes act like a capacitor filled with a dielectric. What is the effective dielectric constant of a membrane that consists of a layer of lipid surrounded by polarized surface layers and water?

SOLVE

As shown in Figure 17-20, the net thickness of water layers associated with the inner and outer surfaces of membrane is 1.1 nm. The dielectric constant of water is 80. The two polar layers together have thickness 4.4 nm and have a dielectric constant of 50. The lipid layer is 2.1 nm thick and has a dielectric constant of 2.2. The total thickness of the membrane is 1.1 nm + 4.4 nm + 2.1 nm, or 7.6 nm, so

$$\kappa = \frac{7.6\ \text{nm}}{\dfrac{1.1\ \text{nm}}{80} + \dfrac{4.4\ \text{nm}}{50} + \dfrac{2.1\ \text{nm}}{2.2}} = 7.2$$

REFLECT

Cell membranes have a relatively high dielectric constant, which accounts for the large difference between the measured value of its capacitance and the value predicted by using $C = \epsilon_0 A/d$ (Equation 17-20) or by considering the membrane of a cell as a capacitor filled with air.

Practice Problem 17-9 Imagine a situation that reduces the thickness of the polarized surface layers in a membrane by one-half. Find the effective dielectric constant of the thinner membrane. Compare your answers to the result of Example 17-9.

> ✳ **What's Important 17-7**
> Filling a capacitor with a dielectric material increases the capacitance. A dielectric is made up of polar substances that tend to align with the field that arises between the plates of a capacitor.

Answers to Practice Problems

17-1 $-ED$

17-2 $-ED$

17-3   220 pF

17-4   11.1 cm

17-5   (a) 0.344 C, (b) 473 J

17-7   (a) 5.83 μF, (b) 2.99 μF

17-8   Effective dielectric constant is found to be κ.

17-9   5.4

Answers to Got the Concept Questions

17-1   We have seen a number of similarities between the electric and gravitational forces, as well as between electric and gravitational potential energies. We have also drawn an analogy between electric field and the concept of a gravitational field. Because the force of gravity is uniform near Earth's surface, the field here would be represented by parallel arrows pointing down. Rotate Figure 17-5 90° clockwise, so that the arrows representing the electric field point down. Point a could represent a point at the bottom of a mountain, and b and c two points at the top. Paths *a* to *b* and *a* to *c* then represent two different ways that a climber could ascend the potential difference, which is a direct measure of the work done by the field.

17-2 Figure 17-21 shows a two-dimensional representation of the electric field and three of the associated equipotential surfaces. In two dimensions, the equipotential surfaces appear as circles, because every point on a circle is perpendicular to a radial line. In three dimensions, the equipotential surfaces themselves are spherical shells.

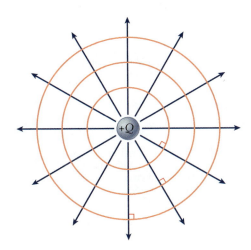

Figure 17-21

17-3 In the limit $y \gg R$, we treat R as negligible when compared to y. So then

$$\lim_{y \gg R} V = \lim_{y \gg R} \frac{2kQ}{R^2}\left(\sqrt{R^2 + y^2} - y\right)$$

$$= \frac{2kQ}{R^2}\left(\sqrt{y^2} - y\right)$$

The term in parentheses is zero, which means that in this limit V approaches zero. Recall that it is the difference in potential between two points that has physical significance, and that implicit in our result for the disk is a choice of reference point for electrical potential. We use the standard convention that V is zero at an infinite distance from a charged object. Because the limit $y \gg R$ is essentially equivalent to the limit $y \to \infty$, we should therefore expect that V goes to zero in this limit. We have essentially programmed it into our result!

17-4 The stored energy must increase. The plates carry equal but opposite charge and therefore attract each other. To pull the plates apart, you must do work on the system; because energy is conserved, the work you do increases the energy stored in the capacitor. You can

verify this conclusion by noting that as d increases, C decreases by using $C = \epsilon_0 A / d$ (Equation 17-20). And by using $\Delta U = \frac{1}{2} Q^2 / C$ (Equation 17-21), as the capacitance decreases, ΔU increases.

17-5 The maximum amount of energy that can be stored decreases. Capacitance decreases as the separation between the plates increases, so the capacitor with the larger plate separation can store less charge (from $q = CV$; Equation 17-19). The potential energy is more affected by a difference in stored charge than a difference in capacitance because, from Equation 17-21, ΔU is proportional to the *square* of the charge but is only proportional to (the inverse of) capacitance to the first power. Although capacitance decreases as the plate separation is increased, the square of the charge stored decreases far more. The maximum amount of energy that can be stored therefore decreases.

17-6 Connecting capacitors in series results in a smaller capacitance. Using Equation 17-27, you can see that each capacitor contributes $1/C$ to the sum in the denominator, so the denominator is then N multiplied by $1/C$, or N/C. Taking the inverse of N/C is C/N, which is certainly smaller than C. The charge that would reside on the equivalent capacitor is the same as the charge on the plates of each of the separate capacitors. The voltage across it, however, is larger than that across any of the separate capacitors. So because capacitance is charge divided by voltage (from $q = CV$, Equation 17-19), the equivalent capacitance must be smaller than that of any of the individual capacitors.

17-7 Connecting capacitors in parallel results in a larger capacitance. As you can see from Equation 17-30, the capacitance of capacitors connected in parallel add. When capacitors are connected in parallel, their combined capacity to store charge is greater than that of each capacitor individually. The voltage difference across each is the same as the voltage difference across an equivalent capacitor, so the combined effect is to create an equivalent capacitance greater than the individual capacitances.

17-8 Capacitors C_2 and C_3 form the simplest group in the circuit. They are connected in parallel; we redraw the circuit in Figure 17-22a showing an equivalent capacitor C_{23} in place of C_2 and C_3 separately. We can now see that C_{23} is connected to C_4 in series. In Figure 17-22b, these have been replaced by an equivalent capacitor C_{234}, making it clear that C_{234} and C_5 are connected in parallel. We replace C_{234} and C_5 by an equivalent capacitor C_{2345} in Figure 17-22c, from which it is clear that C_1 and C_{2345} are connected in series. The circuit containing a capacitor equivalent to all five individual capacitors is shown in Figure 17-22d.

Figure 17-22

C₂ and C₃ were combined in parallel.

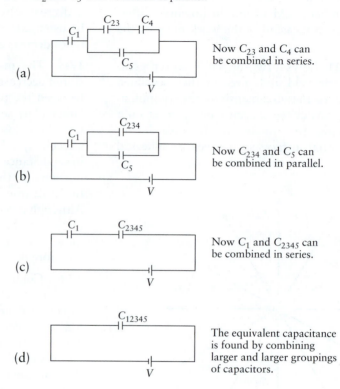

Now C_{23} and C_4 can be combined in series.

Now C_{234} and C_5 can be combined in parallel.

Now C_1 and C_{2345} can be combined in series.

The equivalent capacitance is found by combining larger and larger groupings of capacitors.

SUMMARY

Topic	Summary	Equation or Symbol	
capacitance	Capacitance C characterizes the amount of charge q that can be stored on a capacitor for a given applied voltage V. The SI unit of capacitance is the farad (F).	$q = CV$	(17-19)
capacitor	A capacitor is an electronic device used to store charge, and thereby store energy in the form of an electric field. The capacitance C of two parallel plates is proportional to the area A of the plates and the distance d between them.	$C = \dfrac{\epsilon_0 A}{d}$	(17-20)
dielectric	A dielectric material, or dielectric, is composed of polar molecules on which charge is distributed so that more negative charge is concentrated on one end of each molecule and more positive charge on the other end. Adding a dielectric between the plates of a capacitor increases the capacitance by a factor of the dielectric constant κ.	$C = \kappa C_0$	(17-32)
dielectric constant	The dielectric constant measures how well the polar molecules in a material align in the presence of an electric field, and therefore how much the capacitance increases when that material is placed in a capacitor.	κ	

electric potential difference	Electric potential difference, or electric potential, or voltage, characterizes the work that an electric field does on charged objects, in particular, the acceleration the charge undergoes in the presence of an electric field. Electric potential difference is a measure of the strength of an electric field. The SI unit of electric potential difference is the volt (V).	V	
electric potential energy	Energy stored in an electric field is electric potential energy.	U	
equipotential	The potential is the same everywhere along an equipotential line, equipotential curve, or, simply, an equipotential. Any path that is perpendicular to the lines of field is an equipotential.		
equipotential surface	A surface that is perpendicular to a field everywhere is an equipotential surface.		
equivalent capacitance	The equivalent capacitance of a group of interconnected capacitors is the capacitance of a single capacitor that is the equivalent replacement of the group.		
parallel	When the capacitors are connected in parallel, all of the right plates are connected and (separately) all of the left plates are connected. Connecting capacitors in parallel results in a larger capacitance.	$C_{equiv} = \sum_{i=1}^{N} C_i$	(17-30)
series	When capacitors are connected in series, the right plate of one capacitor is connected to the left plate of the capacitor to its right. Connecting capacitors in series results in a smaller capacitance.	$C_{equiv} = \dfrac{1}{\sum_{i=1}^{N} \dfrac{1}{C_i}}$	(17-27)
voltage	Voltage is equivalent to electric potential difference.	V	

QUESTIONS AND PROBLEMS

In a few problems, you are given more data than you actually need; in a few other problems, you are required to supply data from your general knowledge, outside sources, or informed estimate.

Interpret as significant all digits in numerical values that have trailing zeros and no decimal points.

For all problems, use $g = 9.8$ m/s^2 for the free-fall acceleration due to gravity. Neglect friction and air resistance unless instructed to do otherwise.

- Basic, single-concept problem
- Intermediate-level problem, may require synthesis of concepts and multiple steps
- Challenging problem

SSM *Solution is in Student Solutions Manual*

Conceptual Questions

1. •What is the difference between electric potential and electric field?

2. •What is the difference between electric potential and electric potential energy?

3. •Explain why electric potential requires the existence of only one charge, but a finite electric potential energy requires the existence of two charges.

4. •An electron is released from rest in an electric field. Will it accelerate in the direction of increasing or decreasing potential? Why?

5. •Does it make sense to say that the voltage at some point in space is 10.3 V? Explain your answer. SSM

6. ••Explain why an electron will accelerate toward a region of lower electric potential energy but higher electric potential.

7. •Discuss how a topographical map showing various elevations around a mountain is analogous to the equipotential lines surrounding a charged object.

8. •How much work is required to move a charge from one end of an equipotential path to the other? Explain your answer.

9. •(a) If the electric potential throughout some region of space is zero, does it necessarily follow that the electric field is zero? (b) If the electric field throughout a region is zero, does it necessarily follow that the electric potential is zero? SSM

10. •Explain why capacitance depends neither on the stored charge Q nor on the potential difference V between the plates of a capacitor.

11. •Describe three methods by which you might increase the capacitance of a parallel plate capacitor.

12. •If the voltage across a capacitor is doubled, by how much does the energy stored change?

13. ••The capacitance of several capacitors in series is less than any of the individual capacitances. What, then, is the advantage of having several capacitors in series?

14. •What is the advantage to arranging several capacitors in parallel?

15. ••You charge a capacitor and then remove it from the battery. The capacitor consists of large movable plates with air between them. You pull the plates a bit farther apart. What happens to the stored energy? SSM

16. •Which way of connecting (series or parallel) three identical capacitors to a battery would store more energy?

17. •Qualitatively explain why the equivalent capacitance of a parallel combination of identical capacitors is larger than the individual capacitances.

18. •Does inserting a dielectric into a capacitor increase or decrease the energy stored in the capacitor? Explain your answer.

19. •What are the benefits, if any, of filling a capacitor with a dielectric other than air?

20. ••Capacitors A and B are identical except that the region between the plates of capacitor A is filled with a dielectric. As shown in **Figure 17-23**, the plates of these capacitors are maintained at the same potential difference by a battery. Is the electric field intensity in the region between the plates of capacitor A smaller, the same, or larger than the field in the region between the plates of capacitor B? Explain your answer.

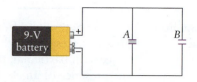

Figure 17-23 Problem 20

Multiple-Choice Questions

21. ••For a positive charge moving in the direction of the electric field,
    A. its potential energy increases and its electric potential increases.
    B. its potential energy increases and its electric potential decreases.
    C. its potential energy decreases and its electric potential increases.
    D. its potential energy decreases and its electric potential decreases.
    E. its potential energy and its electric potential remain constant. SSM

22. ••If a negative charge is released in a uniform electric field, it will move
    A. in the direction of the electric field.
    B. from high potential to low potential.
    C. from low potential to high potential.
    D. in a direction perpendicular to the electric field.
    E. in circular motion.

23. •An equipotential surface must be
    A. parallel to the electric field at every point.
    B. equal to the electric field at every point.
    C. perpendicular to the electric field at every point.
    D. tangent to the electric field at every point.
    E. equal to the inverse of the electric field at every point.

24. •A positive charge is moved from one point to another point along an equipotential surface. The work required
    A. is positive.
    B. is negative.
    C. is zero.
    D. depends on the sign of the potential.
    E. depends on the magnitude of the potential.

25. •The electric potential measured at a point equidistant from two particles that have charges equal in magnitude but of opposite sign is
    A. larger than zero.
    B. smaller than zero.
    C. equal to zero.
    D. equal to the average of the two distances times the charges.
    E. equal to the net electric field.

This page is intentionally left blank.

For complete end of chapter problem sets, please go to
www.whfreeman.com/kestentauck

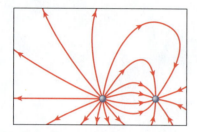

Figure 17-25 Problem 40

41. •In **Figure 17-26**, equipotential lines are shown at 1-m intervals. What is the electric field at (a) point A and (b) point B?

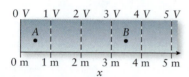

Figure 17-26 Problem 41

42. •Equipotential lines for some region of space are shown in **Figure 17-27**. What is the electric field at (a) point A and (b) point B?

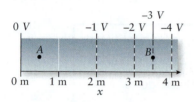

Figure 17-27 Problem 42

43. ••Two charged, parallel plates are 20 cm apart and there is a potential difference of 100 V across them. The plates produce a uniform electric field between them. Referring to **Figure 17-28**, draw in the three equipotential lines (A, B, and C) at the x positions indicated and determine the voltage associated with each. Assume that the potential of the right plate (at $x = 20$ cm) is zero.

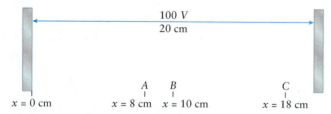

Figure 17-28 Problem 43

44. ••Draw the (a) equipotential lines and (b) electric field lines surrounding a dipole ($+q$ is a distance L from $-q$) (**Figure 17-29**). All charges are "points" of very small (negligible) dimensions.

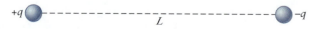

Figure 17-29 Problem 44

45. ••Draw the (a) equipotential lines and (b) electric field lines surrounding pairs of charges with the same sign charge (**Figure 17-30**). SSM

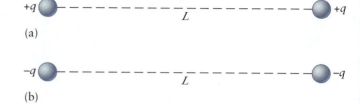

Figure 17-30 Problem 45

46. ••Draw the electric field lines and the electric equipotential lines for the following charge distribution (**Figure 17-31**).

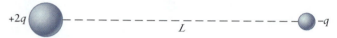

Figure 17-31 Problem 46

17-3: Electrical Potential Due to Certain Charge Distributions

47. •**Chemistry** What is the electric potential due to the nucleus of hydrogen at a distance of 5×10^{-11} m?

48. •(a) What is the electric potential due to a point charge of $+2$ μC at a distance of 0.5 cm? (b) How will the answer change if the charge has a value of -2 μC?

49. •The electric potential has a value of -200 V at a distance of 1.25 m from a point charge. Find that charge.

50. •At point P in **Figure 17-32** the electric potential is zero. (As usual, we take the potential to be zero at infinite distance.) (a) What can you say about the two charges? (b) Are there any other points of zero potential on the line connecting P and the two charges?

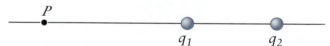

Figure 17-32 Problem 50

51. •Two point charges are placed on the x axis: $+0.5$ μC at $x = 0$ and -0.2 μC at $x = 10$ cm. At what point(s), if any, on the x axis is the electric potential equal to zero? SSM

52. •A charge of $+2.00$ μC is at the origin and a charge of -3.00 μC is on the y axis at $y = 40.0$ cm. (a) What is the potential at point a, which is on the x axis at $x = 40.0$ cm? (b) What is the potential difference $V_b - V_a$ when point b is at (40.0 cm, 30.0 cm)? (c) How much work is required to move an electron at rest from point a to rest at point b?

53. •Two equal point charges, each -0.02 μC, are placed on the x axis as shown in **Figure 17-33**. Find the electric potential (a) at point A and (b) at point B.

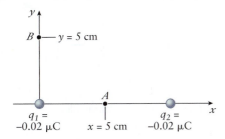

Figure 17-33 Problem 53

54. ••Calculate the electric potential at the origin O due to the point charges in **Figure 17-34**.

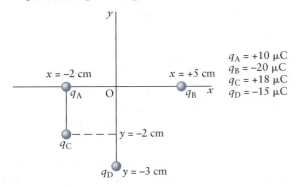

Figure 17-34 Problem 54

55. ••**Calc** Derive an expression for the electric potential at a point $r > R$ on the outside of a uniformly charged, solid sphere that has a radius R and carries a charge of Q. *Hint*: Use Gauss' law first to find $E(r)$.

56. ••Consider a 1.00-m³ cube that has $+2.00$ μC charges located at seven of its corners as shown in **Figure 17-35**. (a) Find the potential at the vacant corner. (b) How much work by an external agent is required to bring an additional $+2.00$ μC charge from rest at infinity to rest at the vacant corner? (c) How does the problem change, if at all, if a -2.00 μC charge is brought in to fill the empty corner rather than a positive charge?

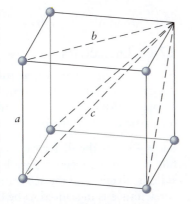

Figure 17-35 Problem 56

57. •••As shown in **Figure 17-36**, three particles, each with charge q, are at different corners of a rhombus with sides of length a and with one diagonal of length a and the other of length b. (a) What is the electric potential energy of the charge distribution? (b) How much work

by an external agent is required to bring a fourth particle, also of charge q, from rest at infinity to rest at the vacant corner of the rhombus? (c) What is the total electric potential energy of the four charges? SSM

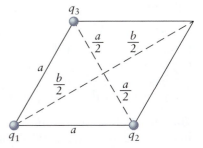

Figure 17-36 Problem 57

58. •••**Calc** The x axis coincides with the symmetry axis of a uniformly charged thin disk that has a radius R and a uniform positive surface charge density σ centered at the origin. (a) Make a sketch of E_x versus x for $-4R < x < +4R$. (b) Make a sketch of $V(x)$ versus x for $-4R < x < +4R$. (c) Let the disk have an infinite radius and redraw the graphs.

17-4: Capacitance

59. •A 2.0 μF capacitor is connected to a 12 V battery. What is the magnitude of the charge on each plate of the capacitor?

60. •Using a single 10 V battery, what capacitance do you need to store 10 μC of charge?

61. •A parallel plate capacitor has square plates that have edge lengths equal to 100 cm and are separated by 1 mm. What is the capacitance of this device?

62. •A parallel plate capacitor has a plate separation of 1 mm. If the material between the plates is air, what plate area is required to provide a capacitance of 2.0 pF?

63. •A parallel plate capacitor has square plates that have edge length equal to 1.0 m. If the material between the plates is air, what separation distance is required to provide a capacitance of 8850 pF? SSM

64. •A parallel plate capacitor has plates measuring 10 cm × 10 cm and a plate separation of 1.0 mm. If you want to construct a parallel plate capacitor of the same capacitance but with plates measuring 5.0 cm × 5.0 cm, what plate separation do you need?

17-5: Energy Stored in a Capacitor

65. •A parallel plate capacitor has square plates that have edge length equal to 100 cm and are separated by 1 mm. It is connected to a battery and is charged to 12 V. How much energy is stored in the capacitor?

66. •Using a single 10 V battery, what capacitance do you need to store 0.0001 J of electric potential energy?

67. •A capacitor has a capacitance of 80 μF. If you want to store 160 J of electric energy in this capacitor, what potential difference do you need to apply to the plates?

68. •You charge a 2.0 μF capacitor to 50 V. How much additional energy must you add to charge it to 100 V?

69. •**Medical** A defibrillator containing a 20 μF capacitor is used to shock the heart of a patient by holding it to the patient's chest. Just prior to discharging, the capacitor has voltage of 10 kV across its plates. How much energy is released into the patient? SSM

70. ••(a) You want to store 10^{-5} C of charge on a capacitor, but you only have a 100 V voltage source with which to charge it. What must be the value of the capacitance? (b) You want to store 10^{-3} J of energy on a capacitor, and you only have a 100 V voltage source with which to charge it. What must be the value of the capacitance?

17-6: Capacitors in Series and Parallel

71. •Three capacitors have capacitances 10 μF, 15 μF, and 30 μF. What is their effective capacitance if the three are connected (a) in parallel and (b) in series?

72. ••How should four 1.0 pF capacitors be connected to have a total capacitance of 0.75 pF?

73. ••Two capacitors provide an equivalent capacitance of 8.00 μF when connected in parallel and 2.00 μF when connected in series. What is the capacitance of each capacitor? SSM

74. ••A series circuit consists of a 0.50 μF capacitor, a 0.10 μF capacitor, and a 220 V battery. Determine the charge on each of the capacitors.

75. •A 0.05 μF capacitor and a 0.10 μF capacitor are connected in parallel across a 220 V battery. Determine the charge on each of the capacitors.

76. ••A 2.0 μF capacitor is first charged by being connected across a 6.0 V battery. It is then disconnected from the battery and connected across an uncharged 4.0 μF capacitor. Find the charge on each of the capacitors.

77. •Calculate the equivalent capacitance between a and b for the combination of capacitors shown in **Figure 17-37**.

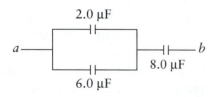

Figure 17-37 Problem 77

78. •What is the equivalent capacitance of the network of three capacitors shown in **Figure 17-38**?

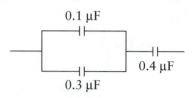

Figure 17-38 Problem 78

79. ••A 10.0 μF capacitor, a 40.0 μF capacitor, and a 100.0 μF capacitor are connected in series across a 12.0 V battery. (a) What is the equivalent capacitance of the combination? (b) What is the charge on each capacitor? (c) What is the potential difference across each capacitor? SSM

80. ••A 10.0 μF capacitor, a 40.0 μF capacitor, and a 100.0 μF capacitor are connected in parallel across a 12.0 V battery. (a) What is the equivalent capacitance of the combination? (b) What is the charge on each capacitor? (c) What is the potential difference across each capacitor?

81. ••For the capacitor network shown in **Figure 17-39**, the potential difference across ab is 75.0 V and all capacitors are accurate to three significant figures. How much charge and how much energy are stored in this system?

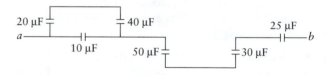

Figure 17-39 Problem 81

17-7: Dielectrics

82. •A parallel plate capacitor has plates of 1.0 cm by 2.0 cm. The plates are separated by a 1.0-mm-thick piece of paper. What is the capacitance of this capacitor? The dielectric constant for paper is 2.7.

83. •What is the dielectric constant of the material that fills the gap between a parallel plate capacitor (area of sides is 20 cm^2 and the separation distance is 1 cm) if the capacitance is measured to be 0.0142 μF?

84. ••A parallel plate capacitor has square plates that have edge lengths equal to 100 cm and are separated by 1 mm. It is connected to a battery and charged to 12 V. How much energy would be stored in the capacitor if a ceramic dielectric material (κ is 5.8) fills the space between the plates?

85. ••A 2800 pF air-gap capacitor is connected to a 16 V battery. If a ceramic dielectric material ($\kappa = 5.8$) fills the space between the plates, how much charge will flow from the battery? SSM

86. ••A parallel plate capacitor that has a plate separation of 0.5 cm is filled halfway with a slab of dielectric material (κ is 5) (**Figure 17-40**). If the plates are 1.25 cm by 1.25 cm in area, find the capacitance of the combination.

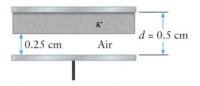

Figure 17-40 Problem 86

87. •••(a) Determine the capacitance of the parallel plate capacitor shown in **Figure 17-41**. The dielectric with constant κ_1 fills up one-quarter of the area, but the full separation of the plates. The materials with constants κ_2 and κ_3 fill the other three-quarters of the area, and divide the separation of the plates in half. (b) What happens to the total capacitance if the material with dielectric constant κ_3 is replaced by air?

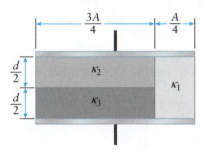

Figure 17-41 Problem 87

General Problems

88. ••A lightning bolt transfers 20 C of charge to Earth through an average potential difference of 30 MV. (a) How much energy is dissipated in the bolt? (b) How much water (at a temperature of 100°C) can be turned into steam with this energy?

89. ••A uniform electric field of 2.00 kN/C points in the $+x$ direction. (a) What is the potential difference $V_b - V_a$ when point a is at $x = -30.0$ cm and point b is at $x = +50.0$ cm? (b) A test charge $q_0 = +2.00$ nC is released from rest at point a. What is its kinetic energy when it passes through point b? (c) If a negative charge instead of a positive charge were used in this problem, qualitatively how would your answers change? SSM

90. ••Calc An electric field is given by $E = (-3.00 \text{ kV/m}^2)\hat{x}$ on the x axis. (a) What is the potential difference $V_b - V_a$ when point a is at $x = -30.0$ cm and point b is at $+50.0$ cm? (b) How much work is done by an external agent in bringing a test charge $q_0 = +2.00$ nC from rest at point a to rest at point b? (c) If point b were moved to $x = +30.0$ cm, how much work would be required from the external agent? (d) Explain how the result is possible by describing the physical system.

91. ••A point charge q_0 that has a charge of 0.5 μC is at the origin. (a) Calculate the potential at $x = 0.8$ m (taking it to be zero at infinite distance). (b) A second particle q that has a charge of 1 μC and a mass of 0.08 g is placed at $x = 0.8$ m. What is the potential energy of this system of charges? (c) If the particle with charge q is released from rest, what will its speed be when it reaches $x = 2$ m?

92. ••In 2004, physicists at the Stanford Linear Accelerator Center (SLAC) in California fired electrons toward each other at very high speeds so that they came within 1.0×10^{-15} m of each other (approximately the diameter of a proton). (a) What was the electrical force on each electron at closest approach? (b) Would this force be large enough to move you? (c) How much energy did it take to get the electrons that close together? (d) How fast must the electrons have been traveling initially, when far from each other, to get that close? (e) Do you notice anything suspicious about your answer?

93. ••Biology Two red blood cells each have a mass of 9.0×10^{-14} kg and carry a negative charge spread uniformly over their surfaces. As described in Example 16-2, the repulsion arising from the excess charge prevents the cells from clumping together. One cell carries -2.50 pC of charge and the other -3.10 pC, and each cell can be modeled as a sphere 7.5 μm in diameter. (a) What speed would they need when very far away from each other to get close enough to just touch? Assume that there is no viscous drag from any of the surrounding liquid. (b) What is the maximum acceleration of the cells in part (a)?

94. ••Biology Under certain circumstances, potassium ions (K^+) move across the 8.0-nm-thick cell membrane from the inside to the outside. The potential inside the cell is -70 mV and the potential outside is zero. (a) What is the change in the electrical potential energy of the potassium ions as they move across the membrane? Does their potential energy increase or decrease as they move from the inside to the outside? (b) What electrical force (magnitude and direction) is exerted on the potassium ions when they are inside the potassium channel that spans the cell membrane, assuming that it behaves electrically like water?

95. ••**Calc** An infinitely long line charge of linear charge density λ equal to 2.00 nC/m lies on the z axis. (a) Find the potential difference $V_c - V_b$ when point b is at (40.0 cm, 30.0 cm, 0) and point c is at (200 cm, 0, 50.0 cm). (b) Does your answer depend on your choice of the reference point at which $V = 0$? (c) Describe the equipotential surfaces. SSM

96. •••**Calc** A charged semicircular ring that has a radius R equal to 7.00 cm and is centered at the origin has a linear charge density that varies with angle according to

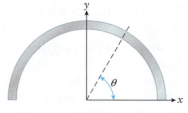

Figure 17-42 Problem 96

$$\lambda = (4.00 \, \mu \times C/m) \sin\theta$$

(**Figure 17-42**). Use direct integration to determine the electric potential at the origin due to this charge distribution.

97. •••**Calc** A uniformly charged sphere has radius R and a total charge Q. (a) Use Gauss' law to find an expression for the electric field E_r for $0 < r < \infty$. (b) Use your result from part (a) to find an expression for the electric potential $V(r)$ everywhere. (c) Sketch a graph for both E_r and V as functions of r.

98. ••**Calc** Using calculus, determine the electric potential at a point P that is 25 cm above the central axis of a solid disk having +200 mC of charge (**Figure 17-43**). The charge is uniformly spread over the area of the disk and the radius of the disk is 50 cm.

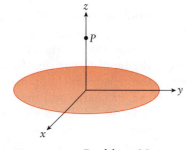

Figure 17-43 Problem 98

99. •••**Biology** We can model a neuron, freshly dissociated from the brain, as a sphere surrounded by a membrane that is 8.0 nm thick. Cell diameter can vary considerably, depending on the cell, but 25 μm is reasonable. The membrane carries equal but opposite charges on its inner and outer surfaces. In its resting state, the potential everywhere on the inner surface of the cell (but not inside the membrane itself) is -70 mV. (a) What is the potential of the outer surface of the cell membrane? (b) Which is positively charged, the inner surface of the membrane or the outer surface? How do you know? (c) What are the magnitude and direction of the electric field inside the membrane? (d) What is the electric field inside the cell?

100. •••**Calc** (a) Calculate the net electric potential at the point $x = +35$ cm due to the 26 point charges (all with a value of $+1$ μC) evenly spaced between $x = 0$ and $x = 25$ cm (**Figure 17-44a**). (b) Now using calculus, calculate the electric potential at the point $x = +35$ cm due to a uniformly charged rod of length 25 cm and charge $+26$ μC (**Figure 17-44b**). Assume the rod is very thin. (c) Compare and comment on your answers to parts (a) and (b).

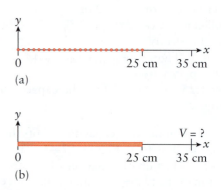

Figure 17-44 Problem 100

101. ••A parallel plate capacitor has a plate separation of 1.5 mm and is charged to 600 V. If an electron leaves the negative plate, starting from rest, how fast is it going when it hits the positive plate?

102. ••Suppose you are supplied with five identical capacitors (each at 10 μF). Determine all of the unique combinations that use all five capacitors and the equivalent capacitance of each.

103. •Calculate the equivalent capacitance of the combination in **Figure 17-45**. SSM

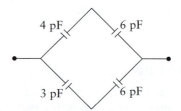

Figure 17-45 Problem 103

104. ••The following four capacitors have an equivalent capacitance of 8 μF (**Figure 17-46**). Calculate the value of C_x.

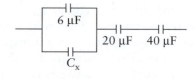

Figure 17-46 Problem 104

105. ••Determine the equivalent capacitance of the combination in **Figure 17-47**.

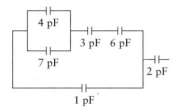

Figure 17-47 Problem 105

106. ••Calculate the charge stored on each capacitor in the circuit shown in **Figure 17-48**.

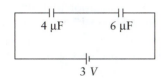

Figure 17-48 Problem 106

107. ••**Calc** What is the approximate capacitance of a cylindrical capacitor (**Figure 17-49**)? Assume that an excess charge of $+Q$ is placed onto an inner cylinder of length L and radius R_1. An excess charge of $-Q$ is placed onto the surface of a thin, concentric cylinder of radius R_2 ($R_2 > R_1$) and length L. Assume that there is air in between the two cylinders.

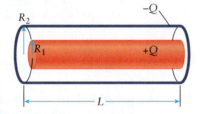

Figure 17-49 Problem 107

108. •••**Calc** One useful tool to detect radiation is the Geiger-Müller (G-M) tube. It consists of a rigid metal cylindrical wire of radius a and length L, surrounded by a larger metal cylinder of the same length and radius b. The wire lies along the central axis of the outer cylinder. A voltage V is placed across the two cylinders, producing an electric field between them. Charged particles of radiation that enter the field can be detected because they temporarily perturb the potential difference between the cylinders. The electric field outside the wire is the same as that of a line of charge. (a) Show (by integration) that the potential difference V between the two cylinders is $V = \lambda/2\pi\epsilon_0 \ln(b/a)$, where λ is the linear charge density on the wire. (b) Measurements of an actual G-M tube show that the diameter of the inner wire is 1.00 mm, the diameter of the inside of the outer cylinder is 30.0 mm, and the length is 6.50 cm. The operating voltage of the tube is 925 V. Approximately how much charge is on the surface of the inner wire? (c) When the G-M tube in part (b) is in operation, find the electric field just outside the surface of the wire and just inside the inner surface of the outer cylinder, but not within the metal of the cylinder. (d) What is the capacitance of the G-M tube in part (b)? (e) How much energy is stored in the G-M tube at its operating potential?

109. •••**Calc** (a) Derive an expression for the capacitance of a concentric, spherical, air-filled capacitor with inner conductor of radius R_1 and outer radius R_2 (**Figure 17-50**). *Hint:* Use Gauss' law to find E and then integrate to find V. (b) If the inner sphere is made larger, will the capacitor store more or less charge per volt? Explain your answer. SSM

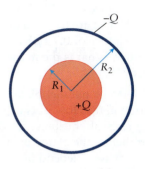

Figure 17-50 Problem 109

110. ••A parallel plate capacitor is made by sandwiching 0.1-mm sheets of paper (dielectric constant 2.7) between three sheets of aluminum foil (A, B, and C in **Figure 17-51**) and rolling the layers into a cylinder. A capacitor that has an area of 10 m² is fabricated this way. (To be practical it would then have to be folded up so as to fit in a small package.) What is the capacitance of this capacitor?

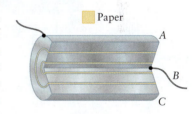

Figure 17-51 Problem 110

111. ••A parallel plate capacitor with a capacitance of 5 μF is charged with a 12 V battery. After fully charging the plates, the battery is removed. How much work is required to increase the separation between the plates by a factor of 3?

112. ••An air-gap, parallel plate capacitor with area A and gap width d is connected to a battery that maintains the plates at potential difference V. (a) The plates are pulled apart, doubling the gap width, while they remain in electrical contact with the battery terminals. By what factor does the potential energy of the capacitor change? (b) If the capacitor is removed from the battery and thus isolated, what happens to the stored potential energy when the gap width is doubled? Explain your answer.

113. ••An air-filled parallel plate capacitor is attached to a battery with a voltage V. While attached to the battery, the area of the plates is doubled and the separation of the plates is halved. During this process, what happens to (a) the capacitance, (b) the charged stored on the positive plate of the capacitor, (c) the potential across the plates of the capacitor, and (d) the potential energy stored in the capacitor, as compared to the original configuration? (e) How would your answers change if, once the capacitor was charged by the battery, it was disconnected from the battery while the area and separation were changed as above?

114. •••You have a bucketful of capacitors, each with a capacitance of 1.00 μF and a maximum voltage rating of 250 V. You are to come up with a combination that has a capacitance of 0.75 μF and a maximum voltage rating of 1000 V. What is the minimum number of capacitors you need?

115. •••A parallel plate capacitor has area A and separation d. (a) How is its capacitance affected if a *conducting* slab of thickness $d' < d$ is inserted between, and parallel to, the plates as shown in **Figure 17-52**? (b) Does your answer depend on where the slab is positioned vertically between the plates? SSM

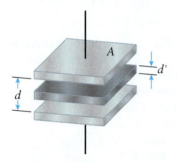

Figure 17-52 Problem 115

116. •••Three 0.18 μF capacitors are connected in parallel across a 12 V battery, as shown in **Figure 17-53**. The battery is then disconnected. Next, one capacitor is carefully disconnected so that it doesn't lose any charge and is reconnected backward, that is, with its positively charged side and its negatively charged side reversed. (a) What is the potential difference across the capacitors now? (b) By how much has the stored energy of the capacitors changed in the process?

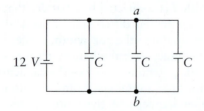

Figure 17-53 Problem 116

117. •(a) Calculate the charge and energy on the 25 μF capacitor when the switch S is placed at position A (**Figure 17-54**). (b) Repeat for both the 25 μF and the 20 μF capacitors after the switch is then placed at position B.

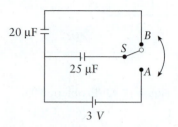

Figure 17-54 Problem 117

118. ••There are three equipotential lines for 30 V, 10 V, and −10 V shown in **Figure 17-55**. A proton's speed as it passes point A is 80 km/s. It follows the path shown in the figure. What is the proton's speed at point B?

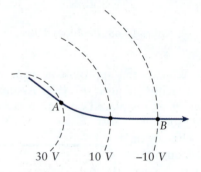

Figure 17-55 Problem 118

119. ••A parallel plate, air-gap capacitor has a charge of 20.0 μC and a gap width of 0.100 mm. The potential difference between the plates is 200 V. (a) What is the electric field in the region between the plates? (b) What is the surface charge density on the positive plate? (c) If the plates of the capacitor are moved closer together while the charge remains constant, how are the electric field, surface charge density, and potential difference going to change, if at all? Explain your answers.

18 Moving Charges

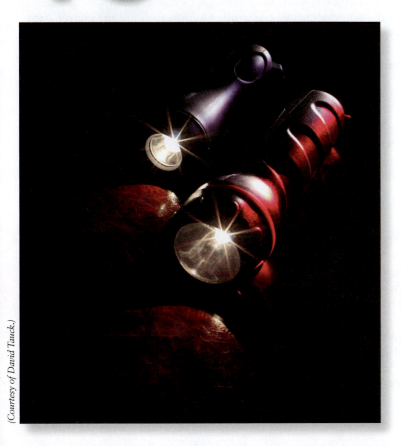

(Courtesy of David Tauck.)

The potential difference between the terminals of the battery in a flashlight causes electrons to move in the filament of the bulb. However, the material making up the filament offers resistance to that flow of charge, causing the electrons to lose energy. That energy goes into making the filament warmer (which in turn makes the bulb warm), and some of the energy is then radiated away in the form of light. The resistance of the filament and the rest of the circuit, together with the potential difference generated by the battery determines the rate at which energy is delivered—the power ("brightness") of the bulb.

While the phenomena associated with static charges are both interesting and important, the really exciting stuff happens when charges move. Although a flash of lightning suddenly brightening a dark sky may inspire poets, the effects of charges flowing through neurons, the heart, and electronic circuits are what excite us. In this chapter, we will study simple circuits and explore some of the things that happen when charges move through them.

18-1 Current

In the previous two chapters, we've examined a variety of topics which fall into the general category of electrostatics. We imagined, for example, adding a certain amount of excess charge to a block of metal, and then determining the electric field it would generate at various points in space by applying Gauss' law. In doing so, we considered the charges to be fixed in place—static. Are the charges really not moving? No! But are they experiencing the kind of motion to which the title of this chapter refers? No, again. Imagine the charges as members of a marching band. As musicians move around on a football field they form one pattern after another, but *as a group* the marching band stays fixed in one place, more or less, on the field. In contrast, when they march in a parade, *as a group* they can move from one end of Main Street to the other. In this chapter, we focus on the latter kind of net motion. (If members of the band were like electrons, however, they would bump into lots of things along the way.)

The net flow of charges is **current**. Although electrons carry charges along wires and through the components of electronic circuits, ions are the charge carriers in biological systems such as neurons and muscle cells. However, the principles governing current are the same. If a net charge Δq moves past a certain point in a time Δt, the current i is defined as the flow rate:

$$i = \frac{\Delta q}{\Delta t} \qquad (18\text{-}1)$$

In much the same way as we treated fluid flow in Chapter 11 we will limit our discussion of currents to a steady flow of charge carriers. This assumption is not unrealistic under most circumstances, and it allows us to equate Equation 18-1 to a more proper definition of current, the derivative of charge with respect to time:

$$i = \frac{dq}{dt} \qquad (18\text{-}2)$$

$\sqrt{x}$ *See the Math Tutorial for more information on Calculus*

This equation is true for any material that conducts electric current, for example, metals, but also for aqueous solutions that conduct ions. As we saw in Section 16-3, metals are especially good electrical conductors because one or more of the electrons associated with a metal atom are only weakly bound to the nucleus, and can therefore move around relatively freely within the material. We will use the term "wire" to refer to a generic metallic conductor that has a circular cross section unless a special situation requires some other particular shape.

The SI unit of current is the ampere, named after the French scientist and mathematician André Ampère. One ampere (abbreviated amp or A) is equivalent to one coulomb of charge passing a given point per second:

$$1\,\text{A} = 1\,\frac{\text{C}}{\text{s}}$$

Although the charge of an electron is clearly a fundamental constant in nature, it has a curious value of about 1.6×10^{-19} C. We humans make up all of the units we use, so why isn't the charge of an electron something simple such as, say 1 C? Our understanding of currents stems from work begun by Ampère and other scientists more than 250 years ago, long before anyone knew the identity of the fundamental charge carriers. The English physicist J. J. Thomson discovered the electron only about 100 years ago, so it's not surprising that the way we quantify current is neat and tidy compared to the way we quantify charge.

Imagine a net flow of electrons in a wire loop, pushed along by a potential difference supplied by a battery (**Figure 18-1**). If the current is 1 A or 1 C/s, the number of electrons moving past any specific point on the wire is

An electric field is created due to the potential difference between the two terminals of the battery. Charges feel a force due to this field, resulting in a current in the wire.

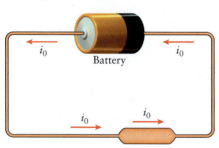

There's no way for electrons to leave or join the flow, and electrons cannot accumulate in one place. So the current is constant around the loop,

Figure 18-1 A net flow of electrons arises in a wire loop due to the potential difference supplied by a battery.

$$1\,\text{A} = 1\,\frac{\text{C}}{\text{s}} = 1\left(\frac{\text{C}}{\text{s}}\right)\left(\frac{1\,e^-}{1.6 \times 10^{-19}\,\text{C}}\right) = 6.0 \times 10^{18}\,\frac{e^-}{\text{s}}$$

If we were to pass an imaginary plane through the wire, even a plane not perpendicular to the direction of the net flow of charge, we would therefore find about 6×10^{18} electrons crossing that plane per second. (Again, this value would be the *net* flow; there can always be some electrons moving randomly in any direction in the wire.) Where can those 6×10^{18} electrons go, except to continue their net flow around the loop? There is no way for them to accumulate at some point or another, no way for more electrons to join the flow, and no way to destroy any electrons. In this regard there is no difference between the regions where the wire is thin and where it is thick, so the value of the current remains the same everywhere in the loop.

How does current arise? We saw in Chapter 16 that electric fields exert a force on charges. In Chapter 17, we discovered a relationship between electric field and the potential difference between two points in the field. Together, these ideas suggest that when a potential difference is applied, for example, between the terminals of a battery or across the membrane of a muscle cell, an electric field is created that exerts a force on any charges in the region. The force causes a net motion of charges that are free to move; that net flow of charge is electric current.

Although we will rarely make use of it, current has a direction as well as a magnitude. Notice, however, that neither quantity used to define current in Equation 18-1 is a vector, so current is not a vector either. We'll need to pay attention to this distinction mostly when we consider mathematical operations that involve current. Electric potential difference can be either positive or negative; it has a polarity. The sign of a potential difference determines the direction of the electric field, which in turn determines the direction of the force experienced by a charged object within the field. By convention, the direction assigned to a current is the direction in which positive charge carriers would move.

Why positive charge carriers? In electrical circuits and many other situations in which currents arise, it is the negatively charged electrons that are free to move. Protons, the positive building blocks of atoms, are locked into the heavy nuclei of atoms and therefore remain relatively fixed in place. So why isn't the direction of current defined by the motion of negative charge carriers? The answer is purely historical. As we noted above, scientists explored phenomena associated with currents long before discovering that current is a net flow of electrons. The convention for current direction, often attributed to the American statesman and scientist Benjamin Franklin, was therefore an arbitrary choice. However, in nearly all of the situations we will consider, the physics of negatively charged objects moving in one direction is the same as the physics of positively charged objects moving in the other direction. So the fact that the **conventional current** is in the opposite direction of the flow of electrons will not affect our work. Moreover, in biological systems, currents arise from the flow of ions.

Example 18-1 Charging Sphere

In our demonstration room we have a Van de Graaff generator, shown in **Figure 18-2**. The device includes a metal sphere of radius $R = 0.15$ m electrically isolated from its surroundings except for a wire which can carry a current. When the current is a steady $i = 5.0$ μA, how long does it take for the sphere to reach its operating potential of 400 kV if it is initially uncharged?

SET UP

The electric potential on the surface of a charged sphere of radius R depends on the amount of excess charge Q on the sphere (Equation 17-10):

$$V = \frac{kQ}{R}$$

(Recall that this equation is shorthand for the difference between the potential at the surface and at an infinite distance from the sphere; we have used the standard convention that $V = 0$ at infinite distance.) We are asked to find how long it takes to charge the sphere. How do we bring time into our solution? Current relates time to charge.

Figure 18-2 When the large metal sphere of a Van de Graaff generator is raised to its operating potential, a spark jumps to a nearby, grounded sphere. *(Sciencephotos/Alamy.)*

SOLVE

We defined current to be the net flow of charge per unit time:

$$i = \frac{\Delta q}{\Delta t} \tag{18-1}$$

We can rearrange the equation as an expression for the time interval:

$$\Delta t = \frac{\Delta q}{i}$$

where Δq represents the total charge transferred to the sphere in time Δt. Because the sphere is initially neutral, Δq is also the charge Q on the sphere at any time, so

$$\Delta t = \frac{Q}{i} = \frac{VR/k}{i} = \frac{VR}{ki} \tag{18-3}$$

Finding a numerical solution is now straightforward:

$$\Delta t = \frac{VR}{ki} = \frac{(4.0 \times 10^5 \text{ V})(0.15 \text{ m})}{(8.99 \times 10^9 \text{ N} \cdot \text{m}^2/\text{C}^2)(5.0 \times 10^{-6} \text{ A})} = 1.3 \text{ s}$$

REFLECT

Although you probably cannot use your experiences to assess whether the answer should be on the order of seconds (or say, tenths of a second or tens of seconds), if you've ever seen a demonstration of a Van de Graaff generator, you'll certainly agree that it starts to throw off sparks relatively quickly. So a result on the order of 1 s is reasonable.

Practice Problem 18-1 The operating potential of the metal sphere of a Van de Graaff generator is 400 kV. A current of 0.20 mA exists when a fully charged sphere, of radius 0.15 m, is connected to ground. What is the potential of the sphere after it has discharged for 0.01 s?

A current is the net flow of charge along the wire.

n charges per unit volume

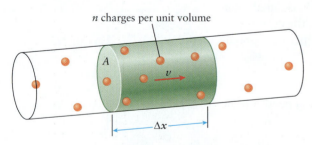

In any volume, the next moving charge is the product of the volume, *n*, the number of moving charges per volume, and the fundamental charge *e*.

Figure 18-3 Charges drift through a volume of cross-sectional area *A* and length Δx in a wire carrying a current.

We defined current as a net flow rate of charge, so it's not unreasonable to ask how fast the individual charges that make up the current actually move. To find an answer, we need a relationship between current and velocity.

Consider a wire that has a cross-sectional area A and carries a current i, as shown in **Figure 18-3**. Charges are moving ("drifting") through the green region at an average velocity of v_{drift}, and at any time, the net moving charge in the region is ΔQ. This charge can be expressed in terms of n (the number of moving charges per volume), the volume of the green region, and the fundamental unit of electric charge:

$$\Delta Q = n(\text{volume})e = nA\,\Delta x e \tag{18-4}$$

where n multiplied by volume is the total number of charge carriers and e is the charge of one of the carriers. Our functional definitions of current and velocity are $\Delta Q/\Delta t$ and $\Delta x/\Delta t$, respectively; both expressions can be found in the previous equation (Equation 18-4) if we divide each side by a time interval Δt:

$$\frac{\Delta Q}{\Delta t} = \frac{nA\,\Delta x e}{\Delta t} = nA\frac{\Delta x}{\Delta t}e$$

So

$$i = n A v_{\text{drift}} e$$

We see, then, that the average rate at which the charge carriers move is related to the current, the cross-sectional area of the wire that carries the current, and the density n of charge carriers:

$$v_{\text{drift}} = \frac{i}{nAe} \tag{18-5}$$

This equation is a compact expression and shows the dependence of the *drift speed* on parameters that describe the current and the wire which carries it. But aren't you curious to know what the actual velocity of electrons is in an actual wire, when you throw a switch to turn on a light? Is the speed fast or slow? Let's do a quick, rough calculation to find out.

Example 18-2 How Fast?

Estimate the drift speed of electrons flowing as a current under normal conditions through a typical copper wire behind the wall of your classroom. Assume a current of 5 A through a 12-gauge wire, which has a radius of about 1 mm. Also, because metals such as copper have a single, loosely bound electron in their outermost atomic orbital, on average, the number of electrons per unit volume can be taken to be the number of atoms per unit volume. The number density of copper is 8×10^{28} atoms/m^3.

SET UP
Equation 18-5 gives the drift speed:

$$v_{\text{drift}} = \frac{i}{nAe}$$

The charge of an electron is 1.6×10^{-19} C.

SOLVE
With the given values, the drift speed of electrons in a 5 A current in the 12-gauge copper wire is

$$v_{\text{drift}} = \frac{5 \text{ A}}{(8 \times 10^{28} \text{ e/m}^3)\pi(10^{-3} \text{ m})^2(1.6 \times 10^{-19} \text{ C/e})} = 1.2 \times 10^{-4} \text{ m/s}$$

With our estimated value we are justified in quoting the answer only to one significant digit, that is, the drift speed is approximately 1×10^{-4} m/s.

REFLECT
The drift speed of the electrons that make up the current is on the order of 10^{-4} m/s. Is that fast or slow? At that rate, the time it would take electrons to make a 10-m trip between the switch and the lights in a classroom is

$$\Delta t = \frac{d}{v_{\text{drift}}} = \frac{10 \text{ m}}{10^{-4} \text{ m/s}} = 10^5 \text{ s}$$

10^5 s—that's more than a day! Because the electrons encounter frequent collisions with atoms in the wire, and because their paths are not straight, each individual electron moves relatively slowly—it drifts down the wire. But because all of the electrons in the wire start to move at nearly the same time, as each electron moves, changes in the electric field propagate at close to the speed of light! So it only takes an instant, not a day, for the lights to come on after the switch is thrown.

✴ What's Important 18-1

An electric field exerts force on charges. In the absence of an applied electric field, the free charges in an electrical conductor move at random; an applied electric field results in a net flow of charge or electric current. The SI unit of current is the ampere, commonly abbreviated amp or A. One ampere is equivalent to one coulomb of charge passing a given point per second. Current has a magnitude and a direction, but it is not a vector. By convention, the direction assigned to a current is the direction positive charge carriers would move, although in general, it is the negative electrons that carry current in non-biological systems.

18-2 Resistance and Resistivity

Resistance is the opposition that an object offers to the flow of charge. To study the effect of resistance in a circuit—either in a biological one or in the kind you find, say, in a cell phone—we need to develop a physical description of it. What determines the resistance of a given piece of material? How does resistance affect the amount of current in a circuit? The answers to these questions come by considering the electric field that drives charges into motion.

Recall that in Chapter 17, we found a relationship between the area A of the plates of a capacitor, the charge q on it, and the electric field:

$$E = \frac{q}{\epsilon_0 A} \tag{17-18}$$

where ϵ_0 is the permittivity of free space, equal to $8.85 \times 10^{-12}\,\mathrm{C^2/(N \cdot m^2)}$. This expression led to a relationship between the charge and the electric potential difference V applied between the plates:

$$q = \frac{\epsilon_0 A}{d} V \tag{18-6}$$

where the plate area A and the plate separation d describe the geometry of the capacitor. The fractional term is capacitance C, so,

$$q = CV \tag{17-19}$$

Can we find a similar relationship between the current in a wire and the applied potential difference?

An electric field is created whenever a potential difference exists between two points in a conducting material, between the ends of a wire, for example. Experimental observations show that for many materials, the current that arises is proportional to both the electric field E created by the potential and the cross-sectional area A of the conductor,

$$i \propto E A = \sigma E A$$

where the proportionality constant σ, or **conductivity**, is a measure of how well or poorly a particular material conducts an electric current. We can change this expression into a relationship between charge and potential, as we did in Chapter 17 for static charge stored on a capacitor, using the relationship between the field and the potential:

$$V = -\int \vec{E} \cdot d\vec{s} \tag{17-5}$$

This integral yields $V = EL$ over a distance L in the direction parallel to a uniform field. So along a length L of a straight wire to which a potential V is a applied,

$$i = \sigma \frac{V}{L} A = \frac{\sigma A}{L} V \qquad (18\text{-}7)$$

The similarity between Equation 18-6 and Equation 18-7 suggests that the fraction in Equation 18-7 is a fundamental characteristic of moving charge in much the same way that capacitance is for stored, static charge. For historical reasons, **resistance** R is defined as the inverse of that term:

$$R = \frac{L}{\sigma A} \qquad (18\text{-}8)$$

Equation 18-7 becomes

$$i = \frac{1}{R} V$$

or

$$V = iR \qquad (18\text{-}9)$$

Equation 18-9 is sometimes incorrectly referred to as Ohm's law. For many conducting materials resistance is constant over a wide range of applied potential. In those cases the relationship between V and i is linear; materials for which this is true are said to obey Ohm's law or to be ohmic. However, there are any number of materials and electronic devices for which the resistance is not constant as the potential applied across it changes. For this reason they are not ohmic, but they nevertheless obey Equation 18-9. It is therefore not strictly correct to label Equation 18-9 as Ohm's law.

As the name suggests, resistance is the characteristic of an object to resist the flow of charge. For that reason, it makes sense to define resistance in terms of **resistivity** ρ, the measure of how well or poorly a material resists an electric current, rather than conductivity σ, the measure of how well or poorly a material conducts current. As with conductivity, resistivity differs from material to material; some typical values are given in Table 18-1. The larger the value of resistivity, the better a material is able to resist the current. As you see from the table, materials that are good insulators, such as rubber and glass, have values of resistivity many orders of magnitude higher than conductors such as copper.

Resistivity is the inverse of conductivity:

$$\rho = \frac{1}{\sigma} \qquad (18\text{-}10)$$

In other words, a good resistor is a poor conductor, and vice versa. In terms of resistivity, the definition of resistance (Equation 18-8) becomes

$$R = \frac{\rho L}{A} \qquad (18\text{-}11)$$

The SI unit of resistance is the ohm or Ω. (The corresponding units for resistivity are then $\Omega \cdot m$.) An ohm is a relatively small resistance—the resistors you might find for sale in a local store will likely be in the range of 1 kΩ to 10 MΩ. The resistance of a human body can be more than 0.5 MΩ.

Note that resistance depends on both the resistivity and the shape of an object. Remember that resistivity is a property of a material; so, two wires drawn from the same piece of copper will have the same

Table 18-1 Resistivity of Some Conductors and Insulators at 21°C

Conductor	ρ ($\Omega \cdot m$)
aluminum	2.733×10^{-8}
copper	1.725×10^{-8}
gold	2.271×10^{-8}
iron	9.98×10^{-8}
nichrome	$150. \times 10^{-8}$
nickel	7.2×10^{-8}
silver	1.629×10^{-8}
titanium	43.1×10^{-8}
tungsten	5.4×10^{-8}

Insulator	ρ ($\Omega \cdot m$)
glass	10^{12}
hard rubber	10^{13}
fused quartz	7.5×10^{17}

resistivity, but they could have very different values of resistance depending on their shapes. For example, a long, narrow wire will have a much higher resistance than a short, thick one.

Finally, we note that the resistivity of materials, and therefore the resistance of an object, depends on temperature. Atoms move more rapidly and are likely to be less well organized at higher temperatures compared to lower temperatures. As a result, flowing charges are more likely to suffer more collisions with the atoms in a material when the temperature is higher. It's not surprising, then, that resistivity tends to increase as temperature increases. The magnitude of the dependence of resistance on temperature is a characteristic of each material. The resistivity measurements listed in Table 18-1 were made at room temperature.

Example 18-3 Stretched Wire

A 10-m-long wire has a radius of 1 mm and a resistance of 50 Ω. The wire is stretched to a length of 100 m. Find the resistance of the wire after it has been stretched.

SET UP

The resistance of a piece of wire depends on the resistivity of the material from which it is made, as well as its length and cross-sectional area. The problem statement doesn't tell us either the material or its resistivity, but because resistivity is a characteristic of a given material it won't change as the wire is stretched. Certainly the length changes from the beginning to the end of the stretching. Does the cross-sectional area change? The key is the realization that because the amount of material doesn't change, the volume of the wire, under most circumstances, will remain the same. This unchanging volume determines how the cross-section area will change (Figure 18-4).

SOLVE

We can set up two general relationships for the resistance of the wire before and after stretching, using the subscript "i" to indicate initial values and "f" to indicate values after the wire has been stretched:

$$R_i = \frac{\rho L_i}{A_i} \tag{18-12}$$

and

$$R_f = \frac{\rho L_f}{A_f} \tag{18-13}$$

By applying relationships between the initial and final lengths and the cross-sectional areas, we can directly compare Equation 18-13 and Equation 18-12. First, we require the volume to remain constant, so

$$A_i L_i = A_f L_f$$

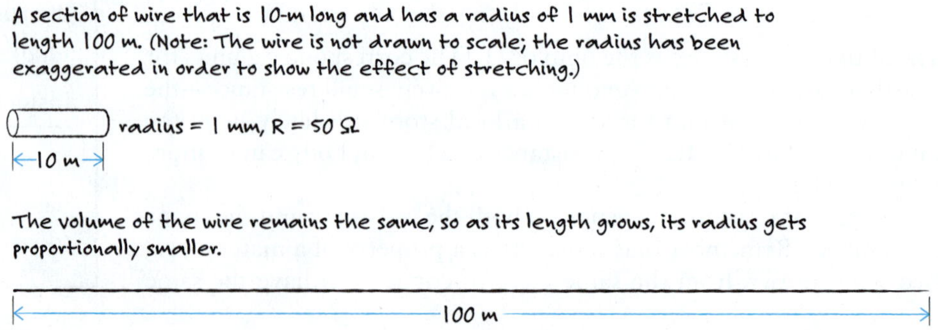

A section of wire that is 10-m long and has a radius of 1 mm is stretched to length 100 m. (Note: The wire is not drawn to scale; the radius has been exaggerated in order to show the effect of stretching.)

radius = 1 mm, R = 50 Ω

$\leftarrow$10 m$\rightarrow$

The volume of the wire remains the same, so as its length grows, its radius gets proportionally smaller.

$\leftarrow$————————————100 m————————————$\rightarrow$

Figure 18-4 What is the resistance of a wire when it is stretched by a factor of 10?

which can be solved for A_f in terms of A_i:

$$A_f = \frac{A_i L_i}{L_f}$$

So, Equation 18-13 becomes

$$R_f = \frac{\rho L_f}{(A_i L_i / L_f)}$$

We know that $L_f = 10 L_i$, so

$$R_f = \frac{\rho(10 L_i)}{(A_i L_i / 10 L_i)} = \frac{\rho(100 L_i^2)}{A_i L_i} = 100\left(\frac{\rho L_i}{A_i}\right)$$

The final term in parentheses is just the initial resistance, as defined in Equation 18-12, so

$$R_f = 100 R_i$$

The resistance of the wire after it has been stretched is 100 times the original resistance, or 5000 Ω.

REFLECT

Stretching the wire makes it longer and thinner. A longer wire has more resistance than a shorter one when both are made from the same material, and a thinner wire has more resistance than a thicker one. So, we should expect that the stretched wire in this problem will have a much higher resistance than the wire had initially.

Practice Problem 18-3 By what factor must a wire be stretched to increase its resistance by a factor of 25?

Example 18-4 One Block, Two Directions

A rectangular block of iron is 5 cm long, 2 cm wide, and 1 cm high. A potential difference is applied either across the width or across the length of the block. Determine the resistance of the block in each case.

SET UP

It might be tempting to take the resistance of the block to be the same regardless of how voltage is applied. However, note that the length and cross-sectional area found in Equation 18-11 must depend on how we apply the potential difference. So let's start by making a quick sketch of the two situations. In **Figure 18-5a** the potential difference is applied across the width of the block. In **Figure 18-5b** the potential difference is applied across the length of the block.

In each case, the effective length of the conductor is along the direction of the applied voltage. So the length required to determine the resistance in the orientation shown in Figure 18-5a is W, but in the orientation shown in Figure 18-5b it is L. Similarly, the cross-sectional area of the conductor is HL in the first case and HW in the second.

SOLVE

We can now express Equation 18-11 as a relationship that includes the length and area of the conductor. For the orientation shown in Figure 18-5a, the length is W so the cross-sectional area is HW and the resistance R is

$$R = \frac{\rho W}{HL}$$

Figure 18-5 What is the resistance of a block of iron when a potential difference is applied either (a) across the width of the block or (b) across the length of the block?

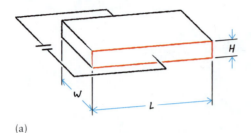

(a)

Case 1, short and wide
The potential is applied across the width of the block, so the effective length of the conductor is W.

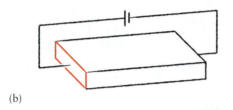

(b)

Case 2, long and narrow
The potential is applied across the length of the block, so the effective length of the conductor is L.

For the orientation shown in Figure 18-5b, the length is L so the cross-sectional area is HW and the resistance R is

$$R = \frac{\rho L}{HW}$$

So by substituting in given values together with the resistivity of iron from Table 18-1, we find the resistance in part (a) is

$$R = \frac{\rho W}{HL} = \frac{(9.98 \times 10^{-8}\ \Omega \cdot \text{m})(0.02\ \text{m})}{(0.01\ \text{m})(0.05\ \text{m})} = 3.99 \times 10^{-6}\ \Omega = 3.99\ \mu\Omega$$

and in part (b) the resistance is

$$R = \frac{\rho L}{HW} = \frac{(9.98 \times 10^{-8}\ \Omega \cdot \text{m})(0.05\ \text{m})}{(0.01\ \text{m})(0.02\ \text{m})} = 2.50 \times 10^{-5}\ \Omega = 25.0\ \mu\Omega$$

REFLECT

Changing the orientation of the block of iron from short and wide (part a) to long and narrow (part b) increases the length and decreases the cross-sectional area of the block; both of these effects increase the resistance. This example once again underscores the fact that although the resistivity of an object is a characteristic of the material from which it is made, the resistance is a function of its shape.

Practice Problem 18-4 A rectangular block of iron is 5 cm long, 2 cm wide, and 1 cm high. Determine the resistance of the block when a potential difference is applied across the sides that are 1 cm apart.

⁕ **What's Important 18-2**

Conductivity is a measure of how well or poorly a particular material conducts an electric current while resistivity is a measure of how well or poorly it inhibits the electric current. Both conductivity and resistivity differ from material to material. Rubber, a typical insulator, has a value of resistivity many orders of magnitude higher than copper, a conductor. The resistance of an object depends on both its shape and the resistivity of the material from which it is made; a long, narrow copper wire has a much higher resistance than a short, thick one. The SI unit of resistance is the ohm or Ω.

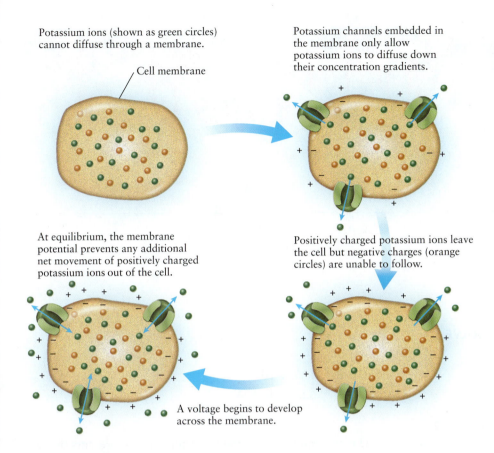

Potassium ions (shown as green circles) cannot diffuse through a membrane.

Cell membrane

Potassium channels embedded in the membrane only allow potassium ions to diffuse down their concentration gradients.

At equilibrium, the membrane potential prevents any additional net movement of positively charged potassium ions out of the cell.

Positively charged potassium ions leave the cell but negative charges (orange circles) are unable to follow.

A voltage begins to develop across the membrane.

Figure 18-6 The flow of K$^+$ ions out of a cell leads to the development of a potential difference across the membrane.

18-3 Physical and Physiological Resistors

Electrical resistance can be found everywhere you look. The circuits in a computer or a cell phone contain hundreds of resistive elements, called **resistors**. The heating element in a toaster is essentially a big resistor. But even in the absence of electrical devices and appliances you aren't far from examples of electrical resistance—your body is filled with them! Consider any of the millions of cells in your body. The concentration of potassium (K$^+$) ions is always higher inside cells than outside, which causes K$^+$ ions to leak out, as shown in **Figure 18-6**. The flow of K$^+$ ions is a current; the channels that allow K$^+$ ions to leak out of the cell behave like resistors through which the charge flows.

Example 18-5 Single Ion Channels

Using an instrument called a *patch clamp*, scientists are able to control the electric potential difference across a patch of cell membrane and measure the current through single ion channels. **Figure 18-7** shows a patch clamp recording of current as a function of time in a tiny piece of membrane sucked into the end of an electrode. When the channel was closed there was no current. When the channel was opened by applying a voltage of 120 mV across the patch of membrane, K$^+$ ions carried 6.6 pA of current. What is the resistance of the K$^+$ channel?

SET UP
Regardless of what carries charge, when a potential difference is applied across a resistive material the current is described by

$$V = iR \qquad\qquad (18\text{-}9)$$

Figure 18-7 A patch clamp recording shows current as a function of time in a tiny piece of cell membrane. The scale on lower left shows the vertical length corresponding to 4 pA of current and the horizontal length corresponding to 40 ms. Lipton, Stuart A. and Tauck, David L., "Voltage-dependent Conductances of Solitary Ganglion Cells Dissociated from the Rat Retina" J. Physiol. (1987), 385, pp. 361–391.

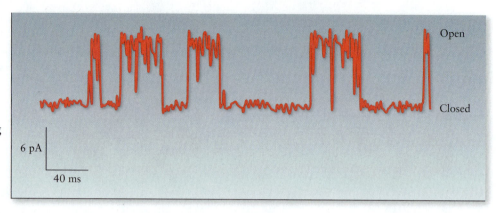

or

$$R = \frac{V}{i}$$

SOLVE

So

$$R = \frac{120 \text{ mV}}{6.6 \text{ pA}} = \frac{120 \times 10^{-3} \text{ V}}{6.6 \times 10^{-12} \text{ A}} = 18 \times 10^{9} \text{ } \Omega = 18 \text{ G}\Omega$$

REFLECT

A resistance of 18 GΩ may seem relatively high, but this value represents the resistance of a single channel made of a complex of four protein molecules embedded in the membrane. Also, because neuroscientists are more interested in how easy it is for an ion to move through a channel rather than how difficult it is, they prefer to describe an ion channel by its conductance, the inverse of resistance. The SI unit of conductance is the siemens (S); the conductance of this ion channel is therefore $(18 \text{ G}\Omega)^{-1} = 55$ pS, which is well within the range of 10 to 300 pS that is typical for a conductance of a single channel.

Practice Problem 18-5 The potential across a cell membrane is 50 mV. What current is carried through an open channel if the resistance is 22 GΩ?

The nervous system encodes information as electrical signals called *action potentials* that propagate along *axons*, the cablelike fibers than extend from nerve cells. The speed of action potential propagation depends in part on the resistance of the axon. As we know from Equation 18-11, the longitudinal resistance depends on both the length and diameter of the axon. The larger the diameter of the axon, for example, the lower the longitudinal resistance and the faster the electrical signal travels. Some axons, such as the giant ones that mediate escape reflexes in squid, for example, have exceptionally large diameters.

Example 18-6 Giant Axons in Squid

Running along each side of the back of a squid is a tubelike structure that can be as large as 1.5 mm in diameter in some species. Originally thought to be blood vessels, in the late 1930s the anatomist J.Z. Young discovered that the structures are connected to the nervous system, not to the heart. Although the vast majority of axons in the squid range in diameter from about 10 μm to 50 μm, as squid develop, axons from about 30,000 neurons fuse together to form the giant axons. Figure 18-8 shows a drawing of a squid and its giant axons. Compare the longitudinal resistance of a giant axon to that of one of the same length but a more typical diameter.

SET UP

How does the diameter of an axon affect its resistance? Equation 18-11 shows how resistance depends on area. Using this equation, we can write

$$R = \frac{\rho L}{\pi (D/2)^2} = \frac{4\rho L}{\pi D^2}$$

Note that $A = \pi (D/2)^2$. So, as diameter increases, resistance decreases. If R_G and D_G represent the resistance and diameter, respectively, of a giant axon, and R_S and D_S represent the resistance and diameter of a small axon, then

$$R_G = \frac{4\rho L}{\pi D_G{}^2}$$

and

$$R_S = \frac{4\rho L}{\pi D_S{}^2}$$

Because both axons are assumed to have the same resistivity and have the same length, these quantities affect the resistance of the two axons in the same way. Because such common terms will cancel in the process of taking a ratio, we can get an answer by taking the ratio of R_G to R_S.

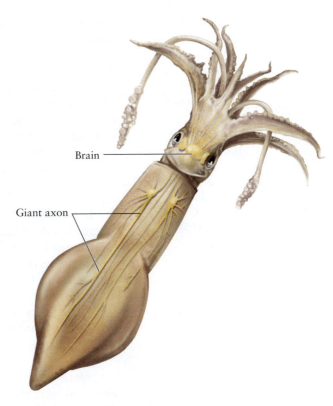

Figure 18-8 A drawing of a giant squid shows the relative size of its giant axons.

SOLVE

The ratio of R_G to R_S is

$$R_G/R_S = \left(\frac{4\rho L}{\pi D_G{}^2}\right) \Big/ \left(\frac{4\rho L}{\pi D_S{}^2}\right) = \left(\frac{4\rho L}{\pi D_G{}^2}\right)\left(\frac{\pi D_S{}^2}{4\rho L}\right) = \frac{D_S{}^2}{D_G{}^2}$$

So the resistance of the giant axon relative to the resistance of a small axon is

$$R_G = \frac{D_S{}^2}{D_G{}^2} R_S$$

Using the numbers provided in the problem statement (and using the high-end value of the range of D_S),

$$R_G = \frac{(50 \times 10^{-6} \text{ m})^2}{(1.5 \times 10^{-3} \text{ m})^2} R_S = 1.1 \times 10^{-3} R_S$$

The longitudinal resistance of the giant axon is about 1000 times smaller than the longitudinal resistance of normal, small diameter axons.

REFLECT

When a squid recognizes danger, it sends electrical signals (action potentials) along the giant axons to trigger muscle contraction. Because the longitudinal resistance along an axon determines the speed at which nerve signals propagate, signals get to the muscle 1000 times more quickly through the giant axon than they would if the squid's nervous system were entirely comprised of nerve cells with smaller diameter axons.

Practice Problem 18-6 We just compared the longitudinal resistance of a giant axon to the longitudinal resistance of an axon of the same length but a more typical diameter, using the high end of the range of typical small axon diameters. Compare the longitudinal resistance of a giant axon to the longitudinal resistance of an axon of the same length but of a diameter 10 μm, the low end of the range for small axons.

In electronic devices, resistors are often small and cylindrical; a code based on colored bands indicates the number of ohms offered by the resistor. Other resistors look more like tiny, black cubes with the number of ohms etched on one face. Some typical resistors are shown in **Figure 18-9**. A typical electronic circuit, such as the one shown in **Figure 18-10**, can contain many resistors. The conventional symbol for resistance is $-\mathcal{W}-$.

What purpose does a resistor serve? As we have seen, a potential difference between two points in a conducting material causes a current; but according to

$$V = iR \tag{18-9}$$

the greater the resistance, the smaller the current. Stated another way, the resistance allows us to control the current at any particular applied voltage. Consider again the flashlights pictured on the first page of this chapter. If the bulb filament offered only a very little resistance, the flashlights would shine quite brightly, but only for a short time. Without resistance to impede the flow of electrons, the battery would quickly exhaust itself.

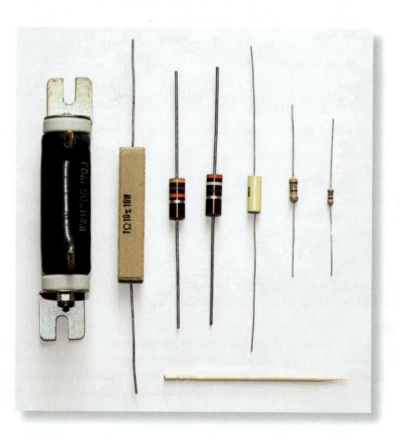

Figure 18-9 Before circuits were built by robots, a code based on colored bands was often used to indicate the number of ohms offered by a resistor. *(Courtesy of David Tauck.)*

Figure 18-10 A typical electronic circuit is likely to contain many resistors. *(Courtesy of David Tauck.)*

★ What's Important 18-3

Resistors are everywhere. For a given applied voltage, the greater the resistance in an electrical system, the smaller is the current. In electronic circuits, resistance allows us to control the current at any particular applied voltage.

18-4 Direct Current Circuits

Let's consider some simple direct current (DC) circuits that contain a source of electrical potential difference and one or more resistors. We'll represent these circuits using the symbols for electrical components. Our goals are first to understand how various resistor configurations affect voltage and current and then to explore how the effects of resistors combine when they are connected together.

Figure 18-11a shows a battery of voltage V connected to a single resistor of resistance R_1. The battery provides an increase in potential, which drives a current. The presence of the resistor results in a drop in potential that is equal to the voltage supplied by the battery. The plot that accompanies the circuit shows the potential changes in the circuit; each colored band corresponds to a different component in the circuit. Although the wires that connect the battery and resistor offer resistance, their resistances are so small compared to the other resistances in a typical circuit that they are negligible. We will therefore assume that there is effectively no drop in voltage along the connecting wires in a circuit.

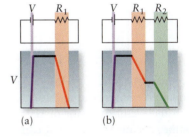

(a)   (b)

Figure 18-11 (a) In a circuit that contains a voltage source and a resistor, the battery provides an increase in potential, which drives a current. The potential drop across the resistor equals the voltage supplied by the battery. We have neglected the typically small resistance present in the connecting wires. (b) In a circuit that contains a voltage source and two resistors connected in series, the sum of the voltage drops across the two resistors equals the voltage supplied by the battery.

Now we add a second resistor, of resistance R_2, as shown in **Figure 18-11b**. In this configuration the resistors are said to be connected in **series**, using the same nomenclature as for connections of capacitors. Again, the battery provides an increase in potential, and the potential drops across each resistor. As you can see from the plot beneath the circuit, the sum of the drops across each resistor separately equals the voltage supplied by the battery. In other words, the sum of all voltage changes around the closed loop is zero. This conclusion was first proposed by the German physicist Gustav Kirchhoff in the mid-nineteenth century and is often referred to as **Kirchhoff's loop rule:**

> The sum of the changes in potential around a closed electrical loop must equal zero.

Kirchhoff's loop rule is a statement only about the sum of all voltage drops around a closed loop; in general, the drop in voltage across each resistor connected in series is different.

How do the currents through the two resistors in Figure 18-11b compare? The current represents the net motion of charge around the loop. Because there is no mechanism in the circuit for electrons to be created, destroyed, or diverted, the same (net) number of charges must be flowing at every point. The currents through R_1 and R_2 are therefore the same.

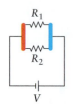

Figure 18-12 How does the voltage drop across each of two resistors connected in parallel compare?

Now consider what happens when we connect two resistors in **parallel**, as in **Figure 18-12**. How does the voltage drop across R_1 compare with that across R_2? In our analysis of a single resistor connected to a voltage source, we concluded that there is no difference in potential between two points in a connecting wire. So all points on the wire connected to the right ends of the resistors, shown in blue, are at the same potential. The same is true for all points on the wire that connects the left ends, shown in red. The potential difference across R_1 must therefore be equal to the potential difference across R_2. This statement is always correct. The change in voltage across each resistor is the same for each of the resistors connected in parallel, regardless of whether the resistances are the same or different.

We concluded that the currents through the two resistors in series in Figure 18-11b were the same, in part because there is no way for electrons to be diverted as they are driven around the loop. That conclusion is not correct for the circuit in **Figure 18-13**, because flowing charge can be diverted at the junctions, marked A, B, C, and D. A **junction** is a point in a circuit at which three or more wires are connected. Any individual electron entering a junction can only flow out along one of the other wires, so the current that comes in can split, perhaps unevenly, as it exits. It must be true, however, that the total current entering a junction must equal the total current leaving, because there is no mechanism in the circuit for electrons to be created or destroyed. This second rule of electronic circuit loops was also proposed by Kirchhoff, and is often referred to as **Kirchhoff's junction rule:**

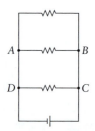

Figure 18-13 The current through each of the resistors in this circuit may not be the same, because the current can split, perhaps unevenly, at the junctions (A, B, C, and D).

> The sum of the currents entering a junction equals the sum of the currents leaving it.

Kirchhoff's two rules can be used to analyze a wide variety of circuits.

Physicist's Toolbox 18-1   Kirchhoff's Rules

KIRCHHOFF'S LOOP RULE

The sum of the changes in potential around a closed electrical loop must equal zero. The rule in mathematical form is

$$\sum \Delta V_i = 0 \qquad (18\text{-}14)$$

where the sum is over all electrical components in the loop.

KIRCHHOFF'S JUNCTION RULE
The sum of the currents entering a junction equals the sum of the currents leaving it. The rule in mathematical form is

$$\sum i_{in,j} = \sum i_{out,k} \tag{18-15}$$

where $i_{in,j}$, the sum on the left, is over all currents entering a junction and $i_{out,k}$, the sum on the right, is over all currents leaving the junction.

✳ What's Important 18-4

In a DC circuit, the potential drop across a single resistor equals the voltage supplied by the battery. If more than one resistor is connected in series, the sum of the voltage drops across each resistor separately equals the voltage supplied by the battery. When resistors are connected in parallel, the voltage drop across each resistor is the same, regardless of whether the resistances are the same or different. A junction is a point in a circuit where three or more wires are connected to each other. The total current entering a junction equals the total current leaving it.

18-5 Resistors in Series and Parallel

A friend of yours is desperately trying to finish building a robot for an engineering class and has asked you to zip over to the local electrical supply store and buy some 200 kΩ resistors. Big trouble: There are none in stock! Is all hope lost for your friend, or is there a way to combine resistors of other values to create a resistance equivalent to the 200 kΩ resistors she needs?

Resistors in Series

Figure 18-14a shows a circuit that has three resistors in series. Imagine we replace the three separate resistors with a single one of a resistance equivalent to that of the series combination, as shown in Figure 18-14b. How do the voltage drops and the currents in the two circuits compare?

According to Equation 18-9, the voltage drop across a resistor, the current through it, and its resistance are related by

$$V = iR \tag{18-9}$$

Our goal is to find an expression that has the same form for the combination of resistors. That is, we want a relationship for the series of resistors that looks like

$$V = i \times \text{something}$$

where that "something" is the resistance of the series of resistors. This **equivalent resistance** in the corresponding circuit is shown in Figure 18-14b.

To include all three resistors in our determination of the equivalent resistance, we start by writing Equation 18-9 separately for each resistor in the first circuit:

$$V_1 = i_1 R_1 \tag{18-16a}$$
$$V_2 = i_2 R_2 \tag{18-16b}$$
$$V_3 = i_3 R_3 \tag{18-16c}$$

There are no junctions in this circuit, so the currents i_1, i_2, and i_3 must all be the same. Using

$$i = i_1 = i_2 = i_3$$

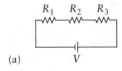
(a)

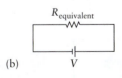
(b)

Figure 18-14 (a) A circuit contains three resistors connected in series to a voltage source. (b) The three resistors have been replaced by a single, equivalent resistor; the equivalent resistance R_{equiv} is the same as the resistance of the three resistors in series.

we can then rewrite Equations 18-16a, b, and c:

$$V_1 = iR_1 \tag{18-17a}$$

$$V_2 = iR_2 \tag{18-17b}$$

$$V_3 = iR_3 \tag{18-17c}$$

Also, according to Kirchhoff's loop rule,

$$V - V_1 - V_2 - V_3 = 0 \tag{18-18}$$

Combining Equations 18-17a, b, and c with Equation 18-18 yields

$$V - iR_1 - iR_2 - iR_3 = 0$$

A simple rearrangement of this equation yields the form $V = i \times$ something:

$$V = i(R_1 + R_2 + R_3)$$

so the equivalent resistance of the three resistors in series is

$$R_{\text{equiv}} = R_1 + R_2 + R_3$$

This equation leads to a general statement for the equivalent resistance of N resistors in series:

$$R_{\text{equiv}} = \sum_{i=1,N} R_i \tag{18-19}$$

By combining resistors in series, we create a circuit with a higher equivalent resistance than that of any of the individual resistors. For example, combining two resistors of the same resistance in series results in an equivalent resistance twice as big:

$$R_{\text{equiv}} = R + R = 2R$$

Resistors in Parallel

Figure 18-15 shows a circuit with three resistors in parallel. Imagine we replace the three separate resistors with a single one of resistance equivalent to that of the parallel combination. How do the voltage drops and the currents in the two circuits compare?

As in the example above of resistors in series, we want to find an equation of the form $V = iR_{\text{equiv}}$, where i is the total current in the circuit. Also, the relationships that we used between voltage, current, and resistance for each resistor separately remain the same as before, that is,

$$V_1 = i_1 R_1 \tag{18-16a}$$

$$V_2 = i_2 R_2 \tag{18-16b}$$

$$V_3 = i_3 R_3 \tag{18-16c}$$

There are four junctions in this circuit, marked A, B, C, and D in Figure 18-15, which means that—unlike in the series combination—the currents i_1, i_2, and i_3 are not required to be all the same. Kirchhoff's junction rule determines how the current splits at each junction. For example, at junction A, current i enters the junction, while currents i_1 and i' leave. So

$$i = i_1 + i' \tag{18-20}$$

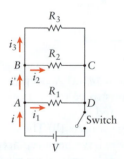

Figure 18-15 A circuit contains a voltage source and three resistors connected in parallel.

In the figure, we show the direction of the currents according to the direction that electrons would flow. Our analysis of the circuit would be the same had we used the direction of conventional current. The only requirement is that the directions are consistent for all currents in the circuit.

Applying the junction rule to junction B gives

$$i' = i_2 + i_3 \tag{18-21}$$

The relationship between the total current and individual currents in each of the three resistors is obtained by combining Equation 18-20 and Equation 18-21:

$$i = i_1 + i_2 + i_3 \tag{18-22}$$

As we saw in Section 18-4, there is no potential difference between different points on a connecting wire. So, the left ends of the three resistors in Figure 18-15 must all be at the same potential, and the right ends of the three resistors must all be at the same potential. The requirement results in the potential difference across each resistor being the same:

$$V = V_1 = V_2 = V_3$$

We can then rewrite Equations 18-16a, b, and c as

$$V = i_1 R_1, \qquad V = i_2 R_2, \qquad \text{and} \qquad V = i_3 R_3$$

Each of the equations can be rearranged as an expression for one of the three currents:

$$i_1 = \frac{V}{R_1} \tag{18-23a}$$

$$i_2 = \frac{V}{R_2} \tag{18-23b}$$

$$i_3 = \frac{V}{R_3} \tag{18-23c}$$

which can be combined with Equation 18-22 to yield

$$i = \frac{V}{R_1} + \frac{V}{R_2} + \frac{V}{R_3}$$

A simple rearrangement of this equation puts it into the form $V = iR_{\text{equiv}}$:

$$i = V\left(\frac{1}{R_1} + \frac{1}{R_2} + \frac{1}{R_3}\right)$$

or

$$V = i\left(\frac{1}{\dfrac{1}{R_1} + \dfrac{1}{R_2} + \dfrac{1}{R_3}}\right)$$

so the equivalent resistance of the three resistors in parallel is

$$R_{\text{equiv}} = \frac{1}{\dfrac{1}{R_1} + \dfrac{1}{R_2} + \dfrac{1}{R_3}}$$

In general, the equivalent resistance of N resistors in parallel is

$$R_{\text{equiv}} = \frac{1}{\displaystyle\sum_{i=1,N} \frac{1}{R_i}} \tag{18-24}$$

By combining resistors in parallel we create a circuit with a smaller equivalent resistance than any of the individual resistors. For example, combining two identical resistors in parallel results in an equivalent resistance half as big:

$$R_{equiv} = \frac{1}{\frac{1}{R} + \frac{1}{R}} = \frac{1}{\frac{2}{R}} = \frac{R}{2}$$

Two Resistors in Parallel—A Special Case

We are likely to encounter many situations in which exactly two resistors are connected in parallel. For two resistors R_1 and R_2, Equation 18-24 becomes

$$R_{equiv} = \frac{1}{\frac{1}{R_1} + \frac{1}{R_2}}$$

So

$$R_{equiv} = \frac{1}{\frac{R_2}{R_1 R_2} + \frac{R_1}{R_1 R_2}} = \frac{1}{\frac{R_1 + R_2}{R_1 R_2}} = \frac{R_1 R_2}{R_1 + R_2} \tag{18-25}$$

For example, when a 3 Ω and a 6 Ω resistor are connected in parallel, the equivalent resistance is

$$R_{equiv} = \frac{R_1 R_2}{R_1 + R_2} = \frac{(3\ \Omega)(6\ \Omega)}{3\ \Omega + 6\ \Omega} = \frac{18\ \Omega}{9\ \Omega} = 2\ \Omega$$

? Got the Concept 18-1
Combinations of Resistors

(a) Rank the four circuits shown in **Figure 18-16a–d** in order of increasing equivalent resistance. (b) Rank the four circuits shown in Figure 18-16a–d in order of increasing current. (c) Why do your answers not depend on the specific value of R_3?

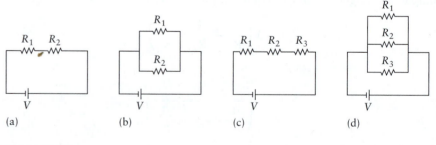

(a)   (b)   (c)   (d)

Figure 18-16

Example 18-7 Resistors in Combination

Go to Picture It 18-1 for more practice dealing with resistors in combination

We are finally in a position to help your friend, the one who desperately needs 200 kΩ resistors to finish her robot. The store stocks only 300 kΩ resistors. What combinations of two or three such resistors can you assemble to create a circuit that has an equivalent resistance of 200 kΩ?

SET UP

Combining resistors in series results in a circuit with larger equivalent resistance; combining them in parallel results in a smaller equivalent resistance. Because we need to end up with a resistance smaller than the individual resistors, we can therefore rule out any combination that does not include some of the resistors connected in parallel. What possible combinations are left? The possibilities are shown in **Figure 18-17**. We will try them all.

SOLVE

The equivalent resistance of two equal resistors in parallel is half the resistance of each separately. So two 300 kΩ resistors in parallel, shown in Figure 18-17a, are equivalent to one 150 kΩ resistor. This value is not our goal of 200 kΩ.

The equivalent resistance of three equal resistors in parallel can be determined from Equation 18-24:

$$R_{equiv} = \frac{1}{\frac{1}{R} + \frac{1}{R} + \frac{1}{R}} = \frac{1}{\frac{3}{R}} = \frac{R}{3}$$

So, three 300 kΩ resistors in parallel, shown in Figure 18-17b, are equivalent to one 100 kΩ resistor. This value is also not our goal of 200 kΩ.

The circuit in Figure 18-17c is a combination of two 300 kΩ resistors in parallel connected in series to a third 300 kΩ resistor. We just determined that the parallel part of the circuit has an equivalent resistance of 150 kΩ. The two resistors in parallel connected to the 300 kΩ resistor in series would result in a total resistance of 450 kΩ. Again, that value is not what we want.

In Figure 18-17d there is also a mix of a series and a parallel connections; two 300 kΩ resistors are in series and this combination is in parallel with a third 300 kΩ resistor. The two in series add to 600 kΩ, so the combination is 600 kΩ in parallel with 300 kΩ. The equivalent resistance can be found using Equation 18-25:

$$R_{equiv} = \frac{(600\,\text{k}\Omega)(300\,\text{k}\Omega)}{600\,\text{k}\Omega + 300\,\text{k}\Omega} = \frac{180000\,\text{k}\Omega}{900\,\text{k}\Omega} = 200\,\text{k}\Omega$$

This configuration of 300 kΩ resistors is equivalent to a single, 200 kΩ resistor. The robot can be finished!

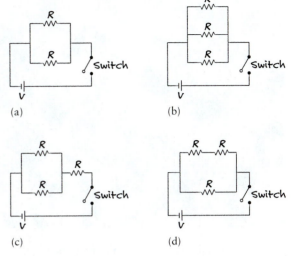

Figure 18-17 Identical resistors are connected in a circuit resulting in an equivalent resistance lower than any one resistor separately. The four parts of this figure show the four possibilities.

REFLECT

There are only a limited number of ways to connect two or three resistors, all of the same value. Using a variety of values and more resistors, we could create a large number of combinations and equivalent resistances. Combination of resistors may be done on purpose, but often separate components in both physical and physiological systems can contribute resistances that combine in unanticipated ways.

Practice Problem 18-7 How can three 200 kΩ resistors be connected so that the equivalent resistance of all three together equals 300 kΩ?

Example 18-8 Giant Axons in Squid, Redux

Approximately 30,000 nerve cells fuse together to form each giant axon in squid. (See Example 18-6.) A representative value of the resistivity of a squid axon is 3100 Ω·m. Taking the diameter of a typical axon of one of these cells to be

Figure 18-18 Two squid swim near the ocean floor.
(Fred Bavendam/Minden Pictures.)

50 μm and the length to be 10 cm, find the longitudinal resistance of a giant squid axon by considering it as 30,000 separate axons in parallel. **Figure 18-18** shows a photograph of a squid.

SET UP

The equivalent resistance of a group of resistors connected in parallel is described by Equation 18-24:

$$R_{equiv} = \frac{1}{\sum_{i=1,N} \frac{1}{R_i}} \tag{18-24}$$

When each of N resistors has the same value R, we can write the sum as

$$R_{equiv} = \frac{1}{\frac{1}{R} + \frac{1}{R} + \cdots + \frac{1}{R}}$$

Because the term $1/R$ appears N times in the denominator, the equation becomes

$$R_{equiv} = \frac{1}{\frac{N}{R}} = \frac{R}{N}$$

SOLVE

The resistance of one axon can be found using Equation 18-11, or

$$R = \frac{\rho L}{A} = \frac{(3100\ \Omega \cdot m)(0.1\ m)}{\pi \left(\frac{50 \times 10^{-6}\ m}{2} \right)^2} = 1.6 \times 10^{11}\ \Omega = 1.6 \times 10^5\ M\Omega$$

The equivalent resistance of 30,000 such axons connected in parallel is

$$R_{equiv} = \frac{R}{N} = \frac{1.6 \times 10^{11}\ \Omega}{30,000} = 5.3 \times 10^6\ \Omega = 5.3\ M\Omega$$

REFLECT

In Example 18-6, we found that the resistance of a giant squid axon is about 1000 times smaller than that of a normal axon. The equivalent resistance of 30,000 small axons connected in parallel is smaller than this. Why? In this example we made an implicit assumption that total cross-sectional area of the giant axon is 30,000 times the cross-sectional area of a small axon. For a resistor of area NA, length L, and resistivity ρ,

$$R_{equiv} = \frac{\rho L}{NA} = \frac{\left(\frac{\rho L}{A} \right)}{N} = \frac{R}{N}$$

However, in the actual biological system (the squid), the cross-sectional area of the giant axon has a diameter of 1.5 mm, which is much smaller than the total cross-sectional area of 30,000 small axons each of diameter 50 μm. Although about 30,000 cells contribute to the formation of the giant axon, the process by which the giant axon forms is not an actual fusing together of 30,000 fully formed smaller axons.

Practice Problem 18-8 A resistive device is made by connecting 1000 small resistors in parallel. Each small resistor provides 55 kΩ of resistance. What is the total resistance of the device?

> ✱ **What's Important 18-5**
>
> The equivalent resistance of resistors connected in series is equal to the sum of the individual resistances and therefore is higher than the resistance of any of the individual resistors. In contrast, the equivalent resistance of resistors connected in parallel is lower than the resistance of any of the individual resistors.

18-6 Power

The coils in a toaster are resistors. Pass a high enough current through them and they get red hot. The filament in an incandescent bulb is a resistor. Pass a high enough current through it and it glows brightly. In both cases, moving charge delivers energy to the resistive load of the circuit, transforming some of it into heat and light energy in the process.

Power is defined as the rate at which energy flows into or out of a system, or changes from one form of energy to another. In a resistor, the flow of charge delivers energy; what is the relationship between power and the quantities that we have been using to describe circuits—current, voltage, and resistance?

The potential V between two locations in an electric field describes the work W required to move a charge q from the initial location to the final location, or the change in potential energy of the charge. From Section 6-5 and Equation 17-3,

$$V = \frac{-W}{q} = \frac{\Delta U}{q}$$

Recall that the minus sign reminds us that W is the work that the field does on the charge; positive work done by the field increases the kinetic energy of the charge, which results in a decrease in the potential energy. Power, the rate at which the potential energy changes, is

$$P = \frac{d}{dt}\Delta U = \frac{d}{dt}Vq = V\frac{dq}{dt} = Vi$$

Using the three possible rearrangements of $V = iR$ (Equation 18-9), we can write an expression for power in terms of any two of the variables voltage, current, and resistance:

$$P = Vi \tag{18-26}$$

$$P = Vi = V\left(\frac{V}{R}\right) = \frac{V^2}{R} \tag{18-27}$$

$$P = Vi = (iR)i = i^2R \tag{18-28}$$

The unit of power is the watt (W); from $P = d(\Delta U)/dt$,

$$1\text{ W} = 1\frac{\text{J}}{\text{s}}$$

What is a watt? You've probably encountered this unit of power before, for example, in reference to audio systems or lightbulbs. A sound system that pumps out, say, 400 W per channel can be turned up quite loud. You might put a 60 W bulb in a reading lamp; you've probably held your hand near a bulb, so you have a sense of how to interpret that much power in terms of heat. Imagine the power used by the mercury vapor street lamps that consume up to 1000 W!

Example 18-9 A Flashlight

A flashlight uses two 1.5 V (AA) batteries to power a bulb rated at 1.0 W. What is the current in the bulb?

SET UP

The resistance of the filament in the bulb results in the delivery of power—the energy of the moving charges causes the filament to heat up, giving off light and making the filament warm. We do not need to know the actual resistance to investigate current and power, however, because we have established a relationship between power, current, and voltage:

$$P = Vi \tag{18-26}$$

which can be rearranged to give

$$i = \frac{P}{V}$$

SOLVE

A 1.0 W bulb connected to a voltage source which supplies a total of $V = 2(1.5) = 3.0$ V draws a current of

$$i = \frac{1.0 \text{ W}}{3.0 \text{ V}} = 0.33 \text{ A}$$

Flashlight bulbs usually operate at between 0.2 and 0.9 A.

REFLECT

Let's check the units of our answer. One watt is equivalent to one joule per second. Potential is defined as the work required to move a charge in an electric field divided by the magnitude of the charge. In terms of the units, then,

$$[i] = \frac{W}{V} = \frac{\text{energy/time}}{\text{work/charge}} = \frac{J/s}{J/C} = \frac{C}{s}$$

This result is what we expect because one ampere is defined as one coulomb per second.

Practice Problem 18-9 Each of a pair of heated gloves uses an 8.6 V battery to produce about 11 W. What is the current through the heating element in the gloves?

Watch Out

Time is embedded in the units of power

The units of power often confuse beginning physics students because it seems that watts requires an additional time measurement. For example, to find the power in watts that a hair dryer draws, a student might ask, "Find the power for what amount of time?" Notice, however, that time is already embedded in watts because 1 W is equal to 1 J *per second*. When a hair dryer is rated at 1500 W, that means it requires 1500 J of energy every second to operate. Time is already taken into account.

Why, then, does the local power company charge based on watts multiplied by hours, usually kilowatt-hours (kW · h)? The power company charges for how much *energy* a customer consumes. A kilowatt-hour is the amount of energy in joules used when, say, a 1000 W device is used for one hour.

Example 18-10 How Long Will the Flashlight Last?

In the previous example, we found that the flashlight bulb draws a current of 0.33 A. If the batteries have a capacity of 1600 mA · h, how long will the flashlight run before the batteries are exhausted?

SET UP

You may have seen a rating in milliampere-hours (mA · h) on batteries. This value is called the *battery's capacity*. What does battery capacity mean? Consider the dimensions of current and time; current is a measure of the number of charges passing a certain point in a given time interval, so the product of current and time just gives us charge:

$$[i \times t] = \frac{\text{charge}}{\text{unit time}} \text{time} = \text{charge} \qquad (18\text{-}29)$$

So capacity is the total amount of excess charge that a battery can drive through a circuit as a current before it stops working. Also, because the capacity is equivalent to current multiplied by time, the time for which a battery can supply current is its capacity divided by the current it supplies, or in units

$$[t] = \frac{[\text{capacity}]}{[\text{current}]} = \frac{\text{mA} \cdot \text{h}}{\text{mA}} = \text{h}$$

SOLVE

Each battery in this problem can supply 1600 mA · h; two batteries therefore supply twice this amount, or 3200 mA · h. When the flashlight operates at 0.33 A, or 330 mA (using the value found in the last example), it will therefore run for

$$t = \frac{3200 \text{ mA} \cdot \text{h}}{330 \text{ mA}} = 9.7 \text{ h}$$

REFLECT

It seems quite plausible that a small flashlight would last for about 10 h before the batteries are exhausted.

Practice Problem 18-10 Each of a pair of heated gloves uses an 8.6 V battery to produce about 11 W. If each battery has a capacity of 2200 mA · h, how long will the gloves provide heat before the batteries are exhausted?

Example 18-11 Drying Your Hair

A hair dryer is rated at 1500 W. If your local power company charges $0.11 per kW · h, and you spend 15 min per day drying your hair, how much does it cost per month to run the hair dryer?

SET UP

The hair dryer consumes 1500 W of power, or 1500 J of energy per second when it is running. Multiplying the power by a time interval yields the total energy consumed in that time. We must take care, of course, to use values with the proper units.

SOLVE

The energy consumed by the hair dryer is

$$E = P\Delta t$$

The power is 1500 W, or 1.5 kW. If the dryer is run for 30 days at 15 min or 0.25 h per day, then

$$E = (1.5 \text{ kW}) (30 \text{ d}) \left(0.25 \frac{\text{h}}{\text{d}} \right) = 11.25 \text{ kW} \cdot \text{h}$$

At a cost of $0.11 per kW·h, the energy costs

$$\text{cost} = E \frac{\$}{\text{kW} \cdot \text{h}} = (11.25 \text{ kW} \cdot \text{h}) \left(\frac{\$0.11}{\text{kW} \cdot \text{h}} \right) = \$1.24$$

REFLECT

The cost to run your hair dryer is not so much, perhaps, but it's not free!

Practice Problem 18-11 You accidentally leave four 100 W bulbs on in your room for one full week over spring break. What is the cost of the energy used, at a rate of $0.11 per kW·h?

Example 18-12 Load Matching

A resistor in your electric socks is connected to a 1.5 V battery in order to keep your toes warm. If the battery has an internal resistance of 0.1 Ω, what resistance, called the *load*, should be connected in series with the battery in order to maximize the power delivered to the socks?

SET UP

Figure 18-19 shows the circuit for the electric socks. Power depends on both current and resistance ($P = i^2 R$). When the resistance gets close to zero, the delivered power is also close to zero. But when the resistance is large, we know from $V = iR$ that the current gets close to zero, so again, power is close to zero. These observations must mean that there is a value of R greater than zero and less than some large number for which the power is a maximum, as suggested by Figure 18-20. We'll refer to this resistance as R_0.

SOLVE

At the value of resistance R_0, the slope of the power curve shown in Figure 18-20 is zero. For that reason, we find R_0 by setting the derivative of power with respect to resistance, or the slope, equal to zero.

The power delivered in a resistor of resistance R is

$$P = i^2 R \tag{18-28}$$

The current in the circuit depends on the voltage and the equivalent resistance:

$$i = \frac{V}{R_{equiv}} = \frac{V}{r_{internal} + R}$$

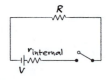

Figure 18-19 What load resistance R should be connected to a battery with an internal resistance $r_{internal}$ in order to maximize the power delivered?

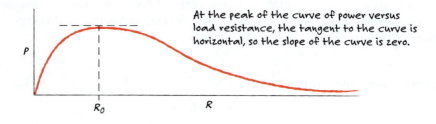

Figure 18-20 A graph of power delivered to the load resistor in Figure 18-19 versus the load resistance shows a peak at resistance R_0.

At the peak of the curve of power versus load resistance, the tangent to the curve is horizontal, so the slope of the curve is zero.

So the power is then

$$P = \left(\frac{V}{r_{\text{internal}} + R}\right)^2 R \qquad (18\text{-}30)$$

Set the derivative of P with respect to R equal to zero:

$$\frac{dP}{dR} = \frac{d}{dR}\left(\frac{V}{r_{\text{internal}} + R}\right)^2 R = 0$$

The derivative is carried out by applying the product rule. To make that easier, we first rewrite the denominator as a factor with a negative exponent:

$$\frac{d}{dR}\left(\frac{V}{r_{\text{internal}} + R}\right)^2 R = \frac{d}{dR} V^2 (r_{\text{internal}} + R)^{-2} R$$

then we expand the expression

$$\frac{d}{dR}\left(\frac{V}{r_{\text{internal}} + R}\right)^2 R = V^2\left[R\frac{d}{dR}(r_{\text{internal}} + R)^{-2} + (r_{\text{internal}} + R)^{-2}\frac{d}{dR}R\right]$$

and rearrange to get

$$\frac{d}{dR}\left(\frac{V}{r_{\text{internal}} + R}\right)^2 R = V^2\left[-2(r_{\text{internal}} + R)^{-3}R + (r_{\text{internal}} + R)^{-2}(1)\right]$$

When this expression is set equal to zero, that is,

$$0 = V^2\left[-2(r_{\text{internal}} + R)^{-3}R + (r_{\text{internal}} + R)^{-2}(1)\right]$$

both sides can be divided by both V^2 and $(r_{\text{internal}} + R)^{-2}$, leaving

$$0 = -2(r_{\text{internal}} + R)^{-1}R + 1$$

Substituting R_0 for R to indicate that the slope is zero only when R equals R_0, we get

$$0 = -2\left(\frac{R_0}{r_{\text{internal}} + R_0}\right) + 1$$

which simplifies to

$$-(r_{\text{internal}} + R_0) = -2R_0$$

or

$$-r_{\text{internal}} = -R_0$$

or, finally,

$$R_0 = r_{\text{internal}}$$

The maximum power is delivered when the external resistance is matched with the internal resistance, or $R_0 = 0.1\ \Omega$ in this problem.

REFLECT

When the resistance is set to $R_0 = 0.1\ \Omega$, the power delivered, according to Equation 18-30, is

$$P = \left(\frac{1.5\ \text{V}}{0.1\ \Omega + 0.1\ \Omega}\right)^2 (0.1\ \Omega) = 5.6\ \text{W}$$

This power should be enough to keep your toes warm. (Commercially available electric socks draw about 10 W.) Also, you can use a calculator or spreadsheet to verify the curve shown in **Figure 18-21** for other

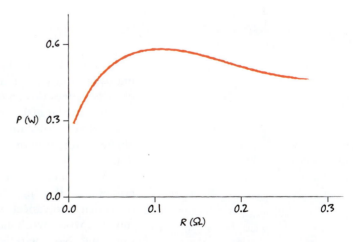

Figure 18-21 A graph of power delivered to the load resistor in Figure 18-19 versus the load resistance, for the values of R and r_{internal} given for the sock warmer of Problem 18-12.

values of resistance, which shows how the delivered power depends on the load resistance.

> ✳ **What's Important 18-6**
> Power is the rate at which energy is delivered to a system. In a resistor, the flow of charge delivers energy. The unit of power is the watt (W); 1 W is equivalent to 1 J per second.

18-7 Series *RC* Circuits

A resistor connected in a closed loop to a voltage source restricts the flow of charge; this relationship is described by $V = iR$. A capacitor connected in a closed loop to a voltage source stores charge; this relationship is described by $q = CV$. Something curious occurs when both a resistor and a capacitor are connected in a circuit with a voltage source.

Charging a Series *RC* Circuit

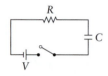

Figure 18-22 A circuit includes a resistor and capacitor connected in series to a battery. The switch enables the time when the loop becomes closed to be set precisely.

We will consider a circuit with a resistor and capacitor connected in series to a battery, as shown in **Figure 18-22**. By using a switch we can precisely set the time when the loop becomes closed. The circuit must obey Kirchhoff's rules; for example, the sum of all voltage changes around a closed loop equals zero:

$$0 = V_{battery} + V_{resistor} + V_{capacitor}$$

or

$$0 = V + (-iR) + \left(-\frac{q}{C}\right)$$

Both the current and the charge on the capacitor can change over time, so to show this change clearly we'll rewrite the previous expression as

$$0 = V - \frac{dq(t)}{dt}R - \frac{1}{C}q(t)$$

or

$$\frac{V}{R} = \frac{dq(t)}{dt} + \frac{1}{RC}q(t) \tag{18-31}$$

This is a first order linear differential equation, which might look (and sound) imposing. But basic math and a little intuition is all we need to find the function $q(t)$ that solves this equation.

Notice that $q(t)$ must have the property that when it is added to its time derivative (with the constants R and C factored in) the result does not depend on time. So whatever function of t is present in $q(t)$, it must cancel out the time dependence of $dq(t)/dt$.

What function of time cancels out the time dependence of its derivative? Not a polynomial. If $q(t) = t^n$, then $(dq(t)/dt) = nt^{n-1}$, and it is not possible for the time dependence to cancel when two different powers of t are added. A sine or cosine function won't work either, because neither one has a derivative equal to the function itself. No, the only function that has a derivative which cancels the function itself is the exponential to a negative power. For example, if $q(t) = A\,e^{-\omega t}$, then

$$\frac{dq(t)}{dt} + \omega q(t) = -\omega A e^{-\omega t} + \omega A e^{-\omega t} = 0$$

We had to multiply the $q(t)$ term by ω in order to make the cancellation work. Comparing the previous expression to Equation 18-31, the variable ω above is equal to $1/RC$, so $q(t)$ must be related to

$$q(t) = e^{-t/RC}$$

How can we get the constant term V/R on the left side of Equation 18-31? The derivative of a constant is zero, which means that adding a constant term to $q(t)$ won't change $dq(t)/dt$. The $e^{-\omega t}$ terms will still cancel, but the extra piece will survive. Using

$$q(t) = \text{constant}_1 + e^{-t/RC}$$

and

$$\frac{dq(t)}{dt} = -\left(\frac{1}{RC}\right)e^{-t/RC}$$

we obtain the expression

$$\frac{dq(t)}{dt} + \left(\frac{1}{RC}\right)q(t) = \text{constant}_2$$

Because $q(t)$ is multiplied by $1/RC$ in Equation 18-31, the constant$_1$ term in $q(t)$ must be CV because

$$\frac{1}{RC}q(t) = \frac{1}{RC}(CV + e^{-t/RC}) = \frac{V}{R} + \frac{1}{RC}e^{-t/RC}$$

and Equation 18-31 is satisfied:

$$q(t) = CV + e^{-t/RC} \qquad \textbf{(18-32)}$$

We must make one last adjustment to have the correct form of $q(t)$. Notice that Equation 18-32 predicts that at $t = 0$ the charge on the capacitor is $CV + 1$. This value can't be correct—the charge on the capacitor must be zero at the moment the switch is closed. Also, the second term on the right side of Equation 18-32 is dimensionless, but the term on the right side—charge—is not. We can fix both problems by multiplying the $e^{-t/RC}$ term by a constant. Our analysis would not be changed by this; setting that constant equal to $-CV$ gives the right value of charge when $t = 0$, and solves the issue of matching dimensions. So all together, the complete solution of Equation 18-31 is

$$q(t) = CV - CVe^{-t/RC}$$

or

$$q(t) = CV(1 - e^{-t/RC}) \qquad \textbf{(18-33)}$$

At $t = 0$, the charge on the capacitor is zero. After a long time has elapsed, so that $e^{-t/RC}$ approaches zero, the charge approaches CV. Because CV is constant we say that the capacitor is fully charged; the maximum charge that can be stored is therefore $q_{max} = CV$. This relationship is the same relationship we discovered in Chapter 17 between charge, capacitance, and applied voltage. Charge on the capacitor as a function of time is shown in **Figure 18-23a**.

The flow of charge (current) causes the capacitor to become charged. To find the current in the circuit we take the derivative of the charge on the capacitor,

$$i(t) = \frac{dq(t)}{dt} = \frac{d}{dt}[CV(1 - e^{-t/RC})]$$

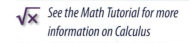

See the Math Tutorial for more information on Calculus

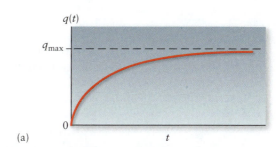

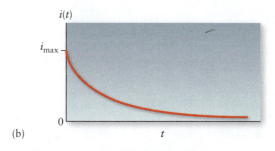

Figure 18-23 (a) The charge on the capacitor in a charging series *RC* circuit increases as a function of time and asymptotically approaches its maximum value q_{max}. (b) The current in a series *RC* circuit falls exponentially as a function of time. By comparing the two curves, notice that as the current decreases, the rate at which charge is deposited on the capacitor decreases.

or

$$i(t) = \frac{V}{R}e^{-t/RC} \qquad (18\text{-}34)$$

At $t = 0$, Equation 18-34 predicts the current should be V/R; the circuit initially obeys the relationship $V = iR$. Current versus time is shown in **Figure 18-23b**. Notice that as the current decreases, the rate at which charge is deposited on the capacitor decreases.

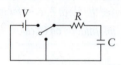

When the switch is closed up, the battery, resistor, and capacitor are in series, and the capacitor charges.

When the switch is closed down, the battery is no longer in the circuit, so the capacitor discharges.

Figure 18-24 The location of the switch in this series RC circuit enables us to choose whether or not to include the battery in the loop with the resistor and capacitor.

Discharging a Series *RC* Circuit

Figure 18-24 shows a series RC circuit. It differs from the series RC circuit shown in Figure 18-22 in that the location of the switch enables us to choose whether or not to include the battery in the loop with the resistor and capacitor. Let's analyze this circuit first by closing the switch up to include the battery; we'll leave the switch up long enough to fully charge the capacitor. Once the capacitor is fully charged we'll move the switch down, so that the battery is no longer part of the circuit.

As in the case of the charging series RC circuit, Kirchhoff's loop rule applies, so

$$0 = V_{\text{resistor}} + V_{\text{capacitor}}$$

or

$$0 = -iR - \frac{q}{C}$$

or

$$0 = \frac{dq(t)}{dt} + \frac{1}{RC}q(t)$$

We can gather all of the $q(t)$ terms on one side and all of the t terms on the other in order to integrate

$$\frac{dq(t)}{q(t)} = -\frac{1}{RC}dt$$

At time $t = 0$, the charge on the capacitor is $q_{\text{max}} = CV$. After some time t has elapsed, the charge is q. We can add up all of the charge which flows in time t by

$$\int_{q_{\text{max}}}^{q} \frac{dq(t)}{q(t)} = \int_{0}^{t} -\frac{1}{RC}dt$$

which leads to

$$\ln(q) - \ln(q_{\text{max}}) = \ln\left(\frac{q}{q_{\text{max}}}\right) = -\frac{1}{RC}t$$

 See the Math Tutorial for more information on Logarithms

To eliminate the natural log functions, make each side of the equation an exponent of e:

$$e^{\ln(q/q_{\text{max}})} = e^{-(1/RC)t}$$

By definition, $e^{\ln x} = x$, so

$$\frac{q}{q_{\text{max}}} = e^{-(1/RC)t}$$

or, going back to our notation in which q is explicitly a function of time,

$$q(t) = q_{\text{max}}e^{-(1/RC)t} \qquad (18\text{-}35)$$

The current $i(t)$ is then

$$i(t) = \frac{dq(t)}{dt} = \frac{d}{dt}\left(q_{max}e^{-(t/RC)}\right) = -\frac{q_{max}}{RC}e^{-(t/RC)}$$

or

$$i(t) = \frac{dq(t)}{dt} = -\frac{CV}{RC}e^{-(t/RC)} = -\frac{V}{R}e^{-(t/RC)}$$

or

$$i(t) = -\frac{V}{R}e^{-(t/RC)} \qquad (18\text{-}36)$$

According to Equation 18-36, at the instant the switch is closed, the current is $i(0) = -V/R$. The value of the initial current shouldn't surprise you, because initially the capacitor is fully charged, and the potential across the fully charged capacitor is exactly equal to the battery potential. Of course, without the battery in the circuit, there is no electric field to hold the excess charge on the capacitor, and as it leaves the plates, the potential across the capacitor drops. The decreasing charge results in a corresponding decrease in current, as seen in **Figure 18-25a** and **Figure 18-25b**. After some long time, the capacitor has become completely discharged.

Did you wonder why $i(t)$ is positive when the capacitor charges (Equation 18-34) but negative when it discharges (Equation 18-36)? Current has a direction, and when the capacitor discharges, the excess charges flow in the direction opposite to the direction of the current that charged it.

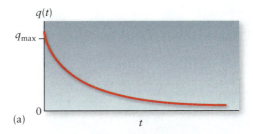

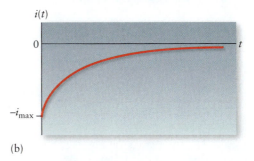

Figure 18-25 (a) The charge on the capacitor in a series *RC* circuit decreases exponentially as a function of time when a fully charged capacitor is discharged. (b) The current in a discharging series *RC* circuit approaches zero exponentially as a function of time. The current is negative to indicate that charges flow in a direction opposite to the current that caused the capacitor to become charged.

Watch Out

An *RC* circuit is not an oscillator.

We've all encountered electric circuits that oscillate, for example, a circuit that makes a light blink on and off. Is an *RC* circuit at the heart of that kind of circuit? After all, we've just seen how a resistor and capacitor connected in series can be made to either charge up or discharge, and together the two processes might be considered a complete cycle. Don't be fooled, however; the processes do not make up an oscillation! To make the circuit switch between the charging and discharging modes, we had to manually change the position of the switch in the circuit in Figure 18-24. An electronic oscillator must cycle on its own, so a series *RC* circuit is not an oscillator.

Time Constant

How can we characterize how quickly or slowly a series *RC* circuit takes to charge up or discharge? It might be tempting to ask how long it takes for the current to fall to zero, but Equation 18-35 makes it clear that in either case this doesn't happen until an infinite amount of time has elapsed. Because this is true regardless of the particular values of resistance and capacitance, the time for the current to fall completely to zero, which is also the time for the circuit to fully charge or discharge, is not a useful way to characterize an *RC* circuit.

Consider the current versus time curves plotted for two different charging *RC* circuits shown in **Figure 18-26a** and **Figure 18-26b**. Given the same time scale in

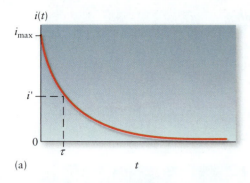

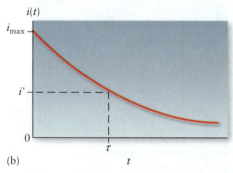

Figure 18-26 Current as a function of time is shown for two charging, series RC circuits. Both curves exhibit exponential behavior, but the circuit that corresponds to part (a) charges more quickly than the circuit that corresponds to part (b) because the time τ it takes for the current to fall to i' (an arbitrarily chosen value) is shorter. The current and time scales are the same in both plots.

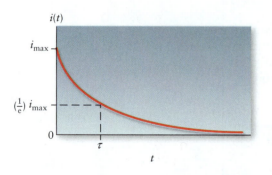

Figure 18-27 The time constant, or decay time, τ of a series RC circuit is defined as the time it takes for the current to fall to $1/e$ of its initial value.

both figures, the circuit that corresponds to Figure 18-26a clearly charges more quickly, because the time τ it takes to decrease to the value i' is shorter than for the circuit shown in Figure 18-26b. The way to characterize an RC circuit, then, is to give the time τ it takes for the current to decrease to a value of i' set to be a standard fraction of i_{max}. Most common for this kind of application is to set

$$i' = \frac{1}{e}i_{max} = e^{-1}i_{max}$$

Figure 18-27 shows the **time constant,** sometimes called the **decay time,** when defined in this way.

The fraction e^{-1} is about 0.37. The reason for using this curious fraction is that the time dependence of current also includes an exponential (Equation 18-34), so treating the exponential is straightforward. At time τ the current is given by

$$i(\tau) = \frac{V}{R}e^{-(\tau/RC)}$$

So requiring $i(\tau) = e^{-1}i_{max} = e^{-1}\frac{V}{R}$ means that

$$\frac{V}{R}e^{-(\tau/RC)} = e^{-1}\left(\frac{V}{R}\right)$$

so

$$e^{-(\tau/RC)} = e^{-1}$$

The exponents must be equal:

$$-\frac{\tau}{RC} = -1$$

or

$$\tau = RC \qquad (18\text{-}37)$$

When resistance is measured in ohms and capacitance is expressed in farads, the units of τ are seconds. After one time constant has elapsed, the current in the charging RC circuit has decreased to 37% of its initial value. You can verify by using your calculator that after 5 time constants the current is less than 1% of its value immediately after the switch is closed; after 10 time constants, the current is less than 0.01% of the starting current. In addition, as you can see by comparing Figure 18-23a and Figure 18-23b, the decrease in current corresponds to an increase in charge stored on the capacitor. So the time constant is also a measure of the rate at which an RC circuit charges.

Because the time constant τ characterizes an RC circuit, it enables us to compare circuits with different resistors and capacitors. For example, as the resistance in an RC circuit is decreased, the time it takes for the circuit to charge gets shorter (as evident from Equation 18-37). This observation is not surprising, because a smaller resistance allows larger current, which means that the capacitor can accumulate charge at a higher rate. Decreasing the capacitance also shortens the charging time; because the capacitor can store less charge, it takes less time to charge it.

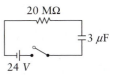

Figure 18-28 A 20 MΩ resistor is connected in series with a 3 μF capacitor in a circuit with a switch and a 24 V battery.

Example 18-13 A Charging Series *RC* Circuit

A 20 MΩ resistor is connected in series with a 3 μF capacitor in a circuit with a switch and a 24 V battery, as shown in **Figure 18-28**. The capacitor is initially uncharged. (a) What is the current in the circuit immediately after the switch is closed? (b) What is the charge on the capacitor once it is fully charged? (c) How long after the switch is closed does it take for the current to decrease to 37% of its value immediately after the switch is closed? (d) How long after the switch is closed does it take for the charge on the capacitor to reach 63% of its maximum value?

SET UP
In a circuit like this one in which the capacitor is initially uncharged, the largest current exists immediately after the switch is closed. At that moment, because there is not yet any excess charge on the capacitor, the circuit—just for an instant—acts as if there were only a battery and resistor, so $i(0) = V/R$. The capacitor eventually becomes fully charged, so no more charge can flow, and the potential difference across the capacitor is identical to that of the battery, so $q_{max} = CV$. In between the time when the switch is closed and when the capacitor is fully charged, the charge and current obey Equation 18-33 and Equation 18-34.

SOLVE
(a) *What is the current in the circuit immediately after the switch is closed?* Both according to the argument above and also Equation 18-34,

$$i(0) = \frac{V}{R} = \frac{24 \text{ V}}{20 \times 10^6 \text{ Ω}} = 1.2 \times 10^{-6} \text{ A} = 1.2 \text{ μA}$$

(b) *What is the charge on the capacitor once it is fully charged?* Both according to the argument above and also Equation 18-33,

$$q_{max} = CV = (3 \times 10^{-6} \text{ F})(24 \text{ V}) = 7.2 \times 10^{-5} \text{ C} = 72 \text{ μC}$$

(c) *How long after the switch is closed does it take for the current to decrease to 37% of its value immediately after the switch is closed?* We have defined the time constant (Equation 18-37) to be the time it takes for the current to decrease to 37% of its initial value:

$$\tau = RC = (20 \times 10^6 \text{ Ω})(3 \times 10^{-6} \text{ F}) = 60 \text{ s}$$

(d) *How long after the switch is closed does it take for the charge on the capacitor to reach 63% of its maximum value?* The charge on the capacitor increases over time, but at a slower and slower rate. This observation can be seen from Equation 18-33 and in Figure 18-23a. To find the time t' such that $q(t')$ equals 63% of q_{max}, set $q(t') = 0.63q_{max}$:

$$q(t') = CV(1 - e^{-t'/RC})$$

and

$$q(t') = 0.63q_{max} = 0.63CV$$

So

$$0.63CV = CV(1 - e^{-t'/RC})$$

or

$$0.63 = 1 - e^{-(t'/RC)}$$

 Go to Interactive Exercise 18-1 for more practice dealing with *RC* circuits

 Go to Interactive Exercise 18-2 for more practice dealing with *RC* circuits

 Go to Interactive Exercise 18-3 for more practice dealing with time constants

$$0.37 = e^{-(t'/RC)}$$

$$t' = -RC \ln(0.37)$$

You should recognize 0.37 as nearly equal to e^{-1}, so (perhaps even without a calculator),

$$t' = RC$$

The time for the excess charge on the capacitor to increase to 63% of its final value is the same as the time for the current to decrease to 37% of its initial value; both times are given by the decay time τ. In this problem, the decay time is 60 s.

REFLECT

The time constant is a direct way to characterize the rate at which charge and current change over time in an RC circuit.

Practice Problem 18-13 A 15 kΩ resistor is connected in series with a 6.0 pF capacitor in a circuit with a switch and a 12-V battery as in Figure 18-22. The capacitor is initially uncharged. (a) What is the current in the circuit immediately after the switch is closed? (b) What is the charge on the capacitor once it is fully charged? (c) After the switch is closed, how long does it take for the current to fall to 37% of its value immediately before the switch is closed?

✱ What's Important 18-7

A series RC circuit contains a resistor and a capacitor connected in series. Connecting the resistor and an uncharged capacitor to a voltage source causes a current and causes the capacitor to charge. Removing the voltage supply (without breaking the circuit) causes a current in the opposite direction as the capacitor discharges. The current in a charging or discharging RC circuit decreases to $1/e$ (37%) of its initial value in one time constant.

18-8 Bioelectricity

To code and transmit information in the brain and throughout the body, animals generate small electrical signals called *action potentials*. The signals are only about 100 mV in amplitude and 2 ms in duration, but they propagate at speeds up to 100 m/s in some cells, about the speed of a race car. In this section, we'll consider the physics of action potentials and the membranes that generate them.

Membranes

Membranes separate cells from their environment. The fabric of a membrane is a thin bilayer of phospholipids, 7.5 to 10 nm thick. One end of a phospholipid molecule is charged and therefore hydrophilic; the other end is uncharged and therefore hydrophobic (**Figure 18-29**). In the aqueous environment of living things, the uncharged lipid tails of the phospholipids face each other, effectively creating an insulating region sandwiched between layers of the charged phosphate heads of the molecules (**Figure 18-30**). As we saw in Chapter 17, because membranes consist of two charged surfaces separated by a thin layer of a nonconducting material, electrically they behave like capacitors.

The cell membrane forms a selective barrier that determines what enters and what leaves the cell. Only small, lipid-soluble substances such as oxygen and carbon

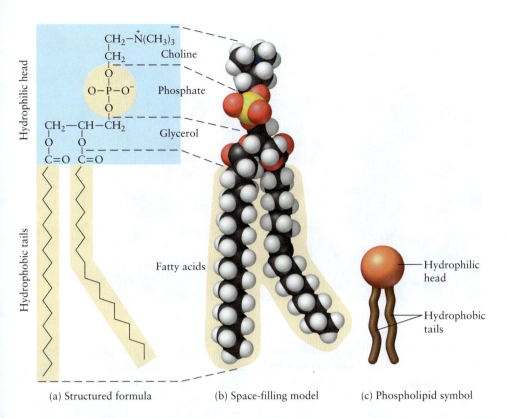

Figure 18-29 Phospholipid molecules are charged on one end and uncharged on the other. Therefore, one end of the molecule is hydrophilic and the other is hydrophobic.

Hydrophilic head

$CH_2-\overset{+}{N}(CH_3)_3$
CH_2 Choline

O
$O-P-O^-$ Phosphate
O

$CH_2-CH-CH_2$ Glycerol
O O
$C=O$ $C=O$

Hydrophobic tails

Fatty acids

Hydrophilic head

Hydrophobic tails

(a) Structured formula   (b) Space-filling model   (c) Phospholipid symbol

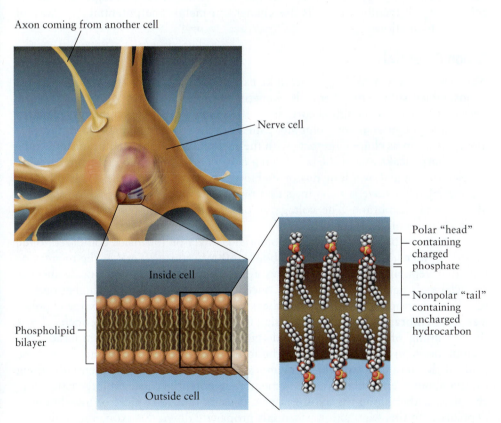

Axon coming from another cell

Nerve cell

Inside cell

Phospholipid bilayer

Outside cell

Polar "head" containing charged phosphate

Nonpolar "tail" containing uncharged hydrocarbon

Figure 18-30 A biological membrane is composed of a bilayer of phospholipid molecules and behaves like a capacitor. The charged, hydrophilic heads of the phospholipids form the surfaces of the membrane while the uncharged, hydrophobic tails of the phospholipids form the interior of the membrane. In other words, a membrane is two charged surfaces separated by a thin layer of a nonconducting material.

dioxide can pass directly through the membrane; everything else can diffuse across a membrane only through a channel created by the presence of one or more specific proteins. For example, some of the proteins embedded in cell membranes

Figure 18-31 The phospholipid bilayer with its embedded protein channels forms a selective barrier that determines what enters and what leaves the cell. Small molecules such as oxygen and carbon dioxide can diffuse through the bilayer, but ions must pass through channels created by the presence of one or more specific proteins.

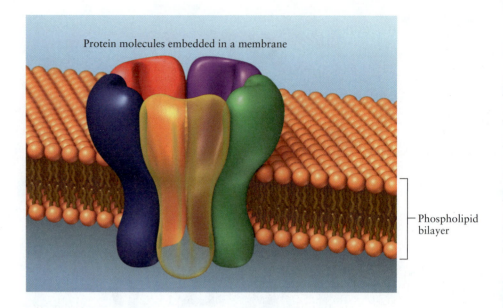

Protein molecules embedded in a membrane

Phospholipid bilayer

form channels that allow certain ions to diffuse across the phospholipid bilayer (**Figure 18-31**). When these channels are open the ions can carry a current across the membrane, and the channels act like resistors. Although the concentration of ions inside and outside the cell remains essentially unchanged, this process changes the balance of charge close to the membrane, which in turn changes the voltage across the cell membrane. Cells use changes in membrane potential as a way of coding information.

Action Potentials

Neurons typically have long, threadlike extensions of their cell membranes, called axons, which stretch to other cells. Sometimes axons can be very long, such as the ones that go from your spinal cord down to your big toe, for example. A change in membrane potential at any point along an axon decays exponentially with distance along the axon as charge interacts with the capacitance of the membrane and some of the charge leaks out of the cell. Such a decay could prevent a cell from carrying the electrical signal over long distances. However, nerve and muscle cell membranes are special. They have ion channels that repeatedly amplify the signal as it spreads along a cell membrane. The voltage across the membrane controls whether the channels open or close.

At rest, the baseline membrane potential is lower on the inside compared to the outside of a cell, typically between about -60 mV and -90 mV. If the membrane potential depolarizes, that is, becomes a little less negative on the inside surface, Na^+ channels open, and because the concentration of Na^+ is always higher outside of cells than inside, Na^+ rushes into the cell. This influx of positive charge depolarizes the membrane even further, causing even more Na^+ channels to open. This is the beginning of an action potential (**Figure 18-32**). At its peak amplitude, the membrane is positively charged on the inside compared to the outside. From the peak of the action potential, the membrane potential decays exponentially along the axon, but even a small change in membrane potential causes voltage-gated Na^+ channels farther along the axon to open, causing that next piece of membrane to depolarize. In this way, action potentials propagate down the axon, amplifying the decaying signal back to its original peak value. The process continues until the action potential reaches the end of the axon (**Figure 18-33**).

How does the axon get back to its resting conditions? First, Na^+ channels automatically close. But in addition, as the potential difference across the cell

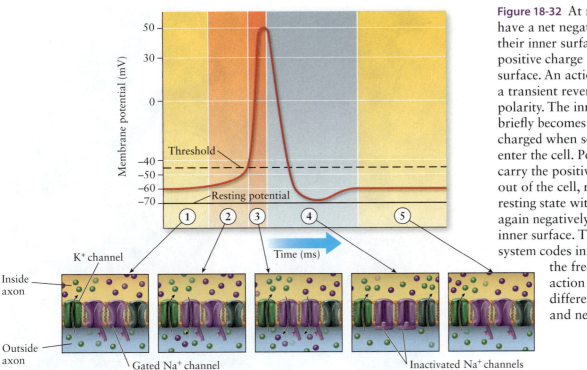

Figure 18-32 At rest, membranes have a net negative charge on their inner surface and a net positive charge on their outer surface. An action potential is a transient reversal of this polarity. The inner surface briefly becomes positively charged when sodium ions enter the cell. Potassium ions carry the positive charge back out of the cell, restoring the resting state with the membrane again negatively charged on its inner surface. The nervous system codes information by the frequency of action potentials in different cells and neural circuits.

membrane changes during the initial phase of an action potential, K^+ channels begin to open. The membrane quickly becomes much more permeable to K^+ than to Na^+, so because the concentration of K^+ ions is always higher *inside* cells, positive charge begins to flow out of the cell. This outward positive current drives the membrane potential back to its resting level. As a result, during an action potential a brief wave of excess charge spreads along a membrane as the polarity of charge on the plates of the capacitor—the inside and outside surfaces of the cell membrane—reverses for an instant.

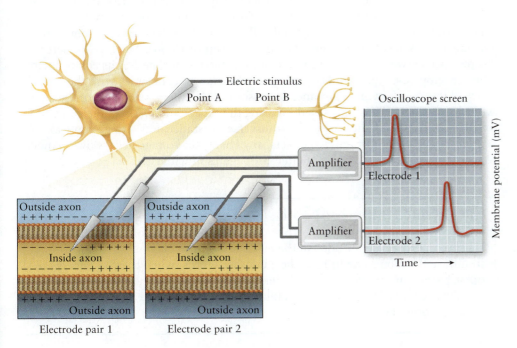

Figure 18-33 From the peak of the action potential, at which time the membrane is positively charged on the inside surface compared to the outside, the membrane potential decays exponentially along the axon. This causes voltage-gated Na^+ channels farther along the axon to open, thus propagating the action potential down the axon. This process continues until the action potential reaches the end of the axon.

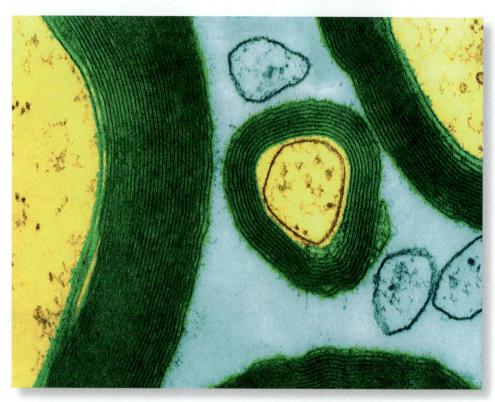

Figure 18-34 This cross section through a nerve shows myelin (colored green) wrapped around axons (colored yellow). By increasing the distance between the inside of the axon and the extracellular fluid (colored blue), myelin reduces the capacitance of the axon. *(Dennis Kunkel Microscopy, Inc./Visuals Unlimited.)*

What determines how fast the action potential can travel? In Example 18-6, we saw that the larger the diameter of the axon, the smaller the longitudinal resistance and the easier it is for charge to move down the axon. So the larger the diameter of an axon, the faster the action potential will propagate. But there's another, more efficient way to make action potentials propagate more quickly, even in small-diameter axons. In vertebrates, special cells wrap themselves around some axons, forming a thick insulating layer called *myelin* (**Figure 18-34**). The myelin increases the distance between the inside surface of the axon membrane and its environment. The capacitance of these additional layers of membrane, in series with the capacitance of the axon's own membrane, decreases the effective capacitance of the axon, while increasing its resistance. By decreasing the capacitance, less charge gets stored on the membrane; by increasing the transverse resistance, less charge can leak out of the cell. As a result of the two effects, any change in voltage decays over a much longer distance, so the action potential propagates much faster.

✱ What's Important 18-8

Biological membranes have the physical properties of capacitance and resistance. Proteins embedded in the membranes form channels that allow ions to diffuse from one side to the other; in other words, when these channels are open, particular ions can carry a current across the membrane and the channels act like resistors. The phospholipid bilayer of the membrane acts like a capacitor and stores charge.

Answers to Practice Problems

18-1 280 kV

18-3 5

18-4 998 nΩ

18-5 2.3 pA

18-6 $R_G = 4.4 \times 10^{-5}\,R_S$

18-7 One resistor in series with the other two connected in parallel.

18-8 55 Ω

18-9 1.3 A

18-10 1.7 h

18-11 $7.39

18-13 (a) 0.80 mA, (b) 72 pC, (c) 90 ns

Answers to Got the Concept Questions

18-1 (a) Combining resistors in series results in a circuit with larger equivalent resistance; combining them in parallel results in a smaller equivalent resistance. Therefore the parallel circuits in Figures 18-16b and d have lower resistances than the series circuits in Figures 18-16a and c. In addition, for the same reason, the equivalent resistance of circuit d is lower than that of b, and the equivalent resistance of circuit a is lower than that of c. So in order of increasing equivalent resistance, the circuits are d < b < a < c. (b) The larger the equivalent resistance, the smaller the current for a fixed voltage; so in order of increasing current, the circuits are c < a < b < d. (c) Adding any resistance in parallel to a group of resistors lowers the equivalent resistance. Adding any resistance in series to a group of resistors increases the equivalent resistance. Therefore the conclusion that the equivalent resistance of the circuit in Figure 18-16d is lower than that of b, and that the equivalent resistance of circuit a is lower than that of c, does not depend on the specific value of R_3.

SUMMARY

Topic	Summary	Equation or Symbol
conductivity	Conductivity is a measure of how well or poorly a particular material conducts an electric current.	σ
conventional current	Although electric current in a wire is the flow of electrons, conventional current defines a current direction as if positive charge were flowing.	
current	Current is the net flow of charge, often in a wire or circuit. Current i is defined as the amount of charge Δq that flows past a selected point in a time Δt, or in the limit of a small time interval, the derivative of charge with respect to time is the instantaneous charge flow rate.	$i = \dfrac{\Delta q}{\Delta t}$ (18-1) or $i = \dfrac{dq}{dt}$ (18-2)
decay time	The decay time (or time contant) τ of a series RC circuit characterizes how quickly or slowly the capacitor charges or discharges. τ is defined as the time it takes for the current to fall to $1/e$ (about 0.37) of its initial value.	$\tau = RC$ (18-37)
drift speed	Drift speed is the speed at which the individual charges that make up a current move. Drift speed is a function of the current i and the cross-sectional area A of the conductor, as well as the number density of free charges n and the charge e of each charge carrier.	$v_{\text{drift}} = \dfrac{i}{nAe}$

equivalent resistance	The equivalent resistance of a group of interconnected resistors is the resistance of a single resistor that is the equivalent replacement of the group.	
junction	A junction is a point in a circuit at which three or more wires are connected.	
Kirchhoff's junction rule	Kirchhoff's junction rule states that the sum of the currents entering a junction equals the sum of the currents leaving it.	
Kirchhoff's loop rule	Kirchhoff's loop rule states that the sum of the changes in potential around a closed electrical loop equals zero.	
parallel	When resistors are connected in parallel, all of the right sides of each are connected and (separately) all of the left sides are connected. Connecting resistors in parallel results in a smaller resistance.	$R_{equiv} = \dfrac{1}{\sum\limits_{i=1,N} \dfrac{1}{R_i}}$ (18-24)
power	Power is defined as the rate at which energy flows P into or out of a system, or changes from one form of energy to another. The SI unit of power is the watt (W). 1 W is equivalent to 1 J per s.	P
resistance	Resistance is the characteristic of an object to resist the flow of charge. The resistance R of an object depends on its length L and cross-sectional area A, and also on the resistivity ρ of the material from which it is made.	$R = \dfrac{\rho L}{A}$ (18-11)
resistivity	Resistivity is the characteristic of a material to resist the flow of charge. The resistivity ρ of a material is the inverse of the material's conductivity σ.	$\rho = \dfrac{1}{\sigma}$ (18-10)
resistor	A resistor is an electronic device employed in electronic circuits to limit the flow of charge. The standard symbol for a resistor is shown to the right. ⟋⟍⟋⟍⟋	
series	When resistors are connected in series, the right side of one resistor is connected to the left side of the resistor to its right. Connecting resistors in series results in a larger resistance.	$R_{equiv} = \sum\limits_{i=1,N} R_i$ (18-19)
time constant	The time constant (or decay time) τ of a series RC circuit characterizes how quickly or slowly the capacitor charges or discharges. τ is defined as the time it takes for the current to fall to $1/e$ (about 0.37) of its initial value.	$\tau = RC$ (18-37)

QUESTIONS AND PROBLEMS

In a few problems, you are given more data than you actually need; in a few other problems, you are required to supply data from your general knowledge, outside sources, or informed estimate.

Interpret as significant all digits in numerical values that have trailing zeros and no decimal points.

For all problems, use $g = 9.8 \text{ m/s}^2$ for the free-fall acceleration due to gravity. Neglect friction and air resistance unless instructed to do otherwise.

• Basic, single-concept problem

•• Intermediate-level problem, may require synthesis of concepts and multiple steps

••• Challenging problem

SSM *Solution is in Student Solutions Manual*

Conceptual Questions

1. •We distinguish the direction of current in a circuit. Why don't we consider it a vector quantity?

2. •Is current dissipated when it passes through a resistor? Explain your answer.

3. •We justified a number of electrostatic phenomena by the argument that there can be no electric field in a

conductor. Now we say that the current in a conductor is driven by a potential difference and thus there is an electric field in the conductor. Is this statement a contradiction? SSM

4. •Two wires, A and B, have the same physical dimensions but are made of different materials. If A has twice the resistance of B, how do their resistivities compare?

5. •The average drift velocity of electrons in a wire carrying a steady current is constant even though the electric field within the wire is doing work on the electrons. What happens to this energy?

6. •Under ordinary conditions, the drift speed of electrons in a metal is around 10^{-4} m/s or less. Why doesn't it take a long time for a lightbulb to come on when you flip the wall switch that is several meters away?

7. •For a given source of constant voltage, will more heat develop in a large external resistance connected across it or a small one? SSM

8. •An ammeter measures the current through a particular circuit element. (a) How should it be connected with that element, in parallel or in series? (b) Should an ammeter have a very large or a very small resistance? Why?

9. •Biology When a bird lands on a high voltage wire and grabs the wire with its feet, will the bird be electrocuted? Explain your answer.

10. •If the only voltage source you have is 36 V, how could you light some 6 V lightbulbs without burning them out?

11. ••Many ordinary strings of Christmas-tree lights contain about 50 bulbs connected in parallel across a 110 V line. Sixty years ago most strings contained 50 bulbs connected in series across the line. What would happen if you could put one of the old-style bulbs into a modern Christmas-tree light set? (The light sockets are made differently to prevent this.) SSM

12. •Biology What causes an action potential?

13. •Biology What causes an action potential to travel along an axon?

14. •Current is always a positive quantity. What, then, is the meaning of a negative current when solving Kirchhoff's equations?

15. •Give a simple physical explanation for why the charge on a capacitor in an RC circuit can't be changed instantaneously.

16. •(a) Does the time required to charge a capacitor through a given resistor with a battery depend on the voltage of the battery? (b) Does it depend on the total amount of charge to be placed on the capacitor? Explain your answers.

Multiple-Choice Questions

17. •If a current carrying wire has a cross-sectional area that gradually becomes smaller along the length of the wire, the drift velocity
   A. increases along length of the wire.
   B. decreases along length of the wire.
   C. remains the same along the length of the wire.
   D. increases along length of the wire if the resistance increases too.
   E. decreases along length of the wire if the resistance decreases too. SSM

18. •Biology What causes an electric shock?
   A. current
   B. voltage
   C. both current and voltage
   D. resistance and current
   E. resistance and voltage

19. •Two copper wires have the same length, but one has twice the diameter of the other. Compared to the one that has the smaller diameter, the one that has the larger diameter has a resistance that is
   A. larger by a factor of 2.
   B. larger by a factor of 4.
   C. the same.
   D. smaller by a factor of 1/2.
   E. smaller by a factor of 1/4.

20. ••A wire has a length L and a resistance R. It is stretched uniformly to a length of $2L$. The resistance of the wire after it has been stretched is
   A. $\frac{1}{4}R$.
   B. $\frac{1}{2}R$.
   C. R.
   D. $2R$.
   E. $4R$.

21. •When a thin wire is connected across a voltage of 1 V, the current is 1 A. If we connect the same wire across a voltage of 2 V, the current is
   A. $\frac{1}{4}$ A.
   B. $\frac{1}{2}$ A.
   C. 1 A.
   D. 2 A.
   E. 4 A. SSM

22. •When a wire that has a large diameter and a length L is connected across the terminals of an automobile battery, the current is 40 A. If we cut the wire to half of its original length and connect the piece that has a length $L/2$ across the terminals of the same battery, the current will be
   A. 10 A.
   B. 20 A.
   C. 40 A.
   D. 80 A.
   E. 160 A.

This page is intentionally left blank.

For complete end of chapter problem sets, please go to
www.whfreeman.com/kestentauck

44. ••An 8-m-long length of wire has a resistance of 4 Ω. The wire is uniformly stretched to a length of 16 m. Find the resistance of the wire after it has been stretched.

45. ••The resistance ratio of two conductors that have equal cross-sectional areas and equal lengths is 1:3. What is the ratio of the resistivities of the materials from which they are made?

46. •A power transmission line is made of copper that is 1.8 cm in diameter. If the resistivity of copper is 1.7×10^{-8} Ω·m, find the resistance of 1 mi of the line.

47. ••When 120 V is applied to the filament of a 75 W lightbulb, the current drawn is 0.63 A. When a potential difference of 3 V is applied to the same filament, the current is 0.086 A. Is the filament made of an ohmic material? Explain your answer.

48. ••A piece of 14-gauge copper wire has a diameter of 0.163 cm and is 14.0 m long. (a) What is its resistance? (b) If a potential difference of 3.00 V is applied across the wire, what is the current in it? (c) What is the electric field in the wire?

49. ••**Calc** Calculate the resistance of a hollow, cylindrical conductor that has an inner radius r_i and an outer radius r_o, for charges that flow radially from the inner radius to the outer radius (**Figure 18-36**). Assume the metal has a resistivity of ρ and a length L. Integrate over concentric cylindrical shells.

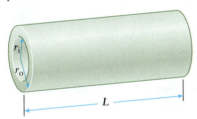

Figure 18-36 Problem 49

50. ••**Calc** Determine the resistance across a thin spherical shell that has an inner radius r_i and an outer radius r_o (**Figure 18-37**). Assume the conductor has a resistivity of ρ and that charge flows from the inner radius to the outer radius.

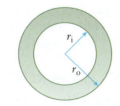

Figure 18-37 Problem 50

18-3: Physical and Physiological Resistors

51. •A common resistor used in a class lab has a color code shown in **Figure 18-38**. Write the numerical value of the resistor and give its *tolerance* (the uncertainty

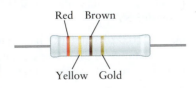

Figure 18-38 Problem 51

in the stated value). You may need to look up the resistor code using your favorite online search engine.

52. •• There is current of 112 pA when a certain potential is applied across a certain resistor. When that same potential is applied across a resistor made of the identical material but 25 times longer, the current is 0.044 pA. Compare the effective diameters of the two resistors.

53. •What are the first three color codes depicting a resistor of that value of 20 GΩ. You may need to look up the resistor code using your favorite online search engine.

54. ••Electrical cables often consist of a large number of strands of conducting wire. Due to variances in manufacturing, cables from a certain company can have between 850 and 950 separate strands, each with a mean diameter of 0.72 ± 0.07 mm. Determine the ratio of the (a) lowest and (b) highest possible resistance of one of these cables to the average resistance of many strands of conducting wire.

55. ••A certain flexible conducting wire changes shape as environmental variables, such as temperature, change. If the diameter of the wire increases by 25% while the length decreases by 12%, by what factor does the resistance of the wire change?

18-4: Direct Current Circuits

56. •If a lightbulb draws a current of 1.0 A when connected to a 12 V circuit, what is the resistance of its filament?

57. •If a flashlight bulb has a resistance of 12.0 Ω, how much current will the bulb draw when it is connected to a 6.0 V circuit?

58. •A lightbulb has a resistance of 8.0 Ω and a current of 0.5 A when it is connected to a voltage. At what voltage is it operating?

59. •Calculate the voltage difference $V_A - V_B$ in each of the following situations (**Figure 18-39**).

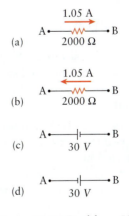

Figure 18-39 Problem 52

18-5: Resistors in Series and Parallel

60. •An 18 Ω resistor and a 6.0 Ω resistor are connected in series. What is the equivalent resistance of the resistors?

61. •An 18 Ω resistor and a 6.0 Ω resistor are connected in parallel. What is the equivalent resistance of the resistors?

62. ••A 9.0 Ω resistor and a 3.0 Ω resistor are connected in series across a 9.0 V battery. Find (a) the current through each resistor and (b) the voltage drop across each resistor.

63. ••A 9.0 Ω resistor and a 3.0 Ω resistor are connected in parallel across a 9.0 V battery. Find (a) the current through each resistor and (b) the voltage drop across each resistor.

64. ••A potential difference of 3.6 V is applied between points a and b in **Figure 18-40**. Find (a) the current in each of the resistors and (b) the total current the three resistors draw from the power source.

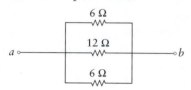

Figure 18-40 Problem 64

65. ••The following four resistors have an equivalent resistance of 8 Ω (**Figure 18-41**). Calculate the value of R_x.

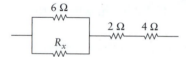

Figure 18-41 Problem 65

66. ••Find the equivalent resistance of the following combination (**Figure 18-42**).

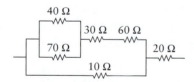

Figure 18-42 Problem 66

67. ••A metal wire of resistance 48 Ω is cut into four equal pieces that are then connected side by side to form a new wire which is one-quarter of the original length. What is the resistance of the new wire?

68. ••Two resistors A and B are connected in series to a 6.0 V battery; the voltage across resistor A is 4.0 V. When A and B are connected in parallel across a 6.0 V battery, the current through B is 2.0 A. What are the resistances of A and B? The batteries have negligible internal resistance.

69. ••A potential difference of 7.50 V is applied between points a and c in **Figure 18-43**. (a) Find the difference in potential between points b and c. (b) Is the current through the 60 Ω resistor larger or smaller than that through the 35 Ω resistor? Why?

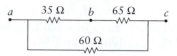

Figure 18-43 Problem 69

18-6: Power

70. •A heater is rated at 1500 W. How much current does it draw when it is connected to a 120 V voltage source?

71. •A 4.0 Ω resistor is connected to a 12 V voltage source. What is the power dissipated by the resistor?

72. •A coffeemaker has a resistance of 12 Ω and draws a current of 15 A. What power does it use?

73. ••When connected in parallel across a 120 V source, two lightbulbs consume 60 W and 120 W, respectively. What powers do the lightbulbs consume if instead they are connected in series across the same source? Assume the resistance of each lightbulb is constant.

74. •A transmission line carries 1200 A and has a resistance of 40 Ω. Calculate the energy lost per second due to *joule heating*, the transformation of electrical energy to thermal energy when current through wires encounters resistance.

75. •A stereo speaker has a resistance of 8 Ω. If the power output is 40 W, calculate the current passing through the speaker wires.

76. •If your local power company charges $0.11/kW · h, what would it cost to run a 1500 W heater continuously during an 8-h night?

77. ••A house is heated by a 24-kW electric furnace using resistance heating, and the local power company charges $0.10/kW · h. The heating bill for January is $218. How long must the furnace have been running on an average January day?

18-7: Series *RC* Circuits

78. •A 4.00 MΩ resistor and a 3.00 μF capacitor are connected in series with a power supply. What is the time constant for the circuit?

79. •A capacitor of 20 μF and a resistor of 100 Ω are quickly connected in series to a battery of 6.0 V. What is the charge on the capacitor 0.0010 s after the connection is made? SSM

80. •A 10.0 μF capacitor has an initial charge of 100.0 μC. If a resistance of 20.0 Ω is connected across it, what is the initial current through the resistor?

81. ••A 10 μF capacitor carries an initial charge of 80 μC. (a) If a resistance of 25 Ω is connected across it, what is the initial current in the resistor? (b) What is the time constant of the circuit?

82. ••A 12.5 μF capacitor is charged to a potential of 50.0 V and then discharged through a 75.0 Ω resistor. (a) How long after discharge begins does it take for the capacitor to lose 90.0% of its initial (i) charge and (ii) energy? (b) What is the current through the resistor at both times in part (a)?

18-8: Bioelectricity

83. ••**Biology** A single ion channel has a resistance of 1.0 GΩ. During an action potential of 80 mV, the channel is open for approximately 1.0 ms for the flow of K$^+$ ions. How many K$^+$ ions travel through the channel?

84. ••**Biology** Cell membranes contain *channels* that allow ions to cross the phospholipid bilayer. A particular potassium channel carries a current of 1.9 pA. How many potassium ions pass through it in 1.0 ms?

85. ••**Biology** Cell membranes contain channels that allow K$^+$ ions to leak out. Consider a channel that has a diameter of 1.0 nm and a length of 10 nm. If the channel has a resistance of 18 GΩ, what is the resistivity of the solution in the channel?

86. ••**Biology** An axon is myelinated so that its effective capacitance is decreased by a factor of $\frac{1}{4}$. If the time constant of the effective RC circuit decreases by $\frac{1}{2}$, by what factor would the resistance of the axon change?

87. ••**Biology** An action potential propagates at 64 m/s down an axon that is 30 cm long. How much energy is transmitted down the axon if the action potential has a maximum value of 75 mV and the resistance is 20,000 Ω?

General Problems

88. ••**Astronomy** The electric charge that is attracted to a wayward satellite in the Van Allen belt is found to be given by

$$q(t) = (100 \ \mu C/s^2)t^2 + (150 \ \mu C/s)t + 28 \ \mu C$$

$$\text{(SI units)}$$

(a) Determine the amount of charge that the satellite will possess at $t = 0$ s.
(b) Determine the amount of charge that the satellite will possess at $t = 2.5$ s.
(c) Calculate the rate at which charge flows into the satellite at $t = 0$ s.
(d) Calculate the rate at which charge flows into the satellite at $t = 12$ s.

89. ••**Biology** Giant electric eels can deliver a voltage shock of 500 V and up to 1.0 A of current for a brief time. For a snorkeler swimming in salt water, the resistance of her skin is negligible, so her body resistance is

about 600 Ω (this can vary). A current of about 500 mA can cause heart fibrillation and death if it lasts too long. (a) What is the maximum power a giant electric eel can delivery to its prey? (b) If the snorkeler is unfortunate enough to be struck by the eel, what current will pass through her body? Is this large enough to be dangerous? (c) What power does the snorkeler in part (b) receive from the eel?

90. •••An electric heater consists of a single resistor connected across a 110 V line. It is used to heat 200.0 g of water in a coffee cup from 20°C to 90°C in 2.70 min. (a) Assuming that 90% of the energy drawn from the power source goes into heating the water, what is the resistance of the heater? (b) Assuming no other heat losses, how much longer will it take to heat your water if you had to power the water heater with your 12.0 V car battery?

91. •Determine the current through each resistor in **Figure 18-44**.

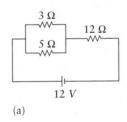

(a)

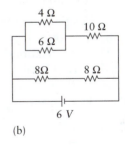

(b)

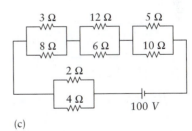

(c)

Figure 18-44 Problem 91

92. ••How much power is dissipated in each resistor (**Figure 18-45**)?

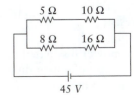

Figure 18-45 Problem 92

93. ••• Measurements made during severe electrical storms reveal that lightning bolts can carry as much as about 30 C of charge and can travel between a cloud and the ground in time intervals of around 100 μs (and sometimes even shorter). Potential differences have been measured as high as 400 million volts. (a) What is the current in such a lightning strike? How does it compare with typical household currents? (b) What is the resistance of the air during such a strike? (c) How much energy is transferred during a severe strike? (d) What mass of water at 100°C could the lightning bolt in part (c) evaporate? (See Table 14-4.)

94. ••• A voltmeter of resistance R_V and an ammeter of resistance R_A are connected in order to measure the unknown resistor R_X in the following circuit (**Figure 18-46**). The actual resistance is $R_X = V/i_X$, where i_X is the current through R_X and V is the reading of the voltmeter. The measured resistance is $(R_X)_{measured} = V/i$, where i is the ammeter reading. Derive a relationship between R_X, $(R_X)_{measured}$, and R_V. If $V = 10.0$ V and $i = 10.0$ mA, by how much do the values for R_X and $(R_X)_{measured}$ differ when $R_V = 1$ MΩ for the voltmeter and $R_A = 10$ Ω for the ammeter?

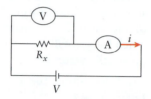

Figure 18-46 Problem 94

95. ••• Repeat problem 101 this time with the voltmeter connected across both the unknown resistor and the ammeter (**Figure 18-47**). Derive a relationship between R_X, $(R_X)_{measured}$, and R_A. If $V = 10.0$ V and $i = 10.0$ mA, by how much do the values for R_X and $(R_X)_{measured}$ differ when $R_V = 1$ MΩ for the voltmeter and $R_A = 10$ Ω for the ammeter?

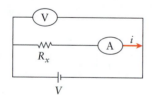

Figure 18-47 Problem 95

96. •• In the circuit shown in **Figure 18-48**, a potential difference of 5.00 V is applied between points a and b. Determine (a) the equivalent total resistance, (b) the current in each resistor, and (c) the power dissipated in each resistor. (d) Comment on the amount of power dissipated in the 20 Ω resistor compared to the other two.

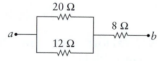

Figure 18-48 Problem 96

97. •• Two circuits, A and B, are shown in **Figure 18-49** with all quantities accurate to 3 significant figures. (a) Which circuit draws the most power from the battery, and (b) what is that maximum power?

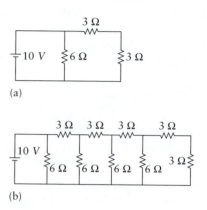

(a)

(b)

Figure 18-49 Problem 97

98. ••Biology The resistance of the body is made up of two basic parts: the resistance of the skin (see the following problem) and the resistance of the interior of the body. The internal resistance varies somewhat, but about 500 Ω measured between the two hands is typical. We can model this part of the body as a cylinder 1.6 m long and 14 cm in diameter. (a) What is the average internal resistivity of the body? (b) How much power will be delivered to the body of a person who accidentally grabs the ends of wires connected across a 110 V electrical outlet?

99. ••Biology Most of the resistance of the human body comes from the skin, as the interior of the body contains aqueous solutions that are good electrical conductors. For dry skin, the resistance between a person's hands is measured at typically 500 kΩ. The skin varies in thickness, but on the average it is about 2.0 mm thick. We can model the body between the hands as a cylinder 1.6 m long and 14 cm in diameter with the skin wrapped around it. (a) What is the resistivity of the skin? (b) Compare your answer with that found in part (a) of the previous problem for the internal resistivity of the body. How do you explain the fact that the resistance of the skin is about 1000 times greater than the internal resistance of the body, even though the resistivity of the skin is only about 60 times that of the interior of the body?

100. •• The capacitor in the flash of a disposable camera has a value of 160 μF. What is the resistance of the filament in the bulb if it takes 10 s to charge the capacitor to 80% of its maximum charge?

101. •• A discharging capacitor starts with a maximum charge and exponentially decays to zero. (a) How much time (in terms of the time constant τ) will it take before a discharging capacitor holds only 50% of the maximum charge? (b) What percentage of the original charge will the capacitor hold at a time that is 3 times the time constant?

102. ••Derive a formula that describes the time dependence of the energy stored in a charging capacitor in a series RC circuit.

103. ••A 3.0 μF capacitor is put across a 12 V battery. After a long time ($\sim$20 time constants), the capacitor is disconnected and placed in series through an open switch, with a 200 Ω resistor. (a) Determine the charge on the capacitor before it is discharged. (b) What is the initial current through the resistor when the switch is closed? (c) At what time will the current reach 37% of its initial value?

104. ••You are working late on a project and find that you need a 75 Ω resistor, but unfortunately you only have a box of 50 Ω resistors on hand. All the electronics shops are closed, so you cannot buy the needed resistor. (a) How can you make a 75 Ω resistor using the resistors you have on hand? (b) How could you use your 50 Ω resistors to make a 60 Ω resistor? (c) Suppose you have only 50 μF capacitors. How could you use them to make a 75 μF capacitor and (d) to make a 60 μF capacitor?

105. ••A circuit probe is any device, such as a meter or an oscilloscope, that gives information about the circuit. Oscilloscopes and voltmeters are designed to measure the voltage across a circuit element, so they are connected across that element. Ideally, the probe should not disturb the circuit in any way by introducing resistance or capacitance, but in reality we cannot avoid some disturbance. A probe should disturb the circuit as little as possible. (a) When it is connected across a circuit element, is the probe in series or in parallel with that element? (b) Should a probe connected in this way have (i) a very large or a very small resistance, (ii) a very large or a very small capacitance? Explain your reasoning and use the properties of series and parallel connections to justify your answer.

106. ••For most purposes, we store electrical energy in batteries. But there are drawbacks to batteries: They release their energy rather slowly and are very damaging environmentally. Capacitors would be much cleaner for the environment and can be quickly recharged. Unfortunately they don't store much energy. (a) A new 1.5 V AAA battery has a "capacity" (*not* capacitance) of 1250 mA·h. What does this "capacity" actually represent? Express it in standard SI units. (b) How many joules of energy can be stored in the AAA battery? (c) At a steady current of 400 mA, how many hours will the AAA battery last? (d) How much energy can be stored in a typical 10-μF capacitor charged to a potential of 1.5 V? How does that compare to the energy stored in the AAA battery?

107. ••An *ultracapacitor* is a very high-capacitance device capable of storing much more energy than an ordinary capacitor. It is designed so that the spacing of the plates is around 1000 times smaller than in ordinary capacitors. Furthermore the plates contain millions of microscopic nanotubes, which increase their effective area 100,000 times. (a) If the plate separation and effective area of a 10 μF capacitor are changed as described above to make an ultracapacitor, what is its new capacitance? (b) How much energy does the ultracapacitor store if charged to 1.5 V? Compare this to the energy in an ordinary 10 μF capacitor at that potential. (c) Compare the energy stored in the ultracapacitor to that of the AAA battery in the previous problem. (d) If the ultracapacitor is to take 1.0 min to decrease to 1/e of its initial maximum charge, what resistance must it discharge through?

19 Magnetism

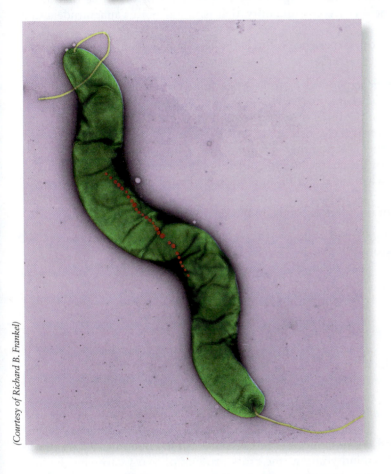

(Courtesy of Richard B. Frankel)

For organisms as tiny as the bacterium shown in the photo, gravity is relatively inconsequential. To find their way back down to the sediments at the bottom of the ponds and lakes in which they live, some bacteria depend on Earth's magnetic field to guide them. These bacterial cells have a row of magnetized iron particles (shown here as red dots) running down their length. The iron acts like a compass needle; in the Northern Hemisphere, for example, it orients the bacteria toward the magnetic north pole. Due to the orientation of Earth's magnetic field, that direction is somewhat downward, so the beating of their flagella propels the bacteria down toward their food supply in the sediments that accumulate at the bottom of lakes and ponds.

Nearly 3000 years ago in a region of Greece called Magnesia, people discovered curious stones that attracted similar stones as well as bits of iron to themselves. The Chinese are often credited with first using these stones, called *lodestones* from the Middle English word *lode* meaning "way" or "path," to fashion compasses to aid in navigation. Lodestone is a naturally occurring magnet.

Lodestones, and also the magnet you might use to hold a favorite photograph to the door of your refrigerator, are permanent magnets. They contain **ferromagnetic** materials, often iron, which become *magnetized* in the presence of another magnet and retain their magnetic properties after the external magnet is removed. Other magnets with which you might come into contact, such as the magnet clips that hold doors open in many college residence halls, are not constructed from permanent magnets but rather from *electromagnets*. Transient magnetic effects arise due to electric currents; in the case of magnetic doorstops, a coil in the system becomes magnetic when electric current passes through it. During a fire drill, the current is turned off in the coil, which deactivates the magnet and allows the door to close. In this chapter, we will explore these transient magnetic effects and the relationship between moving charge and magnetism.

19-1 Magnetic Force and Magnetic Field

In Chapter 16, we defined the electric field in terms of the electrostatic force that arises between charges. Remember that the electrostatic force arises from the interaction of a charged object with an electric field in which it is immersed. In a similar way, we will define a **magnetic field** $\vec{B}$ in terms of the force that arises from it. Magnetic field is a vector quantity, that is, it has both a magnitude and a direction.

Electric charges can experience a force due to a magnetic field. The **magnetic force** exhibits these characteristics:

1. An electric charge experiences a force when immersed in a magnetic field, but *only when the charge is moving.*
2. The magnitude of the force is proportional both to the charge and to the magnitude of the magnetic field.
3. The force is in the direction that is mutually perpendicular to the direction of motion of the charge and the direction of the magnetic field. There is no net force when the direction of motion is aligned with the magnetic field.
4. The direction of the force depends on the sign of the charge.

For a point charge, these characteristics are embodied in the **Lorentz force law,** named after the nineteenth-century Dutch physicist Hendrik Lorentz:

$$\vec{F} = q\vec{v} \times \vec{B} \tag{19-1}$$

Physicists often write the relationship to also include the effect of an electric field on the charge:

$$\vec{F} = q(\vec{E} + \vec{v} \times \vec{B}) \tag{19-2}$$

The presence of velocity $\vec{v}$ in the Lorentz force law accounts for the fact that only moving charges experience a magnetic force. Both charge q and magnetic field $\vec{B}$ appear linearly, that is, with an exponent of 1. This is the mathematical restatement of the fact that the magnitude of the force is proportional both to the charge and to the magnitude of the magnetic field. The characteristic of force being perpendicular to both $\vec{v}$ and $\vec{B}$ is described by the cross product in Equation 19-1. A glance at Figure 8-37 will remind you that the cross product of two vectors results in a third vector that is mutually perpendicular to the first two. We'll return shortly to the remaining characteristic of magnetic force, the relationship between the sign of the charge and the direction of the force.

Go to Interactive Exercise 19-1 *for more practice dealing with magnetic fields*

? Got the Concept 19-1
Electron in a Magnetic Field

An electron is held at rest in the presence of a strong magnetic field and then released. Does the electron experience a force? Would your answer be different if the electron was given a slight push when released?

The SI unit of magnetic field is the **tesla** (T). Magnetic field is also often specified in **gauss** (G); 1 G is 10^{-4} T. But be careful. Although it is common to find magnetic field values given in gauss, SI calculations must be done using teslas.

Rearranging Equation 19-1 clarifies the units of the magnetic field:

$$[B] = \frac{[F]}{[q][v]} = \frac{\text{N}}{\text{C(m/s)}}$$

By replacing C/s in the denominator with A, the units become

$$[B] = \frac{N}{(C/s)m} = \frac{N}{A \cdot m}$$

In other words, 1 T is equal to 1 N/(A·m). We're not surprised at the appearance of units of current; the observation that only moving charge experiences a magnetic force suggests a relationship between current and magnetic field. We will return to this relationship in the next section.

The strength of magnetic fields in our human experience varies over a wide range. Our bodies create a weak magnetic field, on the order of 10^{-10} T. The strength of Earth's magnetic field is about 0.5×10^{-4} T. The magnitude of the field due to the magnet on your refrigerator is 10 or so times stronger, and you can easily buy small neodymium magnets (made from an alloy of neodymium, iron, and boron) that have field strengths of 1 T or more. The strongest magnetic field most of us will likely experience directly is the 2 T (or so) field inside a magnetic resonance imaging (MRI) device. The strong field used in conjunction with radio waves causes hydrogen atoms in your body to align magnetically and then wobble; images of internal regions of the body are created by detecting the wobbles. The strongest magnetic fields in human experience are those fields used in scientific experiments, for which magnets have been built that produce fields of 60 T or more.

Let's return to magnetic force as described by Equation 19-1. First, recall from Chapter 8 that the right-hand rule can be used to determine the result of a cross product of two vectors. We can therefore apply it to determine the direction of the force vector; when you curl the fingers on your right hand from the direction of velocity $\vec{v}$ toward the direction of the magnetic field $\vec{B}$, your thumb points in the direction of the cross product of the two vectors. However, this direction is valid only when the moving charge is positive. The direction of the force on a negatively charged particle is opposite to that predicted by the right-hand rule. In **Figure 19-1a**, a small charged particle moves from left to right in a magnetic field that points into the page

A charge moving through a magnetic field can experience a force.

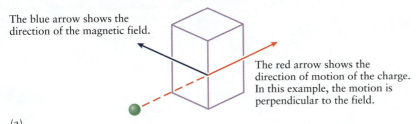

The blue arrow shows the direction of the magnetic field.

The red arrow shows the direction of motion of the charge. In this example, the motion is perpendicular to the field.

(a)

The direction of the magnetic force is mutually perpendicular to the direction of motion and the direction of the field.

A positive charge feels a force in the direction defined by the right-hand rule.

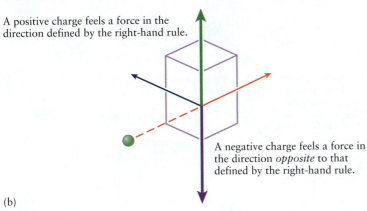

A negative charge feels a force in the direction *opposite* to that defined by the right-hand rule.

(b)

Figure 19-1 (a) Charges moving through a magnetic field experience a force oriented perpendicular to both the direction of motion and the direction of the magnetic field. (b) The direction of the force on a positive charge is determined by the right-hand rule. The force on a negative charge is in the opposite direction.

and to the left. According to the right-hand rule, the cross product of a vector pointing left to right with one pointing into the page is a vector pointing up, shown by the green arrow in **Figure 19-1b**. However, the right-hand rule only gives the direction of the magnetic force on a positive particle. A negatively charged particle experiences a force in the opposite direction, shown by the purple arrow in the figure.

(You may find a slightly different approach to the right-hand rule handy for determining the direction of magnetic force from Equation 19-1. For example, point your right index finger in the direction of the velocity and your other fingers in the direction of magnetic field, then stick your thumb straight out. That's the direction of the resulting force.)

Notice that in Figure 19-1 we need to identify three mutually perpendicular directions, corresponding to the directions of $\vec{v}$, $\vec{B}$, and $\vec{F}$. To represent the *three*, mutually perpendicular directions on a *two*-dimensional surface like this page we drew the vectors along the edges of a sketch of a box. The blue arrow shows the direction of the magnetic field $\vec{B}$. The red arrow shows the direction of the motion of the charge. (In this example, we chose to let $\vec{v}$ be perpendicular to $\vec{B}$, but that is not a requirement.) The magnetic force $\vec{F}$ is mutually perpendicular to the direction of motion $\vec{v}$ and the direction of the field $\vec{B}$. In the figure the green arrow shows the force if the charge is positive and the purple arrow represents the force if the charge is negative. It is usually more convenient, however, to avoid making three-dimensional drawings. To indicate a direction that does not lie in the plane of the page, we can make use of the convention, shown in **Figure 19-2**, that the symbol $\otimes$

When viewed from this perspective, the vector arrow is pointing in. Think of the symbol $\otimes$ as representing the feathers (fletching) of an arrow pointing directly away from you.

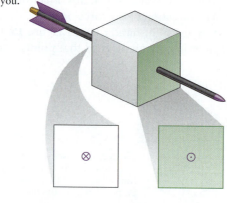

When viewed from this perspective, the vector arrow is pointing out. Think of the symbol $\odot$ as representing the tip of an arrow pointing directly toward you.

Figure 19-2 To represent a vector pointing into or out of the plane of the page, we use the image of an arrow. When it points directly out of the page, you see only the arrow's tip, which we represent by a circle with a dot in the center. When it points directly into the page, you only see the arrow's fletching, which we represent by an "×" enclosed in a circle.

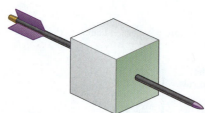

In this region the vector is pointing into the page. You can think of this symbol as representing the feathers (fletching) of an arrow pointing directly away from you.

In this region the vector is pointing out of the page. You can think of this symbol as representing the tip of an arrow pointing directly toward you.

represents a vector pointing into the page, and that the symbol ⊙ represents a vector pointing out of the page. You can think of these two symbols as the tail end and tip, respectively, of the kind of arrow that might be shot from a bow. Using this convention, we can redraw Figure 19-1b as shown in **Figure 19-3**. We have drawn a positive charge and a negative charge separately in the figure.

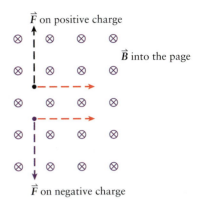

? Got the Concept 19-2
A Proton in a Magnetic Field

A proton is fired horizontally from left to right into a region of constant magnetic field. As shown in **Figure 19-4**, the proton moves from left to right through a field pointing into the page. In which direction does the path of the proton bend? Which of the electrons, labeled A through D, would be bent by the magnetic force in a direction toward the point where the proton enters the field?

Figure 19-3 A positive (green) and negative (purple) charge move through a magnetic field that points into the page. The red arrows show the direction of motion, and the green and purple arrows show the direction of the magnetic force on each. The directions of the magnetic field, the motion of the charges, and the magnetic force on the charges are orthogonal to each other.

Example 19-1 Potassium Ion

When potassium ions (K^+) cross a membrane through an ion channel, they acquire a velocity on the order of 10^{-2} m/s. Approximately what are the minimum and maximum values of the magnetic force on a K^+ due to Earth's magnetic field?

SET UP
We want to apply Equation 19-1. First, let's take the magnitude of the force:

$$F = qvB \sin \varphi$$

where φ is the angle between $\vec{v}$ and $\vec{B}$. Here q is the charge of the K^+, which is one elementary charge unit,

$$q = 1.6 \times 10^{-19} \text{ C}$$

SOLVE
The minimum force on the K^+ must be zero, when the angle between Earth's field and the direction of motion of the K^+ is 0°. The maximum force occurs when φ is 90°, in which case

$$F = (1.6 \times 10^{-19} \text{ C}) (10^{-2} \text{ m/s}) (0.5 \times 10^{-4} \text{ T}) = 8 \times 10^{-26} \text{ N}$$

REFLECT
The force on an ion crossing a cell membrane due to Earth's field is rather small. Nevertheless, the ion's acceleration is not. The mass of a potassium ion is about 6.5×10^{-26} kg, which results in an acceleration:

$$a = F/m_K = 8 \times 10^{-26} \text{ N}/6.5 \times 10^{-26} \text{ kg} = 1.2 \text{ m/s}^2$$

Small magnetic fields can deflect moving ions unless, as in this example, they are constrained to follow a specific path.

Practice Problem 19-1 A number of cellular processes involve the movement of Ca^{2+} ions into and out of a cell's cytoplasm. The magnitude of the positively charged Ca^{2+} ion is $2e$, that is, twice the elementary charge unit. Find the maximum force, due to Earth's magnetic field, on a Ca^{2+} ion moving at 0.01 m/s.

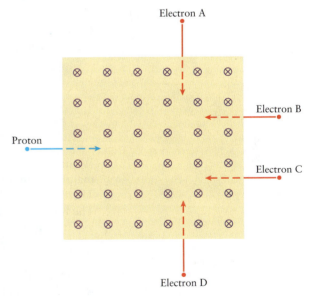

Figure 19-4 A proton enters a magnetic field that points into the page. In which direction will its path be bent by the magnetic force? Which of the electrons that enter the field will be bent toward the point where the proton entered?

★   **What's Important 19-1**
Electric charges moving through a magnetic field experience a magnetic force. The magnitude of the force is proportional to the amount of charge and to the magnitude of the field. The direction of the force depends on the sign of the charge and is mutually perpendicular to the direction of motion of the charge and the direction of the magnetic field. There is no net force when the direction of motion is aligned with the magnetic field. The SI unit of magnetic field is the tesla (T).

19-2 Magnetic Force on a Current

A freely moving charge in the presence of a magnetic field experiences a force according to $\vec{F} = q\vec{v} \times \vec{B}$ (Equation 19-1), causing it to accelerate by changing direction. We would say that the path of a moving charge bends due to the force, and its trajectory will continue to bend until either it leaves the field or it undergoes a collision. What if the particle is confined to remain in a region, for example, an electron constrained to stay within a current-carrying wire? Because individual moving charges experience a force due to a magnetic field, a wire carrying a current will experience a force as well.

Let's convert the Lorentz force law from one that applies to a single, moving charged particle to one that describes the magnetic force on a current. Consider the dimensions of the quantities in $\vec{F} = q\vec{v} \times \vec{B}$:

$$\text{magnetic force} = \text{charge} \left(\frac{\text{distance}}{\text{time}} \right) \text{magnetic field}$$

Current has dimensions of charge divided by time, so we can express the magnetic force in terms of current by moving the units of time together with those of charge:

$$\text{magnetic force} = \frac{\text{charge}}{\text{time}} (\text{distance}) \text{magnetic field}$$

or

$$\vec{F} = i\vec{l} \times \vec{B} \tag{19-3}$$

We have used $\vec{l}$ to represent displacement along the wire carrying the current i. The magnitude of $\vec{l}$ in any specific case is the length of the wire on which the force acts. The direction of the wire (and also the current) gives the direction of the $\vec{l}$ vector. The force depends on the angle between $\vec{l}$ and $\vec{B}$, which we make more apparent by expanding the cross product. The magnitude of the force becomes

$$F = ilB \sin \varphi$$

where φ is the angle between $\vec{l}$ and $\vec{B}$. Notice that this equation only makes sense when i, l, B, and φ are constant—we wouldn't know what specific value to use for any quantity that wasn't constant. Equation 19-3, then, applies most directly to a straight wire carrying a constant current in a constant magnetic field. Wires that are short enough can be treated as straight, which makes Equation 19-3 useful. In addition, we will shortly modify Equation 19-3 to apply it to bent wires.

The direction of the cross product in the Lorentz force is found according to the right-hand rule, but it only applies to positive charges. The laws of physics don't make any such requirement, however; this is simply a convention that physicists have adopted. The convention comes to us from Ben Franklin and his definition of what we now refer to as the *conventional current*. As we saw in Chapter 18, conventional current treats the charge carriers in a current as positive. So here as well, the direction given by the right-hand rule applies only when $\vec{l}$ points in the direction of the conventional current. Said another way, if you take $\vec{l}$ to point in the direction in which the electrons move, applying the right-hand rule to Equation 19-3 gives a result in the opposite direction of that which is correct.

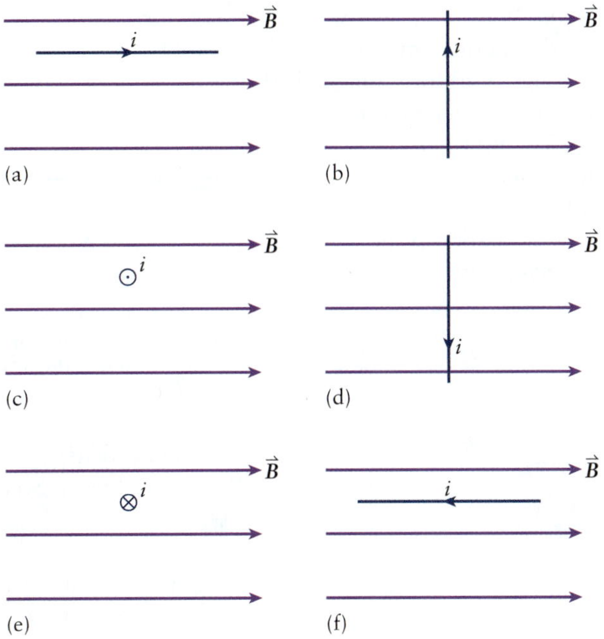

Figure 19-5 A long, straight wire carries a current through a uniform magnetic field. The current direction is (a) from left to right, (b) up, (c) out of the page, (d) down, (e) into the page, and (f) from right to left.

? Got the Concept 19-3
Direction of Force

A long, straight wire carrying a current can be placed in various orientations with respect to a constant magnetic field as shown in **Figure 19-5a** through **Figure 19-5f**. The direction of the conventional current is indicated by the variable i together with the accompanying arrow. Find the direction of the force in each.

Estimate It! 19-1 Desk Lamp Wires

Estimate the maximum force the wires in the cord that connects a desk lamp to the wall socket experience due to Earth's magnetic field.

SET UP
When the wires are oriented in a direction perpendicular to the direction of Earth's field the magnitude of the magnetic force on the wires in the cord is, from Equation 19-3,

$$F = i_{wire} l_{wire} B_{Earth}$$

where l_{wire} is the length of a wire and i_{wire} is the current it carries. We know that Earth's magnetic field is about 0.5×10^{-4} T. We need to determine big, round—but reasonable—values for the current and length in order to complete this estimate.

SOLVE
Let's say the bulb in the lamp is 60 W. From Equation 18-26, the current through it depends on the power P and the voltage V:

$$i_{wire} = \frac{P}{V}$$

We'll discuss wall voltage in Chapter 21, but for now we'll take the voltage to be 120 V, the rms voltage in the United States. Then the current is

$$i_{wire} = \frac{60 \text{ W}}{120 \text{ V}} = 0.5 \text{ A}$$

Let's say the length of the cord is 2 m, which is a reasonable, round number. The magnetic force is then

$$F = (0.5 \text{ A})(2 \text{ m})(0.5 \times 10^{-4} \text{ T}) = 0.5 \times 10^{-4} \text{ N}$$

REFLECT
We estimate the maximum force on the wires in the desk lamp cord to be about 0.5×10^{-4} N. We can, for example, compare it to the weight of the cord. The linear

Figure 19-6 (a) A bent wire carries a constant current through a uniform magnetic field that points from left to right. (b) To find the force, we find the force acting on each infinitesimally small piece of wire and add up the contributions of the pieces to the total force.

A wire carries a constant current through a constant magnetic field.

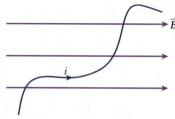

Because the wire isn't straight we cannot directly apply $\vec{F} = i\vec{l} \times \vec{B}$ to find the force on it.

(a)

To find the force, break the wire into an infinite number of infinitesimally small pieces...

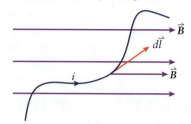

...and add up the force $d\vec{F} = i\,d\vec{l} \times \vec{B}$ each piece contributes to the total. $d\vec{l}$ is tangent to the wire at the location of each piece of the wire.

(b)

mass density μ of "zip cord" commonly used for home electrical devices is about 0.03 kg/m. So a piece of zip cord of length L equal to 2 m has weight

$$W = mg = \mu L g = (0.03 \text{ kg/m})(2 \text{ m})(9.8 \text{ m/s}^2) = 0.6 \text{ N}$$

The weight of the cord is four orders of magnitude larger than the magnetic force it experiences. The magnetic force is relatively small.

We found that

$$\vec{F} = i\vec{l} \times \vec{B} \tag{19-3}$$

applies to a straight wire carrying a constant current in a constant magnetic field. What if one or more of the quantities is not constant? Consider, for example, the wire in **Figure 19-6a**. The wire carries a constant current i in a constant field $\vec{B}$. However, because the wire isn't straight, the direction of the wire and therefore the angle between the direction vector and $\vec{B}$ changes as we follow the current from one end of the wire to the other. To find the force, we model the wire as an infinite number of infinitesimally small pieces of wire, each so small that it can be treated as straight, and then add up the force on each. The force on any one of the pieces is

$$d\vec{F} = i\,d\vec{l} \times \vec{B}$$

so the total force is

To carry out the vector sum of the infinite number of $d\vec{l}$ vectors that make up the wire (blue), we add (separately) the components of $d\vec{l}$ parallel and perpendicular to the line that connects the ends of the wire.

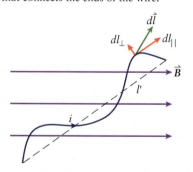

There is no net displacement of the wire from the line connecting the ends, so the sum of all $dl_\perp$ must be zero. Then $\int_{wire} dl = \int_{wire} d\vec{l}_{||} = l'$, where l' is the straight line distance between the ends of the wire.

Figure 19-7 To find the net force on a bent wire carrying a current through a uniform magnetic field, the effective length of the wire is the direct distance from one end of the wire to the other, shown as a dashed line.

$$\vec{F} = \int_{wire} d\vec{F} = \int_{wire} i\,d\vec{l} \times \vec{B} \tag{19-4}$$

where the bounds take the integral from one end of the wire to the other. As shown in **Figure 19-6b**, the $d\vec{l}$ vector is tangent to the wire at each point along it. Because the magnitude and direction of $\vec{B}$ are constant we just need to understand how to add the $d\vec{l}$ vectors along the length of the wire to carry out the integral.

We evaluate a vector sum, such as the integral in Equation 19-4, by adding the vectors' components in mutually perpendicular directions. Although we can choose any coordinate axis, it is convenient to use the directions parallel and perpendicular to the dashed line that connects the ends of the wire, as in **Figure 19-7**. Let's first add the components of $d\vec{l}$ in the direction perpendicular to the dashed line. These components are the displacements either away from or toward the line as we follow the path of the wire from one end to the other. Notice that because we start and end exactly on the dashed line, the net displacement in the direction perpendicular to the dashed line must be zero, in other words, the sum of the perpendicular components of the $d\vec{l}$ contributions equals zero. Therefore the only terms that contribute to the integral are the parallel components. These components are, by definition, aligned with the dashed line, so integrating from one end of the wire to the other simply gives the length of the dashed line. This distance, labeled l', is the length along a straight path from one end of the wire to the other. We refer to this as the end-to-end length.

We can then replace the integral over the infinitesimal elements of the curving path in Equation 19-4 by the end-to-end length:

$$\vec{F} = i\vec{l'} \times \vec{B} \tag{19-5}$$

Example 19-2 Force on a Semicircular Arc of Current

A semicircular arc of wire of radius R is immersed in a magnetic field of magnitude B that is directed into the page as shown in **Figure 19-8a**. The wire is connected to a source of current i_0. Find the force on the semicircular section of the wire (a) by integration using Equation 19-4 and (b) by using the end-to-end length approach of Equation 19-5.

SET UP

We start by considering the direction of the force by applying the right-hand rule. Select any infinitesimal piece of the wire, such as the one shown in **Figure 19-8b**. Point your index finger in the direction of the current there, and point your other fingers in the direction of the field (that is, into the page). Your thumb points in the direction of the force, which as indicated in Figure 19-8b, is radially outward. Of course, radially outward is different for each different infinitesimal contribution around the semicircle, so we'll need vector addition to find the net force.

As suggested by **Figure 19-8c**, for every piece of the wire we select, there is another one such that the y components of the two cancel. (The example shown in the figure is the pair of dF_y and dF'_y.) We conclude that the net force on the wire is to the left. Notice that the same conclusion is drawn by considering the approach of part (b), in which the wire is treated as a straight segment running from one end of the semicircular to the other. To apply the right-hand rule to this equivalent wire, point your index finger in the positive y direction and your other fingers into the page. The net force is in the direction of your thumb, which should point to the left.

To find the magnitude of the net force in part (a), then, we will add together the x components of the force on each infinitesimal section of the semicircle. From **Figure 19-8d** and Equation 19-3 the x components of the force contributions are given by

$$dF_x = i_0\, dl\, B \cos\theta$$

Notice that the cosine term from the cross-product in the preceding equation equals 1, because $\vec{B}$ and $d\vec{l}$ are perpendicular everywhere along the wire.

For part (b), we can apply Equation 19-5 directly.

SOLVE

(a) As suggested in Figure 19-8d, we take each infinitesimal section of the wire to have angular extent $d\theta$. Then the integral over the wire of the x components of the force contributions becomes (from Equation 19-4)

$$F = \int_{\text{wire}} dF_x = \int_{\text{wire}} i_0\, dl\, B \cos\theta = \int_{-\pi/2}^{\pi/2} i_0\, R\, d\theta\, B \cos\theta$$

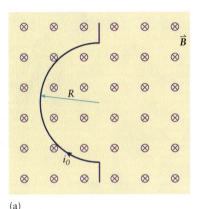

(a)

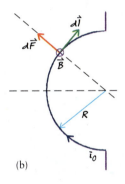

(b)

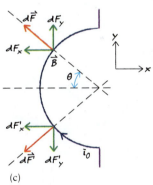

(c)

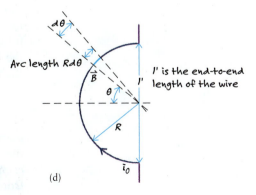

(d)

Figure 19-8 (a) A semicircular arc of wire of radius R carries a constant current i_0 through a uniform magnetic field. (b) To apply the right-hand rule to an infinitesimal piece of the wire, point your right forefinger in the direction of the current and point your other fingers in the direction of the magnetic field (into the page); your thumb points in the direction of the force, shown as a red arrow. (c) For each small element of the wire there is some other element for which the y components of the two forces cancel. For this reason we only need to add the x components of each contribution to the total force. (d) Each infinitesimal piece of wire has an angular extent $d\theta$ and arc length $R\, d\theta$.

The arc length of an infinitesimal section of the wire is $R\,d\theta$, so in the last step we set dl equal to $R\,d\theta$. With the integral now with respect to θ, we also specify the bounds on the integral. Because θ is defined from the middle of the arc, the ends of the semicircle are at θ equal to $-\pi/2$ and $+\pi/2$. To gather up all of the infinitesimal contributions from one end of the wire to the other, then, the bounds on the integral are $-\pi/2$ to $+\pi/2$.

Thus

$$F = i_0 RB \int_{-\pi/2}^{\pi/2} \cos\theta\,d\theta = i_0 RB (\sin\theta)\Big|_{-\pi/2}^{\pi/2}$$

or

$$F = i_0 RB\big[\sin(\pi/2) - \sin(-\pi/2)\big] = i_0 RB\big[1 - (-1)\big] = 2iRB$$

(b) The vectors $\vec{l}'$ and $\vec{B}$ are perpendicular, so the cross product in Equation 19-5 gives 1. So the magnitude of the force is

$$F = i_0 l' B$$

The end-to-end length of the wire is the diameter of the semicircle; that is, the magnitude of $\vec{l}'$ equals $2R$. (The end-to-end length is not πR, the length of the wire. Remember that l' is the direct distance from one end of the wire to the other.) So

$$F = i_0(2R)B = 2i_0 RB$$

REFLECT

Both approaches give the same result, evidence that the argument we made for replacing the integral over path length by the end-to-end distance is correct.

Practice Problem 19-2 Let the semicircular arc of wire in Figure 19-8 be replaced by three straight wire segments each of length R that form three sides of a rectangle. The wire is immersed in a magnetic field of magnitude B that is directed into the page, and the wire is connected to a source of current i_0. Find the magnetic force on the rectangular section of the wire.

According to Equation 19-5, the force on a wire carrying a current in a magnetic field is proportional to the end-to-end length of the wire. What, then, is the force on the current in a wire that has been bent so that the two ends touch? **Figure 19-9a** shows a straight wire bent into a rectangular loop, carrying a current i_0, and immersed in a magnetic field. (We'll soon uncover a way to cause a current through such a loop, even though there are no ends to which to connect a voltage source.) Because we have brought the two ends of the initially straight wire together, the end-to-end distance (the magnitude of $\vec{l}'$ in Equation 19-5) is zero. The net magnetic force on the loop is zero.

Why the interest in loops? In many situations that involve current, charges flow from and eventually return to a source, say, a battery. The loop that the current makes may not look like the one in Figure 19-9, but we can treat it in the same way.

You can verify that the net magnetic force on the loop is zero by applying the right-hand rule to each side. For example, for the left side of the loop, point your index finger up following the current, and point your other fingers into the page to follow the magnetic field. Your thumb gives the direction of the force, which is to the left in this case. **Figure 19-9b** shows the force on all four sides; the forces on each pair of opposite sides of the loop cancel when vector addition is used. We again conclude that the net magnetic force on the loop is zero.

Although the net force on the current-carrying loop in Figure 19-9 is zero, the net torque may not be. When the loop is free to rotate, say around an axis in the plane of the loop, a nonzero torque that arises due to the interaction of the field and the current will give rise to an angular acceleration. The loop, and the forces on each side, are shown in a three-dimensional view in **Figure 19-10a**, and then directly from the side in **Figure 19-10b**. The forces on the near and far sides of the loop,

A straight wire is bent so that the ends touch, forming a rectangular loop.

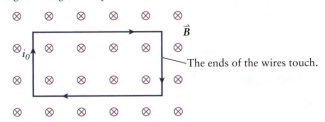

The ends of the wires touch.

(a)   Because the end-to-end distance is zero, the net force on the wire is zero.

Another way to find understand the net force is to apply right-hand rule to each side of the loop.

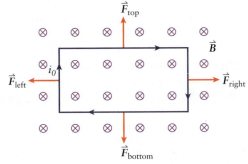

(b)   The magnetic forces on each pair of opposite sides of the loop cancel.

Figure 19-9 (a) A wire is bent into a rectangular loop so that its ends touch. The loop carries a constant current through a uniform magnetic field. Because the direct (end-to-end) distance between the ends of the wire is zero, the net magnetic force on the wire is zero. (b) The net magnetic force on the loop can be determined by applying $d\vec{F} = i\,d\vec{l} \times \vec{B}$ to each side. According to the right-hand rule, the directions of the forces are such that $\vec{F}_{top}$ and $\vec{F}_{bottom}$ cancel, and $\vec{F}_{right}$ and $\vec{F}_{left}$ cancel, resulting in a net force equal to zero.

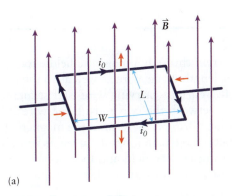

(a)

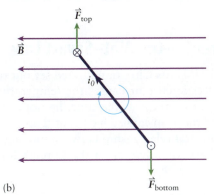

The magnetic forces on the top and bottom sides of the loop both result in a torque in the same rotational direction. The loop, as shown here, will experience an angular acceleration. Even if initially stationary, it will rotate if it is free to do so.

(b)

Figure 19-10 When a current-carrying loop of wire is free to rotate in a magnetic field, a nonzero torque arises due to the interaction of the field and the current. (a) The loop is seen in a three-dimensional view. (b) Here the loop is seen edge-on, so that the left side is seen as a straight wire segment, the top and bottom sides are receding directly into the page, and the right side of the loop is hidden behind the left side. The current goes into the page along the top side and comes out of the page along the bottom side.

which point out of and into the page respectively, cancel, but the forces on the top and bottom sides both give rise to a torque. The two torques add because they are in the same rotational direction, clockwise in this case. The net torque is therefore not zero in general.

<hr>

? Got the Concept 19-4
Torque

Is there an orientation of the loop in Figure 19-10b such that the net torque on the loop is zero? How would the torque on the loop change if the loop were tilting into the magnetic field rather than away from it? Will a current-carrying loop immersed in a magnetic field rotate continually?

<hr>

The magnitude of the torque on the loop is determined from

$$\tau = rF \sin \varphi \tag{8-23}$$

Taking L and W as the lengths of the sides of the loop (see Figure 19-10a), the distance from the application of the force to the axis of rotation (r) is $L/2$. The force comes from Equation 19-5, and φ varies as the loop rotates. Then

$$\tau = (L/2)(i_0 WB) \sin \varphi$$

is the torque due the force on each of the top and bottom sides of the loop, and

$$\tau = 2(L/2)(i_0 WB) \sin \varphi = i_0 LWB \sin \varphi$$

is the net torque. Notice the LW is the area of the loop, so we can write

$$\tau = i_0 AB \sin \varphi \tag{19-6}$$

where A is the area of the loop. This form of the torque applies to a current loop of any shape, not just a rectangular one.

<hr>

✴ What's Important 19-2
A straight wire carrying a current through a magnetic field experiences a magnetic force proportional to the length of the wire as well as to the magnitudes of the current and the field. If a current-carrying wire isn't straight, the force depends on the direct distance between the two ends. If the wire is bent into a closed loop the direct, end-to-end distance is zero, so the magnetic force on it is zero. Although the net magnetic force on a loop of current is zero, if the loop is free to rotate, it is possible to orient the field and loop so that the net torque on it is nonzero. This gives rise to an angular acceleration.

<hr>

19-3 Magnetic Field and Current—the Biot–Savart Law

In the early 19th century Danish physicist Hans Christian Ørsted set out to demonstrate to a group of students that electric current causes the temperature of a wire to increase. Although he could not explain it at the time, he noticed that a nearby compass needle moved whenever the loop carried a current. This accidental discovery provided a clue to the fundamental relationship between moving charge and magnetic field, a relationship we have already seen in the Lorentz force law $\vec{F} = q\vec{v} \times \vec{B}$ (Equation 19-1).

The Biot-Savart law is used to determine the magnetic field at any point...

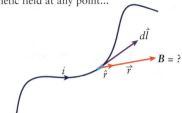

...by adding vectorally the field contribution due to each **small element** of current $id\vec{l}$. The unit vector $\hat{r}$ points from the charge element to the point at which we want to determine the field.

Figure 19-11 We can use the Biot–Savart law to determine the magnetic field at any point. The small light blue rectangle represents an infinitesimally short segment of a current-carrying wire, shown as a curved blue line. The $\vec{r}$ vector represents the distance and direction to the point at which we will determine the field.

Simply put, moving charge gives rise to a magnetic field. Not long after Ørsted's discovery, two French physicists, Jean-Baptiste Biot and Félix Savart, developed a relationship between moving charge and the magnetic field it produces. Referred to as the **Biot–Savart law**, it describes the contribution $d\vec{B}$ to the magnetic field due to an infinitesimally small current element $id\vec{l}$:

$$d\vec{B} = \frac{\mu_0}{4\pi} \frac{id\vec{l} \times \hat{r}}{r^2} \tag{19-7}$$

In **Figure 19-11**, the small light blue piece of the current i represents one current element. The vector $d\vec{l}$ points in the direction of the current at the location of the current element, and both $\vec{r}$ and the unit vector $\hat{r}$ point from the charge element to the point at which we want to determine the field. The value of μ_0, the **permeability of free space**, is $4\pi \times 10^{-7}\ \text{T} \cdot \text{m/A}$. Permeability is a measure of the degree to which a magnetic field forms in a material or a region in response to an applied magnetic field. Permeability is the magnetic analog of permittivity for electric fields.

The total magnetic field is found by adding the contribution from every part of the current. Note that this is a vector addition:

$$\vec{B} = \int_{\text{current}} d\vec{B} = \int_{\text{current}} \frac{\mu_0}{4\pi} \frac{id\vec{l} \times \hat{r}}{r^2} \tag{19-8}$$

In Chapter 16, we saw that stationary charge gives rise to an electric field. The mathematical relationship between a stationary point charge dq and the associated electric field $d\vec{E}$ is represented by what is effectively Coulomb's law (Equation 16-4):

$$d\vec{E} = \frac{1}{4\pi\epsilon_0} \frac{dq}{r^2} \hat{r}$$

Notice the similarities between the Biot–Savart law and Coulomb's law. Both depend on a physical constant associated with the field, ϵ_0 for electric fields and μ_0 for magnetic fields. Both depend, of course, on the charge element that gives rise to the field, an infinitesimally small stationary charge dq for electric field and an infinitesimally small current element $id\vec{l}$ for magnetic field. In addition, both are inversely proportional to $4\pi r^2$. The fundamental difference between the two is the direction of the field. The electric field points either directly away from or directly toward the charge, as indicated by the direct relationship in Coulomb's law between $d\vec{E}$ and $\hat{r}$, the unit vector that points radially from the charge. From the Biot–Savart law, the magnetic field points in the direction of the result of the cross product of $id\vec{l}$ and $\hat{r}$, which is the direction mutually perpendicular to the two vectors. The specific direction is determined by applying the right-hand rule, when $id\vec{l}$ is in the direction of the conventional current.

? Got the Concept 19-5
A Long Straight Wire I

A long, straight wire carries a current from right to left as shown in **Figure 19-12**. In which direction is the magnetic field at the three marked points? Note that P_2 is on the wire.

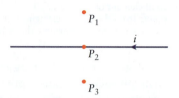

Figure 19-12 A wire carries a current from right to left. Point P_1 is located above the wire, point P_2 is located on the wire, and point P_3 is located below the wire.

? Got the Concept 19-6
A Long Straight Wire II

A long, straight wire carries a current from right to left as shown in Figure 19-12. Describe in words the direction of the magnetic field near the wire.

! Watch Out
It is generally not correct to describe a magnetic field direction as into or out of the page, or even clockwise or counterclockwise.

Consider the previous two Got the Concept questions. It is certainly not correct to describe the magnetic field as "out of the page" or "into the page" because each of those statements is correct at only one point in space. It would be correct, however, to describe the magnetic field as, for example, "out of the page at point P_1 (Figure 19-12)." It is also not correct to describe the field direction as, for example, "counterclockwise." The field is only counterclockwise when viewed from a particular vantage point. In order to use a description like "counterclockwise" or "clockwise" you need to specify your viewing orientation, for example, "counterclockwise when the wire is viewed so that the current is coming toward you."

! Watch Out
At any point in space the magnetic field is represented as a straight arrow, not a curved one.

Magnetic field is a vector because at any point in space it has a specific direction, and therefore must be represented as a straight arrow.

There are electric currents in your body, for example, in your heart and in your brain, and those moving charges give rise to magnetic fields. The currents, and therefore the magnetic fields, are relatively weak. The electrical impulses that drive the contraction and relaxation of the cardiac muscle produce magnetic fields of magnitude about 5×10^{-11} T. The magnetic fields in the brain are even smaller, on the order of 10^{-13} T. Magnetocardiograms (MCGs), magnetic signals from the heart, and magnetoencephalograms (MEGs), magnetic signals from the brain, can be recorded using sensitive medical devices such as the MEG system shown in **Figure 19-13**. MEGs offer a higher spatial and temporal resolution than electroencephalograms (EEGs) because magnetic fields suffer less distortion than electric fields from the resistive properties of the head. **Figure 19-14** show the brain's response to certain visual stimuli recorded in an MEG.

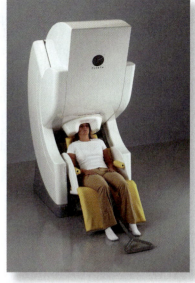

Figure 19-13 This device records a magnetoencephalogram, a map of the magnetic fields induced by electric charges moving through the brain.
(Courtesy of Elekta)

Magnetic Field on the Axis of a Circular Loop of Current

Let's use the Biot–Savart law to find the magnitude and direction of the magnetic field at a point a distance y along the central axis of a circular loop through which a current is carried. As in **Figure 19-15a**, the loop of radius R carries current i_0.

To see the angular dependence of the Biot–Savart law more clearly, we expand the cross product:

$$d\vec{l} \times \hat{r} = |d\vec{l}||\hat{r}| \sin \varphi$$

where φ is the angle between $d\vec{l}$ and $\hat{r}$. Notice that because $\hat{r}$ is a unit vector, it has magnitude equal to 1. Notice also that $d\vec{l}$ and $\hat{r}$ are perpendicular to each other, so φ is 90° and $\sin \varphi$ is 1. The cross product then simply equals the magnitude of $d\vec{l}$. The magnitude of each contribution to the field becomes (from Equation 19-7)

$$dB = \left| \frac{\mu_0}{4\pi} \frac{id\vec{l} \times \hat{r}}{r^2} \right| = \frac{\mu_0}{4\pi} \frac{i_0 dl}{r^2} \qquad (19\text{-}9)$$

Next we add the contribution each current element makes to the total field, as in Equation 19-8. But we need to proceed carefully: Equation 19-8 represents a vector addition, so we need to consider the directions of each $d\vec{B}$. Here, the $d\vec{B}$ vector due to each current element points in a different direction, so to properly sum them, we add the contributions by component. **Figure 19-15b** shows the loop of current from the side and includes the magnetic field contribution from the current element at the point where the current is moving directly toward our vantage point (on the left) and directly away from our vantage point (on the right). These two infinitesimal contributions to the field are labeled $d\vec{B}_\odot$ and $d\vec{B}_\otimes$, respectively. Because of the inherent symmetry, $d\vec{B}_\odot$ and $d\vec{B}_\otimes$ have equal but opposite components in the horizontal direction, so their horizontal components cancel. Thus, when the horizontal components of the infinite number of field contributions are added, the result is zero. The total magnetic field therefore points straight up. To find it, we combine Equation 19-9 with Equation 19-8, and add only the vertical components:

$$|\vec{B}| = \left| \int_{\text{ring}} d\vec{B} \right| = \int_{\text{ring}} \frac{\mu_0}{4\pi} \frac{i_0 \, dl \cos \alpha}{r^2}$$

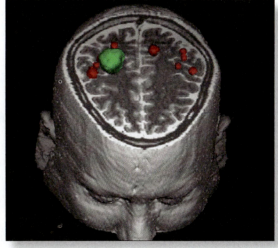

Figure 19-14 The brain responds to visual stimuli with electric currents and associated magnetic fields in particular parts of the brain. In this magnetoencephalogram, colors are used to represent the strength of the magnetic field generated in different regions of the brain.
(Courtesy of Elekta)

The $\cos \alpha$ term, as you can verify from the figure, gives the vertical component of each $d\vec{B}$.

As we go around the loop to complete the integration, both r and α are constant, so the integral simplifies to

$$|\vec{B}| = \frac{\mu_0}{4\pi} \frac{i_0 \cos \alpha}{r^2} \int_{\text{ring}} dl$$

Figure 19-15 (a) A wire loop of radius R carries current i_0. This sketch helps us apply the Biot–Savart law to find the magnitude of the magnetic field at a point a distance y along the central axis of a circular loop. (b) The contributions to the magnetic field from current elements on opposite sides of the loop suggest that the horizontal components of the fields cancel when summed over the entire loop.

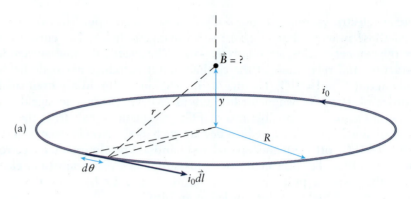

(a)

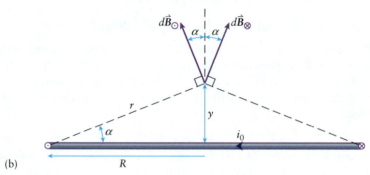

(b)

The integral of dl around the loop is its length, that is, its circumference. So

$$|\vec{B}| = \frac{\mu_0}{4\pi} \frac{i_0 \cos \alpha}{r^2} 2\pi R$$

Note that

$$\cos \alpha = \frac{R}{r}$$

so

$$|\vec{B}| = \frac{\mu_0}{4\pi} \frac{i_0}{r^2} \frac{R}{r} 2\pi R = \frac{\mu_0}{2} \frac{i_0 R^2}{r^3}$$

Finally, because r is the hypotenuse of a triangle with legs y and R,

$$r = (R^2 + y^2)^{1/2}$$

So the magnitude of the magnetic field due to the circular loop of current at a point y along the central axis of the ring is

$$B = \frac{\mu_0}{2} \frac{i_0 R^2}{(R^2 + y^2)^{3/2}}$$

By designating the y direction to be straight up,

$$\vec{B} = \frac{\mu_0}{2} \frac{i_0 R^2}{(R^2 + y^2)^{3/2}} \hat{y} \qquad\qquad (19\text{-}10)$$

 Watch Out

The magnetic field vectors that are used to determine the field which arises from current in a loop do not point along the line that connects the current element with the point of interest.

You may be tempted, based on our study of electric charge, to draw magnetic field vectors along the line connecting a current element with the point at which you want to find the magnetic field. Drawing the vectors this way is not correct; **Figure 19-16** shows the right and wrong way to draw the fields for the loop we just considered. Notice that drawing the point of interest close to the loop helps make the direction clear.

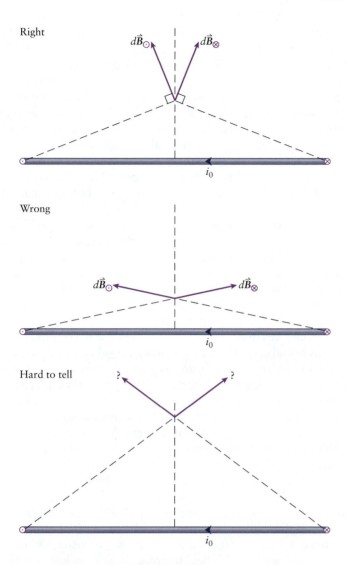

Figure 19-16 The direction of the magnetic field is given by the right-hand rule, so the field at any point is perpendicular to the line which connects that point with the current element that produced the field. It's easy to tell that the top sketch shows the field lines for the current elements on opposite sides of a current ring (shown edge-on, from the side). It's also easy to tell that the middle sketch is not correct. Because the point at which we consider the magnetic field is far from the loop in the bottom sketch, it's harder to tell if the direction of the field lines is correct.

Equation 19-10 describes the magnetic field of the circular loop along its central axis. The shape of the field at all points in space, mapped out in **Figure 19-17**, is similar to the shape of the electric field around an electric dipole. Indeed, the circular loop of current is a **magnetic dipole**. It is customary to label the poles of the magnetic dipole "north" and "south." By convention, the field lines point toward the south pole and away from the north pole. The field lines are loops; each line that points away from the north pole eventually returns to the south pole. (The line that points directly away from the north pole is a circle of infinite radius, so it returns to the south pole in a mathematical sense.)

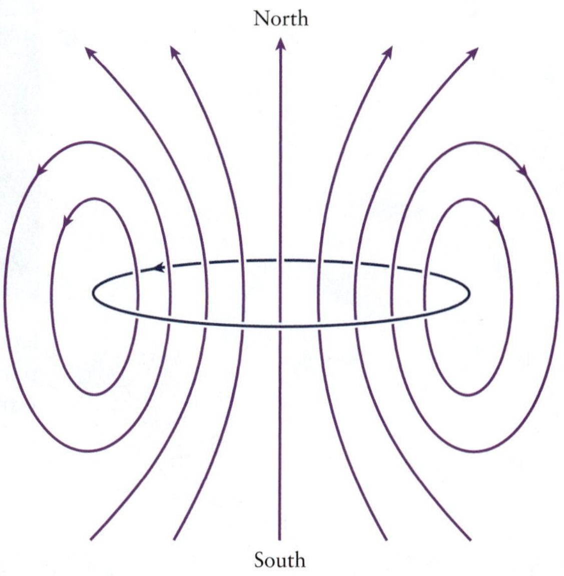

North

South

Figure 19-17 The magnetic field that arises due to a circular loop of current, shown in blue, is a magnetic dipole.

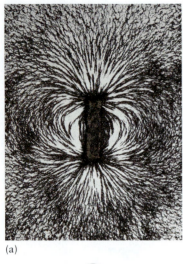

(a)

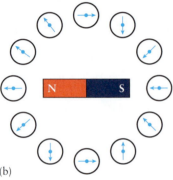

(b)

Figure 19-18 (a) Iron filings line up along the magnetic field produced by a bar magnet. Notice how the photograph looks much like the drawing in Figure 19-17. *(NASA)* (b) We can visualize the direction of the magnetic field lines by placing compasses around the bar magnet.

A bar magnet is a magnetic dipole. In **Figure 19-18a**, the lines of field around a bar magnet are traced out by small bits of iron. The shape of the field looks identical to the drawing in Figure 19-17. We can see the direction of the lines of field by placing a number of compasses around a bar magnet, as illustrated in **Figure 19-18b**.

An electric dipole consists of a positive and negative charge, and the electric field that surrounds it is the superposition of the fields of each charge separately. This is not the case for a magnetic dipole. Even a permanent bar magnet cannot be broken into separate magnetic "charges," properly called *magnetic monopoles*. If you were to break a bar magnet into two pieces, each would itself exhibit a north pole and a south pole; that is, each would be a magnetic dipole. Many experiments have been conducted to look for evidence of a magnetic monopole, but none has been found.

Earth's field is nearly that of a dipole, with the axis of the field tilted about 11° from Earth's rotation axis. (So the magnetic poles, the place on Earth's surface where the field lines point straight and vertical to the surface, are close to but not quite the same as Earth's geographic poles.) Earth cannot, however, be a large permanent bar magnet. All ferromagnetic materials lose the ability to retain the internal order required to sustain a magnetic field as temperature increases. Iron loses its magnetic field at about 800°C, a temperature far lower than Earth's core temperature of some thousands of degrees Celsius. It is generally believed that the rotation of Earth's liquid core, which moves charges in circular currents, gives rise to the dipole magnetic field of our planet. This liquid core, and the field that arises due to it, has played a critical role in the development of life on Earth. Most importantly, Earth's magnetic field shields Earth from solar radiation. More than 10^9 kg of high energy, charged particles stream from the Sun every second; this solar wind would have long ago stripped Earth of its atmosphere if it were not deflected by Earth's magnetic field. In addition, the magnetic force on those particles that do get close to Earth causes them to be drawn to Earth's poles and away from most of the surface. This is the source of the aurora borealis, or "northern lights" (**Figure 19-19**) and the aurora australis, or "southern lights.". Without an atmosphere and with high energy radiation bathing the surface, Earth would be a far less hospitable place.

The dipole nature of Earth's magnetic field provides a clear directionality to the field vectors. The applicability of the directional nature of Earth's field has played an important role in human navigation for thousands of years. Other living organisms also take advantage of Earth's field to guide them from location to location. In addition to bacteria (such as those shown on the opening page of this chapter),

Figure 19-19 The force of Earth's magnetic field on charged particles that stream from the Sun causes them to be drawn to Earth's magnetic poles. This is the source of the aurora borealis, or northern lights, an often spectacular display in the night sky at extreme northern locations. *(NASA)*

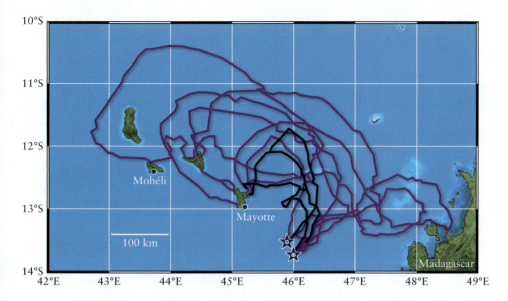

Figure 19-20 Sea turtles navigate
using magnetic cues. When
turtles are transported away
from their nesting sites on
Mayotte, an island between
Madagascar and Mozambique,
and then released (at the points
marked by stars), they take a
relatively direct path back,
marked by the black lines.
Turtles transported after a
magnet has been attached to
their head take wildly
circuitous routes, marked by
the purple lines.

pigeons, honeybees, and sea turtles among others rely to some extent on an internal
magnetic compass to navigate. Although not yet well understood, it is clear that
magnetic phenomena underlie the organisms' ability to navigate. Sea turtles, for
example, have been observed to travel hundreds of kilometers and still find their
way, relatively directly, back to their nesting sites. Yet when they are transported
away from their nests after a magnet has been attached to their heads, the turtles
take wildly circuitous routes back to the nesting site (**Figure 19-20**). The magnet
clearly disrupts the turtles' ability to determine their position using Earth's magnetic field.

Example 19-3 Magnetic Field Due to a Very Long, Straight Wire

A very long, straight wire carries a current i_0. Use the Biot–Savart law to find the
magnetic field a distance d from the wire, far from its ends.

SET UP

We start by making a good picture, carefully labeling all of the variables on which
the Biot–Savart law depends. Notice, in **Figure 19-21**, that we labeled l, the distance along the wire, as well as $d\vec{l}$; we need to understand how l varies in order to
assign bounds to the integration. The unit vector $\hat{r}$ has magnitude equal to 1 and
points from each infinitesimal path element toward the point at which we want to
determine the magnetic field. The cross product depends on φ, the angle between
$d\vec{l}$ and $\hat{r}$.

Next consider the vector aspect of the integral in Equation 19-8. Take any
arbitrarily chosen $d\vec{l}$ along the wire and find the direction of $d\vec{B}$ according to the
right-hand rule. Point your thumb in the direction of the current, to the right in the
figure; your other fingers curl in the direction of the circulation of the field. For the
point shown, the $\vec{B}$ vector points out of the page. This is true for every choice of $d\vec{l}$;
because all field contributions $d\vec{B}$ point in the same direction (out of the page), we
don't need to consider vector components in taking the integral.

Using Equation 19-8, we see that

$$B = \int_{\text{current}} dB = \int_{\text{current}} \frac{\mu_0}{4\pi} \frac{i_0 \, dl \sin\varphi}{r^2} \qquad \text{(19-11)}$$

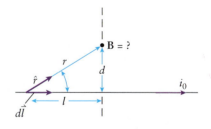

Figure 19-21 This sketch helps
us apply the Biot–Savart law to
find the magnetic field due to a
long, straight wire carrying a
constant current.

Notice that l, φ, and r all vary. To evaluate the integral we need to express two of these in terms of the third. Once we do that, we can also assign bounds to the integral.

SOLVE

Although l, φ, and r all vary, any two of the quantities depend on the third. You are therefore free to choose any one of the three with which to evaluate the integral. We pick l, so we will first express φ and r in terms of l. From the right triangle formed by l, r, and d,

$$\sin \varphi = \frac{d}{r}$$

and

$$r = (d^2 + l^2)^{1/2}$$

Then Equation 19-11 becomes

$$B = \int_{\text{current}} \frac{\mu_0}{4\pi} \frac{i_0 \, dl \, (d/r)}{r^2} = \int_{\text{current}} \frac{\mu_0}{4\pi} \frac{i_0 \, d}{r^3} dl$$

or

$$B = \frac{\mu_0 i_0 \, d}{4\pi} \int_{-\infty}^{+\infty} \frac{dl}{(d^2 + l^2)^{3/2}}$$

Once the integral is in terms of a single variable, l in this case, we set the bounds on the integral to be the smallest and largest values that the variable must take in order to include the contributions from the entire current. Because the wire is very long we take the ends to be infinitely far away in each direction. Evaluating the integral at these bounds gives

$$B = \frac{\mu_0 i_0 d}{4\pi} \frac{l}{d^2 (l^2 + d^2)^{1/2}} \Big|_{-\infty}^{+\infty} = \frac{\mu_0 i_0}{4\pi d} \frac{l}{(l^2 + d^2)^{1/2}} \Big|_{-\infty}^{+\infty}$$

Notice then when at $l = \pm\infty$ we can neglect d^2 compared to l^2, so $(l^2 + d^2)^{1/2}$ is approximately $(l^2)^{1/2}$ or just l. Also, note that because this value is the square root of the square of l it is always positive, so

$$B = \frac{\mu_0 i_0}{4\pi d} \left(\frac{+\infty}{+\infty} - \frac{-\infty}{+\infty} \right) = \left(\frac{\mu_0 i_0}{4\pi d} \right) 2$$

or

$$B = \frac{\mu_0 i_0}{2\pi d} \tag{19-12}$$

REFLECT

The magnitude of the magnetic field is inversely proportional to the distance from the wire. The direction of the field is given by the right-hand rule. When the wire and current are coming straight toward us, the field can be represented as in Figure 19-22. Notice that the lines of field are farther apart farther from the wire, indicating that the strength of the field decreases with distance. Also, we have drawn in the magnetic field vectors at a few specific points in space. Although we represent the field as circles centered on the wire, at any specific point the field is a vector with a specific direction, represented as a straight arrow.

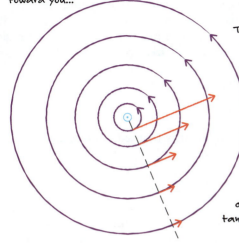

The blue circle and dot represents a long, straight wire carrying current directly toward you...

The purple lines show the magnetic field due to the current. They get farther apart as the distance from the wire increases and the magnitude of the field decreases.

The red arrows represent the magnetic field vectors at five different, specific points in space. The direction of the field at any point is tangent to the circle.

Figure 19-22 The magnetic field around a long, straight wire carrying a current is represented by circular lines of field. The field vector at any given point is represented by an arrow, in the same way that we represent *any* vector quantity. Red arrows show the magnetic field at five specific points along the dashed line.

✳ What's Important 19-3

Current along a wire gives rise to a magnetic field that circulates around the wire. For a long, straight wire this is represented by circular field lines centered on the wire. Because magnetic field is a vector, however, it has a specific direction, so at any particular point in space it must be represented as a straight arrow. The magnetic field points in a direction mutually perpendicular to the direction of the current and the direction radially outward from the wire.

19-4 Magnetic Field and Current—Ampère's Law

The Biot–Savart law for magnetic fields is similar in its approach to our application of Coulomb's law to determine an electric field. In the latter, we break a configuration of charge into infinitesimally small elements and add the contribution each makes to the total electric field. In the former, we break a configuration of current into infinitesimally small elements and add the contribution each makes to the total magnetic field. As we discovered for charges and electric fields, it is often easy to write down the integral required to sum the field contributions but not so easy to evaluate it. Especially in cases that exhibit symmetry, Gauss' law is far simpler to apply than Coulomb's law. In this section, we will explore *Ampère's law*, the magnetic analog of Gauss' law.

Gauss' law relates the charge enclosed by a closed surface and the electric field passing through it:

$$\oint \vec{E} \cdot d\vec{A} = \frac{q_{\text{encl}}}{\epsilon_0} \tag{16-31}$$

We will look for the magnetic analog of Gauss' law by considering the quantities that make up the Gauss' law equation (note that the quantities pertain to electric fields) and finding comparable quantities that pertain to magnetic fields. This determination is not the way André Ampère arrived at the law that bears his name (we encountered Ampère, after whom the SI unit of current is named, in Chapter 18), but it helps to emphasize the similarity between Gauss' and Ampère's laws.

Certainly we should replace electric field $\vec{E}$ with magnetic field $\vec{B}$. Because magnetic field arises from current, we'll also replace charge q with current i. Finally, from a comparison of Coulomb's law, which depends on $1/\epsilon_0$, and the Biot–Savart law, which depends on μ_0, we'll replace $1/\epsilon_0$ in Gauss' law with μ_0. (Recall that permeability of free space μ_0 is related to how a magnetic field affects a region of space or a material and that the permittivity of free space ϵ_0 is related to how an electric field affects a region or material.)

With these substitutions, Gauss' law has become

$$\oint \vec{B} \cdot \_ = \mu_0 i$$

What must the units be of the unknown quantity that will replace $d\vec{A}$? The SI units of μ_0 are $\mathrm{T \cdot m/A}$, so on the right-hand side

$$[\mu_0 i] = \frac{\mathrm{T \cdot m}}{\mathrm{A}} \mathrm{A} = \mathrm{T \cdot m}$$

On the left-hand side, the units of $\vec{B}$ are T, so in order for the units to match, the unknown quantity must have units of meters; the quantity must be length. We will therefore replace the area differential $d\vec{A}$ with a path differential $d\vec{l}$, arriving at

$$\oint \vec{B} \cdot d\vec{l} = \mu_0 i_{\text{through}} \tag{19-13}$$

Dimensional analysis, as we have effectively done here, will not correctly identify constants that might be required in all cases; in this case, we have arrived at the exact form of **Ampère's law**. Ampère's law is the magnetic analog of Gauss' law; Gauss' law provides the relationship between the electric field at the surface of a closed boundary and the charge the boundary encloses, while Ampère's law provides the relationship between the magnetic field on a closed loop in terms of the current that passes through the loop.

The symbol on the left indicates an integral around a closed path. In the same way that we apply Gauss' law by enclosing charge in a closed (Gaussian) surface, we apply Ampère's law by choosing a closed path, or Amperian loop, through which the current passes. We added the subscript "through" to the variable i to remind us of this closed path, in the same way that we added "encl" to the quantity of charge for Gauss' law. The point in space at which we want to determine the magnetic field must lie on the Amperian loop, as in **Figure 19-23**.

The magnetic field that arises from an electric current depends only the current. We must expect, then, that the application of Ampère's law is independent of the shape of the Amperian loop. Nevertheless, some choices of the loop will make the integral in Equation 19-13 easier (or harder) to evaluate. As is the case when applying Gauss' law, symmetries simplify Ampère's law problems. In particular, we look for paths on which the magnetic field is constant, or for which the cosine of the angle between $\vec{B}$ and $d\vec{l}$ (in the dot product) is either 0 or 1, or for which we are able to determine the path length. Any of these factors, or a combination of them, simplifies the integral on the left side of the Ampère's law equation.

Let's reconsider the magnetic field far from the ends of a very long, straight wire carrying a current. We determined the magnetic field in Example 19-3 in the last section by applying the Biot–Savart law.

To use Ampere's law to find the magnetic field at a point in space due to a current...

...define an Amperian loop through which the current passes. The point of interest must lie on the loop.

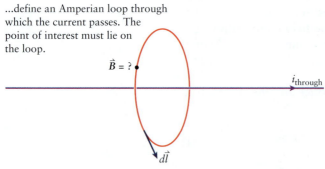

Integrate $\vec{B} \cdot d\vec{l}$ around the loop using Ampere's law

$$\int \vec{B} \cdot d\vec{l} = \mu_0 i_{\text{through}}$$

The vector $d\vec{l}$ is the path differential.

Figure 19-23 This sketch describes the process of applying Ampère's law to find the magnetic field due to a current.

The long, straight wire is an ideal candidate for the application of Ampère's law. Because the wire is long and the current is uniform, at any particular distance from the wire the magnitude of the field must be the same regardless of direction. We can therefore infer from this symmetry that the field lines are circles centered on the wire and therefore choose a circular path as our Amperian loop. As shown in **Figure 19-24**, the magnetic field vectors point in the same direction as the path differentials along a circular path. (The wire is so long that its ends are not shown in the figure, and therefore the source of the current is also not shown.)

We will apply Ampère's law to the current, using the circular Amperian loop shown in Figure 19-24. On the left side of Equation 19-13,

$$\oint \vec{B} \cdot d\vec{l} = \oint B \, dl \cos \varphi$$

where φ, the angle between the field $\vec{B}$ and an infinitesimal element of the path $d\vec{l}$, is 0° based on the argument above. Therefore $\cos \varphi$ equals 1, so we see that the equation simplifies to

$$\oint \vec{B} \cdot d\vec{l} = \oint B \, dl$$

Because the field is constant around any circular path centered on the wire, B can be pulled from the integral:

$$\oint \vec{B} \cdot d\vec{l} = B \oint dl$$

Finally, the integral of the path length around the circular path is the circumference of the circle; let the radius of the circle is r so

$$\oint \vec{B} \cdot d\vec{l} = B \, 2\pi r$$

For the right side of the Ampère's law equation, note that all of the current i_0 passes through the Amperian loop, so

$$\mu_0 i_{\text{through}} = \mu_0 i_0$$

We now set the two sides of the Ampère's law equation equal:

$$\oint \vec{B} \cdot d\vec{l} = \mu_0 i_{\text{through}} \Rightarrow B \, 2\pi r = \mu_0 i_0$$

or

$$B = \frac{\mu_0 i_0}{2\pi r}$$

This expression is, not surprisingly, the same result as we obtained using the Biot–Savart law (Equation 19-12). (In applying the Biot-Savart law we let the distance from the wire be d.) Which method should you pick, then, to find the magnetic field due to a current? It is often the case that while it's easier to set up the integral required for the Biot–Savart law, evaluating the integral can be challenging. Conversely, it takes a bit of analysis and experience to properly take advantage of the symmetry in a problem to apply Ampère's law, and in general problems that do not involve symmetry are poor candidates for the application using Ampère's law. But for problems that present symmetry, such as the long, straight current-carrying wire, once you've set up the problem, working it out tends to be relatively straightforward. For example, the most complicated mathematical manipulation required to complete the wire problem above was finding the circumference of a circle!

Apply Ampère's law to a long, straight wire carrying a current i_0.

Because the wire is long and the current is uniform, at any distance r from the wire the magnitude of the field. The symmetry allows us to choose a circular path as the Amperian loop.

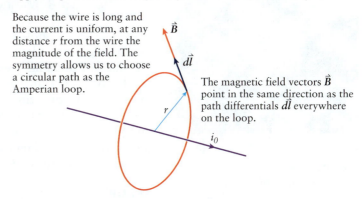

The magnetic field vectors $\vec{B}$ point in the same direction as the path differentials $d\vec{l}$ everywhere on the loop.

Figure 19-24 A circular Amperian loop is a good choice when applying Ampère's law to a long, straight current, because the magnetic field vector points in the same direction as the path differential everywhere along the loop.

Figure 19-25 Coaxial cable like the one shown here carries signals from your service provider to your television. *(Courtesy of David Tauck)*

Example 19-4 Magnetic Field Due to a Coaxial Cable

A coaxial cable, shown in **Figure 19-25**, consists of a conductor surrounded by insulation, surrounded by a conductor made of either fine wire mesh or a thin metallic foil, and finally enclosed in a layer of insulation on the outside. The inner conductor carries current in one direction, and the outer conductor carries it back in the opposite direction. For a coaxial cable that carries a constant current i_0, find the magnetic field (a) inside the inner conductor at a distance r from its central axis and (b) outside the coaxial cable. Assume that the moving charge in the inner conductor is distributed uniformly over the volume of the conductor. Let the radius of the inner conductor be R.

SET UP

Both the inner conductor separately and the coaxial cable as a whole have the same symmetry as the wire (above), so we again choose a circular path as the Amperian loop. The point at which we want to determine the field must lie on the loop, so for part (a), inside the inner conductor, the radius of the loop must be smaller than the radius of the inner conductor (**Figure 19-26a**). Only a fraction of the current in the inner conductor passes through this loop. For part (b), in which we want to find the magnetic field outside the entire coaxial cable, the radius of the loop must be larger than the radius of the outer conductor. In this case, the entire cable passes through the Amperian loop.

The left side of the Ampère's law equation (Equation 19-13) depends on the angle φ between $\vec{B}$ and $d\vec{l}$:

$$\oint \vec{B} \cdot d\vec{l} = \oint B \, dl \cos \varphi$$

and, as argued above, for a straight wire carrying constant current, the angle between $\vec{B}$ and $d\vec{l}$ is zero. The field is constant based on symmetry, so (again, as was the case for the wire above)

$$\oint \vec{B} \cdot d\vec{l} = B \, 2\pi r$$

We only need to determine the current that is carried through the loop for each of the cases to complete the problem.

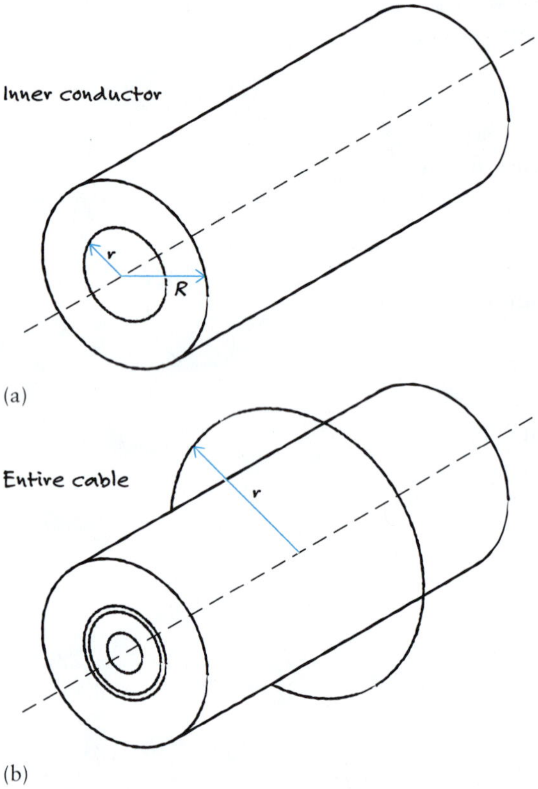

SOLVE

(a) For the point inside the inner conductor, only a fraction of the total current is carried through the loop. That fraction is the same as the ratio of the area of the Amperian loop to the cross-sectional area of the conductor:

$$i_{\text{through}} = \frac{\pi r^2}{\pi R^2} i_0 = \frac{r^2}{R^2} i_0$$

So, setting the two sides of the Ampère's law equation equal,

$$\oint \vec{B} \cdot d\vec{l} = \mu_0 i_{\text{through}} \Rightarrow B \, 2\pi r = \mu_0 \frac{r^2}{R^2} i_0$$

or

$$B = \frac{\mu_0 i_0}{2\pi R^2} r$$

The magnitude of the field is given by a constant multiplied by r; in other words, the field grows linearly as distance increases from the central axis (**Figure 19-27**).

Figure 19-26 (a) Inside the inner conductor we establish a circular Amperian loop of radius r. Only a fraction of the current in the inner conductor passes through the loop. (b) Outside the cable the currents in both the inner and outer conductors pass through a circular Amperian loop. Because the currents in the inner and outer conductors are equal in magnitude but flow in opposite directions, the net current is zero; there is no magnetic field outside the coaxial cable.

Inner conductor

(a)

Entire cable

(b)

(b) For an Amperian loop large enough that the entire cable passes through it (Figure 19-26b), notice that the current in the inner conductor and the current in the outer conductor both pass through the loop. Because the currents in the inner and outer conductors are equal in magnitude but in opposite directions, as much charge flows in one direction through the loop as in the other direction. The net flow of charge through the loop, and therefore the net current, is zero. For part (b), then,

$$i_{through} = 0$$

The right side of the Ampère's law equation is therefore zero:

$$\oint \vec{B} \cdot d\vec{l} = \mu_0 i_{through} \Rightarrow B\,2\pi r = 0$$

so

$$B = 0$$

REFLECT

There is no magnetic field outside the coaxial cable.

The magnetic field inside the inner conductor goes up linearly with distance from the central axis. Notice that while the circumference of whatever Amperian loop we might choose in that region grows proportionally to the radius, the current through the loop increases proportionally to the square of the radius. Because the two terms are on opposite sides of the Ampère's law equation, net effect is that the result is proportional to one power of radius.

Coaxial cables are often referred to as "shielded" cables. The arrangement of the two conductors eliminates the presence of stray magnetic fields outside the cable. The shielding also serves to isolate the inner conductor from external electromagnetic signals, which is the primary reason that both scientists and audio enthusiasts use coaxial cables.

Practice Problem 19-4 In a particular coaxial cable, such as the one shown in Figure 19-25, the inner conductor carries current $2i_0$ in one direction and the outer conductor carries current i_0 in the opposite direction. Find the magnitude of the magnetic field in the region outside the coaxial cable as a function of r, the distance from the central axis of the inner conductor.

A straight, helical coil of wire, called a **solenoid**, produces a relatively uniform magnetic field when a constant current is in the wire. Figure 19-28 shows a simple solenoid. The properties of a solenoid can be characterized by its length L, diameter D, and N, the number of loops or windings that make up the coil (Figure 19-29). We will also use winding density n to characterize a solenoid, where

$$n = \frac{N}{L}$$

The units of n are windings per meter.

To visualize the magnetic field that arises when a solenoid carries a current, we imagine the coil as a series of circular loops of current, as in Figure 19-30. The field due to each loop is a dipole, and the total field is the vector sum of each one separately. The lower part of Figure 19-30 shows three windings of a solenoid, with just the fields close to each loop drawn in. You can see that inside the solenoid the fields add to create a total field that runs parallel to the central axis; this is especially noticeable close to the axis and far from the ends. Outside the solenoid the fields

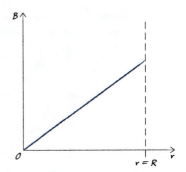

Figure 19-27 The magnitude of the magnetic field inside the inner conductor of a coaxial cable increases linearly with distance from the center of the cable.

▶ **Go to Interactive Exercise 19-2** for more practice dealing with Ampere's law

▶ **Go to Interactive Exercise 19-3** for more practice dealing with Ampere's law

Figure 19-28 A solenoid produces a relatively uniform magnetic field when a constant current is in the wire. (Courtesy of David Tauck)

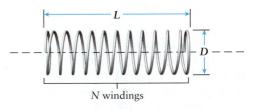

Figure 19-29 The properties of a solenoid can be characterized by its length L, diameter D, and the number N of windings in the coil.

In this short section of a solenoid current comes out toward you...

...here,   ...here,   ...and here,

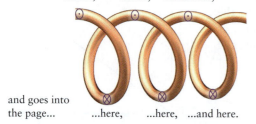

and goes into
the page...   ...here,   ...here,   ...and here.

We treat each winding as a separate, circular loop...

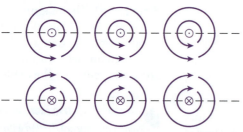

...the fields of the separate loops add constructively inside the solenoid, and destructively outside.

Figure 19-30 To find the magnetic field we can treat the solenoid as a series of separate, circular loops of current.

nearly cancel, especially close to the solenoid and far from the ends. Cancellation of the fields outside the solenoid is more complete when the length of the solenoid is large compared to its diameter, and also when the coils of the solenoid are tightly wound. For an ideal solenoid, one for which L is much larger than D and for which n is large, we can consider the field uniform and parallel to the central axis inside the solenoid and zero outside. This is suggested in **Figure 19-31**.

Let's use Ampère's law to find the magnetic field of a long, narrow, tightly wound (ideal) solenoid of length L, diameter D, and winding density n carrying a current i_0. Because solenoids are so often used to create a uniform magnetic field, we will apply Ampère's law along the axis of the coil and far from the ends, where we expect a relatively uniform field.

The form of Ampère's law

$$\oint \vec{B} \cdot d\vec{l} = \mu_0 i_{\text{through}} \qquad (19\text{-}13)$$

reminds us that to apply it we must first select an Amperian loop *through* which the current is carried and *around* which to carry out the path integral ($\oint$). Remember that our choice of shape and size of the loop won't affect the final answer, although this choice can make the problem either easier or more difficult to solve. Also, recall that the point at which we want to determine the field must lie on the Amperian loop.

Figure 19-32 shows a cutaway view of a section of the solenoid. The view is similar to the lower part of Figure 19-30, in which the current is shown coming out of the page along the top part of the windings and going into the page along the bottom. We picked a rectangle of width w as the Amperian loop and positioned it so that the central axis of the solenoid (along which we want to determine the magnetic field) lies on the rectangular path. The path integral around the rectangle is the sum of the integrals along each side:

$$\oint \vec{B} \cdot d\vec{l} = \int_1 \vec{B} \cdot d\vec{l} + \int_2 \vec{B} \cdot d\vec{l} + \int_3 \vec{B} \cdot d\vec{l} + \int_4 \vec{B} \cdot d\vec{l}$$

Figure 19-31 For a solenoid for which its length is much larger than its diameter, and for which the number of windings per unit length is large, we can consider the field uniform and parallel to the central axis inside the solenoid and zero outside.

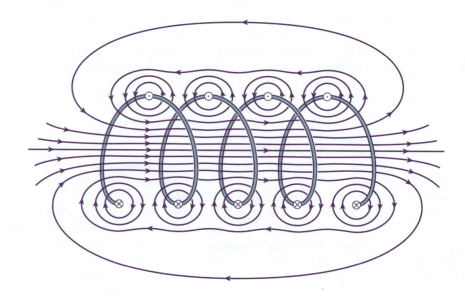

These are the windings of the solenoid. The current comes out toward you along the top and flows into the page along the bottom.

The Amperian loop is a rectangle of width w....

...For side 4 and parts of sides 1 and 3, the magnetic field is zero: no contribution to $\vec{B} \cdot d\vec{l}$ in Ampere's law.

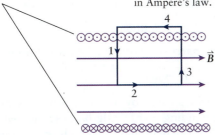

Figure 19-32 A sketch showing a cutaway view of a section of the solenoid helps us apply Ampère's law on a rectangular Amperian loop that encloses some of the windings of the solenoid.

...For the parts of sides 1 and 3 that are inside the solenoid, the magnetic field is perpendicular to the direction of $d\vec{l}$, which follows the path. So $\vec{B} \cdot d\vec{l}$ equals $\vec{B}\, d\vec{l} \cos(90°)$ which is zero.

...For side 2 $\vec{B}$ and $d\vec{l}$ are parallel, so $\vec{B} \cdot d\vec{l}$ equals $B\, dl \cos(0°)$, which is $B\, dl$.

The rectangle is an ideal choice here because side 4 is in a region of no magnetic field, so

$$\int_4 \vec{B} \cdot d\vec{l} = 0$$

Side 1 partially runs through a region of no magnetic field, and inside the solenoid is perpendicular to the field. So

$$\int_1 \vec{B} \cdot d\vec{l} = \int_1 B\, dl \cos 90° = 0$$

The same is true for side 3, so $\int_3 \vec{B} \cdot d\vec{l}$ is also equal to zero. The only side of the Amperian loop that makes a nonzero contribution to the path integral is side 2, on which $\vec{B}$ and $d\vec{l}$ are parallel. So

$$\int_2 \vec{B} \cdot d\vec{l} = \int_2 B\, dl \cos 0° = \int_2 B\, dl$$

Because the field is constant we take B out of the integral, and the integral of dl over the path is just the length of the path:

$$\int_2 \vec{B} \cdot d\vec{l} = B \int_2 dl = Bw$$

We've already done all the hard work! We can put together the four sides of the Amperian loop to evaluate the left side of the Ampère's law equation:

$$\oint \vec{B} \cdot d\vec{l} = \int_1 \vec{B} \cdot d\vec{l} + \int_2 \vec{B} \cdot d\vec{l} + \int_3 \vec{B} \cdot d\vec{l} + \int_4 \vec{B} \cdot d\vec{l} = 0 + Bw + 0 + 0 = Bw$$

On the right side of the equation, we need to evaluate $i_{through}$, the current that passes through the loop. This quantity is *not* the current in the solenoid i_0; notice that every winding of the coil that passes through the loop brings a contribution i_0 to $i_{through}$. So

$$i_{through} = m i_0$$

where m is the number of windings of the solenoid that pass through the Amperian loop. In general,

number of windings through rectangle

$$= \frac{\text{number of windings}}{\text{length}} \times \text{length of rectangle}$$

so m is given by

$$m = nw$$

Figure 19-33 A solenoid used in electronic devices typically has a large number of tightly packed windings. *(Paul Silverman/Fundamental Photographs)*

Then

$$i_{\text{through}} = nwi_0$$

so setting the two sides of the Ampère's law equation equal gives

$$Bw = \mu_0 nwi_0$$

The field along the axis of a long, narrow, tightly packed solenoid is

$$B = \mu_0 ni_0 \tag{19-14}$$

The magnitude of the uniform field created by a long solenoid is proportional to winding density n, the number of windings per length. Because each winding contributes to the total magnetic field, the more windings that can be packed into a length of the solenoid, the larger the field. A typical solenoid used for electronic applications is therefore more likely to look like the one in **Figure 19-33** than the one shown in Figure 19-28. Such solenoids have many layers of tightly packed windings.

! Watch Out

Every winding of a current-carrying wire that passes through an Amperian loop makes a separate contribution to i_{through}.

You may encounter a variety of current configurations, such as the solenoid, in which the wire carrying the current winds around to form loops. Each winding that passes through an Amperian loop you set up to find the magnetic field contributes current to the quantity we label i_{through}, the current that passes through the loop. If the wire of a solenoid carries current i_0, for example, and 10 windings of the solenoid pass through a certain Amperian loop, then i_{through} equals $10i_0$.

Physicists like simplicity! Why do a problem a hard way when an easy way is available? This issue often confronts us when trying to determine the magnetic field due to a current configuration. For many problems, Ampère's law is a better choice than the Biot-Savart law. We'll see that while it might be easier to set up a problem using the Biot-Savart (that is, by defining a small current element and then setting up an integral to add up the contribution each makes to the magnetic field), the integration necessary to do the addition can be challenging. Conversely, it takes a bit of analysis to properly set up a problem using Ampère's law, but once set up, working out the problem tends to be relatively straightforward.

✳ What's Important 19-4

Ampère's law relates the current through a wire to the magnetic field it generates. Current through a solenoid (a straight, helical coil of wire) produces a relatively uniform magnetic field along its axis that is proportional to the winding density as well as to the current. For an ideal solenoid, the magnetic field is uniform and parallel to the central axis inside the solenoid and zero outside.

19-5 Magnetic Force between Current-Carrying Wires

André Ampère suggested that if a wire carrying a current produced a magnetic field, as evidenced by the fact that current could turn a compass needle, then two wires each carrying a current should either attract or repel each other magnetically.

Through a series of experiments, Ampère showed that two parallel, straight wires carrying current in the same direction attract each other, and that two parallel, straight wires carrying current in opposite directions repel.

A wire carrying a current gives rise to a magnetic field. A second current-carrying wire experiences the force described by Equation 19-3:

$$\vec{F}_{1\to2} = i_2 \vec{l}_2 \times \vec{B}_1 \qquad (19\text{-}15)$$

The subscripts indicate that the force of the magnetic field arises from the current in wire 1 and acts on wire 2. Not surprisingly, the interaction of the current in wire 1 and the field due to the current in wire 2 also results in a force. This force,

$$\vec{F}_{2\to1} = i_1 \vec{l}_1 \times \vec{B}_2 \qquad (19\text{-}16)$$

completes the force pair as required by Newton's third law, that for every force there is an equal but opposite reaction force.

Let's consider two long, straight, parallel wires, each carrying a current. Figure 19-34 shows two such wires, separated by distance d and carrying currents i_1 and i_2, respectively, in the same direction. The current in each wire gives rise to a magnetic field that has a circulation around the wire as defined by the right-hand rule. For example, the magnetic field due to current i_1 is shown as clockwise circles from the perspective taken in the figure. At any point in space, however, the magnetic field due to current i_1 is a straight arrow. At any point on wire 2, for example, $\vec{B}_1$ points straight up, as shown. The magnitude of the force on a length l_2 of wire 2 is then given by Equation 19-15:

$$F_{1\to2} = i_2 l_2 B_1 \sin 90° = i_2 l_2 B_1$$

where we have set the angle between $\vec{l}_2$ and $\vec{B}_1$ equal to 90° as described by the right-hand rule. In addition, we have previously found the strength of the magnetic field a distance d from a current-carrying wire. From Equation 19-12,

$$B_1 = \frac{\mu_0 i_1}{2\pi d}$$

So

$$F_{1\to2} = i_2 l_2 \frac{\mu_0 i_1}{2\pi d} = \frac{\mu_0 i_1 i_2 l_2}{2\pi d} \qquad (19\text{-}17)$$

We can apply the same process to the force on wire 1 due to the current in wire 2. At any point on wire 1, $\vec{B}_2$ points straight down according to the right-hand rule, so the angle between $\vec{l}_1$ and $\vec{B}_2$ is 90°. Then, from Equation 19-16,

$$F_{2\to1} = i_1 l_1 B_2 \sin 90° = i_1 l_1 B_2$$

where $F_{2\to1}$ is the force on some length l_1 of wire 1. Also

$$B_2 = \frac{\mu_0 i_2}{2\pi d}$$

from Equation 19-12; so, by substitution we see that

$$F_{2\to1} = i_1 l_1 \frac{\mu_0 i_2}{2\pi d} = \frac{\mu_0 i_1 i_2 l_1}{2\pi d} \qquad (19\text{-}18)$$

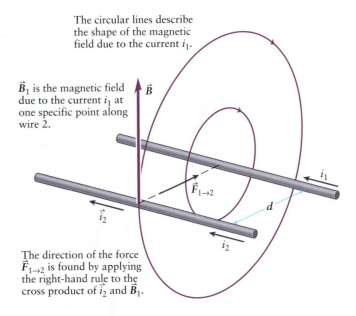

The circular lines describe the shape of the magnetic field due to the current i_1.

$\vec{B}_1$ is the magnetic field due to the current i_1 at one specific point along wire 2.

The direction of the force $\vec{F}_{1\to2}$ is found by applying the right-hand rule to the cross product of $\vec{i}_2$ and $\vec{B}_1$.

Figure 19-34 The force between two long, straight parallel wires carrying currents can be understood as the force that the magnetic field due to one wire exerts on the current in the other wire. When the two currents are in the same direction, the force between the wires attracts each to the other.

To compare $F_{1 \rightarrow 2}$ with $F_{2 \rightarrow 1}$, set the lengths of the two wires on which the force is applied to be the same:

$$l_1 = l_2 = l$$

The force of wire 1 on wire 2 is then

$$F_{1 \rightarrow 2} = \frac{\mu_0 i_1 i_2 l}{2 \pi d}$$ (19-19)

and the force of wire 2 on wire 1 is

$$F_{2 \rightarrow 1} = \frac{\mu_0 i_1 i_2 l}{2 \pi d}$$

That the two forces are equal in magnitude should not be a surprise, because they form a force pair. Notice also that the force on each wire points directly toward the other wire, so $\vec{F}_{1 \rightarrow 2}$ and $\vec{F}_{2 \rightarrow 1}$ point in opposite directions. Thus $\vec{F}_{1 \rightarrow 2}$ and $\vec{F}_{2 \rightarrow 1}$ are equal in magnitude and opposite in direction, exactly as required by Newton's third law.

Watch Out

A wire carrying a current does not experience a force due to its own magnetic field.

Both wires in the previous discussion carry a current, so both give rise to a magnetic field. There is, therefore, a net field at any point in space which is the vector sum of the field due to each wire at that point. This net field is not, however, the field used to determine the force on one of the two wires. A wire does not experience a force due to the field created by the current that it carries.

Got the Concept 19-7
Parallel Wires

For two long, straight, parallel wires carrying current in the same direction, as in Figure 19-34, the force on each wire points toward the other, according to the right-hand rule. Suppose the current in wire 2 in the figure were flowing in the opposite direction. Would that change the direction of the force on wire 2, and if so, how? Would it change the direction of the force on wire 1, and if so, how?

Example 19-5 Wires in a Computer

The two long, straight wires that run along the back of a computer case to drive the cooling fan carry 0.11 A in opposite directions. The wires are separated by 0.5 cm. (a) Find the magnitude of the force per unit length between the wires. (b) Is the force attractive or repulsive? (c) The mass per unit length of the wire is 5.0×10^{-3} kg/m. What acceleration does one of the wires experience due to the force?

SET UP
Taking the wires as long, straight, and parallel, we can rely directly on Equation 19-19. Letting the current in both wires be i_0, the force on each due to the current in the other is

$$F = \frac{\mu_0 i_0 i_0 l}{2 \pi d} = \frac{\mu_0 i_0^2 l}{2 \pi d}$$

where d is the distance between the wires. The force per unit length is then

$$\frac{F}{l} = \frac{\mu_0 i_0^2}{2\pi d}$$

SOLVE

(a) We compute a numeric answer using the given values:

$$\frac{F}{l} = \frac{(4\pi \times 10^{-7}\,\text{T}\cdot\text{m/A})(0.11\,\text{A})^2}{2\pi(0.005\,\text{m})} = 4.8 \times 10^{-7}\,\frac{\text{N}}{\text{m}}$$

(b) Because the wires carry current in opposite directions, the force on each wire pushes it away from the other one.

(c) We don't know the length of the wire, which would be necessary to find the magnitude of the net force on one of the wires. We don't, however, need to know the total force in order to determine the acceleration. The net force equals mass multiplied by acceleration, so

$$a = \frac{F}{m}$$

We found F/l, and were given the mass per unit length (m/l) of the wire, so we divide both numerator and denominator by l:

$$a = \frac{F/l}{m/l}$$

With our result for F/l above and the given value of mass per unit length,

$$a = \frac{4.8 \times 10^{-7}\,\text{N/m}}{5.0 \times 10^{-3}\,\text{kg/m}} = 9.6 \times 10^{-5}\,\text{m/s}^2$$

REFLECT

Both the force on each wire and the acceleration each would experience just due to the magnetic force are quite small. We wouldn't expect to see any effects of the force.

Practice Problem 19-5 The two long, straight parallel wires each carry a current of 0.33 A in the same direction. The wires are separated by 0.5 cm. (a) Find the magnitude of the force per unit length that each wire exerts on the other. (b) Is the force attractive or repulsive? (c) The mass per unit length of the wire is $5.0 \times 10^{-3}\,\text{kg/m}$. What acceleration does one of the wires experience due to the force?

✱ What's Important 19-5

Two wires attract each other when carrying current in the same direction and repel each other when carrying currents in opposite directions. The force on each wire is equal in magnitude and opposite in direction, exactly as required by Newton's third law. A wire does not experience a force due to the field created by the current that it carries.

Answers to Practice Problems

19-1 $1.6 \times 10^{-25}\,\text{N}$

19-2 $i_0 RB$

19-4 $B = \dfrac{\mu_0 i_0}{2\pi r}$

19-5   (a) $4.4 \times 10^{-6}\,\text{N/m}$, (b) attractive, (c) $8.7 \times 10^{-4}\,\text{m/s}^2$

Answers to Got the Concept Questions

19-1 An electron at rest experiences no force due to the presence of a magnetic field. A magnetic field only exerts a force on a moving charge. So, yes, if the electron were given a push to set it in motion, it would experience a magnetic force as long as the direction of the motion is not aligned with the direction of the field. Because the force depends on the component of the field perpendicular to the motion (represented by the cross product in the Lorentz force law), the net force is zero when the motion and the field are completely aligned.

19-2 Protons are positively charged, so a proton obeys Equation 19-1 combined with the right-hand rule. To apply the right-hand rule, start by pointing your index finger from left to right to represent left-to-right velocity. For the magnetic field, point your other fingers into the page. When you stick your thumb out straight it points toward the top of the page. The proton feels a force toward the top of the page, which causes its path to bend upwards in the figure.

Before applying the right-hand rule to Equation 19-1 for electrons, recall that they are negatively charged. The direction of the magnetic force each electron experiences is therefore opposite to that specified by the right-hand rule. Let's apply the rule for the cross product in Equation 19-1 to the electron labeled A. Your index finger points toward the bottom of the page, and your other fingers point into the page. (You'll have to twist your arm around a bit to get into this configuration!) Your outstretched thumb should point to the right, which means the electron feels a force to the left. Electron A therefore feels a force toward the point at which the proton enters the field. For electron B, the right-hand rule has your thumb pointing toward the bottom of the page, so the force on the electron is toward the top of the page. The path of the electron does not bend toward the proton's entry point. The right-hand rule predicts the same direction for the magnetic force on electron C, but in this case a force toward the top of the page does bend the electron's path toward the point where the proton enters the field. Finally, when you point your index finger toward the top of the page for electron D and point your other fingers into the page ($\vec{B}$), your thumb sticks out to the left. For the negatively charged electron the magnetic force is therefore to the *right*. The path of electron D does not bend to the proton's entry point into the field.

19-3 The right-hand rule gives the direction for the force vector. In the case of Figure 19-5b, the direction is into the page; in the case of c, the direction is up; in the case of d, the direction is out of the page; and in the case of e, the direction is down. For the cases of Figure 19-5a and f in which the angle between $\vec{I}$ and $\vec{B}$ is 0° and 180°, respectively, the cross product in Equation 19-3, and therefore the force, is zero. So in these last two cases, because there is no magnetic force vector, we cannot assign a direction.

19-4 If the loop were oriented so that the near and far sides were perpendicular to the magnetic field, the force on the top and bottom sides would be acting at 0° with respect to the line connecting the force to the axis of rotation. (Convince yourself of this conclusion by looking at the sketch in **Figure 19-35**.) Torque depends on the sine of the angle between the force and the line that connects the axis of rotation and the point at which the force is applied. When the angle is 0°, the torque is therefore zero. When the loop is tilted into the magnetic field, the forces are in the same directions, but the torque will be counterclockwise. (Make a sketch to convince yourself of this conclusion.) So no, the loop will not rotate continuously. Instead it will rotate only until the plane of the loop is perpendicular to the direction of the magnetic field, with the side of the loop carrying current into the page at the top.

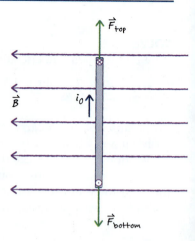

Figure 19-35 If the loop were oriented so that the near and far sides are perpendicular to the magnetic field, the force on the top and bottom sides would be acting at 0° with respect to the line connecting the force to the axis of rotation.

19-5 We use the right-hand rule to evaluate the direction of the cross product in the Biot–Savart law. There are two ways to apply the right-hand rule in this case; choose which ever is more comfortable for you. One way is to point your right index finger in the direction of the current (that is, the direction of $i\,d\vec{l}$) and your other fingers in the direction radially out from the wire (that is, in the direction of $\hat{r}$). Your thumb then points in the direction of the magnetic field. For P_1, for example, your index finger should point to the left, your other fingers toward the top of the page, and your thumb into the page. (You'll have to twist around a bit!) The magnetic field at P_1 therefore points into the page. Note that your knuckles should be positioned over the point P_1, to remind you that you are finding the field *at* P_1. For P_3, your index finger again points to the right, but because $\hat{r}$ points from the wire to P_3, your other fingers will point down the page. So now your thumb should point out of the page. At P_3, then, the field points out of the page. This application of the right-hand rule is demonstrated in **Figure 19-36a**. Finally, because P_2 is on the wire, $\hat{r}$ cannot be defined—the magnetic field is zero.

A second way to apply the right-hand rule to determine the direction of a magnetic field due to a current is to point your right thumb in the direction of the current and curl your other fingers. The direction of your other fingers *at the point of interest* gives the direction of the

Using your right hand, point your...

1. Index finger in the direction of the current
2. Other fingers radially away from the wire
3. Thumb in the direction of the magnetic field

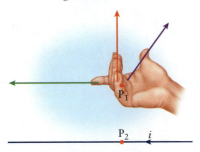

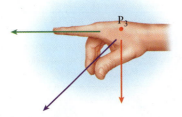

1. Index finger in the direction of the current
2. Other fingers radially away from the wire
3. Thumb in the direction of the magnetic field

(a)

Point your right thumb in the direction of the current and put your knuckles at the point of interest. Your fingers curl in the direction of the magnetic field...

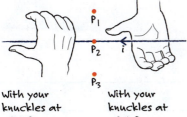

With your knuckles at point P_1 your fingers curl into the page.

With your knuckles at point P_3 your fingers curl out of the page.

(b)

Figure 19-36 (a) The right-hand rule gives the direction of the magnetic field at three points due to a straight wire carrying a current. (b) A second way to apply the right-hand rule is to point your thumb in the direction of the current.

field. At point P_1 your other fingers should be curling into the page, and at point P_3, out of the page (**Figure 19-36b**).

19-6 We can extend the answer to the previous Got the Concept question to see that the magnetic field is represented by circular paths centered on the wire (**Figure 19-37**). Looking head-on into the current, the magnetic field is said to circulate counterclockwise. Unlike your blood, which moves as it circulates, the circulation of a magnetic field is simply meant to suggest the circular, changing direction of the field.

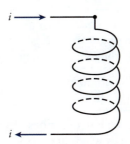

Figure 19-37 The magnetic field due to a wire carrying a current is represented by circular paths centered on the wire. Looking head-on into the current, the magnetic field is said to circulate counterclockwise.

19-7 If the current in wire 2 were in the opposite direction to the current in wire 1, the force on each wire would point away from, rather than toward, the other wire. If the direction of i_2 in Figure 19-34 were reversed, the direction of $\vec{l}_2$ would also be reversed. $\vec{B}_1$ does not change, so it remains as shown in the figure. Apply the right-hand rule to the cross product $\vec{l}_2 \times \vec{B}_1$ in Equation 19-15: Your index finger points left to right (somewhat tilted) along the current, your other fingers point upward in the direction of $\vec{B}_1$, and your thumb (representing the force) points right to left and away from wire 1. For the force on wire 1, notice that although the current in wire 1 continues in the direction as shown in the figure, the magnetic field due to the current in wire 2 changes direction (**Figure 19-38**). So in this case as well, the right-hand rule predicts that the cross product $\vec{l}_1 \times \vec{B}_2$ in Equation 19-16 points away from the other wire.

The circular lines describe the shape of the magnetic field due to the current i_1.

$\vec{B}_1$ is the magnetic field due to the current i_1 at one specific point along wire 2.

The direction of the force $\vec{F}_{1\to2}$ is found by applying the right-hand rule to the cross product of $\vec{i}_2$ and $\vec{B}_1$.

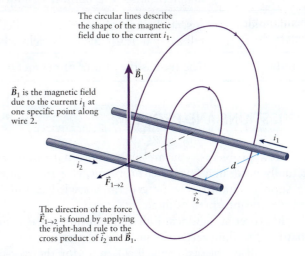

Figure 19-38 When the currents are in opposite directions in two long, straight parallel wires, the force that each wire exerts on the other repels them.

SUMMARY

Topic	Summary	Equation or Symbol
Ampère's law	Ampère's law provides a relationship between the magnetic field B at any point on a closed loop in terms of the current $i_{through}$ that passes through the loop.	$\oint \vec{B} \cdot d\vec{l} = \mu_0 i_{through}$ (19-13)
Biot–Savart law	The Biot–Savart law is a relationship between a moving charge and the magnetic field it produces.	$d\vec{B} = \dfrac{\mu_0}{4\pi} \dfrac{i\,d\vec{l} \times \hat{r}}{r^2}$ (19-7)
ferromagnetic	Ferromagnetic materials, such as iron, become magnetized in the presence of another magnet but retain their magnetic properties after the external magnet is removed.	
gauss	Magnetic field is often specified in units of gauss; 1 G is equivalent to 10^{-4} T.	G
Lorentz force law	The Lorentz force law describes the force a moving, charged object experiences when immersed in a magnetic field $\vec{B}$. The force depends on the charge q of the object and its velocity $\vec{v}$, as well as the angle between $\vec{v}$ and $\vec{B}$.	$\vec{F} = q\vec{v} \times \vec{B}$ (19-1)
magnetic dipole	A magnetic dipole, for example, a bar magnet, is formed from two opposite poles, commonly called "north" and "south." By convention, the field lines of a magnetic dipole point toward the south pole and away from the north pole.	
magnetic field	Moving charge gives rise to a magnetic field. The field represents the force that another moving charge experiences as a result of the first. A ferromagnet also gives rise to a magnetic field, and it can exert a magnetic force on both other ferromagnets and moving charges.	$\vec{B}$
magnetic force	An electric charge experiences a magnetic force when immersed in a magnetic field, but *only when it is moving*.	
permeability of free space	Permeability is related to how a magnetic field affects a region of space or a material. The quantity permeability is the magnetic analog to permittivity for electric fields.	μ_0
solenoid	A straight, helical coil of wire, called a solenoid, produces a relatively uniform magnetic field when a current is in the wire.	
tesla	The tesla is the SI unit of magnetic field strength.	T

QUESTIONS AND PROBLEMS

In a few problems, you are given more data than you actually need; in a few other problems, you are required to supply data from your general knowledge, outside sources, or informed estimate.

Interpret as significant all digits in numerical values that have trailing zeros and no decimal points.

For all problems, use $g = 9.8$ m/s² for the free-fall acceleration due to gravity. Neglect friction and air resistance unless instructed to do otherwise.

• Basic, single-concept problem
•• Intermediate-level problem, may require synthesis of concepts and multiple steps
••• Challenging problem
SSM *Solution is in Student Solutions Manual*

Conceptual Questions

1. •You are given three iron rods. Two of them are magnets but the third one is not. How could you use the two magnets to find that the third rod is not magnetized? SSM

2. •If a magnetic field exerts a force on charged particles, is it capable of doing work on the particles? Explain your answer.

3. •A current-carrying wire is in a region where there is a magnetic field, but there is no magnetic force acting on the wire. How can this be?

4. •How is it possible for an object that experiences no net magnetic force to experience a net magnetic torque?

5. ••A velocity selector consists of crossed electric and magnetic fields, with the magnetic field directed straight up. A beam of positively charged particles passing through the velocity selector from left to right is undeflected by the fields. (a) In what direction is the electric field? (b) The direction of the particle beam is reversed so that it travels from right to left. Is it deflected? If so, in what direction? (c) A beam of electrons (negatively charged) moving with the same speed is passed through from left to right. Is it deflected? If so, in what direction? SSM

6. •Physicists refer to crossed electric and magnetic fields as a *velocity selector*. In the same sense, the deflection of charged particles in a strong magnetic field perpendicular to their motion can be thought of as a *momentum selector*. Why?

7. •A long, straight wire is carrying current and is placed in a cubic region that has a uniform magnetic field as shown in Figure 19-39. Does the force on the wire depend on the width of the magnetic field? Explain your answer.

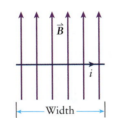

Figure 19-39 Problem 7

8. •A very flexible helical coil is suspended as shown in Figure 19-40. What will happen when a sizable current is sent through the coil? Explain your answer.

9. •What are the similarities and differences between the Biot–Savart law and Coulomb's law?

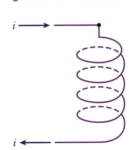

Figure 19-40 Problem 8

10. •In a lightning strike, there is a negative charge moving rapidly from a cloud to the ground. In what direction is a lightning strike deflected by Earth's magnetic field?

11. •In telephone lines, two wires carrying currents in opposite directions are twisted together. How does this reduce the magnetic fields surrounding the wires? SSM

12. •A power cord for an electronic device consists of two parallel straight wires carrying currents in opposite directions. Is there any force between them? Explain your answer.

13. •Parallel wires exert magnetic forces on each other. What about perpendicular wires? Explain your answer.

Multiple-Choice Questions

14. •The magnetic force on a charged moving particle
  A. depends on the sign of the charge on the particle.
  B. depends on the magnetic field at the particle's instantaneous position.
  C. is in the direction which is mutually perpendicular to the direction of motion of the charge and the direction of the magnetic field.
  D. is proportional both to the charge and to the magnitude of the magnetic field.
  E. is described by all of the above options, A through D.

15. •A proton traveling to the right enters a region of uniform magnetic field that points into the page. When the proton enters this region, it will be
  A. deflected out of the plane of page.
  B. deflected into the plane of page.
  C. deflected toward the top of the page.
  D. deflected toward the bottom of the page.
  E. unaffected in its direction of motion. SSM

16. •An electron is moving northward in a magnetic field. The magnetic force on the electron is toward the northeast. What is the direction of the magnetic field?
  A. up
  B. down
  C. west
  D. south
  E. This situation cannot exist, because of the orientation of the velocity and force vectors.

17. •A proton with a velocity along the $+x$ axis enters a region where there is a uniform magnetic field $\vec{B}$ in the $+y$ direction. You want to balance the magnetic force with an electric field so that the proton will continue along a straight line. The electric field should be in the
  A. $+x$ direction.
  B. $-x$ direction.
  C. $+z$ direction.
  D. $-z$ direction.
  E. $-y$ direction.

18. •A circular flat coil that has N turns, is enclosed in area A, and carries a current i has its central axis parallel to a uniform magnetic field $\vec{B}$ in which it is immersed. The net force on the coil is
  A. zero.
  B. $NIAB$.
  C. NIB.
  D. IBA.
  E. NIA.

This page is intentionally left blank.

For complete end of chapter problem sets, please go to
www.whfreeman.com/kestentauck

Problems

19-1: Magnetic Force and Magnetic Field

32. •Convert the units for the following expressions for magnetic fields as directed:

(a) $5.00 \text{ T} = \underline{\hspace{1cm}} \text{G}$ (b) $25,000 \text{ G} = \underline{\hspace{1cm}} \text{T}$

(c) $7.43 \text{ mG} = \underline{\hspace{1cm}} \mu\text{T}$ (d) $1.88 \text{ mT} = \underline{\hspace{1cm}} \text{G}$

33. •Determine the directions of the magnetic forces that act on the positive charges moving in the magnetic fields, as shown in **Figure 19-42**. SSM

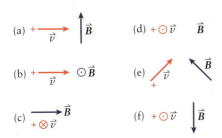

Figure 19-42 Problem 33

34. •Determine the direction of the missing vector, $\vec{v}$, $\vec{B}$, or $\vec{F}$, in the following scenarios (**Figure 19-43**). All moving charges are positive.

(a) $\vec{v} \rightarrow$ $\uparrow \vec{F}$ (d) $\odot \vec{v}$ $\otimes \vec{F}$

(b) $\leftarrow \vec{v}$ $\otimes \vec{F}$ (e) $\uparrow \vec{B}$ $\leftarrow \vec{F}$

(c) $\rightarrow \vec{F}$ (f) $\odot \vec{B}$ $\downarrow \vec{F}$
  $\otimes \vec{v}$

Figure 19-43 Problem 34

35. •A $+1$ C charge moving at 1 m/s makes an angle of $45°$ with a uniform, 1 T magnetic field. What is the magnitude of the magnetic force that the charge experiences?

36. •An electron is moving with a speed of 18 m/s in a direction parallel to a uniform magnetic field of 2.0 T. What are the magnitude and direction of the magnetic force on the electron?

37. •A proton travels with a speed of 18 m/s toward the top of the page through a uniform magnetic field of 2.0 T directed into the page as shown in **Figure 19-44**. What are the magnitude and direction of the magnetic force on the proton? SSM

Figure 19-44 Problem 37

38. •A proton is propelled at 2×10^6 m/s perpendicular to a uniform magnetic field. If it experiences a magnetic force of 5.8×10^{-13} N, what is the strength of the magnetic field?

39. ••An electron moves with velocity $\vec{v} = (1000\hat{x} - 4000\hat{y})$ m/s in a magnetic field $\vec{B} = (-0.10\hat{x} + 0.20\hat{y})$ T. Calculate the magnitude and direction of the magnetic force on the electron.

40. •A proton moves in a circle with a speed of 280 m/s through a uniform magnetic field of 2.0 T directed into the page as shown in **Figure 19-45**. What is the orbital radius R?

41. •An electron moves with a velocity of 10^7 m/s in the x-y plane at an angle of $45°$ to both the $+x$ and $+y$ axes. There is a magnetic field of 3.0 T in the $+y$ direction. Calculate the magnetic force (magnitude and direction) on the electron. SSM

Figure 19-45 Problem 40

42. ••There is a uniform magnetic field of magnitude 2.2 T in the $+z$ direction. Find the force on a particle of charge -1.2 nC if its velocity is (a) 1.0 km/s in the y–z plane in a direction that makes an angle of $40°$ with the z axis and (b) 1.0 km/s in the x–y plane in a direction that makes an angle of $40°$ with the x axis.

43. ••A beam of protons is directed in a straight line along the $+z$ direction through a region of space in which there are crossed electric and magnetic fields. If the electric field is 500 V/m in the $-y$ direction and the protons move at a constant speed of 10^5 m/s, what must be the magnitude and direction of the magnetic field such that the beam of protons continues along its straight-line trajectory?

44. ••A beam of particles (each particle has a charge $q = -2e$ and a kinetic energy of 4.00×10^{-13} J) is deflected by the magnetic field of a bending magnet as shown in **Figure 19-46**. The radius of curvature of the beam is 20.0 cm and the strength of the magnetic field is 1.50 T. (a) What is the mass of the particles making up the beam? (b) Sketch the path for the given ions in the given magnetic field as a reference path. Then sketch a path for a more massive doubly ionized negative ion, and a less massive doubly ionized positive ion, both with the same speed as the given ions, for comparison.

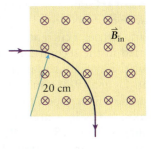

Figure 19-46 Problem 44

19-2: Magnetic Force on a Current

45. •A straight segment of wire 35.0 cm long carrying a current of 1.40 A is in a uniform magnetic field. The segment makes an angle of 53° with the direction of the magnetic field. If the force on the segment is 0.200 N, what is the magnitude of the magnetic field? SSM

46. •A straight wire of length 0.5 m is conducting a current of 2 A and makes an angle of 30° with a 3 T uniform magnetic field. What is the magnitude of the force exerted on the wire?

47. •A wire of length 0.5 m is conducting a current of 8.0 A in the $+x$ direction through a 4.0-T uniform magnetic field directed parallel to the wire. What are the magnitude and direction of the magnetic force on the wire?

48. •A wire of length 0.5 m is conducting a current of 8.0 A toward the top of the page and through a 4.0 T uniform magnetic field directed into the page as shown in **Figure 19-47**. What are the magnitude and direction of the magnetic force on the wire?

Figure 19-47 Problem 48

49. •A straight wire is positioned in a uniform magnetic field so that the maximum force on it is 4.0 N. If the wire is 80 cm long and carries a current that is 2 A, what is the magnitude of the magnetic field?

50. •A 1.5-m length straight wire experiences a maximum force of 2.0 N when in a uniform magnetic field that is 1.8 T. What current must be passing through it?

51. •A long wire stretches along the y axis and carries a 1.0 A current in the $+y$ direction. The wire is in a uniform magnetic field $\vec{B} = (0.10\hat{x} - 0.20\hat{y} + 0.30\hat{z})$. Calculate the components of the force on the wire per meter of length. SSM

52. •••**Calc** A segment of wire that has a total length L and carries a current i is situated in the second quadrant of the x–y plane (as shown in **Figure 19-48**), making an angle θ with respect to the $-x$ axis. It lies in a nonuniform

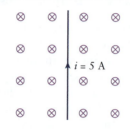

Figure 19-48 Problem 52

magnetic field $\vec{B} = -(B_0/L)x\hat{z}$, which is in the positive z direction since $x < 0$ for the wire. (a) What is the magnetic force on the wire? (b) How would your approach to this problem change if $\vec{B}$ were a function of y instead of x?

53. ••What is the torque on a round loop of wire that carries a current of 100 A, has a radius of 10 cm, and makes an angle of 30° with a magnetic field of 0.244 T? Describe how the answer changes if the angle decreases to 10° and increases to 50°. SSM

54. ••A square loop 10 cm on a side consists of 100 turns of wire that experiences a minimum torque of zero and a maximum torque of 0.045 N · m. If the current in the loop is 2.82 A, calculate the magnetic field strength that the loop rests in.

19-3: Magnetic Field and Current—the Biot–Savart Law

55. ••A long, straight wire carries current in the $+x$ direction. Determine the direction of the magnetic field due to the current at the points O, P, Q, and R (**Figure 19-49**).

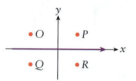

Figure 19-49 Problem 55

56. ••A long, straight wire carries current in the $+z$ direction (out of the page). Determine the direction of the magnetic field due to the current at the points O, P, Q, and R (**Figure 19-50**).

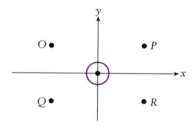

Figure 19-50 Problem 56

57. •••**Calc** Calculate the net magnetic field (magnitude and direction) at the point P due to the two currents in **Figure 19-51**. SSM

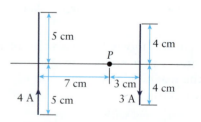

Figure 19-51 Problem 57

58. •••**Calc** Derive an expression for the magnetic field at a point X on the positive x axis due to a long, straight wire that carries a current i straight along the $+y$ direction (**Figure 19-52**). Integrate the current using the variable y ($-\infty < y < +\infty$) and be sure to include the direction of the field.

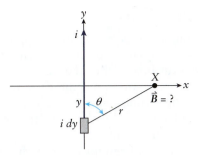

Figure 19-52 Problem 58

59. •••Calc Repeat problem 19-58. However, this time, integrate over the angle θ.

60. •••Calc A straight wire segment that is 20 cm long carries 2 A of current along the $+y$ axis, as shown in **Figure 19-53**. Calculate the magnetic field at the point P on the x axis, 5 cm from the origin.

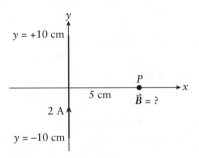

Figure 19-53 Problem 60

61. ••Derive an expression for the magnetic field at the point C at the center of the circular, current-carrying wire segments shown in **Figure 19-54**. SSM

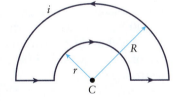

Figure 19-54 Problem 61

62. ••Using Figure 19-55, derive an expression for the magnetic field at the point C located at the center of the two circular, current-carrying arcs and the connecting radial lines. Assume the radii of the small and large arcs are r_1 and r_2, respectively.

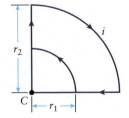

Figure 19-55 Problem 62

19-4: Magnetic Field and Current— Ampère's Law

63. ••Calc There is a current i around the rectangular loop as shown in **Figure 19-56**. Evaluate the integral $\oint \vec{B} \cdot d\vec{l}$ for the paths A, B, C, and D.

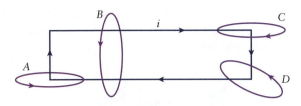

Figure 19-56 Problem 63

64. ••Calc Using Ampère's law, derive an expression for the magnetic field at a point that is a radial distance of R from a long, straight wire carrying a current i. Be sure to specify the direction as well as the magnitude.

65. ••Jerry wants to predict the magnetic field that a high voltage line creates in his apartment. A current of 100 A passes through a wire that is 5 m from his window. Use Ampère's law to calculate the magnetic field. How does the field compare to the magnitude of Earth's magnetic field of about 5×10^{-5} T in New York, New York? SSM

66. •A solenoid with 25 turns per centimeter carries a current of 25 mA. What is the magnetic field in the interior of the coils?

67. •You want to wind a solenoid that is 3.5 cm in diameter, is 16 cm long, and will have a magnetic field of 0.0250 T when a current of 3 A is in it. What total length of wire do you need?

68. ••Calc Using Ampère's law, calculate the magnetic field inside and outside of a toroidal coil with a radius of 25 cm. There are 2 turns per centimeter wrapped around the cylinder that comprises the torus (the cylinder is 2 cm in diameter). The current in each wire is 1.25 A.

69. ••Calc A coaxial cable consists of a solid inner conductor of radius R_i, surrounded by a concentric outer conducting shell of radius R_o. Insulating material fills the space between the conductors. The inner conductor carries current to the right, and the outer conductor carries the same current to the left down the outer surface of the cable, as shown in **Figure 19-57**. Using Ampère's law, derive an expression for the magnetic field in three separate regions of space: inside the inner conductor, between the two conductors, and outside of the outer conductor. SSM

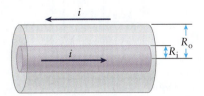

Figure 19-57 Problem 69

70. ••Calc Using Ampère's law, derive an expression for the magnetic field inside the hollow cylinder of radius R

that carries a current i tangentially, on the surface of the cylinder (**Figure 19-58**). Repeat to find the magnetic field outside the cylinder.

Figure 19-58 Problem 70

71. ••**Calc** Using Ampère's law, calculate the magnitude of the magnetic field at a perpendicular distance of 2.2 m from a long, copper pipe that has a diameter of 2 cm and carries a current of 20 A.

19-5: Magnetic Force between Current-Carrying Wires

72. •Using the right-hand rule, verify the assertion that parallel currents attract and opposite currents repel. *Hint:* Find the magnetic field of wire 1 at the location of wire 2, then find the force on wire 2 due to the field.

73. •Find the force (magnitude and direction) acting on a 1-cm section of wire 1 due to wire 2, as shown in **Figure 19-59**. SSM

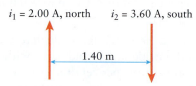

$i_1 = 2.00$ A, north $i_2 = 3.60$ A, south

1.40 m

Figure 19-59 Problem 73

74. •What is the net force on the rectangular loop of wire that is 2 cm wide, is 6 cm long, and is located 2 cm from a long straight wire that carries 40 A of current as shown in **Figure 19-60**? Assume a current of 20 A is in the loop.

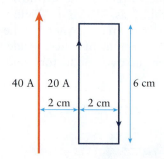

40 A   20 A   6 cm

2 cm   2 cm

Figure 19-60 Problem 74

75. ••The fasteners on overhead power lines are 50 cm long. What force must they be able to withstand if two high voltage lines are 2 m apart, each carrying 2500 A in the same direction?

General Problems

76. ••Horizontal electric power lines supported by vertical poles can carry large currents. Assume that Earth's magnetic field runs parallel to the surface of the ground from south to north with a magnitude of 0.50×10^{-4} T and that the supporting poles are 32 m apart. Find the magnitude and direction of the force that Earth's magnetic field exerts on a 32-m segment of wire carrying 95 A if the current runs (a) from north to south, (b) from east to west, or (c) toward the northwest making an angle of 30° north of east. (d) Are any of the above forces large enough to have an appreciable effect on the power lines?

77. ••A levitating train is three cars long (180 m) and has a mass of 100 metric tons (1 metric ton = 1000 kg). The current that runs in the superconducting wires is about 500 kA and even though the traditional design calls for many small coils of wire, assume for this problem that there is a 180-m-long wire carrying the current. Find the size of the magnetic field needed to levitate the train.

78. ••An electron and a proton have the same kinetic energy on entering a region of constant magnetic field and their velocity vectors are perpendicular to the magnetic field. Suppose the magnetic field is strong enough to allow the particles to circle in the field. What is the ratio of the radii of their circular paths r_p/r_e?

79. •••In **Figure 19-61**, which shows a mass spectrometer, an electron enters a region of magnetic field that has a strength of 0.00242 T, into the page. The velocity of the electron is confirmed with a velocity selector (a region of crossed electric and magnetic fields). The electric field is 90,000 V/m (down) and the magnetic field is 0.0053 T (into page) in the velocity selector. If the radius of curvature of the electron is 4.00 cm, calculate the mass of the electron. SSM

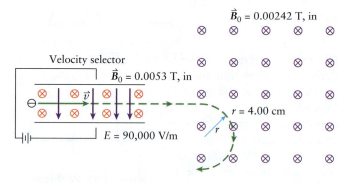

Figure 19-61 Problem 79

80. ••**Biology** The National High Magnetic Field Laboratory holds the world record for creating the largest magnetic field. Their long-pulse magnet produced a magnetic field of 60 T. To see if such a strong magnetic field could pose health risks for nearby workers, calculate the maximum acceleration the field could produce on Na^+

ions (of mass 3.8×10^{-26} kg) in blood traveling through the aorta. The speed of blood is highly variable, but 50 cm/s is reasonable in the aorta. Does your result indicate that it would be dangerous to expose workers to such a large magnetic field?

81. ••During electrical storms, a bolt of lightning can transfer 10 C of charge in 2.0 μs (the amount and time can vary considerably). We can model such a bolt as a very long current-carrying wire. (a) What is the magnetic field 1.0 m from such a bolt? What is the field 1.0 km away? How do the fields compare with Earth's magnetic field? (b) Compare the fields in part (a) with the magnetic field produced by a typical household current of 10 A in a very long wire at the same distances from the wire as in (a). (c) How close would you have to get to the wire in part (b) for its magnetic field to be the same as the field produced by the lightning bolt at 1.0 km from the bolt?

82. ••A straight wire carries a current of 8 A toward the top of the page. What are the magnitude and direction of the magnetic field at point P, which is 8 cm to the right of the wire as shown in Figure 19-62?

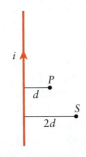

Figure 19-62 Problem 82

83. ••A long straight wire carries a current as shown in Figure 19-63. A charged particle moving parallel to the wire experiences a force of 0.8 N at point P. Assuming the same charge and same velocity, what would be the magnitude of the magnetic force on the charge at point S?

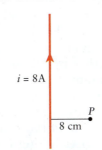

Figure 19-63 Problem 83

84. ••Two long, straight wires parallel to the x axis are at $y = \pm 2.5$ cm (Figure 19-64). Each wire carries a current of 16 A in the $+x$ direction. Find the magnetic field on the y axis at (a) $y = 0$, (b) $y = 1$ cm, and (c) $y = 4$ cm.

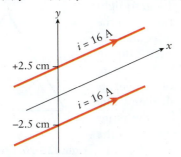

Figure 19-64 Problem 84

85. •••Calc A thin ring of radius R lies with its center at the origin in the $x-y$ plane (Figure 19-65). It contains a total charge of Q uniformly spread over its nonconducting surface. The ring begins to spin at ω rad/s about its center. Calculate the magnetic field due to the rotating ring at the point P a distance of Z above the center of the ring. SSM

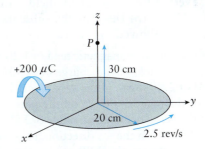

Figure 19-65 Problem 85

86. •••Calc A thin disk with a radius of 20 cm contains +200 μC of charge, uniformly distributed over its surface (Figure 19-66). The disk is rotated at 2.5 rev/s about its central axis. Find the magnetic field at the point P, a height of 30 cm above the center of the rotating disk.

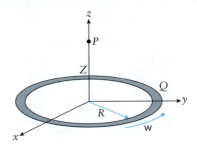

Figure 19-66 Problem 86

87. ••Medical Transcranial magnetic stimulation (TMS) is a noninvasive method to stimulate the brain using magnetic fields. It is used in treating strokes, Parkinson's disease, depression, and other physical conditions. In the procedure, a circular coil is placed on the side of the forehead to generate a magnetic field inside the brain. Although values can vary, a typical coil would be about 15 cm in diameter and contain 250 thin circular windings. The magnetic field in the cortex (3.0 cm from the coil measured along a line perpendicular to the coil at its center) is typically 0.50 T. (a) What current in the coil is needed to produce the desired magnetic field inside the brain? (b) What is the magnetic field at the center of the coil at the forehead? (c) If the current needed in part (a) seems too large, how could you easily achieve the same magnetic field with a smaller current?

88. ••A wire of mass 40 g slides without friction on two horizontal conducting rails spaced 0.8 m apart. A steady current of 100 A is in the circuit formed by the wire and the rails. A uniform magnetic field of 1.2 T, directed into

the plane of the drawing, acts on it. (a) In which direction in **Figure 19-67** will the wire accelerate? (b) What is the magnetic force on the wire? (c) How long must the rails be if the wire, starting from rest, is to reach a speed of 200 m/s? (d) If the magnetic field were directed out of the page, how would your answers differ? (e) What if the magnetic field were directed toward the top of the drawing?

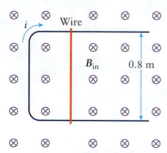

Figure 19-67 Problem 88

89. ••A small 20-turn current loop with a 4.00-cm diameter is suspended in a region with a magnetic field of 1000 G, with the plane of the loop parallel with the magnetic field direction. (a) What is the current in the loop if the torque exerted by the magnetic field on the loop is 4.00×10^{-5} N·m? (b) Describe the subsequent motion of the loop if it is allowed to rotate.

90. ••There is a uniform magnetic field of 1.20 T in the $+z$ direction. (a) A particle with a charge of -2.00 μC moves with a speed of 2.20 km/s in the $y-z$ plane. Calculate the magnitude of the magnetic force on the particle when the direction of the velocity makes an angle of 55° with the $+y$ direction and 35° with the $-z$ direction. (b) A second particle, also with a charge of -2.00 μC, moves with a speed of 2.20 km/s in the $x-y$ plane. Calculate the magnitude of the magnetic force on the particle when the direction of the velocity makes an angle of 55° with the $+y$ direction and an angle of 35° with the $+x$ direction.

91. •••Two straight conducting rods, which are 1.0 m long, exactly parallel, and separated by 0.85 mm, are connected by an external voltage source and a 17 Ω resistance, as shown in **Figure 19-68**. The 0.5 Ω rod "floats" above the 2.5 Ω rod, in equilibrium. If the mass of each rod is 25 g, what must be the potential of the voltage source? SSM

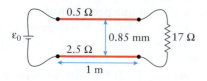

Figure 19-68 Problem 91

92. ••Medical When operated on a household 110-V line, typical hair dryers draw about 1650 W of power. We can model the current as a long straight wire in the

handle. During use, the current is about 3.0 cm from the user's head. (a) What is the current in the dryer? (b) What is the resistance of the dryer? (c) What magnetic field does the dryer produce at the user's head? Compare the field with Earth's magnetic field to decide if we should have health concerns about the magnetic field created when using a hair dryer.

93. ••Three very long, straight wires lie at the corners of a square of side d, as shown in **Figure 19-69**. The magnitudes of the currents in the three wires are the same, but the two diagonally opposite currents are directed into the page while the other one is directed outward. Calculate the magnetic field (magnitude and direction) at the fourth corner of the square.

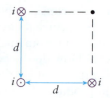

Figure 19-69 Problem 93

94. ••A 2.0-m lamp cord leads from the 110 V outlet to a lamp having a 75 W lightbulb. The cord consists of two insulated parallel wires 4.0 mm apart and held together by the insulation. One wire carries the current into the bulb and the other carries it out. What is the magnitude of the magnetic field the cord produces (a) midway between the two wires and (b) 2.0 mm from one of the wires in the same plane in which the two wires lie? (c) Compare each of the fields in parts (a) and (b) with Earth's magnetic field (0.5×10^{-4} T). (d) What magnetic force (magnitude and direction) do the two wires exert on one another? Is the force large enough to stress the insulation holding the wires together?

95. ••Some people have raised concerns about the magnetic fields produced by current-carrying high voltage lines in residential neighborhoods. Currents in such lines can be up to 100 A. Suppose you have such a line near your house. If the wires are supported horizontally 5.0 m above the ground on vertical poles and your living room is 12 m from the base of the poles, what magnetic field strength does the wire produce in your living room if it carries 100 A? Express your answer in teslas and as a multiple of Earth's magnetic field. Does the magnetic field from such wires seem strong enough to cause health concerns? SSM

96. •••Three long, straight wires are positioned on the vertices of an equilateral triangle of side 2.00 m (**Figure 19-70**). Each wire carries 40 A of current as directed. Calculate the magnetic field at the point P (in the center of the triangle).

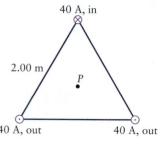

Figure 19-70 Problem 96

97. ••Helmholtz coils are composed of two coils of wire that have their centers on the same axis, separated by a distance that is equal to the radius of the coils (**Figure 19-71**). The coils have N turns of wire that carry a current of i in the same direction. Derive expressions for the net magnetic field due to the coils at the points (a) $x = R/2$ and (b) $x = 2R$.

98. ••Geophysicists may use the gauss unit for magnetic field (10^4 G = 1 T). Earth's magnetic field at the equator can be taken as 0.7 G directed north. At the center of a flat circular coil that has 10 turns of wire and is 1.4 m in diameter, the coil's magnetic field exactly cancels Earth's field. (a) What must be the current in the coil? (b) How should the coil be oriented?

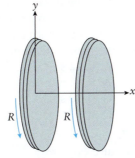

Figure 19-71 Problem 97

99. ••**Biology** Migratory birds use Earth's magnetic field to guide them. Some people are concerned that human-caused magnetic fields could interfere with bird navigation. Suppose that a pair of parallel high voltage lines, each carrying 100 A, are 3.00 m apart and lie in the same horizontal plane. Find the magnitude and direction of the magnetic field the lines produce at a point 15.0 m above them equidistant from both lines in each of the following cases. (a) The lines run in the north–south direction and both currents run from north to south. (b) Both lines run in the north–south direction and the current in the eastern line runs northward while the current in the western line runs southward. (c) The lines run in the east–west direction and both currents run from west to east. (d) Is it reasonable to think that the fields caused by the wires are likely to interfere with bird migration?

100. ••**Medical** Magnetoencephalography (MEG) is a technique for measuring changes in the magnetic field of the brain caused by external stimuli such as touching the body or viewing images of food. Such a change in the field occurs due to electrical activity (current) in the brain. During the process, magnetic sensors are placed on the skin to measure the magnetic field at that location. Typical field strengths are a few femtoteslas (1 femtotesla = 1 fT = 10^{-12} T). An adult brain is about 140 mm wide, divided into two sections (called "hemispheres" although the brain is not truly spherical) each about 70 mm wide. We can model the current in one hemisphere as a circular loop, 65 mm in diameter, just inside the brain. The sensor is placed so that it is along the axis of the loop 2.0 cm from the center. A reasonable magnetic field is 5.0 fT at the sensor. According to this model, (a) what is the current in the brain and (b) what is the magnetic field at the center of the hemisphere of the brain?

101. ••A long, straight wire carries a current of 1.2 A toward the south. A second, parallel wire carries a current of 3.8 A toward the north and is 2.8 cm from the first wire. What is the magnetic force per unit length each wire exerts on the other?

MCAT® Section
The section that follows includes material from previously administered MCAT® items and is reprinted with permission of the Association of American Medical Colleges (AAMC).

Passage II (Questions 74–80)
An electromagnetic railgun is a device that can fire projectiles by using electromagnetic energy instead of chemical energy. A schematic of a typical railgun is shown below.

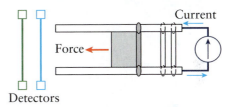

The operation of the railgun is simple. Current flows from the current source into the top rail, through the movable conducting armature into the bottom rail, then back to the current source. The current in the two rails produces a magnetic field directly proportional to the amount of current. This field produces a force on the charges moving through the movable armature. The force pushes the armature and the projectile along the rails.

The force is proportional to the square of the current running through the railgun. For a given current, the force and the magnetic field will be constant along the entire length of the railgun. The detectors placed outside the railgun give off signal when the projectile passes them. This information can be used to determine the exit speed and kinetic energy of the projectile. Projectile mass, rail current, and exit speed are listed in the table below.

Projectile mass (kg)	Rail current (A)	Exit speed (km/s)
0.01	10.0	2.0
0.01	15.0	3.0
0.02	10.0	1.4
0.04	10.0	1.0

74. Which of the following graphs best represents the dependence of exit speed on projectile mass?

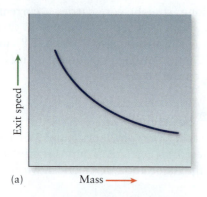

(a)   Mass ⟶

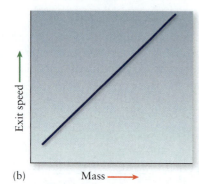

(b)   Mass ⟶

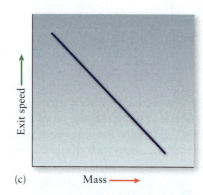

(c)   Mass ⟶

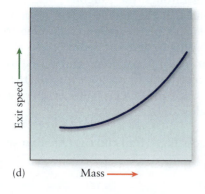

(d)   Mass ⟶

75. What change made to the railgun would reduce power consumption without lowering the exit speeds?
  A. lowering the rail current
  B. lowering the rail resistivity
  C. lowering the rail cross-sectional area
  D. reducing the magnetic field strength

76. Starting from a resting position at the right end of the rail gun, the armature applies a constant force of 3.0 N to a projectile with mass of 0.06 kg. How long will it take for the projectile to move 1.0 m?
  A. 0.02 s
  B. 0.04 s
  C. 0.20 s
  D. 0.40 s

77. Lengthening the rails would increase the exit speed because of
  A. an increased rail resistance.
  B. a stronger magnetic field between the rails.
  C. a larger force on the armature.
  D. a longer distance over which the force is present.

78. For a given mass, if the current were decreased by a factor of 2, the exit speed would
  A. increase by a factor of 2.
  B. increase by a factor of $\sqrt{2}$.
  C. decrease by a factor of $\sqrt{2}$.
  D. decrease by a factor of 2.

79. If a projectile leaves a railgun near the surface of Earth with a speed of 2.0 km/s horizontally, how far will it fall in 0.01 s?
  A. 4.9×10^{-4} m
  B. 4.9×10^{-2} m
  C. 9.8×10^{-2} m
  D. 2.0×10^{1} m

80. If a projectile with a mass of 0.1 kg accelerates from a resting position to a speed of 10 m/s in 2 s, what is the average power supplied by the railgun to the projectile?
  A. 0.5 W
  B. 2.5 W
  C. 5.0 W
  D. 10.0 W

20 Magnetic Induction

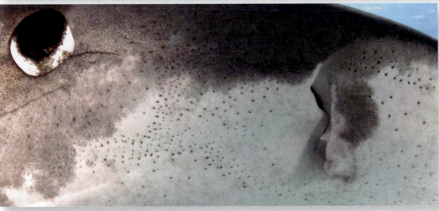

(Pacific Stock / SuperStock)

The shark's large, sharp teeth probably instantly draw your attention, but a physicist might be more interested in the ducts marked by large pores on the shark's snout (see inset). The ducts, called *ampullae of Lorenzini,* are gel-filled, electrically conductive channels leading to sensory cells that enable sharks to detect changes in magnetic field of the order of 0.2×10^{-7} T, or about 0.04% of Earth's field. As sharks move from one place to another, small variations in the magnetic field induce a potential difference across the ampullae of Lorenzini, causing a current. Sensory neurons respond to the current, thus enabling the shark to use small, local changes in Earth's magnetic field as a map by which to navigate.

The nineteenth-century English physicist Michael Faraday first described the phenomenon by which an electric potential arises when the magnetic field in a region changes. The induced potential, which in turn may give rise to a current when the region contains conducting material, provides the means for the magnetic brakes found in electric trains. On a smaller scale, it is possible that sharks, skates, and rays use induced current generated by small variations in magnetic field to navigate. On a smaller scale still, changes in the minute magnetic fields that surround hydrogen atoms can be detected as an induced current in a nearby wire coil. The process by which such changes are created and recorded is magnetic resonance imaging (MRI). All are examples of the application of *magnetic induction*, which we explore in this chapter.

20-1 Faraday's Law of Induction

Let's take a step back to reconsider the magnetic force on a wire carrying a current. The top part of **Figure 20-1** describes what happens when we put a long, straight wire carrying a current into a uniform magnetic field $\vec{B}_0$. Here $\vec{B}_0$ points from left to right, as shown on the left side of the top part of the figure. The current-carrying wire is represented by the blue circle in the middle of the upper right side of the figure; the blue dot within it indicates that the current is coming out of the page. According to the right-hand rule, the current gives rise to a magnetic field $\vec{B}_{\text{wire}}$ that circulates counterclockwise around the wire in a direction as described by the right-hand rule. (To apply the right-hand rule to find the direction of the field, make a fist and then point your right thumb out of the page, in the direction of the current. Notice that your other fingers curl in the direction of the magnetic field.) We have drawn the external, uniform field and the field due to the wire separately to show more clearly how to obtain the net field, shown at the bottom right of the figure.

The net magnetic field in the region near the wire is the vector sum of $\vec{B}_0$ and $\vec{B}_{\text{wire}}$, as suggested by the addition and equal signs in Figure 20-1. In the region above the wire, $\vec{B}_0$ and $\vec{B}_{\text{wire}}$ are in opposite directions, so they partially or completely cancel. (The fields completely cancel at the point marked with a red dot. Below that the field points right to left, and above it left to right.) The cancellation makes the net field lower than $\vec{B}_0$ in the region immediately above the wire. The opposite occurs below the wire, where $\vec{B}_0$ and $\vec{B}_{\text{wire}}$ are in the same direction and therefore reinforce each other. The addition makes the net magnetic field stronger below the wire than above it, as the higher density of field lines below the wire indicates. Although not pertinent to this discussion, note that $\vec{B}_{\text{wire}}$ is relatively large close to the wire, so the net field there is mostly due to $\vec{B}_{\text{wire}}$, but as the distance from the wire increases, $\vec{B}_{\text{wire}}$ gets smaller and smaller, making the sum of $\vec{B}_0$ and $\vec{B}_{\text{wire}}$ approach $\vec{B}_0$ alone. For these reasons the net field is somewhat circular near the wire, and is uniform far from it.

The shape of the net field above and below the wire in the lower part of Figure 20-1 suggests an interaction between the two fields. You can imagine, as did the nineteenth-century English physicist Michael Faraday, that the lines of field are like elastic bands. As shown in the lower part of the figure, below the wire these bands

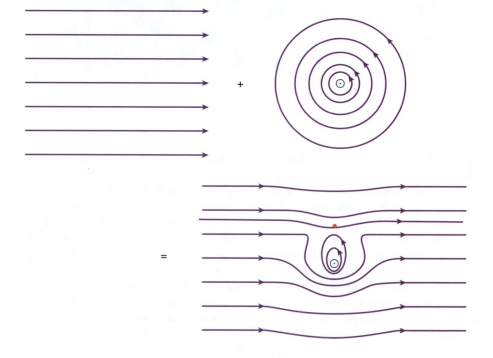

Figure 20-1 A long, straight wire carrying a constant current is immersed in a uniform magnetic field. At every point in space the net field, shown in the lower part of the figure, is the vector sum of the external field (upper left) and the field due to the current in the wire (upper right).

are "stretched" considerably more than above; in order to relax to a shorter, less-stretched length the bands would straighten, resulting in the wire being pushed up. This visualization helps us to see that the magnetic force exerted on the current-carrying wire can be understood as the interaction of two magnetic fields, the external field $\vec{B}_0$ and that of the wire $\vec{B}_{\text{wire}}$. We have previously treated this force as the interaction between a magnetic field and a current; the alternate view suggested by Figure 20-1 is that the force is an interaction between the external field and the magnetic field produced by the current. You can check that the first view gives the same result by using the right-hand rule to find the direction of the force using $\vec{F} = i\vec{l} \times \vec{B}$ (Equation 19-3). The direction of the force is the direction of the cross product of $i\vec{l}$ and $\vec{B}$. So point your index finger out of the page, in the direction of the current, and your other fingers to the right, in the direction of the magnetic field. Your thumb points upward, in the direction of the force.

❗ Watch Out

A wire carrying a current does not experience a force due to its own magnetic field.

A current-carrying wire gives rise to a magnetic field. That field can exert a force on another wire that carries a current. The field does not, however, exert a force on the wire (current) that created it. So, for example, if there were only a single current-carrying wire present, there would be a magnetic field but no magnetic force.

Faraday imagined a magnetic field as the lines we have used to represent it in Figure 20-1; he thought that as the lines extended into a region of space they would have an effect on anything in that region. To test his idea, Faraday wound two coils of wire around an iron ring, one connected to a battery and the other to a device that measures current, as depicted in **Figure 20-2**. Faraday expected that when current flowed in the first coil, the magnetic field it generated would cause a current to flow in the second coil. (It was already known that iron supports magnetic fields better than air, so Faraday included the iron ring to improve the connection of the two coils by the field.) The ring improved the connection—but not in the way you might guess. Current appeared in the second coil at the moment the battery was connected and again when the battery was removed from the first coil, but not in between. In addition, the direction of the current as the battery was removed was opposite to the direction when the battery was first connected.

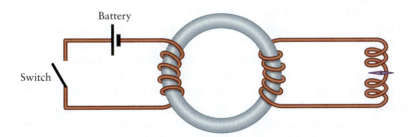

Figure 20-2 Faraday used a device like this to demonstrate that the magnetic field generated by a current in a wire coil (on the left side of the solid ring) would cause current in a nearby coil (on the right side of the ring). A compass placed near a solenoid attached to the second coil indicated the flow of charge. Current appeared in the second coil at the moment the battery was connected to the first one and again when the battery was removed, but not in between.

To understand this phenomenon, we need to define **magnetic flux**. Like electric flux (see Section 16-6), the magnetic flux is a measure of how much magnetic field, or in Faraday's language, how many field lines, pass through a given two-dimensional region. The magnetic flux Φ_B is

See the Math Tutorial for more information on Integrals

$$\Phi_B = \int \vec{B} \cdot d\vec{A} \tag{20-1}$$

Analogous to our definition of electric flux (Equation 16-29), the dot product incorporates the angle φ between the magnetic field vector and the plane of the area. That is, Equation 20-1 can be expanded to

$$\Phi_B = \int B \, dA \cos \varphi \tag{20-2}$$

The flux is largest when the magnetic field vector $\vec{B}$ is perpendicular to the plane of the area, so that $\vec{B}$ and $d\vec{A}$ are parallel, and zero when $\vec{B}$ is parallel to the plane of the area, so that $\vec{B}$ and $d\vec{A}$ are perpendicular.

Recall that the direction of the area vector of a region is perpendicular to the plane of that region. This description is only meaningful, of course, when the region lies in a plane. What if the region through which the magnetic field passes is not flat? That's where the integral comes in. As we have done in other situations, we can break up the region into pieces of area $d\vec{A}$ so infinitesimally small that each can be treated as if it were flat, and then find the flux by adding up the contribution each makes to the total. We choose the bounds of the integral to ensure that all of the infinitesimal pieces that make up the region are included in the addition.

An **induced potential** drives the current that Faraday observed in the second coil. We say the potential is induced because it does not result from some element connected to the second coil, but rather from something happening that is external to it. Here is the key aspect of the relationship between the magnetic field in the first coil and the induced potential in the second: The induced potential is equal to the *rate of change in magnetic flux* through the second coil. This observation explains why there is only current when we connect or remove the battery from the first coil. The current in the first coil, and therefore the magnetic field it creates, only changes at those two times. When the current in the first coil is steady—in between turning the current on and turning it off, or when the battery is disconnected—the field generated is also constant. When the field is constant there is no change in magnetic flux through the second coil, so there is no induced potential. With no potential difference around the coil, no current exists.

The mathematical relationship between induced potential, usually written $\mathscr{E}$, and magnetic flux is

See the Math Tutorial for more information on Calculus

$$\mathscr{E} = -\frac{d\Phi_B}{dt} \tag{20-3}$$

This equation is known as **Faraday's law of induction**.

The minus sign in Faraday's law suggests a polarity of the induced potential, which in turn determines the direction of the current. As the magnetic flux increases, $d\Phi_B/dt$ is positive and $\mathscr{E}$ is negative. In contrast, $d\Phi_B/dt$ is negative and $\mathscr{E}$ is positive as the magnetic flux decreases. We will deal more carefully with the sign in the next section, but certainly it explains Faraday's observation that the induced current has one direction when the current begins in the first coil and in the opposite direction when the current stops. At the instant that the current in the first coil begins to increase, the magnitude of the magnetic field that arises also increases. The magnetic flux through the second coil therefore increases, too, and the induced potential $\mathscr{E}$ is negative. Current is now in the second coil. The current in the first coil quickly becomes steady. At that point neither the field nor the flux through the second coil is changing, so there is no induced potential and no current in the second coil. Finally, when the current in the first coil begins to decrease, the field also decreases, so the magnetic flux through the second coil decreases. The induced potential is now positive (the negative of a negative value), so the current through the second coil is in a direction opposite to the initial current.

When a magnetic flux passes through multiple loops of a conductor, say, the windings of a coil, each separate loop contributes to the total induced potential. In such a case, we must therefore account for the total number of loops. Equation 20-3 becomes

$$\mathscr{E} = -N\frac{d\Phi_B}{dt} \qquad (20\text{-}4)$$

where N is the number of loops in the coil.

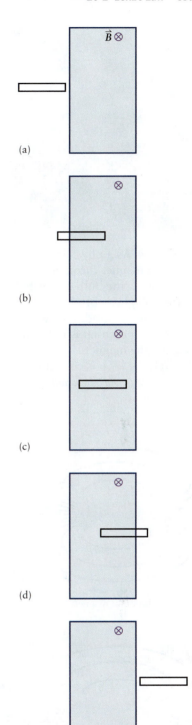

(a)

(b)

(c)

(d)

(e)

Figure 20-3 A rectangular loop moving from left to right approaches, enters, and then leaves a region of constant magnetic field that is perpendicular to the plane of the loop. At each of these moments is the flux decreasing, remaining the same, or increasing? Why?

? Got the Concept 20-1
Changing Flux

A rectangular loop passes first into, then out of, a region of constant magnetic field that points in a direction perpendicular to the plane of the loop. **Figure 20-3** shows the moments when the loop (a) is approaching the field, (b) entering the field, (c) completely inside the field, (d) leaving the field, and (e) completely out of the field. In each of the cases, is the magnetic flux through the loop decreasing, constant, or increasing? Explain your answers.

We note in passing that an electric potential arises when an electric field exists over a certain region. This observation is embodied in $V = -\int_a^b \vec{E} \cdot d\vec{s}$ (Equation 17-5). It was the Scottish physicist James Clerk Maxwell who proposed that the fundamental effect of a changing magnetic flux is the creation of an electric field rather than an electric potential or an induced current. Indeed, while charges need to be present to have a current, a changing magnetic flux will give rise to an electric field even in empty space.

✷ What's Important 20-1
A current-carrying wire gives rise to a magnetic field. That field can exert a force on another wire that carries a current, but it does not exert a force on the wire (current) that created it. A changing magnetic flux in a region gives rise to an induced potential, which in turn can create an induced current. As Maxwell observed, this is fundamentally because a changing magnetic flux gives rise to an electric field.

20-2 Lenz's Law

In this section, we turn our attention to an important detail of Faraday's law of induction, the minus sign in

$$\mathscr{E} = -\frac{d\Phi_B}{dt} \qquad (20\text{-}3)$$

The overall sign suggests that the induced potential $\mathscr{E}$ has a polarity, which in turn determines the direction of the current. More than simply a direction, however, this sign provides a clue to a fundamental aspect of nature. To examine it more closely, let's consider a straightforward application of Faraday's law.

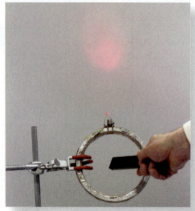

Figure 20-4 As a physicist pushes a magnet through a coil of wire, the bulb connected to the coil glows brightly. According to Faraday's law, an electric potential is induced because the magnetic flux through the loop is changing. (*Courtesy of David Tauck and Ted Szczepanski*)

The photograph in **Figure 20-4** shows a person pushing a magnet through a loop of wire connected directly to a lightbulb. As more of the magnet enters the loop, the strength of the magnetic field through the loop increases. The magnetic flux through the loop is therefore changing, so according to Faraday's law an electric potential is induced. The potential drives an induced current, and as you can see the bulb glows.

The induced current has another effect, too. A current loop gives rise to its own magnetic field. The field is a dipole, the same shape as the field due to a bar magnet, so as the current begins and the field due to the loop of current increases, a second, induced magnetic field begins to pass through the loop. In which direction is the induced field?

There are only two possibilities for the direction of the current in the loop, either clockwise or counterclockwise from the vantage point of the picture. Within the loop, the induced magnetic field points in a direction perpendicular to the plane of the loop, so it either points in the same direction as the external magnetic field or in the direction opposite to it. These two possibilities are shown in **Figure 20-5a** and **b**. We will consider separately the implications of each of the two, noting that only one—the one shown at the bottom—is correct.

Could the first possibility (Figure 20-5a) be correct? No! If it were, as the flux through the loop increased, the induced magnetic field due to the induced current in the loop would point in the same direction as the field of the bar magnet. The field due to the current loop would add to the external field, so that the net field through the loop would be larger than the field due simply to the bar magnet. As a result, the magnetic flux through the loop would increase by an amount greater than that due simply to the motion of the bar magnet. A larger increase in flux, according to Faraday, would result in an even larger induced potential. A larger potential would give rise to an even larger current than first predicted, and a larger current would give rise to an even larger magnetic field through the loop. The change in flux through the loop would then also be larger than simply due to the motion of the magnet, and so on. One small push of the magnet and, were the first possibility correct, the current could light not one, but hundreds or perhaps thousands of bulbs! Do you suspect that the lighting of thousands of bulbs can't be correct?

The second possibility (Figure 20-5b) shows what actually happens. As the flux through the loop increases, the induced magnetic field due to the induced current in the loop points in the opposite direction of the field of the bar magnet. In this case, the field due to the current loop would at least partially cancel the external field and, as a result, oppose the change in magnetic flux. In other words, as the bar magnet moves toward the loop, the flux would not increase at as high a rate as would be the case simply due to the motion of the bar magnet. Only by pushing the magnet at a much higher speed toward the loop would the induced current get appreciably larger.

The second of the possibilities is the correct one, certainly. The requirement that energy must be conserved would never allow the first, in which a small change would result in a runaway reaction that could, in the absurd limit, create an infinite current and an infinite amount of power.

The polarity of the induced potential is embodied in the minus sign in Faraday's law:

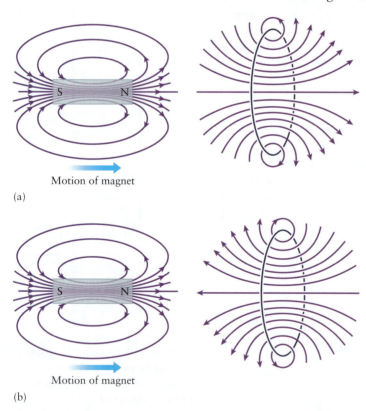

(a) Motion of magnet

(b) Motion of magnet

Figure 20-5 A bar magnet is pushed into a conducting loop. As the magnetic flux increases, does the magnetic field induced in the loop point in the same direction as the field of the bar magnet (a) or in the opposite direction (b)?

$$\mathcal{E} = -\frac{d\Phi_B}{dt} \qquad (20\text{-}3)$$

Because the direction, and therefore the sign, hints at a fundamental aspect of nature, it is appropriate to give it a name separate from the overall law. The sign in Equation 20-3 is **Lenz's law**, named after the nineteenth-century Russian physicist Heinrich Lenz. Lenz's law tells us that the induced potential and the change in flux have opposite sign. A more physical, and therefore more applicable, interpretation is

> The direction of the magnetic field induced in a conducting loop opposes the change in magnetic flux that created it.

Notice that Lenz's law concerns itself with the direction of the change in magnetic flux (is the flux increasing or decreasing?) not with the direction of the field that causes the change in flux. It is not correct to base a determination of the direction of the induced field on the direction of the external field.

Consider the conducting U-shaped wire, along which a conducting bar can slide, as shown in **Figure 20-6a**. Together, the bar and part of the wire to the right of it form a conducting loop. A constant, external magnetic field, which we have arbitrarily chosen to be directed down through the loop, is present in the entire region. When the bar is at rest, there is no current. When the bar is given a small push to the right, however, the area of the loop begins to get smaller, which means that the magnetic flux through the loop decreases. By Faraday's law of induction, the change in flux results in an induced potential, which gives rise to an induced current in the loop. That is, when the bar is moving, there is a current around the loop formed by it and the part of the wire to the right.

What is the direction of the current? To answer the question, we demand that, in keeping with the spirit of Lenz's law, the force exerted on the bar by the external field ($\vec{F} = i\vec{l} \times \vec{B}$, Equation 19-3) oppose the motion, that is, it must point to the left. Conservation of energy, as well as physical reasonability, precludes the possibility that the force points to the right, which would cause the bar to accelerate,

A conducting bar slides to the right on a "U"-shaped wire, forming a conducting loop. A magnetic field is present throughout the region.

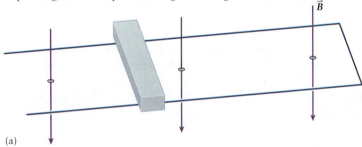

(a)

When the bar slides to the right, the magnetic flux through the area of the loop decreases; this induces a current in the loop. Looking from above, the current is clockwise. This means that the force of the field on the bar opposes, rather than reinforces, the motion.

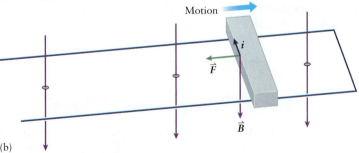

(b)

Figure 20-6 (a) A conducting bar can slide on a U-shaped wire, so that, together, the bar and the part of the wire to the right of it form a conducting loop. (b) An external magnetic field passes through the region, so that when the bar slides to the right, the magnetic flux through the conducting loop changes, which induces an electric potential in the loop. This creates a current in the bar and wire.

which would induce more current, which would result in a greater acceleration, and on and on, resulting in a runaway process that would cause the bar to attain an enormously high speed. No, the force of the field on the current-carrying bar must point to the left. According to the right-hand rule the current must be clockwise as viewed from above (**Figure 20-6b**). (To apply the right-hand rule, point the thumb on your right hand to the left and your non-index fingers down into the loop; your index finger points in the direction of the current.)

The current loop acts like a dipole magnet, so once the bar is moving there is a second, induced magnetic field. You can determine its direction by picking any side of the loop, say the one closest to you in the figure, pointing your thumb in the direction of the current, and curling your other fingers into the region of the loop. Your other fingers, which mark the direction of the induced magnetic field, will be curling into the loop from above. This direction is the *same direction* as the external field. The induced field, in this case, does not point in the direction opposite to the external field. It does, however point so that it opposes the change in flux that created it. In this case, the magnetic flux is decreasing, so the induced field arises in a direction that works to prevent that decrease. The induced field points in the same direction as the external field so that one augments the other.

! Watch Out

The magnetic field due to a current induced in a loop opposes the change in magnetic flux that created it, not the direction of the magnetic field.

When the magnetic flux is decreasing, the induced field will be aligned with the external field. When the flux is increasing, the induced field will point in the direction opposite to that of the external field.

? Got the Concept 20-2
Induced Current

A rectangular loop passes first into, then out of, a region of constant magnetic field that points in a direction perpendicular to the plane of the loop. Refer back to Figure 20-3 to see the moment when the loop is (a) approaching the field, (b) entering the field, (c) completely inside the field, (d) leaving the field, and (e) completely out of the field. In each of these cases, in which direction is the current induced in the loop?

✳ What's Important 20-2

The direction of a potential induced by a changing magnetic flux opposes the change in flux that created it. When the potential is induced in a conducting loop, the induced magnetic field also opposes the change in magnetic flux that created it.

20-3 Applications of Faraday's and Lenz's Laws

A gondola on the Giant Drop ride at Dreamworld on Australia's Gold Coast (Figure 2-13) drops about 120 m in 5 s, reaching a top speed of nearly 40 m/s.

Riders enjoy the thrill, and probably appreciate that every trip ends with a safe, smooth stop. The physics of Faraday's law of induction and Lenz's law guarantee the reliability of the Giant Drop braking system. Even if you don't frequent amusement parks, however, you still encounter this physics on a daily basis. The electricity you use is generated by converting motion in a magnetic field to current. Information is read from a computer's hard drive as the magnetically encoded data rotates past a wire coil and induces a current. We'll look more closely at the physics of Faraday and Lenz in this section.

Let's create a simple generator, a device that converts mechanical motion to electrical energy. **Figure 20-7a** shows a rectangular, flat, conducting wire loop that can be pushed into a region of uniform magnetic field $\vec{B}$. We have arbitrarily set up the loop and field so that the field points into the page from the perspective of the figure and the plane of the loop is perpendicular to the direction of the field. Certainly, the area of the loop immersed in the field increases as the loop is pushed into the field. The magnetic flux ($\Phi_B = \int \vec{B} \cdot d\vec{A}$, Equation 20-1) therefore increases, so an induced electric potential arises ($\mathscr{E} = -d\Phi_B/dt$, Equation 20-3), creating an induced current in the loop. The current could, for example, be used to power a video game or to heat the wires in a toaster.

Let's find the induced current when the loop is pushed into the field at speed v_0. We let the loop have width w and resistance R, and assign B_0 as the magnitude of the magnetic field. First, the current is found from the general relationship between (induced) potential, current, and resistance ($V = iR$, Equation 18-9):

$$i = \frac{\mathscr{E}}{R}$$

or by substituting Equation 20-1 and Equation 20-3,

$$i = -\frac{\dfrac{d\Phi_B}{dt}}{R} = \frac{-\dfrac{d}{dt}\int \vec{B} \cdot d\vec{A}}{R}$$

The area vector $d\vec{A}$ is defined to be perpendicular to the plane of the loop. In the configuration shown in the figure, the field is also perpendicular to the plane of the loop, so $\vec{B}$ is parallel to $d\vec{A}$. The cosine of the angle between them is therefore equal to 1, which gives

$$i = \frac{-\dfrac{d}{dt}\int B_0 \, dA}{R} = \frac{-\dfrac{d}{dt}B_0 wx}{R}$$

where x is the length of the region of the loop inside the field at any instant. Note that both B_0 and w are constant here, and that the rate at which x changes as the loop is pushed into the field is the speed v_0, so

$$i = \frac{-B_0 w \dfrac{dx}{dt}}{R} = \frac{-B_0 w v_0}{R} \tag{20-5}$$

The magnitude of the current is proportional to the speed at which the loop is pushed into the field.

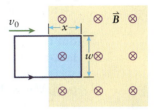

As the loop is pushed into the field...

...the area through which the field passes (shaded in blue) increases.

According to Faraday's law of induction, the changing magnetic flux through the loop induces an electric potential.

(a)

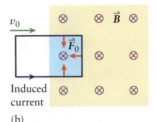

Lenz's law requires that the force $\vec{F}_0$ oppose the motion

Induced current

(b)

Figure 20-7 (a) A rectangular, flat, conducting wire loop is pushed into a region of uniform magnetic field. (b) An induced electric potential arises as the magnetic flux through the loop changes, which creates a current in the loop. Lenz's law dictates that the force exerted on the current in the loop due to the uniform field opposes its motion.

Current is generated as the loop is pushed into the field, but not without a cost. To move the loop, work must be done and power must be supplied. As soon as the loop is pushed into the field and the current begins to flow, the wire feels a force due to the interaction of that current and the magnetic field. In keeping with Lenz's law, this force opposes the motion; so in order for the loop to move at constant speed, the applied force must balance the force exerted on the current loop by the field. When the two forces balance the net acceleration is zero.

The forces on the three sides of the loop that are within the field are shown in **Figure 20-7b**. The forces on the top and bottom sides cancel, so the net force due to the field on the loop is $\vec{F}_0$, the force on the side of length w within the field. The magnitude of the force on this section of the loop, from $\vec{F} = i\vec{l} \times \vec{B}$ (Equation 19-3), is

$$F_0 = \frac{B_0 w v_0}{R} w B_0 \tag{20-6}$$

where we used the result of the induced current (Equation 20-5) and also the fact that the side of length w is perpendicular to the field. To move the loop at constant velocity, the externally applied force must equal F_0.

The power (that is, the energy per unit time) that must be supplied to generate current using this loop and magnetic field is

$$P = \frac{dW}{dt} = \frac{d}{dt} F_0 x = F_0 v_0$$

or

$$P = \frac{B_0 w v_0}{R} w B_0 v_0 = \frac{B_0^2 w^2 v_0^2}{R} \tag{20-7}$$

Notice that the required power goes up as the square of the speed with which the loop is pushed into the field. To generate more power requires moving the loop more and more quickly, which requires more and more power to be supplied to move the loop.

Notice that the direction of the net force $\vec{F}_0$ of the field on the current in the loop is to the left in the figure in order to satisfy Lenz's law. This requires that the induced current is counterclockwise. The counterclockwise current results in an induced magnetic field that points out of the page within the loop. The induced magnetic field is in the opposite direction of the external field when the loop is pushed into the external field. The direction of the induced current is shown in Figure 20-7b.

The power you deliver to the loop is converted to electrical power which, from $P = i^2 R$ (Equation 18-28), is

$$P = \left(\frac{-B_0 w v_0}{R} \right)^2 R = \frac{B_0^2 w^2 v_0^2}{R}$$

Some of the power generated could be used to run an electrical device, and the rest would be lost in the form of heat.

? Got the Concept 20-3
Pulling and Pushing

How would our analysis of the loop and field change if the loop were pulled out of, rather than pushed into, the field?

? Got the Concept 20-4
Toast, the Hard Way

Imagine a (peculiar!) manual toaster, one in which a piece of bread is held close to a coil of wire that is repeatedly pushed into and pulled out of a magnetic field. You want to double the power delivered to cut the toasting time in half. How should you change the speed at which the coil is moved in and out of the field?

Estimate It! 20-1 Are You a Generator?

Estimate the power you could generate by moving a square loop of wire in and out of a magnetic field. For the estimate, assume that you have a 0.4 T magnetic field (which is not hard to find) and that you make the loop using 12-gauge, insulated copper wire (also not hard to find). You could easily look up the mass and resistance per unit length of 12-gauge copper wire, but here are some big, round numbers: 1000 m of 12-gauge copper wire has resistance of about 5 Ω and a mass of about 30 kg.

SET UP

According to Equation 20-7, the power generated when a rectangular loop is moved into or out of a region of constant magnetic field B_0 at a constant speed v_0 is

$$P = \frac{B_0 w v_0}{R} w B_0 v_0 = \frac{B_0^2 w^2 v_0^2}{R}$$

where R is the resistance of the loop and w is the length of the sides in the direction perpendicular to the motion. We have been given a value for the field, and we can determine the resistance of the loop once we determine a good value for its size. So we need to determine reasonable values for the size of the loop and the speed with which you can move it.

SOLVE

For a big, round number, is it reasonable to consider a loop 0.5 m on a side? A loop that size would not be awkward to hold or move. In addition, at about 30 kg per 1000 m, the 2-m-long piece of 12-gauge copper wire from which the loop is formed would have a mass of 60 g—also not difficult to move. So we'll use w equal to 0.5 m for the estimate. At a resistance of about 5 Ω per 1000 m, the loop has a resistance of

$$R = \frac{5\ \Omega}{1000\ \text{m}} 2\ \text{m} = 0.01\ \Omega$$

Finally, for a reasonable estimate of the speed, let's say we can move the loop back and forth once per second. At constant speed, that's the length of one side (0.5 m) every half second, or v_0 equal to 1 m/s. Then

$$P = \frac{B_0^2 w^2 v_0^2}{R} = \frac{(0.4\ \text{T})^2 (0.5\ \text{m})^2 (1\ \text{m/s})^2}{0.01\ \Omega} = 4\ \text{W}$$

REFLECT

We estimate the power generated by moving the loop in and out of the field to be about 4 W. The output is not a lot of power, but with it you could light a typical flashlight bulb, at least until you got tired.

Example 20-1 Terminal Speed

A square loop of wire that has sides equal to 0.25 m falls into a region of constant magnetic field that has a strength equal to 0.40 T, in which the plane of the loop is perpendicular to the lines of field (**Figure 20-8a**). The mass of the loop is 0.10 kg and its resistance is 0.010 Ω. Find the terminal speed of the loop before it has completely entered the field, assuming that it reaches terminal speed in that time.

▶ **Go to Interactive Exercise 20-1** for more practice dealing with applications of Faraday's and Lenz's laws

SET UP

First, let's understand why the loop reaches terminal speed. When the resistive or drag force (Section 5-4) on an object equals the force causing it to accelerate, the net force and therefore the acceleration is zero. In any situation in which the resistive force on an object increases with speed, the object accelerates up to the speed at which the net force is zero, and then continues to move at that terminal speed. To find the terminal speed of the loop in this problem, we therefore need to set the force of the field on the induced current in the loop equal to the force of gravity on the loop (**Figure 20-8b**). Because we are interested here only in the terminal speed, we don't need to find a general expression for the speed of the loop. Instead we simply balance force due to gravity with the net force on the loop due to the interaction between the magnetic field and the induced current.

As the loop falls into the field, the magnetic flux through it increases, inducing a current in the loop.

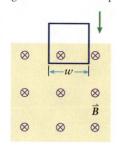

(a)

The force of the external field on the induced current eventually balances the force due to gravity, resulting in constant (terminal) speed.

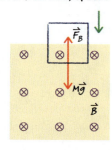

(b)

Figure 20-8 (a) A square loop of wire falls into a region of constant magnetic field that points in a direction perpendicular to the plane of the loop. (b) The magnetic force on the loop increases until it just balances the force on the loop due to gravity. When the net force on it is zero, the loop falls at a constant, terminal velocity.

SOLVE

When a square loop of side W and resistance R moves at speed v while it is partially within a uniform field of magnitude B perpendicular to the plane of the loop, the net force on it is (from Equation 20-6)

$$F = \frac{B^2 W^2 v}{R}$$

The loop moves at its terminal speed when the net force on it is zero, that is,

$$\frac{B^2 W^2 v_{term}}{R} = Mg$$

The terminal speed is then

$$v_{term} = \frac{MgR}{B^2 W^2} \tag{20-8}$$

Using the values given in the problem statement, we calculate

$$v_{term} = \frac{(0.10 \text{ kg})(9.8 \text{ m/s}^2)(0.010 \text{ Ω})}{(0.40 \text{ T})^2 (0.25 \text{ m})^2} = 0.98 \text{ m/s}$$

REFLECT

Although it's hard to have a sense of what result would be physically reasonable, the value for the terminal velocity is well within the range typical for the kinds of coils and fields with which we could carry out this experiment. (Copper wire of this resistance is commonplace, and it's not hard to obtain a magnet that has field strength of 0.40 T close to its surface.) In addition, the terminal speed of about 1 m/s is slow relative to what we would expect if the loop were allowed to fall under gravity in the absence of a magnetic field, which gives us a sense of the power and applicability of using the physics of induction to provide braking to a moving object.

Practice Problem 20-1 The square loop of wire shown in Figure 20-8a is fixed in place, so that the entire loop is within the magnetic field. The top side is detached from the rest of the loop and this conducting "rod" is allowed to slide down while remaining in electrical contact with the sides of the loop. Find the terminal speed of the rod as it falls through the magnetic field. Note that the mass of the rod is one-fourth of the mass of the square loop.

The magnetic flux through a loop of wire ($\Phi_B = \int B\, dA \cos\varphi$, Equation 20-2) depends on the magnitude of the magnetic field B, the area of the loop A, and the orientation of the loop with respect to the direction of the field. Even if the field and the area of the loop remain constant, the magnetic flux can change if the orientation of the loop changes. For example, the magnetic flux through the loop changes when the loop rotates so that the angle φ between the field and the area vector of the loop changes (**Figure 20-9**). According to Faraday's law of induction, a potential and therefore a current will be induced in the loop. The applications of changing the orientation of a conducting loop within a magnetic field include measuring eye movement and generating electricity.

Figure 20-10 shows a subject wearing a device developed to measure and record eye movements. Look carefully and you can see a ring circling his right pupil. The ring is made of gold and therefore conducting. When his eye moves, the orientation of the gold loop with respect to a magnetic field generated by a coil in the glasses changes. The changing magnetic flux induces a current in the loop, which then generates its own magnetic field. That field can be detected by another coil in the glasses, and in this way the movement of the subject's eye can be measured accurately.

How else might we change the orientation of a conducting loop within a magnetic field? Perhaps by using wind, which can turn a loop attached to a windmill (**Figure 20-11**), or by sending water rushing over a waterwheel (**Figure 20-12**) or through the turbines beneath the Grand Coulee Dam (**Figure 20-13**). All of these

As the loop rotates in the magnetic field, the flux through it changes because the angle φ between the field vector $\vec{B}$ and the area vector $\vec{A}$ changes.

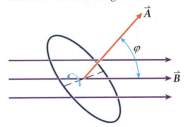

Figure 20-9 The magnetic flux through a conducting loop immersed in a uniform magnetic field changes as the loop rotates.

Figure 20-10 The movement of this person's eye can be measured accurately by detecting the magnetic field induced in the gold loop attached to his right eye. A coil in the special glasses creates a magnetic field, so that when the orientation of the gold loop changes with eye movement, the changing magnetic flux through the loop gives rise to an induced magnetic field. A second coil in the glasses detects the induced field. *(John Van Opstal)*

Figure 20-11 An electric potential arises when wind energy turns a conducting loop immersed in a magnetic field, which causes the magnetic flux through the loop to change. *(Idaho National Laboratory)*

Figure 20-12 Water rushing over a waterwheel can be used to turn a conducting loop immersed in a magnetic field, in order to generate an induced electric potential in the loop. *(age fotostock / SuperStock)*

Figure 20-13 Hydroelectric power is generated as water turns turbines beneath the Grand Coulee Dam. *(Bonneville Power Administration)*

technologies convert linear motion, of air or water, for example, into rotational motion of a conducting coil within a magnetic field. Similarly, the heat released by burning coal can be used to convert water into steam which then turns a turbine. Each of these methods is a commonly used electrical generator.

? Got the Concept 20-5
Rotating Loop

Does the current induced when the conducting loop in Figure 20-9 is rotated always have the same direction around the loop?

Example 20-2 Simple Generator

A conducting coil that consists of 160 circular windings (each winding has a radius of 0.10 m) rotates in a uniform magnetic field of strength 0.20 T. The orientation of the rotation axis of the coil is such that the plane of the coil is alternately parallel to and perpendicular to the direction of the field. What is the maximum induced voltage in the coil when it makes 19 rev/s?

SET UP

The magnetic flux through the coil changes as it rotates. To find the induced potential $\mathcal{E}$ we take the time derivative of the flux according to Faraday's law of induction:

$$\mathcal{E} = -N\frac{d\Phi_B}{dt} \qquad (20\text{-}4)$$

where we have used the form of the equation that includes N, the number of windings in a coil. Each winding makes an equal contribution to the total flux. Using Equation 20-2 we find that

$$\mathcal{E} = -N\frac{d}{dt}\int B\,dA\,\cos\varphi$$

SOLVE

The magnitude of the field is constant so it can be removed from the integral. The cosine term does vary with time but is constant *at any instant of time*, so it can also be removed, leaving us with

$$\mathcal{E} = -N\frac{d}{dt}B\cos\varphi\int dA = -N\frac{d}{dt}BA\cos\varphi$$

or

$$\mathcal{E} = -NBA\frac{d}{dt}\cos\varphi$$

The derivative of the cosine term gives sine, but we must also take the derivative of the argument, so

$$\mathcal{E} = NBA\sin\varphi\frac{d\varphi}{dt} \qquad (20\text{-}9)$$

It is the $d\varphi/dt$ term that describes mathematically the changing potential described above. The maximum induced potential, which occurs when $\sin\varphi$ equals 1, is then

$$\mathcal{E}_{max} = NBA\frac{d\varphi}{dt}$$

The units of $d\varphi/dt$ are rad/s; the rotation rate ω in rev/s is

$$\omega = \frac{1}{2\pi}\frac{d\varphi}{dt}$$

so

$$\mathcal{E}_{max} = NBA(2\pi\omega)$$

As a magnet falls through an aluminum tube, the magnetic flux through any ring-shaped slice increases then decreases.

The change in flux induces an eddy current in the ring, which in turn produces an induced magnetic field.

Figure 20-14 The magnetic flux in any region of an aluminum tube increases and then decreases as a bar magnet falling through the tube passes the region. The changing magnetic flux induces an eddy current in the region of the tube, which in turn gives rise to an induced magnetic field.

Using the values given, the maximum induced potential is

$$\mathcal{E}_{max} = (160)(0.20\,\text{T})\left[\pi(0.10\,\text{m})^2\right]\left[2\pi(19\,\text{rev/s})\right]$$

$$= 120\,\text{V}$$

REFLECT

The values we used result in a maximum induced potential of 120 V. The induced potential is not constant, however, but instead increases and decreases as the orientation of the coil changes. Following the sine term in Equation 20-9, the induced potential varies from -120 V to $+120$ V.

Magnets attract bits of iron. Perhaps due to a lifetime of everyday experiences with magnets, you may expect that *only* metals containing iron can feel a magnetic force, but that is not correct. Indeed, any conducting material can be the source of a magnetic field and can participate in magnetic interactions.

Consider a bar magnet falling through an aluminum tube as in **Figure 20-14**. At any point along the tube, the strength of the magnetic field increases then decreases as the magnet approaches that point and then recedes from it. So the magnetic flux through any small ring-shaped slice of the tube, for example, the shaded one in the figure, also increases and then decreases as the magnet moves through the tube. The changing flux induces a potential, which in turn induces a current around the ring. These **eddy current** loops produce dipole magnetic fields that interact magnetically with the field of the bar magnet. Although the magnet and the aluminum tube exert no magnetic force on one another when the magnet and tube are stationary relative to each other, the eddy currents induced when the magnet moves result in a magnetic interaction between them.

Faraday's law of induction provides the physical explanation for the eddy currents that form as a magnet moves past a conducting material. The *direction* of the eddy currents is determined by Lenz's law; the magnetic field that arises due to the induced current must oppose the change in flux. The net effect of the direction of the induced field is to oppose the motion of the magnet. It is this magnetic, or eddy current, braking that reliably brings gondolas falling from the top of the Giant Drop ride at Dreamworld to a smooth, safe stop. Each gondola is equipped with a copper fin that passes through powerful permanent magnets mounted on the lower part of the tower. As a result, eddy currents arise in the fin, which in turn give rise to a magnetic field. The interaction between the field due to the eddy currents and the field of the permanent magnets causes a deceleration that smoothly brings the gondola to a low terminal speed. The motion of the gondola is completely stopped by springs mounted below it.

? Got the Concept 20-6
Braking Smoothly

Explain why deceleration mediated by the eddy current braking system in an amusement park ride is smooth and gradual.

? Got the Concept 20-7
Eddy Current Braking

A pendulum is allowed to swing so that the bob passes between the pole pieces of a magnet. When the bob consists of a metal ring with a small piece removed, the pendulum swings freely and the motion damps out slowly. When the bob is a complete metal ring, the pendulum quickly comes to a stop as the bob passes through the magnetic field. Why?

Faraday's law of induction applies equally well when a conductor moves past a stationary magnet, for example, if the magnet in Figure 20-14 were stationary and the tube were moving. In general, eddy currents arise whenever a conducting material, even a nonmagnetic one, moves past a magnetic field. Electromagnetic flow meters that rely on this physics can be used to measure the rate of blood flow in an artery. Blood is an electrical conductor; eddy currents are induced in the blood as it flows past magnets in the flow meter. The device records the small but measurable magnetic fields due to the currents and uses them to determine the rate of flow. The obvious advantage of such a device over other methods for measuring blood flow rate is that it is noninvasive—no component of the device need be surgically introduced into the body.

A change in magnetic flux induces an electric potential ($\mathscr{E} = -d\Phi_B/dt$, Equation 20-3), and the magnetic flux ($\Phi_B = \int \vec{B} \cdot d\vec{A}$, Equation 20-1) depends on the field, the area of a region, and the orientation of the field with respect to the region. We have seen that eddy currents arise when the magnitude of the field changes in a region of a conductor due to the movement of an external magnet near the conductor or vice versa. A time-varying magnetic field, say, due to a loop of wire in which the current is controlled by a time-varying voltage, can also induce eddy currents in a conducting material. Magnetic induction tomography (MIT) is a relatively new, experimental technique for medical imaging in which fields created by coils placed near a part of the body induce eddy currents used to monitor brain swelling. Another application of the physics of induction is the controlled, repeated delivery of medication. A capsule containing the drug is implanted in the body; the capsule is made from a gel that heats up slightly when there are eddy currents, opening pores through which the medication is released. The advantage of this approach is that no implanted electronics are required—a recent application of the relatively old physics of induction!

? Got the Concept 20-8
Induction (Hearing) Loop

An induction loop, a wire loop that circles an area or a room, assists people who wear hearing aids. Installed in many public places, such as the commuter train platforms in Melbourne, Australia (**Figure 20-15**), these devices help people hear announcements made over the speaker system in a noisy environment by transmitting sound information directly to the conducting coil that is part of most hearing aids. Can you explain how an induction loop works, given our discussion of Faraday's law of induction?

Figure 20-15 A wire loop that circles an area, or a room, helps people who wear hearing aids to hear announcements made over a speaker system in a noisy environment. (*Alpha Lau/ Flickr*)

Figure 20-16 Transcranial magnetic stimulation initiates electrical activity in the brain. *(Journal of Psychiatric Research 35 (2001) 193– 215, Transcranial magnetic stimulation as a therapeutic tool in psychiatry: What do we know about the neurobiological mechanisms? Anke Post, Martin E. Keck)*

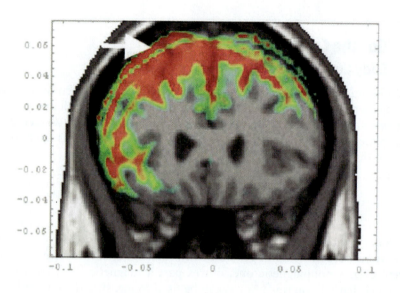

✱ What's Important 20-3

The electrical power generated when a loop is moved into or out of a magnetic field is proportional to the square of the speed. Eddy currents arise whenever a conducting material, even a nonmagnetic one, moves past a magnetic field. Any conducting material can support a loop of current and therefore be the source of a magnetic field.

20-4 Inductance

Is it possible to deliver electrical power without a wired connection? Based on our experience with Faraday's law of induction, the answer must be "yes"—a changing magnetic field through a conducting loop induces a current in that loop without any direct electrical connection. Applications are not uncommon; for example, the chargers in use for electric toothbrushes and other small appliances work inductively by generating a changing magnetic field in close proximity to the battery. A relatively new technique, transcranial magnetic stimulation (TMS), allows physicians to stimulate electrical activity in the brain without sticking electrodes to the scalp or inserting them through the skull! **Figure 20-16** shows currents generated in a human brain using a TMS device. The technology has been used with some success in treating cases of depression that have not responded to more conventional therapy.

The property of **inductance** characterizes the ability of a conducting loop to be coupled to another without a direct electrical connection. To explore inductance, let's start with a simple configuration of two wire coils (coil 1 and coil 2) placed near each other, as in **Figure 20-17**. We allow a current i_1 through the N_1 tightly wound loops of coil 1 and then examine the effect on the N_2 tightly wound loops in coil 2. According to Faraday's law of induction (Equation 20-4), a changing magnetic flux through coil 2 results in an induced electric potential:

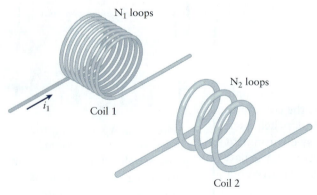

N_1 loops

i_1 Coil 1

N_2 loops

Coil 2

Figure 20-17 A changing current in coil 1 induces a current in the coil 2.

$$\mathcal{E}_2 = -N_2 \frac{d\Phi_2}{dt} \qquad (20\text{-}10)$$

Faraday's law of induction depends on a changing magnetic flux through one of the coils, which in turn depends on a changing current in the other coil. The flux Φ_2 results from the magnetic field that arises due to i_1, the current in coil 1, so for the derivative $d\Phi_2/dt$ to be nonzero, the current in coil 1 must be changing. The field is proportional to the current, so the rates of change are also proportional:

$$\frac{d\Phi_2}{dt} \propto \frac{di_1}{dt}$$

The induced potential in coil 2 is therefore proportional to the rate of change of the current in coil 1:

$$\mathscr{E}_2 \propto \frac{di_1}{dt}$$

or

$$\mathscr{E}_2 = -M\frac{di_1}{dt} \tag{20-11}$$

where the proportionality constant M is the **mutual inductance** of coil 1 with respect to coil 2. Notice that we did not include a subscript on M; the mutual inductance, as the name implies, describes the inductive coupling of two things (coils, in this case) and is the same whether a changing current in coil 1 gives rise to an induced current in coil 2 or vice versa.

An explicit relationship for the mutual inductance M is obtained by setting the right sides of Equations 20-10 and 20-11 equal:

$$-N_2\frac{d\Phi_2}{dt} = -M\frac{di_1}{dt}$$

and then integrating. We find that

$$N_2\Phi_2 = Mi_1$$

Rearranging the equation gives us an expression for mutual inductance:

$$M = \frac{N_2\Phi_2}{i_1} \tag{20-12}$$

The SI unit of mutual inductance is the henry (H), after the American physicist Joseph Henry. Henry discovered the physics of magnetic induction about the same time as Faraday.

An electric toothbrush charger relies on mutual inductance. A time-varying current is created in a coil in the charger base, and the time-varying magnetic field that arises inductively couples the coil to a similar coil in the toothbrush itself. This setup offers the advantage of having no exposed electrical connections, which must be avoided for a device such as a toothbrush that could get wet. As you have likely discovered, however, inductive chargers are inefficient—you have to leave your toothbrush on the stand overnight to recharge the battery.

 Watch Out

Only a *changing* current in one coil in Figure 20-17 results an induced current in the other.

Remember that Faraday's law of induction depends on a changing magnetic flux through one of the coils, which in turn depends on a changing current in the other coil. If a constant current is in one of the coils, the field created is constant and, therefore, so too is the magnetic flux through the other coil. No current is induced in the second coil when the current is constant; there is no inductive coupling of the two coils unless the current through one coil is changing.

? Got the Concept 20-9
Two Coils

Two identical coils can be arranged either with their two planes parallel, as in **Figure 20-18a**, or with one plane perpendicular to the other as in **Figure 20-18b**. In which configuration is the mutual inductance larger, or is the mutual inductance the same for both arrangements? Explain your answer.

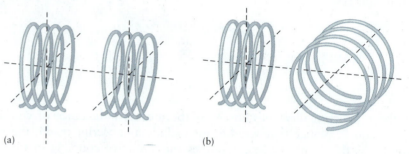

(a)   (b)

Figure 20-18 (a) Two identical coils are arranged so that their planes are parallel. (b) Two identical coils are arranged so that the plane of one is perpendicular to the plane of the other. In which configuration is the mutual inductance larger, or is the mutual inductance the same for both arrangements?

? Got the Concept 20-10
A Coil and a Solenoid

A coil is placed near a solenoid, so that their axes align (**Figure 20-19**). Does the mutual inductance between them get smaller, stay the same, or get larger when the distance between the two is increased? Explain your answer.

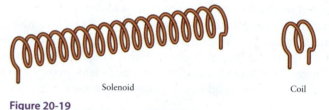

Solenoid   Coil

Figure 20-19

The physics of the mutual induction between two coils applies equally well to the multiple loops in a single coil. Although it's a bit of a simplification, you can think of each turn of the coil as a separate loop; as a changing current passes through any loop, the changing magnetic field that arises induces an electric potential in the other loops. This induced potential, according to Lenz's law, opposes the change in flux that created it. So if the current is increasing in the coil, the induced potential will oppose an increase in current. If the current is decreasing, the induced potential will be directed so that the current is augmented. This **self-inductance** has the net effect of opposing a change in current in a coil.

For a coil of N windings through which there is a current i and through which a magnetic flux Φ_B passes, the self-inductance, commonly represented by the variable L and often referred to simply as inductance, is defined in analogy to mutual inductance as

$$L = \frac{N\Phi_B}{i} \tag{20-13}$$

The unit of self-inductance is the henry, abbreviated as H.

Although the form of L varies based on the geometry of the coil, because the field and therefore the flux depends on the current, both the numerator and the denominator of Equation 20-13 are proportional to the current i. The inductance L therefore does not depend on i; L depends only on geometrical factors, such as the dimensions of the coil and physical constants. For example, the magnetic field along the axis of an ideal solenoid of n windings per unit length through which a current i_0 flows is

$$B = \mu_0 n i_0 \tag{19-14}$$

The magnetic flux through the cross-sectional area A of the solenoid is then (from Equation 20-2 and setting φ equal to 0)

$$\Phi_B = BA = \mu_0 n i_0 A$$

So

$$L = \frac{N\mu_0 n i_0 A}{i_0} = N\mu_0 n A$$

To simplify this expression a bit, note that $n = N/l$, where l is the length of the solenoid, so

$$L = N\mu_0 \frac{N}{l} A = \frac{\mu_0 N^2 A}{l} \tag{20-14}$$

For the solenoid, as for any coil, we see that the inductance depends only on its dimensions and a physical constant; inductance does not vary with current. Also notice that, from Equation 20-14, we can find an alternate set of units for μ_0, the permeability of free space. Because this relationship can be expressed as

$$\mu_0 = \frac{Ll}{N^2 A}$$

the units of μ_0 are

$$[\mu_0] = \frac{[L][l]}{[N^2][A]} = \frac{\text{H} \cdot \text{m}}{\text{m}^2} = \frac{\text{H}}{\text{m}}$$

We have used $4\pi \times 10^{-7}\,\text{T} \cdot \text{m/A}$ as the value of μ_0, which is useful when addressing magnetic force and magnetic field problems. For problems that involve magnetic inductance, expressing μ_0 in henrys per meter (H/m) makes more sense.

Example 20-3 How Large Is 1 H?

To get an intuitive feeling for the size of the henry, you decide to create a solenoid with inductance of 1 H by tightly wrapping wire around a plastic pipe. The wire has a diameter of 2.6 mm and the pipe has an outside radius of 3.0 cm. How long a pipe will you need? Is 1 H a large or a small amount of inductance?

SET UP

Winding the wire tightly around the pipe will create a solenoid. The length of the solenoid may likely be large compared to its diameter, so we'll treat it as ideal. (If it turns out not to be long, we'll have to make some adjustments to our approach.) Using Equation 20-14 the inductance of the solenoid is given by

$$L = \frac{\mu_0 N^2 \pi R_p^2}{l_p}$$

where R_p is the outside radius of the pipe, l_p is the length of the pipe, and N is the number of windings that form the solenoid. The expression can be rearranged to get l_p in terms of the other variables.

SOLVE

Before we rearrange the equation, note that the number of windings N depends on how many widths (diameters) of the wire fit along the length of the pipe,

$$N = \frac{l_p}{D_w}$$

where D_w is the diameter of the wire (**Figure 20-20**). Noting this dependence is a critical step, because otherwise we have no way to know a value of N. Substituting for N in the equation for inductance, we find that

$$L = \frac{\mu_0 l_p^2 \pi R_p^2}{D_w^2 l_p} = \frac{\mu_0 l_p \pi R_p^2}{D_w^2}$$

so the length of the pipe can be expressed as

$$l_p = \frac{L D_w^2}{\mu_0 \pi R_p^2}$$

We can now substitute the given values to arrive at an answer:

$$l_p = \frac{(1\,\mathrm{H})(0.0026\,\mathrm{m})^2}{(4\pi \times 10^{-7}\,\mathrm{H/m})\,\pi\,(0.030\,\mathrm{m})^2} = 1.9 \times 10^3\,\mathrm{m}$$

Each width of wire contributes one loop to the total number of windings...

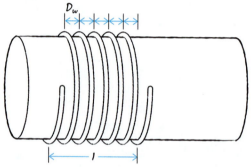

...so in any length l, the number of windings is l/Dw.

Figure 20-20 In winding a solenoid around a pipe, the number of windings depends on how many widths (diameters) of the wire fit along the length of the pipe.

REFLECT

According to the specifications given, we need a pipe nearly 2 km long to construct a 1 H inductor. It is reasonable to assume that the solenoids used to introduce inductance in electronic devices are not that long, so we conclude the typical solenoid found in such devices is much smaller than 1 H. This conclusion is true enough; 1 H is a fairly large inductance. We could make the 1 H solenoid smaller by using finer wire and by wrapping more than one layer around the pipe, but even if our imaginary solenoid had, say, 100 layers and 10 times as many windings per meter, it would still need to be about 2 m long in order to provide an inductance of 1 H. Finally, note that the length we determined is certainly long enough so that we were justified in treating this as an ideal solenoid.

Practice Problem 20-3 A solenoid is created by winding a single layer of wire that has a diameter equal to 0.26 mm tightly around a cylinder of outer radius 0.0050 m and length 0.10 m. What is the inductance of the solenoid?

The definition of self-inductance (Equation 20-13) leads to an expression for the induced electric potential similar to Equation 20-11 for mutual inductance:

$$N\Phi_B = Li$$

Taking the derivative of the equation with respect to time gives

$$N\frac{d\Phi_B}{dt} = L\frac{di}{dt}$$

The left side of the equation is the negative of the induced electric potential according to Faraday's law of induction (Equation 20-4), so

$$\mathscr{E} = -L\frac{di}{dt} \tag{20-15}$$

Let's stop for a moment and reflect on this result. We have seen two other simple relationships between electric potential and a quantity related to charge. Specifically, the potential difference across a capacitor, from Equation 17-19,

$$V = \frac{1}{C}q$$

is proportional to the charge q on it. Similarly, the potential difference across a resistor, from Equation 18-9,

$$V = Ri$$

is proportional to the current i through it. In addition, the proportionality constants for both capacitors and resistors depend only on the dimensions, the shape, and physical constants of the components. For example, the capacitance of a parallel plate capacitor of plate area A and plate separation d is given by $C = \epsilon_0 A/d$ (Equation 17-20) where ϵ_0 is the permittivity of free space. The resistance of a cylindrical resistor of cross-sectional area A and length L, for example, is given by $R = \rho L/A$ (Equation 18-11), where ρ is the resistivity of the material. So Equation 20-15 for self-inductance has the same form as our fundamental relationships for capacitance and resistance, in which electric potential is proportional to a quantity related to charge, and for which the proportionality constant depends only on dimensions and a physical constant. Surely, then, inductance must characterize some electrical component, as capacitance does for capacitors and resistance does for resistors. Yes, we have uncovered the **inductor**, the third basic electrical component.

Like a capacitor, an inductor stores energy. The capacitor stores energy in the form of an electric field; in contrast, the inductor stores energy in the form of a magnetic field. The energy associated with current in the inductor delivers power, which results in energy stored in the inductor. Power is the rate at which energy is delivered to, or stored in, the inductor:

$$P = \frac{dU}{dt}$$

or

$$dU = P\,dt$$

We find the total stored energy by adding up all of the infinitesimal contributions dU over the time it takes the current in the inductor to rise to its final value:

$$\int dU = \int P\,dt$$

To carry out the integrations we need to express P in terms of i, and apply the appropriate bounds. For the first part, the power delivered due to a potential $\mathcal{E}$ and current i (from Equation 18-26) is

$$P = \mathcal{E}i$$

Substituting the right side of the equation for power in the integral, we find that

$$\int dU = \int \mathcal{E}i\, dt$$

or applying $\mathcal{E} = L(di/dt)$ (taking the magnitude of the induced potential from Equation 20-15),

$$\int dU = \int L\frac{di}{dt}i\, dt = \int Li\, di$$

As the current increases from 0 to its final value i_{final}, the energy in the magnetic field increases from 0 to the total stored energy U_L. So

$$\int_0^{U_L} dU = \int_0^{i_{final}} Li\, di$$

or

$$U_L = \frac{1}{2}Li_{final}^2$$

More generally, the energy stored in the magnetic field of an inductor carrying a current i is

$$U_L = \frac{1}{2}Li^2 \tag{20-16}$$

The relationship for the energy stored in an inductor—in the form of a magnetic field—is similar to Equation 17-21 for the energy stored in the electric field of a parallel plate capacitor. To emphasize the similarities, we rewrite Equation 17-21 as

$$U_E = \frac{1}{2}\frac{1}{C}q^2 \tag{20-17}$$

We will explore the similarity between the energy stored in an inductor and in a capacitor in the next section.

Let's also compare the relationship between electric potential and inductance, and between potential and resistance. For a resistor, from Equation 18-9,

$$V = Ri$$

For an inductor, using Equation 20-15 and writing electric potential as V:

$$V = -L\frac{di}{dt}$$

The forms of the two equations reinforce a conclusion we drew above, that self-inductance has the net effect of opposing a change in current in a coil. For a resistor, V is proportional to i; a resistor resists the current in a circuit. For an inductor, V is proportional to di/dt; an inductor resists a *change* of current in a circuit.

✳ What's Important 20-4

An inductor stores energy in the form of a magnetic field. The electric potential that arises in the inductor causes a current, and the current delivers power to the inductor. The property of inductance characterizes the ability of a conducting loop to be coupled to another without a direct electrical connection.

20-5 *LC* Circuits

Energy in the form of an electric field is associated with a capacitor. The electric field arises from the charge stored on the plates of the capacitor. Energy in the form of a magnetic field is associated with an inductor, but in this case the field arises from a current, that is, from the motion of charges. The energy in the capacitor is potential energy, in that it is associated with the stored charge at any instant of time. The energy in the inductor, on the other hand, is associated with motion; in other words, it is in some way analogous to kinetic energy. We have studied systems in which gravitational or spring potential energy is transferred to kinetic energy and vice versa. What happens if we connect a capacitor and an inductor?

Imagine a capacitor and an inductor connected in series by a switch and nothing else, as in **Figure 20-21**. We treat this idealized series *LC* circuit as if it has no resistance. (All conducting parts of any real circuit, such as the wires, will offer some but typically small resistance.) At the instant shown in the figure, the capacitor is charged; we have drawn some number of electrons to the lower plate, leaving a deficit of electrons on the upper plate. The field inside the capacitor is at its maximum, as is the energy stored in the electric field. The field points down, toward the negatively charged plate, as indicated by the red arrows in the figure. This is not a stable configuration, however; the excess electrons on the lower plate are each repelled by the other electrons, and they are also attracted to the upper, positively charged plate. Once the switch is closed electrons flow from the negative plate to the positive plate, or clockwise in this case. **Figure 20-22** follows the flow of the electrons in the circuit, by showing the circuit at four different times. The drawing of the circuit on the left of the figure shows the moment immediately after the switch is closed, so that the capacitor is still fully charged.

At some point in time all of the excess charge has left the negatively charged plate of the capacitor. The electric field and therefore the energy stored in the capacitor are both zero. Energy must be conserved, so all of the energy that had been stored in the form of an electric field is, at this moment, associated with the

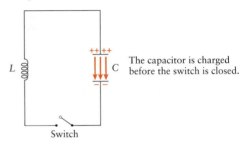

A capacitor and an inductor are connected in series...

The capacitor is charged before the switch is closed.

Switch

Figure 20-21 A capacitor, an inductor, and a switch are connected in series.

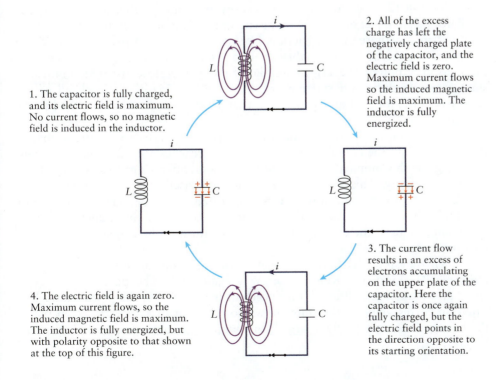

1. The capacitor is fully charged, and its electric field is maximum. No current flows, so no magnetic field is induced in the inductor.

2. All of the excess charge has left the negatively charged plate of the capacitor, and the electric field is zero. Maximum current flows so the induced magnetic field is maximum. The inductor is fully energized.

3. The current flow results in an excess of electrons accumulating on the upper plate of the capacitor. Here the capacitor is once again fully charged, but the electric field points in the direction opposite to its starting orientation.

4. The electric field is again zero. Maximum current flows, so the induced magnetic field is maximum. The inductor is fully energized, but with polarity opposite to that shown at the top of this figure.

Figure 20-22 The capacitor in an *LC* circuit is initially charged (1). Such a configuration is not stable, so charge leaves the negative plate, which gives rise to a current in the circuit; the inductor becomes energized (2). As current continues to flow, the capacitor becomes charged again (3) and eventually discharges in the opposite direction around the circuit (4). The process repeats; the circuit oscillates.

magnetic field arising from the current in the inductor. The magnetic field, and therefore the current, must be at their maximum values. (This observation is true even though no negative charge can be drawn from the capacitor, because it takes a finite time for the charges to make their way from one plate to the other. There is current in the circuit even at the moment when no more charges are flowing from the capacitor.)

The state of the circuit when the current is maximum is represented by the circuit shown at the top of Figure 20-22. We say that the inductor is fully energized at this moment. Note that the application of the right-hand rule to the direction of the current gives a magnetic field directed upward through the inductor at this moment, as indicated by the purple, oval arrows in the figure.

The electrons that make up the current eventually find themselves on the upper plate of the capacitor. When as many electrons as possible have accumulated on that plate, the capacitor is once again fully charged, although the electric field now points toward the upper plate, which is in the direction opposite to its starting orientation. This moment in the motion of the charges in the series LC circuit (often just "LC circuit") is shown at the right in Figure 20-22. Once again, the electric field in the capacitor, and therefore the energy stored in the field, is at its maximum. The energy in the form of the magnetic field is zero, because there is no current. (If there were current, we could not claim that the maximum possible number of electrons had accumulated on the plate.)

You can probably guess what happens next. Just as at the start of the process we are describing, each electron on the negatively charged plate of the capacitor will experience a repulsive force. Charge will begin to flow (current will begin). Because the top plate is now negatively charged, the current is counterclockwise, opposite to the direction of the current at the start. Once again, the system reaches a moment when the capacitor has no charge and when current and therefore the energy in the magnetic field is at a maximum.

The entire process repeats, and because we assumed no resistance in the circuit, the process repeats without any loss. The electrons slosh back and forth, with energy continually transforming between the potential energy associated with an electric field and the kinetic energy associated with a magnetic field. We have seen a system like this before—an object at the end of an ideal spring.

In Section 12-3, we considered the energy associated with a simple harmonic oscillator or an object of mass M at the end of a spring of constant k. (In this case, "simple" means there are no frictional or other nonconservative effects.) When the spring is stretched and the object held fixed, all of the energy in the system is in the form of potential energy. Potential energy is therefore at a maximum. This state corresponds to our starting point of the LC circuit, in which all of the energy is in the form of an electric field that arises from the charge stored in the capacitor.

After the object is released, the spring force causes it to accelerate, so that as it passes through the equilibrium point its speed reaches a maximum. The potential energy at this moment is zero, and all of the oscillator's energy is in the form of kinetic energy. This state of the mechanical oscillator corresponds to the moment in the LC circuit, shown in top part of Figure 20-22, when the capacitor is completely discharged. At this moment the current in the inductor is at its maximum, so all of the circuit's energy is in the form of the magnetic field due to moving charge.

When the spring is fully stretched in the opposite direction, the object is momentarily stationary. Kinetic energy is again zero. The potential energy is again at a maximum, although displacement is opposite to the initial state. This situation corresponds to the state shown to the right in Figure 20-22, in which the capacitor is again fully charged, but with polarity opposite to that of the initial state.

The transfer between potential energy and kinetic energy in the mechanical oscillator, summarized in Figure 12-10, repeats over and over without loss, in a way analogous to the transfer between U_E and U_B in the *LC* circuit. It would appear that the series *LC* circuit is an electrical version of a simple harmonic oscillator.

In addition to the qualitative similarities, we can find other direct connections between the mechanical object and spring system and the electrical inductor and capacitor system. At any moment the spring potential energy (from Equation 12-11) is proportional to the square of the displacement of the system from equilibrium:

$$U \propto x^2$$

and the kinetic energy of the object attached to the spring is proportional to the square of its speed:

$$K \propto v^2$$

Notice that speed is the time derivative of displacement: $v = dx/dt$.

For the *LC* circuit, the energy stored in the capacitor (from Equation 20-17) is proportional to the square of the stored charge:

$$U_E \propto q^2$$

and the energy associated with the inductor (from Equation 20-16) is proportional to the square of the current:

$$U_L \propto i^2$$

Current, the parameter that governs the energy associated with the motion of charge, is the time rate of change of charge, the parameter that governs the potential energy in the circuit:

$$i = \frac{dq}{dt} \tag{20-18}$$

The dependence of the energies on q and dq/dt in the *LC* circuit and on x and dx/dt for the object and spring reinforces our conclusion that the *LC* circuit is the electrical analog of the mechanical simple harmonic oscillator.

We found an expression for the displacement from equilibrium of the simple harmonic oscillator by setting the net force on the object, $F = -kx$, equal to mass multiplied by acceleration, $m\, d^2x/dt^2$. This expression leads to the differential equation

$$\frac{d^2x}{dt^2} + \frac{k}{m}x = 0 \tag{12-1}$$

A solution is

$$x(t) = A \cos(\omega_0 t + \phi) \tag{12-4}$$

where

$$\omega_0 = \sqrt{\frac{k}{m}} \tag{12-2}$$

Let's compare the simple harmonic oscillator equation to the equation that describes the *LC* circuit. Kirchhoff's loop rule (Section 18-4) states that the sum of the potential drops around any closed circuit must be zero, so here

$$V_L + V_C = 0$$

Using expressions for V_L and V_C from Equation 20-15 and Equation 17-19, respectively,

$$-L\frac{di}{dt} - \frac{1}{C}q = 0 \tag{20-19}$$

The minus signs simply indicate that both terms represent drops in voltage. Using the definition of current (Equation 20-18),

$$\frac{di}{dt} = \frac{d^2q}{dt^2}$$

so Equation 20-19 becomes

$$-L\frac{d^2q}{dt^2} - \frac{1}{C}q = 0$$

or

$$\frac{d^2q}{dt^2} + \frac{1}{LC}q = 0$$

The form of the equation is identical to the equation for the simple harmonic oscillator. The series LC circuit is indeed an *electrical oscillator*.

We can now write a description of the charge on the capacitor as a function of time, and also the current in the circuit as a function of time. By analogy with Equations 12-4 and 12-2,

$$q(t) = Q_0 \cos(\omega_0 t + \phi) \tag{20-20}$$

where Q_0 is the initial, maximum charge on the capacitor, the natural frequency ω_0 is

$$\omega_0 = \sqrt{\frac{1}{LC}} \tag{20-21}$$

and the phase ϕ allows us to select that moment at which we want to set time equal to zero. The current in the circuit is

$$i = \frac{d}{dt}Q_0 \cos(\omega_0 t + \phi) = -Q_0\omega_0 \sin(\omega_0 t + \phi) \tag{20-22}$$

Both the charge and current oscillate. Because sine and cosine are 90° out of phase with each other, so too are the capacitor charge and the current. When the charge is at its maximum, the current is zero, and when the current is at its maximum, the stored charge is zero. This pattern is exactly what we would expect from a simple harmonic oscillator.

The total energy of the mechanical object and spring system is constant ($E = \frac{1}{2}kA^2$, Equation 12-10). We can compare this energy to the total energy in the LC circuit by writing the energy stored in the electric field of the capacitor and the energy stored in the magnetic field of the inductor as a function of time.

Combining the electric field energy equation (Equation 20-17) with the equation of charge as a function of time (Equation 20-20) gives

$$U_E = \frac{1}{2}\frac{1}{C}Q_0^2 \cos^2(\omega_0 t + \phi)$$

Combining the magnetic field energy equation (Equation 20-16) with the equation of current as a function of time (Equation 20-22) gives

$$U_L = \frac{1}{2}LQ_0^2\omega_0^2 \sin^2(\omega_0 t + \phi)$$

The total energy is the sum of the two equations; we can combine the terms by noticing that ω_0 is defined in terms of L and C (Equation 20-21). Then

$$U_E + U_L = \frac{1}{2}\frac{1}{C}Q_0^2 \cos^2(\omega_0 t + \phi) + \frac{1}{2}LQ_0^2\omega_0^2 \sin^2(\omega_0 t + \phi)$$

$$= \frac{1}{2}\frac{1}{C}Q_0^2 \cos^2(\omega_0 t + \phi) + \frac{1}{2}LQ_0^2\frac{1}{LC} \sin^2(\omega_0 t + \phi)$$

$$= \frac{1}{2}\frac{1}{C}Q_0^2 [\cos^2(\omega_0 t + \phi) + \sin^2(\omega_0 t + \phi)]$$

or

$$E_{\text{total}} = \frac{1}{2}\frac{1}{C}Q_0^2 \qquad\qquad (20\text{-}23)$$

This equation makes sense when considered in terms of the energy stored in the electric field at any moment:

$$U_E = \frac{1}{2}\frac{1}{C}q^2 \qquad\qquad (20\text{-}17)$$

When the charge on the capacitor is Q_0, the maximum value, then U_E equals $\frac{1}{2}\frac{1}{C}Q_0^2$. Also, when the energy in the electric field is largest, no energy can be in the form of the magnetic field. That is, E_{total} equals U_E, which gives Equation 20-23. However, although Equation 20-17 looks similar to Equation 20-23, they are different in an important way. The first equation describes the energy stored in the capacitor at some moment in time, while the second equation describes the total, constant energy of the entire circuit.

Whereas the total energy of the mechanical oscillator depends on the characteristics of the spring (k) and how far we initially stretch it (A), the total energy of the electrical oscillator depends on the characteristics of the capacitor (C) and how much we initially charge it (Q_0). This, again, reinforces our conclusion that the series *LC* circuit is the electrical analog of the mechanical oscillator.

Example 20-4 Shared Energy

The total energy stored in an oscillating *LC* circuit is shared between the energy stored in the electric field of the capacitor and the energy stored in the magnetic field of the inductor. For example, when one-quarter of the total energy is in the form of electric field, three-quarters of the energy is in the form of magnetic field. When the energy is shared in this way, what is the charge on the capacitor? Give an answer as a fraction of the capacitor's charge when it stores *all* of the energy of the circuit.

SET UP

To answer, it is important for us to distinguish between the energy stored in the electric field of the capacitor at any time and the maximum energy that the capacitor stores. At a moment when the charge on the capacitor is q, the energy stored in the electric field is given by Equation 20-17:

$$U_E = \frac{1}{2}\frac{1}{C}q^2$$

The energy changes as the charge q changes. The total energy in the circuit is

$$E_{total} = \frac{1}{2}\frac{1}{C}Q_0^2 \tag{20-23}$$

This energy does not change; Q_0 is the maximum charge placed on the capacitor. Although the two equations look similar, they are fundamentally different: The first describes the energy stored in the capacitor at some moment in time, while the second describes the total, constant energy of the entire circuit.

SOLVE

The requirement that one-quarter of the total energy be in the form of the electric field can then be written mathematically as

$$U_E = \frac{1}{4}E_{total}$$

or

$$\frac{1}{2}\frac{1}{C}q^2 = \frac{1}{4}\left(\frac{1}{2}\frac{1}{C}Q_0^2\right)$$

Canceling common terms in the equation gives

$$q^2 = \frac{1}{4}Q_0^2$$

or

$$q = \frac{1}{2}Q_0$$

When the charge on the capacitor is half of the maximum charge, the energy stored in the capacitor is one-quarter of the total energy.

REFLECT

The fraction of the total energy stored in the electric field is not the same as the fraction of the charge to the total charge. Rather, the energy stored in the field is proportional to the *square* of the charge. So for the energy stored in capacitor to be one-quarter of the maximum, it must be charged to *one-half* the maximum. The same kind of statement can be made about the mechanical oscillator—when the object at the end of a spring is at a displacement half of the maximum (that is, half of the amplitude of motion), only one-quarter of the total energy is in the form of potential energy.

We can look at this physics another way, by asking what fraction of the total energy is stored in the capacitor when it is charged to one-quarter of the maximum. In that case, using Equation 20-17,

$$U_E = \frac{1}{2}\frac{1}{C}\left(\frac{1}{4}Q_0\right)^2 = \frac{1}{16}\left(\frac{1}{2}\frac{1}{C}Q_0^2\right)$$

So when the charge on the capacitor is one-quarter Q_0, the energy stored in the capacitor is just one-sixteenth of the total.

Practice Problem 20-4 In terms of the capacitor's charge when it stores all of the energy in an oscillating *LC* circuit, what is the charge on the capacitor when one-half of the total energy is in the form of electric field and one-half in the form of magnetic field?

> ★ **What's Important 20-5**
> The series *LC* circuit is the electrical analog of the mechanical oscillator. In an *LC* circuit, current varies as a function of time as energy continually transforms between the potential energy associated with an electric field and the kinetic energy associated with a magnetic field. Both the charge and current oscillate.

20-6 *LR* Circuits

In the last section, we saw that current in a series *LC* circuit varies as a function of time. At some instants, the current is maximum, while at others there is no current in the circuit. In this section, we will uncover another explicit way to include time in a description of circuits that include inductors.

The presence of the inductor in an *LC* circuit acts to oppose the current through it from changing. As a result, even in the idealized circuit, which we treat as having no resistance, although there are times when the current increases it does not rise instantly to its maximum. Similarly, the current does not decrease instantly to zero during times when the current is decreasing. Both the capacitor and the inductor provide a mechanism for energy storage, which allows the energy in an *LC* circuit to be shared and exchanged between the two. This sharing and exchange naturally leads to an oscillation, a flow of energy back and forth between the inductor and the capacitor, in the same way energy flows back and forth between potential and kinetic in an object–spring system.

Let's now consider a circuit in which an inductor is connected in series with a resistor, as shown in **Figure 20-23a**. Because the resistor is not an energy-storage device, we should not expect oscillations in this series **LR circuit**. *There can be no oscillation without two mechanisms for energy storage.* There must be some dependence on time, however, because the potential across the inductor ($\mathscr{E} = -L(di/dt)$, Equation 20-15) changes over time.

As we did in considering the *LC* circuit in the last section, we take the *LR* circuit to be initially energized. In this case, that means that there is current through the inductor. We can accomplish an energized circuit using the circuit shown in **Figures 20-23b** and **20-23c**. We begin with the switch set as in Figure 20-23b, so

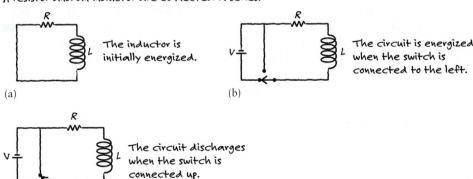

A resistor and an inductor are connected in series.

(a) The inductor is initially energized.

(b) The circuit is energized when the switch is connected to the left.

(c) The circuit discharges when the switch is connected up.

Figure 20-23 (a) A resistor and an initially energized inductor are connected in series. The addition of a battery and a switch allows the circuit to be (b) energized or (c) discharged.

that the battery drives a current in the circuit. Once the current reaches our desired value (whatever that might be), we throw the switch into the position shown in Figure 20-23c so that the circuit includes only the inductor and resistor in series. The inductor is energized at this instant because there is current.

Kirchhoff's loop rule (Section 18-4) demands that the sum of the potential drops around any closed circuit be zero, so here

$$V_L + V_R = 0$$

Using expressions for V_L and V_R from Equations 20-15 and 18-9, respectively, we see that

$$-L\frac{di}{dt} - Ri = 0$$

or

$$\frac{di}{dt} + \frac{R}{L}i = 0 \tag{20-24}$$

The minus signs in the first equation indicate that both terms represent drops in voltage.

We have encountered an equation of this form before, when examining the discharge of a series RC circuit (Section 18-7). In that case, although the differential equation is in terms of charge rather than current,

$$\frac{dq}{dt} + \frac{1}{RC}q = 0$$

the form of the solution is the same (see Discharging a Series RC Circuit in Section 18-7). The solution of Equation 20-19 gives current as a function of time; by analogy to our solution to the RC circuit,

$$i(t) = i_0 e^{-(R/L)t} \tag{20-25}$$

where i_0 is the initial current in the inductor. You might want to refer back to Section 18-7 for the details of how the solution of this form was obtained.

Equation 20-25 tells the story of the series LR circuit. When the value of t is set to zero, the exponential term is equal to 1, so

$$i(0) = i_0 e^{-(R/L)(0)} = i_0$$

This equation is exactly the initial condition we created. Of more interest, perhaps, is what happens after the initial moment. As t gets larger the exponential term $e^{-(R/L)t}$ gets R/L, so the current decreases (exponentially) as shown in Figure 20-24a.

The decay time τ, defined as the time it takes for the current to decrease to $1/e$ of its initial value, characterizes the decreasing current. To find the decay time, we substitute τ for t in Equation 20-25 and set the new expression equal to $1/e$ of i_0. These manipulations yield

$$i_0 e^{-(R/L)\tau} = \frac{1}{e}i_0$$

Note that $1/e$ equals e^{-1}, so

$$e^{-(R/L)\tau} = e^{-1}$$

or

$$-(R/L)\tau = -1$$

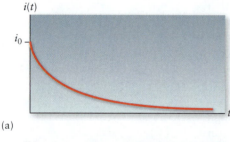

(a)

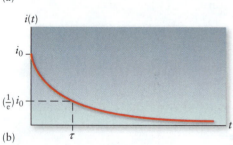

(b)

Figure 20-24 (a) The current in a discharging RL circuit decreases exponentially. (b) The decay time τ is the time that it takes the current in a discharging RL circuit to fall to $1/e$ of i_0, the maximum current in the circuit before discharging begins.

Finally,

$$\tau = L/R$$

This definition of decay time is shown in **Figure 20-24b**.

To complete the picture of the *LR* circuit, let's examine the process by which the current increases to i_0 after the switch is connected as in Figure 20-23b. Once again, the sum of the potential drops around the circuit is zero (Kirchhoff's loop rule), so

$$V - V_L - V_R = 0$$

or, using Equations 20-15 and 18-9,

$$V - L\frac{di}{dt} - Ri = 0 \qquad (20\text{-}26)$$

We make use of the work we did in analyzing the charging *RC* circuit in Section 18-7 by rearranging Equation 20-26 into the form of Equation 18-31:

$$\frac{V}{L} = \frac{di(t)}{dt} + \frac{R}{L}i(t)$$

The solution then takes the form of Equation 18-33:

$$i(t) = \frac{V}{R}(1 - e^{-(R/L)t}) \qquad (20\text{-}27)$$

As you see in **Figure 20-25**, when the value of t is set to zero in Equation 20-27, the current is zero:

$$i(0) = \frac{V}{R}(1 - e^{-(R/L)(0)}) = \frac{V}{R}(1 - 1) = 0$$

The result is not surprising, the current only starts at the instant the switch is closed. As t gets larger, the exponential term $e^{-(R/L)t}$ gets smaller, so the current increases. After a long time, when t is so large that we can consider the $e^{-(R/L)t}$ term to be zero, the current tends to V/R. This situation is the starting point for the case when the switch is flipped so that the inductor is energized but the battery is excluded from the loop of the circuit, as shown in Figure 20-23c,

$$i_0 = \frac{V}{R}$$

Look closely! This relationship is simply

$$V = i_0 R$$

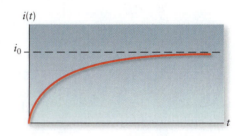

Figure 20-25 In a charging *RL* circuit the current rises asymptotically to the maximum value of i_0.

which is what we would expect if the inductor were not present in the circuit. Indeed, once the current has reached its maximum value, the inductor acts like a passive element in the circuit. Until the time when we try to change the current again, for example, by changing the position of the switch to exclude the battery from the loop, the current in the circuit acts as if the inductor is not present.

? Got the Concept 20-11

Bulb Brightness

Two identical lightbulbs, which can be treated as resistors, are connected to an inductor and a battery as shown in **Figure 20-26**. Compare the brightness of the bulbs immediately after the switch is closed, and after the switch has been closed for a long time.

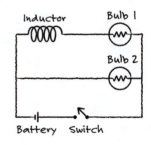

Figure 20-26 Two lightbulbs B_1 and B_2 are connected to an inductor and a battery.

Example 20-5 Let There Be Light

Two lightbulbs B_1 and B_2 are connected to an inductor R/L and a battery of voltage $V = 12$ V as shown in Figure 20-26. The bulbs can be treated as resistors; B_1 has resistance 6 Ω and B_2 has resistance 3 Ω. (a) Find the power output of each bulb immediately after the switch is closed. (b) Find the power output of each bulb a long time after the switch is closed. (c) Find the power output of each bulb immediately after the switch is reopened after having been closed for a long time.

SET UP

The primary effect of the inductor is to act against a change in the current through it. So when a potential is applied across it, current does not instantly rise to its maximum value; when the potential is removed, the current does not instantly decrease to zero. When the switch is first closed, there is no current through the inductor or bulb B_1; bulb B_1 does not immediately begin to glow and bulb B_2 glows brightly. The inductor eventually becomes fully energized, so the two bulbs share the total current. When the switch is reopened, the battery is no longer part of the circuit; if the circuit didn't include an inductor the current would immediately decrease to zero. The current in the inductor, however, decreases exponentially:

$$i(t) = i_0 e^{-(R/L)t} \tag{20-25}$$

There will be current around the loop that includes the two bulbs, so both bulbs will continue to glow, growing dimmer with time.

In all of the cases, we find the power output of the bulbs by determining the current and applying Equation 18-28:

$$P = i^2 R$$

SOLVE

(a) The power output of B_1 is initially zero, because there is no current through the inductor at the instant after the switch is closed. The circuit is effectively a battery in series with B_2 (**Figure 20-27a**), so the current through the bulb (from Equation 18-9) is

$$i = \frac{V}{R} = \frac{12 \text{ V}}{3\Omega} = 4 \text{ A}$$

and the power (Equation 18-28) is

$$P_2 = (4 \text{ A})^2 (3\Omega) = 48 \text{ W}$$

That is, initially $P_1 = 0$ W and $P_2 = 48$ W.

(b) After a long time, the current in the inductor is no longer changing, so by using

$$\mathscr{E} = -L\frac{di}{dt} \tag{20-15}$$

you can conclude there is no longer a potential drop across it. At that time, the inductor plays no active role in the circuit, so we can treat the circuit as two resistors in parallel with the battery (**Figure 20-27b**). The total current through each resistor (bulb) is then

$$i_1 = \frac{V}{R_1} = \frac{12 \text{ V}}{6\Omega} = 2 \text{ A}$$

and

$$i_2 = \frac{V}{R_2} = \frac{12 \text{ V}}{3\Omega} = 4 \text{ A}$$

▶ Go to Interactive Exercise 20-2 for more practice dealing with LR Circuits

When the switch is first closed, no current flows through the inductor. The circuit is effectively one resistor in series with the battery.

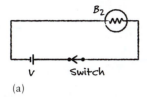

(a)

Figure 20-27 These sketches show how the circuit shown in Figure 20-26 responds after the switch is (a) just closed, (b) closed for a long time, and (c) reopened.

Figure 20-27 These sketches show how the circuit shown in Figure 20-26 responds after the switch is (a) just closed, (b) closed for a long time, and (c) reopened.

After a long time has elapsed, the inductor has become a positive element, so the circuit is a battery with two resistors in parallel.

When the switch is reopened the battery is no longer part of the circuit, so the current loop includes only the two bulbs.

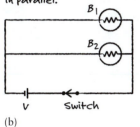

(b)

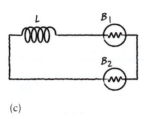

(c)

The power output of the bulbs is

$$P_1 = (2\text{ A})^2(6\,\Omega) = 24\text{ W}$$

and

$$P_2 = (4\text{ A})^2(3\,\Omega) = 48\text{ W}$$

(c) When the switch is reopened, so that the battery is not part of the loop, the circuit is effectively just the two bulbs in series as in **Figure 20-27c**. At that instant the current in the loop is 2 A, the current in the inductor. Because the two bulbs are in series, the current is the same through both, so their power outputs are

$$P_1 = (2\text{ A})^2(6\,\Omega) = 24\text{ W}$$

and

$$P_2 = (2\text{ A})^2(3\,\Omega) = 12\text{ W}$$

REFLECT

When the switch is first closed it takes some time for bulb B_1 to turn on but bulb B_2 begins to glow immediately. After enough time has elapsed, however, both bulbs draw as much current as the battery will supply, so both glow as brightly as possible. In this particular example, the resistance of bulb B_2 is half that of bulb B_1, so its power output at that time is twice that of B_1. (For part b we found $P_2 = 48$ W compared to $P_1 = 24$ W.) Once the switch is reopened the only current in the circuit is that through the inductor, so initially bulb B_1 continues to glow at the same level of brightness ($P_1 = 24$ W in part c). However, in the absence of the battery, bulb B_2 can no longer draw more current, so its power output drops ($P_2 = 12$ W in part c) and it gets dimmer after the switch is opened. Both bulbs grow dimmer with time after the switch is reopened, as the current through the inductor decreases exponentially to zero. After a long time, both bulbs will be dark, which is not surprising because with the switch open there is no ongoing source of potential in the circuit.

Practice Problem 20-5 A lightbulb of resistance 6.0 Ω is connected with a switch, a 20 H inductor, and a 12 V battery. The switch is initially open and the inductor not energized. How long does it take, after the switch is closed, for the bulb to reach one-half of its maximum brightness?

✳ What's Important 20-6

Because a resistor is not an energy-storage device, we should not expect oscillations in a series *LR* circuit. There can be no oscillation without two mechanisms for energy storage. Once the current has reached its maximum value, the inductor acts like a passive element in the circuit.

Answers to Practice Problems

20-1   0.25 m/s

20-3 1.5×10^{-4} H

20-4 $Q_0/\sqrt{2}$

20-5   0.10 s

Answers to Got the Concept Questions

20-1 (a) The magnetic flux through the loop is constant and equal to zero, because the magnitude of the field is zero. (b) The flux is increasing. Although the area of the loop is not changing, the area that is immersed in the field is increasing. Only the region of the loop through which the magnetic field passes experiences a nonzero flux, so as the loop enters the field, the area through which the field passes increases. (c) The flux is constant. The field, the area of the loop through which the field passes, and the angle between the field vector and the area vector are all constant. (d) The flux is decreasing. As the loop leaves the field the area through which field passes decreases. (e) The flux through the loop is constant and equal to zero, because the magnitude of the field is zero.

20-2 First, as noted in the answer to Got the Concept 20-1, the magnetic flux through the loop is not changing in cases a, c, and e. For those three cases, then, there is no induced current. When the loop is entering the field, as in Figure 20-3b, the area of the loop immersed in the field, and therefore the magnetic flux, is increasing. The induced field opposes the increase, so it points out of the page as it passes through the loop. According to the right-hand rule, the direction of the induced current is counterclockwise to produce an induced magnetic field pointing out of the page. When the loop is leaving the field, as in Figure 20-3d, the area of the loop immersed in the field, and therefore the magnetic flux, is decreasing. To oppose the

decrease, the induced field must reinforce the external field, that is, it also points into the page. According to the right-hand rule, the direction of the induced current must be clockwise in order to produce an induced magnetic field pointing into the page in the region inside the loop.

20-3 The magnitude of the induced current, and therefore the magnitude of the induced magnetic field, would remain the same, but the direction of both would be reversed. The magnitude of the induced current depends on the magnitude of the change in magnetic flux, which in turn depends on the rate at which the area of the loop through which field passes is changing. If the loop is pulled out of the field at the same speed as it was pushed in (v_0), the change of magnetic flux is the same, so the induced current is the same. The induced magnetic field depends on the current, so it too is the same regardless of whether the loop is pulled out or pushed in. However, Lenz's law requires that the induced field oppose the change in flux that created it. When the loop is pulled out, the change in flux is negative because the area of the loop through which the field passes is getting smaller. The induced magnetic field will therefore point in the same direction as the external field. This direction is opposite to the direction of the induced field when the loop is pushed into the external magnetic field. For a similar reason, the induced current is clockwise, also opposite to the direction we found when the loop is pushed in.

20-4 The electrical power generated when a loop is moved into or out of a magnetic field is proportional to the square of the speed (Equation 20-7). To double the power, the speed must be increased by a factor of $\sqrt{2}$.

20-5 The direction of the induced current in the loop is alternately clockwise and then counterclockwise as the loop rotates. The loop in the figure is immersed in a uniform magnetic field; when the loop rotates, the angle between the direction of the field and the area vector of the loop changes. As the loop rotates the magnetic flux through it alternatively gets larger and smaller; in order to oppose the changing flux as required by Lenz's law, the direction of the induced magnetic field will point opposite to the external field when the flux is increasing and be aligned with it when the flux is decreasing. The direction of the induced field is determined, from the right-hand rule, by the direction of the induced current. So, the direction of current reverses whenever the flux switches from increasing to decreasing and vice versa.

20-6 The magnitude of the induced magnetic field that arises due to eddy currents depends on the rate at which the magnetic flux in a region is changing. The faster the flux changes, the larger the induced current and induced magnetic field. In other words, the braking force, which is due to the induced magnetic field, is larger when the velocity of the gondola is large, and it gradually decreases as the speed of the gondola decreases. In a sense, the strength of the braking force is automatically adjusted to the speed of the gondola, which results in a gradual, smooth change in speed.

20-7 In the second case, the changing magnetic flux as the ring passes through the field causes eddy currents. The eddy currents are around the ring, resulting in an induced magnetic field, and the interaction between the induced field and the external field resists the motion according to Lenz's law. The ring quickly slows as a result. In the first case, however, the gap in the ring prevents eddy currents. Eddy currents can still be in other directions, but the gap minimizes the braking effect of induced magnetic fields.

20-8 Announcements on the train platform are typically broadcast over a speaker system. Most people on the platform hear the announcements as sound waves traveling through air, and must therefore separate out the important sounds, such as the announcer's voice, from background noise. The induction loop, however, broadcasts the announcements directly to any hearing aids in use on the platform. The control unit of the induction loop converts a sound signal, say, someone's voice, to an electrical signal. The electrical signal is used to create a current in the loop, which in turn gives rise to a magnetic field—the induction loop becomes a large dipole magnet. As the current in the loop varies in concert with the sound signal, the magnetic field also changes, and this

changing field results in a changing magnetic flux through the coil in the hearing aid. In this way a changing current is induced in the coil; circuitry in the hearing aid converts the changing current back into sound. Sounds of interest, such as the platform announcements, are delivered directly to the ear, minimizing background noise.

20-9 The mutual inductance is largest when the planes of the two coils are parallel. The mutual inductance depends on the amount of magnetic flux that passes through one of the coils when there is a current, and a magnetic field arises, in the other. When the two coils are parallel, the field lines of the dipole field due to a current in one coil pass as directly as possible through the other coil (**Figure 20-28**).

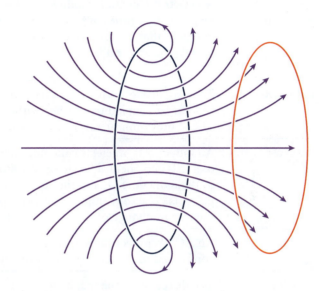

Figure 20-28

20-10 Mutual inductance characterizes the coupling of two electrical elements through magnetic flux. Regardless of whether there is current in the solenoid or the coil in this problem, as the distance between them is increased, the field of one that passes through the other must decrease, so the magnetic flux through it decreases as well. The coupling of the solenoid and the coil, that is, their mutual inductance, will therefore be diminished.

20-11 An inductor resists changes to the current in a circuit. When a potential is placed across an inductor, as occurs when the switch is closed, it takes time before the current increases to its maximum value. Immediately after the switch is closed, all of the charge flows through bulb 2; bulb 2 glows while bulb 1 is dark. The current rises through the part of the circuit that includes the inductor, however, so after a long time the current in the inductor reaches its maximum value. Because the current through the inductor is no longer changing, it becomes a passive element in the circuit—it's as if it weren't there at all! So now current is equal in both bulbs, and they both glow with the same brightness.

SUMMARY

Topic	Summary	Equation or Symbol
eddy current	An eddy current is induced by a change in magnetic flux in a conductor, for example, due to a magnet moving past a conducting material.	
Faraday's law of induction	Faraday's law of induction describes the relationship between the change in magnetic flux $d\Phi_B/dt$ and the electric potential $\mathscr{E}$ it induces.	$\mathscr{E} = -\dfrac{d\Phi_B}{dt}$ (20-3)
induced potential	An electric potential is induced in a material or region when the magnetic flux through that material or region changes. Unlike the potential that arises from connecting a battery or other voltage source to, say, a circuit, an induced potential is due to something external to the circuit.	$\mathscr{E}$
inductance	The property of inductance characterizes the ability of a conducting loop to be coupled to another without a direct electrical connection. The SI unit of inductance is the henry (H).	L
inductor	An inductor is an electrical component that stores energy in the form of a magnetic field. The energy U_L stored in an inductor is proportional to the inductance L and the square of the current i. The electric potential $\mathscr{E}$ across an inductor is proportional to the inductance and the rate of change of the current passing through it.	$U_L = \dfrac{1}{2}Li^2$ (20-16) $\mathscr{E} = -L\dfrac{di}{dt}$ (20-15)
LC circuit	An LC circuit contains a capacitor and an inductor. The LC circuit considered in this chapter is a capacitor and an inductor in series.	
Lenz's law	The direction of the magnetic field induced in a conducting loop opposes the change in magnetic flux that created it. Lenz's law can be thought of as the minus sign in Faraday's law of induction.	
LR circuit	An LR circuit contains a resistor and an inductor. The LR circuit considered in this chapter is an inductor and a resistor in series.	
magnetic flux	Magnetic flux is a measure of how much magnetic field passes through a given two-dimensional region. The magnetic flux Φ_B depends on the magnetic field $\vec{B}$, the area of the region $\vec{A}$, and the angle between the field vector and the vector that represents the area of the region. The angular dependence is described by the dot product between the field and area, $\vec{B} \cdot d\vec{A}$.	$\Phi_B = \int \vec{B} \cdot d\vec{A}$ (20-1)
mutual inductance	Mutual inductance characterizes the coupling of two electrical elements through magnetic flux. The mutual inductance M arises between two coils when current in one (i_1) gives rise to a magnetic flux (Φ_2) in the other. In Equation 20-12, N_2 is the number of loops in the second coil.	$M = \dfrac{N_2\Phi_2}{i_1}$ (20-12)
self-inductance	Self-inductance L is a measure of the mutual inductance between different loops in a single coil. Self-inductance depends on the number of loops N in the coil, the magnetic flux Φ_B through one loop due to the others, and the current i in the coil. The self-inductance of a coil has the net effect of opposing a change in current in that coil.	$L = \dfrac{N\Phi_B}{i}$ (20-13)

QUESTIONS AND PROBLEMS

In a few problems, you are given more data than you actually need; in a few other problems, you are required to supply data from your general knowledge, outside sources, or informed estimate.

Interpret as significant all digits in numerical values that have trailing zeros and no decimal points.

For all problems, use $g = 9.8 \text{ m/s}^2$ for the free-fall acceleration due to gravity. Neglect friction and air resistance unless instructed to do otherwise.

• Basic, single-concept problem
•• Intermediate-level problem, may require synthesis of concepts and multiple steps
••• Challenging problem
SSM *Solution is in Student Solutions Manual*

Conceptual Questions

1. •A common physics demonstration is to drop a small magnet down a long, vertical aluminum pipe. Describe the motion of the magnet and the physical explanation for the motion. SSM

2. ••The following diagram depicts an electron in between the poles of an electromagnet (**Figure 20-29**). Explain how the electron is accelerated if the magnetic field is gradually being increased.

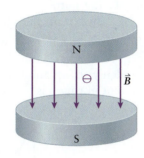

Figure 20-29 Problem 2

3. ••In a popular demonstration of electromagnetic induction, a metal plate is suspended in midair above a large electromagnetic coil, as shown in **Figure 20-30**. (a) How does this work? (b) If your professor does the demonstration, one thing you'll notice is that the plate gets quite hot. (In fact, you can end the demonstration by frying an egg on the plate!) Why does the plate become hot? (c) Would the trick work if the plate were made of an insulating material?

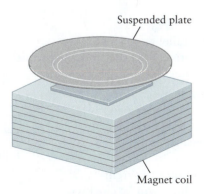

Suspended plate

Magnet coil

Figure 20-30 Problem 3

4. •In hospitals with magnetic resonance imaging facilities and at other locations where large magnetic fields are present, there are usually signs warning people with pacemakers and other electronic medical devices not to enter. Why?

5. •Two conducting loops with a common axis are placed near each other, as shown in **Figure 20-31**, and initially the currents in both loops are zero. If a current is suddenly set up in loop a, is there also a current in loop b? If so, in which direction? What is the direction of the force that loop a exerts on loop b? Explain your answer.

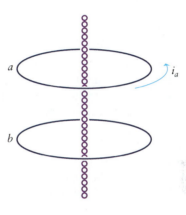

Figure 20-31 Problem 5

6. •A conducting rod slides without friction on conducting rails in a magnetic field as shown in **Figure 20-32**. It is given an initial velocity $\vec{v}$ to the right. Describe its subsequent motion and justify your answer.

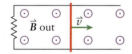

Figure 20-32 Problem 6

7. •Explain why the mutual inductance of two circuits does not depend on the current in either circuit. SSM

8. •What change must you make in the current in an inductor to double the energy stored in it?

9. •Compare the expressions for the energy stored in an inductor and the energy stored in a capacitor and explain the similarities.

10. •A motor will sometimes burn out when its load is suddenly increased. Why? What burns out?

11. •A given length of wire is wound into a solenoid. How will its self-inductance be changed if it is rewound into another coil of (a) twice the length or (b) twice the diameter? SSM

12. •Do a careful units analysis and convince yourself that the units of henrys per ohm (H/Ω) are in fact seconds.

13. •When the switch S is opened in the RL circuit shown in **Figure 20-33**, a spark jumps between the switch contacts. Why?

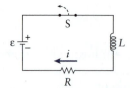

Figure 20-33 Problem 13

Multiple-Choice Questions

14. •On which variable does the magnetic flux depend?
A. the magnetic field
B. the area of a region through which magnetic field passes
C. the orientation of the field with respect to the region through which is passes
D. all of above
E. none of above

15. •**Figure 20-34** shows a sequence of a rectangular loop passing from left to right through a region of constant magnetic field. The field points out of the page and perpendicular to the plane of the loop. In which one of the sequences is the magnetic flux through the loop decreasing? SSM

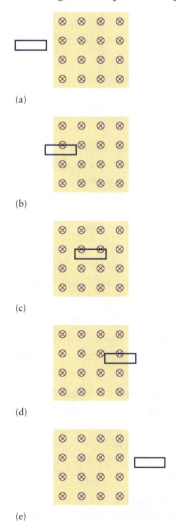

Figure 20-34 Problem 15

A. from left to right approaching the magnetic field
B. entering the magnetic field
C. inside the magnetic field
D. leaving the magnetic field
E. from left to right moving away from the magnetic field

16. •**Figure 20-35** shows two coils wound around an iron ring. Current appears in the second coil the moment

Figure 20-35 Problem 16

A. the battery is connected by closing the switch.
B. the battery is connected continually with the switch closed.
C. the battery is disconnected by opening the switch.
D. the battery is connected by closing the switch or the battery is disconnected by opening the switch.
E. the battery is disconnected continually with the switch open.

17. •Two metal rings with a common axis are placed near each other, as shown in **Figure 20-36**. If current i_a is suddenly set up and is increasing in ring a as shown, the current in ring b is

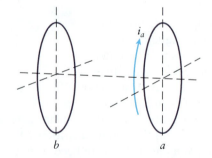

Figure 20-36 Problem 17

A. zero.
B. parallel to i_a.
C. antiparallel to i_a.
D. alternatively parallel and antiparallel to i_a.
E. perpendicular to i_a. SSM

18. •A conducting loop moves at a constant speed parallel to a long, straight, current-carrying wire, as shown in **Figure 20-37**.

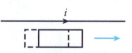

Figure 20-37 Problem 18

A. The induced current in the loop will be clockwise.
B. The induced current in the loop will be only parallel to the current i.

This page is intentionally left blank.

For complete end of chapter problem sets, please go to
www.whfreeman.com/kestentauck

is the magnitude of the flux through the ring? (b) What is the magnitude of the magnetic flux through the half of the circle where $x > 0$? (c) What is the magnitude of the magnetic flux through the half of the circle where $y > 0$? Explain your answers. SSM

32. •A 30-turn coil with a diameter of 6.00 cm is placed in a constant, uniform magnetic field of 1.00 T directed perpendicular to the plane of the coil. Beginning at time $t = 0$ s, the field is increased at a uniform rate until it reaches 1.30 T at $t = 10.0$ s. The field remains constant thereafter. What is the magnitude of the induced potential in the coil at (a) $t < 0$ s, (b) $t = 5.00$ s, and (c) $t > 10.0$ s? (d) Plot the magnetic field and the induced potential as functions of time for the range -5.00 s $< t < 15.0$ s.

20-2: Lenz's Law

33. •Determine the direction of the induced current in the loop for each case shown in **Figure 20-39**.

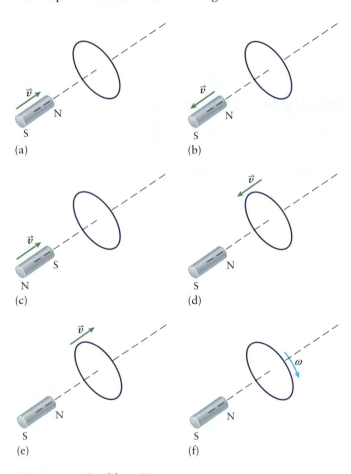

Figure 20-39 Problem 33

34. •A bar magnet is moved steadily through a wire loop as shown in **Figure 20-40**. Make a qualitative sketch of the induced voltage in the loop as a function of time (be sure to include the times t_1, t_2, and t_3).

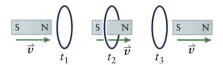

Figure 20-40 Problem 34

35. •Repeat problem 34, but this time the leading edge of the magnet is the south pole instead of the north pole. SSM

36. •A single-turn circular loop of wire that has a radius of 5.0 cm lies in the plane perpendicular to a spatially uniform magnetic field. During a 0.12-s time interval, the magnitude of the field increases uniformly from 0.2 T to 0.4 T. Determine the electric potential induced in the loop during the time interval.

20-3: Applications of Faraday's and Lenz's Laws

37. ••An electromagnetic generator consists of a coil that has 100 turns of wire, has an area of 400 cm², and rotates at 60 rev/s in a magnetic field of 0.25 T. What potential is induced in the coil? SSM

38. ••Calc A rectangular loop of wire with a length a and width b lies in the x–y plane, as shown in **Figure 20-41**. Within the loop, there is a time-dependent and spatial-dependent magnetic field given by

$$\vec{B}(t) = B_0\left[x \cos \omega t\, \hat{x} + y \sin \omega t\, \hat{z}\right] \quad \text{SI units}$$

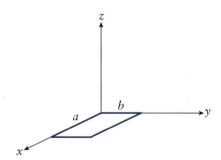

Figure 20-41 Problem 38

(a) Derive an expression for the induced potential in the loop as a function of time. (b) Calculate the numerical value of the induced potential at $t = 2.0$ s, if $a = 20$ cm, $b = 10$ cm, and $B_0 = 0.575$ T/m.

39. •••Calc A long, straight wire and a single-turn rectangular loop both lie in the plane of the page (**Figure 20-42**). The wire is parallel to the long side of the rectangular loop and is 0.5 m away from the closer side. At an instant when the induced potential in the loop is 2.0 V, what is the time rate of change of the current in the wire?

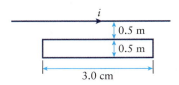

Figure 20-42 Problem 39

40. ●●A 50-turn square coil with a cross-sectional area of 5.00 cm² has a resistance of 20.0 Ω. The plane of the coil is perpendicular to a uniform magnetic field of 1.00 T. The coil is suddenly rotated about the axis shown in **Figure 20-43** through an angle of 60° over a period of 0.200 s. (a) What charge flows past a point in the coil during that time? (b) If the loop is rotated a full 360° around the axis, how much total charge passes the point in the loop? Explain your answer.

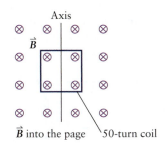

Figure 20-43 Problem 40

41. ●●A pair of parallel conducting rails that are 12 cm apart lies at right angles to a uniform magnetic field of 0.8 T directed into the page, as shown in **Figure 20-44**. A 15 Ω resistor is connected across the rails. A conducting bar is moved to the right at 2 m/s across the rails. (a) What is the current in the resistor? (b) What direction is the current in the bar (up or down)? (c) What is the magnetic force on the bar? SSM

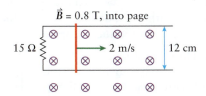

Figure 20-44 Problem 41

42. ●●●Calc As shown in **Figure 20-45**, a coil that is square, has 30 turns, is 0.82 Ω, and is 10.0 cm on a side is between the poles of a large electromagnet that produces a constant, uniform magnetic field of 0.6 T directed into the page. As suggested by the figure, the field drops sharply to zero at the edges of the magnet. The coil moves to the right at a constant velocity of 2.00 cm/s. What is the current through the coil wire (a) before the coil reaches the edge of the field, (b) while the coil is leaving the field, and (c) after the coil leaves the field? (d) What is the total charge that flows past a given point in the coil as it leaves the field? (e) Plot the induced current in the loop as a function of the horizontal position of the right side of the current loop. Let the right-hand edge of the magnetic field region be x = 0. Your plot should be in the range of −5.00 cm < x < 20.0 cm.

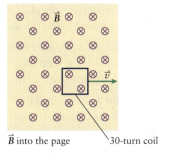

Figure 20-45 Problem 42

43. ●●(a) Determine the magnitude and direction of the force on each side of the coil in problem 42 for situations (a) through (c). (b) As the loop enters the field region from the left, what is the direction of the induced current and the resulting force on each segment of the coil?

20-4: Inductance

44. ●How much voltage is produced by an inductor of value 25 μH if the time rate of change of the current is 58 mA/s?

45. ●●Uncle Leo tunes an old-fashioned radio that has an antenna that is made from a 3.0-cm-long solenoid with a cross-sectional area of 0.50 cm², composed of 300 turns of fine copper wire. Calculate the inductance of the coil assuming it is air-filled. SSM

46. ●What energy is stored in a 250 mH inductor with a current of 0.055 A?

47. ●●The current in an inductor is given by the following equation: $i(t) = i_0 \cos \omega t$. If the inductor has a value of $L = 28$ mH, the maximum current is $i_0 = 125$ mA, and the angular frequency is $\omega = 25$ rad/s, calculate the induced voltage in the inductor when $t = 1.5$ s.

48. ●●A tightly wound solenoid of 1600 turns, cross-sectional area of 6.00 cm², and length of 20.0 cm carries a current of 2.80 A. (a) What is its inductance? (b) How much energy is stored in the solenoid? (c) If the cross-sectional area is doubled, does anything happen to the values of L and U? Explain your answer.

20-5: LC Circuits

49. ●An LC circuit is made up of a 14.4 mH inductor connected in series to a fully charged, 225 μF capacitor. If the maximum charge on the capacitor is 300 μC at $t = 0$ s, find the charge on the capacitor 7.5 ms after the circuit is complete. SSM

50. •Repeat the previous problem if the maximum charge on the capacitor of 300 μC now occurs at $t = 2.5$ ms.

51. •An LC circuit is formed with a 150 mH inductor and a 1000 μF capacitor. Calculate the frequency of oscillation f for the circuit.

52. •You have a 1.0 mH inductor. What size capacitor should you choose to make an oscillator with a natural frequency of 980 kHz?

53. •An LC circuit is built with a 100 μF capacitor and a 400 mH inductor. The current has its maximum value at $t = 0$ s. How long does it take to fully charge the capacitor? SSM

54. •A 200 pF capacitor is charged to 120 V and then quickly connected to an inductor. Calculate the maximum energy stored in the magnetic field of the inductor as the circuit oscillates.

55. •LC circuits are used for filters in electronics. The selectivity of an LC circuit is defined as the ratio L/C. Calculate the selectivity of the bandwidth for an LC circuit composed of an inductor ($L = 0.250$ H) and a capacitor ($C = 875$ μF).

56. ••An LC circuit is composed of a 150 μH inductor in series with a 2.20 pF capacitor. How long will it take before the charge on the capacitor reaches 75% of its maximum value?

57. ••When the charge on the capacitor in an LC circuit is one-half of the maximum stored charge, calculate the ratio of the energy stored in the capacitor compared to the total energy in both the inductor and the capacitor.

58. ••If the ratio of the energy stored in a capacitor compared to the total energy stored in an LC circuit is 0.5, calculate the ratio of the charge stored on the capacitor compared to the maximum charge stored on the capacitor in that circuit.

20-6: LR Circuits

59. •What is the time constant for an LR circuit if the inductor has a value of $L = 22$ mH and the resistor has a value of $R = 360$ Ω? SSM

60. •An LR circuit has a time constant of 256 ms. If the resistance is increased by a factor of 6, what is the new time constant?

61. •Determine the decay time constant for an LR circuit with a 100 Ω resistor and 10 H inductor.

62. •A 12 H inductor of negligible resistance is placed in series with a 12 V battery, a 3.0 Ω resistor, and a switch. What is the current 2.0 s after the switch is closed?

63. •A 12 H inductor of negligible resistance is placed in series with a battery, a 3.0 Ω resistor, and a switch. How long will it take for the current to reach 50% of its maximum value after the switch is closed? SSM

64. ••A 50 mH inductor with negligible resistance is placed in series with a 3.0 V battery, a 5.0 Ω resistor, and a switch. Once the switch is closed calculate the current (a) at a time of 4 s after the switch is closed and (b) when $t \rightarrow \infty$.

65. ••After the switch S in Figure 20-46 is closed, charge begins to flow. At the instant at which one-half the power supplied by the battery is being dissipated in the resistor, what is the current and how fast is it changing?

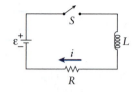

Figure 20-46 Problem 65

66. •••A tightly wound solenoid 18.0 cm long with a diameter of 2.00 cm is made of 1500 turns of 22-gauge copper wire. Calculate the (a) inductance, (b) resistance, and (c) time constant of the solenoid.

67. ••Calc The switch shown in Figure 20-47 is closed and the current through the resistor at some instant is 6.00 A to the left. (a) What is the rate of change of the current at that instant? (b) Is the current increasing or decreasing? (c) What is the potential difference $V_a - V_b$? (d) What would have to be true about the current for us to deduce that it is increasing? SSM

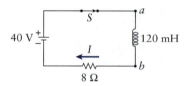

Figure 20-47 Problem 67

General Problems

68. •A 40.0 mH inductor is connected with a resistance of 5.00 Ω and placed across a battery of voltage 36.0 V. How much energy is stored in the inductor once the steady state is reached?

69. •••Calc A copper wire of length L is formed into a circular coil with N turns. When a magnetic field through the coil changes with time, determine the value of N that results in the maximum induced current in the coil.

70. ••Calc The current in the long, straight wire shown in Figure 20-48 is given by $i = i_0 \sin \omega t$, where i_0 equals 15 A and ω equals 120π rad/s. What is the induced current in the rectangular loop at $t = 0$ s and at $t = 2.09$ ms? Assume the resistance of the loop is 2.0 Ω.

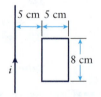

Figure 20-48 Problem 70

71. ••A magnetic field of 0.45 G is directed straight down, perpendicular to the plane of a circular coil of wire that is made up of 250 turns and has a radius of 20 cm. (a) If the coil is stretched, in a time of 15 ms, to a radius of 30 cm, calculate the voltage induced across the coil during the process. (b) Assuming the resistance of the coil is a constant 25 Ω, what is the induced current in the coil during the process? (c) What is the direction of the induced current in the coil (clockwise or counter-clockwise, as viewed from above)? SSM

72. •••Calc A long, rectangular loop of width w, mass m, and resistance R is being pushed into a magnetic field by a constant force $\vec{F}$ (Figure 20-49). Calculate the speed of the loop as a function of time. Assume that $t = 0$ s at the instant that the left edge of the loop enters the field.

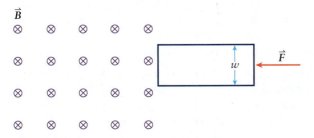

Figure 20-49 Problem 72

73. ••Calc Perhaps it has occurred to you that we could tap Earth's magnetic field as a source of electrical potential from which we could then generate energy. One way to do this would be to spin a metal loop about an axis perpendicular to Earth's magnetic field. Suppose that the metal loop is a square that is 45.0 cm on each side and that we want to generate a potential in the loop of amplitude 120 V at a place where Earth's magnetic field is 0.50×10^{-4} T. At what angular speed (in rev/s) would we have to spin the coil? Does this appear to be a feasible method to extract energy from Earth's magnetic field?

74. ••Astronomy, Calc Activity on the Sun, such as solar flares and coronal mass ejections, hurl large numbers of charged particles into space. When the particles reach Earth, they can interfere with communications and the power grid by causing electromagnetic induction. In one example, a current of millions of amps (known as the *auroral electrojet*) that runs about 100 km above Earth's surface can be perturbed. The change in the current causes a change in the magnetic field it produces at Earth's surface, which induces a potential difference along Earth's surface and in the power grid (which is grounded). Induced electric fields as high as 6.0 V/km have been measured. We can model the circuit at Earth's surface as a rectangular loop made up of the power lines and completed by a path through the ground beneath, and we can treat the magnetic field as being uniform (but not constant). The height of the power lines is 5.0 m, they

are 1.0 km long, and the potential induced in the loop is 6.0 V. At what rate must the magnetic field be changing to induce that potential difference?

75. •Medical During transcranial magnetic stimulation (TMS) treatment, a magnetic field typically of magnitude 0.50 T is produced in the brain using external coils. During the treatment, the current in the coils (and hence the magnetic field in the brain) rises from zero to its peak in about 75 μs. Assume that the magnetic field is uniform over a circular area of diameter 2.0 cm inside the brain. What is the magnitude of the average induced potential in the brain during the treatment?

76. ••Calc The Faraday disk dynamo is a conducting disk rotated about its axis while in a magnetic field directed perpendicularly to the surface of the disk, as shown in Figure 20-50. As the disk rotates, a radial line of the disk sweeps out an area, and a potential is induced between the center of the disk and its outside edge. When connected in a circuit, charges will flow. (a) If the disk has radius R_{disk}, the magnitude of the field has a magnitude B, the resistance is R, and the disk is rotated at an angular speed ω, find an algebraic expression for the current induced in the circuit. (b) Is the center of the disk or the outside edge at a higher potential in this case? Explain your answer.

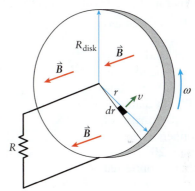

Figure 20-50 Problem 76

77. •••Calc A newly designed kitchen pot has a round metal coil embedded in its base. The coil is 12.0 cm in diameter and has a resistance of 22.5 mΩ. The surface of the stove contains an electromagnet that produces a uniform vertical magnetic field that oscillates sinusoidally with amplitude 0.850 T and frequency 60.0 Hz. (a) At what average rate does the system produce heat in the coil of the pot? *Hint:* The average of $\sin^2\theta$ over one period is equal to $\frac{1}{2}$. (b) How many minutes would it take the stove to heat 0.500 kg of water from 20°C to 50°C if all the heat generated goes into the water? (See Table 14-3.) SSM

78. •A permanent bar magnet with the north pole pointing downward is dropped into a solenoid. (a) Determine the direction of the induced current that would be

measured in the ammeter shown in **Figure 20-51**. (b) If the magnet is suddenly pulled upward through the solenoid, what is the direction of the induced current that would be measured in the ammeter?

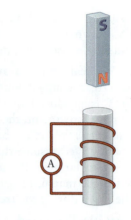

79. ••A tightly wound solenoid is 18.0 cm long, has a 2.00 cm diameter, and is made of 1500 turns of 22-gauge copper wire. It is surrounded by a 20-turn circular coil (3.00 cm in diameter) that is coaxial with the solenoid, as shown in **Figure 20-52**. The circular coil is connected across a resistor of very high resistance. (a) What is the magnitude of the induced potential in the coil when the current in the solenoid is changing at a rate of 100 A/s? (b) Consider the reverse situation and explain how you would find the induced potential in the solenoid when the current in the coil is changing. SSM

Figure 20-51 Problem 78

Figure 20-52 Problem 79

80. ••A circular coil that has 20 turns and is 1.60 cm in diameter is inside and coaxial with a tightly wound solenoid that is 18.0 cm long and has a 2.00 cm diameter made of 1500 turns of 22-gauge copper wire, as shown in **Figure 20-53**. The circular coil is connected across a resistor of very high resistance. What is the magnitude of the induced potential in the circular coil when the current in the solenoid changes at a rate of 100 A/s?

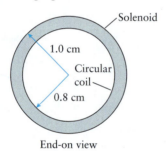

End-on view

Figure 20-53 Problem 80

81. ••A long solenoid that is 2 cm in diameter and is 20 cm long consists of 5000 turns of wire. (a) What is the field inside the solenoid when a current of 5 A is in it? (b) What is the self-inductance of the solenoid? (c) If the current in the solenoid drops to 0 in 0.1 s, what potential is generated in it?

82. •A coaxial cable consists of an inner conductor of radius r_i and an outer conductor of radius r_o (**Figure 20-54**). Assume that the two conductors

Figure 20-54 Problem 82

carry equal currents in opposite directions. Derive an algebraic expression for the inductance per unit length for the cable.

83. ••In addition to inductance, any real solenoid has some resistance due to its metal windings. Suppose that a cylindrical solenoid is 35.0 cm long and 4.50 cm in diameter. The windings are made of 22-gauge copper wire (diameter equal to 0.6438 mm) coated with an insulating varnish of negligible thickness. The windings are wrapped so adjacent loops touch each other but do not overlap. (a) How many loops are there to the solenoid? (b) What are the resistance and self-inductance of the solenoid? (See Table 18-1.) SSM

84. ••Suppose a 5.0-kg object is connected to a spring that has a constant k equal to 1.34 kN/m. The system is then modeled by an LC circuit with a 4.0 H inductor. What should C be in order for the LC circuit to have the same resonant frequency as the object–spring system?

85. ••A 25.0 μF capacitor and a 10.0 mH inductor are connected in series with an open switch between them. The capacitor is initially charged to a potential difference of 50.0 V when the switch is suddenly closed. How long after closing the switch will it take (a) for the energy in the system to first be equally shared between the capacitor and the inductor and (b) for the inductor to first have all the energy?

86. ••Calc The time rate of change of the current in an LR circuit is given by the following:

$$\frac{di}{dt} = J_0 \cos(\omega t) \quad \text{(SI units)}$$

(a) If $J_0 = 35$ mA/s and $\omega = 2$ s^{-1}, calculate the value of the current at $t = 0.4$ s. (b) The resistance of the circuit is 50 Ω. Calculate the inductance.

87. •••In the following circuit (**Figure 20-55**), take $V_0 = 12$ V, $R_1 = 4.0$ Ω, $R_2 = 8.0$ Ω, $R_3 = 2.0$ Ω, and $L = 2.0$ H. Switch S has been open for a very long time to start. (a) Redraw the circuit as it would function at the moment the switch is closed. (b) Redraw the circuit as it would function a very long time after the switch is closed. (c) After the current has reached a steady state with the switch closed, it is reopened. Calculate the currents in the three resistors. (d) A long time after the switch is reopened, what are the currents in the three resistors?

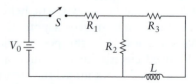

Figure 20-55 Problem 87

21 AC Circuits

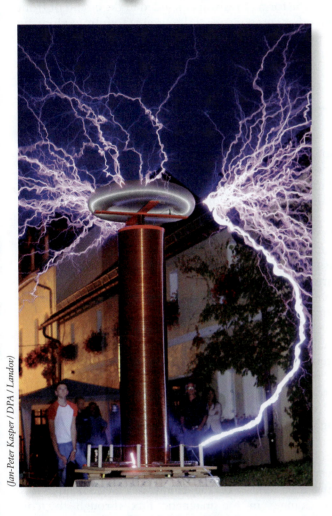

(Jan-Peter Kasper / DPA / Landov)

A Tesla coil can produce enormous sparks by charging a conducting sphere—up to 100,000 V or more—and then allowing the sphere to discharge through the air. At the heart of the device are two coils: a flat one (on top of the white base) and a cylindrical one formed by winding copper wire around and around a long, vertical pipe. When the current changes in the flat coil, the magnetic field it produces changes, which in turn creates a changing magnetic flux in the cylindrical coil. According to Faraday's law, the changing magnetic flux induces an electric potential in the cylindrical coil, which charges the conducting sphere. Although there are only about 10 windings in the flat coil, there may be thousands in the cylindrical one. The induced potential is proportional to the number of windings, so the voltage in the cylindrical coil is much higher than in the flat one.

Interesting phenomena arise when we apply an external driving force to a system that can oscillate freely. For example, when a mechanical system that is free to oscillate experiences a periodic driving force, and when the frequency of that force matches the natural frequency of the system, the amplitude of oscillations becomes large (Section 12-8). This phenomenon, known as resonance, can also occur in electrical systems, as we saw in the last chapter (Section 20-5). There we introduced the *LC* circuit as an oscillating system, although we treated it in an idealized way by ignoring the fact that it dissipates energy. We also simplified our description by not considering the response of the circuit to an external, time-dependent driving voltage. In this chapter, we will extend our discussion of the electrical oscillator to include both dissipative and driving terms.

For a mechanical oscillator such as a block and spring, the total energy is $E_{\text{total}} = \frac{1}{2}kA^2$ (Equation 12-10), where k is the spring constant and A is the amplitude of oscillations. The total energy E_{total} is constant because both k and A are constant. For the *LC* circuit, we found that the total energy can also be expressed in terms of the capacitance C and the maximum charge Q_0 on the capacitor, $E_{total} = \frac{1}{2}\frac{1}{C}Q_0^2$ (Equation 20-23). Both C and Q_0 are constant, so here again, E_{total} does not change with time. Constant energy implies, however, that the amplitude of oscillations would not diminish over time, which we know can't be true in a real system. For the block and spring, damping always causes the amplitude of oscillations to die away. In an electrical circuit, the dissipation of energy results from resistance, either in the form of resistors included in the circuit or from resistance offered by wires and other components of the circuit. To fully explore the electrical oscillator, then, we need to include resistance as well as inductance and capacitance.

21-1 Alternating Current

Current arises in a circuit as a result of electric potential. Until now, we have only considered sources of potential that are fixed at a constant voltage. The batteries that power your laptop or your flashlight are sources of constant voltage. A 9 V battery, for example, introduces a constant 9 V potential difference between the two points at which it connects to a circuit. By convention, we refer to a circuit driven by a fixed voltage source, or one that does not change direction, as a **direct current**, or **DC**, **circuit**. To discuss the full range of electric circuits, we will now turn our attention to those driven by an electric potential that not only varies but also "alternates," that is, it changes direction over time. A varying potential, often a periodically varying potential, gives rise to a time-varying current. For that reason, the alternating voltage source is referred to as **alternating current** or **AC**. AC voltage is what you get when you plug an electronic device into a wall outlet in your room.

Electric generators, such as those run by water flowing through the Grand Coulee Dam (**Figure 21-1**), convert linear motion into the rotational motion of a conducting coil within a magnetic field. As the coil turns, the angle φ between the fixed direction of the magnetic field and the rotating area vector of the coil varies periodically; this causes changes in the magnetic flux through the coil $\Phi_B = \int B \, dA \cos \varphi$ (Equation 20-2). Therefore, according to Faraday's law ($\mathscr{E} = -d\Phi_B/dt$, Equation 20-3), the induced or generated potential will also vary periodically. In other words, today's electric generators produce alternating current. In the early days of electrical power distribution, DC was used exclusively, but as we will see, the advantages of AC over DC far outweigh the disadvantages.

Perhaps the most important advantage of AC over DC is that it is relatively straightforward to change the voltage delivered by an AC source. Most common electrical devices used by consumers operate at a low and relatively safe voltage. But the power generated by, say, the hydroelectric generators at the Grand Coulee Dam must be transmitted over long wires to reach your room and run your laptop and desk lamp. As we will see, significant losses can occur in the process of transmitting power from one place to another over wires.

Figure 21-1 Water flowing through electric generators inside the Grand Coulee Dam convert the linear motion of water into the rotational motion of a conducting coil within a magnetic field. Located in central Washington on the Columbia River, this dam generates 6.5 million kilowatts of power. *(Christopher J. Morris / Corbis)*

In Chapter 18, we saw that when there is current in a wire, the power dissipated by the wire is proportional to the resistance of the wire and to the square of the current it carries. Thus, the loss of power in a transmission wire is

$$P_{\text{loss}} = i^2 R \qquad (21\text{-}1)$$

The resistance of a wire increases with length, so the power loss, mostly in the form of thermal heating, increases as the length of the wires grows. In addition, we see from

$$P = Vi \qquad (18\text{-}26)$$

that to transmit a significant amount of power P, either the current, the voltage, or both must be high. Consider the consequences of delivering power at high current. In our equation for power loss (Equation 21-1) the value of i is squared, so given that R will be large for long wires, delivering power at high current over any appreciable distance results in significant power loss. This method of delivery would be inefficient and costly. However, while the resistance of the wires over which electrical power is transmitted is fixed by the choice of wire diameter, length, and composition, we can keep the current low by choosing to deliver the power at a high voltage. From Equation 21-1, $P_{\text{loss}} = i^2 R$, we know that smaller current results in less power loss. It is far more efficient, then, to transmit power on high voltage power lines, such as the ones in **Figure 21-2**, and then, before supplying power to the consumer, to reduce the voltage with a *transformer* such as the ones we will explore in the next section.

We describe AC voltage $V(t)$ mathematically as

$$V(t) = V_0 \sin(\omega t) \qquad (21\text{-}2)$$

where V_0 is the peak value, or amplitude, of the sinusoidally varying voltage and ω is its angular frequency. The angular frequency is related to the frequency f of the time-varying voltage by a factor of 2π:

$$\omega = 2\pi f$$

In **Figure 21-3**, we have labeled the period T equal to $1/f$ and $2\pi/\omega$. The voltage supplied to homes, *wall voltage*, oscillates at a frequency of 60 Hz in North America and at 50 Hz in Europe, Australia, and other parts of the world.

Figure 21-2 Transmitting power on high voltage power lines is much more efficient than delivering power at high current. (*Peter & Georgina Bowater / Stock Connection / Aurora*)

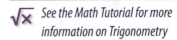 *See the Math Tutorial for more information on Trigonometry*

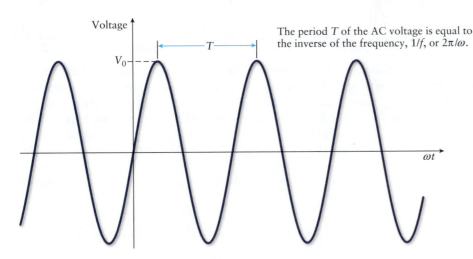

The period T of the AC voltage is equal to the inverse of the frequency, $1/f$, or $2\pi/\omega$.

Figure 21-3 V_0 is the peak amplitude of the sinusoidally varying voltage and ω is its angular frequency. The period T is equal to $1/f$ or $2\pi/\omega$.

Wall voltage in the United States and Canada is often referred to as "120 volts AC." This value is *not* the amplitude of the voltage V_0, but rather an average value of the voltage as it varies with time. This average is not, however, the arithmetic mean, which is computed by adding values and then dividing the sum by the number of values included. AC voltages vary sinusoidally and are just as often negative as positive, so the arithmetic mean of the AC voltage described by Equation 21-2 is zero. Instead of taking the arithmetic mean, we will use the **root mean square (rms) average**, which is not influenced by whether the values are positive or negative.

Although the arithmetic mean of a sinusoidally varying voltage is zero, the arithmetic mean of the power delivered by that voltage is not, because it is proportional to the square of the voltage, V^2, a quantity that is always positive. From Equations 18-27 and 21-2, we see that power delivered at any instant of time is related to the voltage by

$$P(t) = \frac{V^2(t)}{R} \tag{21-3}$$

or

$$P(t) = \frac{V_0^2 \sin^2(\omega t)}{R}$$

The arithmetic mean of power for N instants of time is

$$P_{\text{avg}} = \frac{1}{N} \sum_{i=1}^{N} \frac{V_0^2 \sin^2(\omega t_i)}{R}$$

This expression is not zero, in general, because squaring the sine term allows for no negative values. Of course, the voltage varies in a continuous fashion so we shouldn't pick out discrete values of time, but instead average over all times. To do this we convert the sum to an integral. Over one full cycle of the voltage, the average power is then

$$P_{\text{avg}} = \frac{1}{T} \int_0^T \frac{V_0^2 \sin^2(\omega t)}{R} \, dt \tag{21-4}$$

In Equation 21-4, V_0 and R are both constant, so both can be pulled out of the integral (Math Box 21-1).

$$P_{\text{avg}} = \frac{V_0^2}{R} \frac{1}{T} \int_0^T \sin^2(\omega t) \, dt$$

Math Box 21-1 Integrals and Averages

Although expressed in integral form Equation 21-4 is still an add-them-up, divide-by-N average, which means we can pull any terms that are constant out of the integral. As a simple example, consider the average of three numbers, 6, 8, and 16, which is $(6 + 8 + 16)/3$ or 10. Now express the three numbers as 2×3, 2×4, and 2×8. The average is $(2 \times 3 + 2 \times 4 + 2 \times 8)/3$. The value 2 can be pulled from the sum, which gives $2((3 + 4 + 8)/3)$ or 10, the same result as before.

A straightforward way to carry out the integral is to use the trigonometric substitution

$$\sin^2 \alpha = \frac{1 - \cos(2\alpha)}{2}$$

so

 See the Math Tutorial for more information on Integrals

$$P_{avg} = \frac{V_0^2}{R} \frac{1}{T} \int_0^T \frac{1 - \cos(2\omega t)}{2} dt$$

and

$$P_{avg} = \frac{V_0^2}{R} \frac{1}{T} \left[\int_0^T \frac{1}{2} dt - \int_0^T \frac{\cos(2\omega t)}{2} dt \right]$$

The integral on the left yields $T/2$. The second integral must be zero, because any periodic function is positive half of the time and is negative the other half over one full cycle. (We encourage you to evaluate the integral and prove it to yourself!) Thus,

$$P_{avg} = \frac{V_0^2}{R} \frac{1}{T} \left[\frac{T}{2} - 0 \right] = \frac{1}{2} \frac{V_0^2}{R} \qquad (21\text{-}5)$$

Power as a function of time is described by

$$P(t) = \frac{V^2(t)}{R} \qquad (21\text{-}3)$$

The average of the left side of the equation equals the average of the right side, and this must equal the average given in Equation 21-5. So

$$\left(\frac{V^2(t)}{R} \right)_{avg} = \frac{1}{2} \frac{V_0^2}{R}$$

or

$$(V^2(t))_{avg} = \frac{V_0^2}{2}$$

We define the rms average of the voltage by taking the square root of both sides:

$$V_{rms} = \sqrt{(V^2(t))_{avg}} = \sqrt{\frac{V_0^2}{2}} = \frac{V_0}{\sqrt{2}} \qquad (21\text{-}6)$$

As you can see, we have taken the square *root* of the average (*mean*) of the *square* of the voltage in order to define the root mean square value.

The rms value of the AC voltage we get from a standard wall outlet in North America is 120 V. Thus, the amplitude of the voltage in Equation 21-2 is

$$V_0 = \sqrt{2} V_{rms} = \sqrt{2}(120 \text{ V}) = 170 \text{ V}$$

The *peak* voltage from an AC wall outlet in North America is about 170 V. In other parts of the world, for example, Europe, Australia, and most of South America, Africa, and Asia, the rms voltage is 230 V. The peak voltage is $\sqrt{2}$ times the rms voltage or about 325 V. In Japan, the rms voltage is 100 V and the peak voltage about 140 V.

? Got the Concept 21-1
Power, Voltage, and Current

For a circuit driven by an AC voltage, rewrite the fundamental relationships for power, Equations 18-26, 18-27, and 18-28, in terms of the rms values of voltage and current, V_{rms} and i_{rms}.

Example 21-1 Going Abroad? Be Careful with Your Hair Dryer!

Seasoned travelers know that an electrical device that works in one country may not work in another. Imagine taking a 1500 W (the average maximum power) hair dryer made in the United States, where the rms wall voltage is 120 V, to Australia, where the rms wall voltage is 230 V. Find the power output of the hair dryer when connected to a wall outlet in Australia. *Hint:* Treat the resistance of the hair dryer as the same regardless of the wall voltage. In reality, the resistance depends somewhat on the temperature of the dryer's heating elements.

SET UP

Because the resistance of the hair dryer is the same regardless of the wall voltage, resistance provides a way to compare the power output of the hair dryer for different values of V_{rms}. The average power (Equation 21-5) can be expressed in terms of V_{rms} through the relationship between V_{rms} and V_0. From Equation 21-6,

$$V_0^2 = 2V_{rms}^2$$

So, from Equation 21-5,

$$P_{avg} = \frac{1}{2}\frac{2V_{rms}^2}{R} = \frac{V_{rms}^2}{R}$$

Rearranging the expression we find that

$$R = \frac{V_{rms}^2}{P_{avg}}$$

The resistance of the hair dryer in the United States is the same as in Australia, so

$$\left.\frac{V_{rms}^2}{P_{avg}}\right|_{US} = \left.\frac{V_{rms}^2}{P_{avg}}\right|_{Australia}$$

SOLVE

To arrive at a solution, we rearrange the above expression:

$$P_{avg,\,Australia} = V_{rms,\,Australia}^2 \frac{P_{avg,\,US}}{V_{rms,\,US}^2}$$

We then substitute in the values given in the problem statement:

$$P_{avg,\,Australia} = (230\text{ V})^2 \frac{1500\text{ W}}{(120\text{ V})^2} = 5500\text{ W}$$

REFLECT

A power of 5500 W is *considerably* higher than the intended output of the hair dryer. You may know how hot the air from a 1500 W hair dryer can be, or perhaps you've felt how hot a 100 W bulb can get after it's been on for a while. So you can probably imagine how hot 5500 W would make the heating element of the hair dryer. To radiate that much power, the required temperature of the heating element might well exceed the melting point of the material!

Practice Problem 21-1 Many desk lamps are made to be used with a 60 W bulb in the United States where the rms wall voltage is 120 V. If you take such a lamp to Australia, where the rms wall voltage is 230 V, will the lamp be brighter or dimmer, and by how much?

★ What's Important 21-1

The arithmetic mean of a sinusoidally varying voltage is zero because the voltage is negative as often as it is positive. The average power is not zero because it is proportional to the square of the voltage, which is always positive. For this reason it is common to characterize AC voltage by its root mean square (rms) value. The peak voltage is $\sqrt{2}$ times the rms voltage.

21-2 Transformers

For reasons of safety, most of the electrical devices you use everyday operate at low voltages. It is not efficient, however, to transmit power over long distances, say, from the hydroelectric plant to your home, at low voltage. Transmitting power over long distances at low voltage necessarily requires high current, and as we saw in the previous section, power loss increases as the square of the current ($P_{\text{loss}} = i^2 R$, Equation 21-1). To deliver power efficiently as well as economically, most long-distance power lines operate at low current and high voltage. In this section, we examine the **transformer**, a device that increases the voltage of an electric current while decreasing the current, or vice versa. Such a device allows power to be transmitted efficiently at high voltage but used at safer, low voltages.

The transformer shown in **Figure 21-4** consists of two coils formed by winding separate wires around a piece of iron. The coil on the left, called the **primary coil**, is connected to the input voltage. The coil on the right, the **secondary coil**, is connected to the output. Notice that the two coils have a different number of windings, which we have labeled N_p and N_s.

When an input voltage is applied to the primary coil in the transformer, the current loops give rise to a magnetic field. Because the field changes as the AC voltage changes, the magnetic flux in the region is also continually changing, and the changing flux induces an electric potential in the secondary coil. The magnetic flux and the change in magnetic flux are the same in both coils.

According to Faraday's law of induction (Equation 20-4), the applied voltage V_p together with the self-inductance of the primary coil gives rise to a changing magnetic flux $d\Phi_B$ in the primary coil:

$$V_p = -N_p \frac{d\Phi_B}{dt} \tag{21-7}$$

This equation is true at any instant of time and also when averaged over a period of time, so we can choose to take V_p as the rms average of the input voltage. The changing flux $d\Phi_B$ induces an electric potential in the secondary coil:

$$V_s = -N_s \frac{d\Phi_B}{dt} \tag{21-8}$$

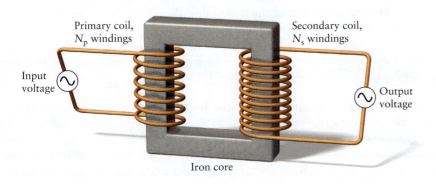

Primary coil, N_p windings

Secondary coil, N_s windings

Input voltage

Output voltage

Iron core

Figure 21-4 Transformers consist of two coils formed by winding separate wires around a piece of iron. Application of an input voltage to the primary coil gives rise to a magnetic field and induces an electric potential in the secondary coil. Notice that the two coils have a different number of windings, which we have labeled N_p and N_s.

where, again, we will take V_s as the rms value of the output voltage. Rearranging both Equation 21-7 and Equation 21-8 yields

$$\frac{d\Phi_B}{dt} = -\frac{V_p}{N_p} = -\frac{V_s}{N_s}$$

The voltage across the secondary coil is therefore

$$V_s = \frac{N_s}{N_p} V_p \tag{21-9}$$

In other words, the ratio of the number of windings in the primary coil to the number of windings in the secondary coil determines the output voltage relative to the voltage input to the transformer. In a *step-up transformer*, N_s is greater than N_p, so the output voltage is greater than in the input voltage. In a *step-down transformer*, N_s is less than N_p, which results in an output voltage less than the input voltage.

How does the current in the secondary coil compare to the current in the primary coil? If we take the transformer to be ideal by neglecting thermal losses and by assuming that there are no losses in the inductive coupling between the primary and secondary coils, conservation of energy requires that the rate at which energy is delivered to the primary coil by the input voltage must equal the rate at which energy is transferred from the primary to the secondary coil. The rate of energy delivery is power, so

$$P_p = P_s$$

Power is the product of voltage and current (Equation 18-26), so

$$V_p i_p = V_s i_s$$

Here i_p and i_s are the rms values of the currents in the two coils. The two currents are related by

$$i_s = \frac{V_p}{V_s} i_p$$

or, by applying the relationship between the voltages (Equation 21-9),

$$V_s = \frac{N_s}{N_p} V_p$$

$$i_s = \frac{N_p}{N_s} i_p \tag{21-10}$$

Compare Equation 21-10 with Equation 21-9. Because the voltage in the secondary coil varies as N_s/N_p but the current varies as the inverse N_p/N_s, the increase in voltage in a step-up transformer, for which N_s is greater than N_p, is accompanied by a corresponding decrease in current. The reverse is true for a step-down transformer.

Recall that for long-distance transmission of power, we want to reduce losses in the wires by driving a low current at high voltage. A step-up transformer does exactly that. At the delivery end of the power line, step-down transformers lower the voltage for use in the electric devices that you commonly use at home and at school.

Example 21-2 High Voltage Transformer

Each of the 17 generators employed in the hydroelectric power plant at Hoover Dam in Colorado can generate up to 133 MW of power. The power is delivered at about 8 kV to the transformers that connect to long-distance transmission lines. Consider a transmission line that operates at 500 kV. What is the ratio of the number of windings in the secondary coil to the number in the primary coil in a step-up transformer that transforms 133 MW of power from 8 kV to 500 kV? What is the current in the high voltage transmission line?

SET UP

The relationship between the output and input voltages of an ideal transformer depends only on the ratio of the number of windings in the two coils:

$$V_s = \frac{N_s}{N_p} V_p \qquad (21\text{-}9)$$

Because we know both the transmission voltage (500 kV) and the power delivered (133 MW) we can find the current from

$$P = Vi \qquad (18\text{-}26)$$

SOLVE

Using Equation 21-9,

$$\frac{N_s}{N_p} = \frac{V_s}{V_p} = \frac{500 \times 10^3 \text{ V}}{8 \times 10^3 \text{ V}} = 62.5$$

So for every winding in the primary coil there must be 62.5 windings in the secondary coil. Or, because windings come in whole numbers, there must be 125 windings in the secondary for every 2 in the primary.

The current, by using Equation 18-26, is

$$i = \frac{P}{V} = \frac{133 \times 10^6 \text{ W}}{500 \times 10^3 \text{ V}} = 266 \text{ A}$$

REFLECT

A step-up transformer that has 125 windings in the secondary for every 2 windings in the primary takes AC voltage from 8 kV to 500 kV. In the process of delivering power to the end user, the voltage must ultimately be stepped down to 120 V. This is not normally done with a single transformer. One reason is that if only one transformer were used for the stepping-down process, there would need to be more than 4000 windings in the primary coil for every one in the secondary. (The ratio of 500 kV to 120 V is 4167:1.)

We found a current of 266 A in the transmission line. Although that sounds large, it is only about a factor of 10 more than the current in household wiring. In addition, the power loss in the transmission wire is relatively low at such a current. A few hundred kilometers of transmission wire might have a resistance of, say, 25 Ω. Using Equation 21-1, the loss in such a line would be about

$$P_{loss} = (266 \text{ A})^2 (25 \text{ }\Omega) = 1.8 \times 10^6 \text{ W}$$

The loss of about 2 MW is only 2/133 or 1.5% of the total power transmitted.

> ### ✳ What's Important 21-2
> In a step-up transformer, there are more windings in the secondary coil than in the primary coil, so the output voltage is greater than the input voltage. In a step-down transformer, there are fewer windings in the secondary coil than in the primary coil, so the output voltage is less than the input voltage. The increase in voltage in a step-up transformer is accompanied by a corresponding decrease in current. The reverse is true for a step-down transformer.

21-3 The Series *LRC* Circuit

In Section 12-8, we saw that when an oscillating system is driven at its natural frequency, the amplitude of the oscillations becomes large. A similar resonance phenomenon arises when we drive a series *LC* circuit with a current that varies in time with a frequency equal to the circuit's natural frequency ω_0:

$$\omega_0 = \sqrt{\frac{1}{LC}} \tag{20-21}$$

Causing an *LC* circuit to go into resonance can be useful. For example, if the inductance is fixed and known and the capacitance can change, the resonance response of the *LC* circuit tells us about whatever is causing that change. A common application employs a capacitor in which the plate separation grows and shrinks in response to external pressure; the frequency at which resonance occurs in a series *LC* circuit built with such a capacitor provides a measurement of the pressure.

Our goal for the bulk of the rest of this chapter is to understand the electrical analog of a forced, damped harmonic oscillator. This analog is the circuit shown in **Figure 21-5**, in which an inductor, a resistor, and a capacitor are connected in series to each other and to an AC voltage source. The inductor and capacitor together are the electrical equivalent of a simple harmonic oscillator (Section 20-5), the resistor provides the electrical equivalent of damping, and the periodic AC voltage source is the electrical equivalent of the driving force for a mechanical oscillator. We refer to the circuit as a series *LRC* circuit. It is the fundamental building block of a wide variety of electronic devices, including devices used to perform magnetic resonance imaging (MRI).

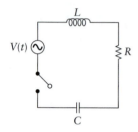

Figure 21-5 *LRC* circuits are used in many electronic devices.

In the course of the next three sections, we will analyze the series *LRC* circuit connected to an AC voltage source, primarily to understand how the current in the circuit varies with time. That is, for a voltage that varies like

$$V(t) = V_0 \sin(\omega t) \tag{21-2}$$

where V_0 is the peak voltage and ω is the angular frequency at which the voltage varies, we want to know how the current varies with time. In general,

$$i(t) = i_0 \sin(\omega t - \varphi) \tag{21-11}$$

so we need to determine i_0, the peak current (or current amplitude), and φ, the phase difference between the voltage and the current. Recall that φ tells us how one periodic function rises and falls with respect to another. When φ equals zero, the voltage and current rise and fall together; other values of the phase difference indicate whether the rise and fall of current precedes or follows the voltage, and by how much.

We have analyzed other circuits by requiring that the sum of all voltage drops around a closed loop equals zero, according to Kirchhoff's loop rule (Section 18-4). For example, in a circuit with only an inductor and a capacitor connected in series, we satisfied Kirchhoff's loop rule by setting the sum of the voltage drops ($V_L + V_C$) equal to zero, and substituting Equation 20-15 and Equation 17-19 respectively, for V_L and V_C, or

$$-L\frac{di}{dt} - \frac{1}{C}q = 0 \tag{20-19}$$

The solution to this differential equation gives the current in the circuit as a function of time. Summing the voltage drops around the circuit in Figure 21-5 gives

$$V(t) - V_L - V_R - V_C = 0 \tag{21-12}$$

or by using $V(t)$ from Equation 21-2 and by using V_L, V_R, and V_C from Equation 20-15, Equation 18-9, and Equation 17-19 respectively,

$$V_0 \sin (\omega t) - L\frac{di}{dt} - Ri - \frac{1}{C}q = 0 \qquad \textbf{(21-13)}$$

The current is the time rate of change of the charge that passes any cross-sectional area of the circuit and ends up on the capacitor, so

$$i = \frac{dq}{dt} \qquad (18\text{-}2)$$

and

$$\frac{di}{dt} = \frac{d^2q}{dt^2}$$

So Equation 21-13 becomes

$$L\frac{d^2q}{dt^2} + R\frac{dq}{dt} + \frac{1}{C}q = V_0 \sin (\omega t) \qquad \textbf{(21-14)}$$

This equation has the same form as the equation that describes the damped, forced mechanical oscillator:

$$m\frac{d^2x}{dt^2} + b\frac{dx}{dt} + kx = F(t) \qquad (12\text{-}29)$$

The series *LRC* circuit is an electrical damped, forced oscillating system.

Equation 21-14 may appear a bit daunting, but don't worry. Just as we did for the mechanical oscillator, we can learn a lot about the series *LRC* circuit without directly solving the differential equation. Once we've done that, we'll combine all the pieces into one, complete solution.

By first finding the angular frequency ω and the phase difference φ, we can then determine the current in the series *LRC* circuit (Equation 21-11). To accomplish this we employ a method that relies on rotating vectors, called **phasors**, to represent the voltage and current, both of which vary periodically over time. **Figure 21-6a** shows a phasor A. Here A is both the label of the phasor and also its magnitude. By convention, phasors rotate counterclockwise; the curved arrow shows the rotation of phasor A at angular frequency ω. The changing angle between the phasor and the positive x axis is therefore ωt. The value of the phasor at any time (its instantaneous value) is defined as the projection of the phasor onto the y axis. (Taking the instantaneous value of the phasor as the projection on the y axis rather than the x axis is arbitrary but standard.) In the figure, the right triangle formed by the phasor and the angle ωt has been enlarged to make it more clear that

$$A(t) = A \sin (\omega t)$$

You can see that $A(t)$ ranges from $-A$, when the phasor points straight down, to $+A$, when the phasor points straight up. At the time shown in **Figure 21-6b**, for example, the phasor has just passed its most positive value. At the time shown in **Figure 21-6c** the phasor is approaching its most negative value.

The phase difference between two sinusoidal functions can be read directly from a phasor diagram such as the one in **Figure 21-7a**. Phasors A and B are separated by angle φ. At the moment shown in **Figure 21-7b**, when $A(t)$ has its maximum positive value, the value of $B(t)$ is still rising toward its maximum. The instantaneous values of the two phasors are shown in **Figure 21-7c** as a function of time, although note that because the argument of a sine function must be an angle, we use the angle ωt as the horizontal axis. In Figure 21-7c, the times at which $A(t)$ and $B(t)$ are maximum are marked by a red vertical dashed line and a blue vertical dashed line, respectively.

Figure 21-6 Phasors, shown here as blue arrows, are rotating vectors and represent a quantity that varies periodically over time. (a) The value of a phasor at any instant in time is its projection onto the y axis. (b) The value of the phasor is largest when it points straight up. (c) The value of the phasor is smallest (and negative) when it points straight down.

The phasor A rotates counterclockwise at angular frequency ω.

The value of the phasor at any time is its projection onto the y axis, $A(t) = A \sin(\omega t)$.

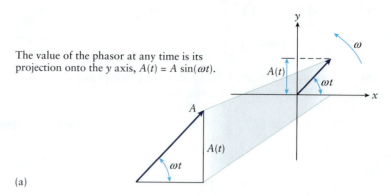

(a)

At this moment the phasor has just passed its most positive value.

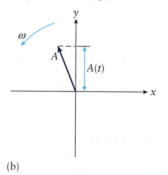

(b)

At this moment the phasor is approaching its most negative value.

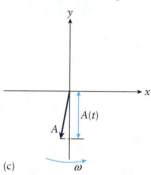

(c)

You can see that one of the sine functions plotted in Figure 21-7c is shifted from the other by the same angle φ that separates the phasors in Figure 21-7a and Figure 21-7b. We could represent the time-varying values of the phasors as

$$A(t) = A \sin(\omega t) \tag{21-15}$$

and

$$B(t) = B \sin(\omega t - \varphi) \tag{21-16}$$

In this case, because A reaches its peak before B we say that A leads B by φ. It is equivalent to say that B lags A by φ.

 Watch Out

When phasor B reaches its peak after phasor A, the phase difference is a negative value.

We represented the time-varying values of two phasors A and B as sine functions in Figure 21-7c. Time increases to the right in the figure, so $A(t)$ reaches its peak *earlier* than $B(t)$, which is why we say that A leads B. This results in a negative value of φ when $A(t)$ and $B(t)$ are represented as sine functions as in Equations 21-15 and 21-16. You can verify that φ has the proper sign by considering the value of $B(t)$ when t equals 0. At that instant, when $A(t)$ is equal to 0, $B(t)$ should have value of $B \sin(-\varphi)$ according to Equation 21-16. Because the value of φ is between 0° and 90° (by inspection of Figure 21-7a), we therefore expect $\sin(-\varphi)$ to be negative, which it is in Figure 21-7c.

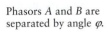

Phasors *A* and *B* are separated by angle φ.

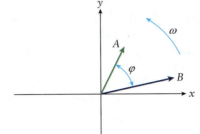

(a)

When *A* is at its maximum positive value, *B* must still rotate by angle φ to reach its maximum value.

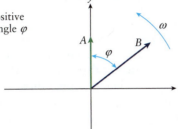

(b)

A(*t*) and *B*(*t*) vary sinusoidally with time. When *A*(*t*) is at its largest positive value (the red dashed line), *B*(*t*) is still rising toward its maximum value.

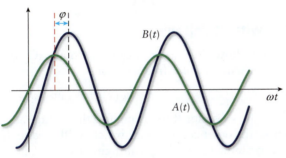

A reaches its peak before B, so A leads B by φ, or equivalently, B lags A by φ.

(c)

Figure 21-7 (a) The phase difference between two sinusoidal functions can be read directly from a phasor diagram. (b) You can think of the phase difference between A and B as the angle that B must rotate before it reaches its maximum at the moment when A is at its peak. (c) Plotting phasors A and B as sinusoidal functions gives the same information as the phasor diagram.

? Got the Concept 21-2

Phasor Diagrams

Make a phasor diagram that shows both phasors in each of these cases: (a) $A(t) = 5 \sin(\omega t)$ and $B(t) = 3 \sin(\omega t + \pi/2)$, (b) $A(t) = 2 \sin(\omega t)$ and $B(t) = \sin(\omega t)$, and (c) $A(t) = 5 \sin(\omega t)$ and $B(t) = 3 \sin(\omega t - \pi)$. In your diagrams, show the relative magnitudes and the phase differences. Notice that where the phase difference is not zero, the value is given in radians.

To determine the current amplitude and the phase difference between the current and voltage in the series *LRC* circuit (that is, to find i_0 and φ in Equation 21-11), we will consider separately the effect of a resistor, a capacitor, and an inductor on a circuit driven by an AC voltage. In the next section, we will draw and then analyze a phasor diagram that represents the voltage and current in each case.

✳ What's Important 21-3

Phasors are rotating vectors that represent voltage and current as they vary periodically over time. By convention, phasors rotate counterclockwise. The instantaneous value of a phasor at any moment of time is given by the projection of the phasor onto the y axis. The phase difference between two sinusoidal functions can be read directly from a phasor diagram.

21-4 *L, R, C* Separately with AC

The voltage difference across electrical components connected in series is the sum of the voltage differences across each one separately. Moreover, each component has the same current when they are connected in series. We can therefore examine the voltage across and the current through each component in the series *LRC* circuit separately (Figure 21-5), then combine those pieces into the complete picture of the full circuit. In this section, we will consider three separate circuits, one with a resistor connected to an AC voltage source, one with a capacitor, and one with an inductor. We start with the resistor circuit because it is mathematically more straightforward than the other two.

Resistor with AC Voltage

Figure 21-8 shows a resistor in series with an AC voltage source $V_R(t)$ that varies sinusoidally with angular frequency ω:

$$V_R(t) = V_0 \sin(\omega t) \tag{21-17}$$

We added the subscript R to distinguish the voltage across the resistor from the voltages across the other components, which we'll need to do when we combine the separate resistor, inductor, and capacitor circuits. Note, however, that there is no need to label V_0 with a subscript R. In the combined circuit, there is only one voltage source, which means that the applied voltage $V_0 \sin \omega t$ will be the same in all three circuits.

The current through the resistor comes directly from $V = iR$ (Equation 18-9), or

$$i_R(t) = \frac{V_R(t)}{R} \tag{21-18}$$

Substituting Equation 21-17 into Equation 21-18 we see that

$$i_R(t) = \frac{V_0}{R} \sin(\omega t) \tag{21-19}$$

Both voltage (Equation 21-17) and current (Equation 21-19) are proportional to $\sin \omega t$, which means that the current and voltage in the circuit are in phase. Phasors representing the voltage and current therefore lie on top of each other as in Figure 21-9.

We drew the current phasor in Figure 21-9 with a smaller amplitude than the voltage phasor. This assignment was a completely arbitrary choice on our part. The current amplitude is defined by V_0/R, by using Equation 21-19. The resistance can be more than, less than, or equal to 1 Ω, so the current amplitude can be larger than, smaller than, or equal to any numerical value of V_0.

Capacitor with AC Voltage

Figure 21-10 shows a capacitor in series with an AC voltage source that varies sinusoidally with angular frequency ω. As we did for the circuit with a resistor

Figure 21-8 A resistor is connected in series with an AC voltage source.

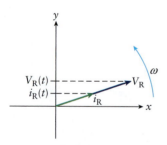

Figure 21-9 Because the current and voltage are in phase in a circuit with a resistor connected to an AC voltage source, phasors representing the voltage (shown in blue) and the current (shown in green) lie on top of each other. Although we drew the amplitude of the current phasor smaller than that of the voltage phasor, the current amplitude can be larger than, smaller than, or equal to any numerical value of voltage.

connected to an AC voltage source, we begin by writing the voltage difference across the capacitor:

$$V_C(t) = V_0 \sin(\omega t) \quad\quad (21\text{-}20)$$

and then find the current in the circuit. Again, V_0 does not carry an added subscript because the same voltage source is common to all three components in the series *LRC* circuit.

The current in the circuit determines the rate at which charge accumulates on the capacitor. The amount of charge stored on a capacitor depends on the voltage between the two plates as well as the capacitance:

$$q = CV \quad\quad (17\text{-}19)$$

Because current is defined as the time rate of change of charge (Equation 18-2), we can find the current by

$$i(t) = \frac{dq(t)}{dt}$$

or, by substituting $CV(t)$ for $q(t)$,

$$i_C(t) = \frac{d}{dt}CV_C(t) \quad\quad (21\text{-}21)$$

Note that although no current is conducted *across* the capacitor, charge flows in the rest of the circuit in order to charge the capacitor.

Combining Equation 21-21 with the expression for voltage (Equation 21-20) gives

$$i_C(t) = \frac{d}{dt}CV_0 \sin(\omega t) = \omega CV_0 \cos(\omega t)$$

To make the phase difference between $i_C(t)$ and $V_C(t)$ easier to see, note that

$$\cos\theta = \sin\left(\theta + \frac{\pi}{2}\right)$$

(For example, the cosine of 0 equals 1, as does the sine of $\pi/2$.) So the current in the circuit with a capacitor connected to an AC voltage source is

$$i_C(t) = \omega CV_0 \sin\left(\omega t + \frac{\pi}{2}\right) \quad\quad (21\text{-}22)$$

The phase difference between the current and voltage in the circuit is $\pi/2$ rad. Because the phase difference is positive, the current leads the voltage, as shown in **Figure 21-11**. Notice that we have drawn the phasors so the current points in the same direction as in Figure 21-9, the phasor diagram for the resistor plus AC circuit.

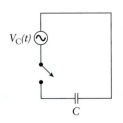

Figure 21-10 A capacitor is connected in series with an AC voltage source.

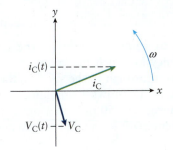

Figure 21-11 The current in a circuit with a capacitor connected to an AC voltage source leads the voltage by 90°.

This designation will serve to remind us that when we combine our results for each of the three components separately connected to an AC voltage source, the current in that series LRC circuit is the same as for each separately.

> ### ? Got the Concept 21-3
> **Change the Phase?**
>
> In Figure 21-11, the phase angle between the current and voltage in an AC circuit containing a capacitor is $\pi/2$ rad. Is it possible to adjust the phase difference between the current and voltage in such a circuit? Explain your answer.

From Equation 21-22, the current amplitude i_C is

$$i_C = \omega C V_0$$

or

$$V_0 = i_C \frac{1}{\omega C} \tag{21-23}$$

This equation has the same form as the relationship between the voltage across and current through a resistor, $V = iR$ (Equation 18-9). Indeed, $1/\omega C$ acts as a resistive term. So here's a bit of a surprise—when a capacitor is connected to an AC voltage, it impedes the flow of charge in the circuit. The property of a capacitor (or an inductor) to offer resistance to the flow of charge in an AC circuit is called **reactance**; the **capacitive reactance** X_C of a capacitor C driven by an AC voltage of angular frequency ω is

$$X_C = \frac{1}{\omega C} \tag{21-24}$$

The SI unit of capacitive reactance is the ohm.

By using the definition of capacitive reactance (Equation 21-24), the relationship between the peak voltage V_0 and the peak current i_C (Equation 21-23) can be rewritten as

$$V_0 = i_C X_C \tag{21-25}$$

This relationship holds for the peak values of voltage and current, or for the rms average values:

$$V_{rms} = i_{rms,C} X_C \tag{21-26}$$

However, because the voltage and current vary and are not in phase, the relationship between voltage, current, and reactance (Equation 21-25 or 21-26) does not hold for *instantaneous* values of voltage and current.

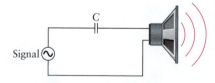

Figure 21-12 This simple circuit filters out low frequencies but lets high frequencies pass through to the speaker.

Example 21-3 High Pass Filter

Audio systems convert electrical signals to sound by moving the cone of a speaker back and forth at the same frequency at which the voltage varies. If the low frequency variation of wall voltage passes through to the speaker, the result is an unpleasant, low frequency sound. **Figure 21-12** shows a simplified version of a circuit that filters out low frequencies. By using $C = 50$ nF and a signal of $V_{rms} = 120$ V, compare the rms current in the speaker circuit for a 60 Hz signal and a more musical 880 Hz signal.

SET UP

The rms current is proportional to the rms voltage:

$$V_{rms} = i_{rms,C} X_C \qquad (21\text{-}26)$$

or

$$i_{rms,C} = \frac{1}{X_C} V_{rms}$$

We introduce the frequency of the signal into our analysis through the definition of capacitive reactance (Equation 21-24):

$$X_C = \frac{1}{\omega C}$$

so

$$i_{rms,C} = \omega C V_{rms}$$

or

$$i_{rms,C} = 2\pi f C V_{rms}$$

SOLVE

For the 60 Hz signal, the rms current is

$$i_{rms,C} = 2\pi (60\,\text{Hz}) (50 \times 10^{-9}\,\text{F}) (120\,\text{V}) = 2.3 \times 10^{-3}\,\text{A}$$

For the 880-Hz signal, the rms current that drives the speaker is

$$i_{rms,C} = 2\pi (880\,\text{Hz}) (50 \times 10^{-9}\,\text{F}) (120\,\text{V}) = 3.3 \times 10^{-2}\,\text{A}$$

REFLECT

Because the current in the speaker circuit is so much smaller for the low frequency signal compared to the high frequency signal, this simple "high pass filter" suppresses the 60-cycle hum while allowing the higher frequency electrical signals that carry the musical information to pass through to the speaker.

Inductor with AC Voltage

Figure 21-13 shows an inductor in series with an AC voltage source that varies sinusoidally with angular frequency ω. As we already did for the resistor and capacitor, we first write the voltage difference across the inductor:

$$V_L(t) = V_0 \sin(\omega t) \qquad (21\text{-}27)$$

As before, V_0 is the amplitude of the voltage that will be supplied in common to the inductor, resistor, and capacitor in the series *LRC* circuit, so we will not add the subscript *L* to it.

The current in the circuit depends on the voltage difference across the inductor. By using Equation 20-15, the magnitude of the voltage difference is

$$V = L\frac{di}{dt}$$

or

$$V_L(t) = L\frac{di_L(t)}{dt}$$

We obtain a relationship for the current by solving the equation for $di(t)$:

$$di_L(t) = \frac{1}{L} V_L(t)\, dt$$

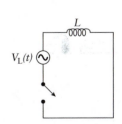

Figure 21-13 An inductor is connected in series with an AC voltage source.

and then applying Equation 21-27 in order to integrate,

$$i_L(t) = \int di_L(t) = \frac{1}{L}\int V_L(t)\,dt = \frac{1}{L}\int V_0 \sin(\omega t)\,dt$$

so

$$i_L(t) = -\frac{V_0}{\omega L}\cos(\omega t)$$

To see the phase difference between the current and the voltage more clearly, note that

$$-\cos\theta = \sin\left(\theta - \frac{\pi}{2}\right)$$

So the current in the circuit that contains only an inductor and an AC voltage source is

$$i_L(t) = \frac{V_0}{\omega L}\sin\left(\omega t - \frac{\pi}{2}\right) \tag{21-28}$$

The phase difference between the current and applied voltage (Equation 21-27) is $-\pi/2$, in other words, the voltage leads the current by 90°. The voltage and current phasors for the inductor plus AC circuit are shown in **Figure 21-14**.

 Watch Out

In the inductor plus AC circuit, the voltage leads the current.

Because the relative phase angle between the voltage (Equation 21-27) and the current is negative in the inductor plus AC circuit (Equation 21-28), the voltage reaches its peak value before the current peaks. For this reason, we say that the voltage leads the current.

 Got the Concept 21-4

Change the Phase, Again?

As shown in Figure 21-14, the current leads the voltage in an AC circuit containing an inductor by $\pi/2$ rad. Is it possible to adjust the phase difference between the current and voltage in such a circuit? Explain your answer.

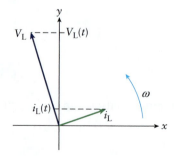

Figure 21-14 The voltage and current phasors for an inductor plus AC circuit are not in phase; the voltage leads the current by 90°.

Relying on our realization that a capacitor impedes current when connected to an AC voltage, notice that the current amplitude i_L, from Equation 21-28, is

$$i_L = \frac{V_0}{\omega L}$$

or

$$V_0 = i_L \omega L \qquad (21\text{-}29)$$

This equation, too, has the same form as the relationship between the voltage across and current through a resistor, $V = iR$ (Equation 18-9). We conclude that ωL is the **inductive reactance** in a circuit in which an inductor with inductance L is connected to an AC voltage of angular frequency ω. In analogy to the capacitive reactance, we define the inductive reactance as

$$X_L = \omega L \qquad (21\text{-}30)$$

The SI unit of inductive reactance is the ohm.

As with capacitive reactance, the inductive reactance relates the peak voltage V_0 and the peak current i_L in the AC plus inductor circuit. Using Equation 21-30, we rewrite Equation 21-29:

$$V_0 = i_L X_L \qquad (21\text{-}31)$$

The rms average values of voltage and current in the AC plus inductor circuit are also related by the inductive reactance:

$$V_{\text{rms}} = i_{\text{rms}, L} X_L \qquad (21\text{-}32)$$

As with capacitive reactance, the voltage does not reach its peak value at the same instant in time that the current reaches its peak value. In addition, the voltage and current don't rise and fall together. For these reasons there is no one instant in time when Equation 21-31 or Equation 21-32 holds.

? Got the Concept 21-5
Another Filter

In Example 21-3, we discovered that a capacitor can be used as a simple high pass filter to suppress low frequency noise in an audio system. **Figure 21-15** shows a speaker system consisting of a small speaker for high frequencies (a tweeter) and a larger speaker for low frequencies (a woofer). What purpose do the capacitor and inductor play in the simple "crossover" circuit shown?

★ What's Important 21-4
The current and voltage are in phase in a circuit containing a resistor in series with an AC voltage source. The phase angle between the current and applied voltage in an AC circuit containing a capacitor is $\pi/2$ rad. Because the phase difference is positive, the current leads the voltage. The phase difference between the current and applied voltage in an AC circuit containing an inductor is $-\pi/2$ rad; in other words, the voltage leads the current by 90°.

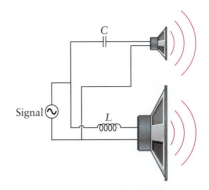

Figure 21-15 This sound system consists of a small tweeter speaker for high frequencies and a larger woofer speaker for low frequencies.

21-5 *L, R, C* in Series with AC

Our primary goal for this chapter is to understand how the current acts in a series *LRC* circuit connected to an AC voltage source, such as the one shown in Figure 21-5. The general form of the current as a function of time, $i(t)$,

$$i(t) = i_0 \sin(\omega t - \varphi) \qquad (21\text{-}11)$$

depends on the amplitude i_0 and the phase φ. Our task is to find relationships for the two parameters in terms of the quantities that define the circuit: V_0, ω, L, R, and C.

Our strategy relies on the use of phasors. In the previous section, we developed a phasor diagram for a resistor, inductor, and capacitor connected separately to an AC voltage source (Figure 21-9, Figure 21-11, and Figure 21-14). We are now ready to combine those results into a single phasor diagram that represents the complete series *LRC* circuit. When we're done, we will know the current amplitude i_0 and the phase angle φ in terms of V_0, ω, L, R, and C.

Kirchhoff's loop rule requires that the sum of the voltage drops around the circuit in Figure 21-5 equals zero. By Equation 21-12,

$$V(t) - V_L(t) - V_R(t) - V_C(t) = 0$$

or

$$V(t) = V_L(t) + V_R(t) + V_C(t) \qquad (21\text{-}33)$$

We have been careful to evaluate the loop rule at a specific instant of time t because the voltage differences across the inductor, resistor, and capacitor are not in phase.

 Watch Out

The total voltage in the series *LRC* circuit connected to an AC source is equal to the sum of the *instantaneous* voltage across the inductor, the resistor, and the capacitor.

In the series *LRC* circuit, the voltage differences across the inductor, the resistor, and the capacitor vary sinusoidally with time. However, the voltage differences are not in phase. According to Kirchhoff's loop rule, the voltage applied to the circuit by the AC source is equal to the sum of the voltage differences across each component separately *at every instant of time*. But because the voltages across the three components do not rise and fall together, they don't reach their peak values at the same time. **Figure 21-16a** shows the voltage differences across the three components in a series *LRC* circuit as a function of time. The instantaneous values of $V_L(t)$, $V_R(t)$, and $V_C(t)$ are marked with circles at two selected values of time. Their sum is clearly not the sum of the peak values, marked with triangles. $V(t)$, the time-varying sum of the three voltages, is shown in **Figure 21-16b**. Certainly the sum of the peak values, which is constant because the peak values are constants, cannot equal $V(t)$. So although it might be tempting to set the sum of the *peak* voltages equal to the amplitude of the AC voltage, it is not correct to do so! Similarly, the sum of the rms values of the voltage differences across the three components is not equal to the rms value of the voltage in the whole circuit.

The phasor diagrams we developed in the last section show the phase relationship between voltage and current in three circuits in which an inductor, a resistor, and a capacitor are separately connected to an AC voltage source. To see the

In a series LRC circuit, $V_L(t)$, $V_R(t)$, and $V_C(t)$ vary with time but they do not reach their peak values at the same instant of time.

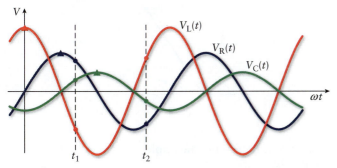

Triangles mark the peak values of the voltages. Circles show the instantaneous values of voltage at times t_1 and t_2.

(a)

Figure 21-16 (a) In *LRC* circuits, the voltages across the three components do not rise and fall together, so they don't reach their peak values at the same time. (b) Here we see the voltage differences across the three components as a function of time.

$V(t)$, the sum of the instantaneous values of the three voltages $V_L(t)$, $V_R(t)$, and $V_C(t)$, varies sinusoidally.

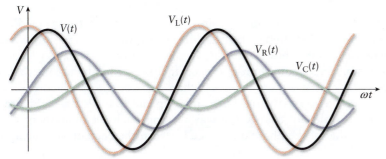

$V(t)$ can not be equal to the sum of the peak values of $V_L(t)$, $V_R(t)$, and $V_C(t)$, because the peak values are constants!

(b)

relationship between the current in the circuit and the *total* applied voltage, we must combine the three separate voltage phasors into a single, total voltage phasor. We accomplish this by recognizing that the current is the same through electrical components connected in series. So in **Figure 21-17**, the V_L, V_R, and V_C phasors are each drawn relative to a single current phasor. The phase of each relative to the current is the same as shown in Figures 21-14, 21-9, and 21-11, respectively.

How should we determine the phasor that represents the total applied voltage in the series *LRC* circuit? To answer, first recall that in the phasor representation, the value of a voltage at any specific instant is defined as the projection of a rotating vector on the y axis (Figure 21-6a). The values of the voltage across each component at a specific instant of time t, $V_L(t)$, $V_R(t)$, and $V_C(t)$, are then the y projections of the three phasors, and these projections can be added directly to get the value of the applied voltage at time t, according to Equation 21-33. This is vector addition; the sum of vectors is determined by adding their projections onto orthogonal axes. The phasor that represents the applied voltage is the vector sum of the V_L, V_R, and V_C phasors.

In a series LRC circuit connected to an AC voltage the same current is everywhere. The phase differences between the phasors V_L, V_R, and V_C (representing the voltage differences across each component) are set by considering simple circuits, each with an AC source and one of the three electrical components.

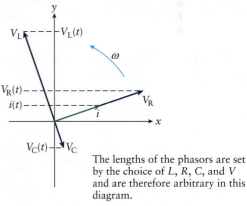

The lengths of the phasors are set by the choice of L, R, C, and V and are therefore arbitrary in this diagram.

Figure 21-17 To see the relationship between the current in an *LRC* circuit and the *total* applied voltage, we must combine the three separate voltage phasors into a single, total voltage phasor. Because the current is the same through electrical components connected in series, the V_L, V_R, and V_C phasors are each drawn relative to a single current phasor.

Because phasors V_L and V_C point in opposite directions, their vector sum is $V_L - V_C$.

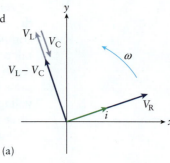

(a)

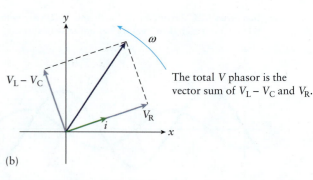

The total V phasor is the vector sum of $V_L - V_C$ and V_R.

(b)

Figure 21-18 The phasor that represents the applied AC voltage is the vector sum of the $V_L - V_C$ and V_R phasors.

Phase difference φ, which describes how the AC voltage rises and falls with respect to the rise and fall of the resulting current, is the angle between the V and i phasors.

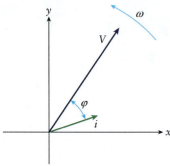

Figure 21-19 We redrew the phasor diagram shown in Figure 21-18 to show only the total voltage and current phasors. The overall phase difference between the time-varying AC voltage and the current is the angle φ between the V phasor and the i phasor.

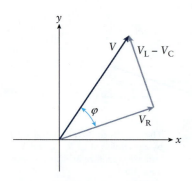

Figure 21-20 In carrying out the vector sum of $V_L - V_C$ and V_R (Figure 21-18b), we formed a right triangle with hypotenuse V and sides $V_L - V_C$ and V_R.

Because V_L and V_C point in opposite directions, their vector sum is $V_L - V_C$, as shown in **Figure 21-18a**. Remember that the lengths of the voltage and current phasors depend on specific choices of L, R, C, and V, so the fact that V_L is longer than V_C in these diagrams is an arbitrary choice. It could certainly happen that in a specific circuit the magnitude of V_L is less than that of V_C, in which case $V_L - V_C$ would point down and to the right instead of up and to the left as in Figure 21-18a. This instance would not change the way we approach the analysis of the circuit, but only the value of the phase difference between the current and the applied AC voltage.

The phasor that represents the applied AC voltage is the vector sum of the $V_L - V_C$ and V_R phasors, as shown graphically in **Figure 21-18b**. We redrew this diagram in **Figure 21-19** to show only the total voltage and current phasors. The overall phase difference between the time-varying AC voltage and the current is the angle φ between the V phasor and the i phasor. All that remains, then, is to determine the magnitude of the current i (i_0) and the value of the phase angle φ in terms of V_0, ω, L, R, and C.

We now take advantage of the power of the phasor approach. Notice that in carrying out the vector sum of $V_L - V_C$ and V_R (Figure 21-18b), we formed a right triangle (**Figure 21-20**) with hypotenuse V and sides $V_L - V_C$ and V_R. Therefore,

$$V_0 = \sqrt{(V_{L,0} - V_{C,0})^2 + V_{R,0}^2} \qquad \text{(21-34)}$$

This expression is written in terms of the *magnitudes* of the voltage phasors. Using the magnitudes allows us to use the relationships we developed in the last section between the current and $V_{L,0}$, $V_{R,0}$, and $V_{C,0}$ separately. From Equation 21-19 the magnitude of $V_{R,0}$ is

$$V_{R,0} = i_R R$$

and

$$V_{C,0} = i_C \frac{1}{\omega C} \qquad \text{(21-23)}$$

$$V_{L,0} = i_L \omega L \qquad (21\text{-}29)$$

Because the same current is everywhere in the series *LRC* circuit, the equations become

$$V_{R,0} = i_0 R$$

$$V_{C,0} = i_0 \frac{1}{\omega C} \qquad (21\text{-}25)$$

$$V_{L,0} = i_0 \omega L$$

where i_0 is the current amplitude. Equation 21-34 then gives V_0 in terms of i_0:

$$V_0 = \sqrt{\left(i_0 \omega L - i_0 \frac{1}{\omega C}\right)^2 + (i_0 R)^2}$$

We factor i_0^2 from each term, giving

$$V_0 = i_0 \sqrt{\left(\omega L - \frac{1}{\omega C}\right)^2 + R^2} \qquad (21\text{-}36)$$

One of our goals for the examination of the series *LRC* circuit driven by an AC voltage is to find i_0 in terms of V_0, ω, L, R, and C. Done!

$$i_0 = \frac{V_0}{\sqrt{\left(\omega L - \dfrac{1}{\omega C}\right)^2 + R^2}} \qquad (21\text{-}37)$$

Notice that Equation 21-36 has the same form as the relationship between the voltage across and current through a resistor ($V = iR$, Equation 18-9). Thus, the term $\sqrt{(\omega L - 1/\omega C)^2 + R^2}$ acts as a resistive term. This result isn't a surprise, because a resistor impedes the flow of charge and we've seen that in circuits driven by AC voltage both a capacitor and an inductor do as well. We define the denominator of Equation 21-37 as the **impedance Z**:

$$Z = \sqrt{\left(\omega L - \frac{1}{\omega C}\right)^2 + R^2} \qquad (21\text{-}38)$$

Impedance describes the overall opposition to the flow of charge in a circuit driven by a time-varying voltage. The SI unit of impedance is the ohm (Ω).

The characteristic of an inductor and a capacitor to impede the flow of charge in an AC circuit is reactance, so it is helpful to express the impedance in terms of reactance. Recall that capacitive reactance X_C is defined as

$$X_C = \frac{1}{\omega C} \qquad (21\text{-}24)$$

and inductive reactance X_L as

$$X_L = \omega L \qquad (21\text{-}20)$$

The terms in parentheses in Equation 21-38 are therefore inductive reactance and capacitive reactance, so impedance can be expressed as

$$Z = \sqrt{(X_L - X_C)^2 + R^2} \qquad (21\text{-}39)$$

Finally, because impedance contributes to the series LRC circuit driven by an AC voltage in a way similar to resistance, we combine Equations 21-37 and 21-38 to obtain

$$i_0 = \frac{V_0}{Z} \tag{21-40}$$

The triangle we formed in carrying out the vector sum of the V_L, V_C, and V_R phasors (Figure 21-20) enables us to find the phase angle φ. We defined the angle as the angle between the voltage phasor V and the current phasor i. Because the V_R phasor is aligned with i, φ is also the angle between V and V_R in the triangle, so

$$\tan \varphi = \frac{V_{L,0} - V_{C,0}}{V_{R,0}}$$

where the lengths of the sides of the triangle are written in terms of the magnitudes of the three separate voltage phasors. We rewrite the equation in terms of R, C, and L using the expressions above for the voltage magnitudes:

$$V_{R,0} = i_0 R$$

$$V_{C,0} = i_0 \frac{1}{\omega C} \tag{21-35}$$

$$V_{L,0} = i_0 \omega L$$

so

$$\tan \varphi = \frac{i_0 \omega L - i_0 \dfrac{1}{\omega C}}{i_0 R}$$

or

$\sqrt{x}$ *See the Math Tutorial for more information on Trigonometry*

$$\tan \varphi = \frac{\omega L - \dfrac{1}{\omega C}}{R} \tag{21-41}$$

In this expression, we have found φ in terms of the parameters that define a particular series LRC circuit driven by an AC voltage. Equation 21-41 together with Equation 21-37 provides the complete description of the current in the circuit. That is, when an inductor L, a resistor R, and a capacitor C are connected in series to an AC voltage $V(t)$,

$$V(t) = V_0 \sin(\omega t) \tag{21-2}$$

a time-varying current

$$i(t) = i_0 \sin(\omega t - \varphi) \tag{21-11}$$

is in the circuit. The current amplitude i_0 is

$$i_0 = \frac{V_0}{\sqrt{\left(\omega L - \dfrac{1}{\omega C}\right)^2 + R^2}} \tag{21-37}$$

and the phase angle is

$$\tan \varphi = \frac{\omega L - \dfrac{1}{\omega C}}{R} \tag{21-41}$$

Example 21-4 A Series *LRC* Circuit Driven by an AC Voltage—Impedance

An inductor, a resistor, and a capacitor are connected in series with a 120 V (rms) AC voltage source that varies with a frequency of 60 Hz. (A generic series *LRC* circuit is shown in Figure 21-5.) Let *L* equal 78.0 mH, *R* equal 25.0 Ω, and *C* equal 53.0 μF. Find the capacitive reactance, the inductive reactance, and the impedance in the circuit.

SET UP

Reactance and impedance are the property of electrical components to resist the current in a circuit driven by a time-varying voltage. For a voltage varying at angular frequency ω, the capacitive reactance X_C is

$$X_C = \frac{1}{\omega C} \tag{21-24}$$

the inductive reactance X_L is

$$X_L = \omega L \tag{21-30}$$

and the impedance Z is

$$Z = \sqrt{(X_L - X_C)^2 + R^2} \tag{21-39}$$

Note that the time variation of the voltage is given in terms of frequency, so in order to make numeric calculations we need to convert frequency to angular frequency by

$$\omega = 2\pi f$$

SOLVE

The capacitive reactance X_C is

$$X_C = \frac{1}{2\pi f C} = \frac{1}{2\pi (60 \text{ Hz})(53.0 \times 10^{-6} \text{ F})} = 50.0 \text{ }\Omega$$

the inductive reactance X_L is

$$X_L = 2\pi f L = 2\pi (60 \text{ Hz})(78.0 \times 10^{-3} \text{ H}) = 29.4 \text{ }\Omega$$

and the impedance Z is

$$Z = \sqrt{(29.4 \text{ }\Omega - 50.0 \text{ }\Omega)^2 + (25.0 \text{ }\Omega)^2} = 32.4 \text{ }\Omega$$

REFLECT

In this circuit, X_C is greater than X_L. The capacitor will therefore have a greater effect on the current than the inductor. Such a circuit is said to be more capacitive than inductive. X_L is greater than X_C in a circuit that is more inductive than capacitive.

Practice Problem 21-4 An inductor, a resistor, and a capacitor are connected in series with a 120 V (rms) AC voltage source that varies with a frequency of 60 Hz. (A generic series *LRC* circuit is shown in Figure 21-5.) In the circuit, the capacitive reactance X_C is 36.8 Ω, the inductive reactance X_L is 17.7 Ω, and the impedance Z is 30.0 Ω. Find the inductance of the inductor, the capacitance of the capacitor, and the resistance of the resistor.

Example 21-5 A Series *LRC* Circuit Driven by an AC Voltage—Current

For the series *LRC* circuit described in Example 21-4, find the peak current i_0 in the circuit.

SET UP

The peak current is a function of L, R, C, the peak voltage V_0, and the angular frequency ω at which the voltage varies:

$$i_0 = \frac{V_0}{\sqrt{\left(\omega L - \dfrac{1}{\omega C}\right)^2 + R^2}} \tag{21-37}$$

We can make use of the results of the previous problem, in which we found the impedance in the circuit, by writing the expression in the form of Equation 21-40:

$$i_0 = \frac{V_0}{Z}$$

Also, notice that the problem statement gives the rms voltage, not the peak voltage V_0. Using Equation 21-6, the relationship between the two is

$$V_0 = \sqrt{2}\, V_{rms}$$

SOLVE

In terms of the rms value of the voltage, Equation 21-40 becomes

$$i_0 = \frac{\sqrt{2}\, V_{rms}}{Z}$$

We obtain a numeric answer by substituting the values provided:

$$i_0 = \frac{\sqrt{2}\,(120\text{ V})}{32.4\ \Omega} = 5.24\text{ A}$$

REFLECT

We have found the peak current i_0, but it is important to recognize that the current in the circuit varies with time, according to

$$i(t) = i_0 \sin(\omega t - \varphi) \tag{21-11}$$

To find the current at any specific instant, we now need only determine the phase angle φ. We do this in the following example.

Practice Problem 21-5 Determine the peak current i_0 in a circuit similar to the series LRC circuit described in Example 21-4, but in which the inductance, resistance, and capacitance have all been reduced to one-half of the values given in Example 21-4.

Example 21-6 A Series *LRC* Circuit Driven by an AC Voltage—Phase Angle

For the series LRC circuit described in Example 21-4, find the phase angle between the time-varying current and voltage in the circuit.

SET UP

The phase angle φ is given by

$$\tan\varphi = \frac{\omega L - \dfrac{1}{\omega C}}{R} \tag{21-41}$$

To take advantage of our results in Example 21-4, we can write the equation in terms of the inductive and capacitive reactances using Equation 21-30 and Equation 21-24:

$$\tan\varphi = \frac{X_L - X_C}{R}$$

▶ *Go to Interactive Exercise 21-1* for more practice dealing with *LRC* Circuits

SOLVE
Then

$$\tan \varphi = \frac{29.4 \; \Omega - 50.0 \; \Omega}{25.0 \; \Omega} = -0.824$$

or

$$\varphi = \tan^{-1}(-0.824) = -39.5°$$

REFLECT
The voltage and current in a series *LRC* circuit obey

$$V(t) = V_0 \sin(\omega t) \qquad\qquad (21\text{-}2)$$

and

$$i(t) = i_0 \sin(\omega t - \varphi) \qquad\qquad (21\text{-}11)$$

In this problem, we found φ to be negative, so the overall sign of the phase term in Equation 21-11 is positive. Specifically,

$$i(t) = i_0 \sin(\omega t + 39.5°)$$

in this problem. So the current reaches its peak on every cycle before the voltage, that is, the current leads the voltage in the circuit. This is made clear by the phasor diagram shown in **Figure 21-21**. How does the voltage phasor end up lagging the current? We found the voltage phasor by taking the vector sum of the V_L, V_R, and V_C phasors. In Figure 21-18a and Figure 21-18b, we had arbitrarily chosen the length of V_L to be greater than V_C, which resulted in $V_L - V_C$ and therefore the total voltage V leading the current. For the voltage to *lag* the current, V_L must be *smaller* than V_C. In that case, $V_L - V_C$ points in the same direction as V_C rather than opposite to it, which brings the vector sum (V) to the lagging side of the current. Recall that in the simple circuit in which a capacitor is connected to an AC source, the current leads the voltage (Figure 21-11).

We will examine the voltages in the circuit in the next example; based on the conclusions in the previous paragraph, we expect to find that V_L is smaller than V_C. However, notice that in Example 21-4 we found that in a circuit in which X_C is greater than X_L, the capacitor has a greater effect on the current than the inductor. This observation also suggests that the $V_L - V_C$ phasor is influenced more by the capacitor than the inductor, that is, that V_L is smaller than V_C.

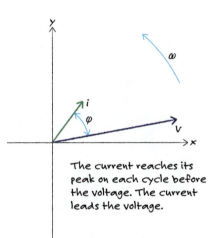

The current reaches its peak on each cycle before the voltage. The current leads the voltage.

Figure 21-21 In this problem the current reaches its peak on every cycle before the voltage. In other words, the current leads the voltage in the circuit.

Practice Problem 21-6 For the series *LRC* circuit described in Example 21-4 we found the capacitive reactance X_C equal to 50.0 Ω and the inductive reactance X_L equal to 29.4 Ω. Find the value of the phase angle φ if the values of the capacitor and inductor were changed so that X_C equaled 29.4 Ω and X_L equaled 50.0 Ω.

Example 21-7 A Series *LRC* Circuit Driven by an AC Voltage—Peak Voltages

For the series *LRC* circuit described in Example 21-4, find the peak voltages across each component in the circuit.

SET UP

The peak voltages are determined from the peak current, as in the equations labeled Equation 21-35:

$$V_{R,0} = i_0 R$$

$$V_{C,0} = i_0 X_C$$

$$V_{L,0} = i_0 X_L$$

SOLVE

R is given as 25.0 Ω. We found X_C and X_L in Example 21-4, $X_C = 50.0\ \Omega$ and $X_L = 29.4\ \Omega$. From Example 21-5, the peak current i_0 is 5.24 A. Then

$$V_{R,0} = (5.24\ \text{A})(25.0\ \Omega) = 131\ \text{V}$$

$$V_{C,0} = i_0 X_C = (5.24\ \text{A})(50.0\ \Omega) = 262\ \text{V}$$

$$V_{L,0} = i_0 X_L = (5.24\ \text{A})(29.4\ \Omega) = 154\ \text{V}$$

REFLECT

As we expected, the peak voltage across the capacitor in the circuit is greater than the peak voltage across the inductor. The circuit is more capacitive than inductive, which results in the current leading the voltage.

Also, notice that the sum of the peak voltages, $131 + 262 + 154\ \text{V} = 547\ \text{V}$, is not equal to the peak voltage in the circuit, which is 170 V. Not even close! Because the rise and fall of the voltages across the three components are not in phase with each other, when the voltage across one component is at its peak, the voltages across the others are not, so the sum of the peak values has no physical meaning. However, the sum of the values of the voltages at any instant of time must equal the total voltage at that time; we explore this conclusion in the next example.

Example 21-8 A Series *LRC* Circuit Driven by an AC Voltage—Instantaneous Voltages

For the series *LRC* circuit described in Example 21-4, find the instantaneous voltages across each component in the circuit at 0.926 ms after the start of a cycle of the applied AC voltage, and compare the sum of the voltages to the applied voltage at that instant.

SET UP

We found the peak voltages $V_{R,0}$, $V_{C,0}$, and $V_{L,0}$ in Example 21-7 to be 131 V, 262 V, and 154 V, respectively. The phasors that represent the voltages are shown in Figure 21-22. Also shown in the figure are the current phasor, the total voltage phasor, and the phase angle φ between them, which we found to be $-39.5°$ in Example 21-6. From the diagram you can see that with the time-varying value of the applied AC voltage described as

$$V(t) = V_0 \sin(\omega t) \tag{21-2}$$

the voltages across the three electrical components are

$$V_R(t) = V_{R,0} \sin(\omega t + |\varphi|)$$

$$V_L(t) = V_{L,0} \sin\left(\omega t + |\varphi| + \frac{\pi}{2}\right)$$

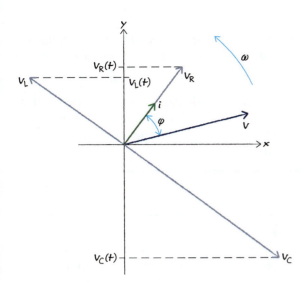

Figure 21-22 We found the peak voltages $V_{R,0}$, $V_{C,0}$, and $V_{L,0}$ in Example 21-7 to be 131 V, 262 V, and 154 V, respectively. The phasors that represent the voltages are shown here, as well as the current phasor, the total voltage phasor, and the phase angle φ between them, which we found to be $-39.5°$ in Example 21-6.

$$V_C(t) = V_{C,0} \sin\left(\omega t + |\varphi| - \frac{\pi}{2} \right)$$

Here the value of the phase angle must be in radians, that is

$$\varphi = -39.5° \frac{\pi \text{ rad}}{180°} = -0.689 \text{ rad}$$

Because φ is negative, we use the absolute value of φ in the voltage relationships in order to make the angular direction clear.

The angle between the positive x axis and the V phasor is ωt at any time t. At the time specified in the problem the angle is

$$\omega t = 2\pi f t = 2\pi (60 \text{ Hz}) (0.926 \times 10^{-3} \text{ s}) = 0.349 \text{ rad}$$

We will use this value in the subsequent computations.

SOLVE
We apply Equation 21-6 to convert the magnitude of the AC voltage V_0 from the rms voltage given:

$$V_0 = \sqrt{2} \, V_{\text{rms}} = \sqrt{2} (120 \text{ V}) = 170 \text{ V}$$

Notice that the vector sum of the phasors representing the voltages across the resistor, inductor, and capacitor (from Equation 21-34) gives the same result:

$$V_0 = \sqrt{(V_{L,0} - V_{C,0})^2 + V_{R,0}^2} = \sqrt{(154 \text{ V} - 262 \text{ V})^2 + (131 \text{ V})^2} = 170 \text{ V}$$

The instantaneous applied voltage is then

$$V(t) = V_0 \sin(\omega t) = (170 \text{ V}) \sin(0.349 \text{ rad}) = 58.1 \text{ V}$$

Using the known values of $V_{R,0}$, $V_{C,0}$, $V_{L,0}$, and φ, at t equal to 0.926 ms the voltages across the resistor, inductor, and capacitor are

$$V_R(t) = (131 \text{ V}) \sin(0.349 \text{ rad} + 0.689 \text{ rad}) = 113 \text{ V}$$

$$V_L(t) = (154 \text{ V}) \sin(0.349 \text{ rad} + 0.689 \text{ rad} + \pi/2) = 78.2 \text{ V}$$

$$V_C(t) = (262 \text{ V}) \sin(0.349 \text{ rad} + 0.689 \text{ rad} - \pi/2) = -133 \text{ V}$$

So at this instant in time, the total voltage is

$$V_R(0.926 \text{ ms}) + V_L(0.926 \text{ ms}) + V_C(0.926 \text{ ms}) = 113 \text{ V} + 78.2 \text{ V} - 133 \text{ V} = 58 \text{ V}$$

REFLECT

As we expected, the sum of the voltage differences across the resistor, inductor, and capacitor equals the magnitude of the AC voltage at this, or indeed any, instant of time.

Practice Problem 21-8 For the series LRC circuit described in Example 21-4, find the instantaneous voltages across each component in the circuit at 0.750 ms after the start of a cycle of the applied AC voltage, and compare the sum of these voltages to the applied voltage at this instant.

★ What's Important 21-5

In this section we found the current as a function of time in a series LRC circuit driven by a sinusoidal voltage. In $i(t) = i_0 \sin(\omega t - \varphi)$ (Equation 21-11) we found $i_0 = V_0/\sqrt{(\omega L - 1/\omega C)^2 + R^2}$ (Equation 21-37) and $\tan\varphi = (\omega L - 1/\omega C)/R$ (Equation 21-41). The denominator of Equation 21-37 is the impedance Z, which describes the overall opposition to the flow of charge in a circuit driven by a time-varying voltage. The SI unit of impedance is the ohm (Ω).

21-6 Applications of a Series *LRC* Circuit

When a mechanical system that is free to oscillate (such as an object on the end of a spring) experiences a periodic driving force and when the frequency of that force matches the natural frequency of the system, the amplitude of oscillations becomes large. This phenomenon, which we explored in Section 12-8, is resonance. The series LRC circuit driven by an AC voltage is the electrical analog of the mechanical oscillator, and just like the object on the end of a spring, the circuit also exhibits the phenomenon of resonance. In the case of an electrical oscillator, it is the amplitude of the current that becomes large when the resonance occurs. This characteristic has practical applications; for example, as the natural frequency of the tuning circuit in an FM radio is swept past the frequency of a radio station's signal, the increase in current is used to tune in the station.

The current amplitude i_0 in a series LRC circuit driven by an AC voltage depends on the angular frequency ω at which the voltage varies as well as L, R, C, and the peak voltage V_0:

$$i_0 = \frac{V_0}{\sqrt{\left(\omega L - \dfrac{1}{\omega C}\right)^2 + R^2}} \tag{21-37}$$

For a circuit of fixed L, R, and C, and for which the peak voltage doesn't change, i_0 is largest when the denominator of Equation 21-37 is as small as possible. The resonance condition occurs when the term in parentheses equals zero, that is,

$$\omega L - \frac{1}{\omega C} = 0$$

or

$$\omega L = \frac{1}{\omega C}$$

Resonance therefore occurs when the driving AC voltage varies with angular frequency

$$\omega = \sqrt{\frac{1}{LC}}$$

This quantity is the natural frequency of the *LC* circuit; in other words, resonance occurs when the circuit is driven at its natural frequency. We should have expected this! The resistor affects the amplitude but not the frequency of the oscillations; only the inductor and capacitor affect the resonance characteristics of the circuit.

What if we select a specific inductance, resistance, and capacitance for a circuit, and then sweep the angular frequency of the driving AC voltage over a range starting from below the natural frequency and going to a frequency above it? For very small values of ω, the $1/\omega C$ term in Equation 21-37 becomes large, resulting in a relatively small peak current. Similarly, the ωL term in Equation 21-37 becomes large for large values of ω, again resulting in a relatively small peak current. At some value of ω in between the extremes, however, the peak current in the circuit grows large; this result is shown in **Figure 21-23**. The specific value of ω for which the peak current is largest is the natural frequency ω_0. To make that clear in the figure, we plot the peak current on the vertical axis versus ω/ω_0 horizontally. The largest value of peak current therefore corresponds to 1 on the horizontal axis.

When the angular frequency ω equals $\sqrt{1/LC}$, the term in parentheses in Equation 21-37 equals zero, which minimizes the denominator and results in the maximum peak current:

$$i_{0,\text{max}} = \frac{V_0}{\sqrt{0 + R^2}} = \frac{V_0}{\sqrt{R^2}} = \frac{V_0}{R} \qquad (21\text{-}42)$$

The maximum value of the peak current is therefore inversely proportional to the resistance, as is the case whenever voltage is applied across resistance. **Figure 21-24** shows peak current versus driving frequency for three series *LRC* circuits that differ only by the value of the resistance. As Equation 21-42 suggests, the maximum peak current increases as resistance decreases.

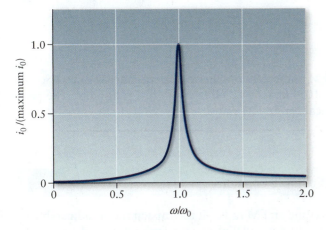

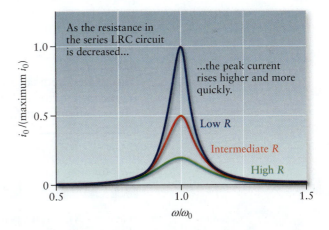

Figure 21-23 This figure plots peak current at each value of angular frequency ω as a fraction of the maximum possible peak current. The horizontal axis represents the angular frequency of the applied AC voltage, plotted as a fraction of the natural frequency. The peak of the curve occurs where the angular frequency ω equals the natural frequency ω_0, or 1 on the horizontal axis.

Figure 21-24 Whenever voltage is applied across a resistance, the maximum value of the peak current is inversely proportional to the resistance. Here we see peak current versus driving frequency for three series *LRC* circuits that differ only by the value of the resistance. The maximum peak current increases as resistance decreases.

Example 21-9 Greatest Hits 99.7

The tuner knob on an FM radio moves the plates of an adjustable capacitor which is in series with a 0.130 μH inductor and a net resistance of 750 Ω. The peak current induced in the circuit by a radio wave of a specific carrier frequency becomes large when the natural frequency of the circuit matches the carrier frequency. What is the frequency of the FM station that is tuned in when the capacitor in the radio is adjusted to 19.6 pF? If the peak operating voltage in the tuning circuit is 9 V, what is the peak current?

SET UP

The tuning circuit resonates when driven by a radio wave—an alternating wave that induces a potential in the circuit—of frequency equal to the natural frequency f_0 of the circuit. To tune a radio station of frequency f_{radio}, then, the tuning circuit must be adjusted so that

$$f_0 = f_{radio}$$

where, according to Equation 20-21,

$$f_0 = \frac{\omega_0}{2\pi} = \frac{1}{2\pi}\sqrt{\frac{1}{LC}}$$

SOLVE

So, to determine the radio station frequency, we use

$$f_{radio} = \frac{1}{2\pi}\sqrt{\frac{1}{LC}} \tag{21-43}$$

and the values given in the problem statement:

$$f_{radio} = \frac{1}{2\pi}\sqrt{\frac{1}{(0.130 \times 10^{-6}\,\text{H})(19.6 \times 10^{-12}\,\text{F})}} = 99.7 \times 10^6\,\text{Hz}$$

The peak current, by using Equation 21-42, is

$$i_{0,max} = \frac{V_0}{R} = \frac{9\,\text{V}}{750\,\Omega} = 0.012\,\text{A}$$

REFLECT

The carrier frequencies of FM radio stations lie in the megahertz range. So the frequency we found, 99.7×10^6 Hz, is 99.7 MHz on an FM radio dial. The peak current, about 10 mA, is typical for an FM radio.

Estimate It! 21-1 An Adjustable Capacitor

A series LRC circuit is used to tune an FM radio. If the capacitance is adjustable by changing the effective plate area of a parallel plate capacitor, estimate the factor by which the area needs to be adjusted in order to tune in the full range of FM radio stations.

SET UP

We need three pieces of information to carry out the estimation: the frequency range of FM stations, the relationship between radio carrier frequency and the capacitance of the tuner circuit, and the relationship between capacitance and plate area for a parallel plate capacitor.

If you've used radio to listen to music, you probably know that the frequency range of FM stations is about 88 MHz to 107 MHz. Using Equation 21-45, the capacitance must be adjusted to

$$C = \frac{1}{4\pi^2 L f_{radio}^2}$$

in order to be sensitive to a station of carrier frequency f_{radio}. Finally, from Equation 17-20 for a parallel plate capacitor, plate area A and plate separation d gives rise to capacitance:

$$C = \frac{\varepsilon_0 A}{d}$$

SOLVE

Our goal is to determine the factor between the smallest and largest plate area required, so we only need to know how plate area depends on the frequency to be tuned. Because the inductance is constant, from the first equation above, the capacitance C required is proportional to the inverse of the square of the frequency to be tuned:

$$C \propto \frac{1}{f_{radio}^2}$$

Because the plate separation is constant, from the second equation above, the capacitance is proportional to the plate area:

$$C \propto A$$

So the plate area required to tune in a station is proportional to the inverse of the square of the frequency to be tuned:

$$A \propto \frac{1}{f_{radio}^2}$$

So A_{max} is proportional to $1/f_{min}^2$ and A_{min} is proportional to $1/f_{max}^2$. Then the ratio of the largest to the smallest plate area required is

$$\frac{A_{max}}{A_{min}} = \frac{f_{max}^2}{f_{min}^2}$$

Notice that because the proportionality constant between A and $1/f_{radio}^2$ is the same for all values of f_{radio} (it's a constant, after all!), it cancels in the *ratio* of areas. So A_{max}/A_{min} is *equal* to f_{max}^2/f_{min}^2.

With f_{min} equal to 88 MHz and f_{max} equal to 107 MHz, the ratio of the areas is

$$\frac{A_{max}}{A_{min}} = \frac{(107 \text{ MHz})^2}{(88 \text{ MHz})^2} = 1.5$$

REFLECT

The plate area must be adjustable so that the largest area is about one and a half times the smallest area.

Filters

In Example 21-3, we considered a simple circuit that filters the unpleasant sound generated when the low frequency variation of wall voltage passes through the speakers of an audio system. Circuits composed of inductors, resistors, and capacitors connected in series create filters that pass or block electrical signals over both wide and narrow ranges of frequencies.

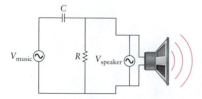

Figure 21-25 A time-varying voltage V_{music} that carries a music signal would likely be composed of a wide range of frequencies. The series LRC circuit (here we let L equal 0) filters out the low frequency hum due to the wall voltage. This circuit is a high pass filter.

Music is a superposition of many frequencies, so an electrical signal that carries musical information is as well. In the circuit shown in **Figure 21-25**, for example, a time-varying voltage V_{music} that carries a music signal would likely be composed of a wide range of frequencies. This series LRC circuit (here we let L equal 0) filters out the low frequency hum due to the wall voltage. We can see how that happens by comparing the magnitude of the voltage that drives the speaker to the magnitude of V_{music}.

A current loop is formed by the voltage source V_{music}, the capacitor, and the resistor. According to Equation 21-37, the peak current in that loop is

$$i_0 = \frac{V_{music,0}}{\sqrt{\left(0 - \dfrac{1}{\omega C}\right)^2 + R^2}} = \frac{V_{music,0}}{\sqrt{\dfrac{1}{\omega^2 C^2} + R^2}} \tag{21-44}$$

where we set L equal to 0 and let the peak voltage of the music signal be $V_{music,0}$. The resistor and the speaker form a second current loop. Only one current can be in the resistor, so the peak current through the resistor, and therefore around the second loop, must be the same as the peak current in the first loop, given by Equation 21-44. According to $V = iR$ (Equation 18-9) the current is also given by

$$i_0 = \frac{V_{speaker,0}}{R} \tag{21-45}$$

where $V_{speaker,0}$ is the peak value of the voltage driving the speaker. To compare the two voltages, we solve Equations 21-44 and 21-45 for voltage and then take the ratio:

$$\frac{V_{speaker,0}}{V_{music,0}} = \frac{i_0 R}{i_0\sqrt{\dfrac{1}{\omega^2 C^2} + R^2}} = \frac{R}{\sqrt{\dfrac{1}{\omega^2 C^2} + R^2}}$$

Notice that when ω is small, the fraction $1/\omega^2 C^2$ gets large, so the ratio $V_{speaker,0}/V_{music,0}$ is small. In the limit as ω goes to zero,

$$\lim_{\omega \to 0} \frac{V_{speaker,0}}{V_{music,0}} = \frac{R}{\sqrt{\dfrac{1}{0^2 C^2} + R^2}} = \frac{R}{\infty} = 0$$

For larger values of ω, the fraction $1/\omega^2 C^2$ gets smaller, so in the limit of large values of ω the ratio is

$$\lim_{\omega \to \infty} \frac{V_{speaker,0}}{V_{music,0}} = \frac{R}{\sqrt{\dfrac{1}{\infty^2 C^2} + R^2}} = \frac{R}{\sqrt{R^2}} = 1$$

As you can see in the graph of the ratio of $V_{speaker,0}$ to $V_{music,0}$ shown in **Figure 21-26**, the higher frequencies in the music signal are passed to the speaker while the lower frequencies are not. The circuit in Figure 21-25 is a high pass filter.

For small values of frequency, $V_{speaker,0}$ is a small number multiplied by $V_{music,0}$, so very little of the low frequency signals get through to the speaker.

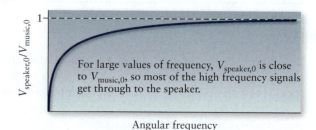

For large values of frequency, $V_{speaker,0}$ is close to $V_{music,0}$, so most of the high frequency signals get through to the speaker.

Figure 21-26 Higher frequencies in the music signal are passed to the speaker but lower frequencies are not.

Example 21-10 Another Filter Circuit

The circuit shown in **Figure 21-27** is similar to that in Figure 21-25, except that the capacitor and resistor have been switched. Find the ratio of the peak voltage driving the speaker to the peak voltage in the music signal, $V_{speaker,0}/V_{music,0}$, and graph it versus the angular frequency of the music signal.

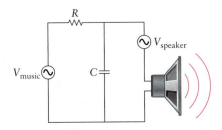

SET UP

There are two current loops in the circuit shown. V_{music}, the capacitor, and the resistor form one loop. This circuit is a series *LRC* circuit driven by an AC voltage in which L is equal to 0. The capacitor and the speaker form the second loop. This loop can also be treated as a series *LRC* circuit; but in this loop both the resistance and inductance are 0. The peak current must be the same in both loops, because they share a common element on which there can be only one current.

Figure 21-27 This circuit is similar to that in Figure 21-25, except that the capacitor and resistor have been switched.

SOLVE

According to Equation 21-37, the peak current in the first loop is

$$i_0 = \frac{V_{music,0}}{\sqrt{\left(0 - \dfrac{1}{\omega C}\right)^2 + R^2}} = \frac{V_{music,0}}{\sqrt{\dfrac{1}{\omega^2 C^2} + R^2}} \qquad (21\text{-}46)$$

where we set L equal to 0 and let the peak voltage of the music signal be $V_{music,0}$. The current in the second loop is also found from Equation 21-37, setting both L and R equal to 0:

$$i_0 = \frac{V_{speaker,0}}{\sqrt{\left(0 - \dfrac{1}{\omega C}\right)^2 + 0^2}} = \omega C V_{speaker,0} \qquad (21\text{-}47)$$

$V_{speaker,0}$ is the peak value of the voltage driving the speaker. Solving Equations 21-46 and 21-47 for voltage and then taking the ratio gives

$$\frac{V_{speaker,0}}{V_{music,0}} = \frac{i_0/\omega C}{i_0\sqrt{\dfrac{1}{\omega^2 C^2} + R^2}} = \frac{1}{\omega C\sqrt{\dfrac{1}{\omega^2 C^2} + R^2}}$$

or

$$\frac{V_{speaker,0}}{V_{music,0}} = \frac{1}{\sqrt{1 + \omega^2 C^2 R^2}}$$

A graph of this ratio versus ω is shown in **Figure 21-28**.

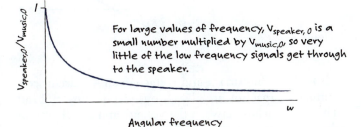

For small values of frequency. $V_{speaker,0}$ is close to $V_{music,0}$, so most of the low frequency signals get through to the speaker.

For large values of frequency, $V_{speaker,0}$ is a small number multiplied by $V_{music,0}$, so very little of the low frequency signals get through to the speaker.

Figure 21-28 In a low pass filter, most of the low frequency signals get through to the speaker but high frequency signals do not.

REFLECT

For small values of frequency, the ratio $V_{speaker,0}/V_{music,0}$ is close to 1, so $V_{speaker,0}$ is close to $V_{music,0}$. Most of the low frequency signals therefore get through to the speaker. When ω is large, $V_{speaker,0}/V_{music,0}$ is small. In this situation, $V_{speaker,0}$ is a small number multiplied by $V_{music,0}$ and very little of the high frequency signals get through to the speaker. This configuration of a resistor and a capacitor is a low pass filter.

✳ What's Important 21-6

The series LRC circuit driven by an AC voltage is the electrical analog of the mechanical oscillator and, just like the object on the end of a spring, the circuit also exhibits the phenomenon of resonance. In the case of an electrical oscillator, the amplitude of the current becomes large when the resonance occurs. The resistor affects the amplitude but not the frequency of the oscillations; only the inductor and capacitor affect the resonance characteristics of the circuit.

Answers to Practice Problems

21-1 About four times as bright

21-4 $L = 47.0$ mH, $C = 72.1$ μF, $R = .1$ Ω

21-5 1.97 A

21-6 $+39.5°$

21-8 $V(t) = 47.4$ V, $V_R(t) = 108$ V, $V_C(t) = -148$ V, and $V_L(t) = 86.8$ V. The sum of $V_R(t)$, $V_C(t)$, and $V_L(t)$, when computed before rounding the three values, equals $V(t)$.

Answers to Got the Concept Questions

21-1 We can combine Equation 21-5 and Equation 21-6 to form

$$P_{avg} = \frac{V_{rms}^2}{R}$$

which—because we started our study of the average voltage using Equation 18-27 ($P = V^2/R$)—is equivalent to Equation 18-27. In a similar way, we can rewrite Equations 18-26 and 18-28:

$$P = Vi \text{ becomes } P = V_{rms}i_{rms} \quad (21\text{-}50)$$
$$P = i^2R \text{ becomes } P = i_{rms}^2R \quad (21\text{-}51)$$

21-2

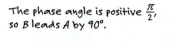
The phase angle is positive $\frac{\pi}{2}$, so B leads A by 90°.

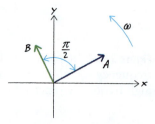

A and B are in phase.

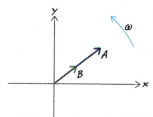

A and B are 180° out of phase. The sign of the phase doesn't matter in this case.

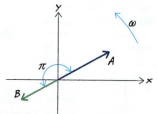

Figure 21-29

21-3 No. The current and voltage in an AC circuit containing a capacitor are always out of phase by $\pi/2$ rad. The relationship between the voltage and the charge on the capacitor (Equation 17-19) determines the phase difference, so no phase other than $\pi/2$ rad is possible.

21-4 No. The current and voltage in an AC circuit containing an inductor are always out of phase by $\pi/2$ rad. The relationship between the voltage and the current in the inductor (Equation 20-14) determines the phase difference.

21-5 Inductive reactance is directly proportional to the frequency (or angular frequency) of the voltage signal (Equation 21-30) while capacitive reactance is inversely proportional to the frequency. So while X_C is large for low frequencies, X_L is large for high frequencies. In this circuit,

SUMMARY

Topic	Summary	Equation or Symbol
AC (alternating current) circuit	An AC circuit is one driven by an alternating voltage source, so that the current oscillates. It is common for the voltage and therefore the current to vary sinusoidally in an AC circuit.	
capacitive reactance	The property of a capacitor to offer resistance to the flow of charge in an AC circuit is the capacitive reactance X_C. Capacitive reactance depends on the capacitance C of the capacitor and the angular frequency ω of the time-varying voltage.	$X_C = \dfrac{1}{\omega C}$ (21-24)
DC (direct current) circuit	A DC circuit is one driven by a voltage source that does not polarity, so that the current always flows in one direction.	
impedance	Impedance Z describes the overall opposition to the flow of charge in a circuit driven by a time-varying voltage. Impedance depends on the capacitance C, the inductance L, and the resistance R, as well as the angular frequency ω of the time-varying voltage.	$Z = \sqrt{\left(\omega L - \dfrac{1}{\omega C}\right)^2 + R^2}$ (21-38)
inductive reactance	The property of an inductor to offer resistance to the flow of charge in an AC circuit is the inductive reactance X_L. Inductive reactance depends on the inductance L of the inductor and the angular frequency ω of the time-varying voltage.	$X_L = \omega L$ (21-30)
phasor	A phasor is a rotating vector that represents a time-varying quantity, such as voltage or current, as it varies periodically over time. The instantaneous value of a phasor is given by its projection onto the y axis.	
primary coil	The primary coil in a transformer is connected to the input voltage.	
reactance	Reactance is the property of a capacitor or an inductor to offer resistance to the flow of charge in an AC circuit.	
root mean square (rms) average	The rms average is found by taking the square root of the average of the square of a function. The rms average is a function that varies sinusoidally between positive and negative values is not zero. (The arithmetic mean of such a function is zero.)	
root mean square (rms) voltage	The rms average of the time-varying voltage in an AC circuit V_{rms} is the peak voltage V_0 divided by $\sqrt{2}$.	$V_{rms} = \sqrt{(V^2(t))_{avg}}$ $= \sqrt{\dfrac{V_0^2}{2}} = \dfrac{V_0}{\sqrt{2}}$ (21-6)
secondary coil	The secondary coil in a transformer is the one connected to the output.	
transformer	A transformer is an electrical device that increases the voltage of an electric current while decreasing the current or vice versa. A transformer can be used to transmit power efficiently at high voltage and then used to transform the power to a safer, low voltage. A transformer consists of two coils; an alternating current in one (the primary coil P of N_p windings) induces a current in the other (the secondary coil S of N_s windings). The voltages V_p and V_s and the currents i_p and i_s in the two coils are related by the ratio of the number of windings in the two coils.	$V_s = \dfrac{N_s}{N_p} V_p$ (21-9) $i_s = \dfrac{N_p}{N_s} i_p$ (21-10)

QUESTIONS AND PROBLEMS

then, the capacitor serves as a simple high pass filter, suppressing low frequencies from getting to the tweeter, and the inductor functions as a simple low pass filter, suppressing high frequencies signals from getting to the woofer.

In a few problems, you are given more data than you actually need; in a few other problems, you are required to supply data from your general knowledge, outside sources, or informed estimate.

Interpret as significant all digits in numerical values that have trailing zeros and no decimal points.

For all problems, use $g = 9.8 \text{ m/s}^2$ for the free-fall acceleration due to gravity. Neglect friction and air resistance unless instructed to do otherwise.

• Basic, single-concept problem

•• Intermediate-level problem, may require synthesis of concepts and multiple steps

••• Challenging problem

SSM Solution is in Student Solutions Manual

Conceptual Questions

1. •What does this statement mean: The voltage drop across an inductor leads the current by 90°? SSM

2. •Why does a transformer whose primary coil has 10 times as many turns as its secondary coil normally delivers about 10 times more power than it receives?

3. •Does Equation 18-9 apply to alternating currents? If not, is there any manner in which it can be amended so that it will be valid?

4. •Explain why AC voltage is often described using the root mean square value rather than the average voltage.

5. •Search the Internet to determine the root mean square value of voltage in a common wall receptacle for five countries that are not mentioned in this chapter.

6. •Give a simple explanation as to why an electric appliance, which is supposed to be operated with a certain root mean square voltage, operated with a slightly increased root mean square voltage will lead to catastrophic results.

7. •Examine the label on the power converter for a laptop computer, digital camera, or cell phone. If you took the converter to a country where the rms voltage is 240 V, would you need a power transformer? Explain your answer.

8. •(a) What is a transformer? (b) How does it change the voltage input to some different voltage output? (c) Will a transformer work with a DC input?

9. •Chemistry Some of the most toxic substances that have been widely used in the United States (and many other countries in the world) are known as polychlorinated biphenyls or PCBs. (PCBs are a class of organic compounds that have 2 to 10 chlorine atoms attached to biphenyl, a molecule composed of two benzene rings.) Liquid PCBs were often used to fill the interior of transformers. Research PCBs and explain why they were used in transformers.

10. •Why is electric power for domestic use in the United States, Canada, and most of the Western Hemisphere transmitted at very high voltages and stepped down to 120 V by a transformer near the point of consumption?

11. •Discuss the storage of energy in an ideal LC circuit with no losses. SSM

12. •Why is the total or equivalent reactance of a series LC combination the difference between the inductive and capacitive reactances rather than the sum?

13. •The reactance of a capacitor decreases with increased frequency, whereas the reactance of an inductor increases. Why?

14. •Is the current through a resistor in an AC circuit always in phase with the potential applied to the circuit? Why or why not?

15. •In a coil that has both resistance and inductance, does the phase angle between the current through the coil and the voltage drop across it vary with the frequency? Explain your answer. SSM

16. •Define the concept of reactance for a capacitor and an inductor. What are the units of reactance?

17. •Define the concept of impedance for an LRC circuit. What are the units of impedance?

18. •What is a phasor and how is it used in the mathematical solution of LRC circuits?

19. •Describe how a crossover filter routes the high frequency notes to the tweeter and the low frequency notes to the woofer.

Multiple-Choice Questions

20. •The most important advantage of AC over DC is that
    A. electric power can't be delivered by a DC source.
    B. DC could only be used in the early days of electrical power distribution.
    C. DC results in more power loss in wire than AC.
    D. it is relatively straightforward to change the voltage delivered by an AC source.
    E. AC is safer.

21. •In a sinusoidal AC circuit with rms voltage V, the peak-to-peak voltage equals SSM
    A. $\sqrt{2}V$
    B. $2V$
    C. $2V/\sqrt{2}$
    D. $2\sqrt{2}V$
    E. $V/2$

22. •The common electrical receptacle voltage in North America is often referred to as "120 volts AC." One hundred twenty volts is
    A. the arithmetic mean of the voltage as it varies with time.
    B. the root mean square (rms) average of the voltage as it varies with time.
    C. the peak voltage from an AC wall receptacle.
    D. the average voltage over many weeks of time.
    E. one-half the peak voltage.

23. •If a power utility were able to replace an existing 500 kV transmission line with one operating at 1 MV, it would change the amount of heat produced in the transmission line to
    A. $\frac{1}{4}$ of the previous value.
    B. $\frac{1}{2}$ of the previous value.
    C. 2 times the previous value.
    D. 4 times the previous value.
    E. 0.

24. •For long-distance transmission of power, in order to reduce losses in the wires by driving a low current at high voltage, we should
    A. use a step-up transformer.
    B. use a step-down transformer.
    C. not use a transformer at all.
    D. change the length of the wires between transmission towers.
    E. change the type of conducting material.

25. •Two phasors A and B vary periodically over time according to the expressions $A(t) = A \sin(\omega t)$ and $B(t) = B \sin(\omega t - \pi)$.
    A. A leads B by π.
    B. A lags B by π.
    C. A and B are in phase.
    D. A and B are out of phase.
    E. A and B are out of phase by $\pi/2$.

26. •When the instantaneous voltage and current in an AC circuit are in phase, we know that
    A. the capacitive reactance is zero.
    B. the inductive reactance is zero.
    C. the resistance is zero.
    D. the impedance equals the resistance.
    E. all three, capacitive reactance, inductive reactance, and resistance, are zero.

27. •In an LRC circuit, when the inductive and capacitive reactances are equal,
    A. the phase angle is zero.
    B. the resistance is zero.
    C. the current is zero.
    D. the voltage is zero.
    E. the phase is $\pi/2$. SSM

28. •The voltage leads the current
    A. in resistor plus AC circuits.
    B. in capacitor plus AC circuits.
    C. in inductor plus AC circuits.
    D. in any AC circuit.
    E. in only resistive AC circuits.

29. •Which reactance will increase when the angular frequency in the circuit increases?
    A. resistive
    B. capacitive
    C. inductive
    D. both capacitive and inductive
    E. both capacitive and resistive

Estimation/Numerical Analysis

30. •Estimate the power lost in a length of wire when it is transmitted by DC compared to AC.

31. •Estimate the AC current in an average household appliance. How much current does an average U.S. household draw? SSM

32. •Estimate the number of turns of the secondary compared to the number of turns of the primary for a transformer that is used to power your hair dryer that you purchased in the United States and used in Europe.

33. •The voltage applied to a series LRC circuit is shown in the table below. Predict and plot the current as a function of time if $f = 60$ Hz, $L = 100$ mH, $C = 133 \ \mu C$, and $R = 50 \ \Omega$.

t (s)	V (V)	t (s)	V (V)
1	12	14	12
2	12	15	12
3	12	16	−12
4	12	17	−12
5	12	18	−12
6	−12	19	−12
7	−12	20	−12
8	−12	21	12
9	−12	22	12
10	−12	23	12
11	12	24	12
12	12	25	12
13	12		

Problems

21-1: Alternating Current

34. •A sinusoidally varying voltage is represented by $V(t) = (75 \text{ V})\sin(120\pi t)$. What are its frequency and peak voltage?

35. •Write an expression for the instantaneous voltage delivered by an AC generator supplying 120 V (rms) at 60 Hz.

36. •What is the rms current provided to a 60 W lightbulb that is plugged into a 120 V wall receptacle?

This page is intentionally left blank.

For complete end of chapter problem sets, please go to
www.whfreeman.com/kestentauck

magnitudes and the phase differences. (a) $A(t) = 3 \sin(\omega t)$ and $B(t) = 5 \sin\left(\omega t - \frac{\pi}{2}\right)$, (b) $A(t) = 3 \sin(\omega t)$ and $B(t) = 2 \sin(\omega t)$, (c) $A(t) = 5 \sin(\omega t)$ and $B(t) = 3 \sin(\omega t + \pi)$

56. • An *LRC* circuit contains a 1.00 μF capacitor, a 5.00 mH inductor, and a 100 Ω resistor. What is its resonant frequency?

57. •An *LRC* circuit contains a 500 Ω resistor, a 5.00 H coil, and an unknown capacitor. The circuit resonates at 1000 Hz. Find the value of the capacitor. SSM

58. •The resonant frequency of an *LRC* circuit is 250 Hz. If the resistor is 200 Ω and the capacitor is 125 nF, find the value of the inductor.

21-4: *L, R, C* Separately with AC

59. •A sinusoidal voltage with a frequency of 60 Hz is applied to a 0.20 H inductor. What is the reactance of the inductor?

60. •A sinusoidal voltage that is 120 V (rms) and has a frequency of 60 Hz is applied to a 0.20 H inductor. Calculate the peak value of the current.

61. •A sinusoidal voltage with a frequency of 60 Hz is applied to a 50.0 μF capacitor. What is the reactance of the capacitor? SSM

62. •A capacitor has a capacitive reactance of 160 Ω at 60 Hz. What is the reactance at 600 Hz?

63. •A sinusoidal voltage of 120 V (rms) and a frequency of 60 Hz are applied to a 50.0 μF capacitor. Calculate the peak value of the current.

64. •Calculate the reactance of a 5.0 μF capacitor at (a) 60 Hz, (b) 6000 Hz, and (c) 6 MHz.

65. •Calculate the reactance of a 5.0 mH inductor at (a) 60 Hz, (b) 6000 Hz, and (c) 6 MHz. SSM

66. ••A 100 Ω resistor is connected across a 120 V rms, 60 Hz AC power line. Calculate (a) i_{rms}, (b) i_{max}, and (c) the average power dissipated in the resistor.

67. ••A sinusoidal voltage of 50.0 V (peak) at a frequency of 400 Hz is applied to a capacitor of unknown capacitance. The current in the circuit is 400 mA (rms). (a) What is the capacitance? (b) If the frequency of the voltage is increased, what, if anything, will happen to the rms value of the current in the circuit? Why?

68. ••A potential of 40 V (rms) and a frequency of 100 Hz is applied to (a) a 0.2 H inductor and (b) a 50 μF capacitor. In each case, find the peak value of the current.

69. ••A sinusoidal voltage of 40.0 V rms and a frequency of 100 Hz is applied to (a) a 100 Ω resistor, (b) a 0.200 H inductor, and (c) a 50.0 μF capacitor. Find the peak value of the current and the average power delivered in each case. SSM

21-5: *L, R, C* in Series with AC

70. ••A 35 mH inductor with 0.20 Ω resistance is connected in series to a 200 μF capacitor and a source that is 60 Hz, AC, and 45 V. Calculate the (a) rms current and (b) phase angle for the circuit.

71. ••In an *LRC* series circuit, the inductance is 250 mH, the resistance is 20 Ω, and the capacitance is 350 μF. If the AC voltage applied across the circuit is given by

$$V(t) = (10 \text{ V}) \sin 12\pi t$$

calculate the voltage across each element at a time of $t = 0.04$ s. SSM

72. ••The circuit in **Figure 21-32** consists of a 1.5 μF capacitor and a 100 Ω resistor in series with negligible inductance. (a) Draw a vector diagram showing the voltage drops across *C* and *R*, and find the phase difference between the potential and the current at a frequency of 300 Hz. (b) Repeat for a frequency of 5000 Hz.

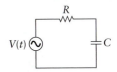

Figure 21-32 Problem 72

73. ••Assume that the circuit in **Figure 21-33** has *L* equal to 0.60 H, *R* equal to 250 Ω, and *C* equal to 3.5 μF. At a frequency of 60 Hz, what are the impedance and the phase angle between the current and voltage?

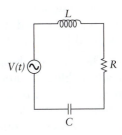

Figure 21-33 Problems 73, 74

74. ••Assume that the circuit in **Figure 21-33** has *L* equal to 0.60 H, *R* equal to 280 Ω, and *C* equal to 3.5 μF. The amplitude of the driving voltage is 150 V (rms). At a frequency of 60 Hz, what is the rms current in the circuit?

75. •In a standard *LRC* series circuit, the inductive reactance is 2500 Ω, the capacitive reactance is 3400 Ω, and the resistance is 1000 Ω. What is the impedance? SSM

76. •Suppose the total reactance of an *LRC* circuit is 1400 Ω and the total resistance is 600 Ω. What is the impedance?

77. •An inductor has a reactance of 10 Ω and a resistance of 8.0 Ω. When connected across a 120 V rms AC receptacle, how much current will it draw?

78. ••In an *LRC* circuit, the voltage amplitude and frequency of the source are 100 V and 500 Hz, respectively. The resistor has a value of 500 Ω, the inductor has a value of 0.20 H, and the capacitor has a value of 2.0 μF. (a) What is the impedance of the circuit? (b) What is the amplitude of the current from the source? (c) If the voltage of the source is given by

$$V(t) = (100 \text{ V}) \sin 1000\pi t$$

how does the current vary with time?

21-6: Applications of a Series *LRC* Circuit

79. ••The circuit in **Figure 21-34** consists of a 250 Ω resistor and a 0.04 H inductor in series with negligible capacitance. Find the voltage across the inductor (a) at a frequency of 50 Hz and (b) at a frequency of 5000 Hz. (c) An *LR* circuit is often used in electronic circuits as a filter. What do you suppose it filters? SSM

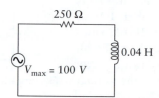

250 Ω

0.04 H

V_{max} = 100 V

Figure 21-34 Problem 79

80. ••Suppose your lab partner wants to build a high pass filter using only a 75 nF capacitor. Compare the rms current that is created if the frequency of the 120 V rms source is 60 Hz versus 100 Hz.

81. ••The tuner knob on an FM radio moves the plates of an adjustable capacitor that is in series with a 0.400 μH inductor and a net resistance of 1000 Ω. The peak current induced in the circuit by a radio wave of a specific carrier frequency becomes large when the natural frequency of the circuit matches the carrier frequency. What is the frequency of the FM station that is tuned in when the capacitor in the radio is adjusted to 5.80 pF? If the peak operating voltage in the tuning circuit is 9 V, what is the peak current?

82. ••A series *LRC* tuning circuit in a TV receiver resonates at 58 MHz. The circuit uses an 18 pF capacitor. What is the inductance?

83. ••An *LRC* series circuit contains a 500 Ω resistor, a 5.0 H choke coil, and a capacitor. What value of capacitance will cause the circuit to resonate at 1000 Hz? SSM

84. ••A 2.5 H choke coil has a negligible resistance and is in series with a 1200 Ω resistor. The circuit is plugged into a transformer that outputs 50 V (rms, AC) at 60 Hz. (a) What voltage drop will appear across the resistor? (b) What about the voltage drop across the inductor?

General Problems

85. •••**Calc** Determine the rms voltage associated with an AC circuit that has a voltage versus time as shown in **Figure 21-35**. Express your answer in terms of the period, *T*, and the maximum voltage, V_0. SSM

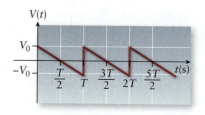

Figure 21-35 Problem 85

86. ••A 150 Ω resistor connected across a 60 cycle AC power supply of voltage amplitude 75 V produces heat in the resistor at a certain rate. If you want to replace the AC source with a DC power supply and still produce the same rate of heating, what should be the voltage of the DC source?

87. •••An immersion coil resistance heater, designed to heat individual cups of water for coffee or tea, has a resistance of 4.50 Ω and operates from an ordinary household outlet (120 V rms, 60 cycle). (a) How long will it take the coil to heat 250 g of water originally at 20°C to the boiling point? (See Table 14-3.) (b) What are the current amplitude and the rms current through the coil while it is heating the water?

88. ••In Europe, the standard voltage is 240 V (rms), AC, and 60 cycles/s. Suppose you take your 5.00 W electric razor to Rome and plug it into the receptacle (with an adapter to fit the receptacle but not to change the voltage). (a) What power will it draw in Rome? (b) What rms current will run through the razor in the United States and in Rome? Is the razor in danger of being damaged by using it in Rome without a voltage adapter? (c) If you want to use the razor in Rome without damaging it, what type of transformer would you need? Be as quantitative as you can. (d) What is the resistance of your razor?

89. ••**Medical** A DC current of 60 mA can cause paralysis of the body's respiratory muscles and hence interfere with breathing, but only 15 mA (rms) of AC current will do the same thing. Suppose a person is working with electrical power lines on a warm humid day and therefore has a low body resistance of 1000 Ω. What DC and what AC (amplitude and rms) potentials would it take to cause respiratory paralysis?

90. •A power cord has a resistance of 8.00×10^{-2} Ω and is used to deliver 1500 W of power. (a) If the power is delivered at 12.0 V (rms), how much power is dissipated in the power cord (assuming the current and voltage are in phase)? (b) If the power is delivered at 120 V rms, how much power is dissipated in the power cord (again assuming the current and voltage are in phase)? (c) Which voltage would you prefer to use to power your electrical device? Why?

91. •••An AC electrical generator is made by turning a flat coil in a uniform constant magnetic field of 0.225 T.

The coil consists of 33 square windings and each winding is 15.0 cm on each side. It rotates at a steady rate of 745 rpm about an axis perpendicular to the magnetic field passing through the middle of the coil and parallel to two of its opposite sides. An 8.50 Ω lightbulb is connected across the generator. (a) Find the voltage and current amplitudes for the lightbulb. (b) At what average rate is heat generated in the lightbulb? (c) How much energy is consumed by the lightbulb every hour? SSM

92. ••A small AC heating coil consisting of 750 windings is 8.50 cm long. It has a diameter of 1.25 cm and a resistance of 2.15 Ω. The coil is connected in series with a 2240 μF ultracapacitor and the combination is plugged into a household outlet (60.0 Hz and 120 V rms). (a) What is the impedance of the circuit? (b) What is the rms current through the resistor? (c) What is the maximum (or peak) current through the resistor?

93. ••Medical An electrician is working with high voltage AC lines having a voltage amplitude of 25 kV. The resistance of his body between his hands and feet could be as low as 1000 Ω if he is wet or as high as 500 kΩ if he is dry. T protect himself against possible electrical shock, the worker wears protective shoes having insulating soles. Each sole measures 8.0 cm by 20.0 cm, and the shoes are designed to limit the AC rms current through the worker to no more than 1.0 mA. (a) If the worker decides not to wear the protective shoes, is he in significant danger when wet? What if he is dry? (b) For maximum protection with the protective shoes, should the worker stand on one foot or on both feet? Why? (c) If the worker is standing on two feet and is wet, what should be the minimum resistivity of the insulating material to provide adequate protection if the soles are to be 2.0 cm thick?

94. ••Using the series LRC circuit in Figure 21-36, determine the maximum voltages V_R, V_C, and V_L across the resistor, capacitor, and inductor, respectively, (a) at a frequency of 100 Hz and (b) at resonance. (c) Explain why the sum of the maximum voltages in each case is larger than the maximum voltage of the applied potential. (d) Why is the maximum voltage across the inductor equal to the maximum voltage across the capacitor at resonance?

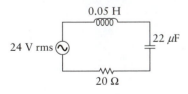

Figure 21-36 Problem 94

95. ••For the circuit shown in Figure 21-37, determine (a) the total impedance and (b) the phase angle between the current and the voltage for each of the current elements (R, L, and C). The current in the circuit is given by $i(t) = (10 \text{ A}) \sin 120\pi t$. SSM

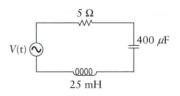

Figure 21-37 Problem 95

96. •••Calc (a) With regard to the charge $q(t)$ on the capacitor in an LRC circuit, show that the maximum amplitude will be equal to:

$$Q_0 = \frac{V_0}{\sqrt{(\omega R)^2 + (\omega^2 L - 1/C)^2}}$$

(b) For what angular frequency, ω_0, will the value of charge have a maximum?

97. ••You construct an LRC AC circuit using a 125 Ω resistor, a 12.5 mH inductor, and a parallel plate capacitor having a plate separation of 2.10 mm with a plastic material completely filling the region between the plates. The rectangular capacitor plates each measure 4.25 cm by 6.20 cm. (a) If you want the maximum rms current through the circuit at an AC frequency of 55.0 Hz, what should be the dielectric constant of the plastic in the capacitor? (b) Consult Table 17-2 to see if it appears feasible to achieve the desired results for the circuit, and explain why or why not. Would the use of an ultracapacitor (which can have a capacitance of up to several farads) allow you to achieve the desired results?

98. ••An LRC circuit consists of a 15.0 μF capacitor, a resistor, and an inductor connected in series across an AC power source of variable frequency having a voltage amplitude of 25.0 V. You observe that when the power source frequency is adjusted to 44.5 Hz, the rms current through the circuit has its maximum value of 65.0 mA. What will be the rms current if you change the frequency of the power source to 60.0 Hz?

99. ••Calc Suppose an AC power supply is in series with a capacitor. If the current in the circuit is found to be $i(t) = i_0 \sin \omega t$, derive an expression for the charge on the capacitor as a function of time. Assume that the charge is zero at $t = 0$ s. SSM

100. ••Calc Suppose an AC power supply is in series with a capacitor. The voltage is

$$V(t) = V_0 \sin(\omega t + \varphi)$$

Derive an expression for the current in the circuit.

101. •••Calc An inductor that has an inductance L and is in series with a resistor of resistance R is powered by a sinusoidal voltage source. The differential equation that results from Kirchhoff's loop rule around the circuit is

$$V_0 \sin(\omega t) = L\frac{di}{dt} + Ri$$

(a) Prove by direct substitution that the current in the circuit is of the form $i(t) = i_0 \sin(\omega t - \varphi)$. Derive expressions for (b) the amplitude (i_0) and (c) the phase angle (φ).

102. ••The circuit in **Figure 21-38** is known as a high pass filter because it allows high frequency signals to pass and attenuates lower frequencies. Suppose it is input an AC signal that is composed of a broad range of frequencies. The voltage across the resistor is the output that is detected. Show that the ratio of the output voltage to the input voltage is

$$\frac{V_o}{V_i} = \frac{1}{\sqrt{1 + 1/(2\pi fRC)^2}}$$

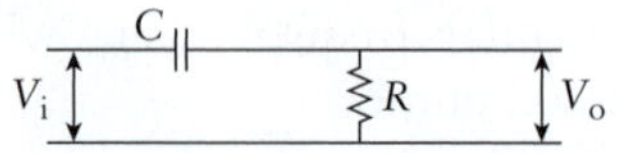

Figure 21-38 Problem 102

103. ••The circuit in **Figure 21-39** is known as a low pass filter because it allows low frequency signals to pass and attenuates higher frequencies. Suppose its input is an AC signal that is composed of a broad range of frequencies. In this case, the voltage across the capacitor is the output that is detected. Show that the ratio of the output voltage to the input voltage is SSM

$$\frac{V_o}{V_i} = \frac{1}{\sqrt{1 + (2\pi fRC)^2}}$$

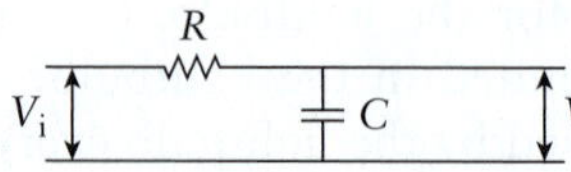

Figure 21-39 Problem 103

104. •••The Q factor of an LRC resonance circuit can be defined as the ratio of the voltage across the capacitor (or inductor) to the voltage across the resistor at resonance. The larger the Q factor, the sharper the resonance curve will be and the sharper the tuning. (a) Show that the Q factor is given by the equation

$$Q = \frac{1}{R}\sqrt{L/C}$$

(b) At a resonant frequency of 1.0 MHz, what must be the values of L and R to produce a Q factor of 1000? Assume that $C = 0.0010 \; \mu\text{F}$.

22 Electromagnetic Waves

(Siebeck et al. 2010. "A Species of Reef Irish that Uses Ultraviolet Patterns for Covert Face Recognition." Current Biology)

The brightly colored feathers of some male hummingbirds help them attract the attention of potential mates, and the orange, red, green, and blue markings on some poisonous frogs say "Don't eat me!" to potential predators. Light is an electromagnetic wave; colors such as the bright blue of a poisonous frog fall within the visible spectrum that can be seen by humans. Some animals see electromagnetic waves outside our visible range. Damselfish, for example, can see ultraviolet waves. Where you see only dark and light bands on the fish's face, damselfish see an intricate pattern. This enables the territorial male Ambon damselfish (top) to attack another Ambon in order to defend its territory, but ignore a male lemon damselfish (bottom).

W hy do you see a flash of bright red when sunlight hits the neck of a male ruby-throated hummingbird at just the right angle? Properties of waves—the way they reflect and the way they interfere with each other—enable you to detect the red flash. When you think of a wave, you might conjure up an image of undulations on the surface of the ocean, wiggles moving along a stretched string, or the pressure variations that travel as sound propagates down a pipe. We recognize waves by these and other characteristics and properties. Light exhibits the same properties because it is an **electromagnetic wave**. We experience a host of other electromagnetic waves in our daily lives, including x-rays, microwaves, and radio waves. In this chapter, we examine phenomena associated with light and other electromagnetic waves.

Figure 22-1 The energy spectrum of electromagnetic waves shows radio waves at the low energy end of the spectrum; at the high energy end of the spectrum are x-rays and gamma rays.

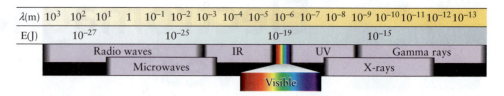

| λ(m) | 10^3 | 10^2 | 10^1 | 1 | 10^{-1} | 10^{-2} | 10^{-3} | 10^{-4} | 10^{-5} | 10^{-6} | 10^{-7} | 10^{-8} | 10^{-9} | 10^{-10} | 10^{-11} | 10^{-12} | 10^{-13} |

22-1 Electromagnetic Waves

Radio waves are electromagnetic waves. They are all around us and, in fact, pass right through us with no ill effect. X-rays are electromagnetic waves that can pierce soft tissue but not bone, allowing doctors to diagnose a broken bone and dentists to see a cavity in your tooth. Some electromagnetic waves, such as high energy gamma rays, can damage DNA, cause cancer, and even kill cells. The difference between radio waves, x-rays, and gamma rays is their energy. Although all electromagnetic waves exhibit the same general properties and obey the same physical laws, the x-rays used in a dentist's office carry about 10^{11}—one hundred billion—times as much energy as the radio waves that deliver signals to your cell phone. The most energetic electromagnetic waves carry about 10^{18} as much energy as the least energetic ones. Visible light is somewhere near the middle of the range.

The energy spectrum of electromagnetic waves (**Figure 22-1**) shows that radio waves are at the low energy and long wavelength end of the spectrum; at the high energy and short wavelength end of the spectrum are x-rays and, beyond them, gamma rays. Other waves can extend over a wide range of energy, too. A water wave, for example, doesn't carry any particular or fixed amount of energy; the ripples that extend outward behind a duck swimming on a pond (**Figure 22-2a**) carry relatively low energy, while the waves at Pe'ahi on Maui, Hawaii (**Figure 22-2b**), pound the shore with enormous energy. There is, however, a fundamental difference between electromagnetic waves and mechanical waves such as the ripples on a pond and the pounding surf at Pe'ahi. The waves of the electromagnetic spectrum exhibit some particle-like properties! The energy of an electromagnetic wave propagates as small, individual packets of energy. These packets are called **photons**, even for waves outside the range of visible light. When we say that a dental x-ray carries 10^{11} times as much energy as a radio wave, we mean that one x-ray photon is 10^{11} times as energetic as one radio photon.

Figure 22-2 Not all waves carry the same amount of energy. (a) The ripples that extend outward behind a duck swimming on a pond carry relatively low energy. *(Cedric Pieterse/Fotolia)* (b) Ocean waves pound the beach with enormous energy. *(Ron Dahlquist/ SuperStock)*

(a)

(b)

An x-ray can potentially deliver far more energy to an object than a radio wave. Especially for biological tissue, the effect of electromagnetic waves depends on whether enough energy is transferred to ionize atoms by pulling off electrons. **Ionizing radiation** breaks apart molecules by pulling electrons from the chemical bonds that hold atoms together. Although ionizing radiation can directly break DNA molecules in living tissue, it's more likely to disrupt some other, more common molecule, such as water, to create highly reactive free radicals that then damage DNA. Depending on the severity of the damage, the cell may be able to recover. If the damage cannot be repaired, however, or if it is repaired incorrectly, the resulting mutations may be lethal. These effects generally go unnoticed until the next time the cell tries to divide. Because cancerous cells divide more frequently than most other cells in the body, they are more susceptible to radiation damage than most healthy cells.

All photons move at the speed of light. In empty space, photons propagate at c, the **speed of light in a vacuum**, where

$$c = 2.997\ 924\ 58 \times 10^8 \text{ m/s}$$

We will use 3.0×10^8 m/s as the value of c for most calculations. Note that in a medium other than a vacuum, light travels at a speed less than c. In water, for example, photons travel at about 2.2×10^8 m/s ($0.73c$), and in diamond, only about 1.2×10^8 m/s ($0.4c$). The speed of photons in air is about $0.9998c$, a small but easily measurable difference from the speed of light in a vacuum.

The energy of a photon does not depend on its speed; all photons travel at the same speed in a given medium. Regardless of its position in the electromagnetic spectrum, a photon's energy is inversely proportional to the wavelength of the electromagnetic wave. In a vacuum, in which the speed is c,

$$E = \frac{hc}{\lambda} \tag{22-1}$$

where **Planck's constant** h has a value of 6.63×10^{-34} J · s. Recall that the propagation speed of a wave is given by the product of wavelength and frequency ($v_p = f\lambda$, Equation 13-8). So for electromagnetic waves in a vacuum, for which the propagation speed (v_p) is c, the right side of Equation 22-1 can be written in terms of the frequency of the wave

$$E = hf \tag{22-2}$$

! Watch Out

The energy that a photon carries does not depend on its speed.

All photons propagate at the same speed in a given medium; for example, in a vacuum, all photons propagate at c. Yet an x-ray photon carries more energy than a visible light photon. As both Equations 22-1 and 22-2 make clear, a photon's energy does not depend on speed, but only on its frequency. You might protest that it could also depend on its wavelength, but as Equation 13-8 makes clear, the wavelength of a photon also depends on its frequency.

? Got the Concept 22-1

X-Rays and Radio Waves

X-rays carry far more energy than radio waves. Which has the longest wavelength? Which has the highest frequency?

Example 22-1   Visible Light

The lowest energy photons that our eyes can detect have wavelengths around 750 nm. The highest energy photons that are visible to us have wavelengths around 380 nm. (a) What colors are these wavelengths? (b) What are the range of energies and the range of frequencies of visible light?

SET UP

(a) Take a look at the visible spectrum shown in **Figure 22-3**.
(b) To find the energies we can use Equation 22-1 directly, that is,

$$E = \frac{hc}{\lambda}$$

We could use Equation 22-2 to find frequency from the energy. Because we have been given wavelengths, we choose to employ Equation 13-8:

$$c = f\lambda$$

so that frequency can be expressed in terms of wavelength:

$$f = \frac{c}{\lambda}$$

SOLVE

(a) From the spectrum shown in Figure 22-3, we see that the longest wavelengths of the visible spectrum are in the red. Red photons therefore carry the lowest energy of all visible light, and also oscillate at the lowest frequency. Photons that are slightly lower in energy than red are termed infrared (IR). Violet light carries the most energy of visible photons. Light of energies above that lies in the ultraviolet (UV) part of the spectrum.

(b) Using Equation 22-1, the energy of the red and violet photons are

$$E_{\text{red}} = \frac{(6.63 \times 10^{-34}\,\text{J}\cdot\text{s})(3.0 \times 10^{8}\,\text{m/s})}{750 \times 10^{-9}\,\text{m}} = 2.7 \times 10^{-19}\,\text{J}$$

and

$$E_{\text{violet}} = \frac{(6.63 \times 10^{-34}\,\text{J}\cdot\text{s})(3.0 \times 10^{8}\,\text{m/s})}{380 \times 10^{-9}\,\text{m}} = 5.2 \times 10^{-19}\,\text{J}$$

The corresponding frequencies are

$$f_{\text{red}} = \frac{3.0 \times 10^{8}\,\text{m/s}}{750 \times 10^{-9}\,\text{m}} = 4.0 \times 10^{14}\,\text{Hz}$$

and

$$f_{\text{violet}} = \frac{3.0 \times 10^{8}\,\text{m/s}}{380 \times 10^{-9}\,\text{m}} = 7.9 \times 10^{14}\,\text{Hz}$$

REFLECT

Red light is the lowest energy visible light, while violet is the highest. The energy of violet light is nearly double that of red light.

Practice Problem 22-1 According to the electromagnetic spectrum shown in Figure 22-1, microwaves range in wavelength from about 10 m to 0.0005 m. Find the range of energies and the range of frequencies of microwaves.

Figure 22-3 Our eyes detect wavelengths in the visual spectrum, between around 380 and 750 nm.

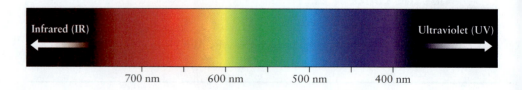

Infrared (IR)   Ultraviolet (UV)

700 nm   600 nm   500 nm   400 nm

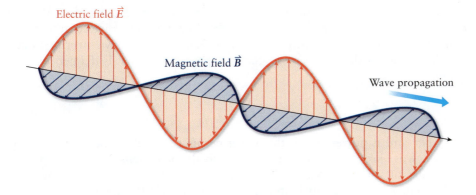

Electric field $\vec{E}$

Magnetic field $\vec{B}$

Wave propagation

Figure 22-4 As an electromagnetic wave travels, an electric field $\vec{E}$ and a magnetic field $\vec{B}$ rise and fall together. The $\vec{E}$ and $\vec{B}$ vectors are perpendicular to each other at any point along the direction in which an electromagnetic wave propagates.

We use wavelength and frequency to describe electromagnetic waves, in the same way as for mechanical waves. For a mechanical wave, such as a transverse wave on a string or a sound wave emanating from the bell of a trumpet, wavelength is the distance over which a disturbance in the medium repeats itself, and frequency is the inverse of the time interval required for any particular region of the medium to be displaced and then return to its initial position. In contrast, an electromagnetic wave is not a disturbance of a medium through which the wave passes. Indeed, electromagnetic waves do not require any medium at all to get from one place to another; light, for example, travels from the Sun to Earth through the approximate vacuum of space. If an electromagnetic wave is not a disturbance of a medium through which the wave propagates, what, then, does the wavelength and frequency represent for an electromagnetic wave?

An electromagnetic wave is the propagation of an electric field and a magnetic field. The two fields vary in phase with each other as the wave travels; that is, $\vec{E}$ and $\vec{B}$ rise and fall together. In addition, at any point along the direction in which an electromagnetic wave propagates, the $\vec{E}$ and $\vec{B}$ vectors are perpendicular to each other. The electric and magnetic fields of an electromagnetic wave are shown in **Figure 22-4**. We will explore the electric and magnetic aspects of electromagnetic waves in the next section.

At what rate does an electromagnetic wave transport energy across a region of space? **Figure 22-5** shows the oscillating electric field and magnetic field components of an electromagnetic wave passing through a boundary of area A. The energy dU

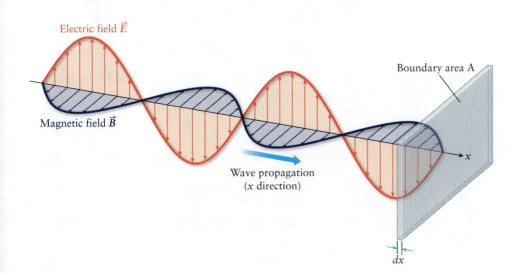

Electric field $\vec{E}$

Boundary area A

Magnetic field $\vec{B}$

Wave propagation
(x direction)

x

dx

Figure 22-5 The oscillating electric and magnetic field components of an electromagnetic wave pass through a boundary of area A. The energy transported across the boundary in a short time is the energy that occupied the volume of the rectangular prism during that time.

transported across A in a short time dt is the energy that occupied the volume of the rectangular prism formed by A and the distance the wave traveled in that time. To find the rate of energy transport across a region of space, we must first determine the energy of the wave crossing area A per time dt, that is dU per unit area per unit time, or

√x̄ *See the Math Tutorial for more information on Calculus.*

$$S = \frac{dU}{A\, dt} \tag{22-3}$$

Energy per time is power, so S is the power per unit area of the boundary through which the wave propagates. We have previously defined *intensity* as power per unit area, so S is intensity. The watt is defined as 1 J/s, so the SI units of S are W/m^2.

We have seen that energy is stored in an electric field (Chapter 17) and a magnetic field (Chapter 20). The energy density of an electromagnetic wave must therefore depend on the magnitudes of the electric and magnetic fields, E and B. In Chapter 17, we found the energy stored U_E in the electric field in a parallel plate capacitor of plate area A and plate separation d from Equation 17-21:

$$U_E = \frac{1}{2}\frac{q^2}{C} \tag{22-4}$$

where q is the magnitude of the charge stored on each plate and C is the capacitance. This can be cast into a form that shows the dependence on the electric field E directly, by using Equation 17-18 for q

$$E = \frac{q}{\epsilon_0 A}$$

and Equation 17-20 for C

$$C = \frac{\epsilon_0 A}{d}$$

where the constant ϵ_0 is the permittivity of free space. Using these two equations to substitute for q and C in Equation 22-4 gives

$$U_E = \frac{1}{2}\frac{(\epsilon_0 A E)^2}{(\epsilon_0 A/d)} = \frac{1}{2}\epsilon_0 E^2 A d$$

This holds for any uniform electric field aligned between parallel regions of area A and separated by d, so it describes the energy stored in the electric field in the infinitesimally thin rectangular prism shown in Figure 22-5. The distance in this case is dx; we let the wave travel through a vacuum, so the speed is c and therefore the infinitesimally small distance dx it travels in time dt is equal to $c\, dt$. The amount of energy dU_E stored in the electric field in the volume is then

$$dU_E = \frac{1}{2}\epsilon_0 E^2 A\, dx = \frac{1}{2}\epsilon_0 E^2 A c\, dt$$

The total amount of energy in the volume must be twice this, because the magnetic and electric fields are equal components of an electromagnetic wave. Thus the total energy dU in the volume is

$$dU = \epsilon_0 E^2 A c\, dt$$

From Equation 22-3, the energy of the wave crossing area A per time dt is then

$$S = \frac{dU}{A\, dt} = \frac{\epsilon_0 E^2 A c\, dt}{A\, dt} = \epsilon_0 c E^2 \tag{22-5}$$

It is standard to express S in terms of B as well as E. These magnitudes are related by

$$E = cB \tag{22-6}$$

By rewriting one factor of E in Equation 22-5, S becomes

$$S = \epsilon_0 c E c B = \epsilon_0 c^2 E B$$

Later in this chapter we will discover that c, ϵ_0, and μ_0, the permeability of free space are related according to $c = \sqrt{1/\mu_0 \epsilon_0}$ (Equation 22-27), so

$$S = \epsilon_0 \frac{1}{\mu_0 \epsilon_0} EB = \frac{EB}{\mu_0} \qquad (22\text{-}7)$$

This magnitude can be combined with the direction in which the electromagnetic wave propagates to form the **Poynting vector**, named after the nineteenth century English physicist John Henry Poynting. The direction of propagation of an electromagnetic wave is mutually perpendicular to the directions in which the electric and magnetic fields oscillate. Recall the direction of the cross product of two vectors is the direction that is mutually perpendicular to those vectors, so

$$\vec{S} = \frac{\vec{E} \times \vec{B}}{\mu_0} \qquad (22\text{-}8)$$

Notice that because $\vec{E}$ and $\vec{B}$ are perpendicular to each other, the magnitude of $\vec{E} \times \vec{B}$ equals EB, so adding the cross product does not change the magnitude of S described in Equation 22-7.

The magnitude of the Poynting vector gives the energy transported by an electromagnetic wave per unit area per unit time at any instant. It is often more useful to know the average value over a time interval. The rms average of the Poynting vector is found from the rms average values of the electric and magnetic fields. From our discussion of the rms average (for example, in obtaining Equation 21-6) these values are given by $E_{\text{rms}} = E_0/\sqrt{2}$ and $B_{\text{rms}} = B_0/\sqrt{2}$, where E_0 and B_0 are the maximum values of the electric field and magnetic field components of the electromagnetic wave. So the rms average of the energy transported by an electromagnetic wave per unit area per unit time is

$$S_{\text{rms}} = \frac{E_{\text{rms}} B_{\text{rms}}}{\mu_0} = \frac{E_0 B_0}{2\mu_0} \qquad (22\text{-}9)$$

 Watch Out

The Poynting vector does more than "point."

If you've only heard someone talking about the Poynting vector, say, in a class, you might be tempted to think of it as the "pointing vector." Not so! Although it does have a direction (it is a vector after all), the most important aspect of the Poynting vector is that its magnitude is the energy transported by an electromagnetic wave per unit area per unit time at any instant.

Example 22-2 Units of the Poynting Vector

Show that when the quantities E, B, and μ_0 are combined to form the Poynting vector (for example, in Equation 22-7), the resulting SI units are watts per square meter (W/m^2).

SET UP
The SI units of electric field are N/C, the SI unit of magnetic field is T, and the SI units of the permeability of free space μ_0 are $\text{T} \cdot \text{m/A}$. Applying these units to Equation 22-7 serves as a starting point; we'll look for equivalences between units to guide us the rest of the way.

SOLVE

From Equation 22-7, the units of S are

$$[S] = \frac{[E][B]}{[\mu_0]} = \frac{(N/C)(T)}{T \cdot m/A}$$

This expression simplifies to

$$[S] = \frac{N \cdot A}{C \cdot m}$$

This doesn't look like our goal of W/m^2. But note that 1 A is defined as 1 C/s, so

$$[S] = \frac{N(C/s)}{C \cdot m} = \frac{N}{s \cdot m}$$

Now recall that work can be determined by the product of force and the distance over which it is applied. (The details of this relationship are not important because any physical variable has the same units and dimensions regardless of the expression in which it is used.) One joule of energy is equivalent to 1 N·m, so 1 N is equivalent to 1 J/m. The units of S can then be written as

$$[S] = \frac{J/m}{s \cdot m} = \frac{J}{s \cdot m^2}$$

As 1 J/s is defined to be 1 W, finally,

$$[S] = \frac{W}{m^2}$$

REFLECT

The units of S are those of intensity, W/m^2.

Example 22-3 Sunlight on Earth's Atmosphere

The rms average electric field strength of sunlight striking the top of Earth's atmosphere is about 7.2×10^2 N/C. The rms average of the magnetic field strength is about 2.4×10^{-6} T. How much power per unit area does sunlight deliver to Earth's atmosphere?

SET UP

Power per unit area is intensity, the magnitude of the Poynting vector. We can use the first form of Equation 22-9 to find it.

SOLVE

Using the given values of E_{rms} and B_{rms},

$$S_{rms} = \frac{E_{rms}B_{rms}}{\mu_0} = \frac{(7.2 \times 10^2 \text{ N/C})(2.4 \times 10^{-6} \text{ T})}{4\pi \times 10^{-7}(\text{T} \cdot \text{m/A})} = 1.4 \times 10^3 \text{ W/m}^2$$

REFLECT

Earth is approximately 1.5×10^{11} m from the Sun. Imagine a sphere of that radius with the Sun at its center; based on our answer each square meter of the surface area of the sphere receives 1.4×10^3 W in the form of electromagnetic radiation.

Practice Problem 22-3 The rms average electric field strength of sunlight striking the surface of Mars is about $4.7 \times 10^2 \, \text{N/C}$. The rms average of the magnetic field strength is about $1.6 \times 10^{-6} \, \text{T}$. How much power per unit area does sunlight deliver to the surface of Mars?

Because electromagnetic waves transport energy, we should expect that they also carry linear momentum. The momentum of an electromagnetic wave is directly related to the energy it carries, and is expressed as

$$p = \frac{E}{c} \tag{22-10}$$

To match the language we used in discussing the Poynting vector, the change in momentum Δp imparted to an object that fully absorbs the energy ΔU transferred by an electromagnetic wave is given by

$$\Delta p = \frac{\Delta U}{c}$$

If we let the energy and momentum transfer occur in an infinitesimally short time dt, then from Equation 7-23, the electromagnetic wave exerts a force on the object given by

$$F = \frac{dp}{dt} = \frac{1}{c}\frac{dU}{dt}$$

This force results in a force per unit area, or **radiation pressure**, of

$$P = \frac{1}{Ac}\frac{dU}{dt}$$

where A is the area of the surface that the electromagnetic wave strikes. We've seen the term $(1/A)(dU/dt)$ before; it is the magnitude S of the Poynting vector (Equation 22-3). Over a longer period of time, the radiation pressure due to the full absorption of an electromagnetic wave by an object is then

$$P = \frac{\langle S \rangle}{c} \tag{22-11}$$

where $\langle S \rangle$ is the average of the Poynting vector over that time.

If an electromagnetic wave is completely reflected by an object, the net change in momentum is Δp as the wave impinges minus $(-\Delta p)$ as the wave is reflected away, or twice Δp. So in this case,

$$\Delta p = \frac{2\Delta U}{c}$$

and the radiation pressure is

$$P = \frac{2\langle S \rangle}{c} \tag{22-12}$$

Estimate It! 22-1 Sunshine on Your Shoulders

In Example 22-3, we determined that at the top of the atmosphere Earth receives about $1.4 \times 10^3 \, \text{W/m}^2$ in the form of electromagnetic radiation from the Sun. About 70% of the power makes it down to the surface on a clear day. Estimate the force on your shoulders due to the radiation pressure from sunlight.

SET UP

For simplicity we assume that all of the sunlight is absorbed, so we can apply Equation 22-11 or

$$P = \frac{\langle S \rangle}{c}$$

Force is the product of pressure and area, so

$$F = PA = \frac{\langle S \rangle A}{c}$$

What is $\langle S \rangle$? Recall that the Poynting vector is itself intensity, so the value of $\langle S \rangle$ comes directly from the Sun's intensity given in the problem statement.

SOLVE

We need a reasonable number for the area of your shoulders. If yours are about the same size as ours, each one is about 15 cm long by 10 cm across, so both together present about 300 cm² or 0.03 m² to the Sun.

Take the power per unit area reaching your shoulders to be 70% of the intensity that strikes the top of the atmosphere from the Sun; 70% of $1.4 \times 10^3 \, \text{W/m}^2$ is about $1 \times 10^3 \, \text{W/m}^2$, so the force is

$$F = \frac{(1 \times 10^3 \, \text{W/m}^2)(0.03 \, \text{m}^2)}{(3 \times 10^8 \, \text{m/s})} = 1 \times 10^{-7} \, \text{N}$$

REFLECT

A force of $1 \times 10^{-7} \, \text{N}$ is imperceptibly small.

✴ What's Important 22-1

Radio waves, visible light, and x-rays are all examples of electromagnetic waves. The energy of a single electromagnetic wave packet is called a photon; the energy of a photon depends on its wavelength and frequency. In a vacuum, all photons move at the speed of light regardless of their energy. An electromagnetic wave is the propagation of an oscillating electric field and an oscillating magnetic field.

22-2 Maxwell's Equations

By the middle of the nineteenth century, scientists understood a great deal about electricity and magnetism. The work of Gauss, Ampère, and Faraday had established fundamental relationships that describe electrical and magnetic phenomena; for example, electric fields arise from the presence of charge and magnetic fields arise from current. It was the Scottish physicist James Clerk Maxwell who brought this understanding into a unified theory. Most importantly he saw that electric and magnetic fields are traveling waves, and that phenomena such as radio waves, x-rays, and visible light are all electromagnetic waves consisting of a time-varying electric field and a time-varying magnetic field. By applying relationships between electric and magnetic quantities to an electromagnetic wave, Maxwell discovered something exciting, that the speed of light in empty space is constant. In this section, we will look back at the physics of electric and magnetic fields in order to cast them into Maxwell's model, and then rediscover his conclusion about the speed of light.

Our starting point is the physics that relates electric and magnetic fields to each other. According to Faraday's law, a changing magnetic flux Φ_B gives rise to an electric potential $\mathscr{E}$,

$$\mathscr{E} = -\frac{d\Phi_B}{dt} \tag{20-3}$$

where Φ_B is defined in terms of the magnetic field $\vec{B}$ and the area A of a surface through which it passes:

$$\Phi_B = \int \vec{B} \cdot d\vec{A} \qquad (20\text{-}1)$$

We can cast Faraday's law in terms of electric field through Equation 17-5 which relates the electric field E to electric potential. Around a closed path Equation 17-5 becomes

$$\oint \vec{E} \cdot d\vec{l} = -\frac{d\Phi_B}{dt} \qquad (22\text{-}13)$$

In this form, Faraday's law suggests that a changing magnetic flux gives rise to an electric field.

Maxwell knew that electric and magnetic fields are intertwined, so he expected that just as a changing magnetic flux gives rise to an electric field, a changing electric field must give rise to a magnetic field. To uncover this relationship Maxwell needed to solve a puzzle. Ampère's law relates the magnetic field $\vec{B}$ and the current i that gives rise to it:

$$\oint \vec{B} \cdot d\vec{l} = \mu_0 i_{\text{through}} \qquad (19\text{-}13)$$

where μ_0 is the permeability of free space. But here's the puzzle. Maxwell imagined an experiment in which a parallel plate capacitor is in the process of being charged. **Figure 22-6** shows such a capacitor, as well as two closed loops around which we will apply Ampère's law. One of the loops is outside the capacitor, so that the current i_0 that is charging the plates passes through it. The other loop is inside the capacitor. Because no charge flows between the plates of a capacitor, no current passes through the second loop.

The Amperian loop we use to evaluate the integral in Equation 19-13 serves only to allow us to determine the magnetic field. Also, Ampère's formulation requires that the result not depend on the choice of loop. But the application of Ampère's law around each of the two loops in Figure 22-6 leads to an apparent contradiction. The current passing through the loop outside the plates gives rise to a magnetic field, and the field extends out in all directions. However, no current flows through the loop which is between the plates, so Ampère's law predicts that there is no magnetic field in this region. Either there is or is not a field everywhere in the region. This is the puzzle that Maxwell addressed.

Maxwell resolved this apparent failing of Ampère's law by recognizing that the changing electric field between the plates gives rise to a magnetic field, in analogy to the process described by Faraday's law. To account for this, Maxwell proposed adding a second term to Ampère's law to account for a changing electric field. This additional term, known as the **displacement current**, is proportional to the rate of change in electric flux, that is, proportional to $d\Phi_E/dt$, where

$$\Phi_E = \int \vec{E} \cdot d\vec{A} \qquad (16\text{-}29)$$

The proportionality constant is ϵ_0, the permittivity of free space; with the addition of the displacement current, Ampère's law becomes

$$\oint \vec{B} \cdot d\vec{l} = \mu_0 \left(i_{\text{through}} + \epsilon_0 \frac{d\Phi_E}{dt} \right) \qquad (22\text{-}14)$$

This extended form of Ampère's law works equally well both inside and outside the charging capacitor. And as Maxwell expected, it suggests that, analogously to Faraday's law, a changing electric flux gives rise to a magnetic field.

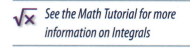
√x *See the Math Tutorial for more information on Integrals*

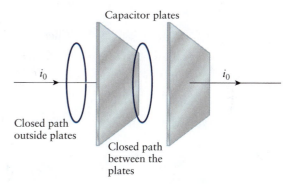

Capacitor plates

i_0 i_0

Closed path outside plates

Closed path between the plates

Figure 22-6 Maxwell imagined an experiment in which a parallel plate capacitor such as the one shown here is in the process of being charged. The current i_0 that charges the plates of the capacitor passes through a closed conducting loop that lies completely outside of the capacitor. A second closed loop lies between the plates of the capacitor. Because no charge flows between the plates, no current passes through the second loop.

To convince ourselves that Maxwell's extension of Ampère's law is correct, we evaluate the right-hand side of Equation 22-14 for both of the loops shown near the charging capacitor in Figure 22-6.

As the charge on the plates increases, the electric field between them increases. In the limit of large plates and a narrow gap between them, however, there is no electric field *outside* the plates. Thus for the loop outside of the gap between the plates the displacement current term in Equation 22-14 is zero and the right-hand side of Ampère's law for this loop is

$$\mu_0\left(i_{\text{through}} + \epsilon_0\frac{d\Phi_E}{dt}\right) = \mu_0\left(i_{\text{through}} + 0\right) = \mu_0 i_{\text{through}}$$

Between the plates there is no current, so the first term in the right-hand side of Ampère's law is zero. To evaluate the displacement current, we use the definition of Φ in the context of Gauss' law, which relates the electric flux, expressed in terms of the electric field $\vec{E}$ and the static charge q that gives rise to that field:

$$\Phi_E = \oint \vec{E} \cdot d\vec{A} = \frac{q_{\text{encl}}}{\epsilon_0} \tag{16-31}$$

Notice that in Equation 16-29 we wrote the flux using the notation $\int$ for the integral, but here we use $\oint$, which indicates that the integral is over a *closed* surface. When the time-varying charge $q(t)$ on one of the plates is enclosed within a Gaussian surface,

$$\Phi_E = \frac{q(t)}{\epsilon_0}$$

The right-hand side of Equation 22-14 around the loop in between the plates is then

$$\mu_0\left(i_{\text{through}} + \epsilon_0\frac{d\Phi_E}{dt}\right) = \mu_0\left(0 + \epsilon_0\frac{d}{dt}\frac{q(t)}{\epsilon_0}\right)$$

The rate at which charge accumulates on the plate is exactly the current i_0 on the plate, so

$$\oint \vec{B} \cdot d\vec{l} = \mu_0\left(\frac{dq(t)}{dt}\right) = \mu_0 i_0$$

With Maxwell's addition of the displacement current term, Ampère's law yields the same result when evaluated on both the loop outside the plates and the loop between the plates of the charging capacitor.

Faraday's law describes the integral of the electric field around a closed path. Ampère's law is the magnetic equivalent of Faraday's law, in that it describes the integral of the magnetic field around a closed path. Gauss' law provides a description of the electric flux through a closed surface. You might wonder, as Maxwell did, whether there is a magnetic equivalent of Gauss' law, a law that describes the magnetic flux through a closed surface.

In uncovering Gauss' law we recognized that the lines of electric field point either radially toward or radially away from a point charge, so that when the charge is enclosed by a Gaussian surface there is a net flux of field lines across the boundary. This is suggested in **Figure 22-7a**, in which a positive charge is enclosed in a spherical surface. (Only a two-dimensional cross section is shown.) Every line of field is leaving the enclosed region as it crosses the boundary. When a source of magnetic field is enclosed, however, there can be no net magnetic flux across the boundary. As we saw in Chapter 19, there are no magnetic monopoles; the simplest magnetic source is a dipole. As suggested by **Figure 22-7b**, in which a bar magnet is contained within a closed boundary, for every line of magnetic field that passes through the boundary pointing outward, indicated by a red dot, another line of

A red dot indicates a line of field leaving the closed boundary. A green dot indicates a line of field entering the closed boundary.

All of the electric field lines due to a positive point charge q leave the blue Gaussian surface. The net flux is non-zero. (Both the field and the surface are three dimensional; only a two-dimensional cross-section is shown.)

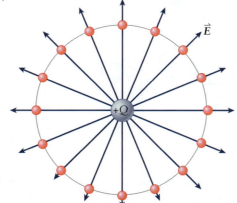

(a)

The same number of magnetic field lines, due to a magnetic dipole, leave the blue closed surface as enter it. The net flux is zero. (Both the field and the surface are three-dimensional; only a two-dimensional cross-section is shown.)

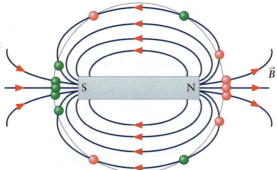

(b)

Figure 22-7 (a) A positive charge is enclosed in a spherical surface, shown here in cross section. Every line of field is leaving the enclosed region as it crosses the boundary. (b) When a source of magnetic field is enclosed, however, there can be no net magnetic flux across the boundary. Here a bar magnet is contained within a closed boundary. For every line of magnetic field that passes through the boundary pointing outward, indicated by a red dot, another line of field passes through pointing inward, shown as a green dot. Because flux is a signed quantity, the net magnetic flux across the closed boundary is zero.

field passes through pointing inward, shown as a green dot. Because flux is a signed quantity, the net magnetic flux across the closed boundary is zero:

$$\oint \vec{B} \cdot d\vec{A} = 0 \tag{22-15}$$

This equation is sometimes referred to as Gauss' law for magnetism.

It was Maxwell who first recognized that electric and magnetic fields underlie all electric, magnetic, and electromagnetic phenomena. It is fitting, then, that the four fundamental relationships above, Gauss' law, Gauss' law for magnetism, Faraday's law, and the modified version of Ampère's law, are known together as **Maxwell's equations**. We summarize them here

$$\oint \vec{E} \cdot d\vec{A} = \frac{q_{\text{encl}}}{\epsilon_0} \tag{16-31}$$

$$\oint \vec{B} \cdot d\vec{A} = 0 \tag{22-15}$$

$$\oint \vec{E} \cdot d\vec{l} = -\frac{d\Phi_B}{dt} \tag{22-13}$$

$$\oint \vec{B} \cdot d\vec{l} = \mu_0 \left(i_{\text{through}} + \epsilon_0 \frac{d\Phi_E}{dt} \right) \tag{22-14}$$

This form of Maxwell's equations applies in the absence of dielectric and magnetic materials.

Although Maxwell's equations provide the fundamental underpinning of electromagnetic waves, none directly shows how the waves move. To make the motion of the wave clear we need a relationship between the strength of either the electric field or the magnetic field, and both position and time. Such a relationship will, for example, enable us to determine the speed at which the waves move. We can obtain

Figure 22-8 (a) To apply Faraday's law or Ampère's law to an electromagnetic wave moving through free space, we need to define a loop around which to carry out the path integral in each. One such loop that lies in the same plane as the electric field component of the wave is shown as a pink rectangle of height h and width Δx. (b) This is an expanded view of the infinitesimally narrow loop shown in part a. The sides of the loop are numbered 1 to 4.

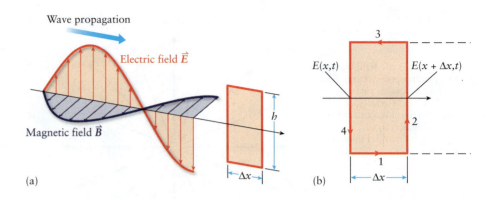

(a)   (b)

√x   See the Math Tutorial for more information on Calculus.

such a relationship by applying Maxwell's equations to an electromagnetic wave. We will apply Faraday's law and Ampère's law, the two of Maxwell's four equations that address time-varying fields.

Figure 22-8a shows a bit of the electric and magnetic components of an electromagnetic wave moving through free space. Here, "free space" implies that there are no charges or currents present. To apply Faraday's law or Ampère's law to the wave, we need to define a loop around which to carry out the path integral. One such loop, which lies in the same plane as the electric field component of the wave, is shown in Figure 22-8a, and expanded in **Figure 22-8b**. This loop is infinitesimally narrow.

Let's apply Faraday's law (Equation 22-13) around the loop at some fixed time t. The path integral on the left-hand side of Faraday's law (Equation 22-13), gives a different result for each of the four sides of the loop:

$$\oint \vec{E} \cdot d\vec{l} = \int_1 E_1\, dl \cos\theta_1 + \int_2 E_2\, dl \cos\theta_2 + \int_3 E_3\, dl \cos\theta_3 + \int_4 E_4\, dl \cos\theta_4$$

The sides of the loop are numbered 1 to 4 in the figure. Because the electric field $\vec{E}$ points up everywhere, the differential path vector $d\vec{l}$ is parallel to $\vec{E}$ along side 2, antiparallel to $\vec{E}$ along side 4, and perpendicular to $\vec{E}$ along sides 1 and 3. Thus, $\cos\theta_2$ equals 1, $\cos\theta_4$ equals -1, and $\cos\theta_1$ and $\cos\theta_3$ both equal 0. The integral reduces to

$$\oint \vec{E} \cdot d\vec{l} = \int_2 E_2\, dl - \int_4 E_4\, dl$$

The electric field is a function of position and time. At the point where side 4 crosses the wave, the magnitude of the electric field at time t is $E(x,t)$. At the point where side 2 crosses the wave, the electric field is slightly different, $E(x + \Delta x,t)$. Because we are considering a specific time t, $E(x,t)$ and $E(x + \Delta x,t)$ can be pulled out of the integrals:

$$\oint \vec{E} \cdot d\vec{l} = E_2 \int_2 dl - E_4 \int_4 dl$$

The integral $\int dl$ gives the length of the path, which is h for both side 2 and side 4 of the loop. All together, then, the left-hand side of Faraday's law gives

$$\oint \vec{E} \cdot d\vec{l} = E_2 h - E_4 h = E(x + \Delta x,t)h - E(x,t)h \qquad (22\text{-}16)$$

For infinitesimally small values of Δx, $E(x + \Delta x,t)$ is directly related to $E(x,t)$. To see how, consider the generic function $f(x)$ plotted in **Figure 22-9**. When we expand a small region between the values of x equal to x_0 and $x_0 + \Delta x$, notice that because Δx is so small, $f(x)$ is essentially a straight line in the expansion. The slope of the line df/dx is

$$\frac{df}{dx} = \frac{\Delta y}{\Delta x}$$

and because the side we labeled Δy is equal to the difference between the function evaluated at $x_0 + \Delta x$ and the function evaluated at x_0,

$$f(x_0 + \Delta x) - f(x_0) = \Delta y = \Delta x \frac{df}{dx}$$

The value of the function at $x_0 + \Delta x$ is then

$$f(x_0 + \Delta x) = f(x_0) + \Delta x \frac{df}{dx} \qquad (22\text{-}17)$$

$E(x + \Delta x, t)$ and $E(x,t)$ for the electromagnetic wave are related in the same way, that is,

$$E(x + \Delta x, t) = E(x,t) + \Delta x \frac{\partial E}{\partial x} \qquad (22\text{-}18)$$

The partial derivative $\partial/\partial x$ indicates that although the electric field is a function of both position and time, we are holding time fixed while carrying out the integral in Faraday's law.

We can use Equation 22-18 to simplify the left-hand side of Faraday's law as applied to the loop in Figure 22-8. From Equation 22-16 we get

$$\oint \vec{E} \cdot d\vec{l} = E(x,t)h + \Delta x \frac{\partial E}{\partial x}h - E(x,t)h = \frac{\partial E}{\partial x}h\,\Delta x$$
$$(22\text{-}19)$$

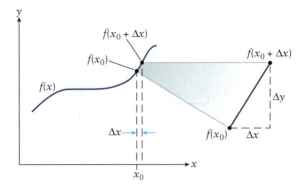

Figure 22-9 For infinitesimally small values of Δx, $f(x + \Delta x, t)$ is directly related to $f(x,t)$. When we expand a small region of the generic function $f(x)$ between the values of x equal to x_0 and $x_0 + \Delta x$, notice that because Δx is so small, $f(x)$ is essentially a straight line in the expansion. The slope of the line is $df/dx = \Delta y/\Delta x$.

The right-hand side of Faraday's law (Equation 22-13) requires us to determine the time derivative of the magnetic flux through the loop:

$$\frac{d\Phi_B}{dt} = \frac{d}{dt}\oint \vec{B} \cdot d\vec{A}$$

First, notice that because $d\vec{A}$ is perpendicular to the area of the loop (by definition), $d\vec{A}$ is parallel to $\vec{B}$. The cosine in the dot product is therefore equal to 1, so the integral becomes $\oint B\,dA$. Because the loop is infinitesimal in width, we can take B to be the average field strength over the width of the loop, so

$$\frac{d\Phi_B}{dt} = \frac{d}{dt}B\oint dA = \frac{d}{dt}BA_{\text{loop}} = \frac{d}{dt}Bh\,\Delta x$$

Because we used an average value for B, suggesting a specific value of position x, to be mathematically correct we must use a partial derivative, so

$$\frac{d\Phi_B}{dt} = \frac{\partial}{\partial t}Bh\,\Delta x = \frac{\partial B}{\partial t}h\,\Delta x \qquad (22\text{-}20)$$

We can now set equal the right-hand (Equation 22-20) and left-hand (Equation 22-19) sides of Faraday's law as applied to the loop in Figure 22-8. From

$$\oint \vec{E} \cdot d\vec{l} = -\frac{d\Phi_B}{dt} \qquad (22\text{-}13)$$

we get

$$\frac{\partial E}{\partial x}h\,\Delta x = -\frac{\partial B}{\partial t}h\,\Delta x$$

or

$$\frac{\partial E}{\partial x} = -\frac{\partial B}{\partial t} \qquad (22\text{-}21)$$

This equation is a compact expression that involves position and time. Our goal, however, was an expression that relates either the electric field or the magnetic field to position and time, so we're not there yet.

To complete the picture of an electromagnetic wave we apply Ampère's law to the wave; from Equation 22-14,

$$\oint \vec{B} \cdot d\vec{l} = \mu_0 \epsilon_0 \frac{d\Phi_E}{dt} \tag{22-22}$$

Because there are no currents present, we have dropped the first term on the right-hand side in Ampère's law. This version of Ampère's law has a structure parallel to Faraday's law. We expect that when it is applied to an appropriately defined loop relative to an electromagnetic wave, the result will be an expression parallel in structure to Equation 22-21:

$$\frac{\partial B}{\partial x} = -\mu_0 \epsilon_0 \frac{\partial E}{\partial t} \tag{22-23}$$

The minus sign arises from the dot product on the left-hand side of Equation 22-22; when the loop is traversed in the same directional sense as the one we chose to evaluate Faraday's law (see Figure 22-8), the magnetic field points in the same direction as the path at x and the opposite direction as the path at $x + \Delta x$. To achieve our goal of finding an expression that relates position and time to one of the two fields, we can use Equation 22-21 or Equation 22-23 to eliminate either E or B from the other equation. For example, we take the second derivative with respect to x of Equation 22-23:

$$\frac{\partial}{\partial x} \frac{\partial B}{\partial x} = -\mu_0 \epsilon_0 \frac{\partial}{\partial x} \frac{\partial E}{\partial t}$$

or

$$\frac{\partial^2 B}{\partial x^2} = -\mu_0 \epsilon_0 \frac{\partial}{\partial t} \frac{\partial E}{\partial x}$$

The order of the partial derivatives on the right-hand side doesn't matter, so we are free to reverse $\partial/\partial t$ and $\partial/\partial x$. We now substitute for $\partial E/\partial x$ from Equation 22-21:

$$\frac{\partial^2 B}{\partial x^2} = -\mu_0 \epsilon_0 \frac{\partial}{\partial t} \left(-\frac{\partial B}{\partial t} \right)$$

or

$$\frac{\partial^2 B}{\partial x^2} = \mu_0 \epsilon_0 \frac{\partial^2 B}{\partial t^2} \tag{22-24}$$

We have achieved our goal of finding a relationship between either E or B and both x and t! Had we eliminated B from the two equations, resulting in an equation for E, the result would look the same:

$$\frac{\partial^2 E}{\partial x^2} = \mu_0 \epsilon_0 \frac{\partial^2 E}{\partial t^2} \tag{22-25}$$

Equations 22-24 and 22-25 are two versions of the **one-dimensional wave equation,** as applied to an electromagnetic wave. Do the equations do what we claimed, that is, do they show the motion of an electromagnetic wave? In particular, do they tell us the speed of an electromagnetic wave?

To answer, we look at the dimensions of the variables in Equation 22-24:

$$\frac{[B]}{[x^2]} = [\mu_0 \epsilon_0] \frac{[B]}{[t^2]}$$

or

$$\frac{\text{magnetic field}}{\text{distance}^2} = \left[\frac{1}{1/\mu_0 \epsilon_0} \right] \frac{\text{magnetic field}}{\text{time}^2}$$

We wrote the $\mu_0 \epsilon_0$ term in that curious way in order to look more closely at the denominators. We have

$$\text{distance}^2 = \left[\frac{1}{\mu_0 \epsilon_0}\right]\text{time}^2$$

or

$$\text{distance} = \left[\sqrt{\frac{1}{\mu_0 \epsilon_0}}\right]\text{time}$$

Distance is given by speed multiplied by time, so we conclude that the dimensions of $\sqrt{1/\mu_0 \epsilon_0}$ must be those of speed. The speed of the wave described by Equation 22-24 is

$$v = \sqrt{\frac{1}{\mu_0 \epsilon_0}} \tag{22-26}$$

The same is true for Equation 22-25. We know the values of both μ_0 and ϵ_0, so we can immediately find the speed of an electromagnetic wave:

$$v = \sqrt{\frac{1}{[4\pi \times 10^{-7}\,(\text{T}\cdot\text{m})/\text{A}]\,[8.85 \times 10^{-12}\,\text{C}^2/(\text{N}\cdot\text{m}^2)]}}$$

We address the units in Example 22-4 below, but because we have used values of μ_0 and ϵ_0 in SI units, we know the result must be in meters per second. So

$$v = \sqrt{\frac{1}{1.11 \times 10^{-17}}} = \sqrt{\frac{1}{1.11 \times 10^{-17}}} = 3.00 \times 10^8\,\text{m/s}$$

This result is the well-known value of c, the speed of light in a vacuum! The speed of an electromagnetic wave defined by the wave equation is the speed of light. We can therefore write

$$c = \sqrt{\frac{1}{\mu_0 \epsilon_0}} \tag{22-27}$$

To be clear, the quantity c is the speed of all electromagnetic waves, not just visible light.

Watch Out

The speed of light in a medium other than a vacuum is less than c.

As we noted earlier in the chapter, light travels at c (approximately 3.0×10^8 m/s) only in a vacuum. We determined the value of c above using ϵ_0, the permittivity of free space, and μ_0, the permeability of free space; "free space" is equivalent to "in a vacuum."

Example 22-4 The Units of c

Show that when μ_0 and ϵ_0 are combined in our expression for the speed of an electromagnetic wave, Equation 22-26, the resulting units of the speed are indeed meters per second (m/s).

SET UP

We start by combining the units of μ_0 and ϵ_0 as in Equation 22-26:

$$[v] = \sqrt{\frac{1}{[\mu_0][\epsilon_0]}} = \sqrt{\frac{1}{[(T \cdot m)/A][C^2/(N \cdot m^2)]}}$$

So

$$[v] = \sqrt{\frac{A \cdot N \cdot m}{T \cdot C^2}} \tag{22-28}$$

The trick to working with units (or dimensions) is to find relationships that relate one or more parameters to others of interest. Any one will do, because any physical variable has the same units and dimensions regardless of the expression in which it is used. A glance at Equation 22-28 suggests we look for an expression that helps us express current (A), charge (C), or magnetic field (T) in terms of some of the other parameters (units).

SOLVE

The force on a charge q moving at velocity v in a magnetic field B is given by

$$\vec{F} = q\vec{v} \times \vec{B} \tag{19-1}$$

The units of magnetic field can therefore be expressed as

$$[B] = T = \frac{[F]}{[q][v]} = \frac{N}{C \cdot m/s}$$

Then Equation 22-28 becomes

$$[v] = \sqrt{\frac{C \cdot m}{N \cdot s}\frac{A \cdot N \cdot m}{C^2}} = \sqrt{\frac{A \cdot m^2}{C \cdot s}}$$

We now employ the definition of the ampere, the SI unit of current; from

$$i = \frac{dq}{dt} \tag{18-2}$$

$$1\,A = 1\,\frac{C}{s}$$

So

$$[v] = \sqrt{\frac{C}{s}\frac{m^2}{C \cdot s}} = \sqrt{\frac{m^2}{s^2}} = \frac{m}{s}$$

REFLECT

The units of the speed of electromagnetic waves, as determined from the one-dimensional wave equation for such waves, are indeed meters per second.

Practice Problem 22-4 Confirm that in the SI system the units of the right-hand sides of Equations 22-1 and 22-2 are joules (J), the SI unit of energy.

Estimate It! 22-2 Just One Foot

Approximately how long does it take light to travel one foot in a vacuum? (For this problem, suspend our exclusive use of the SI system for scientific calculations.)

SET UP

At a speed of c, the distance d light travels in any time t is

$$d = ct$$

or

$$t = \frac{d}{c}$$

Noting that c equals approximately 3×10^8 m/s, we'll take 1 ft to be approximately 0.3 m, to make the calculation easy!

SOLVE

So

$$t = \frac{0.3 \text{ m}}{3 \times 10^8 \text{ m/s}} = 1 \times 10^{-9} \text{ s}$$

REFLECT

It takes light about 10^{-9} s, or 1 ns, to travel 1 ft. This convenient number is worth remembering.

What's Important 22-2

Electric and magnetic fields are traveling waves that propagate at the speed of light. Phenomena such as radio waves, x-rays and visible light are all electromagnetic waves consisting of a time-varying electric field and a time-varying magnetic field.

Answers to Practice Problems

22-1 $E_{min} = 2.0 \times 10^{-26}$ J, $f_{min} = 30$ MHz,
$E_{max} = 4.0 \times 10^{-22}$ J, $f_{max} = 600$ GHz

22-3 6.0×10^2 W/m²

22-4 $[E] = \dfrac{[h][c]}{[\lambda]} = \dfrac{(\text{J} \cdot \text{s})(\text{m/s})}{\text{m}} = \text{J}$,

$[E] = [h][f] = (\text{J} \cdot \text{s})(1/s) = \text{J}$

Answer to Got the Concept Question

22-1 The higher the energy of an electromagnetic wave, the higher the frequency and the shorter the wavelength. Using Equation 22-1, longer (larger values of) wavelengths result in smaller energy, so radio waves have longer wavelengths than x-rays. Using Equation 22-2, waves of higher (larger values of) frequency carry larger energy so x-rays are of higher frequencies than radio waves.

SUMMARY

Topic	Summary	Equation or Symbol
displacement current	To account for a changing electric field Maxwell added a second term to Ampère's law. Known as the displacement current, the additional term is proportional to the rate of change of electric flux Φ_E.	$\oint \vec{B} \cdot d\vec{l} = \mu_0\left(i_{\text{through}} + \epsilon_0 \dfrac{d\Phi_E}{dt}\right)$ (22-14)
electromagnetic wave	Light and other electromagnetic waves contain an oscillating electric field coupled with an oscillating magnetic field.	
ionizing radiation	Ionizing radiation delivers enough energy to strip atoms in an object of one or more electrons.	
Maxwell's equations	Gauss' law, Gauss' law for magnetism, Faraday's law, and the modified version of Ampère's law are known together as Maxwell's equations.	$\oint \vec{E} \cdot d\vec{A} = \dfrac{q_{\text{encl}}}{\epsilon_0}$ (16-31) $\oint \vec{B} \cdot d\vec{A} = 0$ (22-15) $\oint \vec{E} \cdot d\vec{l} = -\dfrac{d\Phi_B}{dt}$ (22-13) $\oint \vec{B} \cdot d\vec{l} = \mu_0\left(i_{\text{through}} + \epsilon_0 \dfrac{d\Phi_E}{dt}\right)$ (22-14)
one-dimensional wave equation	The one-dimensional wave equation applied to electromagnetic waves describes the motion of the oscillating electric field and magnetic field components of an electromagnetic wave.	$\dfrac{\partial^2 B}{\partial x^2} = \mu_0\epsilon_0 \dfrac{\partial^2 B}{\partial t^2}$ (22-24) $\dfrac{\partial^2 E}{\partial x^2} = \mu_0\epsilon_0 \dfrac{\partial^2 E}{\partial t^2}$ (22-25)
photon	A photon is a small, individual packet of electromagnetic energy.	
Planck's constant	Planck's constant is a fundamental physical constant that, for example, relates the energy of an electromagnetic wave to the speed of light.	$h = 6.63 \times 10^{-34}\,\text{J} \cdot \text{s}$
Poynting vector	The magnitude of an electromagnetic wave can be combined with the direction in which the electromagnetic wave propagates to form the Poynting vector. The magnitude of the Poynting vector gives the energy transported by an electromagnetic wave per unit area per unit time at any instant.	$\vec{S} = \dfrac{\vec{E} \times \vec{B}}{\mu_0}$ (22-8)
radiation pressure	The force per unit area of an electromagnetic wave is called the radiation pressure.	$P = \dfrac{\langle S \rangle}{c}$ (22-11)
speed of light in a vacuum	The speed of light in a vacuum is constant, that is, it is independent of the energy of the light. This is true for all electromagnetic radiation.	$c = 2.997\,924\,58 \times 10^8\,\text{m/s}$

QUESTIONS AND PROBLEMS

In a few problems, you are given more data than you actually need; in a few other problems, you are required to supply data from your general knowledge, outside sources, or informed estimate.

Interpret as significant all digits in numerical values that have trailing zeros and no decimal points.

For all problems, use $g = 9.8 \text{ m/s}^2$ for the free-fall acceleration due to gravity. Neglect friction and air resistance unless instructed to do otherwise.

- Basic, single-concept problem
- •• Intermediate-level problem, may require synthesis of concepts and multiple steps
- ••• Challenging problem

SSM *Solution is in Student Solutions Manual*

Conceptual Questions

1. •(a) Rank the following electromagnetic waves from the lowest to the highest wavelength: (a) microwaves, (b) red light, (c) ultraviolet light, (d) infrared light, and (e) gamma rays. (b) Which wavelength has the highest energy? Which has the lowest?

2. •James Clerk Maxwell is credited with compiling the three laws of electricity and magnetism (Gauss' law, Ampère's law, and Faraday's law), adding his own law (also called Gauss' law for magnetism), and understanding the connections between electricity, magnetism, and optics. Were Maxwell's efforts more important than the individual discoveries of Gauss, Ampère, and Faraday? Explain your answer.

3. •Changing electric fields create changing magnetic fields. These vibrating fields propagate at the speed of light. Describe the orientation of the directions of the fields and the velocity of the electromagnetic wave. SSM

4. •Does a wire connected to a DC source, such as a battery, emit an electromagnetic wave?

5. •Describe how the frequency at which the changing electric and magnetic fields oscillate in electromagnetic waves is related to the speed of light.

6. •Describe how the frequency is related to the energy of the electromagnetic waves (photons).

7. •The energy of an ultraviolet light photon is unrelated to the speed of the fundamental electromagnetic waves that make up such radiation. Explain how this is possible. SSM

8. •Calc Match the formal, integral–differential equations that were conceived of by Carl Friedrich Gauss, André Ampère, Michael Faraday, and James Clerk Maxwell with the corresponding written statements:

(i) $\oint \vec{B} \cdot d\vec{l} = \mu_0 i_{\text{through}}$

(ii) $\oint \vec{E} \cdot d\vec{l} = -\dfrac{d\Phi_B}{dt}$

(iii) $\oint \vec{B} \cdot d\vec{A} = 0$

(iv) $\oint \vec{E} \cdot d\vec{A} = \dfrac{q_{\text{encl}}}{\epsilon_0}$

(a) The source of an electric field is an electric charge.

(b) The source of a magnetic field is an electric current.

(c) Changing magnetic fields induce changing electric fields.

(d) Changing electric fields induce changing magnetic fields.

9. •Calc Define the concept of a partial derivative. For example, define these operations:

$$\frac{\partial E}{\partial t} \quad \text{or} \quad \frac{\partial E}{\partial x}$$

10. •Calc Explain the dimensions of the following formula that expresses a magnetic field as a solution to the wave equation:

$$\frac{\partial^2 B}{\partial x^2} = \frac{1}{c^2} \frac{\partial^2 B}{\partial t^2}$$

11. •Explain why the wave equation, the formula that predicts the fundamental wavelike behavior of all oscillating sources of energy, *cannot* have a minus sign on one side. For example:

$$\frac{\partial^2 B}{\partial x^2} \neq -\frac{1}{c^2} \frac{\partial^2 B}{\partial t^2}$$

12. •Name three types of electromagnetic energy that you used today.

Multiple-Choice Questions

13. •X-rays and gamma rays
  A. have same frequency.
  B. have same wavelength.
  C. have same speed.
  D. have the same "color."
  E. None of the above SSM

14. •In comparison to x-rays, visible light has
  A. a speed that is faster.
  B. wavelengths that are longer.
  C. wavelengths that are equal.
  D. wavelengths that are shorter.
  E. frequencies that are equal.

15. •Which of the following requires a physical medium through which to travel?
  A. radio waves
  B. light
  C. x-rays
  D. sound
  E. gamma rays SSM

16. •In comparison to radio waves, visible light has
    A. speed that is faster.
    B. wavelengths that are longer.
    C. wavelengths that are equal.
    D. wavelengths that are shorter.
    E. frequencies that are equal.

17. •In comparison to x-rays, visible light has
    A. a speed that is faster.
    B. wavelengths that are longer.
    C. wavelengths that are equal.
    D. wavelengths that are shorter.
    E. frequencies that are equal.

18. •In an *RC* circuit, the capacitor begins to discharge. In the region of space between the plates of the capacitor,
    A. there is an electric field but no magnetic field.
    B. there is a magnetic field but no electric field.
    C. there are both electric and magnetic fields.
    D. there are no electric and magnetic fields.
    E. there is an electric field whose strength is one-half that of the magnetic field.

19. •Maxwell's equations apply
    A. to both electric fields and magnetic fields that are constant over time.
    B. only to electric fields that are time-dependent.
    C. only to magnetic fields that are constant over time.
    D. to both electric fields and magnetic fields that are time-dependent.
    E. to both time-independent and time-dependent electric and magnetic fields. SSM

20. •The phase difference between the electric and magnetic fields in an electromagnetic wave is
    A. 90°.
    B. 180°.
    C. 0°.
    D. alternately 90° and 180°.
    E. alternately 0° and 90°.

Estimation/Numerical Analysis

21. •Estimate how long it would take for a bird to circumnavigate Earth traveling at 0.1c. (This would be one fast bird!)

22. •Estimate the average wavelength of radio waves that are received by (a) an AM radio and (b) an FM radio.

23. •Estimate the energy associated with the visible light (green) to which the human eye is most sensitive. SSM

24. •Estimate a simple conversion factor between joules and electron volts.

25. •Estimate the wavelength of electromagnetic radiation that would potentially be classified as ionizing radiation.

26. •Estimate the number of photons that are emitted in the 1000-h lifetime of a 100 W lightbulb.

27. •An important news announcement is transmitted by radio waves to people who are 300 km away and sitting next to their radios and also by sound waves to people sitting 3 m from the newscaster in a newsroom. Who receives the news first? Explain your answer. SSM

28. •••Use a computer-based, graphical program to plot the graphs of the electric field (E) versus time (t) and the magnetic field (B) versus time (t) on a three-dimensional graph (see the table below). Let t be on the x axis, E be on the y axis, and B be on the z axis. Explain the shape of the graph.

E (N/C)	B ($\times 10^{-9}$ T)	t (s)
100	333	0
70.7	236	$\pi/4$
0	0	$\pi/2$
−70.7	−236	$3\pi/4$
−100	−333	π
−70.7	−236	$5\pi/4$
0	0	$3\pi/2$
70.7	236	$7\pi/4$
100	333	2π
70.7	236	$9\pi/4$
0	0	$5\pi/2$
−70.7	−236	$11\pi/4$
−100	−333	3π
−70.7	−236	$13\pi/4$
0	0	$7\pi/2$
70.7	236	$15\pi/4$
100	333	4π
70.7	236	$17\pi/4$
0	0	$9\pi/2$
−70.7	−236	$19\pi/4$
−100	−333	5π
−70.7	−236	$21\pi/4$
0	0	$11\pi/2$
70.7	236	$23\pi/4$
100	333	6π

Problems

22-1: Electromagnetic Waves

29. ••Calculate the wavelength of the photons with the following frequencies and tell which type of electromagnetic radiation that each is.

    A. $f = 4.14 \times 10^{15}$ Hz   E. $f = 9.00 \times 10^{12}$ Hz

    B. $f = 7.00 \times 10^{14}$ Hz   F. $f = 3.44 \times 10^{17}$ Hz

    C. $f = 8.00 \times 10^{16}$ Hz   G. $f = 8.23 \times 10^{15}$ Hz

    D. $f = 3.00 \times 10^{13}$ Hz   H. $f = 6.00 \times 10^{15}$ Hz

This page is intentionally left blank.

For complete end of chapter problem sets, please go to
www.whfreeman.com/kestentauck

D. $\dfrac{\partial^2 f}{\partial x^2} = ?$ $f(x,t) = A\,e^{i(kx-\omega t)}$

E. $\dfrac{\partial^2 f}{\partial x\,\partial t} = ?$ $f(x,t) = At^2 x^3$

F. $\dfrac{\partial^2 f}{\partial t\,\partial x} = ?$ $f(x,t) = At^2 x^3$

49. ••Calc Calculate the partial derivative of the following function. SSM

$$\dfrac{\partial^2 f}{\partial t^2} = ?\qquad f(x,t) = A\,e^{\,\alpha t}\sin(kx-\omega t)$$

50. •••Calc Starting with Ampère's law and no conduction current ($\oint \vec{B}\cdot d\vec{l} = \mu_0\epsilon_0\,d\Phi_E/dt$), derive the formula

$$\dfrac{\partial B}{\partial x} = -\mu_0\epsilon_0\dfrac{\partial E}{\partial t}$$

Hint: Start with an excerpt of the transverse electromagnetic wave that typifies the problem and integrate around a small rectangle in the plane of the magnetic field that is Δz long and Δx wide.

51. •••Calc (a) Continue the derivation that was begun in problem 50 to demonstrate that the electric field obeys the wave formula with a speed that equals that of light. (b) Repeat the previous derivation to demonstrate that the magnetic field also obeys the wave equation.

52. ••Calc Starting with the wave equation for plane electromagnetic waves, which are

$$\dfrac{\partial^2 E}{\partial x^2} = \mu_0\epsilon_0\dfrac{\partial^2 E}{\partial t^2}\quad \text{and}\quad \dfrac{\partial^2 B}{\partial x^2} = \mu_0\epsilon_0\dfrac{\partial^2 B}{\partial t^2}$$

show that the following functions solve the second order partial differential equations:

$$E(x,t) = E_0\sin(kx-\omega t)\quad B(x,t) = B_0\sin(kx-\omega t)$$

53. ••Calc Show the function $E(x,t) = E_0\sin(kx-\omega t) + \cos(kx-\omega t)$ satisfies the one-dimensional wave equation:

$$\dfrac{\partial^2 E}{\partial x^2} = \mu_0\epsilon_0\dfrac{\partial^2 E}{\partial x^2}$$

given that $\omega = kc$ and $\mu_0\epsilon_0 = 1/c^2$. SSM

54. ••Calc Show the function $B(x,t) = B_0\sin(kx)\cos(\omega t)$ satisfies the wave equation:

$$\dfrac{\partial^2 B}{\partial x^2} = \mu_0\epsilon_0\dfrac{\partial^2 B}{\partial x^2}$$

given that $\omega = kc$ and $\mu_0\epsilon_0 = 1/c^2$.

General Problems

55. ••Biology A recent study found that electrons having energies between 3.0 and 20 eV can cause breaks in a DNA molecule even though they do not ionize the molecule. If the energy were to come from light, (a) what range of wavelengths (in nanometers) could cause DNA breaks, and (b) in what part of the electromagnetic spectrum does the light lie?

56. ••Medical A dental x-ray typically affects 200 g of tissue and delivers about 4.0 μJ of energy using x-rays that have wavelengths of 0.025 nm. What is the energy (in electron volts) of such x-rays and how many photons are absorbed during the dental x-ray?

57. ••A He–Ne laser is made up of a cylindrical beam of light with a diameter of 0.750 cm. The energy is pulsed, lasting for 1.50 ns, and each burst contains an energy of 2.00 J. (a) What is the length of each pulse of laser light? (b) What is the average energy per unit volume for each pulse? SSM

58. ••(a) What is the energy of a photon of blue-green light that has a wavelength of 525 nm? Give your answer in joules and electron volts. (b) What is the wave number of the photon?

59. ••Calc Show that the quantity $\epsilon_0\,d\Phi_E/dt$ has units of current.

23 Wave Properties of Light

(Courtesy David Tauck)

Nocturnal animals see much better at night than we do, partly because of a reflective structure called the *tapetum lucidum* that lines the back of their eyes. Because of this biological reflector, light that isn't absorbed by photoreceptors on its first pass through the retina reflects straight back in the direction from which it came. Some of the reflected photons are detected by photoreceptors on the second pass, thereby enhancing visual sensitivity at low light levels. Other photons continue out of the eye, leading to the phenomenon of *eyeshine*.

23-1 Refraction

You've probably seen a diagram such as the one in **Figure 23-1**, in which light enters the eye and is focused on the retina. Or you may have experienced the disorienting phenomenon of reaching into a pool of water to pick up a shiny object and discovering that it isn't where you think it is, much like the submerged part of the chopstick in the right-hand image in **Figure 23-2**. Anytime a wave, such as light, passes across the boundary between two different materials at an incident angle other than perpendicular to the boundary, it changes direction. The bending is known as **refraction**. In this section, we will confine our discussion to the refraction of visible light, but all electromagnetic waves, sound waves, water waves, and any other waves that can travel in more than one direction may refract when crossing from one medium to another under the right circumstances.

The phenomenon of wave refraction results fundamentally from change in the propagation speed of a wave as it moves from one medium to another. Often the speed at which a wave travels in a medium is a function of the medium's density. For example, because the concentration of dissolved protein determines the density of plasma, veterinarians can estimate the amount of

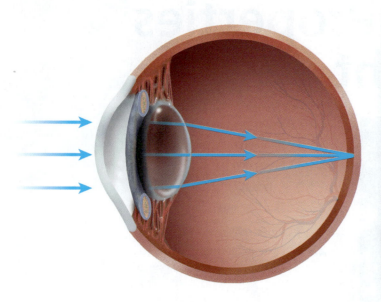

Figure 23-1 Light enters the eye and is focused on the retina.

Figure 23-2 The submerged part of the chopstick on the right isn't where it appears. *(Courtesy David Tauck)*

protein present in an animal's plasma by measuring how much light refracts as it passes through a sample. Similarly, winemakers determine the amount of sugar in their grapes by using the same technique.

To see how the change in speed at which light travels in two media causes light to bend as it crosses the boundary between them, we turn to a theory of light waves introduced by the 17th-century Dutch scientist Christiaan Huygens. He suggested that every point along the crest, sometimes called the *front*, of a two-dimensional or three-dimensional wave be treated as a separate source of a tiny *wavelet* that itself moves at the speed of the wave. Each wavelet propagates from its source much like the ripples spreading out from a pebble dropped in a pond, and the moving wave front itself is then the superposition of all of the wavelets. This phenomenon is **Huygens' principle**.

Figure 23-3 shows a source of light. We treat the light as a wave; each wave front is the surface of a sphere expanding at the propagation speed v of the wave. One section of one wave front is shown in blue. According to Huygens' principle, we imagine the wave front as made up of many tiny sources, shown as brown and red dots in the figure. Each one acts as the source of a tiny wavelet, shown as a brown or red arc, which also propagates at v. The expanding wave front, shown at a later time in black, is the superposition of all of the wavelets. As such, the expanding wave front is tangent to all of the wavelets.

Let's apply Huygens' principle to a beam of light of a fixed energy (and therefore a fixed wavelength and frequency) crossing a boundary from air into glass. **Figure 23-4** shows light entering a plate of glass from air, moving from left to right. The beam is wide enough to strike the surface at many points along the boundary. The blue lines perpendicular to the direction of travel represent the crests of the light waves. According to Huygens' principle, we

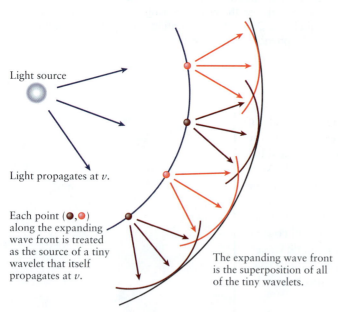

Light source

Light propagates at v.

Each point (●,●) along the expanding wave front is treated as the source of a tiny wavelet that itself propagates at v.

The expanding wave front is the superposition of all of the tiny wavelets.

Figure 23-3 According to Huygens' principle, each point on each expanding wave front of light is treated as the source of a wavelet. The expanding wave front is the superposition of all of the wavelets.

can treat each crest as an assemblage of many sources of tiny wavelets. Two such sources are labeled "1" and "3."

The wavelet generated at point 1 travels through air, so it propagates at the speed of light in air. In one period of the wave, a time T, this wavelet reaches point 2. Point 1 lies on a wave crest, so because the time interval is one period, point 2 lies on the next wave crest. In one period T the wave travels one wavelength λ_{air}, so

$$\lambda_{air} = v_{air} T$$

or

$$T = \frac{\lambda_{air}}{v_{air}} \qquad (23\text{-}1)$$

The wavelet generated at point 3 travels through glass. In the same time T this wavelet reaches point 4, traveling a distance equal to the wavelength of the light in glass:

$$T = \frac{\lambda_{glass}}{v_{glass}} \qquad (23\text{-}2)$$

Setting the two relationships for T equal gives

$$\frac{\lambda_{air}}{v_{air}} = \frac{\lambda_{glass}}{v_{glass}}$$

or

$$\lambda_{glass} = \frac{v_{glass}}{v_{air}} \lambda_{air} \qquad (23\text{-}3)$$

The energy of the photon is the same in air and glass. However, from $\lambda = v/f$ (from Equation 13-8) together with $E = hf$ (Equation 22-2), wavelength is proportional to the speed of the wave and inversely proportional to its energy. Because the speed of light in air is not the same as the speed of light in glass, the wavelength of the photon in air is not the same as in glass. The speed of light in glass is lower than the speed of light in air, so according to Equation 23-3, the wavelength of the light in glass is smaller than the wavelength in air. Although it is often convenient to associate specific wavelength ranges with specific colors, that association only holds for light traveling at c. As we noted in Section 22-1, what we perceive as color depends directly on the energy of the photons in a beam of light. So in this case, although the wavelength of the light changes as it crosses from air to glass, the energy and therefore the color of the photon remain the same when light is refracted at the boundary between two media.

Using Equation 23-3, we see that the fronts of the light waves are closer together in glass than air. The light waves must therefore leave the boundary at an angle different from the one at which they encountered the boundary. The specific angle at which the light refracts depends on the incident angle and the speed of light in the two regions.

Let's analyze the angles in Figure 23-4 by following a narrow beam (or ray) of light and noting that the two right triangles (see **Figure 23-5**) share a common hypotenuse D. We have defined θ_{air}, the angle at which the light approaches the boundary, and θ_{glass}, the angle at which the light refracts with respect

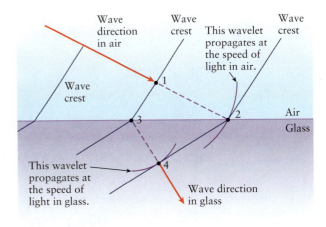

Figure 23-4 The speed of light is lower in glass than in air, so light slows down as it crosses a boundary from air to glass. The wavelength is therefore shorter in glass, which causes the direction of the light to change.

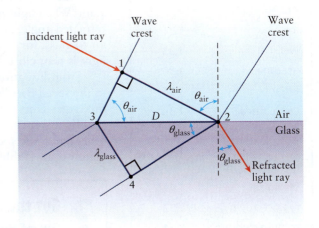

Figure 23-5 The two right triangles formed by the light rays crossing a boundary from air to glass in Figure 23-4 are emphasized in this drawing.

to the normal. As you can verify, however, each is also one of the angles in the triangles. So

$$\sin \theta_{air} = \frac{\lambda_{air}}{D}$$

and

$$\sin \theta_{glass} = \frac{\lambda_{glass}}{D}$$

By solving for D and setting the two resulting equations equal to each other we see that

$$D = \frac{\lambda_{air}}{\sin \theta_{air}} = \frac{\lambda_{glass}}{\sin \theta_{glass}}$$

Using Equations 23-1 and 23-2, the equation becomes

$$\frac{T v_{air}}{\sin \theta_{air}} = \frac{T v_{glass}}{\sin \theta_{glass}}$$

or

$$\frac{1}{v_{glass}} \sin \theta_{glass} = \frac{1}{v_{air}} \sin \theta_{air} \tag{23-4}$$

It is common to express this relationship in terms of the **index of refraction** n of the materials, a measure of the speed of light in each material relative to the speed of light in a vacuum. The definition of n is

$$n = \frac{c}{v} \tag{23-5}$$

Equation 23-4 becomes

$$n_{glass} \sin \theta_{glass} = n_{air} \sin \theta_{air}$$

More generally, for light crossing the boundary between medium 1 and medium 2, Equation 23-4 becomes

$$n_2 \sin \theta_2 = n_1 \sin \theta_1 \tag{23-6}$$

where again θ_1 and θ_2 are defined with respect to the normal to the boundary surface. This equation is known as **Snell's law of refraction**. Snell's law tells us that when a ray of light crosses from one medium to another, the product of the index of refraction and the sine of the angle the ray makes to the normal remains constant. When light passes into a material of higher index of refraction, the sine of the refracted angle, and therefore the angle itself, decreases. So when light passes from air to glass, for example, the refracted angle is smaller than the incident angle. The light is bent closer to the normal.

From Equation 23-5, the index of refraction of a material is inversely related to the speed that light travels in the material. The slower light travels in a medium, the larger is its index of refraction. In air, for example, the speed of light is only slightly lower than c, and the index of refraction is close to 1. In diamond, on the other hand, the speed of light is about $0.4c$, and the index of refraction is approximately 2.42. Table 23-1 lists the index of refraction of some common materials.

Table 23-1
Indices of Refraction

Material	n
vacuum	1.00000
air at 20 °C, 1 atm pressure	1.00029
ice	1.31
water at 20 °C	1.33
acetone	1.36
ethanol	1.36
eye, cornea	1.38
eye, lens	1.41
sugar water (high concentration)	1.49
plexiglas	1.49
typical crown glass	1.52
sodium chloride	1.54
sapphire	1.77
diamond	2.42

? Got the Concept 23-1

Out of the Water and into the Air

A ray of light from an underwater swimming pool lamp exits the pool into the surrounding air. Does the ray bend closer to the normal or farther away from it?

? Got the Concept 23-2
Is It Really That Shallow?

Looking out from the shore you can see the bottom of a shallow pond. Does the pond look deeper or more shallow than it really is?

Example 23-1 Seeing Under Water

A surveyor looking at an aqueduct is just able to see the underwater edge where the far wall meets the bottom (Figure 23-6). If the aqueduct is 4.2 m wide and her line of sight to the near, top edge is 25° above the horizontal, find the depth of the aqueduct.

SET UP

In Figure 23-6 we labeled the angles, with respect to the normal, at which the light ray approaches and leaves the water–air boundary. The labels are helpful in applying Snell's law of refraction, and also to make clear that the angle given in the problem statement is not one of the angles in Snell's law. However, notice that the sum of the given angle (25°) and θ_2, the angle the light ray makes to the normal in air, is 90°, so

$$\theta_2 = 90° - 25° = 65°$$

Also, note that the light ray coming from the bottom, far edge of the aqueduct forms the hypotenuse of a right triangle. The other two legs of the triangle are the depth D and width W (4.2 m) of the aqueduct, so

$$\tan \theta_1 = \frac{W}{D}$$

By solving Snell's law for θ_1, this relationship will allow us to find D:

$$D = \frac{W}{\tan \theta_1} \qquad (23\text{-}7)$$

SOLVE

Snell's law is

$$n_2 \sin \theta_2 = n_1 \sin \theta_1 \qquad (23\text{-}6)$$

> Go to Interactive Exercise 23-1 for more practice dealing with refraction

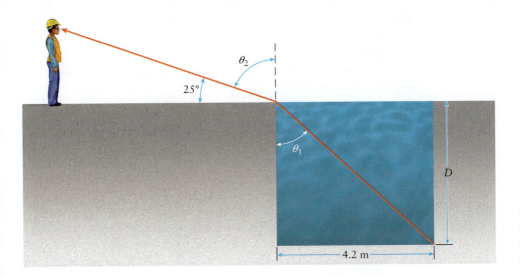

25°

θ_2

θ_1

D

4.2 m

Figure 23-6 A surveyor looking at an aqueduct is just able to see the underwater edge where the far wall meets the bottom.

so

$$\theta_1 = \sin^{-1}\left(\frac{n_2}{n_1} \sin \theta_2\right)$$

Then Equation 23-7 becomes

$$D = \frac{W}{\tan\left(\sin^{-1}\left(\frac{n_2}{n_1} \sin \theta_2\right)\right)}$$

Using the known values, the depth of the aqueduct is

$$D = \frac{4.2 \text{ m}}{\tan\left(\sin^{-1}\left(\frac{1}{1.33} \sin 65°\right)\right)} = 4.5 \text{ m}$$

REFLECT

The definition of Snell's law depends on angles measured with respect to the direction normal to the boundary between two media. It is common, however, especially in problems that involve an observed angle, for one or both angles to be measured from some other direction. In situations like this one, for example, it is often easier to measure the angle from the horizontal. It's straightforward to find the angle with respect to the normal, as long as you pay attention to how the angles in a problem are defined.

Practice Problem 23-1 A surveyor looking at an aqueduct is just able to see the underwater edge where the far wall meets the bottom. If the aqueduct is 4.2 m wide and 4.2 m deep, find her line of sight, above the horizontal, to the near, top edge of the aqueduct. Use Figure 23-6, but note that for this problem the angle is not 25°.

Watch Out

The angles in Snell's law of refraction are defined from the direction normal to the boundary between one light-transmitting medium and another.

Snell's law of refraction is a relationship between the indices of refraction of two light-transmitting media and the angles at which a ray of light enters and leaves the boundary between them. The angles must be taken from the normal to the boundary, but you may be given, or find convenient to use, an angle defined in some other way. That's okay, as long as you find the angle from the normal before applying Snell's law.

✳ What's Important 23-1

Waves refract (change direction) when crossing from one medium to another when the wave speed is different in the two media. When a wave enters a second medium in which the wave speed is lower that the first, it bends toward the direction perpendicular to the surface, and vice versa. Light travels more slowly in glass than in air, so when light enters a piece of glass from air, it is bent closer to the normal. The index of refraction is a measure of the speed of light in a medium relative to the speed of light in a vacuum; the slower light travels in a medium, the larger is its index of refraction.

23-2 Total Internal Reflection

Our experiences tell us that light passes through some materials such as the pane of glass in a window. Does all of the light that strikes a window pass though it? From the photo in Figure 23-7, taken from inside a restaurant looking out, you might guess that it does not. Were you on the sidewalk you could certainly see the tables and chairs inside the restaurant, as well as the reflection of a flag in the window. But notice that the tables and chairs are visible (in the reflection) even when looking through the restaurant's window from the inside. We will consider the phenomenon of reflection more thoroughly in the next chapter, but certainly, the image of the tables and chairs in this photo indicates that some of the light striking the window has been reflected; therefore, not all of the light that strikes the inside surface of the window passes through it. Some of the light is bent—refracted— as it crosses the boundary between air and glass, and some of the light is reflected.

What happens to the light that makes it into the glass? Eventually it strikes the back surface of the glass, and we expect it to be refracted again as it crosses the boundary between glass and air. Is some of the light also *reflected* back into the glass? The answer contains a surprise.

Light travels more slowly in glass than air, so the index of refraction (Equation 23-5) of glass is higher than the index of refraction of air. As light crosses the boundary between glass and air, it is bent away from the normal as in Figure 23-8a. As the incident angle of the light striking the inside surface of the glass increases (Figure 23-8b), the outgoing light ray gets farther from the normal and closer to the surface. There is some angle, called the **critical angle** θ_c, at which the light leaving the glass lies exactly in the plane of the surface of the glass, as shown in Figure 23-8c. For incident angles greater than the critical angle, such as the one shown in Figure 23-8d, the light is completely reflected back into the glass. This phenomenon is **total internal reflection**.

Total internal reflection is possible only when light leaving a boundary between two media is bent away from the normal to the surface. Total internal reflection can therefore only occur when light encounters a boundary from a medium of higher index of refraction to one of a lower index of refraction.

Total internal reflection is a special case of refraction. Snell's law of refraction can therefore be used to find the critical angle θ_c beyond which light is reflected back from a boundary between two media. As in Figure 23-8c, light incident on a boundary at θ_c is refracted along the surface of the boundary, so the refracted angle (with respect to the normal) is 90°. From Equation 23-6, when light goes from a medium of index of refraction n_1 to one of index of refraction n_2:

$$n_2 \sin 90° = n_1 \sin \theta_c$$

Figure 23-7 Some of the light striking a window passes through it, and some is reflected. *(Courtesy of David Tauck)*

(a) A ray of light refracts as it passes from glass to air. The index of refraction of glass is higher than the index of refraction of air.

The dashed line is the normal to the surface.

(b) As the angle at which the ray approaches the boundary increases, the outgoing ray is bent farther from the normal, so closer and closer to the surface...

(c) ...until the incident angle reaches the critical angle θ_c. When the incident angle is θ_c, the outgoing ray lies in the surface of the boundary.

(d) When the incident angle is greater than θ_c, the outgoing ray is completely reflected back into the higher index material. This is total internal reflection.

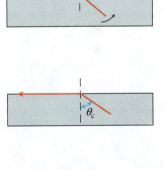
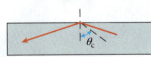
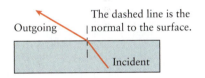

Figure 23-8 (a, b) As the angle of light crossing a boundary between glass and air gets farther from the normal, it is refracted closer and closer to the surface. (c) When the incident angle equals the critical angle, the light ray is refracted along the surface. (d) For incident angles greater than the critical angle, the light ray is completely reflected back into the glass.

or

$$\theta_c = \sin^{-1}\left(\frac{n_2}{n_1}\right) \tag{23-8}$$

Notice that the critical angle depends on the index of refraction of the media on both sides of a boundary. In other words, critical angle is not a characteristic of a material or medium, but only of the boundary between *two* materials.

? Got the Concept 23-3
A Dome of Light

A diver looking straight up toward the surface in clear, still water sees a circle of light surrounded by darkness. What explains this phenomenon?

Figure 23-9 A diver looking straight up toward the surface in clear, still water sees a circle of light surrounded by darkness. (*U.S. Navy photo by Mass Communication*)

Example 23-2 Critical Angle for a Glass–Water Boundary

A beam of light traveling through the piece of crown glass in **Figure 23-10** strikes the internal surface on the right side of the glass and is totally internally reflected. Find the critical angle for the glass–air interface.

SET UP

The critical angle depends on the index of refraction of the media on both sides of a boundary:

$$\theta_c = \sin^{-1}\left(\frac{n_2}{n_1}\right) \tag{23-8}$$

Here n_1 is the index of refraction of glass and n_2 is the index of refraction of air, so

$$\theta_c = \sin^{-1}\left(\frac{n_{air}}{n_{glass}}\right)$$

SOLVE

Using Table 23-1, the index of refraction of typical crown glass is 1.52, and the index of refraction of air is approximately 1, so

$$\theta_c = \sin^{-1}\left(\frac{1}{1.52}\right) = 41.1°$$

REFLECT

The critical angle for total internal reflection when light is moving from glass to air is about 42°.

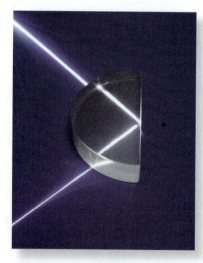

Figure 23-10 A beam of light traveling through the piece of glass strikes the internal surface on the right side of the glass and is totally internally reflected. (*GIPhotoStock / Photo Researchers*)

Practice Problem 23-2 Imagine that the piece of glass in Figure 23-10 is replaced by a polished sapphire crystal, and that the beam of light traveling through it strikes the internal surface on the right side of the crystal and is totally internally reflected. Find the critical angle for the sapphire–air interface.

Example 23-3 Critical Angles

(a) A laser is aimed from under the water toward the surface, as in **Figure 23-11**. Find the critical angle beyond which total internal reflection occurs. (b) Find the critical angle if the liquid in the tank were replaced by water containing a high concentration of dissolved sugar.

SET UP
The critical angle depends on the index of refraction of the media on both sides of a boundary:

$$\theta_c = \sin^{-1}\left(\frac{n_2}{n_1}\right) \tag{23-8}$$

For both parts of this problem n_2 is the index of refraction of air, which we will take as equal to 1 for the purposes of calculations.

SOLVE
In part a, light encounters a boundary between water and air, so n_1 is the index of refraction of water, 1.33 from Table 23-1. Using Equation 23-8, the critical angle for a water–air boundary is

$$\theta_c = \sin^{-1}\left(\frac{1}{1.33}\right) = 48.8°$$

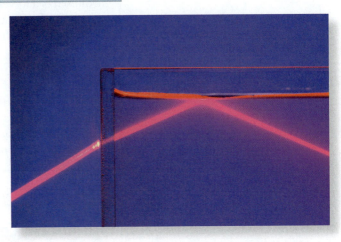

Figure 23-11 Light from a laser aimed from under the water toward the surface is totally internally reflected. (*GIPhotoStock/ Photo Researchers/ Getty Images*)

In part b, light hits a boundary between a solution of sugar dissolved in water, so by using Table 23-1 we find the index of refraction is 1.49. The critical angle when the tank is filled with sugar water is then

$$\theta_c = \sin^{-1}\left(\frac{1}{1.49}\right) = 42.2°$$

where again, n_2 equals 1, the index of refraction of air.

Go to Picture It 23-1 for more practice dealing with practice dealing with refraction and reflection

REFLECT
The addition of sugar to water results in a lower speed of light, which increases the index of refraction. The angle beyond which total internal reflection occurs is therefore smaller for sugar water than for pure water. If a source of light that radiated light uniformly in all directions were placed in the tank so that light strikes the surface at all possible angles, a larger fraction of the light would experience total internal reflection when the tank is filled with sugar water compared to when it is filled with pure water. The smaller the critical angle, the larger the range of angles at which light experiences total internal reflection.

Practice Problem 23-3 A laser is aimed from within a tank of ethanol toward the surface, as in Figure 23-11. Find the critical angle beyond which total internal reflection occurs.

! Watch Out

Total internal reflection can only occur when light passes from a medium of higher index of refraction to one of lower index of refraction.

For total internal reflection to occur, light must be bent away from the normal as it passes from one medium to another. This only happens when the second medium has a lower index of refraction. Try to compute a critical angle when light crosses the boundary from a material of lower index of refraction n_1 to one of higher index of refraction n_2. In the equation for critical angle (Equation 23-8),

$$\theta_c = \sin^{-1}\left(\frac{n_2}{n_1}\right)$$

the ratio n_2/n_1 will be greater than 1 because n_2 is greater than n_1. The sine function is always less than or equal to 1, however, so the inverse sine is not defined. The physics of total internal reflection therefore does not apply when n_2 is greater than n_1.

? Got the Concept 23-4
Mirage on the Road

Look down the highway on a hot day and you're likely to see what appear to be puddles of water on the road. You might even see the reflection of cars and other objects in the "puddles," as in **Figure 23-12**. Given that the speed of light is higher in the hot air near the road than in the cooler air up above the road, can you come up with an explanation for the road mirages?

Figure 23-12 What causes the appearance of the shimmering patches on the road that look like puddles? *(Kent Wood/ Photo Researchers)*

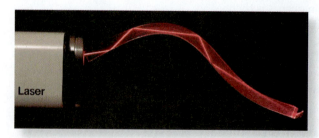

Figure 23-13 Light propagates along a curved bar of Plexiglas by a series of total internal reflections. *(Courtesy of David Tauck)*

Endoscopy is a medical procedure used to see inside the body. It relies on light fiber, an optical device used to carry light, and sometimes data encoded in pulses of light, from one place to another. A beam of light sent down a light fiber experiences total internal reflection at the surface of the fiber, which results in multiple reflections that keep the beam inside the fiber. The curved bar of Plexiglas in **Figure 23-13** carries light in a similar way. Because light experiences total internal reflection when it encounters the surface, no light is lost as the beam is reflected back. Notice in Figure 23-13 that no light leaks out of the block at the points where the beam hits the surface. Most light fiber is made by surrounding a central core with either one or two layers of a "cladding" made from a material of lower index of refraction than the core. The index of refraction of the core and the cladding is chosen to adjust the critical angle at the core–cladding boundary in order to optimize the light-carrying properties of the light fiber.

Example 23-4 Light Fiber

Consider a straight light fiber of length L made from material with index of refraction equal to 1.50. Light propagating along the direction of the axis of the fiber travels a distance L from end to end. (a) If the fiber is surrounded by air, and light moves along the fiber by repeated total internal reflections, how far does light travel as it goes from one end of the fiber to the other? (b) How far does light travel if the fiber is encased in cladding made of material of index of refraction 1.45?

SET UP

Light propagates down the fiber by repeated reflections at slightly greater than the critical angle. A section of one light ray is shown in **Figure 23-14a**. The path the light ray follows forms the hypotenuse of a series of right triangles, one of which is shown in **Figure 23-14b**. For the ray to travel a distance l lengthwise along the fiber (the base of the triangle in Figure 23-14b) it must actually cover a distance l', the hypotenuse of the triangle. To find the total path length, we will essentially add all of the hypotenuses together.

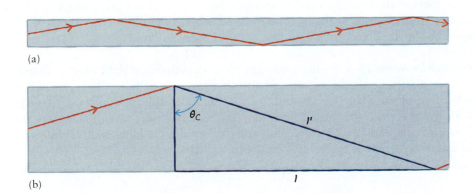

(a)

(b)

Figure 23-14 (a) Light propagates down a light fiber by repeated reflections at the critical angle. (b) The path the light ray follows forms the hypotenuse of a series of right triangles, one of which is shown.

SOLVE

The hypotenuse l' and base l in Figure 23-14b are related by

$$l' = \frac{l}{\sin \theta_c}$$

The variable l' gives the distance for one "bounce." For a fiber of total length L, the distance L' that a light ray travels from one end to the other is the sum of all of the bounces, or

$$L' = \frac{L}{\sin \theta_c}$$

The critical angle θ_c, the angle at which total internal reflection occurs at the boundary between a medium of index of refraction n_1 and a medium of index of refraction n_2, is given by

$$\theta_c = \sin^{-1}\left(\frac{n_2}{n_1}\right) \tag{23-8}$$

so

$$\sin \theta_c = \frac{n_2}{n_1}$$

and then

$$L' = \frac{L}{n_2/n_1} = \frac{n_1 L}{n_2}$$

For total internal reflection to occur n_1 must be greater than n_2, so clearly L' is greater than L. Here n_1 equals 1.50. In part a, the light fiber is surrounded by air, so n_2 equals 1.00. In part b, the light fiber consists of a core encased in a cladding that has an index of refraction n_2 equal to 1.45. So, when the fiber is surrounded by air, the distance the light travels down the fiber is

$$L'_{air} = \frac{1.50L}{1.00} = 1.50L$$

When the fiber is surrounded by the cladding material, the light travels a total distance of

$$L'_{cladding} = \frac{1.50L}{1.45} = 1.03L$$

REFLECT

The real distance light travels to go from one end of the light fiber to the other is far shorter when the fiber is encased in cladding. Light travels only marginally farther than the end-to-end length of the light fiber when it is encased in cladding, but 50% more than the end-to-end length otherwise. The time it takes a light signal to be sent from one end of the light fiber to the other is therefore significantly shorter for a fiber that consists of a cladding-encased core.

Practice Problem 23-4 A short, straight piece of cylindrical light fiber of diameter 1.00 mm is to be used to connect two other sections of light fiber. The index of refraction of the fiber is 1.50, and it is encased in cladding of index of refraction 1.45. What is the longest the connecting piece can be so that light that enters at the center of the end of the connector and strikes the cladding halfway along the connector will exit at the center of the other end?

The ray of light colored blue strikes the surface at an angle greater than the critical angle, and experiences total internal reflection.

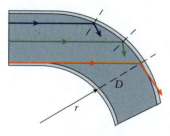

The ray of light colored green strikes the surface at an angle greater than, but closer to, the critical angle and experiences total internal reflection.

The ray of light colored red strikes the surface at an angle less than the critical angle, so it does not experience total internal reflection and leaves the light fiber.

Figure 23-15 What is the minimum radius of curvature at which the fiber can be bent before light, initially traveling parallel to the axis of the fiber, will begin to leak out?

Example 23-5 Laser Lithotripsy

Laser lithotripsy is an invasive procedure used to fragment and disintegrate kidney stones. Light from a powerful laser is directed down a light fiber inserted into a patient's body through a catheter. A commonly used light fiber consists of a glasslike core that has a diameter equal to 0.002 m and an index of refraction equal to 1.55; it is surrounded by a cladding that has an index of refraction equal to 1.46. What is the minimum radius of curvature at which the fiber can be bent before light, initially traveling parallel to the axis of the fiber, will begin to leak out of the fiber (**Figure 23-15**)?

SET UP

Light leaks out of the fiber when it strikes the surface of the fiber at an angle less than the critical angle. As you can see from Figure 23-15, this leakage is more likely to happen the deeper into the bend that light strikes the surface of the fiber. So when the fiber is bent at a radius such that if the radius of the bend were smaller light *would* leak out, the light that strikes the surface farthest into the bend will hit at an angle just at the critical angle. This light is the light ray colored red in the figure. To understand the geometry of this ray, we consider the triangle formed by it and the radius of the bend, as shown in **Figure 23-16**. We labeled the radius r_{min} to indicate that light would leak out were the light fiber bent at a smaller radius.

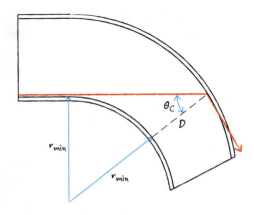

Figure 23-16 The angle in the right triangle formed by the light ray that strikes deepest into the bend of the light fiber can be no smaller than the critical angle if no light is to leak out.

SOLVE

From the triangle we see that

$$\sin \theta_c = \frac{r_{min}}{r_{min} + D}$$

In addition, by using Equation 23-8,

$$\sin \theta_c = \frac{n_{cladding}}{n_{core}}$$

so

$$\frac{r_{min}}{r_{min} + D} = \frac{n_{cladding}}{n_{core}}$$

The equation can be rearranged for r_{min} in terms of the other variables:

$$r_{min} = \frac{n_{cladding}}{n_{core} - n_{cladding}} D$$

Using the values provided in the problem statement, the minimum radius of curvature is

$$r_{min} = \frac{1.46}{1.55 - 1.46} 0.002 \text{ m} = 0.032 \text{ m}$$

REFLECT

The parameters given are typical for light fibers used in biomedical applications. Our calculation suggests that such a light fiber can be bent with a radius of curvature as small as 3 cm before light leaks out. Such a radius is more than adequate to accommodate the twists and turns a fiber would make under normal circumstances, say, to direct laser light onto a kidney stone.

> ### ✱ What's Important 23-2
> As light crosses the boundary from a medium of higher index of refraction to one of lower index of refraction (such as glass to air), some of the light is refracted away from the direction normal to the surface and therefore closer to the surface. There is some incident critical angle θ_c at which the light is refracted exactly in the plane of the surface; for incident angles greater than the critical angle, the light is reflected back into the first medium. The phenomenon is called total internal reflection.

23-3 Dispersion

Sunlight striking a glass crystal hanging in a window throws rainbows of color in every direction. You might see the same brilliant colors in the sky after a rainfall

Figure 23-17 The index of refraction of water, for example, water droplets in the atmosphere, varies with the color of light. This causes dispersion, the spreading out of the colors of visible light, and gives rise to phenomena such as a rainbow. *(iStockphoto/ Thinkstock)*

(**Figure 23-17**). Although we aren't usually directly aware of it, the spectrum of visible light includes all of the colors in these rainbows.

We defined the index of refraction n in terms of the speed of light in a vacuum c and the speed v at which it propagates in a particular medium:

$$n = \frac{c}{v} \tag{23-5}$$

However, because the speed at which light travels varies as a function of the wavelength, the index of refraction in a given medium is normally measured using yellow light. Red light travels slightly faster than yellow light, so the index of refraction of red light in any given medium is slightly lower than the index of refraction of yellow light. Blue light travels slightly slower than yellow light, so the index of refraction of blue light in any given medium is slightly higher than the index of refraction of yellow light. Note that all colors of light travel at c in a vacuum and nearly c in air. We will not, therefore, consider the index of refraction of air to be wavelength dependent.

The angle at which light refracts as it crosses the boundary between two light-transmitting materials depends on their indices of refraction (Snell's law of refraction, Equation 23-6), so different colors of light refract at different angles. For example, when light crosses a boundary into a medium of higher index of refraction, it bends toward the direction normal to the surface (**Figure 23-18**). The higher the index of refraction of the second medium, the more the light is bent. Because in a given medium the index of refraction of blue light is higher than the index of refraction of red light, blue light is therefore bent *more* than red light in the second medium. This causes the different colors of light to be spread out, a phenomenon known as **dispersion**.

White light is composed of light across all wavelengths (colors) in the visible spectrum. The colors spread apart when light crosses the boundary between two light transmitting media.

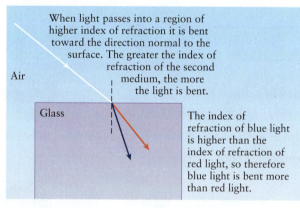

When light passes into a region of higher index of refraction it is bent toward the direction normal to the surface. The greater the index of refraction of the second medium, the more the light is bent.

Air

Glass

The index of refraction of blue light is higher than the index of refraction of red light, so therefore blue light is bent more than red light.

Figure 23-18 The angle at which light refracts as it crosses the boundary between two light-transmitting materials depends on their indices of refraction, so because the index of refraction varies with the color of light, different colors refract at different angles.

The high index of refraction of diamond results in a small critical angle for total internal reflection when light strikes an inside surface.

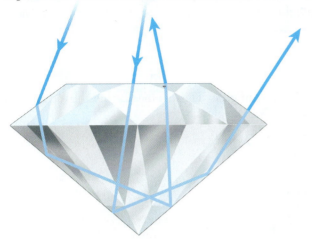

Most of the light experiences multiple total internal reflections before exiting the diamond through a top surface, resulting in the property of cut diamonds called "brilliance."

Figure 23-19 Diamond has a relatively high index of refraction, so the critical angle for light exiting a diamond back into the surrounding air is small. As a result, most of the light entering a cut diamond experiences multiple total internal reflections before exiting through a top surface, which contributes to the property of cut diamonds called brilliance.

The amount of dispersion that occurs in a medium depends on how strongly the index of refraction varies with wavelength. In common glass, yellow light has an index of refraction of 1.52 (Table 23-1). The index of refraction of red light is about 1.51, and the index of refraction of blue light is about 1.53. Because of the differences, the colors spread out as light passes through a glass crystal. Water droplets suspended in or falling through the air disperse white light, forming rainbows such as the one in Figure 23-17.

The range over which the index of refraction of a material varies as a function of color characterizes the dispersion of light in that material. For glass, for example, the difference is about 0.02 (1.53 minus 1.51), while for diamond, in which blue light has an index of refraction of 2.45 and red light 2.41, the spread in index of refraction is about 0.04. Therefore, when white light passes from air into diamond, colors spread out over a wider angle than when it passes into glass. This high value of dispersion contributes to the sparkly character of a cut diamond. In addition, the high index of refraction results in a small critical angle for total internal reflection when light exits a diamond back into the surrounding air. This causes most of the light entering a cut diamond to experience multiple total internal reflections before exiting the diamond through a top surface, as shown in **Figure 23-19**. Both the high dispersion and the multiple internal reflections result in the property of cut diamonds called *brilliance*.

Estimate It! 23-1 Split Apart

A narrow beam of white light enters the narrow edge of a rectangular pane of glass at an angle of 60° from the direction perpendicular to the edge. The pane is 0.50 m in width. Use the data in **Figure 23-20** to estimate how far apart the red and blue parts of the visible spectrum will be when the light leaves the glass, first if the pane is made from crown glass and then if it is made from dense flint glass.

SET UP

The refraction of the light is shown in **Figure 23-21**. According to Snell's law (Equation 23-6),

$$\theta_{\text{color}} = \sin^{-1}\left(\frac{n_{\text{air}}\sin\theta_{\text{air}}}{n_{\text{color}}}\right)$$

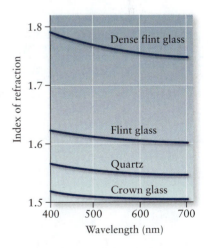

Figure 23-20 The index of refraction in glass varies with the wavelength of light and the type of glass.

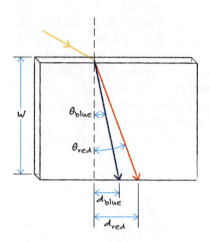

Figure 23-21 Light disperses as it passes through a pane of glass.

where θ_{color} is the angle at which light of a given color is refracted from the direction normal to the edge. The index of refraction n_{color} is different for each type of glass as well as for each light color. The distance d_{color} from the dashed line perpendicular to the edge that each color of light exits the pane is therefore also different; from the figure,

$$d_{color} = W \tan \theta_{color}$$

where W is the width of the pane of glass. The distance Δ that the red and blue light rays are separated when they exit the glass is then

$$\Delta = d_{red} - d_{blue}$$

SOLVE

For each type of glass, then

$$\Delta = W \tan \theta_{red} - W \tan \theta_{blue}$$

or

$$\Delta = W \tan\left(\sin^{-1}\left(\frac{n_{air} \sin \theta_{air}}{n_{red}} \right) \right) - W \tan\left(\sin^{-1}\left(\frac{n_{air} \sin \theta_{air}}{n_{blue}} \right) \right)$$

To calculate numeric values, we need to use Figure 23-20 to estimate the index of refraction for red and blue light, for crown glass and for dense flint glass. First, we pick 700 nm and 400 nm as the wavelengths of red and blue light, respectively. Then we use a ruler to roughly divide each interval of 0.1 on the vertical axis into 10 pieces. In this way we get big, round, but reasonable, values for the index of refraction, by color, for crown glass:

$$n_{crown,red} = 1.51$$

$$n_{crown,blue} = 1.52$$

and for dense flint glass:

$$n_{flint,red} = 1.75$$

$$n_{flint,blue} = 1.79$$

Using n_{air} equal to 1, W equal to 0.50 m, and $\theta_{air} = 60°$, the separation between red and blue light on exiting the crown glass is

$$\Delta_{crown} = (0.50 \text{ m}) \tan\left(\sin^{-1}\left(\frac{1 \sin 60°}{1.51} \right) \right) - (0.50 \text{ m}) \tan\left(\sin^{-1}\left[\frac{1 \sin 60°}{1.52} \right] \right)$$

$$= 0.0034 \text{ m}$$

Similarly, the separation between red and blue light on exiting the dense flint glass is

$$\Delta_{flint} = (0.50 \text{ m}) \tan\left(\sin^{-1}\left(\frac{1 \sin 60°}{1.75} \right) \right) - (0.50 \text{ m}) \tan\left(\sin^{-1}\left(\frac{1 \sin 60°}{1.79} \right) \right)$$

$$= 0.0083 \text{ m}$$

REFLECT

Flint glass, sometimes called "lead glass" because lead is often added during the glass-making process, has a relatively high refractive index. In addition, the range of values of index of refraction versus color is much wider for flint glass than crown glass, the kind of glass commonly used to make windows, for example. The spectrum of visible light is spread out twice as wide in the flint glass compared to crown glass.

✳ What's Important 23-3

Visible light includes all of the colors of the rainbow. When light crosses a boundary into a medium of higher index of refraction, the different colors experience different amounts of bending (refraction). This causes them to spread out, a phenomenon called dispersion. The amount of dispersion depends on how strongly the index of refraction varies with wavelength.

23-4 Polarization

When a honeybee finds nectar, it communicates the location to other bees in the hive. Over 60 years ago, Austrian ethologist Karl Ritter von Frisch established that if bees can see a small patch of blue sky, they can use the position of the Sun to describe the path back to the food from the hive. (For his work on animal behavior, von Frisch received the Nobel Prize in Physiology or Medicine in 1973.) How is it that bees can know the position of the Sun, especially from only a small patch of sky?

The physics behind the scattering of sunlight in the atmosphere underlies the honeybee's ability to use the position of the Sun to navigate between its hive and the location of nectar. Water molecules as well as tiny particles, or aerosols, suspended in the atmosphere absorb light and then reradiate it. Because water molecules and very fine aerosols tend to preferentially scatter light in the shorter wavelengths, blue light ends up being distributed fairly uniformly across the sky, which is why our sky is blue.

Each photon of light coming from the Sun has an associated oscillating electric field $\vec{E}$ and magnetic field $\vec{B}$. Although the fields are perpendicular to each other for a given photon (look back to Figure 22-4), the orientations of the planes for different photons are completely random, until the photons undergo scattering in the atmosphere. During that process, photons with fields aligned in directions tangent to circles centered on the Sun are more likely to be scattered, so that the photons end up with their fields slightly aligned. Honeybees can see the alignment, or **polarization**, in the blue light from the sky and use it to navigate. The photograph in **Figure 23-22**, taken through a polarizing lens, shows the polarization of sunlight A

Figure 23-22 The photograph shows the polarization of sunlight. (© *Harald Edens, reproduced with permission www. weather-photography.com*)

Figure 23-23 The image on the left was taken through a linear polarizing filter, the one on the right was not. Some of the light from the sky is polarized in a direction perpendicular to the filter and does not pass through it, making the sky darker in the image on the left. *(Emmanuel Lattes / Alamy)*

variety of insects and other animals, such as homing pigeons, can see it, too. Even we humans are weakly sensitive to the polarization direction of light, although for us it's easier to notice the polarization of sunlight by using a filter that only allows light of a certain polarization to pass. Compare the images in **Figure 23-23**; the one on the left was taken through a polarizing filter.

Imagine a ray of light in which all photons have an electric field oscillating up and down in the same direction. Such a ray is **linearly polarized**. Most of the sources of light we encounter in our daily lives, such as the lightbulb in your desk lamp, are not polarized. A lightbulb produces unpolarized light with a random distribution of polarization directions. A common way to create linearly polarized light is to shine an unpolarized beam through a polarizing filter. Such a filter is made from a special material in which the molecules act like slits, so that only photons that have an electric field aligned with the slits pass through. This is suggested in **Figure 23-24**, in which photons polarized in the direction of the slits pass through but those polarized in a direction perpendicular to the slits are stopped. (In common polarizing materials such as that found in sunglasses, the photons which pass through are those polarized in the direction *perpendicular* to the molecular slits. This is a consequence of the way light interacts with the electrons in the molecules. The net result is the same: Light passing through the filter is linearly polarized.)

What happens if a polarized beam of light encounters a polarizing filter that is neither aligned with nor perpendicular to its polarization direction? Some of the light can pass through the filter, unlike a transverse wave on a string which will only pass through a slit that is completely aligned with the plane in which the oscillations occur. Remember that the electric field, which defines the polarization

Figure 23-24 A linear polarizing filter allows only light with an electric field component aligned in a certain direction to pass through it.

Two waves of different polarizations encounter a polarizing filter.

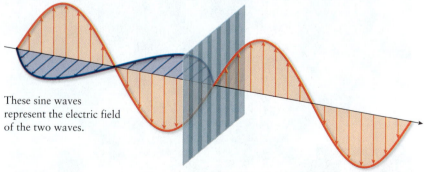

These sine waves represent the electric field of the two waves.

Only the wave with polarization aligned with the direction of the filter passes through it. The light is now linearly polarized.

direction of the photons, is a vector. Except in the case that the polarization direction of a photon is perpendicular to the transmission direction of the filter, there will therefore be a component of its electric field that is aligned with the filter. This component of the wave's electric field therefore passes through the filter. As shown in **Figure 23-25**, if the electric field has magnitude E_0, the component aligned with the filter is

$$E_{aligned} = E_0 \cos \theta \qquad (23\text{-}9)$$

Although some light can pass through, the brightness, or intensity, of the light is reduced because not all of it is transmitted. Intensity is proportional to the square of the amplitude of the field, so

$$I = E_0^2 \cos^2 \theta = I_0 \cos^2 \theta \qquad (23\text{-}10)$$

where I_0 is the intensity of the incoming light and I is the intensity of the light transmitted by the filter. When a beam of light that contains photons of a random mixture of polarization directions passes through a linear polarizing filter, each photon separately obeys Equation 23-10. The average of $\cos^2 \theta$ (and also $\sin^2 \theta$; see Section 21-1) over a range from 0 to 360° is $1/2$, so the net effect of a linear polarizing filter on an unpolarized beam of light is to reduce the overall intensity by one-half.

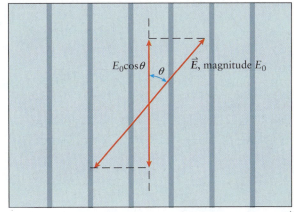

The electric field vector, the polarization direction, of a photon is at an angle θ with respect to the polarization direction of a filter...

$E_0 \cos \theta$ θ $\vec{E}$, magnitude E_0

...the filter only passes the component of $\vec{E}$ ($E_0 \cos \theta$) that is aligned with the filter.

Figure 23-25 The electric field, which defines the polarization direction of the photons, is a vector, so the component of a light wave's electric field aligned with a linear polarizing filter will pass through it.

? Got the Concept 23-5
Polarizing Filters I

Two linear polarizing filters are placed one behind the other, so that their transmission directions are perpendicular to one another. A beam of unpolarized light is directed at the two filters. What fraction of the light will pass through both?

? Got the Concept 23-6
Polarizing Filters II

Two linear polarizing filters are placed one behind the other, so that their transmission directions are perpendicular to one another. A third polarizer is placed in between them, so that its transmission direction is not aligned with either of the first two. A beam of unpolarized light is directed at the three filters. Explain how some of the light can pass through all three.

The polarizing lenses often used for sunglasses can dramatically reduce glare from light reflected off the surface of water or off the glass of a car's windows (**Figure 23-26**), for example. When light strikes a nonmetallic surface, some is reflected and some may be refracted, depending on the angle at which the light hits. In addition, at any incident angle other than perpendicular to the surface, the reflected light is linearly polarized to some extent.

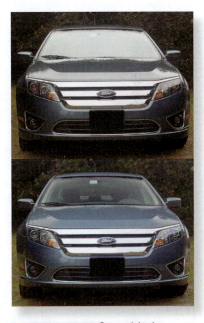

Figure 23-26 Reflected light is linearly polarized. The polarizing lenses often used for sunglasses can dramatically reduce glare from reflected light. (*Courtesy of David Tauck*)

Unpolarized light strikes the surface between air and water. The dots represent polarization in a direction parallel to the surface. The arrows represent polarization in a direction perpendicular to the surface and also to the direction in which the light is traveling.

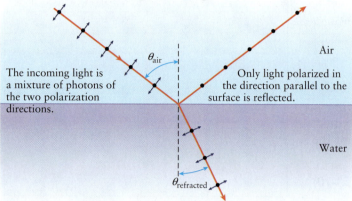

Figure 23-27 The electric fields of reflected photons cannot oscillate in the direction perpendicular to the surface, so this component of the incident wave is suppressed or eliminated when light is reflected. Therefore the outgoing light is, to some degree, linearly polarized.

Figure 23-27 shows light striking the boundary between air and water. Some of the light is polarized in a direction perpendicular to the plane of incidence (and to the page) (the blue dots) and some is polarized in a direction perpendicular to that direction (the small blue arrows). The electric fields of the reflected photons cannot oscillate in the direction perpendicular to the surface, however, because the electrons in the surface cannot oscillate in that direction. This component of the incident electromagnetic wave is therefore suppressed or eliminated when photons are reflected, and the outgoing light is, to some degree, linearly polarized. Because reflection eliminates light polarized in a direction perpendicular to the surface, reflected light is polarized in a direction parallel to the surface.

The amount of polarization depends on the incident angle; the reflected beam is completely polarized, as if it had passed through a linear polarizing filter, when the sum of the incident angle and the angle at which light is refracted equals 90°. For the boundary surface between any two light-transmitting media, the angle of incidence that results in this special condition is called **Brewster's angle** θ_B. Using the angles defined in Figure 23-27,

$$\theta_{\text{incident}} + \theta_{\text{refracted}} = 90° \qquad (23\text{-}11)$$

Using Snell's law of refraction (Equation 23-6), we know the relationship between θ_{incident} and $\theta_{\text{refracted}}$:

$$n_2 \sin \theta_{\text{refracted}} = n_1 \sin \theta_{\text{incident}}$$

where n_1 is the index of refraction of the medium from which the light strikes the surface and n_2 is the index of refraction of the medium into which the light is partially refracted. When θ_{incident} equals Brewster's angle θ_B and using Equation 23-11,

$$\theta_{\text{refracted}} = 90° - \theta_{\text{incident}} = 90° - \theta_B$$

So Snell's law gives

$$n_2 \sin (90° - \theta_B) = n_1 \sin \theta_B$$

$\text{Sin}(90° - \theta_B)$ equals $\cos \theta_B$ (**Figure 23-28**), so

$$n_2 \cos \theta_B = n_1 \sin \theta_B$$

To determine θ_B, we first rearrange the equation to get

$$\frac{\sin \theta_B}{\cos \theta_B} = \frac{n_2}{n_1} = \tan \theta_B$$

Finally, we solve for Brewster's angle:

$$\theta_B = \tan^{-1}\left(\frac{n_2}{n_1}\right) \qquad (23\text{-}12)$$

$\sin(90° - \varphi) = \cos\varphi.$

When one angle of a triangle is 90°, the sum of the other two angles is 90°.

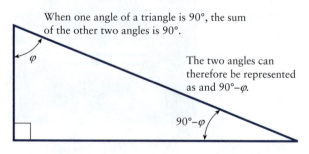

The two angles can therefore be represented as and 90°−φ.

The side opposite to the angle 90°− φ is the side adjacent to the angle φ. so $\sin(90° - \varphi) = \cos\varphi$.

Figure 23-28 The sine of $90° - \varphi$ equals the cosine of φ.

Again, when light strikes the boundary between two light-transmitting media at this angle the reflected light is completely linearly polarized. If the angle of incidence is something other than θ_B, the reflected ray is partly polarized. (Light that strikes the surface perpendicular to it is not polarized.)

Example 23-6 Brewster's Angle for Air to Water

At what angle, with respect to the normal to the surface, must light strike the wet surface of a road so that the reflected light is completely polarized?

SET UP

The reflected light is completely polarized when the incident light strikes the air–water boundary at Brewster's angle. The angle depends on the index of refraction of air and water; from Equation 23-12,

$$\theta_B = \tan^{-1}\left(\frac{n_{water}}{n_{air}}\right)$$

SOLVE

From Table 23-1, the index of refraction of air is about 1 and the index of refraction of water is 1.33, so

$$\theta_B = \tan^{-1}\left(\frac{1.33}{1}\right) = 53.1°$$

Go to Interactive Exercise 23-2 and Interactive Exercise 23-3 for more practice dealing with polarization

REFLECT

Light from the sky that strikes the surface at about 53° will be completely polarized on reflection, and light striking at close to that angle will be strongly polarized.

Practice Problem 23-6 Surfactants, agents that reduce the surface tension of fluids, play an important role in a number of physiological systems. For example, surfactant makes it easier to inflate the lungs by reducing surface tension of the fluid that lines the gas exchange surfaces. The polarization of light reflecting off the surface of a liquid containing a surfactant can be used to measure certain properties of the liquid. What is the index of refraction of a liquid containing a surfactant if the angle (with respect to the normal to the surface) at which reflected light is completely polarized is 57.2°?

✱ What's Important 23-4

Every photon has an associated oscillating electric field and magnetic field. The fields are perpendicular to each other for a given photon, but the orientations of the fields for different photons from most light sources are likely to be random. This randomness is characteristic of unpolarized light. A ray of light in which the electric fields of all photons oscillate in a single direction is linearly polarized.

23-5 Thin Film Interference

The striking, iridescent wings of the butterfly *Morpho menelaus* (Figure 23-29) absorb a significant fraction of energy in the sunlight that falls on them. Do the brilliant, almost glowing colors remind you of the eyeshine we see when light reflects directly back from an animal's eyes, as in the photos on the opening page of this chapter? Both phenomena result from the interference of light, a process similar to the interference of sound waves we investigated in Chapter 13.

Light has wave properties; for example, it has a wavelength and frequency, two parameters we associate with the color of the light. Both the wavelength λ and the frequency f depend on the energy E of the light, specifically,

$$E = \frac{hc}{\lambda} \tag{22-1}$$

Figure 23-29 The striking, iridescent colors on the wings of a *Morpho menelaus* butterfly result from the interference of light that reflects from them. *(Michael Gray/ Dreamstime.com)*

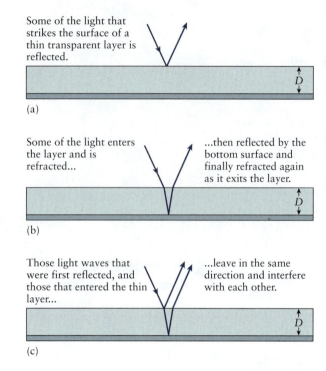

Some of the light that strikes the surface of a thin transparent layer is reflected.

(a)

Some of the light enters the layer and is refracted...   ...then reflected by the bottom surface and finally refracted again as it exits the layer.

(b)

Those light waves that were first reflected, and those that entered the thin layer...   ...leave in the same direction and interfere with each other.

(c)

Figure 23-30 (a) Some of the light waves that strike the outer surface of the thin film are reflected. (b) Some of the light enters the thin layer, refracting as it does so. (c) After the light that entered the film is reflected back up and out, it interferes with the light that was reflected off the outer surface.

and

$$E = hf \qquad (22\text{-}2)$$

As we observed for sound waves in Section 13-4, light waves constructively or destructively interfere depending on how the peaks and troughs of two waves align.

Start with two waves of the same wavelength that are initially in phase, so that they are rising and falling in sync with each other, and follow them along different paths to a common point. When they combine, the extent to which they either constructively or destructively interfere depends on the difference in the number of wavelengths that fit into the difference in the lengths of the two paths. If one wave has undergone an integer number of wavelengths more than (or less than) the other before they combine, then the two waves will still be in phase at that point. In this case, the two waves reinforce each other, resulting in *constructive interference*. If the number of oscillations one wave undergoes is an odd number of half wavelengths more than (or less than) the other, the waves cancel each other out; this is *destructive interference*.

Keeping this description of interference in mind, consider the process that occurs when light of a single wavelength strikes a layer of a transparent material, for example, the thin tapetum lucidum ("shining carpet") that lines the back of some animals' eyes, behind the retina. Behind the tapetum lucidum is a vascular layer and the protective, outer layer of the eye. Light is not transmitted through these last two layers, so for this discussion we restrict ourselves to a thin layer against an opaque backing. The thin layer, often called a *thin film*, is of thickness D.

As shown in **Figure 23-30a**, some of the light waves that strike the outer surface of the thin film are reflected. As we expect, the angle of reflection equals the angle of incidence. Not all of the incident light is reflected, however. As shown in **Figure 23-30b**, some light enters the thin layer, and it is refracting according to Snell's law (Equation 23-6) as it does so. This light eventually strikes the boundary between the film and the opaque backing, where it is reflected back up. (Some light is likely absorbed in the process as well.) Again, the angle of incidence equals the angle of reflection, so the light strikes the inside of the outer surface at the same angle at which it was refracted. The light is refracted again on exiting the thin film, and by Snell's law the angle at which it leaves the thin layer is equal to the angle at which it entered. So this second ray of light therefore leaves the upper surface of the thin film at the same angle as the light which was initially reflected from the upper surface, as shown in **Figure 23-30c**.

But look more closely at what happened! What was originally a single beam of light striking a thin layer of a transparent material has been split in two. Notice that the light that enters the layer and reflects off the inner surface travels farther than the light reflected off the outer surface. For this reason the two light waves, which had to be in phase initially because they were part of the same wave before striking the outer surface, may not be in phase when they recombine. The **thin film interference** that occurs depends on the number of wave cycles that fit into the extra distance traveled by the light that enters the thin layer. If an integer number of cycles fit into that extra distance, then the two outgoing waves are in phase and constructively interfere. The surface appears bright. If an odd number of half cycles fit into the extra distance, however, the two outgoing waves are 180° out of phase and destructively interfere. (We specify an *odd* number of half cycles because an even number of half cycles is the same as an integer number of full cycles, which is the previous case.) When destructive interference occurs the two outgoing waves cancel each other out and the surface appears dark.

Whether constructive interference, destructive interference, or partial interference occurs depends on the number of wavelengths of the incident light that fit into the difference in the length of the path followed by the reflected light and the length of the path followed by the light that enters and then exits the thin film. Because both types of light follow the same path while outside the material, the path length difference is simply the distance traveled by the light while inside the thin film. Each leg of the distance is the hypotenuse L of a triangle, as shown in Figure 23-31, that includes the angle at which the light is refracted inside the thin film. The path length difference Δ_{pl} is then equal to $2L$, where

$$\Delta_{pl} = 2L = 2\left(\frac{D}{\cos \theta_2}\right) \tag{23-13}$$

The angle θ_2 depends on the incident angle θ_1 and the indices of refraction of the thin film and the material that surrounds it, n_2 and n_1, respectively. Using Snell's law (Equation 23-6),

$$\theta_2 = \sin^{-1}\left(\frac{n_1 \sin \theta_1}{n_2}\right)$$

However, for light incident on the thin film close to the normal, both θ_1 and θ_2 are small, and therefore $\cos \theta$ is approximately equal to 1. So in this special case, from Equation 23-13 we see that the path length difference Δ_{pl} is approximately equal to twice the thickness of the thin film:

$$\Delta_{pl} = 2L \approx 2D \tag{23-14}$$

For light incident close to the normal on a thin film, constructive interference occurs when an integer number of wavelengths of the light exactly fits into the path length difference of $2D$. Destructive interference occurs when an odd number of half wavelengths fit into the distance $2D$.

The wavelength of the light inside the thin film is not the same as the wavelength of the light when traveling through the material that surrounds the film. Recall that it is the energy of the light that remains constant when the light wave crosses the boundary between two materials. The energy depends on both the wavelength of the light and the speed of light in the particular medium. Using Equation 22-1, we see that

$$E = \frac{hv}{\lambda}$$

where we have replaced c with v in order to represent the energy of the light wave in a medium other than a vacuum. Using Equation 23-5, the speed of light in a medium of index of refraction n is

$$v = \frac{c}{n}$$

So for light of incident wavelength λ_1,

$$E = \frac{h(c/n_1)}{\lambda_1} \tag{23-15}$$

Inside the thin film, which has index of refraction n_2,

$$E = \frac{h(c/n_2)}{\lambda_2} \tag{23-16}$$

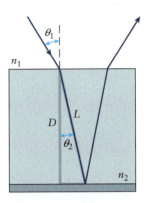

Figure 23-31 The distance light travels inside the thin film is twice the length of the hypotenuse L of this right triangle.

Combining Equations 23-15 and 23-16 yields

$$\frac{1}{n_1 \lambda_1} = \frac{1}{n_2 \lambda_2}$$

or

$$\lambda_2 = \frac{n_1 \lambda_1}{n_2} \qquad (23\text{-}17)$$

We now know everything we need to determine how the light reflected from the outer surface of the thin film interferes with the light that enters the film and is reflected from the inner surface. Constructive interference occurs when

$$2D = m\lambda_2 = m\frac{n_1 \lambda_1}{n_2}, m = 1, 2, 3, \ldots \qquad (23\text{-}18)$$

The minimum thickness of a thin film that results in constructive interference when light of wavelength λ_1 strikes the surface close to the normal is then

$$D_{\min} = \frac{n_1 \lambda_1}{2n_2} \qquad (23\text{-}19)$$

Destructive interference occurs when

$$2D = (2m - 1)\frac{\lambda_2}{2} = (2m - 1)\frac{n_1 \lambda_1}{2n_2}, m = 1, 2, 3, \ldots \qquad (23\text{-}20)$$

The minimum thickness of a thin film that results in destructive interference when light of wavelength λ_1 strikes the surface close to the normal is then

$$D_{\min} = \frac{n_1 \lambda_1}{4n_2} \qquad (23\text{-}21)$$

? Got the Concept 23-7
The Wavelength of Light in Soapy Film

The index of refraction of a soapy film, such as the one in **Figure 23-32**, is about 1.4. Is the wavelength of a wave of light longer, shorter, or the same in the soapy film as it is in air?

Figure 23-32 A thin film was made by dipping an open ring into soapy water, and then holding it vertically. The horizontal bands result from the increasing thickness of the film from top to bottom. *(Carol and Mike Werner / Phototake, Inc.)*

? Got the Concept 23-8
Glowing Eyes

The colors reflected in the eyes of the bobcat shown on the opening page of this chapter are identical. For some animals, however, the eyeshine of each eye can be a different color. Explain why different animals' eyes glow with different colors, and why even the two eyes on one animal can display different colors.

Example 23-7 Soapy Film

Monochromatic light of wavelength 560 nm strikes a thin layer of soap, which has an index of refraction of 1.4 and rests on a bathroom tile. (a) If the layer of soap is 700 nm thick, does constructive interference, destructive interference, or partial interference occur when the light strikes the surface close to the normal? (b) What is the minimum thickness of soap that would result in no (or minimal) reflection from the surface?

SET UP
Interference occurs between the light that is reflected from the top surface of the soap layer and the light that enters the soap and is eventually reflected back out. How the light interferes depends on how many cycles of the waves fit into the difference in lengths of the paths the two rays travel. To answer the questions about interference, then, we need to compare the path length difference and the wavelength of the light as it propagates through the soap.

 Go to Interactive Exercise 23-4 for more practice dealing with thin films

SOLVE
(a) The path length difference between the two light rays is approximately twice the thickness D of the soap layer, that is, from Equation 23-14,

$$\Delta_{pl} \approx 2D = 2(700 \text{ nm}) = 1400 \text{ nm}$$

The wavelength of the light in air is 560 nm; however, because the speed of light is not the same in soap as it is in air, the wavelength of the light is not 560 nm as it propagates through the soap. We can, however, apply Equation 23-17, which we developed to account for the effect of wave speed in a medium on wavelength. When we consider

$$\lambda_2 = \frac{n_1 \lambda_1}{n_2} \tag{23-17}$$

n_1 and n_2 are the indices of refraction of the two media, and n_1 equals 1, for air, and n_2 equals 1.4 for the soap film in this problem. Therefore

$$\lambda_2 = \frac{560 \text{ nm}}{1.4} = 400 \text{ nm}$$

To determine how many wavelengths fit into the path length difference, divide the Δ_{pl} by λ_2:

$$\frac{\Delta_{pl}}{\lambda_2} = \frac{1400 \text{ nm}}{400 \text{ nm}} = 3.5$$

Ah ha! The path length difference is an odd number of half wavelengths. The result is destructive interference between the light reflected from the top surface of the soap and the light that enters the soap before being reflected back out.

(b) As suggested by Equation 23-20, the minimum thickness required for destructive interference is such that the light that enters the soap layer travels an extra distance, compared to the light reflected from the top surface of the soap, equal to one half of a wave cycle. That is, for a layer of thickness D_{min},

$$2D_{min} = \frac{\lambda_2}{2}$$

where λ_2 is the wavelength of the light while propagating in the soap layer. Using Equation 23-17, the equation becomes

$$D_{min} = \frac{n_1 \lambda_1}{4 n_2}$$

which is just Equation 23-21. So

$$D_{min} = \frac{560 \text{ nm}}{4(1.4)} = 100 \text{ nm}$$

REFLECT
When light of wavelength equal to 560 nm strikes the soap layer close to the normal, destructive interference occurs both when the layer is 100 nm and 700 nm thick. For these thicknesses, reflections from the surface would be minimized, and the surface would look dark. Destructive interference always occurs when twice the thickness of the layer is an odd multiple of one-half of the wavelength.

Practice Problem 23-7 Monochromatic light of wavelength 560 nm strikes a layer of cinnamon oil, which has an index of refraction of 1.60 and rests on a kitchen counter. (a) If the layer of cinnamon oil is 700 nm thick, does constructive interference, destructive interference, or partial interference occur when the light strikes the surface close to the normal? (b) What is the minimum thickness of cinnamon oil that would result in no (or minimal) reflection from the surface?

Take another look at the photo of a soap film in Figure 23-32. This film was made by dipping an open ring into soapy water, and then holding it vertically. The horizontal bands result from a combination of the mixture of wavelengths in the light incident on the film and the increasing thickness of the film from top to bottom.

Some of the light that strikes the soap film is reflected from the front surface, while some passes into the film before being reflected at the back surface. When two such light waves recombine, the wavelength of light (the color) that results in constructive interference appears bright. However, whether or not constructive interference occurs also depends on the difference in lengths of the paths traveled by each component of the incident light. This path length difference, which is approximately twice the thickness of the soap film, varies from the top to bottom of the film, because the soap flows downward under the pull of gravity as the ring is held upright. For this reason, as the thickness changes from the top to the bottom of the soap film, the brightest color you see does as well. The thickness of the film is relatively constant *across* the film, however, so the colors appear in bands. All wavelengths of visible light are present in the light illuminating the soap film, so that all colors appear in the light reflected from it.

Did you notice that the region at the very top of the soap film in Figure 23-32 appears dark? To understand why we need to go back to our discussion in Chapter 13 of how waves are reflected from a boundary. In Chapter 13 we saw that a wave pulse on a rope is inverted as it reflects from a fixed boundary. This inversion is equivalent to a phase shift of one-half of a wavelength; the position along the wave that arrived at the boundary as a peak has been reflected as a trough. In general, any wave is inverted when reflected, either partially or completely, from a boundary going from a material of higher wave speed to one of lower wave speed. (For the rope, the wave speed is zero on the other side of the boundary, which is definitely lower than the speed along the rope.) So when light traveling in air is partially reflected from the surface of the soap film, the reflected light wave is inverted. This does not happen, however, to the light that enters the soap film and is reflected from the back surface. At that boundary, light is moving from a medium of lower speed (soap) to one of higher speed (air), so no inversion occurs. So even if there were no path length difference between the light wave that reflects from the front surface and the one that reflects from the back surface, the two waves would be 180° out of phase because one is inverted on reflection but the other is not. This would result in destructive interference.

Is it possible that light reflecting from the front surface travels the same distance as light reflecting from the back surface? Yes, or at least approximately so, when the thickness of the film is very small compared to the wavelength of the light. This explains the dark band across the top of the soap film in Figure 23-32. As the soap flows down toward the lower part of the ring, the film at the top becomes increasingly thin. It is eventually so thin compared to the wavelengths of the incident light that destructive interference occurs for all incident wavelengths. The canceling out of the waves results in a surface that looks dark.

? Got the Concept 23-9
Inversion on Reflection

Figure 23-33 shows light shining on a piece of glass that has an index of refraction of 1.5 and is covered in a thin layer of oil that has an index of refraction of 1.4. As the light reflects off each boundary (air-to-oil, oil-to-glass, and glass-to-air), identify whether or not the reflected light undergoes an inversion.

Example 23-8 Reducing the Reflection

The glass in an LCD display is sometimes given a thin coating to reduce glare. Reflection of light incident on the display particularly light that strikes close to the normal to the plane of the display is minimized by the phenomenon of thin film interference. Zirconium acrylate, a plastic doped with zirconium, which has an index of refraction of about 1.54, is often used for this purpose. What is the minimum thickness of zirconium acrylate that will accomplish the desired reduction in reflection for light of wavelength λ_0 equal to 560 nm? Note that the index of refraction of zirconium acrylate is higher than that of glass.

SET UP

We want the light reflected from the surface of the coating to destructively interfere with the light reflected from the acrylate-to-glass boundary. This requires that when the two light waves recombine they are shifted by an odd number of half wavelengths.

You might be tempted to use Equation 23-21 to solve this problem, but that result doesn't hold here. Equation 23-21 was obtained for a thin layer with an opaque backing, unlike the zirconium acrylate layer backed by transparent glass in this problem. When the backing is opaque no phase shift occurs between the light wave that reflects from the top surface and the light wave that reflects from the bottom surface of the thin film. This is because both waves are inverted on reflection; we built this into Equation 23-21. In the case of zirconium acrylate on glass, however, while incident light is inverted as it reflects from the air-to-acrylate boundary,

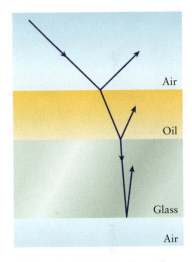

Figure 23-33 As the light reflects off each boundary (air–oil, oil–glass, and glass–air), identify whether the reflected light undergoes an inversion.

no inversion occurs at the acrylate-to-glass boundary because the index of refraction of glass is lower than that of acrylate. This introduces a 180°, or one-half of a wavelength, phase shift between the two light waves. The net phase shift between the two light waves therefore includes this one-half of a wavelength as well as the shift that occurs as a result of the path length difference between the two.

For destructive interference to occur, relative to the light reflected at the outer surface, the light that passes into the zirconium acrylate before coming back out must therefore differ in cycles by an integer multiple of a full cycle. The inversion at the acrylate-to-glass boundary gives a one-half cycle difference, so the minimum distance light must travel in the zirconium acrylate must be one full cycle, or one wavelength.

SOLVE

The path length difference for light incident close to the normal on the coating and glass is approximately twice the thickness of the coating. The minimum thickness D_{min} must therefore be one-half of a wavelength so that the path length difference is one full wavelength. Recall that the wavelength of the light changes when it crosses the boundary from one medium to another; from Equation 23-17 the wavelength $\lambda_{acrylate}$ of the light in the coating is

$$\lambda_{acrylate} = \frac{\lambda_0}{n_{acrylate}}$$

So then

$$D_{min} = \frac{1}{2}\lambda_{acrylate} = \frac{1}{2}\frac{\lambda_0}{n_{acrylate}}$$

The minimum thickness of the coating that results in minimizing the reflection of light of wavelength 560 nm is then

$$D_{min} = \frac{1}{2}\frac{560\ nm}{1.54} = 182\ nm$$

REFLECT

For most people, light sensitivity peaks in the range of 555 nm to 565 nm, which we perceive as yellow. Antireflective coatings are usually optimized for yellow light for that reason.

Practice Problem 23-8 A thin coating of zirconium acrylate which has an index of refraction equal to 1.54, is to be added to eyeglass lenses as an antireflective coating. What is the minimum thickness of zirconium acrylate that will minimize reflection of red light of wavelength 675 nm that strikes close to the normal to the plane of the lenses?

✱ What's Important 23-5

Like sound waves, light waves constructively or destructively interfere depending on how the peaks and troughs of two waves align when they combine. When a single beam of light strikes a thin layer of a transparent material, some of the light is immediately reflected while some enters the layer and is reflected off its inner surface. (When the light strikes a horizontal surface from above, this is the inner, lower surface.) Because the two light rays travel a different distance they may not be in phase when they recombine, resulting in interference. Whether destructive interference or constructive interference, or neither, occurs depends on the number of wave cycles that fit into the extra distance traveled by the light that enters the thin layer.

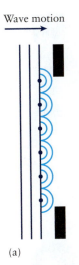

Wave motion →

We follow the front of a wave approaching an opening in an object. According to Huygen's principle we treat each point along the front as the source of a tiny "wavelet."

(a)

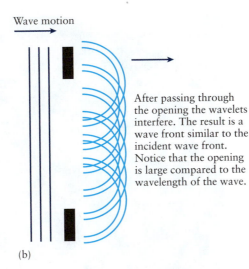

Wave motion →

After passing through the opening the wavelets interfere. The result is a wave front similar to the incident wave front. Notice that the opening is large compared to the wavelength of the wave.

(b)

Figure 23-34 (a) A wave front passes through an opening that is not small compared to the wavelength. Because so many wavelets along each wave front (only a three of which are shown as vertical lines) pass through, (b) their crests pile up along straight lines. This recreates the incident wave front, and the wave does not spread out as it passes through the opening.

23-6 Diffraction

Christiaan Huygens suggested that the points along the front of a wave be treated as many separate sources of tiny wavelets that themselves move at the speed of the wave (Section 23-1). Each wavelet propagates from its source much like the ripples spreading out from a pebble dropped in a pond, and the moving wave front itself is then the superposition of all of the wavelets. This phenomenon is Huygens' principle, shown in Figure 23-3. This approach to the propagation of waves led us to refraction and to Snell's law. Huygens' principle is also at the heart of **diffraction**, in which waves passing through a narrow opening or near the sharp edge of an object tend to spread out.

Consider the wave front passing through an opening in **Figure 23-34a**. The opening is not narrow in this case, where we define "narrow" to be approximately the same size, or smaller than, the wavelength of the wave. Three wave crests have been drawn as vertical lines to show the wavelength of the incident wave. A few of the large number of wavelets along the wave front are drawn, and many of them pass through the opening, as in **Figure 23-34b**. The wavelets interfere with each other, and because there are so many, at any modest distance past the opening, the crests pile up along straight lines. That is, the incident wave front is recreated due to the interference. So this wave does *not* spread out as it passes through the opening. Something rather different happens when the opening is on the same size as the wavelength, however.

Compared to the situation illustrated in Figure 23-34a, in **Figure 23-35a** the opening is much narrower, about one wavelength. (It's not necessary for the opening

Wave motion →

We follow the front of a wave approaching an opening in an object. According to Huygen's principle we treat each point along the front as the source of a tiny "wavelet."

(a)

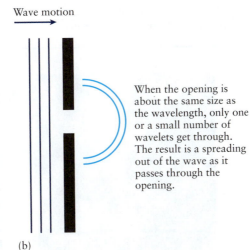

Wave motion →

When the opening is about the same size as the wavelength, only one or a small number of wavelets get through. The result is a spreading out of the wave as it passes through the opening.

(b)

Figure 23-35 (a) A wave front passes through an opening that is small compared to the wavelength. Because relatively few wavelets along each wave front pass through (only one is shown), (b) constructive interference cannot recreate the incident wave front. After passing through the narrow opening the wave spreads out; it has undergone diffraction.

Figure 23-36 Ocean waves diffract through the opening between two rocks. *(© 2012 Google Earth/ image © 2012 DigitalGlobe)*

to be exactly the same size as the wavelength, or smaller than the wavelength, or to bear any particular relationship to the wavelength other than the two being *about the same size*.) Because the opening is narrow, only a relatively few of the wavelets that make up each wave front pass through it; we drew only one in **Figure 23-35b** to exaggerate the process. Unlike in the case in which the opening is large, because so few wavelets get through, the constructive interference that occurs on the expanding wave crests cannot reproduce the straight wave front that was incident on the opening. After passing through the narrow opening, the wave spreads out; it has undergone diffraction. **Figure 23-36** shows ocean waves diffracting through the opening between two rocks. The separation between the rocks looks to be about 10 wavelengths, which is narrow enough to cause the waves to diffract (although the diffraction pattern is more obvious farther from the opening).

Water waves passing directly through a wide opening in **Figure 23-37a** show almost no effects of diffraction. In contrast, the wave spreads as it passes through a narrow opening in **Figure 23-37b**.

? Got the Concept 23-10
Hear through a Door?

The wavelengths of sound in human speech lie in the range of a few meters down to a few tenths of a meter. Why can you hear someone talking on the other side of an open door, even if you're not directly in front of it?

Figure 23-37 (a) Water waves passing directly through a wide opening show almost no effects of diffraction. (b) The same waves diffract through a narrow opening. *(George Resch/ Fundamental Photographs)*

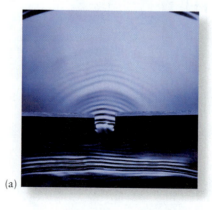

(a)

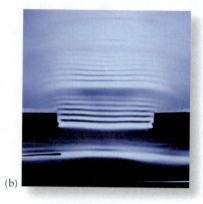

(b)

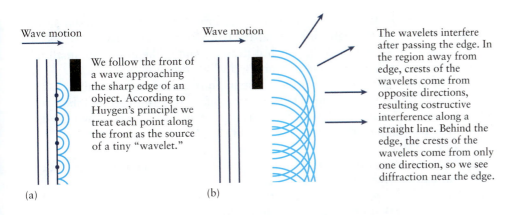

Wave motion →

We follow the front of a wave approaching the sharp edge of an object. According to Huygen's principle we treat each point along the front as the source of a tiny "wavelet."

(a)

Wave motion →

The wavelets interfere after passing the edge. In the region away from edge, crests of the wavelets come from opposite directions, resulting costructive interference along a straight line. Behind the edge, the crests of the wavelets come from only one direction, so we see diffraction near the edge.

(b)

Figure 23-38 (a) A wave moves past a sharp edge of an object. (b) In the open region away from the sharp edge (lower right), the crests of the wavelets come from opposite directions, resulting in constructive interference along a straight line. Behind the edge (upper right), the crests of the wavelets come from only one direction, so the wave spreads—diffracts—as it passes the edge.

When the wave in Figure 23-34b passes through the wide opening, the crests of the many wavelets that get through the opening constructively interfere to recreate the straight, incident wave front. We were careful, however, not to show this interference outside of the region generally directly behind the opening in our sketch, because even for a wide opening, the wave spreads at the edges of the opening. You can see this in the photograph of waves passing through a wide opening in Figure 23-37a. Near the edges of the opening the expanding wavelets come from only one direction, so constructive interference cannot reproduce a straight wave front. So even for a wide opening, spreading of the wave occurs near the edges. In the same way, diffraction occurs when a wave moves past a sharp edge of an object, as in **Figure 23-38a**. In the open region away from the sharp edge (in the lower part of **Figure 23-38b**), the crests of the wavelets come from opposite directions, resulting in constructive interference along a straight line. Behind the edge (in the upper part of Figure 23-38b), however, the crests of the wavelets come from only one direction, so the wave spreads—diffracts—as it passes the edge.

Diffraction at the edge of objects is a significant issue when making observations through a telescope. In **Figure 23-39**, you can see the pattern of concentric circles that forms due to diffraction of the star light around the circular edge of the

Figure 23-39 The pattern of concentric circles around each star is due to diffraction of light around the circular edge of the tube that forms the telescope. Light diffracting at the edges of the thin, crossed struts that support the mirror inside the telescope gives rise to the bright vertical and horizontal lines around each star. *(NASA, ESA and AURA/Caltech)*

Figure 23-40 An image of the stars Castor and Pollux taken using a small telescope. The overlapping diffraction rings limit the ability to make careful measurements of each star separately. *(Randy Culp)*

tube that forms the telescope. You can also see bright vertical and horizontal lines around each star as the light diffracts at the edges of the thin, crossed struts that support a mirror inside the tube. Such diffraction limits the ability of small telescopes to clearly observe objects that are optically close. For example, Castor and Pollux, the two brightest stars in the constellation Gemini, are optically close, so that when viewed through a small telescope the diffraction rings around them overlap. An image of Castor and Pollux taken using a small telescope is shown in **Figure 23-40**. The overlapping diffraction rings limit the ability to make careful measurements of each star separately.

Diffraction is a wave phenomenon. The diffraction of light, then, is evidence that light can be treated as a wave. Scientists did not always accept this as true, however. For example, when the 19th century French physicist Augustin-Jean Fresnel proposed that diffraction of light could be explained by treating light as a wave, Siméon-Denis Poisson, one of the leading mathematicians and physicists of the time, labeled the idea absurd. As proof, Poisson noted that due to constructive interference of light diffracted around the edges of a small object, the theory predicted that a bright spot would form directly in the center of the shadow cast by the object. Imagine Poisson's surprise when the experiment was done and a bright spot was indeed observed directly in the center of the shadow cast by a small disk! **Figure 23-41** shows the result of a similar experiment; the "Poisson spot" is clearly visible in the center of the shadow cast by a ball bearing supported at the end of a needle. Other diffraction effects can be seen in the image as well.

> **? Got the Concept 23-11**
> **Out, Darn Spot!**
>
> Figure 23-41 shows the bright spot that forms at the center of a shadow cast by a small disk. Draw a sketch that accounts for the formation of the Poisson spot due to the constructive interference of light waves diffracted at the edges of the disk.

Figure 23-41 A bright spot appears at the center of the shadow cast by a small disk as a result of the interference of light diffracting around the sharp edge of the disk. *(Omikron / Photo Researchers)*

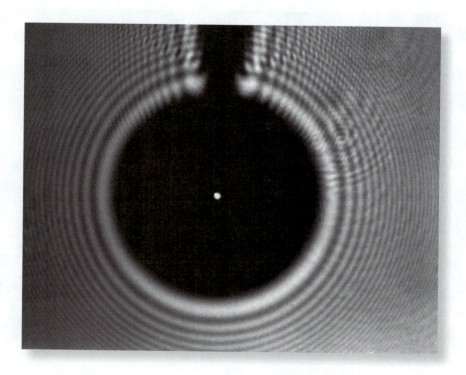

Light diffracts as it passes through an opening, and also at the edges of an object. To study the phenomenon quantitatively, we consider parallel rays of light of a single wavelength λ passing through a long, narrow opening or *slit*. The monochromatic light falls on a screen after passing through the slit. This setup is shown in **Figure 23-42**, in which we have labeled the width of the slit w and the distance between the slit and the screen L.

Our experience with the Poisson spot suggests that as light passes through the slit, it diffracts, forming an interference pattern on the screen. At positions where light rays interfere constructively the screen is light, and at positions where light rays interfere destructively the screen is dark. In other words, where rays from different regions of the slit reach a point on the screen so that one travels an integer number of wavelengths more than the other, we'll see a bright region. Where rays from different regions reach a point on the screen so that one travels an odd number of half waves more than the other we'll see a dark region due to destructive interference.

Light diffracting from the two edges of the slit travels the same distance to reach the point on the screen directly behind the center of the slit. In the same way that the Poisson spot forms directly behind a solid object, there must therefore be a bright region directly behind the center of the slit. Moving to either side of the central maximum along the screen we encounter darker and darker regions until we reach a point at which light diffracting from different regions of the slit interfere completely destructively. This is the location of a dark band or fringe on either side of the central maximum on the screen. Farther out from the dark fringe, where light again interferes constructively on the screen, we find another bright fringe—and then another dark fringe, and another bright fringe, and so on. A typical diffraction pattern arising from monochromatic light passing through a narrow slit is shown in **Figure 23-43**.

The first dark fringe is found where rays of light from the two sides of the slit interfere destructively. We can find the location by demanding that pairs of rays from the two sides reach the same location of the screen having traveled paths

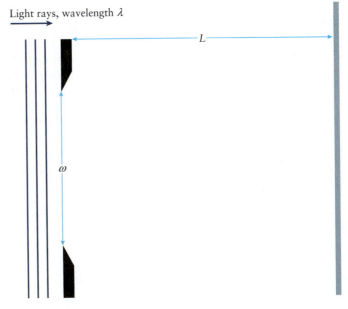

Light rays, wavelength λ

Figure 23-42 Light diffracts as it passes through an opening and also at the edges of an object. We consider parallel rays of light of a single wavelength λ passing through an opening of width w and falling on a screen a distance L from the opening.

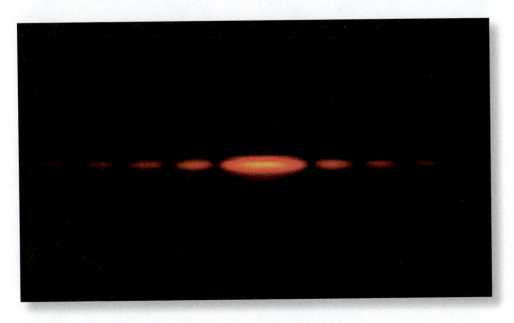

Figure 23-43 A diffraction pattern arises when monochromatic light passes through a narrow slit. *(Edward Kinsman/Science Photo Library)*

that are different in length by one-half of a wavelength. In **Figure 23-44a** we show one such pair of rays, defined by dividing the slit into two equal regions and taking the rays that begin at the upper edge of each section. The rays converge and interfere at a point on the screen an angle θ from the line that connects the center of the slit directly to the screen. (We'll refer to this as the central line.) Yes, angle θ is not quite the same for the two rays, but we can simplify the geometry of the slit, rays, and screen in Figure 23-44a by making the distance from the slit to the screen large compared to the width of the slot ($L \gg w$). In that case the angles each ray makes to the direction from slit to screen are approximately the same.

The two rays in Figure 23-44a travel different distances to arrive at the common point shown on the screen; we have identified the path length difference Δ_{pl} by drawing a line perpendicular to the longer ray and between the points where the two rays start. The point on the screen where two rays interfere is a minimum when Δ_{pl} equals $\lambda/2$:

$$\Delta_{pl} = \frac{\lambda}{2}$$

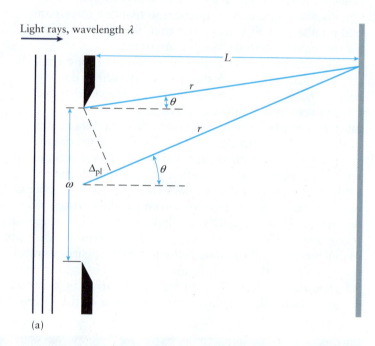

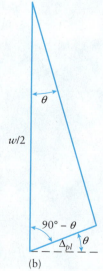

Figure 23-44 (a) Dark fringes are found where rays of light travel paths that differ in length by one-half of a wavelength and therefore interfere destructively. The path length difference Δ_{pl} is shown for a ray that comes from the center of the slit and one that comes from the upper edge. (b) The distance between the origin of the two light rays shown in (a) forms the hypotenuse of a right triangle.

Notice that half the opening of the slit is the hypotenuse of a triangle formed by Δ_{pl} and the perpendicular line we used to identify Δ_{pl}. This triangle is shown in **Figure 23-44b**. The angle opposite to the side of length Δ_{pl} is θ, so

$$\sin \theta = \frac{\Delta_{pl}}{w/2} = \frac{\lambda/2}{w/2}$$

The first dark fringe is therefore found at angle θ from the central line, where θ is defined by

$$\sin \theta = \frac{\lambda}{w} \qquad (23\text{-}22)$$

In addition, although we found this relationship by considering a specific pair of rays that interfere destructively when they reach the screen, for any ray that reaches the screen at that same location, we can always find another ray such that the two interfere destructively. Equation 23-22 is therefore a general equation for the position of the first dark fringe. Also, by repeating the analysis above we would find that all of the dark fringes are similarly defined by

$$\sin \theta = m\frac{\lambda}{w} \qquad (23\text{-}23)$$

where m is a counting integer, that is, m equals 1, 2, 3, and so on. Each bright fringe is found halfway between two successive dark fringes.

The locations of the dark and bright fringes in the interference pattern formed when light of wavelength λ passes through a slit of width w are determined by the ratio λ/w, as evident from Equation 23-23. For example, when the width of the slit is approximately the same as the length of one wave cycle, so that the ratio λ/w is approximately equal to 1, the angle at which the first dark fringe forms is close to $\sin^{-1}(1)$ or $90°$. Even when the screen is placed far from the slit, $90°$ from the center line would put the first dark fringe off the screen on either side. In effect, diffraction through a narrow slit results in light being maximally spread out, as suggested by the water wave diffracting through a narrow opening in Figure 23-37b.

What about when the width of the slit is large compared to the wavelength, that is, when the ratio λ/w is small? We expect little or no effect due to diffraction in this case, even though for a small value of λ/w Equation 23-23 offers many solutions for θ that are under $90°$. The reason there are *not* more and more fringes as the slit is widened is found by combining the mathematical description of the location of the dark fringes (Equation 23-23) with the intensity of the fringes as a function of position. Regardless of the value of λ/w, the central maximum is the brightest, and each fringe on either side is less and less bright as the angle (from the central line) of that fringe increases. The graphs in **Figure 23-45a, 23-45b**, and **23-45c** show the intensity of the bright fringes as a function of θ, the angle between the center of the fringe and the central line, for three values of λ/w. The wavelength is held constant, however, so the progression of the three curves shows how the brightness of the fringes changes as the slit is widened. Notice that although there are more fringes when the width of the slit is eight times the wavelength, compared to the number when the width is, say, four times the wavelength, the brightness of the fringes drops off rapidly away from the central maximum. In the limit when the width is much greater than the wavelength, the central maximum is so bright compared to all of the other fringes that we observe light essentially only passing directly through the slit. In other words, no effects of diffraction are observed.

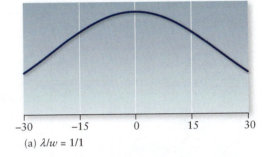

(a) $\lambda/w = 1/1$

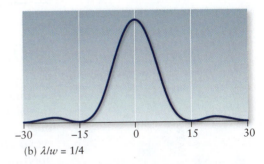

(b) $\lambda/w = 1/4$

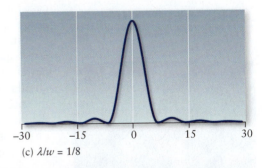

(c) $\lambda/w = 1/8$

Figure 23-45 The intensity of the bright fringes formed when monochromatic light passes through a narrow slit varies as the ratio of the wavelength of the light to the slit width. Intensity is shown for (a) λ/w equal to 1, (b) λ/w equal to 1/4, and (c) λ/w equal to 1/8.

? **Got the Concept 23-12**
Three Slits

Figure 23-46 shows the diffraction pattern that is created when red laser light passes through a narrow slit. The wavelength of the light is the same in all three images, but the width of the slit is different. Order the images from the widest slit to the narrowest one.

(a)   (b)   (c)

Figure 23-46 A diffraction pattern is created when red laser light passes through a narrow slit. The wavelength of the light is the same in all three images, but the width of the slit is different. Can you order the images (a), (b), and (c) in order from the widest slit to the narrowest one? *(Richard Megna/ Fundamental Photographs)*

✱ **What's Important 23-6**

Huygens' principle states that every point along the front of a wave can be treated as a separate source of tiny "wavelets" that themselves move at the speed of the wave. Waves passing through a narrow opening or near the sharp edge of an object tend to spread out, or diffract. Wavelets originating from different regions of an opening can constructively or destructively interfere with each other.

23-7 Circular Apertures

Perhaps without realizing it, you have observed light passing through circular openings on a regular basis—your pupils! Light diffracts through a circular aperture, similar to the way it diffracts through a rectangular slit. The difference, as you might expect, is that the resulting diffraction pattern will have a circular symmetry. As you can see in the pattern formed when red laser light passes through a circular aperture (Figure 23-47), the bright and dark fringes are concentric circles. In this case the laser serves as a distant, pointlike source, so the diffraction pattern is clean and regular. In this section, we consider how light diffracts through a circular opening.

We describe the location of the bright and dark fringes in a circular diffraction pattern as a function of angle θ. Here, as in the case of light passing through a rectangular slit, θ measures the angle between the line that passes through the center of the circular opening, perpendicular to the plane of the circle (the central axis), and the line connecting the center of a fringe to the center of the opening The definition of θ is shown in Figure 23-48. The bright fringes are shown in red and the dark fringes as white, the absence of red light.

As in the case of a rectangular slit, the location of the fringes when light diffracts through a circular aperture depends on the ratio of the wavelength λ of the light and the size of the opening. We characterize the size of a circular aperture by the diameter D of the opening; the location of the fringes depends on the ratio λ/D. In particular, the location of the center of the first dark fringe is given by

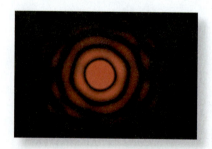

Figure 23-47 The interference of light diffracting through a circular aperture forms bright and dark fringes. *(sciencephotos / Alamy)*

$$\sin \theta = 1.22 \frac{\lambda}{D}$$

(23-24)

Compare this to the location of the first dark fringe formed by light passing through a rectangular slit (Equation 23-22); the only difference is the factor of 1.22, which results from the different geometry of a circular versus a rectangular opening.

What if the light from two distant, pointlike objects passes through a circular aperture? Each gives rise to a diffraction pattern like the one shown in Figure 23-47. The central part of the pattern is repeated in **Figure 23-49a** so that you can see how the brightness varies over the area of the central maximum. When the light from *two* objects passes through the aperture, the patterns may either partially or more completely overlap. In **Figure 23-49b**, the patterns of the two objects overlap but are distinct. In **Figure 23-49c**, however, the two objects are close enough optically so that their diffraction patterns overlap and it becomes impossible to tell them apart. Is Figure 23-49c an image of two point objects or one single, elongated object? We have run into the limit of **resolution**, that is, our ability to resolve, or optically distinguish, the two objects.

Two distant, pointlike objects observed through a circular aperture can be resolved when the central maximum of one coincides with the center of the first dark fringe of the other. This phenomenon is known as Rayleigh's criterion for resolvability, after the 19th century English physicist Lord Rayleigh (John William Strutt, 3rd Baron Rayleigh). The angle θ_R that separates two point objects that are just barely resolved through a circular aperture, known as the **angular resolution** of the aperture, is then

$$\sin \theta_R = 1.22 \frac{\lambda}{D} \qquad (23\text{-}25)$$

This comes directly from Equation 23-24, which describes the angular separation of the central maximum and the first dark fringe in the diffraction pattern of a single point object viewed through a circular aperture.

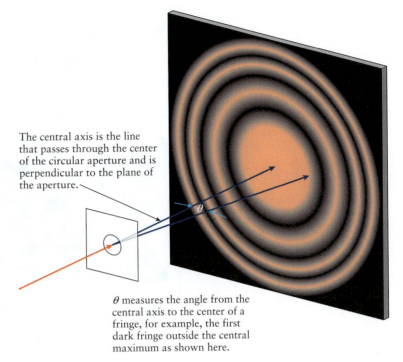

The central axis is the line that passes through the center of the circular aperture and is perpendicular to the plane of the aperture.

θ measures the angle from the central axis to the center of a fringe, for example, the first dark fringe outside the central maximum as shown here.

Figure 23-48 The location of the fringes in a circular diffraction pattern can be described as a function of angle θ.

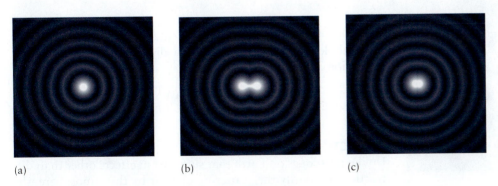

(a)   (b)   (c)

Figure 23-49 (a) Light from a single, distant point source gives rise to a diffraction pattern when it passes through a circular aperture. (b) The diffraction patterns formed as light from two distant point sources pass through a circular aperture partially overlap, but are distinct. (c) At the limit of our ability to resolve two distant point sources, the diffraction patterns formed as their light passes through a circular aperture overlap and are not distinct.

? Got the Concept 23-13
The Whirlpool Galaxy, Twice

The two images of the Whirlpool galaxy in Figure 23-50 were recorded using different telescopes. Which of the two telescopes has the wider aperture? What evidence can you give to support your answer?

Figure 23-50 Two images of the Whirlpool galaxy were recorded using different telescopes. Which of the two telescopes has the wider aperture? (NASA, ESA, S. Beckwith (STScI), and The Hubble Heritage Team (STScI/AURA), Todd Boroson (NOAO), AURA, NOAO, NSF)

Example 23-9 The Hubble Space Telescope

The Hubble Space Telescope has a circular aperture 2.4 m in diameter. What is the theoretical angular resolution of the Hubble Space Telescope for light of wavelength 500 nm?

SET UP

The angular resolution, the smallest angular separation of two objects that can be resolved as a result of diffraction effects, is given by

$$\sin \theta_R = 1.22 \frac{\lambda}{D} \qquad (23\text{-}25)$$

SOLVE

When light of wavelength 500 nm enters the 2.4-m aperture of the Hubble Space Telescope from a distant object, the angular resolution is therefore

$$\theta_R = \sin^{-1}\left(1.22 \frac{500 \times 10^{-9} \text{ m}}{2.4 \text{ m}}\right) = 1.5 \times 10^{-5} \text{ degree}$$

REFLECT

Unlike telescopes on Earth that suffer from effects other than diffraction, primarily the scattering of light by the atmosphere, the Hubble Space Telescope's resolution is essentially diffraction limited. As you can imagine, the ability to resolve two objects that are separated by 1.5×10^{-5} degrees makes it quite sensitive. Figure 23-51 shows an image of Pluto and its moon Charon taken by the Hubble Space Telescope. Pluto and Charon are separated by about 25×10^{-5} degrees in this image, so are easily resolved.

Figure 23-51 Pluto and its moon Charon are separated by about 25×10^{-5} degree in this image, so are easily resolved by the Hubble Space Telescope. The angular resolution of the Hubble Space Telescope is about 1.5×10^{-5} degree. (NASA, ESA, and STScI)

Practice Problem 23-9 The telescopes used in our astronomy laboratory have a circular aperture 15 cm in diameter. What is the theoretical angular resolution of one of these telescopes for light of wavelength 500 nm?

Estimate It! 23-2 Maximum Altitude

Probably the longest line-of-sight view you can get on or near Earth is from the window of a commercial airplane. Assuming that the resolution of your eyes is limited by the diffraction that occurs as light passes through your pupils, estimate the maximum altitude from which you identify cars on a highway below. In bright light the diameter of your pupils is probably about 3 mm.

SET UP
We can estimate the angular resolution of your eye from Rayleigh's criterion (Equation 23-25):

$$\sin \theta_R = 1.22 \frac{\lambda}{D}$$

using big, round, but reasonable values for D, the diameter of your pupil, and λ, the wavelength of visible light. D is on the order of 10^0 (1) mm, and we can take 500 nm, close to the peak of the visible spectrum, as λ to one significant digit.

To find the maximum altitude from which you identify cars on the ground, imagine that from the plane the car below forms a very small arc of a circle of radius equal to your height H above the ground. The length of the car L then forms the length of this arc, so recalling that the length of the arc of a circle of radius R and angular extent θ equals $R\theta$,

$$L = H\theta_R$$

or

$$H = \frac{L}{\theta_R}$$

To calculate H with this equation θ_R must be in radians. To get a reasonable estimate we take the length of a car to be on the order of 1 m.

SOLVE
The angular resolution of your eye is

$$\theta_R = \sin^{-1}\left(1.22 \frac{500 \times 10^{-9}\,\text{m}}{1 \times 10^{-3}\,\text{m}}\right) = 6 \times 10^{-4}\,\text{rad}$$

So

$$H = \frac{L}{\theta_R} = \frac{1\,\text{m}}{6 \times 10^{-4}} \approx 2 \times 10^3\,\text{m}$$

REFLECT
You might be able to see cars from about 2000 m up. But the resolution of the eye is not limited only by diffraction. The spacing of photoreceptors in the retina, chromatic and spherical aberrations, and the ability to maintain focus on a particular spot of the visual field also contribute to the resolving power of the eye. So 2000 m is probably somewhat of an overestimate.

> ✴ **What's Important 23-7**
> The location of the fringes of the diffraction pattern that occurs when light passes through a circular aperture depends on the ratio of the wavelength of the light and the size of the opening. If two objects are close enough optically so that their diffraction patterns overlap, it becomes impossible to resolve, or optically distinguish, one from the other.

Answers to Practice Problems

23-1 20°

23-2 34.4°

23-3 47.3°

23-4 3.78 mm

23-6 1.55

23-7 (a) constructive, (b) 87.5 nm

23-8 219 nm

23-9 2.3×10^{-4} degrees

Answers to Got the Concept Questions

23-1 The index of refraction of water is greater than that of air. According to Snell's law of refraction, the angle that the light makes to the normal in air is larger than the angle it makes as it approaches the boundary while still in water. Thus the light is bent farther away from the normal as it crosses the boundary.

23-2 The pond looks more shallow. In **Figure 23-52** you can follow a light ray that bounces off the bottom of the pond and ends up entering your eyes. Because the index of refraction of water is greater than that of air, the light exits the water at an angle from the normal larger than the angle at which it arrived at the water–air boundary. Your brain assumes that light travels in a straight line, so you *interpret* the origin of the ray on the floor of the pond to be along the line of the light ray as it enters your eyes. Your brain traces the light back along the dashed blue line. The distance you measure to the origin of the light ray is fixed by your stereo vision, so the pond floor appears to be higher than it really is.

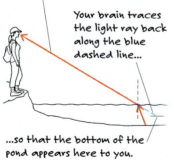

The red line represents a ray of light that bounces off the bottom of the pond and arrives at your eyes.

Your brain traces the light ray back along the blue dashed line...

...so that the bottom of the pond appears here to you.

Figure 23-52

23-3 Refraction, but in particular total internal reflection, is responsible. Because light travels more slowly in water than in air, when light from above enters the water it is refracted closer to the normal. Light that just barely skims along the surface before entering the water, in other words, light that enters the water at nearly 90° to the normal direction, is refracted at the critical angle for the water–air boundary. No light from above the surface can enter the water at an angle greater than 90° so, as you can see from **Figure 23-53**, the diver sees no light at angles greater than the critical angle from vertical.

Light from the sky is refracted toward the normal when it enters the water.

Light entering the water at 90° to the normal is refracted at the critical angle for the air-water boundary.

θ_C

Light cannot enter the water at angles greater than 90° from the normal, so no light reaches the camera at angles greater than the critical angle.

Figure 23-53

23-4 Air sits in layers above the road, each layer a bit cooler than the one below it. Light from the blue sky that is propagating down toward the road is refracted as it encounters the boundary between one layer of air and the next. Because the index of refraction of the layer of air closer to the road is lower the light is bent farther from the normal. The normal direction in these refractions is vertical, so the light is refracted closer to horizontal, as in the left part of **Figure 23-54**. Light that strikes the boundary between layers at a large angle with respect to the normal (a grazing angle as measured from the air layer boundary) can experience total internal reflection and reflect back up into the higher layer, as at the right of Figure 23-54. In either case, your eyes trace the light rays back along straight lines, so that the image is formed close to the road. What looks like a puddle is actually a refracted image of the blue sky.

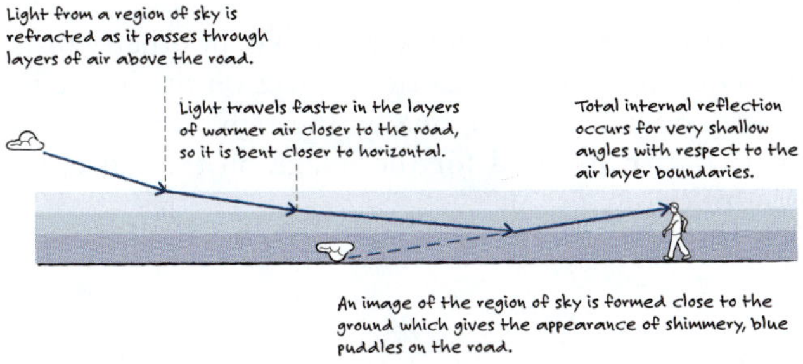

Light from a region of sky is refracted as it passes through layers of air above the road.

Light travels faster in the layers of warmer air closer to the road, so it is bent closer to horizontal.

Total internal reflection occurs for very shallow angles with respect to the air layer boundaries.

An image of the region of sky is formed close to the ground which gives the appearance of shimmery, blue puddles on the road.

Figure 23 -54

23-5   None. Once the light has passed through the first filter it will be completely polarized in the direction defined by the transmission direction of that filter. The second filter is set so that its transmission direction is perpendicular to the first. So the angle between that direction and the polarization direction of the beam that has already passed through the first filter is 90°. In Equation 23-10, then, the cosine term is cos 90°, which equals zero (**Figure 23-55**).

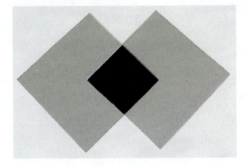

Two linear polarizing filters are placed one behind the other, so that their transmission directions are perpendicular to one another.

No light passes through the two filters.

Figure 23-55   *(Courtesy of David Tauck)*

23-6   Light passing through the first filter is polarized. However, because the transmission direction of the filter placed in between the first two is not perpendicular to the first one, some light will pass through. The light is polarized in the transmission direction of the in-between filter; because this is not perpendicular to the transmission direction of the final filter some of the light will pass it (**Figure 23-56**).

Two linear polarizing filters are placed one behind the other, so that their transmission directions are perpendicular to one another. A third filter is placed in between them, at an angle.

First filter

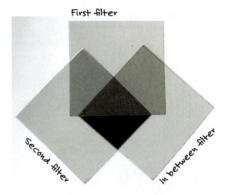

Second filter

In between filter

The polarization direction of light passing through the first filter is not perpendicular to the in between filter, so some light passes through both. The polarization of this light is not perpendicular to the second filter, so again, some light passes through!

Figure 23-56   *(Courtesy of David Tauck)*

23-7   The index of refraction of air is approximately equal to 1, which is lower than the index of refraction of a soapy film. In other words, light travels more slowly in the soapy film than in air. From Equation 23-17, the wavelength in the film is therefore shorter than in air. This phenomenon is true in general. The wavelength of a wave gets shorter when it passes through a medium in which light travels more slowly.

23-8   The thickness of a thin film determines the wavelength at which constructive interference occurs.

23-9   When light is reflected from a boundary going from a material of higher wave speed to one of lower wave speed, the wave is inverted. No inversion occurs when light is reflected from a boundary where the transition is from a medium of lower speed to one of higher speed. The higher a material's index of refraction, the lower the speed of light is in that material. Another way to state the rule is that a light wave is inverted when it is reflected from a boundary going from a material that has a lower index of refraction to one that has a higher index of refraction. So here, the speed of light is highest in air, lower in the oil, and lowest in glass. Light striking the air-to-oil boundary is therefore inverted on reflection, as is light striking the oil-to-glass boundary. In both cases, the boundary separates a material of higher light speed from one of lower light speed. No inversion occurs at the glass-to-air boundary, however, because the speed of light in glass is lower than the speed in air.

23-10   A doorway is on the order of 1 m wide, which is on the same size scale as the sound waves in human speech. The sound waves therefore diffract and spread as they pass through the opening of a door, so you don't have to be directly outside an open door to hear sounds produced inside a room.

23-11 The sketch is shown in **Figure 23-57**.

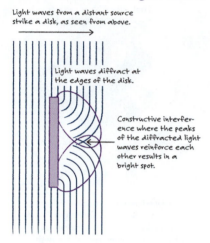

Light waves from a distant source strike a disk, as seen from above.

Light waves diffract at the edges of the disk.

Constructive interference where the peaks of the diffracted light waves reinforce each other results in a bright spot.

Figure 23-57

23-12 The narrower the slit, the more the diffracted light is spread out, which is most obvious by observing the width of the central maximum. The central maximum is widest in image a, so this is the narrowest slit. Of the other two images, c has the wider central maximum, so the slit width is narrower in image c than b. The order from the widest to the narrowest slit is b, c, a.

23-13 The image on the left was recorded with the telescope with the wider aperture. The structure of the galaxy is far better resolved in that image, and the nearby stars are far more pointlike. In other words, diffraction has a less deleterious effect on the image on the left compared to the one on the right which means that the aperture used for the image on the left is the wider of the two.

SUMMARY

Topic	Summary	Equation or Symbol
angular resolution	Angular resolution is the angle θ_R that separates two point objects that can just barely be distinguished as separate objects when viewed through an aperture. For a circular aperture, angular resolution depends on diameter of the aperture D and the wavelength λ of light that enters it.	$\sin \theta_R = 1.22 \dfrac{\lambda}{D}$ (23-25)
Brewster's angle	When light strikes the boundary between two light-transmitting media at Brewster's angle θ_B, the reflected light is completely linearly polarized. Brewster's angle depends on the indices of refraction of the two media, n_1 and n_2.	$\theta_B = \tan^{-1}\left(\dfrac{n_2}{n_1}\right)$ (23-12)
critical angle	When light strikes a boundary between a medium of index of refraction n_1 and a second medium of index of refraction n_2, and n_1 is greater than n_2, total internal reflection occurs when the incident angle is greater than the critical angle θ_c.	$\theta_c = \sin^{-1}\left(\dfrac{n_2}{n_1}\right)$ (23-8)
diffraction	When waves, such as light, pass through a narrow opening or pass near the sharp edge of an object, they tend to spread out, a phenomenon known as diffraction.	
dispersion	The index of refraction of a material or medium depends on the wavelength of light passing through it. For this reason, the angle at which light is refracted when passing from one medium to another also depends on wavelength, so light of different colors is bent at different angles during refraction. This dispersion results in a spreading out of different colors of light, in, say, a beam of white light. The amount of dispersion that occurs in a medium depends on how strongly the index of refraction varies with wavelength.	

Huygens' principle	Huygens' principle states that the front of a wave can be treated as many separate sources of tiny wavelets that themselves move at the speed of the wave.		
index of refraction	The index of refraction n is a measure of the speed of light v in a material relative to the speed of light in a vacuum.	$n = c/v$	(23-5)
linearly polarized	A light ray is linearly polarized when the directions of oscillation of the electric field components of all photons in the ray are aligned. The oscillation directions of the magnetic field components are also aligned when the ray is linearly polarized.		
polarization	Polarization is the level to which the oscillating electric field and magnetic field components of the photons in a beam are aligned.		
refraction	A wave such as light is refracted when it passes across a boundary between materials in which the wave speed is different. The refracted wave leaves the boundary at an angle that depends on the incident angle of the wave as well as the wave speed in the two media. Refraction of light follows Snell's law of refraction.		
resolution	When two objects are viewed through an aperture, such as the eye or a telescope, diffraction causes their images to overlap when the two are optically close. Resolution is the ability to distinguish two objects that are optically close together.		
Snell's law of refraction	When light strikes a boundary between a medium of index of refraction n_1 and a second medium of index of refraction n_2, the incident angle θ_1 and the outgoing, or refracted, angle θ_2 are related by Snell's law of refraction.	$n_2 \sin \theta_2 = n_1 \sin \theta_1$	(23-6)
thin film (interference)	When a beam of light strikes the surface of a transparent medium, some is reflected and some enters the medium, is refracted, reflected from the lower surface, and refracted again as it exits the medium. In this way the single beam is split into two, and the two beams interfere with each other when they recombine. Constructive interference occurs when the difference in the length of paths the two travel is an integer number of wavelengths. Destructive interference occurs when the difference in the length of the paths the two travel is an odd number of half wavelengths.		
total internal reflection	When light strikes a boundary between a medium of index of refraction n_1 and a second medium of index of refraction n_2, if n_1 is greater than n_2, and the incident angle is greater than the critical angle, all light is reflected back into the first medium.		

In a few problems, you are given more data than you actually need; in a few other problems, you are required to supply data from your general knowledge, outside sources, or informed estimate.

Interpret as significant all digits in numerical values that have trailing zeros and no decimal points.

For all problems, use $g = 9.8$ m/s² for the free-fall acceleration due to gravity. Neglect friction and air resistance unless instructed to do otherwise.

• Basic, single-concept problem

•• Intermediate-level problem, may require synthesis of concepts and multiple steps

••• Challenging problem

SSM *Solution is in Student Solutions Manual*

Conceptual Questions

1. •What is Huygens' principle and why is it necessary to understand Snell's law of refraction?

2. •Why do you expect the last color of the sunset to be on the red end of the visible spectrum?

3. •Explain why the Moon appears to change colors during a total lunar eclipse (when Earth's shadow completely blocks the light coming from the Sun). SSM

4. •Does the depth of a pool determine the critical angle that a light ray will have as it travels from the bottom of the pool and heads toward the air above the water? Explain your answer.

5. •Does the refraction of light make a swimming pool seem deeper or shallower? Explain your answer.

6. •Give two common uses of total internal reflection.

7. •In your own words, explain why the phenomenon of total internal reflection only occurs when light moves from a medium with larger index of refraction toward a medium with smaller index of refraction.

8. •Describe the physical interactions that take place when unpolarized light is passed through a polarizing filter. Be sure to describe the electric field of the light before and after the filter as well as the incident and transmitted intensities of the light source.

9. •Recently, researchers have created materials with a negative index of refraction. Explain what happens to the angle of refraction if light enters such a material from air. SSM

10. •Describe how polarized sunglasses work. Why do such sunglasses have *vertically* polarized lenses (as opposed to *horizontally* polarized lenses)?

11. •Give two or three examples of thin film interference.

12. •A thin layer of gasoline floating on water appears brightly colored in sunlight. From where do the colors come?

13. •Does the phenomenon of diffraction apply to wave sources other than light? Give an example if it does.

14. •Why was the discovery of diffraction in the lab slightly later than the theory that explains it?

15. •Sunlight striking a diamond throws rainbows of color in every direction. From where do the colors come? SSM

16. •Linearly polarized light is incident at Brewster's angle on the surface of an optical medium. What can be said about the refracted and reflected beams if the incident beam is polarized (a) parallel to the plane of incidence and (b) perpendicular to the plane of incidence?

Multiple-Choice Questions

17. •Which kind of wave can refract when crossing from one medium to another with the wave having a different speed in the two media?

    A. electromagnetic waves

    B. sound waves

    C. water waves

    D. electromagnetic, sound, and water waves

    E. only electromagnetic and sound waves

18. •When light enters a piece of glass from air with an angle of θ with respect to the normal to the boundary surface,

    A. it bends with an angle larger than θ with respect to the normal to the boundary surface.

    B. it bends with an angle smaller than θ with respect to the normal to the boundary surface.

    C. it does not bend.

    D. it bends with an angle equal to two times θ with respect to the normal to the boundary surface.

    E. it bends with an angle equal to one-half θ with respect to the normal to the boundary surface.

19. •Which phenomenon would cause monochromatic light to enter the prism and follow along the path as shown in **Figure 23-58**?

    A. reflection

    B. refraction

    C. interference

    D. diffraction

    E. polarization

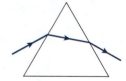

Figure 23-58 Problem 19

20. •Which color of light, red or blue, travels faster in crown glass?
- A. red
- B. blue
- C. their speeds are the same
- D. it depends on the index of refraction
- E. blue if the glass is thin

21. •Two linear polarizing filters are placed one behind the other, so that their transmission directions are parallel to one another. A beam of unpolarized light of intensity I_0 is directed at the two filters. What fraction of the light will pass through both filters?
- A. 0
- B. $\frac{1}{2} I_0$
- C. I_0
- D. $\frac{1}{4} I_0$
- E. $2 I_0$ SSM

22. •Two linear polarizing filters are placed one behind the other, so that their transmission directions form an angle of 45°. A beam of unpolarized light of intensity I_0 is directed at the two filters. What fraction of the light will pass through both?
- A. 0
- B. $\frac{1}{2} I_0$
- C. I_0
- D. $\frac{1}{4} I_0$
- E. $2I_0$

23. •A monochromatic light passes through a narrow slit and forms a diffraction pattern on a screen behind the slit. As the wavelength of the light decreases, the diffraction pattern
- A. shrinks with all the fringes getting narrower.
- B. spreads out with all the fringes getting wider.
- C. remains unchanged.
- D. spreads out with all the fringes getting alternately wider and then narrower.
- E. becomes dimmer.

24. •A monochromatic light passes through a narrow slit and forms a diffraction pattern on a screen behind the slit. As the slit width increases, the diffraction pattern
- A. shrinks with all the fringes getting narrower.
- B. spreads out with all the fringes getting wider.
- C. remains unchanged.
- D. spreads out with all the fringes getting alternately wider and then narrower.
- E. becomes dimmer.

25. •Figure 23-59 shows two single-slit diffraction patterns. The distance between the slit and the viewing screen is the same in both cases. Which

(a)

(b)

Figure 23-59 Problem 25

of the following is true about the width of the slits, w_a and w_b?
- A. $w_a > w_b$
- B. $w_a < w_b$
- C. $w_a = w_b$
- D. $w_a = \frac{1}{2} w_b$
- E. $w_a = \frac{1}{4} w_b$ SSM

Estimation/Numerical Analysis

26. •Estimate the speed of light in (a) air, (b) water, and (c) glass.

27. •Estimate the range of values for the index of refraction that you will use in this class.

28. •Estimate the time required for light to travel from (a) the Moon to Earth, (b) the Sun to Earth, and (c) Earth to Alpha Centauri (the nearest star to our solar system).

29. •Estimate the distance that light travels in (a) one second, (b) one minute, and (c) one year. SSM

30. •Estimate the range of wavelengths of visible light in water.

31. •A beam of light travels from medium 1 to medium 2. Medium 1 is found in quadrants 2 and 3; medium 2 is in quadrants 1 and 4. The beam touches each of the points in the x–y plane given in the table below. Calculate the ratio of the index of refraction of medium 2 to medium 1.

x (cm)	y (cm)	x (cm)	y (cm)
−4	−2.00	+1	+0.296
−3	−1.52	+2	+0.595
−2	−1.02	+3	+0.901
−1	−0.514	+4	+1.20
0	0		

Problems

23-1: Refraction

32. •The speed of light in a newly developed plastic is 1.97×10^8 m/s. Calculate the index of refraction.

33. ••The index of refraction for a vacuum is 1.00000. The index of refraction for air is 1.00029. Determine the ratio of time required for light to travel through 1000 m of air to the time required for light to travel through 1000 m of vacuum.

This page is intentionally left blank.

For complete end of chapter problem sets, please go to
www.whfreeman.com/kestentauck

fiber optic pipe made of acrylic ($n = 1.50$) (**Figure 23-62**)?

Figure 23-62 Problem 42

43. ••At what angle with respect to the vertical must a scuba diver look in order to see her friend standing on the very distant shore? Take the index of refraction of the water to be $n = 1.33$. SSM

44. ••A point source of light is 2.50 m below the surface of a pool. What is the volume of the "cone of light" that is formed in the water? Assume the water has an index of refraction of $n = 1.33$.

45. ••A block of glass that has an index of refraction of 1.55 is completely immersed in water ($n = 1.33$). What is the critical angle for light traveling from the glass to the water?

23-3: Dispersion

46. ••For a certain optical medium, the speed of light varies from a low value of 1.90×10^8 m/s for violet light to a high value of 2.00×10^8 m/s for red light. (a) Calculate the range of the index of refraction for the material for visible light. (b) A white light is incident on the medium from air, making an angle of 30° with the normal. Compare the angles of refraction for violet light and red light. (c) Repeat the previous part when the incident angle is 60°.

47. •••A beam of light shines on an equilateral glass prism at an angle of 45° to one face (**Figure 23-63a**). (a) What is the angle at which the light emerges from the opposite face given that $n_{glass} = 1.57$? (b) Now consider what happens when dispersion is involved (**Figure 23-63b**). Assume the incident ray of light spans the spectrum of visible light between 400 nm and 700 nm (violet to red, respectively). The index of refraction for violet light in the glass prism is 1.572 and it is 1.568 for red light. Find the distance along the right face of the prism between the points where the red light and violet light emerge back in to air.

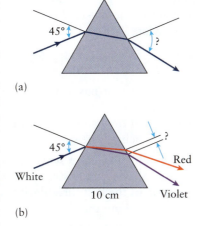

Figure 23-63 Problem 47

Assume the prism is 10 cm on a side and the incident ray hits the midpoint of the left face.

48. ••A light beam strikes a piece of glass with an incident angle of 45°. The beam contains two colors: 450 nm and an unknown wavelength. The index of refraction for the 450 nm light is 1.482. Determine the index of refraction for the unknown wavelength if the angle between the two refracted rays is 0.275°. Assume the glass is surrounded by air.

49. •••Blue light (500 nm) and yellow light (600 nm) are incident on a 12-cm-thick slab of glass as shown in **Figure 23-64**. In the glass, the index of refraction for the blue light is 1.545 and for the yellow light it is 1.523. What distance along the glass slab (side AB) separates the points at which the two rays emerge back into air? SSM

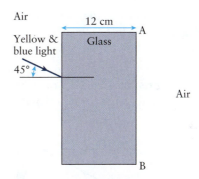

Figure 23-64 Problem 49

23-4: Polarization

50. •Unpolarized light is passed through an optical filter that is oriented in the vertical direction. If the incident intensity of the light is 78 W/m², what are the polarization and intensity of the light that emerges from the filter?

51. ••Vertically polarized light that has an intensity of 400 W/m² is incident on two polarizing filters. The first filter is oriented 30° from the vertical while the second filter is oriented 75° from the vertical. Predict the intensity and polarization of the light that emerges from the second filter.

52. ••**Calc** When unpolarized light passes through a filter, it emerges with one-half the intensity. Prove this statement by finding the average intensity over all angles (in other words, integrate $I = I_0 \cos^2 \theta$ over all angles and divide by 2π):

$$I_{average} = \frac{1}{2\pi} \int_0^{2\pi} I_0 \cos^2\theta \, d\theta$$

53. •What angle(s) does vertically polarized light make relative to a polarizing filter that diminishes the intensity of the light by 25%? SSM

54. ●●Light that passes through a series of three polarizing filters emerges from the third filter horizontally polarized with an intensity of 250 W/m². If the polarization angle between the filters increases by 25° from one filter to the next, find the intensity of the incident beam of light, assuming it is initially unpolarized.

55. ●The critical angle between two optical media is 60°. What is Brewster's angle at the same interface between the two media?

56. ●(a) What would Brewster's angle be for reflections off the surface of water when the light source is beneath the surface? (b) Compare that answer to the angle for total internal reflection when light starts in water and reflects off air.

57. ●What is Brewster's angle when light in air is reflected off a plastic surface ($n_{plastic} = 1.49$)?

58. ●What is Brewster's angle when light in water is reflected off a glass surface? Assume $n_{water} = 1.33$ and $n_{glass} = 1.55$.

59. ●●At what angle θ above the horizontal is the Sun when a person observing its rays reflected off water finds them linearly polarized along the horizontal (**Figure 23-65**)? SSM

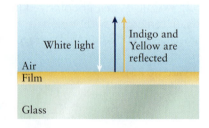

Figure 23-65 Problem 59

23-5: Thin Film Interference

60. ●●A ray of light is reflected from a thin film back into air. If the film is actually a coating on a slab of glass ($n_{film} < n_{glass}$), describe the phase changes that the reflected ray undergoes (a) as it reflects off the front surface of the film and (b) as it reflects off the back surface of the film (the film/glass interface).

61. ●What is the wavelength of red light (700 nm) when it is inside a glass slab with $n = 1.55$?

62. ●●When white light illuminates a thin film with normal incidence, it strongly reflects both indigo light (450 nm) and yellow light (600 nm) (**Figure 23-66**). Calculate the minimum thickness of

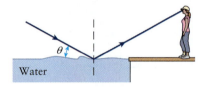

Figure 23-66 Problem 62

the film if it has an index of refraction of 1.28 and it sits atop a slab of glass that has $n = 1.50$.

63. ●●When white light illuminates a thin film normal to the surface, it strongly reflects both blue light (500 nm) and red light (700 nm) (**Figure 23-67**). Calculate the minimum thickness of the film if it has an index of refraction of 1.35 and it "floats" on water with $n = 1.33$. SSM

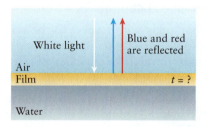

Figure 23-67 Problem 63

64. ●●A soap bubble is suspended in air. If the thickness of the soap is 625 nm and both blue light (500 nm) and red light (700 nm) are *not* observed to reflect from the soap film, what is the index of refraction of the thin film?

65. ●●A thin film of cooking oil ($n = 1.38$) is spread on a puddle of water ($n = 1.33$). What are the minimum and the next three thicknesses of the oil that will strongly reflect blue light having a wavelength in air of 518 nm?

66. ●●What is the minimum thickness of a nonreflective coating of magnesium chloride ($n = 1.39$) so that no light centered around 550 nm will reflect back off a glass lens ($n = 1.56$)?

67. ●●Water ($n = 1.33$) in a shallow pan is covered with a thin film of oil that is 450 nm thick and has an index of refraction of 1.45. What visible wavelengths will *not* be present in the reflected light when the pan is illuminated with white light and viewed from straight above? SSM

23-6: Diffraction

68. ●Light that has a wavelength of 550 nm is incident on a single slit that is 10 μm wide. Determine the angular location of the first three dark fringes that are formed on a screen behind the slit.

69. ●Light that has a wavelength of 475 nm is incident on a single slit that is 800 nm wide. Calculate the angular location of the first three dark fringes that are formed on a screen behind the slit.

70. ●What is the highest order dark fringe that is found in the diffraction pattern for light that has a wavelength of 633 nm and is incident on a single slit that is 1500 nm wide?

71. ●The highest order dark fringe found in a diffraction pattern is 6. Determine the wavelength of light that is used with the single slit that has a width of 3500 nm.

72. •When blue light ($\lambda = 500$ nm) is incident on a single slit, the central bright spot has a width of 8.75 cm. If the screen is 3.55 m distant from the slit, calculate the slit width.

73. ••A He–Ne laser illuminates a narrow, single slit that is 1850 nm wide. The first dark fringe is found at an angle of 20.0° from the central peak. Determine the wavelength of the light from the laser. **SSM**

74. ••Yellow light that has a wavelength of 625 nm produces a central maximum peak that is 24 cm wide on a screen that is 1.58 m from a single slit. Calculate the width of the slit.

23-7: Circular Apertures

75. •Light that is from a helium–neon laser and has a wavelength of 633 nm passes through a 0.18-mm-diameter hole and forms a diffraction pattern on a screen 2.0 m behind the hole. Calculate the diameter of the central maximum.

76. •**Biology** The average pupil is 5.0 mm in diameter and the average normal-sighted human eye is most sensitive at a wavelength of 555 nm. What is the eye's angular resolution in radians?

77. •**Astronomy** The telescope at Mount Palomar has an objective mirror that has a diameter of 508 cm. What is the angular limit of resolution for 560 nm light in degrees and radians?

78. •**Astronomy** The Hubble Space Telescope has a diameter of 2.4 m. What is the angular limit of resolution due to diffraction when a wavelength of 540 nm is viewed?

79. ••The distance from the center of a circular diffraction pattern to the first dark ring is 15,000 wavelengths on a screen that is 0.85 m away. What is the size of the aperture? **SSM**

80. ••**Biology** Assume your eye has an aperture diameter of 3.00 mm at night when bright headlights are pointed at it. At what distance can you see two headlights separated by 1.50 m as distinct? Assume a wavelength of 550 nm, near the middle of the visible spectrum.

General Problems

81. ••One way of describing the speed of light in an optical material is to specify the ratio of the time that is required for light to travel through vacuum to the time required for light to travel through the same length of the optical material. For example, if light travels through a material at 150% of the time for light to travel through a vacuum, the speed of light in the material would be 2/3 (1/1.5) that in a vacuum. Complete the table by giving the speed of light and the index of refraction for each of the following optical materials, listed with the corresponding percentage.

Optical material with percentage of time required for light to pass through compared to an equal length of vacuum	Speed of light	Index of refraction
100%		
125%		
150%		
200%		
500%		
1000%		

82. ••(a) Determine the index of refraction for medium 2 if the distance between points B and C in **Figure 23-68** is 0.75 cm. Assume the index of refraction in medium 1 is 1.00. (b) Suppose $n_2 = 1.55$. Calculate the distance between points B and C.

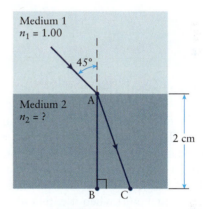

Figure 23-68 Problem 82

83. ••Prove that in the case where there are more than two optically different media sandwiched together, with air on the left and air on the right, the angle at which light returns to air is independent of the indices of refraction of the interior media (**Figure 23-69**). In other words, the refraction angle, θ_n, is *only* dependent on n_1, θ_1, and n_i (not $n_2, n_3, n_4, \ldots$).

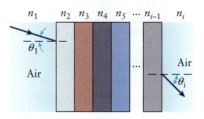

Figure 23-69 Problem 83

84. •••Calc Fermat's principle states that light travels in a path between two points so as to minimize the time of flight. Using this basic principle derive (a) the law of reflection for a single optical medium ($\theta_i = \theta_r$) and (b) Snell's law of refraction for two adjacent media ($n_1 \sin \theta_1 = n_2 \sin \theta_2$). *Hint:* Choose two points so that a ray between the two points can undergo reflection [for part (a)] or refraction [for part (b)]. Sketch in a path that connects the two points, calculate the time required for the light to make the trip, and then minimize the expression to find the optimal path.

85. ••The object in **Figure 23-70** is a depth $d = 0.850$ m below the surface of clear water. (a) How far from the end of the dock, distance D in the figure, must the object be if it cannot be seen from any point on the end of the dock? The index of refraction of water is 1.33. (b) If you could change the index of refraction of the water, how would you change it so that the object could be seen at any distance from the dock?

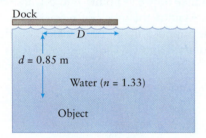

Figure 23-70 Problem 85

86. ••One of the world's largest aquaria is the Monterey Bay Aquarium. The viewing wall is made of acrylic and is 0.33 m thick, and the tank holds 1.2 million gallons of water (**Figure 23-71**). If a ray of light is directed into the plastic from air at an angle of 40°, calculate the angle that the ray will make (a) when it enters the plastic and (b) when it enters the seawater. The indices of refraction for air, acrylic and seawater are listed on the figure.

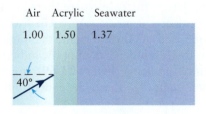

Figure 23-71 Problem 86

87. ••A flat glass surface ($n = 1.54$) has a layer of water ($n = 1.33$) of uniform thickness directly above the glass. At what minimum angle of incidence must light in the glass strike the glass–water interface for the light to be totally internally reflected at the water–air interface? **SSM**

88. ••Find the radius of the circle of light made when a small source of light at the bottom of a 3-m-deep pool of freshwater is directed straight up, as shown in **Figure 23-72**.

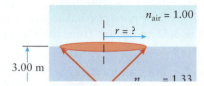

Figure 23-72 Problem 88

89. •••A baseball is hit into a round pool of water that is 4.00 m deep and 17.0 m across (**Figure 23-73**). It lands right in the center of the pool. A large round raft shaped like a lily pad is floating in the pool, concentrically on top of the location of the ball. What minimum diameter must the lily pad have in order to completely obscure the ball from sight? Assume that water has an index of refraction of 1.33.

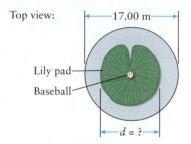

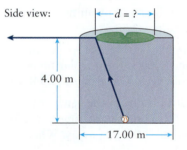

Figure 23-73 Problem 89

90. •••Light rays fall normally on the vertical surface of a glass prism ($n = 1.55$) as shown in **Figure 23-74**. (a) What is the largest value of θ such that the ray is totally internally reflected at the slanted face? (b) Repeat the calculation if the prism is immersed in water with $n = 1.33$.

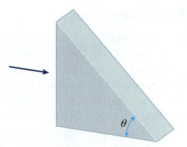

Figure 23-74 Problem 90

91. •••Light that has a wavelength of 500 nm is incident onto a glass slab at an angle of 35°, as shown in **Figure 23-75**. The index of refraction for this blue light in the glass slab is 1.521. A second ray of red light that has a wavelength of 680 nm is incident at the same angle of 35°. The index of refraction for red light in the glass slab is 1.514. Calculate the distance between the red ray and blue ray as they each emerge from the right side of the glass slab. The index of refraction for both red and blue light in air is 1.00. Start by labeling which color is on top and which color is on the bottom.

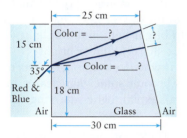

Figure 23-75 Problem 91

92. ••The polarizing angle for light that passes from water ($n = 1.33$) into a certain plastic is 61.4°. What is the critical angle for total internal reflection of the light passing from the plastic into air?

93. ••Unpolarized light of intensity 100 W/m² is incident on two ideal polarizing sheets that are placed with their transmission axes perpendicular to each other. An additional polarizing sheet is then placed between the two, with its transmission axis oriented at 30° to that of the first. (a) What is the intensity of the light passing through the stack of polarizing sheets? (b) What orientation of the middle sheet enables the three-sheet combination to transmit the greatest amount of light? **SSM**

94. ••A glass lens that has an index of refraction equal to 1.57 is coated with a thin layer of transparent material that has an index of refraction equal to 2.10. If white light strikes the lens at near-normal incidence, light of wavelengths 495 nm and 660 nm are absent from the reflected light. What is the thinnest possible layer of material for which this can be accomplished?

95. ••Unpolarized light that has an intensity of 850 W/m² is incident on a series of polarizing filters as shown in **Figure 23-76**. If the intensity of the light after the final filter is 75 W/m², what is the orientation of the second filter relative to the x axis? *Hint:* $\cos(90 - \theta) = \sin \theta$ and $2 \sin \theta \cos \theta = \sin 2\theta$.

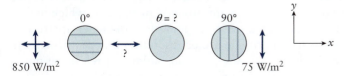

Figure 23-76 Problem 95

96. •••Find the ratio of the intensities of light that will emerge from a series of polarizing filters in **Figure 23-77a** and **Figure 23-77b** (compare I_a to I_b). Assume that, in both cases, the incident light is unpolarized and has an intensity of 100 W/m² to start.

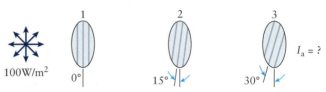

(a) The vertical polarizer (1) is followed by a second filter that is tilted at 15° from the vertical, and then by a third filter tilted at 30° from the vertical.

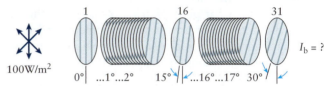

(b) The vertical polarizer (1) is followed by 30 polarizing filters, each one tilted 1° more than the previous one...the figure is meant to be illustrative only

Figure 23-77 Problem 96

97. ••A thin film of soap solution ($n = 1.33$) has air on either side and is illuminated normally with white light. Interference minima are visible in the reflected light only at wavelengths of 400, 480, and 600 nm. What is the minimum thickness of the film?

98. ••A thin brass sheet has a thin slit scratched in it. At room temperature (22.0 °C) a laser beam is shined on the slit and you observe that the first dark diffraction spot occurs at ±25.0° on either side of the central maximum. The brass sheet is then immersed in liquid nitrogen at 77.0 K until it reaches the same temperature as the nitrogen. It is removed and the same laser is shined on the slit. At what angle will the first dark spot now occur? Consult Table 14-1 for the thermal properties of brass.

99. ••A wedge-shaped air film is made by placing a small slip of paper between the edges of two thin plates of glass 12.5 cm long. Light of wavelength 600 nm is incident normally on the glass plates. If interference fringes with a spacing of 0.200 mm are observed along the plate, how thick is the paper? This form of interferometry is a very practical way of measuring small thicknesses. **SSM**

100. ••A thin layer of SiO, having an index of refraction of 1.45, is used as a coating on certain solar cells. The refractive index of the cell itself is 3.5. (a) What is the minimum thickness of the coating needed to cancel visible light of wavelength 400 nm in the light reflected from the top of the coating in air? Are any other visible wavelengths also canceled? (b) Suppose that technological limitations

require you to make the coating 3.0 times as thick as in part (a). Which, if any, visible wavelengths in the reflected light will be canceled and which, if any, will be reinforced?

101. ••You want to coat a pane of glass that has an index of refraction of 1.54 with a 155-nm-thick layer of material that has an index of refraction greater than that of the glass. The purpose of the coating is to cancel light reflected off the top of the film having a wavelength (in air) of 550 nm. The coating material is very expensive, so you want to use the thinnest possible layer. (a) What should be the index of refraction of the coating? (b) If, due to technological difficulties, you cannot achieve a uniform coating at the desired thickness, what are the thicknesses of the next three layers you could use?

102. ••Light of wavelength 500 nm falls at near normal incidence on two flat glass plates 8.00 cm long that are separated at one end by a wire with a diameter of 5 μm as shown in **Figure 23-78**. Calculate the spacing of the dark interference fringes along the length of the plates.

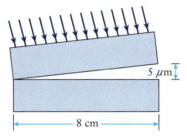

5 μm

8 cm

Figure 23-78 Problem 102

103. ••In 2009, researchers reported on evidence that a giant tsunami had hit the eastern coast of the Mediterranean Sea (present-day Lebanon and Israel) around 1600 BCE, causing huge damage to the civilizations located there. It is believed that the tsunami was caused by the eruption of the Thera volcano near the island of Crete. The waves would have passed through the 100-mile wide opening between Crete and Rhodes, which would cause them to diffract and spread out. Satellite observations of tsunamis show that the waves measure about 250 mi from a crest to the adjacent trough, and the time between successive crests is typically 60 min. (a) How fast do tsunami waves travel? (b) How long after the eruption of Thera would the tsunami reach the eastern shore of the Mediterranean Sea, 600 mi from Thera? (c) For these waves, could we apply the formula $w \sin \theta = m\lambda$ to find the angles at which the waves cancel after passing through the 100-mile "slit" between Crete and Rhodes? Explain why or why not. SSM

104. ••Two point sources of light that each have a wavelength of 500 nm are photographed from a distance of 100 m using a camera with a 50.0-mm focal length lens. The camera aperture is 1.05 cm in diameter. What is the minimum separation of the two sources if they are to be resolved in the photograph, assuming the resolution is diffraction limited?

105. ••If you peek through a 0.75-mm-diameter hole at an eye chart, you will notice a decrease in visual acuity. Calculate the angular limit of resolution if the wavelength is taken as 575 nm. Compare the result to a 4-mm pupil of the eye that has an angular resolution of 1.75×10^{-4} rad.

106. ••**Biology** The pupil of the eye is the circular opening through which light enters. Its diameter can vary from about 2.0 mm to about 8.0 mm to control the intensity of the light reaching the interior. (a) Calculate the angular resolution, θ_R, of the eye for light that has a wavelength of 550 nm in both bright light and dim light. In which light can you see more sharply, dim or bright? (b) You probably have noticed that when you squint, objects that were a bit blurry suddenly become somewhat clearer. In light of your results in part (a), explain why squinting helps you see an object more clearly.

107. ••**Biology** Under bright light, the pupil of the eye (the circular opening through which light enters) is typically 2.0 mm in diameter. The diameter of the eyeball is about 25 mm. Suppose you are viewing something with light of wavelength 500 nm. Ignore the effect of the lens and the vitreous humor in the eye. (a) At what angles (in radians and degrees) will the first three diffraction dark rings occur on either side of the central bright spot on the retina at the back of the eye? (b) Approximately how far (in millimeters) from the central bright spot would the dark rings in part (a) occur? (c) Explain why we do not actually observe such diffraction effects in our vision.

108. ••**Biology** The pupil (the opening through which light enters the lens) of a house cat's eye is round under low light but ciliary muscles narrow it to a thin vertical slit in very bright light. Assume that bright light of wavelength 550 nm is entering the eye parallel to the principal axis of the lens and that the pupil has narrowed to a slit that is 0.500 mm wide. What are the three smallest angles on either side to the central maximum at which no light will reach the cat's retina (a) if we assume the eyeball is filled with air, and (b) if we take into consideration that in reality the eyeball is filled with a fluid having index of refraction of approximately 1.4? (c) Would the cat be aware of the pattern of alternating dark and light fringes? Why?

109. ••**Astronomy** Under the best atmospheric conditions at the premium site for land-based observing (Mauna Kea, Hawaii, elevation ~4.27 km), an optical telescope can resolve celestial objects that are separated by one-fourth of a second of arc (arcsec). The viewing never gets any better than this because of atmospheric turbulence, which makes the images jitter. (a) What minimum diameter aperture is necessary to provide arcsec resolution due to diffraction? (b) Is there ever any point in building a telescope much bigger than this? Explain your answer. SSM

110. ••**Astronomy** The Herschel infrared telescope, launched in 2009, began observations in January of 2010. Its primary mirror is 3.5 m in diameter, and the telescope focuses infrared light in the range of 55 μm to 672 μm. (a) What wavelength in its observing range will give the maximum angular resolution? What is that maximum resolution (in radians and seconds, $1° = 60'$ and $1' = 60''$)? (b) To achieve the same resolution as in part (a) using visible light of wavelength 550 nm, what should be the diameter mirror of an optical telescope? (c) What is the smallest infrared source that the Herschel infrared telescope can resolve at a distance of 150 light-years?

111. ••**Astronomy** The world's largest refracting telescope is at Yerkes Observatory in Williams Bay, Wisconsin. Its objective is 1.02 m in diameter. Suppose you could mount the telescope on a spy satellite 200 km above the ground. (a) Assuming that the resolution is diffraction limited, what minimum separation of two objects on the ground could it resolve? Take 550 nm as a representative wavelength for visible light. (b) Because of atmospheric turbulence, objects on the surface of Earth can be distinguished only if their angular separation is at least 1.00 arcsec. How far apart would two objects on Earth's surface be if they subtended an angle of 1.00 arcsec? Compare this with your answer to part (a).

112. •••**Astronomy** Sometime around 2017, astronomers at the European Southern Observatory hope to begin using the E-ELT (European Extremely Large Telescope), which is planned to have a primary mirror 42 m in diameter. Let us assume that the light it focuses has a wavelength of 550 nm. (a) What is the most distant Jupiter-sized planet the telescope could resolve, assuming it operates at the diffraction limit? Express your answer in meters and light-years. (b) The nearest known exoplanets (planets beyond the solar system) are around 20 light-years away. What would have to be the minimum diameter of an optical telescope to resolve a Jupiter-sized planet at that distance using light of wavelength 550 nm? (1 light-year = 9.461×10^{15} m)

24 Geometrical Optics

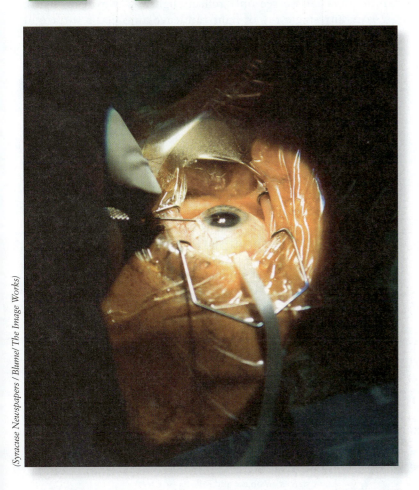

(Syracuse Newspapers / Blume/ The Image Works)

Two structures focus images on the back of our eye allowing us to see clearly: the cornea, which makes up the outer surface of the eye, and the lens, which is suspended inside the eye behind the pupil. Ophthalmologists correct some vision problems, such as nearsightedness, by surgically changing the shape of the cornea. As we age, however, many people develop cataracts as the normally clear lens gradually becomes cloudy. Cataracts eventually impair vision. The only treatment, shown here, is to surgically remove the damaged lens and replace it with an artificial one. More than 1.5 million cataract surgeries are performed in the United States annually.

You use optical devices every time you see something. We don't only mean glasses or contact lenses, although they certainly are optical devices. Your eye itself is an optical device. So is the mirror you use when you shave or apply makeup. The simplest **optical device** is a piece of either transparent or reflective material shaped so that it changes the direction of rays of light in a regular way.

A **mirror** is an optical device that changes the direction of rays of light as they reflect from its surface. A mirror can be made from polished metal or a piece of glass with a layer of shiny material on one surface, but biological mirrors help nocturnal animals see at night. A layer of cells behind the retina reflects light back, giving the light-sensitive cells in the retina a second chance to absorb and respond to the light.

A **lens** is an optical device that changes the direction of rays of light as they refract on entering and then exiting it. Both the cornea and the crystalline lens (just "lens") in your eyes are lenses.

In this chapter we will examine, both qualitatively and quantitatively, the properties of mirrors and lenses.

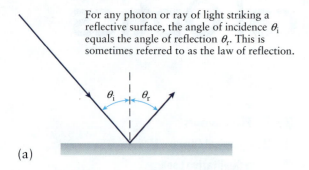

For any photon or ray of light striking a reflective surface, the angle of incidence θ_i equals the angle of reflection θ_r. This is sometimes referred to as the law of reflection.

(a)

Light that reflects from an object's surface in many, random directions is called diffuse or scattered.

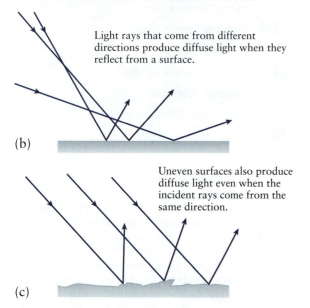

Light rays that come from different directions produce diffuse light when they reflect from a surface.

(b)

Uneven surfaces also produce diffuse light even when the incident rays come from the same direction.

(c)

Figure 24-1 (a) The angle of incidence equals the angle of reflection. So when light strikes a surface from different directions (b) or when light strikes an uneven surface (c) the reflected light is diffuse.

24-1 Plane Mirrors

Our visual system only detects objects that emit or reflect light. We can't see in the dark! Although it's easy to find examples of luminous things, such as lightbulbs, the Sun, or a firefly, most of what we see only reflects light. As shown in **Figure 24-1a**, any single ray of light can leave an object's surface in only one specific direction, such that the angle of incidence θ_i equals the angle of reflection θ_r. This is sometimes referred to as the **law of reflection**. However, reflected light is usually diffuse or scattered, because it reflects from an object's surface in many, random directions. This occurs mainly for two reasons: light shining on a surface generally comes from a range of directions (**Figure 24-1b**) and most surfaces are uneven (**Figure 24-1c**).

The surface of some objects is smooth enough that light rays do not scatter as they are reflected. For example, when light rays hit a flat, **plane mirror**, they tend to be reflected all in the same general direction. A plane mirror changes the direction in which light is traveling in a uniform way. When you look at a mirror you see the light scattered from an object only indirectly; we say that an image has formed as a result of the mirror changing the direction of the scattered light rays.

To study how mirrors (and lenses) form an image, physicists rely on the way our brains interpret information generated when light is absorbed by our eyes. First consider **Figure 24-2**, in which light shining on a small ball is scattered in many directions. No optical device is present, but some of the scattered light, represented in the figure as single light rays, goes directly

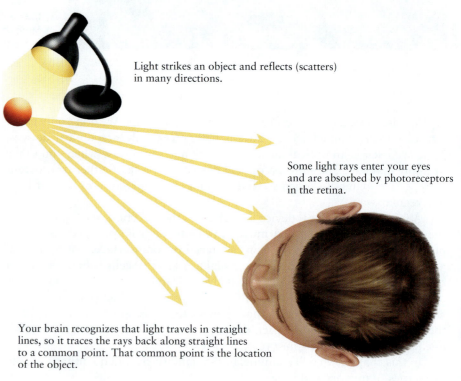

Light strikes an object and reflects (scatters) in many directions.

Some light rays enter your eyes and are absorbed by photoreceptors in the retina.

Figure 24-2 When light strikes our eyes, we trace the rays of light back along straight lines to a common source.

Your brain recognizes that light travels in straight lines, so it traces the rays back along straight lines to a common point. That common point is the location of the object.

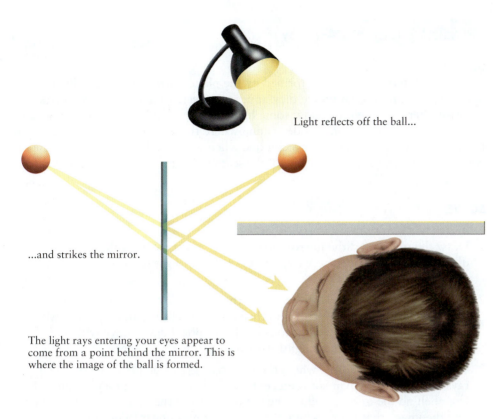

Light reflects off the ball...

...and strikes the mirror.

The light rays entering your eyes appear to come from a point behind the mirror. This is where the image of the ball is formed.

Figure 24-3 When light strikes our eyes after reflecting from a mirror, we trace the rays of light back along straight lines to an apparent common source.

to the observer's eyes. The brain traces those rays backward to a common origin, and interprets that point as the location of the ball.

Now imagine the configuration in **Figure 24-3**. Here a wall between the ball and the observer prevents light scattered by the ball from traveling directly to the observer's eyes. Some of the reflected light strikes a plane mirror, however, and is reflected toward the observer. As indicated by the faint yellow lines, those rays appear to be coming from a point behind the mirror. Because the light rays appear to originate at that common point, the observer sees the image back behind the mirror.

As you can see in Figure 24-3, no light rays actually meet where the image forms. For this reason, the image formed by a plane mirror is said to be a **virtual image**. We will shortly encounter some optical devices that cause light rays to bend toward each other, so that the image forms where light rays do actually meet. An image formed by light rays coming together is called a **real image**.

The ray trace diagram in **Figure 24-4** shows the distance from the arrow to the mirror is the **object distance** d_O; similarly, the **image distance** d_I is the distance of the image from the mirror. The geometry of the light rays requires that the object distance equals the image distance. In addition, the **object height** h_O and the **image height** h_I are also the same, as shown by the dashed gray line in Figure 24-4. Had we traced a light ray from the top of the arrow directly toward the mirror, it would be reflected directly back toward the tip of the arrow. The extension of that ray, traced back behind the mirror, would follow the dashed gray line behind the mirror, leaving no doubt that the height of the image equals the height of the object.

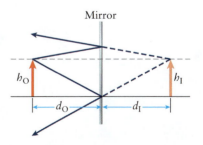

Figure 24-4 A ray trace diagram for an object placed in front of a plane mirror shows that the image forms behind the mirror.

Physicist's Toolbox 24-1   Ray Tracing

SET UP

Our brains interpret the image formed by the plane mirror by tracing the reflected light rays back to a point where they appear to meet. To find the image formed by a mirror, then, we need to follow some light rays scattered from an object as they are reflected by the mirror, and determine where those rays either meet or appear to meet. We will apply a similar technique when we study lenses. To determine the common origin, or apparent origin, of the scattered light, we need to follow at least two light rays.

SOLVE

To create a ray trace diagram,

1. Draw the object and the mirror. It is not important to make the dimensions and distances exact, but it is helpful to get the scale approximately correct.

2. Draw two rays that are scattered from the same point on the object, and follow them first to the mirror and then after reflection. Although any two rays will do, as we consider various kinds of optical devices, we will note some rays that are more straightforward to follow than others.

3. Find the image that forms where the reflected rays meet. Or, if the reflected rays diverge rather than converge, extend the rays back as if they originated from behind the mirror. Follow those extensions to the point where they cross; find the image that forms where the reflected rays *appear* to meet.

REFLECT

Figure 24-4 is an example of a ray trace diagram. An arrow, which is often used to represent a generic object in ray trace diagrams, is placed in front of a plane mirror. We trace two randomly chosen rays coming from the top of the arrow as they first hit and then reflect off the mirror. The rays appear to come from a point behind the mirror, which is where the image forms.

 Watch Out

The image formed by a plane mirror appears *behind* the mirror.

The image does not form "on" a plane mirror but, rather, behind it. This is evident from the ray trace diagrams in Figure 24-4. But if this diagram doesn't convince you, you can prove it to yourself by taping a bit of paper to a plane mirror, then looking at your reflection in it from about 1 m away. You'll find it difficult, likely impossible, to focus your eyes on both your image in the mirror and the piece of paper at the same time. That's because your image in the mirror is twice as far from your eyes as the paper on the surface of the mirror. The image is behind the mirror, not on it.

The light reflected from a ball is diffuse; in other words, the ball scatters light in many, random directions. As suggested by **Figure 24-5**, this means that the reflection of the ball can be seen in the mirror from many vantage points.

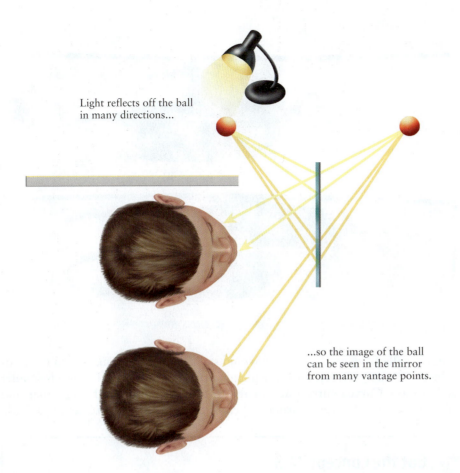

Light reflects off the ball in many directions...

...so the image of the ball can be seen in the mirror from many vantage points.

Figure 24-5 Light reflected from an object such as a ball is diffuse, that is, it scatters in many, random directions. For this reason the reflection of the ball in a mirror can be seen from many vantage points.

? Got the Concept 24-1
See You, See Me I

You look into a mirror hanging near the corner of a hallway and see the eyes of someone standing around the corner. Is she able to see you?

? Got the Concept 24-2
See You, See Me II

You look into a mirror hanging near the corner of a hallway, and see the right hand of someone standing around the corner but not his eyes. Is he able to see you?

Next, let's consider the size of the image compared to that of the object. We define the **lateral magnification m** as the ratio of the image height to the object height:

$$m = \frac{h_I}{h_O} \tag{24-1}$$

(In this text, we will often refer to m simply as the **magnification**.) For a plane mirror, h_O equals h_I, so the magnification is equal to 1. For an optical device that results in magnification greater than 1, the image is larger than the object. Such a device is

Figure 24-6 Curved mirrors can form an image both larger than and smaller than an object placed in front of it. *(Courtesy of David Tauck)*

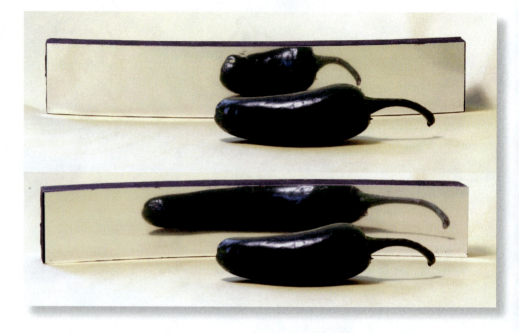

commonly called a *magnifier* and we would say that it forms a magnified image of the object. When *m* is less than 1, the image formed by the optical device is smaller than the object. Curved mirrors are capable of forming images both larger than and smaller than an object placed in front of them. **Figure 24-6** shows an example of both.

? Got the Concept 24-3
A Thick Mirror

Imagine a mirror made by adding a layer of shiny metal to the back surface of a thick plate of glass. Is the image distance equal to the object distance for this mirror? Is the image height equal to the object height? Use the ray tracing technique to find the answer.

✱ What's Important 24-1

The law of reflection says that for any photon or ray of light hitting a reflective surface, the angle of incidence θ_i equals the angle of reflection θ_r. When light rays hit a flat, plane mirror they tend to all be reflected in the same general direction. A real image forms where reflected light rays come together, but no light rays actually meet where a virtual image forms.

24-2 Spherical Concave Mirrors, a Qualitative Look

The eyes of the bobcat shown on the opening page of Chapter 23 appear to glow. This striking feature is the result of the layer of cells at the back of eye called the *tapetum lucidum* (Section 23-5). In addition to its properties as a thin film, the tapetum lucidum acts as a mirror at the back of the eye. Because the photoreceptor cells of the retina are positioned in front of the tapetum lucidum, they have two chances to absorb each photon. The result is improved low-light vision. Many nocturnal animals have a tapetum lucidum. Cats, for example, have the same visual acuity as humans in light levels about 15% of what we require; because of their tapetum lucidum, cat eyes reflect about 130 times more light than human eyes. The key to the enhanced low-light sensitivity is that the tapetum lucidum reflects light generally back in the direction from which it originated.

The tapetum lucidum can be considered a retroreflector, an optical device that reflects light back toward its source. A retroreflector's special property results from its having more than one reflective surface or a reflective surface which is curved. (A plane mirror by itself is not a retroreflector because it reflects light back to the source only when the light strikes the mirror in a direction normal to the surface.) A simple retroreflector, shown in **Figure 24-7**, consists of two plane mirrors set at a 90° angle from one another. Over a wide range of directions from which light can enter such an optical device, light is reflected back in the direction from which the ray of light enters. (For a proof of this claim, see Example 24-1.) Although there are only plane mirrors in the retroreflector shown in the figure, there is more than one reflective surface. A curved mirror is effectively an optical device constructed from many tiny plane mirrors each at a slight angle to the ones adjacent to it. In this and the next few sections we will consider the optical properties of curved reflective surfaces.

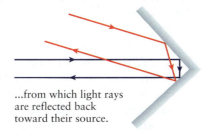

Two plane mirrors set at 90° from each other form a retroreflector...

...from which light rays are reflected back toward their source.

Figure 24-7 A simple retroreflector consists of two plane mirrors set at a 90° angle from one another. Light that enters the device is reflected back in the opposite direction.

▶ *Go to Interactive Exercise 24-1 for more practice dealing with mirrors*

Example 24-1 Retroreflector

Light enters a pair of plane mirrors arranged at 90° from each other (a simple retroreflector) at an arbitrary angle θ_1, as in **Figure 24-8**. Show that the light is reflected back toward its source.

SET UP

The key to most problems that involve light interacting with optical devices is to carefully trace one or more rays of light from its source until it leaves the system.

SOLVE

In the first part of **Figure 24-9**, the incoming ray reflects off the lower mirror. The angle of incidence θ_1 equals the angle of reflection θ_2. In the second part of the figure, the reflected ray strikes the second mirror. Because the second mirror is parallel to the normal direction of the first mirror, the angle labeled θ_3 equals θ_2.

The light ray reflects off the second mirror in the third part of Figure 24-9. As is always the case, the angle of incidence equals the angle of reflection. We normally measure these angles from the normal, but the statement is also true when the angles are measured from the plane of the mirror. Therefore θ_4 equals θ_3. Working back through the three parts of the figure, θ_3 equals θ_2, and θ_2 equals θ_1. So θ_1 equals θ_4. Because both θ_1 and θ_4 were measured from the same direction (vertical in the figure), we conclude that the light ray leaves the retroreflector in the same direction at which it entered.

REFLECT

Reflection was the critical aspect of this problem!

Practice Problem 24-1 Light enters a pair of plane mirrors arranged so that their reflective surfaces are parallel and face each other. Show that when the light strikes one mirror at an angle θ from the normal, it hits that mirror at the same angle θ on subsequent strikes.

Figure 24-8

A light ray enters a simple retroreflector at an arbitrary angle θ_1.

The incoming ray reflects off the lower mirror. The angle of incidence θ_1 equals the angle of reflection θ_2.

The angle θ_3 at which the reflected ray strikes the second mirror, as measured from the mirror (not the normal) equals θ_2.

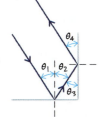

The light ray reflects off the second mirror. Because the angle of incidence equals the angle of reflection, the outgoing angle θ_4 equals θ_3.

Angles θ_4 and θ_3 are equal, θ_3 equals θ_2, and θ_2 equals θ_1. So θ_1 equals θ_4, and both are measured from the same direction. The light ray leaves the retroreflector in the same direction at which it entered.

Figure 24-9 These three sketches show a light ray entering a retroreflector, and then (in sequence) reflecting off one mirror, then the other mirror, and finally exiting the retroreflector in a direction toward the source of the light ray.

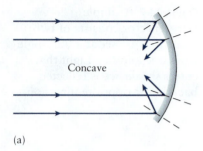

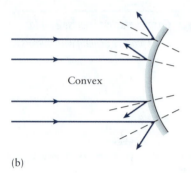

(a)

(b)

Figure 24-10 (a) A concave mirror causes parallel light rays to converge, while (b) a convex mirror causes parallel light rays to diverge.

We extend the application of more than one planar reflective surface with a spherical mirror, a reflective surface in the shape of a section of the surface of a sphere. We imagine such a curved surface as an assemblage of an infinite number of infinitesimally small plane mirrors, each at a slight angle with respect to the ones adjacent to it. Because the orientation of each of these small segments of a spherical mirror is different from its neighbors, for a given angle of incident light, the angle at which the light is reflected depends on where on the mirror the ray strikes.

Two spherical mirrors are shown in **Figures 24-10a** and **24-10b**. If you imagine a spherical mirror as a section cut from a complete sphere, then the reflective side can be either facing the inside or the outside of the sphere. A mirror is called **concave** when the inner surface of the spherical section is reflective, as in Figure 24-10a. The reflective surface of a **convex mirror** is the outer surface of the spherical section, as shown in Figure 24-10b.

Notice that when parallel light rays strike a concave mirror the rays converge, and when parallel light rays strike a convex mirror they diverge. It is important to consider parallel light rays as we examine various types of mirrors and lenses because light rays that come from distant objects are either parallel or nearly so. The Sun's rays, for example, are essentially parallel when they strike Earth. This is evident from the crisp shadows cast by an object placed in the Sun, such as the hanging picture frame in **Figure 24-11a**. In contrast, the shadow of the frame cast by a desk lamp in **Figure 24-11b** is fuzzy because the light rays from the light bulb are not parallel.

Figure 24-6 shows the same jalapeño as reflected from a concave and a convex mirror. The convex mirror must be the one on top in the photograph; because light rays spread out after reflection from a convex mirror, our eyes trace those rays

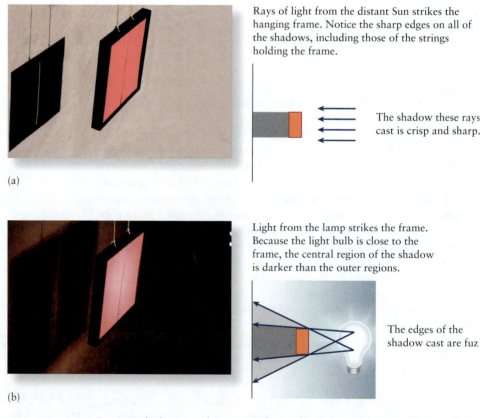

Rays of light from the distant Sun strikes the hanging frame. Notice the sharp edges on all of the shadows, including those of the strings holding the frame.

The shadow these rays cast is crisp and sharp.

(a)

Light from the lamp strikes the frame. Because the light bulb is close to the frame, the central region of the shadow is darker than the outer regions.

The edges of the shadow cast are fuz

(b)

Figure 24-11 (a) Because light rays that come from the Sun are essentially parallel, they cast a crisp, sharp shadow. (b) That is not the case for light from a source close to an object. *(Courtesy of David Tauck)*

back to a region smaller than the size of the object. The photo on the bottom of the figure shows that the concave mirror produces a magnified image, although we'll uncover circumstances in which the image formed by a concave mirror is smaller than the object placed in front of it. We will concentrate on concave mirrors in this and the next section, and explore the properties of convex mirrors in the two sections that follow.

We start by considering what happens when a series of parallel light rays strike a concave mirror formed from a fairly large section of a sphere, as shown in **Figure 24-12**. The incoming rays are parallel to each other and also parallel to the **principal axis** of the mirror, the axis that runs through the center of the sphere and also the center of the mirror. The law of reflection tells us that the angles from the normal to the mirror of the incident and reflected rays are equal at the point where each ray strikes. Notice that while the reflected rays generally converge along the principal axis of the mirror, they do not converge to the exact same point. This characteristic of spherical mirrors is referred to as *spherical aberration*. Only those incoming rays that are relatively close to the axis reflect through a point close enough together for us to say that they are focused on a point. Therefore, we limit the reflective surface of the mirrors we consider to a relatively small section of a sphere. Here "relatively small" means that the size of the reflective surface is small compared to the radius of the sphere. There is no specific cutoff value; rather, the smaller the mirror compared to the radius, the more tightly focused the reflected rays will be. For the rest of this chapter we will deal with mirrors small enough that we will treat them as if all rays parallel to the principal axis are focused to a single point.

Figure 24-13 defines many of the variables we use to describe the physics of a concave, spherical mirror. The point labeled F, the **focal point** of the mirror, is the point along the principal axis at which incident rays parallel to the principal axis converge when they reflect off the mirror. We might also say that parallel rays come to a common focus at the focal point, or that parallel rays are focused at the point. Although the mirror is only a small section of a full sphere, the point C corresponds to the center of that sphere, and the distance from C to any point on the mirror is r, the radius of the sphere. Point C is often referred to as the **center of curvature** of the mirror. The distance from the focal point to the center of the mirror is f, the **focal length**. In the case of a concave mirror small enough that all rays parallel to the principal axis are focused at the focal point, the focal length is half of the radius.

$$f = \frac{r}{2} \qquad (24\text{-}2)$$

To see how a concave mirror forms an image, let's put our standard arrow at a point far from the mirror, as in **Figure 24-14a**. "Far" in this case means that the object distance d_O is greater than the radius r, in other words, the base of the arrow is farther from the mirror, along the principal axis, than the center of the sphere C. Now let's trace two rays of light from the tip of the arrow, as they strike and then are reflected from the mirror. The image of the arrow's tip forms where those two rays intersect. Although any two light rays that originate at the tip of the arrow and that strike the mirror will work, we chose the two used in Figure 24-14 that are particularly convenient. In **Figure 24-14b**, we trace a ray that starts parallel to the principal axis because all light rays parallel to the principal axis are reflected through the focal point. In **Figure 24-14c** we trace a ray that strikes the center of the mirror. Because the normal to the surface at this point lies along the principal axis, it is easy to apply the law of reflection; the incident and reflected rays are symmetric around the axis of the mirror.

Where does the image of the base of the arrow form? The base of the arrow was placed on the principal axis, which is by definition the normal to the mirror's surface at the center of the mirror. So a light ray coming from the base of the arrow is reflected straight back from the center of the mirror, which means the image of the base of the arrow forms along the principal axis. The image of the arrow, then,

A concave mirror causes parallel light rays to nearly converge along the axis of the mirror. When only rays close to the axis are considered, or if the mirror is only a small arc of the complete sphere, we consider the rays of light to be focused, that is, to converge to a single point.

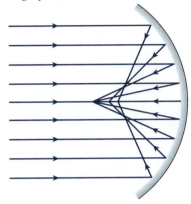

Figure 24-12 Only light rays that are parallel to the principal axis of a concave mirror, and that strike close to its center are reflected so that they converge at a single point.

Describing a spherical, concave mirror

r is the radius of the full sphere f is the focal length

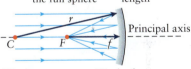

Principal axis

C is the center of the full sphere F is the focus, or focal point of the mirror

Figure 24-13 This drawing defines many of the variables we use to describe the physics of a concave, spherical mirror.

An arrow is placed in front of a spherical, concave mirror. We will trace two light rays from the tip of the arrow to the mirror and as they reflect from it.

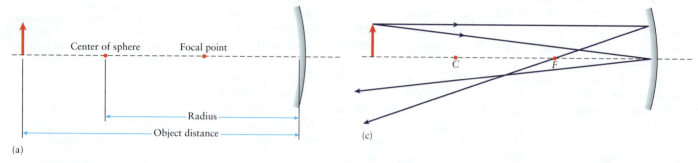

(a)

The angle of incidence of a ray on the mirror always equals the angle of reflection. This is easy to draw for this light ray.

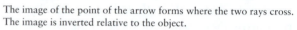

(c)

First, a light ray goes from the tip of the arrow to the mirror, parallel to the principal axis, and is reflected so that it passes through the focal point.

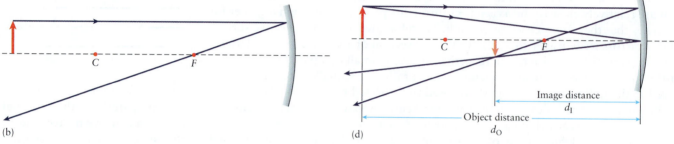

(b)

The image of the point of the arrow forms where the two rays cross. The image is inverted relative to the object.

(d)

Figure 24-14 A sequence of ray trace images shows how an image is formed by a concave mirror. (a) We set up the diagram by placing an upright arrow in front of the mirror. Two rays are traced from the tip of the arrow, (b) one that starts parallel to the principal axis of the mirror and (c) one that strikes the mirror at its center. (d) The image of the arrow's tip forms where the two rays converge.

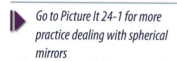

Go to Picture It 24-1 for more practice dealing with spherical mirrors

is as shown in **Figure 24-14d**. It is smaller than the object and its orientation is upside down relative to the object, that is, it is an **inverted image**. Finally, the image is real, that is, it forms in front of the mirror where light rays reflected from different parts of the mirror's surface meet. The object distance d_O and the image distance d_I are shown in the figure.

We now move the object closer and closer to the mirror and in each case consider where the image is formed, whether it is real or virtual and inverted or not. **Figure 24-15a** shows the standard arrow object placed at the center of curvature. We trace the same two light rays as in the previous example; now the two rays meet so that the image forms farther from the mirror, as in **Figure 24-15b**. The image is the same size as the object, and as in the previous case, is both real and inverted. Notice that when the object is placed at the center of curvature, the image forms at the center of curvature.

In **Figure 24-16** the arrow has been moved closer still to the mirror, so that it sits between the center of curvature and the focal point. The two light rays we considered in the previous two cases are traced again. The image has moved farther still from the mirror, with the net result that it is now larger than the object. This image is both real and inverted.

The image is clearly getting larger and larger as we move the object closer to the mirror. The limiting case occurs when the object is placed at the focal point where, after reflection, light rays that come from the tip of the arrow are parallel, as shown in **Figure 24-17**. Because the reflected rays are parallel they never meet, so no image forms when the object is placed at the focal point. When the object is slightly farther from the mirror than the focal point, the image formed by the mirror is nearly infinitely large and nearly infinitely far from the mirror.

What happens when the object is placed between the focal point and the mirror as shown in **Figure 24-18a**? Notice that the reflected rays never actually meet,

An arrow is placed at the center of curvature of a spherical, concave mirror.

(a)

We trace two light rays from the tip of the arrow in order to find the location of the image.

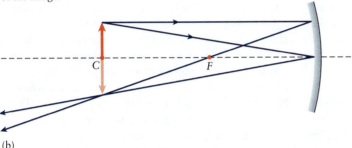

(b)

Figure 24-15 (a) A standard arrow is placed at the center of curvature of a concave mirror. (b) The image of the arrow's tip forms where the two rays we trace converge.

An arrow is placed in between the focal point of a spherical, concave mirror and the mirror itself.

(a)

Two light rays are traced from the tip of the arrow in order to find the location of the image. The two reflected rays appear to meet behind the mirror. The image that forms is upright and virtual.

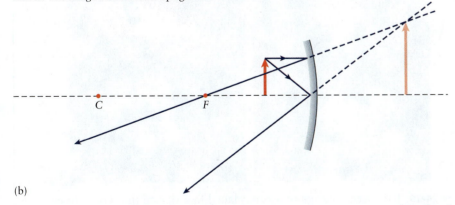

(b)

Figure 24-18 When an arrow is placed between a concave mirror and its focal point the reflected rays never actually meet, but instead they appear to meet at a point behind the mirror. This forms an upright, virtual image that is larger than the arrow.

Two light rays are traced from the tip of the arrow in order to find the location of the image.

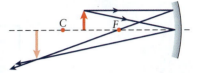

Figure 24-16 An arrow has been placed between the center of curvature and the focal point of a concave mirror. The image is both real and inverted, and is larger than the arrow.

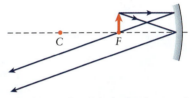

An arrow is placed at the focal point of a spherical, concave mirror. The two reflected rays are parallel, so they never meet and no image is formed.

Figure 24-17 An arrow has been placed at the focal point of a concave mirror. The reflected rays are parallel, so they never meet. No image forms when the arrow is placed at the focal point.

but rather appear to meet at a point behind the mirror, as in **Figure 24-18b**. Like the image formed by the plane mirror of Section 24-1, this image is therefore virtual. The image is not inverted, that is, it is an **upright image**, and it is larger than the object.

! Watch Out

An image can really form in front of a concave mirror.

You are probably comfortable with the idea that the reflection you see in a mirror can appear to be behind it. That's the case for a plane mirror, the type of mirror with which most of us have the most experience. But what about the image that forms when an object is placed farther from a spherical, concave mirror than the focal point? We say the image forms in front of the mirror. Is that just random terminology or is the image somehow floating in front of the reflective surface? The best way to answer is to look at a reflection in a concave mirror, but if you don't have one handy, look at the photos in **Figure 24-19**. Here we have placed a gold ring inside an enclosure that consists of two concave mirrors. The top one has a hole in its center to allow light to go in and come out. The reflection of the ring appears to float at the level of the top surface of the enclosure in part (a). The position of this image is so convincing that you might be tempted to reach out and touch it. But notice in part (b) that no shadow is cast when we shine a light directly across the top of the enclosure, convincing evidence that it is only the image, not the ring itself, which we see there. Part (c) shows the ring sitting on a circular scrap of paper at the bottom of one of the concave mirrors. The other mirror rests on its side in the background. Part (d) shows the entire enclosure. Although the ring remains in exactly the same position as it was in part (c), notice that the front of the ring shows in part (d) but not in part (c). Nature's engineer is a virtual image that casts no shadow!

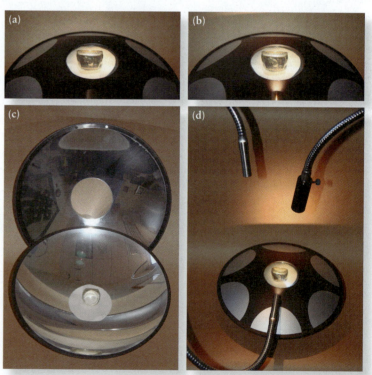

Figure 24-19 The virtual image of a ring placed in a device that consists of two, facing concave mirrors appears to float above it. *(Courtesy of David Tauck)*

To summarize, when the object is placed outside the focal point, the image is large, real, and inverted. As the object is moved closer and closer to the focal point, the image is larger and larger and forms farther and farther from the mirror. And from Figure 24-18b we see that when the object is placed inside the focal point, the image, while still large, is now virtual and upright. Something peculiar clearly happens as an object is moved from beyond the focal point to within the focal point.

When the object is placed exactly at the focal point, we can't even ask whether the image is real or virtual, or inverted or upright, because the light rays from any point on the object neither meet nor appear to meet after being reflected by the mirror. Although not strictly correct, we can imagine that the image is infinitely large and infinitely far away from the mirror, and either real and inverted or virtual and upright. The transition from one to the other is shown in the sequence of ray tracing diagrams shown on the left side of **Figure 24-20**. The sequences on the right show the image of a physiologist's face reflected in the mirror at each corresponding position relative to the mirror.

> **? Got the Concept 24-4**
> **A Soup Spoon**
>
> A shiny spoon is not so different in shape from a spherical mirror, concave on one side and convex on the other. Your reflection from the inside of the spoon, when held at arm's length, is upside down and appears to float in front of the spoon. When you hold the spoon about 6 cm from your eye you see only a blur reflected in the spoon, but when you hold it about 4 cm from your eye your reflection is right side up and appears to be behind the spoon. Explain these phenomena. About where is the focal length of the spoon?

> **✳ What's Important 24-2**
> Parallel rays come to a common focus at the focal point of a concave mirror, a mirror created when the inner surface of a spherical section is reflective. When an object is placed farther from a concave mirror than the focal point, the image is real and inverted. As an object is moved closer and closer to the focal point, the image becomes larger and larger, and forms farther and farther from the mirror. When an object is placed inside the focal point, the image is virtual and upright.

24-3 Spherical Concave Mirrors, a Quantitative Look

In the previous section, we examined the focusing properties of a spherical, concave mirror from a qualitative perspective, answering questions such as "Is the image real or virtual?" and "Is the image inverted or upright?" We now turn our attention to the mathematical details of reflections formed by a concave mirror. Here we will consider questions such as exactly how far from the mirror the image forms and how much magnification the mirror provides.

Knowing the location of an object in front of a mirror, we often want to determine the position of its image. Our first goal is to find a relationship for the distance of the image from the mirror d_I in terms of the distance of the object from the mirror d_O and the radius of the mirror r. To do so, we place a point object O at

Figure 24-20 This sequence of ray trace diagrams shows how the image formed by a concave mirror changes as the person walks from beyond the focal point to a location closer to the mirror than the focal point. *(Courtesy of David Tauck)*

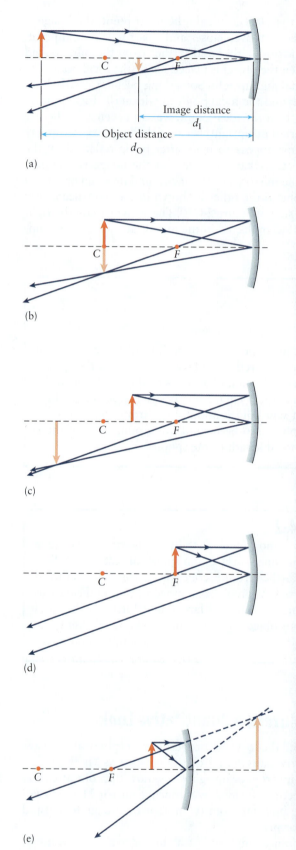

(a)

(b)

(c)

(d)

(e)

distance d_O along the principal axis from the reflective surface and use ray tracing to determine the location of the image. A point object is a good choice, because all light rays emanating from the object come from a single point. In addition, because the principal axis is normal to the surface at the center of the mirror, a ray of light coming from the object along the axis is reflected directly back, so the image forms on the principal axis. This setup is shown in **Figure 24-21**. Also shown is an arbitrarily chosen ray of light coming from the object and reflecting off the mirror. The image I forms where this reflected ray intercepts the principal axis, at image distance d_I from the mirror.

Recall that we have chosen to consider only spherical concave mirrors that represent a tiny piece of the surface of a sphere. To make the drawing in Figure 24-21 to scale would require the mirror to be drawn as a straight vertical line, with the center of curvature C far off the left side of the page. Such a figure wouldn't fit in this book, so instead, the ray trace diagram is not drawn to scale. So while the dashed line drawn from C to the point where the light ray reflects represents the normal to the mirror's surface at that point, it does not appear to be perpendicular in the drawing.

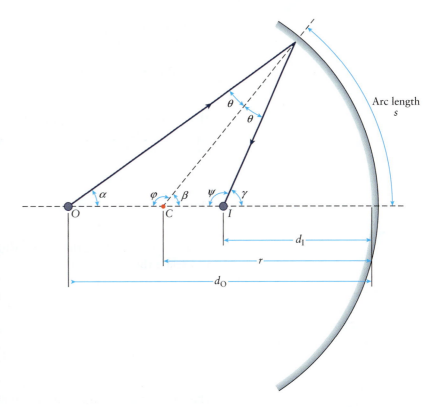

Figure 24-21 The variables we use to study a concave mirror are shown. The object is placed at point O, a distance d_O along the principal axis from the reflective surface. The image forms at point I a distance d_I from the mirror. Point C is the center of curvature of the mirror.

Our goal is to find a relationship for d_I in terms of d_O and r. To do so, we will make use of the angles and triangles formed by combinations of the dashed line, the principal axis, and the incident and reflected light rays in Figure 24-21. The angle between the dashed line and the principal axis is labeled β. The angle of incidence equals the angle of reflection, both labeled θ. We have also labeled angles α and γ, between the principal axis and the incident and reflected light rays, respectively, and φ and ψ, the angles supplementary to β and γ, respectively. Supplementary angles add to 180°.

We need to find a relationship between the angles α, β, and γ. Because we are only considering mirrors composed of a very small section of a full sphere, we can simplify the problem by recognizing that in this approximation α, β, and γ are all small angles. For that reason we can treat each of the three regions formed by one of these angles and the arc length s of the mirror as a sector (a pielike slice) of a circle. What's the same? The arc length is common to all three sectors. This allows us to relate the image distance, the radius of curvature of the mirror, and the object distance. The length of the arc of a circle of radius r subtended by an angle w equals rw (see Figure 3-32), so in Figure 24-21

$$s = d_O\alpha; \quad s = r\beta; \quad s = d_I\gamma \tag{24-3}$$

We can relate these by noting the relationship between the angles. The sum of the angles in any triangle equals 180°, so, for example,

$$\alpha + \theta + \varphi = 180°$$

However, φ and β are supplementary, so

$$\beta + \varphi = 180°$$

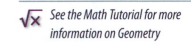

√x See the Math Tutorial for more information on Geometry

and therefore

$$\alpha + \theta = \beta \qquad (24\text{-}4)$$

Using the same method we find that

$$\alpha + 2\theta = \gamma \qquad (24\text{-}5)$$

The relationships in Equation 24-3 are in terms of α, β, and γ, so we eliminate θ in Equations 24-4 and 24-5. Solving Equation 24-4 for θ and substituting the result into Equation 24-5 gives

$$\alpha + 2(\beta - \alpha) = \gamma$$

or

$$\alpha + \gamma = 2\beta \qquad (24\text{-}6)$$

We now solve each of the relationships in Equation 24-3 for the respective angles and relate the distances through Equation 24-6, yielding

$$\frac{s}{d_O} + \frac{s}{d_I} = 2\frac{s}{r}$$

To simplify we divide by s and find that

$$\frac{1}{d_O} + \frac{1}{d_I} = \frac{2}{r}$$

The focal length is one-half the radius of the mirror (Equation 24-2), so this relationship can also be written as

$$\frac{1}{d_O} + \frac{1}{d_I} = \frac{1}{f} \qquad (24\text{-}7)$$

This expression, sometimes called the **mirror equation**, provides a relationship between the focal length of the concave mirror and the locations of the object and image. Most often, though, we want to determine the image position given the position of the object, so first we rearrange Equation 24-7 to put d_I in terms of d_O and f:

$$\frac{1}{d_I} = \frac{1}{f} - \frac{1}{d_O} = \frac{d_O - f}{d_O f}$$

and then solve for d_I:

$$d_I = \frac{d_O f}{d_O - f} \qquad (24\text{-}8)$$

Both d_I and d_O are allowed to be either positive or negative. The sign convention is to take positive values to mark positions on the reflective side of the mirror and negative values behind the mirror. The object distance d_O is always positive because we won't get a reflection if the object is placed behind the mirror. The image distance, however, can be either positive or negative, depending on whether the image is real (positive, in front of the mirror) or virtual (negative, behind the mirror). We explore these two cases below, along with the special case in which an object is placed at the focal point of a concave mirror.

In the previous section, we found that the image of an object placed farther from a concave mirror than the focal point forms in front of the mirror and therefore is real. In contrast, the image of an object placed inside the focal point is virtual because it forms behind the mirror. When the object is placed exactly at the focal point no image forms and we can consider the image to be infinitely far from the mirror. Each of these cases is shown in Figure 24-20. What does Equation 24-8 predict in these cases?

Object placed beyond the focal point (d_O greater than f). In this case, both d_O and f are positive. For values of d_O greater than f, the denominator in Equation 24-8 is positive, so d_I is then also positive. The image therefore forms on the reflective side of the mirror. The reflected light rays can cross on this side of the mirror, so this is a real image. The first three cases in Figure 24-20 (Figures 24-20a, 24-20b, and 24-20c) show an object placed beyond the focal point of a concave mirror.

Object placed inside the focal point (d_O less than f). Here, both d_O and f are positive. For values of d_O less than f, however, the denominator in Equation 24-8 is negative, so d_I is also negative. The image therefore forms behind the mirror. Reflected light rays cannot actually cross on this side of the mirror, but can only appear to cross. This is a virtual image. Figure 24-20e shows an object placed inside the focal point of a concave mirror.

Object placed at the focal point (d_O equals f). When d_O equals f, the denominator in Equation 24-8 equals zero, resulting in an infinitely large value for d_I. We consider this to suggest that the image is infinitely far from the mirror. (Strictly speaking, however, because the reflected light rays neither cross nor appear to cross when an object is placed at the focal point of a spherical, concave mirror, no image forms.) Figure 24-20d shows an object placed at the focal point of a concave mirror.

We conclude that predictions using Equation 24-8, and therefore Equation 24-7, are in agreement with the qualitative analysis of the spherical, concave mirror we made in the previous section. In addition, the analysis using Equation 24-8 supports the claim that positive values of d_I indicate that the image forms in front of the mirror, on the *real side* of the mirror. Conversely, the image forms behind the mirror when d_I has a negative value.

How large is the image formed by a spherical, concave mirror relative to the size of the object? The height of the object h_O and the height of the image h_I are defined in the direction perpendicular to the principal axis, as in **Figure 24-22a**. We can find the image by tracing two light rays reflecting from the tip of the arrow, one that strikes the mirror parallel to the principal axis and a second that strikes the center of the mirror. Because the principal axis is normal to the surface at the center of the mirror, the angles that the incident and reflected rays make to the axis are the same, labeled θ in the figure. The geometry of this ray creates two right triangles, as shown in **Figure 24-22b**. We use these triangles to determine h_I in terms of h_O.

Using the lower triangle in Figure 24-22b, we can obtain an equation relating θ, h_I, and d_I:

$$\tan \theta = \frac{h_I}{d_I}$$

Using the upper triangle in Figure 24-22b, we can obtain an equation relating θ, h_O, and d_O:

$$\tan \theta = \frac{h_O}{d_O}$$

Because both fractions equal $\tan \theta$, the fractions are equal, or

$$\frac{h_O}{d_O} = \frac{h_I}{d_I}$$

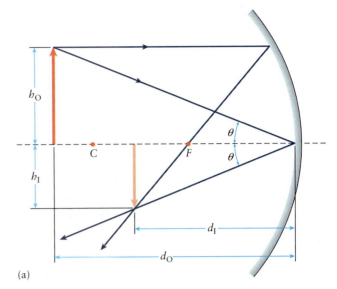

(a)

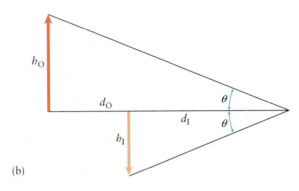

(b)

Figure 24-22 (a) The height of an image formed by a concave mirror can be found in terms of the height of the object and the focal length using a ray trace diagram. (b) We analyze the diagram by relating the two right triangles formed.

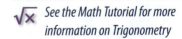 *See the Math Tutorial for more information on Trigonometry*

This equation can be rearranged to be

$$h_I = \frac{d_I}{d_O} h_O \qquad (24-9)$$

We have already seen that the image and object distances are signed quantities; a positive value indicates a location in front of the mirror and a negative value a location behind it. It is also standard to assign a positive value for the height of an upright image and a negative value for the height of an inverted one.

Equation 24-9 makes it clear that when the image forms farther from the mirror than the object, the magnitude of d_I is greater than d_O and therefore the image is larger than the object.

To make the sign conventions clear, let's return to the lateral magnification, defined as

$$m = \frac{h_I}{h_O} \qquad (24-1)$$

The image of an object placed inside the focal point forms behind the mirror and is upright (for example, Figure 24-18b). The sign of h_I is therefore positive, so m is also positive. Besides telling us the size of an image relative to the size of the object, the sign of the magnification m tells us whether the image is upright or inverted. By convention a positive magnification indicates that the image is upright. The image of an object placed outside the focal point forms in front of the mirror and the image is inverted (for example, Figure 24-16). In this case, the sign of h_I is negative, so m is also negative. By convention a negative magnification indicates that the image is inverted.

We can express m in terms of the object and image distances. From Equation 24-9,

$$m = -\frac{d_I}{d_O} \qquad (24-10)$$

where we *inserted* a minus sign in order to match the sign convention applied to object and image distance.

? Got the Concept 24-5

Got the Signs?

Verify that the lateral magnification determined by Equation 24-10 obeys the same sign convention as magnification determined using Equation 24-9.

Example 24-2 An Object Far from a Concave Mirror

Go to Interactive Exercise 24-2 for more practice dealing with concave mirrors

An object is 1.0 cm tall and is placed 14 cm in front of a spherical concave mirror that has a radius of curvature equal to 20 cm. Find the image distance and height.

SET UP

It might be tempting to start by substituting values into the mirror equation. As is almost always the case for physics problems, however, it is helpful to get a qualitative feel for the answer before seeking a quantitative result so that we know whether or not our answer makes sense. Is the image real or virtual? Is the image upright or inverted? We start, therefore, by tracing a few rays from a sample object, as in **Figure 24-23**. In this figure, we trace one ray that leaves the tip of the arrow (object)

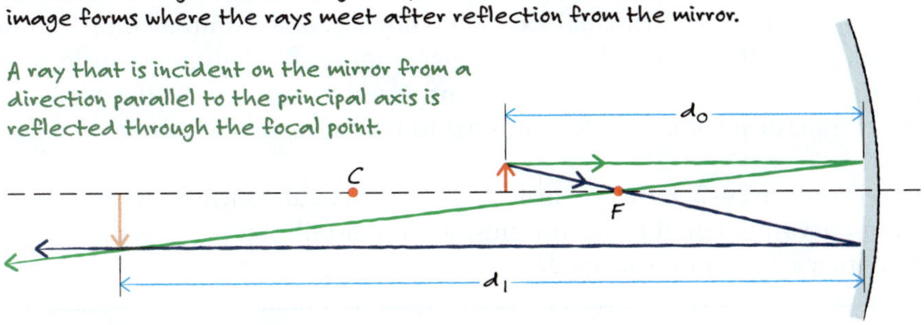

To find the image trace two light rays from the tip of the arrow. The image forms where the rays meet after reflection from the mirror.

A ray that is incident on the mirror from a direction parallel to the principal axis is reflected through the focal point.

A ray that passes through the focal point before striking the mirror is reflected in a direction parallel to the principal axis.

Figure 24-23 A ray trace diagram helps us determine where the image forms, and how high it is, when an object is in front of a spherical concave mirror.

traveling parallel to the principal axis; this ray passes through the focal point after reflection from the mirror. A second ray passes through the focal point before striking the mirror and is then reflected in a direction parallel to the principal axis. The image of the tip of the arrow forms where the two rays cross. Note that we did not attempt to make this drawing to scale, but rather simply ensured that the distances and sizes are generally correct. For example, the radius of curvature of the mirror is 20 cm, making the focal point 10 cm from the mirror (f equals one-half r, Equation 24-2). The object distance is 14 cm, so we marked the focal point (F) halfway between the mirror and the center of curvature (C), and drew the object slightly short of halfway between F and C.

SOLVE

Using Figure 24-23, we now know that the image is real (it forms on the reflective side of the mirror) and inverted. We also expect from the figure that the image forms 10 cm or so beyond the center of curvature of the mirror. This will help us decide whether the numeric answer we get from the mirror equation is reasonable.

The mirror equation, expressed as image distance d_I in terms of object distance d_O and focal length f, is

$$d_I = \frac{d_O f}{d_O - f} \tag{24-8}$$

Here d_O is 14 cm and f is 10 cm, so

$$d_I = \frac{(14\ \text{cm})(10\ \text{cm})}{14\ \text{cm} - 10\ \text{cm}} = 35\ \text{cm}$$

The image forms 35 cm in front of the mirror. The image, as determined by the ray trace diagram in Figure 24-23, is inverted.

The height of the image is found from the magnification, given by Equation 24-10:

$$m = -\frac{d_I}{d_O} = -\frac{35\ \text{cm}}{14\ \text{cm}} = -2.5$$

So by using Equation 24-1,

$$h_I = m h_O = (-2.5)(1.0\ \text{cm}) = -2.5\ \text{cm}$$

REFLECT

Both the position and size of the image are reasonable based on our ray trace diagram. In addition, the negative value of image height confirms that the image is inverted. More interesting, perhaps, is a consideration of our reason for selecting a relatively small object. Because the object is small compared to the radius of curvature, we can be assured that the portion of the mirror used to form the image is relatively close to the principal axis. This satisfies our requirement, noted in the previous section, that to avoid spherical aberration, either light must strike a spherical mirror close to the point at which the principal axis intercepts the mirror or the size of the mirror must be small compared to the radius of curvature.

Practice Problem 24-2 How far in front of a spherical concave mirror that has a radius of curvature equal to 20 cm must a 1.0-cm-tall object be placed so that an inverted image 5 cm tall is formed?

Example 24-3 An Object Close to a Concave Mirror

An object is 1.0 cm tall and is placed 6.0 cm in front of a spherical concave mirror with a radius of curvature equal to 20 cm. Find the image distance and height.

SET UP

As in the previous example, we start by making a ray trace diagram to determine whether the image is real or virtual and upright or inverted. In **Figure 24-24**, we trace one ray that leaves the tip of the arrow (object) traveling parallel to the principal axis; this ray passes through the focal point after reflection from the mirror. A second ray, which can be traced back to the focal point, strikes the mirror and is reflected in a direction parallel to the principal axis. These two rays will never cross; however, to an observer in front of the mirror they would appear to cross in the region behind the mirror. As shown in the figure, the image forms at this location behind the mirror.

SOLVE

From Figure 24-24, we now know that the image is upright; because it forms behind the mirror, it is virtual. In addition, although we did not attempt to make a scale drawing, from the approximate dimensions in the figure we expect that the image forms 10 cm or so beyond the center of curvature of the mirror. This will help us decide whether the numeric answer we get from the mirror equation is reasonable.

The mirror equation, expressed as image distance d_I in terms of object distance d_O and focal length f, is

$$d_I = \frac{d_O f}{d_O - f} \tag{24-8}$$

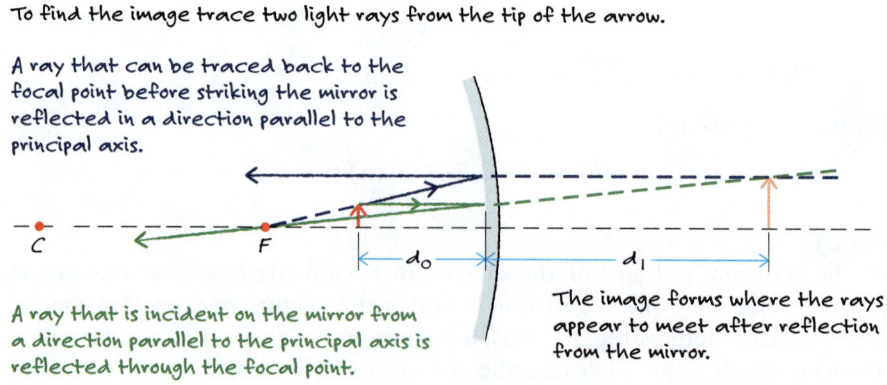

To find the image trace two light rays from the tip of the arrow.

A ray that can be traced back to the focal point before striking the mirror is reflected in a direction parallel to the principal axis.

Figure 24-24 This ray trace diagram helps us determine where the image forms, and how high it is, when an object is placed inside the focal point of a spherical concave mirror.

A ray that is incident on the mirror from a direction parallel to the principal axis is reflected through the focal point.

The image forms where the rays appear to meet after reflection from the mirror.

Here d_O is 6.0 cm and f is 10 cm, so

$$d_I = \frac{(6.0 \text{ cm})(10 \text{ cm})}{6.0 \text{ cm} - 10 \text{ cm}} = -15 \text{ cm}$$

The image forms 15 cm from the mirror; the minus sign indicates that it forms behind the mirror. The image, as determined by the ray trace diagram in Figure 24-24, is upright.

The height of the image, from Equation 24-9, is

$$|h_I| = \frac{|d_I|}{|d_O|}|h_O| = \frac{15 \text{ cm}}{6.0 \text{ cm}}(1.0 \text{ cm}) = 2.5 \text{ cm}$$

Because the image is upright the image height is positive, so $h_I = 2.5$ cm.

REFLECT
Both the position and size of the image are reasonable based on our ray trace diagram. By coincidence, the magnitude of the size of the image, 2.5 cm, is the same in this and the previous example. Were you standing in front of the mirror, however, you would see rather different images in the two cases. In the previous example, an inverted image would appear to hover in front of the mirror. In this example, the image would be upright, and appear to form behind the mirror.

Practice Problem 24-3 How far in front of a spherical concave mirror that has a radius of curvature equal to 20 cm must a 1.0-cm-tall object be placed so that an upright image 5 cm tall is formed?

✳ What's Important 24-3
The mirror equation provides a relationship between the focal length of a concave mirror and the locations of the object and image. By convention, a positive value of the image distance d_I indicates that the image is real; a real image forms in front of the mirror where reflected rays cross. Conversely, a negative value of d_I indicates that the image is virtual; a virtual image forms where reflected rays appear to cross when traced back behind the mirror.

24-4 Spherical Convex Mirrors, a Qualitative Look

Rays of light coming from a relatively distant object are parallel, but they diverge when they reflect from a spherical convex mirror (Figure 24-10b). We'll see in this section that the divergence results in an image that appears squeezed behind the reflective surface of the mirror. In other words, the image formed by convex mirrors encompasses a wide field of view. Do we find convex mirrors in nature? Consider the photograph in **Figure 24-25** of a Ferris Wheel reflected from the surface of someone's eye. Because the eye is relatively spherical, it's not surprising that its outer surface acts like a convex mirror. (This is not, of course, the purpose of the outer surface of the eye.)

Parallel light rays that reflect off a convex mirror appear to be coming from nearly the same point behind the mirror (Figure 24-10b). That point is the virtual focal point, where "virtual" indicates that the rays do not actually come to a common focus but only appear to do so. In addition, if we limit the size of the reflective surface so that it is small relative to the radius of the sphere, we can neglect spherical aberration. This allows us to treat the reflection of all incoming rays that

Figure 24-25 Look closely and you will see a Ferris Wheel reflected from the surface of someone's eye. The eye is relatively spherical so its outer surface acts like a convex mirror. *(Melissa Gaskell / Alamy)*

are parallel to the principal axis as if they emanated from a single, virtual focal point. This focal point is located halfway between the surface of the mirror and the center of the sphere. To indicate that the point is behind the mirror, we will assign a negative value to the focal length *f* of a convex mirror.

As before, we locate the image formed in a convex mirror using our ray tracing technique. In **Figure 24-26a,** an upright arrow has been placed in front of a convex mirror. A light ray that emanates from the tip of the arrow traveling parallel to the principal axis is reflected so that it can be traced back to the virtual focal point,

To locate the image formed by a spherical, convex mirror, we place an arrow in front of it. Then we follow two light rays from the tip of arrow to the reflective surface and then as they reflect from it.

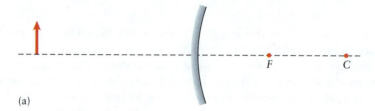

(a)

A light ray that starts off parallel to the principal axis is reflected so that when we trace it back behind the mirror it appears to pass through the *virtual* focal point.

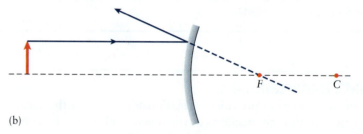

(b)

The angle of incidence of a ray striking the mirror always equals the angle of reflection. This is easy to draw for a light ray that strikes the center of the mirror.

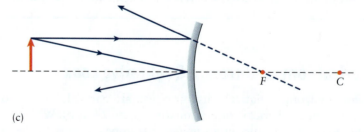

(c)

By tracing the two reflected rays behind the mirror, we find the point where they appear to cross. This is where the upright image forms. It is a virtual image because it forms behind the mirror.

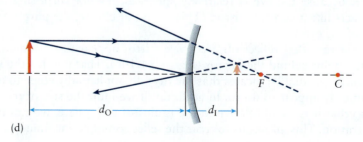

(d)

Figure 24-26 A sequence of ray trace diagrams shows how an image is formed by a convex mirror. (a) We set up the diagram by placing an upright arrow in front of the mirror. Two rays are traced from the tip of the arrow, (b) one that starts parallel to the principal axis of the mirror and (c) one that strikes the mirror at its center. (d) The image of the arrow's tip forms where the two rays appear to converge.

as in **Figure 24-26b**. A second ray, one that strikes the center of the mirror, is shown in **Figure 24-26c**. Because the principal axis is normal to the mirror at its center, the reflected ray is found by simply reflecting the incident ray around the principal axis, as shown. The image of the tip of the arrow forms where the two reflected rays appear to meet. The image of the base of the arrow must lie along the principal axis. A light ray coming from the base of the arrow and traveling along the principal axis strikes the mirror normal to the surface, so it is reflected straight back. The image of the base of the arrow must therefore form somewhere on the principal axis. The image, as you see in **Figure 24-26d**, is thus upright and behind the mirror. It is a virtual image because it forms not where reflected rays actually cross but rather where they appear to cross. For this reason, the image formed by a spherical, convex mirror can never be larger than the object.

When one end (the base) of the object is placed on the principal axis, the height of the image is determined by the point where the reflections of light rays emanating from the point on the object farthest from the principal axis appear to cross. (As we've seen, the image of the base must lie along the principal axis.) For example, one of the rays is initially parallel to the principal axis; when the ray is reflected it traces back to the virtual focal point as shown by the dashed line in Figure 24-26b. The tip of the image of the arrow must lie along this dashed line, as shown in Figure 24-26d. In addition to partly determining where the image of the tip forms, this requirement demands that the image will always be upright, always be smaller than the object, and always be closer to the mirror than the virtual focal point.

If the object distance were infinite, then the image would form *at* the virtual focal point. Consider an object and the image formed by a spherical, convex mirror as the object is moved from a position far from the mirror (**Figure 24-27a**), to a position closer to the mirror (**Figure 24-27b**), and finally to a position very close to the mirror (**Figure 24-27c**). In the first case the image forms close to the virtual focal point, but it's small. As you can see from Figure 24-27c, the image gets larger and larger as the object is moved closer to the mirror. If the object were located at the surface of the mirror, the image would be exactly the same size as the object.

These three ray trace diagrams follow the reflection of an object in a convex mirror as it is moved from a position far from the mirror to one close to it.

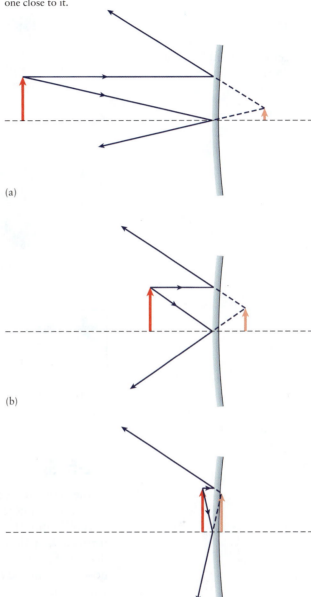

Figure 24-27 Three ray trace diagrams show how the height and position of an image change as an object is moved closer and closer to a convex mirror.

? Got the Concept 24-6
Fire?

The Sun is so far from Earth that rays of sunlight are effectively parallel. Sunlight reflected from a spherical, concave mirror is therefore focused to a single point. This can increase the temperature of whatever might be at that point high enough to cause it to burn (**Figure 24-28**). Is it possible to start a fire using a spherical, convex mirror?

Figure 24-28 Sunlight reflected from a spherical, concave mirror is focused to a point. The temperature of a scrap of paper placed at that point can get high enough to cause it to burn.

✳ What's Important 24-4

A convex mirror, one created when the outer surface of a spherical section is reflective, causes parallel rays to diverge. Reflected parallel rays trace back to a common focus behind the mirror; a convex mirror therefore always forms a virtual, upright image.

24-5 Spherical Convex Mirrors, a Quantitative Look

In the previous section, we examined the properties of spherical, convex mirrors from a qualitative perspective, answering questions such as "Is the image real or virtual?" and "Is the image inverted or upright?" (In all cases, a convex mirror forms a virtual, upright image.) We now turn our attention to the mathematical details of reflections formed by convex mirrors. Once again we want to determine how far from the mirror the image forms and what magnification the mirror provides. Our goal is to find a relationship for image distance d_I in terms of the object distance d_O and parameters that characterize the shape of the spherical mirror, either the radius of curvature r or the focal length f.

To do so, we place a point object O at distance d_O along the principal axis from the reflective surface and use ray tracing to determine the location of the image. The setup is shown in **Figure 24-29**. A point object is a good choice, because all light rays emanating from it are guaranteed to come from a single point. In addition, because the principal axis is normal to the surface at the center of the mirror, a ray of light coming from the object along the axis (not shown in the figure) would be reflected directly back, which means that the image forms on the principal axis. Also shown in the figure is an arbitrarily chosen ray of light coming from the object and reflecting off the mirror. The image forms where the reflected ray, when traced back behind the mirror, appears to intercept the principal axis. This occurs at image distance d_I from the mirror.

In Figure 24-29, r is the distance from the reflective surface of the mirror to the center of curvature C. The red dashed line in Figure 24-29, drawn from the center of curvature C to the point at which the ray of light strikes the mirror, represents

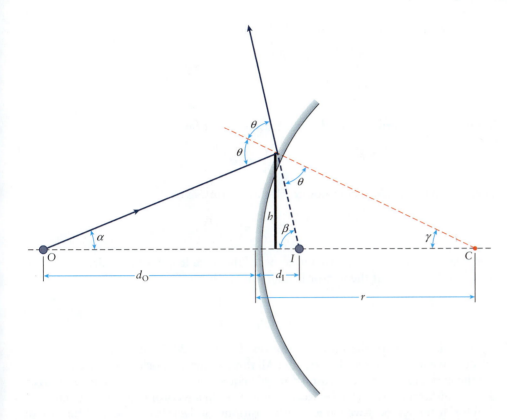

Figure 24-29 The variables we use to study a convex mirror are shown. The object is placed at point O, a distance d_O along the principal axis from the reflective surface. The image forms at point I a distance d_I from the mirror. Point C is the center of curvature of the mirror.

the normal to the mirror's surface at that point. It does not appear to be perpendicular to the surface because the sketch is not drawn to scale; the mirror represents such a small section of a large sphere that if we drew it to scale it would appear nearly straight. Because the dashed line represents the normal, the angle between it and the incident ray, and the angle between it and the reflected ray, are equal. They are labeled θ in the figure.

As in the earlier case of the concave mirror, we can find a relationship for d_I in terms of r and d_O by first finding a relationship between α, the angle between the principal axis and the light ray incident on the convex mirror, β, the angle between the principal axis and the reflected ray, and γ, the angle between the principal axis and red dashed line that defines the normal at the point where the light ray is reflected. As earlier, we ask "What's the same?" to relate the angles.

Notice that each of the three angles forms a right triangle with the thick gray line labeled h as the opposite side. Because we are only considering mirrors made from a very small section of a full sphere so that the reflective surface is nearly flat, we can treat d_O, d_I, and r as the bases of these triangles. Because h is common to all three triangles we can use it to relate the lengths of the bases:

$$h = d_O \tan \alpha; \quad h = d_I \tan \beta; \quad h = r \tan \gamma$$

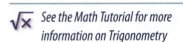

See the Math Tutorial for more information on Trigonometry

All three angles are small in our approximation, and $\tan \theta$ is approximately θ for small values of θ, so

$$h = d_O \alpha; h = d_I \beta; h = r\gamma \tag{24-11}$$

Notice that the rule we established in exploring supplementary angles (for example, Equation 24-4), requires that

$$\theta + \gamma = \beta \tag{24-12}$$

and also

$$\alpha + \beta = 2\theta \tag{24-13}$$

Combining Equations 24-12 and 24-13 yields

$$\alpha + \beta = 2(\beta - \gamma)$$

or

$$\alpha - \beta = -2\gamma$$

We write this in terms of the distances using Equation 24-11:

$$\frac{h}{d_O} - \frac{h}{d_I} = -\frac{2h}{r}$$

Dividing out h which is common to all three terms yields

$$\frac{1}{d_O} - \frac{1}{d_I} = -\frac{2}{r}$$

The relationship is often written in terms of the focal length of the mirror, which is one-half the radius of the mirror (Equation 24-2):

$$\frac{1}{d_O} - \frac{1}{d_I} = -\frac{1}{f} \tag{24-14}$$

This is similar to the mirror equation (Equation 24-7) that we developed for the concave mirror, but look carefully! All three terms are positive in the expression for the concave mirror; in contrast, in this equation describing the convex mirror, the d_I and f terms are negative. Recall, however, that according to our sign convention for distances, positive values mark positions on the reflective side of the mirror and negative values represent locations behind the mirror. So while the object distance is always positive (the object must be in front of the mirror), both the image distance and the focal length are negative for the convex mirror. Therefore, with care taken to use only negative values for d_I and f, Equation 24-14 is identical to the relationship we developed for a spherical, concave mirror:

$$\frac{1}{d_O} + \frac{1}{d_I} = \frac{1}{f} \tag{24-7}$$

This equation applies to both convex as well as concave mirrors—it is indeed the mirror equation.

To determine how far from the mirror the image forms, we rearrange Equation 24-7 to write d_I in terms of d_O and f:

$$d_I = \frac{d_O f}{d_O - f} \tag{24-8}$$

Notice first that because f will always be negative for a convex mirror, the numerator in Equation 24-8 will also always be negative while the denominator will always be positive. Thus we are guaranteed that d_I is negative. A convex mirror can only form an image behind the reflective surface, that is, a virtual image.

Equation 24-8 can also be written as a fraction times the focal length:

$$d_I = \frac{d_O}{d_O - f} f \tag{24-15}$$

Because f is negative, the denominator in the fraction will always be greater than d_O, so the fraction itself will always be less than 1. We conclude, therefore, that the image distance is always less than the focal length, meaning that the image always forms somewhere between the mirror and the focal point. The sequence of ray trace diagrams in Figure 24-27, which takes an object from a point far from a convex mirror to a point close to it, supports this claim.

> ### ❓ Got the Concept 24-7
> ### A Convex Mirror Near and Far
> We saw in the last section that when an object is placed very far from a convex mirror, the image forms close to the focal point, and that when an object is placed very close to a convex mirror, the image forms close to (but still behind) the mirror. Convince yourself that both are true by considering Equation 24-15.

How large is the image formed by a spherical, convex mirror relative to the size of the object? We define both the height of the object h_O and the height of the image h_I in the direction perpendicular to the principal axis, as shown in **Figure 24-30a**. In the figure the image has been located by tracing two light rays that scatter from the tip of the arrow and strike the mirror. The two rays create two right triangles, as shown in **Figure 24-30b**. We use the triangles to determine h_I in terms of h_O. First,

$$\tan \theta = \frac{h_I}{d_I} = \frac{h_O}{d_O}$$

so

$$\frac{h_I}{d_I} = \frac{h_O}{d_O}$$

How large is the image formed by a spherical, convex mirror relative to the size of the object?

(a)

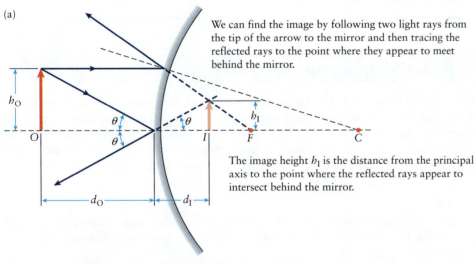

We can find the image by following two light rays from the tip of the arrow to the mirror and then tracing the reflected rays to the point where they appear to meet behind the mirror.

The image height h_I is the distance from the principal axis to the point where the reflected rays appear to intersect behind the mirror.

(b)

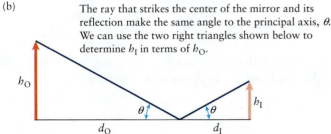

The ray that strikes the center of the mirror and its reflection make the same angle to the principal axis, θ. We can use the two right triangles shown below to determine h_I in terms of h_O.

Figure 24-30 (a) The height of an image formed by a convex mirror can be found in terms of the height of the object and the focal length using a ray trace diagram. (b) We analyze the diagram by relating the two right triangles formed.

or

$$h_{\mathrm{I}} = \frac{d_{\mathrm{I}}}{d_{\mathrm{O}}} h_{\mathrm{O}}$$

This is identical to the relationship between the heights and distances that we found for the spherical concave mirror (Equation 24-9). The magnification is the same as well:

$$m = -\frac{d_{\mathrm{I}}}{d_{\mathrm{O}}} \tag{24-10}$$

Because d_{I} is always negative for a convex mirror, the magnification will always be positive. In other words, the image formed by a convex mirror is always upright.

Watch Out

Positive magnification indicates that an image is upright.

The sign of the magnification of a mirror does not indicate whether the image is larger or smaller than the object, only whether the image is upright or inverted. An upright image, for example, is always associated with a positive magnification, regardless of whether the image is larger than or smaller than the object.

Got the Concept 24-8

Can a Convex Mirror Magnify?

Is it possible for the image of an object in front of a convex mirror to be larger than the object itself?

Estimate It! 24-1 Ornament Reflection

Walking along a street you catch your reflection in a shiny spherical ornament (**Figure 24-31**). If you're about 2 m from the ornament, estimate how large your image appears. Because your reflection occupies only the region of the ornament close to its center, it is approximately correct to treat the reflective surface as if it were only a small part of a larger sphere.

SET UP
We can combine Equations 24-1 and 24-10, which define magnification in terms of either object and image heights (h_{I} and h_{O}) or object and image distances from the mirror (d_{I} and d_{O}):

$$h_{\mathrm{I}} = m h_{\mathrm{O}} = -\frac{d_{\mathrm{I}}}{d_{\mathrm{O}}} h_{\mathrm{O}}$$

For a convex mirror the distance from the mirror at which the image forms depends on the object distance and the focal length f of the mirror (Equation 24-8):

$$d_{\mathrm{I}} = \frac{d_{\mathrm{O}} f}{d_{\mathrm{O}} - f}$$

Figure 24-31 How big is your image reflected from a shiny spherical ornament? (*Marcus Fuehrer/dpa /Landov*)

All we need is some big, round, but reasonable values for the focal length of the reflective surface of the ornament and your height, and we're ready to go. Let's take 10 cm as the diameter of the ornament, which seems about right. The distance to the focal point is then −2.5 cm, which is one-half the radius of curvature, which itself is one-half the diameter. Notice that f is negative because the focal point is inside the sphere, behind the mirrored surface. We'll use 2 m as your height, a nice round number that can't be too far off.

SOLVE

First, the distance the image forms from the surface is

$$d_I = \frac{(200 \text{ cm})(-2.5 \text{ cm})}{200 \text{ cm} - (-2.5 \text{ cm})} = -2.47 \text{ cm}$$

The height of your image is then

$$h_I = mh_O = -\frac{-2.47 \text{ cm}}{200 \text{ cm}}200 \text{ cm} = 2.47 \text{ cm}$$

With our broad estimates of the sizes and distances, a reasonable way to state the answer is that the height of your image in the ornament is between 2 and 3 cm.

REFLECT

Our estimate is equivalent to about one-third the diameter of the ornament. That seems about right based on the photo in Figure 24-31.

Example 24-4 An Object Far from a Convex Mirror

An object is 8.0 cm high and is placed 68 cm in front of a spherical convex mirror with a radius of curvature equal to 48 cm. Find the image distance and height.

SET UP

We start by tracing a few rays from a sample object, as in **Figure 24-32**. We trace one ray that leaves the tip of the arrow (object) traveling parallel to the principal axis; this ray traces back to the focal point after reflection from the mirror. A second ray, which strikes the center of the mirror, is reflected so that the angles between

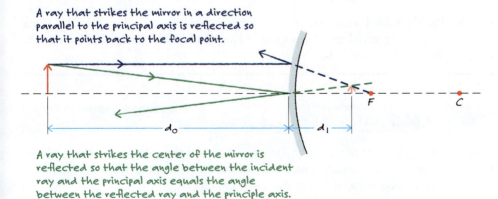

To find the image trace two light rays from the tip of the arrow. The image forms where the rays appear to meet after reflection from the mirror.

A ray that strikes the mirror in a direction parallel to the principal axis is reflected so that it points back to the focal point.

A ray that strikes the center of the mirror is reflected so that the angle between the incident ray and the principal axis equals the angle between the reflected ray and the principle axis.

When we trace this reflected ray back behind the mirror, we clearly see that it cannot intersect the principal axis behind the reflective surface. Therefore, d_I must be less than F.

Figure 24-32 A ray trace diagram helps us determine where the image forms, and how high it is, when an object is placed far from the surface of a spherical convex mirror.

the principal axis and both the incident and reflected rays are the same. This must be the case because the principal axis is normal to the mirror at its center. The image of the tip of the arrow forms where the two rays appear to cross. Note that we did not attempt to make the drawing to scale, but rather simply ensured that the distances and sizes are generally correct. For example, the radius of curvature of the mirror is 48 cm, making the focal point 24 cm from the mirror (f equals one-half r, Equation 24-2). The object distance is 68 cm. So we marked the focal point (F) halfway between the mirror and the center of curvature (C), and drew the object about 1.5 times farther from the mirror than C.

SOLVE

The mirror equation, expressed as image distance d_I in terms of object distance d_O and focal length f, is

$$d_I = \frac{d_O f}{d_O - f} \tag{24-8}$$

Here d_O is 68 cm and f is -24 cm. Note that the value of f is negative because the center of curvature and therefore the focal point is behind the reflective side of the mirror. So

$$d_I = \frac{(68 \text{ cm})(-24 \text{ cm})}{68 \text{ cm} - (-24 \text{ cm})} = -18 \text{ cm}$$

The negative value indicates that the image forms behind the mirror, as shown in the ray trace diagram (Figure 24-32). This virtual image is about 18 cm behind the mirror.

From Equation 24-9, the height of the image is

$$|h_I| = \frac{|d_I|}{|d_O|}|h_O| = \frac{18 \text{ cm}}{68 \text{ cm}}(8.0 \text{ cm}) = 2.1 \text{ cm}$$

The image is 2.1 cm high. In addition, as we see from the ray trace diagram, it is upright.

REFLECT

Both the position and size of the image are reasonable based on our ray trace diagram. Also, note that as we expect from our qualitative analysis of the convex mirror in the previous section, because the object is far from the mirror, the image forms relatively close to the focal point. The ray that traces back to the focal point in part determines the image height, so the closer the image is to the focal point, the smaller it is. In this example the image is close to the focal point and about one-fourth the size of the object.

Practice Problem 24-4 How far in front of a spherical convex mirror that has a radius of curvature equal to 48 cm must a 1.0-cm-tall object be placed so that an upright image 0.5 cm tall is formed?

Example 24-5 An Object Close to a Convex Mirror

This time we place our object only 3.0 cm in front of the spherical convex mirror. Find the image distance and height. As in the previous example, the object height is 8.0 cm and the mirror has a 48 cm radius of curvature.

SET UP

Once again, we start by making a ray trace diagram such as the one in **Figure 24-33**. In the figure, we trace one ray that leaves the tip of the arrow (object) traveling

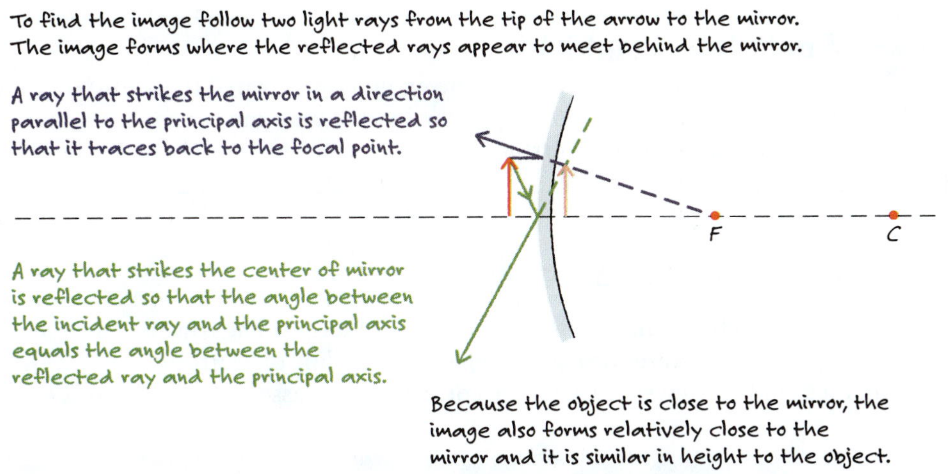

To find the image follow two light rays from the tip of the arrow to the mirror. The image forms where the reflected rays appear to meet behind the mirror.

A ray that strikes the mirror in a direction parallel to the principal axis is reflected so that it traces back to the focal point.

A ray that strikes the center of mirror is reflected so that the angle between the incident ray and the principal axis equals the angle between the reflected ray and the principal axis.

Because the object is close to the mirror, the image also forms relatively close to the mirror and it is similar in height to the object.

Figure 24-33 This ray trace diagram helps us determine where the image forms, and how high it is, when an object is placed close to the surface of a spherical convex mirror.

parallel to the principal axis; the ray traces back to the focal point after reflection from the mirror. A second ray, which strikes the center of the mirror, is reflected so that the angles between the principal axis and both the incident and reflected rays are the same. This must be the case because the principal axis is normal to the mirror at its center. By tracing the reflected rays back behind the mirror, we find the image of the tip of the arrow where the two rays appear to cross.

SOLVE

The mirror equation, expressed as image distance d_I in terms of object distance d_O and focal length f, is

$$d_I = \frac{d_O f}{d_O - f} \tag{24-8}$$

Here d_O is 3.0 cm and f is −24 cm. Note that the value of f is negative because the center of curvature and therefore the focal point is behind the reflective side of the mirror. So

$$d_I = \frac{(3.0\ \text{cm})(-24\ \text{cm})}{3.0\ \text{cm} - (-24\ \text{cm})} = -2.7\ \text{cm}$$

As in the previous example, the negative value indicates that the image forms behind the mirror; in other words, it is a virtual image. The distance between the mirror and the image is 2.7 cm.

We calculate the height of the image using Equation 24-9:

$$|h_I| = \frac{|d_I|}{|d_O|}|h_O| = \frac{2.7\ \text{cm}}{3.0\ \text{cm}}(8.0\ \text{cm}) = 7.2\ \text{cm}$$

The image, as we see from the ray trace diagram, is upright, and 7.2 cm high.

REFLECT

Both the position and size of the image are reasonable based on our ray trace diagram. Also, note that as we expect from our qualitative analysis of the convex mirror in the previous section, because the object is close to the mirror, the image also forms relatively close to the mirror. In general, this has the effect of forming an image that is nearly the same height as the object, and our calculation confirms this.

⁕ **What's Important 24-5**
    The focal length of a spherical, convex mirror is related to the distances of an object and its image from the mirror by the same mirror equation that describes a spherical concave mirror.

24-6 Lenses, a Qualitative Look

Take a look through the magnifying glass in Figure 24-34. The rodent's furry nose appears larger when viewed through the glass. The skyline of Seattle in Figure 24-35 is not in focus unless viewed through the corrective eyeglasses. And in Figure 24-36,

Figure 24-34 A rodent's furry nose appears larger when viewed through a magnifying glass. *(Monika Graff / The Image Works)*

Figure 24-35 A pair of corrective eyeglasses brings the skyline of Seattle into focus. *(Hackfish/ Wikipedia)*

Figure 24-36 The image of a physicist through a lens appears squeezed down to fit within the disk of the lens. *(Courtesy of Rachael Lynn Beaton/ University of Virginia)*

the image of a physicist shining a laser through a piece of glass is squeezed down to fit within the disk of the glass. You may have noticed one or more of these effects while looking through a piece of glass of a particular shape. Our experience with the way mirrors form images probably suggests to you that the pieces of glass used to create such photographs are bending the light rays that come from the rodent, the skyline, and the physicist. In this section we'll explore the *lens* and the way it bends light rays in a regular, predictable way to form images.

In the previous chapter, we learned that light rays bend when crossing the boundary between two transparent media of different indices of refraction, as in Figure 24-37. The angle that the outgoing ray makes to the normal (to the boundary) obeys Snell's law ($n_2 \sin \theta_2 = n_1 \sin \theta_1$, Equation 23-6), so this angle (θ_2, the refracted angle) depends in part on the angle of the incoming ray θ_1. Until now we have only explored the physics of refraction by considering what happens to single rays of light when they cross a flat boundary such as a pane of window glass. We now consider how a series of parallel rays are refracted at a boundary that is not flat.

Consider the parallel light rays striking a glass sphere surrounded by air in Figure 24-38a. At the surface of the sphere, the angle each ray makes with respect to the normal to the surface depends on the point at which the ray strikes the surface. The dashed lines coming from the center of the sphere, labeled C in the figure, represent the normal to the surface at each point where a light ray strikes it. This helps us to apply Snell's law for each ray. Because the index of refraction of glass is higher than that of air, each ray is bent closer to the normal as it enters the glass. (The exception, of course, is the ray that enters the sphere along its principal axis, for which both the incident and refracted angles are zero.) The angle that each refracted ray makes to the normal is smaller than the angle made by the associated incident ray, and the light rays are, as a group, refracted toward nearly the same point on the principal axis. Although the spherical piece of glass does not focus the parallel incident rays to a single point, the rays do converge to some degree.

In Figure 24-38b parallel rays of light *exit* a spherical piece of glass. Here again each ray is refracted at the surface according to Snell's law. Because the index of refraction of

A light ray that crosses the boundary from one transparent medium to another...

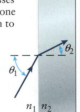

...changes direction according to Snell's law:

$$n_1 \sin \theta_1 = n_2 \sin \theta_2$$

n_1 and n_2 are the indices of refraction of the two media.

Figure 24-37 A ray of light is refracted according to Snell's law when it crosses the boundary from one medium to another.

Figure 24-38 Snell's law is applied separately to a number of light rays as they (a) enter and (b) exit a glass sphere.

Each dashed line represents the normal to the surface at the point where a light ray enters the sphere.

The incident angle θ_1 and the refracted angle θ_2 obey Snell's law.

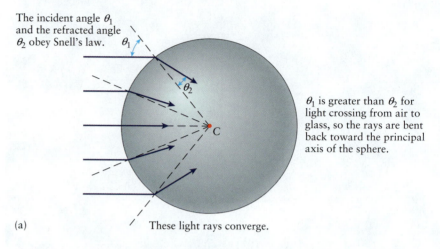

θ_1 is greater than θ_2 for light crossing from air to glass, so the rays are bent back toward the principal axis of the sphere.

(a)   These light rays converge.

Light rays approaching the surface of the sphere parallel to its principal axis and from within the sphere...

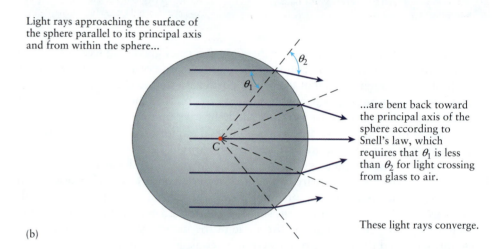

...are bent back toward the principal axis of the sphere according to Snell's law, which requires that θ_1 is less than θ_2 for light crossing from glass to air.

These light rays converge.

(b)

glass is higher than that of air, the angle that each refracted ray makes to the normal to the surface (shown by a dashed line) is larger than the incident angle. This has the net effect of refracting the rays toward nearly the same point along the principal axis. As in the case of light rays entering a glass sphere, the parallel rays nearly converge after they cross the boundary, this time from glass to air.

Considering the shape of the glass in both figures (not the air that surrounds it) and in keeping with the nomenclature we use for mirrors, we describe both of the surfaces in Figure 24-38 as convex.

Now consider a spherical region filled with air and surrounded by glass. What happens when parallel rays of light enter the glass (**Figure 24-39a**) and exit (**Figure 24-39b**). Again, the center of the sphere is labeled C, and the dashed lines show the normal to the boundary at each point where a light ray crosses into the glass. Because the index of refraction of glass is higher than that of air, the angle each ray makes to the normal is larger in air than glass; this has the net effect of causing the rays to be bent toward the normal as they enter the glass (Figure 24-39a) and away from the normal as they exit back into the air (Figure 24-39b). In addition, as you can convince yourself by following a few of the light rays in both Figure 24-39a and Figure 24-39b, the rays are all bent away from the principal axis. In addition, the diverging rays all trace back to nearly the same point on the principal axis. Considering the shape of the glass in both figures (not the air that surrounds it) and in keeping with the nomenclature we use for mirrors, we describe both of the surfaces in Figure 24-39 as concave.

Light rays approaching the glass surface parallel to the principal axis of the spherical region of air and from within the air...

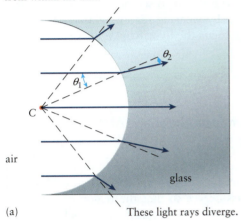

(a) These light rays diverge.

...are bent away from the principal axis according to Snell's law, which requires that the refracted angle θ_2 be smaller than the incident angle θ_1 for light crossing from air to glass.

Parallel light rays that exit the glass into a spherical region of air...

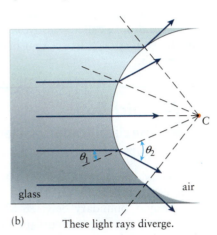

(b) These light rays diverge.

...are bent away from the principal axis according to Snell's law, which requires that the refracted angle θ_2 be larger than the incident angle θ_1 for light crossing from glass to air.

Figure 24-39 Snell's law is applied separately to a number of light rays as they (a) exit and (b) enter a spherical region of air surrounded by glass.

None of the four cases in Figures 24-38 and 24-39 exactly represents the lenses we considered at the beginning of this section. For all of those (the magnifying glass, the corrective eyeglasses, and the lens that forms a small image of the physicist) and, indeed, in most of the applications of lenses that we're likely to experience, light both enters and exits a material. Said another way, the lenses we will study have *two* optical surfaces. A variety of lenses can be created by combining the surfaces in Figures 24-38 and 24-39 as well as flat surfaces; some examples are shown from the side in Figure 24-40. Of particular interest to us are the two on the left side, one made from two convex surfaces and the other from two concave surfaces. The first is called a *convex lens*, sometimes referred to as a biconvex or double convex lens. A convex lens is thicker in the center than at its edges. The second is a *concave lens*, sometimes referred to as a biconcave or double concave lens. A concave lens is thinner in the center than at its edges. As you might guess from Figures 24-38 and 24-39, and as we'll see more clearly in the next section, a convex lens causes light rays to converge while a concave lens causes light rays to diverge.

As we get older, most of us come to rely on convex lenses to see small things such as the font in a newspaper or paperback book. By the time we reach middle

Four examples of lenses created by combining convex, concave, and flat optical surfaces

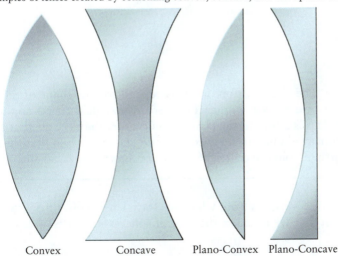

Convex   Concave   Plano-Convex   Plano-Concave

Figure 24-40 Four ways that two optical (spherical glass) surfaces can be combined to form a lens.

age, the lenses in our eyes eventually lose some of their elasticity. As a result, the front and back surfaces of the lenses cannot become as curved as they need to be in order to focus light on the retina, making it difficult to see objects at close range. This condition, called *presbyopia*, is treated by wearing reading glasses. These are made from convex lenses that help the eye focus the light from a nearby object on the retina.

> ### ❗ Watch Out
>
> #### The classification of a lens depends on the shape of the material, not what surrounds it.
>
> Because lenses, unlike mirrors, have two active surfaces, you might find it hard to remember the nomenclature we use to identify lenses. As shown in **Figure 24-41a**, a convex boundary between air and glass curves so that the region of glass is closer to the center of curvature than the air that surrounds it. Conversely, a concave boundary between air and glass curves so that the region of glass is *farther* from the center of curvature than the air that surrounds it. A concave surface is shown in **Figure 24-41b**. Both surfaces of the lens on the far left in Figure 24-40 are convex, and both surfaces of the lens to the right in Figure 24-40 are concave.
>
>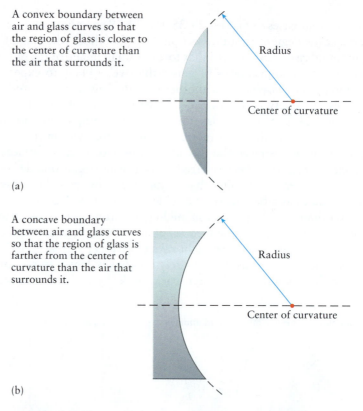
>
> **Figure 24-41** The classification of a lens depends on the shape of the material. (a) A convex boundary between air and glass curves so that the region of glass is closer to the center of curvature than the air that surrounds it. (b) A concave boundary between air and glass curves so that the region of glass is farther from the center of curvature than the air that surrounds it.

The spherical surfaces from which we form lenses suffer from the same spherical aberration that we noted in the case of spherical mirrors, that is, parallel light rays are not focused down to a single point. (Or, in the case of a concave lens, the refracted light does not appear to originate at a single point.) As is the case for mirrors, however, we can neglect the effect by constraining the rays of light to be close to the principal axis. This is assured by using only a small section of a large spherical surface to form each surface of the lens, as in **Figure 24-42**. In doing so we create a **thin lens**, so called because when each surface is a small part of a sphere of relatively large radius, the distance across the lens along the principal axis is negligibly small. We will consider only thin lenses for the rest of this chapter, so that we can neglect spherical aberration and treat each lens as if parallel rays were focused to a single point.

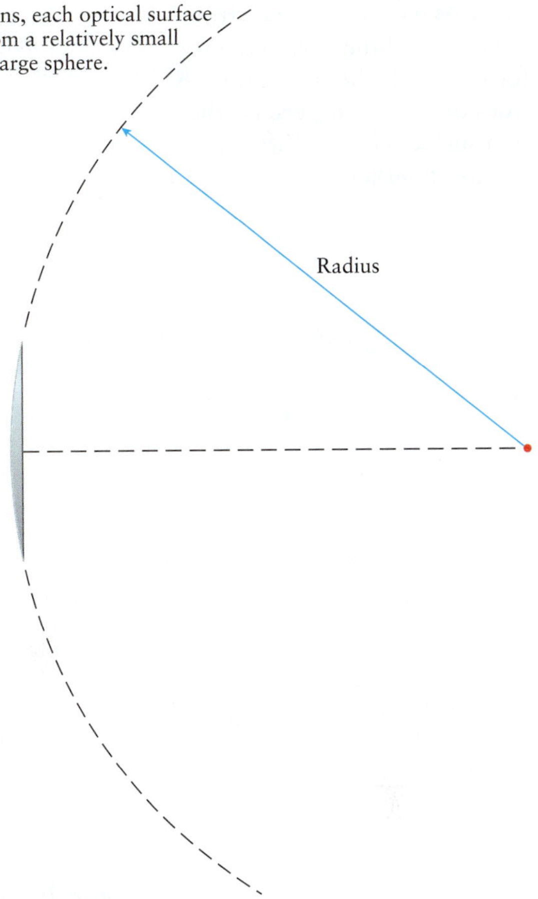

Figure 24-43 shows a thin convex lens, created by placing two thin spherical sections such as the one shown in Figure 24-42 back-to-back. Notice that we have marked two points as a focal point *F*. Light can strike the lens from either direction, so we need to mark a focal point on both sides. As suggested by **Figure 24-44**, for a thin lens the focal length, the distance from the center of the lens to the focal point, is the same on both sides of the lens, even if the two surfaces have a different radius of curvature.

As with mirrors, a ray that arrives at the lens traveling parallel to the principal axis is refracted so that it passes through one of the focal points, in this case, the focal point on the other side of the lens. Likewise a ray that passes through the focal point on its way to the lens exits the lens traveling parallel to the principal axis. In addition, a ray that strikes the lens directly at its center continues in a straight line as it exits. (If you wonder whether by "center" we mean the center of the surface of the lens or the point midway between the two surfaces along the principal axis, remember that for a thin lens, the width of the lens is negligible. So for a thin lens, *both* of these are the "center" of the lens.) A ray of each description, coming from the tip of the arrow that serves as the standard object, is shown in Figure 24-43. The arrow in the figure is farther from the lens than the focal point.

A careful look at Figure 24-43 reveals that it is an approximation. Any light ray that passes through a lens (or any transparent medium) will be refracted first as it enters and again as it exits. In the figure, however, each ray is shown to refract only

For a thin lens, each optical surface is formed from a relatively small section of a large sphere.

Radius

Figure 24-42 A thin lens, one in which spherical aberration is minimized, is formed by using optical surfaces that are only a small section of a large sphere. In this way the rays of light that pass through it are constrained to be close to the principal axis.

Light coming from an object strikes a thin convex lens...

A ray traveling parallel to the principal axis is refracted so that it passes through the focal point.

The focal length, the distance from the lens to the focal point, is the same on both sides of the lens.

F

F

A ray that passes through the focal point on its way to the lens exits the lens traveling parallel to the principal axis.

A ray that strikes the lens directly at its center continues in a straight line as it exits.

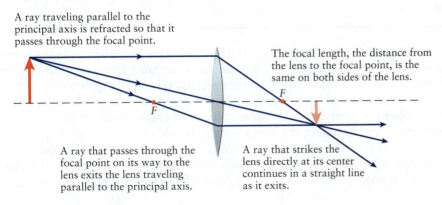

Figure 24-43 A ray trace diagram shows how an image is formed when light passes through a thin, convex lens.

Figure 24-44 The distance from the center of the lens to the focal point is the same on both sides of a thin lens, even if the two surfaces have a different radius of curvature.

For a thin convex lens, the distance from the center of the lens to the point where parallel rays are focused is the same regardless of the direction that light passes through the lens, even if the curvatures of the two surfaces of the lens are not the same.

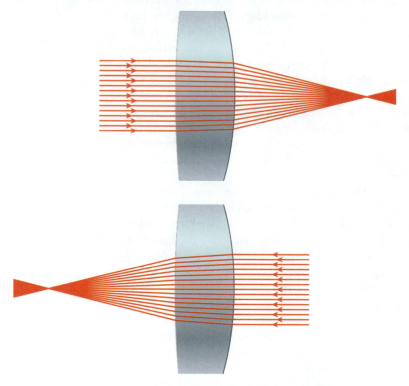

once. In addition, that refraction is shown to occur along the center line of the lens, where no refraction can occur because no boundary exists! What we have done to make the properties of the lens clearer is to draw a single refraction to represent the net change in direction of each light ray and draw that single refraction midway between the two surfaces of the lens because the lens is so thin.

Physicist's Toolbox 24-2 Rules for Tracing Light Rays through a Lens

Here is a summary of the rules we use to trace light rays from a point on an object through a lens:

- A ray that arrives at the lens traveling parallel to the principal axis is refracted so that it passes through one of the focal points
- A ray that passes through the focal point on its way to the lens exits the lens traveling parallel to the principal axis.
- A ray that strikes the lens directly at its center continues in a straight line as it exits.

The image of the tip of the arrow in Figure 24-43 forms where the rays coming from the tip cross. The image of the base of the arrow must form along the principal axis, because the ray that comes from the base and travels along the axis strikes the center of the lens and therefore continues in the same direction. The image of the arrow is seen to be inverted, and because the refracted rays actually cross (as opposed to only appearing to cross), the image is real. In addition, as shown in the figure, when an object is farther than the center of curvature from a convex lens, the image formed is smaller than the object. **Figure 24-45** shows an example of the image formed when an object is farther from a convex lens than the center of curvature.

Figure 24-45 When an object is farther than the focal point from a convex lens, the image formed is inverted, and smaller than the object.
(Derrick Alderman/Alamy)

> ## ⚠ Watch Out
> ### If a lens forms a real image, it must be on the side of the lens opposite to the object.
>
> We have defined a real image as one that forms where reflected or refracted light rays converge. A concave mirror can cause light rays to converge only on the reflective side of the mirror, which is where we would put an object, so this is the only side on which a real image can form. A lens, however, can only cause light rays to converge on the side of the lens *opposite* to the object. Whether or not rays passing through the lens actually converge depends on the shape of the lens and the location of the object, however, if a real image forms it must be on the side of the lens opposite to the object.

In **Figure 24-46**, our standard object has been placed *between* the lens and the focal point. To locate the image, we have traced two light rays from the tip of the arrow. One ray moves parallel to the principal axis of the lens and therefore passes through the focal point after being refracted by the lens. The other ray we have

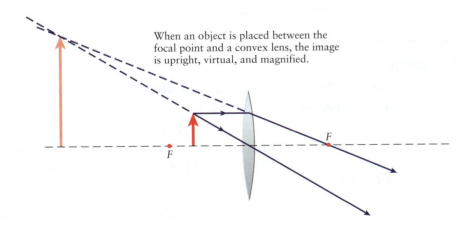

When an object is placed between the focal point and a convex lens, the image is upright, virtual, and magnified.

Figure 24-46 As seen by a ray trace diagram, when an object is placed closer to a thin, convex lens than the focal point, the image formed is upright, virtual, and magnified.

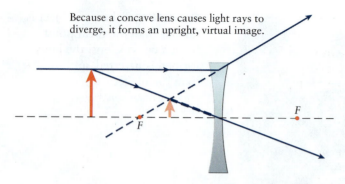

Because a concave lens causes light rays to diverge, it forms an upright, virtual image.

Figure 24-47 The image formed by a thin, concave lens is upright and virtual.

drawn strikes the center of the lens, so it continues in a straight line. Notice that the two rays do not meet. As the dashed lines suggest, however, they appear to meet on the same side of the lens as the object. This apparent crossing of the refracted rays determines the location of the image. As in the previous example, the base of the arrow must lie on the principal axis of the lens, so here, the image is upright. Because it forms not where light rays meet but only where they appear to meet, the image is virtual. Finally, because we placed the object closer to the lens than the focal point, the image is magnified.

We consider one other example, that of a thin, concave lens. To understand the properties of such a lens, we start by making a ray trace diagram. In **Figure 24-47**, we have placed an arrow so that some of the light that scatters from it passes through a lens formed from two thin, concave surfaces. Two light rays have been drawn to determine where the image forms, and as we did for the thin, convex lens, we have simplified the diagram by drawing the rays as if they refract only once, along the center line of the lens. The center line of the lens is the line perpendicular to the principal axis that runs through the center of the lens. Each of the two concave surfaces that comprise the lens cause parallel light rays to diverge (Figure 24-39), so the thin, concave lens does as well. For that reason the refracted rays cannot meet, so the image forms where the refracted rays appear to meet. The ray that leaves the tip of the arrow traveling parallel to the principal axis of the lens is refracted so that it traces back to one of the focal points. The ray that strikes the center of the lens continues in the same direction. The two rays are traced back (the dashed lines) to a common, meeting point. (The dashed line associated with the second light ray lies on top of the ray as it approaches the lens.) The image of the tip of the arrow forms at this point; the image is upright and virtual.

Go to Picture It 24-2 for more practice dealing with lenses

 Got the Concept 24-9
Passing Through the Focal Point

A light ray that passes through the focal point on its way to a thin, convex lens exits the lens traveling parallel to the principal axis, as in Figure 24-43. The same is true for a ray that approaches the lens on a line directed at the focal point on the other side of the lens. Convince yourself that such a ray is part of the formation of the image by a thin, concave lens, by copying Figure 24-47 and adding a light ray that leaves the tip of the arrow and passes through the focal point before striking the lens.

 Watch Out
There's no need to memorize which type of lens is converging and which is diverging.

You shouldn't feel compelled to memorize how images are formed by concave and convex lenses. If you understand the three rules for tracing light rays through a lens (see Physicist's Toolbox 24-2), you can easily reproduce sketches like those in Figures 24-43, 24-46, and 24-47. Such a ray trace figure tells you quickly and easily how a particular lens forms an image.

> ### ⭐ What's Important 24-6
> Each surface of a thin lens is a small part of a sphere of relatively large radius, so the distance across the lens along the principal axis is negligibly small. A thin, convex lens refracts parallel light rays to a common focus, so it can form a real image on the side of the lens opposite that of the object. A thin, concave lens causes parallel light rays to diverge, so the image formed must be virtual, that is, it is formed by tracing the rays back to the same side of the lens as the object.

24-7 Lenses, a Quantitative Look

Light bends as it passes from one transparent medium to another, and as we saw in the last section, this causes light rays to either converge or diverge when the boundary surface is curved. Our eyes bend light rays, too, because both the cornea and the lens have curved surfaces that work together to focus light on the retina (Figure 24-48). But as suggested in the previous section, for example, by comparing Figures 24-43 and 24-46, the location of the image formed depends partly on the location of the object. How can we see both near and far objects, if the location of the image in our eyes depends on our distance from the object? The drawings of reflections on the pupil of an eye shown in Figure 24-49 provide a clue.

In the 1860s the German physicist Hermann von Helmholtz noticed three reflections of a single candle in a person's eye. The front and back surfaces of the lens as well as the front surface of the eye act like spherical mirrors that give rise to the reflections. As Helmholtz moved the candle closer to the eye, the position and size of the reflections changed. The picture on the left in Figure 24-49 shows the reflections when the candle is relatively far from the eye, while the one on the right records the reflections when the candle is close by. The inverted image of the candle, labeled C and on the left in each panel, is due to light reflected from the back

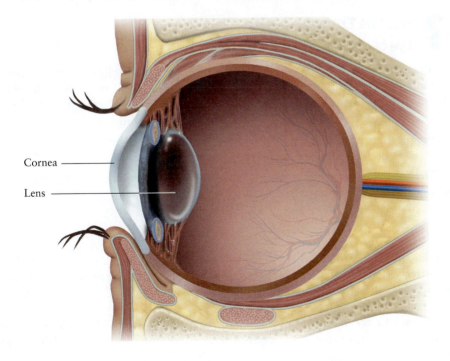

Cornea

Lens

Figure 24-48 Both the cornea and the lens have curved surfaces, both of which cause light to refract. Together, these parts of the eye cause light to be focused on the retina which lines the back of the eye.

Figure 24-49 These drawings are copies of photographs taken of the reflections of a single candle on the pupil of a person's eye. The front and back surfaces of the lens as well as the front surface of the eye act like spherical mirrors, giving rise to three separate reflections of the candle. The image on the right was formed after the candle was moved closer to the eye; the change in the images shows that the shapes of the surfaces of the lens change in order to bring objects of different distances from the eye into focus.

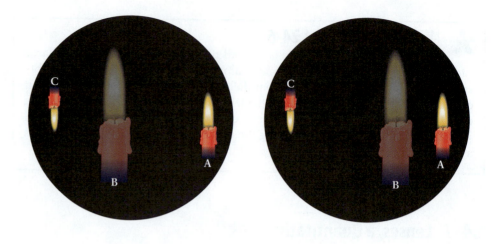

surface of the lens. The larger, dimmer image B in the middle of each diagram is the reflection of the candle off the front surface of the lens, and the brighter, smaller image of the candle on the right in each A is the result of light reflected from the front surface of the eye.

Notice that in Figure 24-49, the reflection of the candle from the front surface of the lens of the eye changes as a function of the distance between the candle and the eye. This suggests that the shape of the surface of the lens changes. The drawing of a pupil focused on an object far from the eye on the left of Figure 24-49 and closer to the eye on the image on the right side of the figure confirms that the lens changes shape to adjust focal length. Indeed, it is through such a change in shape that the image formed by the lens is made to focus on the retina regardless of whether the candle is far from the eye or near it. Unlike a lens made from glass, the lenses in your eyes are flexible, so that as tiny muscles contract and relax the lens becomes more or less round. As the shape of the lens changes, the radius of curvature of its surfaces changes, so that the image focuses on the retina, no matter how far the object is from the eye.

> ### ❓ Got the Concept 24-10
> #### Eye of Mirrors
>
> In the experiment devised by Helmholtz, three surfaces in the eye act as spherical mirrors: the front of the cornea and the front and back surfaces of the lens. Is each of the surfaces convex or concave? Can you tell whether the candle is closer to or farther from the focal point in each case, and if so, where?

For a thin lens made of material of index of refraction n and surrounded by air, the focal length is given by the **lens maker's equation**

$$\frac{1}{f} = (n - 1)\left(\frac{1}{R_1} - \frac{1}{R_2}\right) \qquad (24\text{-}16)$$

Here R_1 and R_2 are the radii of curvature of the front and back surfaces of the lens, respectively. The front surface of the lens is the one closer to the object. Equation 24-16 comes directly from applying Snell's law (Equation 23-6) to a ray of light refracted by both surfaces of a lens.

To see how the image distance is related to the focal length of a thin, convex lens (as well as the object distance), we turn back to Figure 24-43, in which we traced

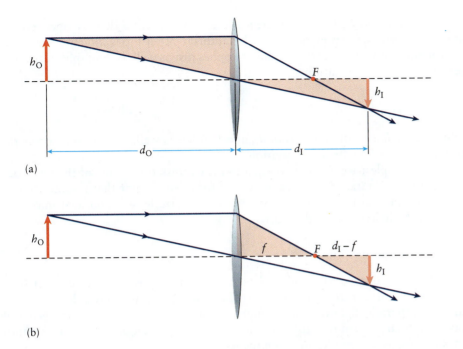

(a)

(b)

Figure 24-50 To study how the image distance is related to the object distance and the focal length of a thin, convex lens, we highlight pairs of right triangles formed in the ray trace diagram of Figure 24-43.

certain light rays from an object, through the lens, to the place where the image forms. Some of those rays, reproduced in **Figures 24-50a** and **24-50b**, form right triangles that we can use to find relationships between the distances of interest.

The shaded regions in Figure 24-50a are similar triangles because the angle opposite to the object height in the leftmost triangle is the same as the angle opposite to the image height in the rightmost triangle. The ratio of the heights of the two triangles is therefore equal to the ratio of their bases, that is,

$$\frac{h_I}{h_O} = \frac{d_I}{d_O} \qquad (24\text{-}17)$$

The two shaded regions in Figure 24-50b are similar triangles for the same reason as those in Figure 24-50a. So the ratio of their heights is equal to the ratio of their bases:

$$\frac{h_I}{h_O} = \frac{d_I - f}{f} \qquad (24\text{-}18)$$

Combining Equations 24-17 and 24-18 we find a relationship between the image distance d_I, the object distance d_O, and the focal length f:

$$\frac{d_I}{d_O} = \frac{d_I - f}{f}$$

Rearranging gives

$$\frac{d_I}{d_O} = \frac{d_I}{f} - 1$$

or

$$\frac{1}{d_O} = \frac{1}{f} - \frac{1}{d_I}$$

or

$$\frac{1}{d_O} + \frac{1}{d_I} = \frac{1}{f} \qquad (24\text{-}7)$$

This equation should look familiar; it is the same equation that we uncovered for spherical mirrors. We previously noted that this relationship is sometimes called the mirror equation, but now we see that it applies just as well to thin, spherical convex lenses. The related equation

$$d_I = \frac{d_O f}{d_O - f} \tag{24-8}$$

which gives d_I directly in terms of d_O and f, holds as well. We will sometimes refer to Equation 24-7 as the **lens equation**.

In its simplest form, the lens equation is expressed in terms of the reciprocals of the object distance, image distance, and focal length. For that reason it is common, especially among optometrists, to characterize lenses (and also mirrors) in terms of the inverse of focal length

$$P = \frac{1}{f} \tag{24-19}$$

Here P is the **power** of the optical device. The unit of P is the inverse meter (m^{-1}); $1\ m^{-1}$ is known as *1 diopter*. The larger the power of a thin, convex lens, the closer to the lens parallel light rays are brought to a focus. A 2-diopter lens brings parallel rays to a focus $1/2$ m (0.5 m) from the lens, while a 5-diopter lens brings parallel rays to a focus $1/5$ m (0.2 m) from the lens.

? Got the Concept 24-11

Where's the image?

Based on the relationship between the image distance, the object distance, and the focal length of a thin, convex lens, where does the image form when the object is placed very far from the lens? Where does the image form when the object is placed at the focal point of the lens?

Physicist's Toolbox 24-3 Signs of Distances

To correctly make calculations with Equations 24-7 and 24-8 for thin lenses, physicists define a sign convention for d_I and f.

- d_I is positive when the image is real, that is, the image forms where light rays cross.
- d_I is negative when the image is virtual, that is, the image forms where light rays appear to cross.
- f is positive for a thin convex (converging) lens.
- f is negative for a thin concave (diverging) lens.

Because a thin, concave lens causes light rays to diverge, it always forms a virtual image, so d_I is always negative. A thin, convex lens can form either a real or virtual image, depending on how far the object is from the lens, relative to the focal point. So d_I can be either positive or negative for a thin, convex lens.

Equation 24-7 holds for both thin concave and thin convex lenses. To make calculations for a concave lens, however, we take the image distance and the focal length to be negative. This is similar to the sign convention we use for a spherical convex mirror, which also causes parallel light rays to diverge. Diverging rays can only form a virtual image (diverging rays can never cross, the requirement for the formation of a real image), so the image distance is negative. In addition, the effective focal point of the concave lens must be on the virtual side of the lens, so we take focal length to be negative as well.

The relationship between the ratio of image and object heights and image and object distances (Equation 24-17) leads to the same definition of magnification that we found for spherical mirrors:

$$m = -\frac{d_I}{d_O} \qquad (24\text{-}10)$$

A positive value of m indicates an upright image. For example, when an object is placed closer than the focal point to a thin, convex lens, an upright, virtual image forms, as in Figure 24-46. In this case, d_I is negative because the image is virtual, and m is therefore positive.

Example 24-6 Magnification

A thin, convex lens has a focal length of 0.15 m. How far from the lens does the image form when an object is placed 0.09 m from the center of the lens? Is the image virtual or real? Is the image inverted or upright? By what factor is the image magnified relative to the object?

SET UP

As with all lens and mirror problems, we start by making a ray trace diagram. We don't worry about making a picture which is exactly to scale, but getting the dimensions approximately correct will help us test numeric answers for reasonability. In **Figure 24-51**, we trace two rays coming from the tip of a standard arrow object placed about 9 cm in front of a thin, convex lens of focal length 15 cm. One of the rays leaves the tip of the arrow traveling parallel to the principal axis of the lens; it is refracted so that it passes through the focal point. The other ray strikes the center of the lens, so it continues on in the same direction after passing through the thin lens. As shown in the figure, although the two rays never meet, they do appear to meet on the same side of the lens as the image.

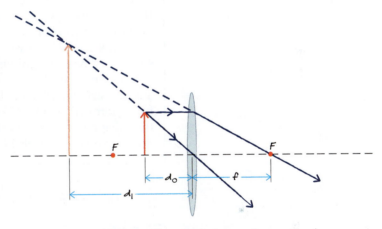

Figure 24-51 A ray trace diagram helps us determine the magnification of a thin, convex lens.

SOLVE

The ray trace diagram provides some of the answers we seek. The image formed by the lens in this case is upright. It is also virtual, because it forms not where light rays meet but instead where they appear to meet.

We find the image distance by applying Equation 24-8:

$$d_I = \frac{d_O f}{d_O - f} = \frac{(0.09 \text{ m})(0.15 \text{ m})}{0.09 \text{ m} - 0.15 \text{ m}} = -0.225 \text{ m} = -0.23 \text{ m}$$

The magnification, from Equation 24-10, is then

$$m = -\frac{d_I}{d_O} = -\frac{-0.225 \text{ m}}{0.09 \text{ m}} = 2.5$$

REFLECT

The image forms about 23 cm from the lens. Because the image distance is negative, we know that the image is virtual and forms on the same side of the lens as the object. The height of the image is 2.5 times the height of the object. Because the magnification is positive, we know that the image is upright. Both of these conclusions are consistent with those we made by examining the ray trace diagram. In addition, although we didn't try to make dimensions in the ray trace diagram exact, by eye the ratio of the focal length to the object distance looks to be about 2:1, certainly consistent with our result of 15:9. We can therefore expect that the object distance in the diagram is about one and one-half times the focal length (23 compared to 15); it is. We can also expect that height of the image in the diagram is about 2.5 times the height of the object. It is!

Practice Problem 24-6 A thin convex lens has a focal length of 0.15 m. How far from the lens does the image form when an object is placed 0.12 m from the center of the lens? Is the image virtual or real? Is the image inverted or upright? By what factor is the image magnified relative to the object?

> **? Got the Concept 24-12**
> **A Magnifying Glass**
>
> By placing an object closer to your eyes it looks larger and you can see it in more detail. The object has an angular size θ_0, as shown in **Figure 24-52a**. If you're typical, however, you can focus on an object only if it's about 25 cm or farther from your eyes, which may not allow you to see small details clearly. A convex lens with a relatively short focal length, a magnifying glass, is the solution. When a magnifying glass is placed in between the object and your eye and the object is closer to the lens than the focal point, the image forms with angular size θ_M, which is larger than θ_0. In addition, the image forms farther from your eye than the 25 cm limit so you can focus clearly on it. Draw a ray trace diagram that shows the principle of a magnifying glass, starting from the sketch shown in **Figure 24-52b**.

When an object is placed close to the eye it spans an angular size θ_0.

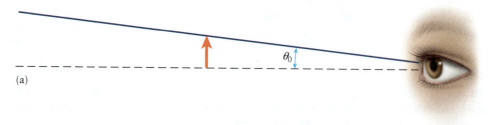

(a)

A magnifying glass is placed between the object and the eye, in order to see more detail without moving the object closer to the eye. Where does the image form?

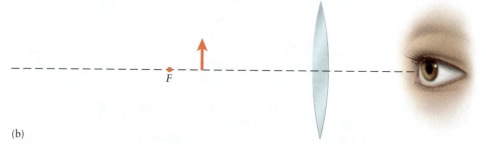

Figure 24-52 (a) The angular size of an object depends on the angle it subtends. (b) Use this setup to draw a ray trace diagram that demonstrates the principle of a magnifying glass.

(b)

? Got the Concept 24-13
Microscopes

You've used microscopes many times, but do you know how they work? By applying what you've learned about lenses, you can figure it out for yourself! A simple compound microscope has two lenses, the objective and the eyepiece. An object is placed just beyond the focal point of the objective in order to form a magnified image. Using the sketch in **Figure 24-53**, determine the location and size of the image by tracing rays from the tip of the arrow to the eye. Is the image real or virtual?

Go to Interactive Exercise 24-3 for more practice dealing with lenses

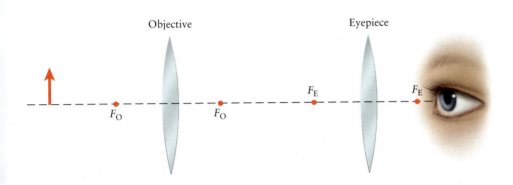

Figure 24-53 Use this setup to draw a ray trace diagram that demonstrates the principle of a simple compound microscope.

✳ What's Important 24-7

The focal length of a thin lens is related to the distances of an object and its image from the lens by the same equation that describes spherical mirrors. A similar sign convention applies. Image distance d_I is positive when the image is real and negative when the image is virtual. Also, the focal length f is positive for a thin, convex lens and negative for a thin, concave lens.

Answers to Practice Problems

24-1 Because the mirrors are parallel, the directions normal to the mirrors are also parallel. The angle from the normal at which the light leaves the first mirror and the angle of incidence on the second mirror together form alternate interior angles and are therefore equal. The same is true as the light reflects back to the first mirror, and therefore the second strike occurs at the same angle θ as the first.

24-2 12 cm

24-3 8.0 cm

24-4 24 cm

24-5 0.60 m, virtual, upright, 5

Answers to Got the Concept Questions

24-1   If you see someone's eyes in a mirror, she can see you as well. This must be true because if light rays can reflect off the mirror from her eyes to yours, then light can follow that path in the opposite direction and reflect from your eyes to hers, too.

24-2   From the information given it's not possible to know whether the person around the corner can see you. The situation is sketched in Figure 24-54. Light rays, shown in red, that reflect off the mirror from his right hand to your eyes enable you to see his hand. You can't see his eyes because the path that light would need to follow for this to happen, shown by the gray, dashed line, does not intercept the mirror. Whether the other person can see some part of you, however, depends on how far out you are holding your right hand. If it is far enough away from your body, light rays can follow the path drawn in blue, from your hand to his eyes.

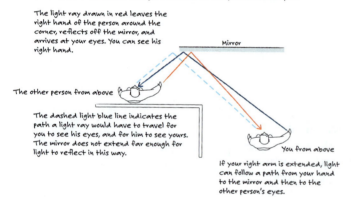

You and someone else stand around a corner from each other (as seen from above). You can see a reflection of his right hand, but not his eyes. Can he see you?

The light ray drawn in red leaves the right hand of the person around the corner, reflects off the mirror, and arrives at your eyes. You can see his right hand.

Mirror

The other person from above

The dashed light blue line indicates the path a light ray would have to travel for you to see his eyes, and for him to see yours. The mirror does not extend far enough for light to reflect in this way.

You from above

If your right arm is extended, light can follow a path from your hand to the mirror and then to the other person's eyes.

Figure 24-54

24-3   Figure 24-55 shows two light rays traced from the tip of an arrow placed in front of the mirror, through the glass and then reflecting off the shiny metal layer. Notice that the rays refract on entering the glass plate, reflect off the metal, and then refract again as they leave the glass. The reflections obey the law of reflection, that is, the angle of incidence equals the angle of reflection. The refractions obey Snell's law, so light is bent closer to the normal on entering the glass and bent farther from the normal on exiting. The outgoing rays appear to be coming from a point behind the mirror; however, the image forms closer to the mirror than the distance of the object from it. This is similar to the phenomenon we explored in the last chapter, in which a pool of water appears shallower than it really is. Although the image distance does not equal the object distance in this case, as you can verify

from Figure 24-55, the heights of the image and object are the same.

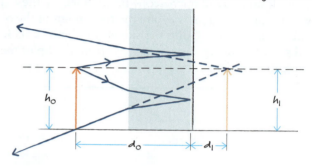

Light reflecting from a plane mirror with a thick glass front.

Figure 24-55

24-4   When an object is placed farther from a concave mirror than the focal point, the image is real and inverted. Note that a real image forms in front of a mirror, that is, it appears to be in front of the mirror. So when you hold the spoon at arm's length, the object (you!) must therefore be farther from the reflective surface than the focal point. When you hold the spoon 4 cm from your eye, however, you (your eye) must be closer to the spoon than its focal point. An object placed closer to a reflective surface than the focal point forms a virtual image. Because the virtual image is formed at the point where light rays from any point on an object appear to meet (but don't actually), the image forms behind the mirror, and is upright. The focal length of the spoon must be about 6 cm from the spoon. At the focal point no image forms because light rays from any point on your face are reflected parallel to each other.

24-5   The image of an object placed inside the focal point forms behind the mirror, as in, for example, Figure 24-18. The sign of d_I is therefore negative, so the fraction in Equation 24-10 is also negative. However, the fraction is negated to find m, so magnification is positive. The sign convention we established using Equation 24-9 indicates that m is positive for an upright image. The image of an object placed outside the focal point (for example, Figure 24-16) forms in front of the mirror and is inverted. In this case, the sign of d_I is positive, so m is negative. Again, this follows the sign convention we established using Equation 24-9.

24-6   It is not possible to start a fire with a spherical, convex mirror. Unlike a concave mirror, a convex mirror does not focus parallel light rays. The image formed by a convex mirror is a result of the apparent, but not actual, crossing of reflected rays behind the mirror. Light must actually cross to focus a large amount of energy in a small region of space.

24-7 For an object "very far" from the mirror we can treat d_O as overwhelmingly large compared to f, so that f can be neglected in the denominator of the fraction in Equation 24-15. The fraction is then approximately equal to 1, so that d_I is approximately equal to f. When the object is "very close" to the mirror, d_O is approximately equal to zero in Equation 24-15. In that case, d_I is also approximately equal to zero. Because f is negative, d_I is also negative so the image forms just behind the mirror.

24-8 It is not possible to create an image larger than the object using a convex mirror. From Equation 24-15, d_I equals d_O multiplied by the fraction $f/(d_O - f)$. Because f is negative the absolute value of this fraction will always be less than 1, regardless of the value of the object distance. The absolute value of d_I will therefore always be less than d_O (the image forms closer to the mirror than the object); from Equation 24-10 the absolute value of the magnification, then, will always be less than or equal to 1.

24-9 See Figure 24-56, in which a light ray that passes through the focal point has been drawn in green.

The light ray drawn in green leaves the tip of the arrow in the direction of one of the focal points, so it must be refracted in a direction parallel to the principal axis. The refracted ray traces back to the top of the image of the arrow.

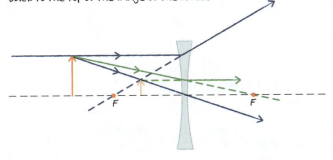

Figure 24-56

24-10 Common sense probably tells you that the front surfaces of the cornea and the lens act as convex mirrors, or you might guess this is the case from the diagram in Figure 24-48. Either way, the reflections from those two surfaces, shown as A and B in Figure 24-49, support that conclusion. The images formed fit into the eye, so we imagine they are smaller than the candle that serves as the object, and the images are upright. As shown in Figure 24-32, when an object is farther from a convex mirror than the focal point, an upright image, smaller than the object is formed. We conclude, therefore, that the candle is farther from the eye than the focal point of

those two mirrorlike surfaces. The back surface of the lens must act like a concave mirror. The reflection from the surface is inverted, and only a concave spherical mirror can form an inverted image. The candle must be farther from the back surface of the lens than the focal point of that reflective surface, because a concave mirror only forms an inverted image when the object is farther from the mirror than the focal point.

24-11 When the object is very far from the lens, the image forms at the focal point. This is the definition of the focal point! To see this, notice in Equation 24-8, that when d_O is very large compared to f, the denominator on the right side is approximately equal to d_O. In this case, the fraction reduces to f. Conversely, when the object is placed at the focal point so that d_O equals f, the denominator on the right side of Equation 24-8 becomes zero, so that the fraction, and therefore d_I, becomes infinitely large. This is also part of the definition of the focal point.

24-12 The ray trace diagram is shown in **Figure 24-57**.

The effect of the magnifying glass is to increase the angular size of the image to θ_M.

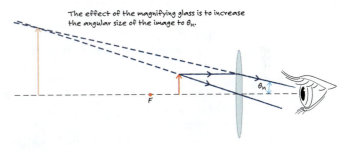

Figure 24-57

24-13 The ray trace diagram is shown in **Figure 24-58**.

In tracing the rays through the two lenses notice that the ray that passes through the focal point of the objective exits parallel to the principal axis of the objective, so that this ray enters the eyepiece parallel to the principal axis of the eyepiece. For that reason, this ray passes through the focal point of the eyepiece when it exits.

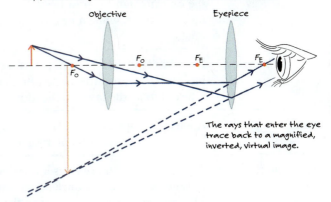

The rays that enter the eye trace back to a magnified, inverted, virtual image.

Figure 24-58

SUMMARY

Topic	Summary	Equation or Symbol
center of curvature	Spherical mirrors and lenses are made from a small section of a full sphere. The center of that sphere is the center of curvature.	
concave mirror	A mirror is concave when the inner surface of the spherical section is reflective.	
convex mirror	A mirror is convex when the outer surface of the spherical section is reflective.	
focal length, lens	The focal length f of a lens is the distance from the focal point to the center of the lens.	
focal length, mirror	The focal length f of a mirror is the distance from the focal point to the center of the mirror. In the case of a mirror small enough that all rays parallel to the principal axis are brought to a single real or virtual focus, the focal length is half of the radius.	$f = \dfrac{r}{2}$ $\quad$ (24-2)
focal point	The focal point of a mirror or lens is the point along its principal axis at which incident rays parallel to the principal axis converge after reflecting (from a mirror) or refracting (through a lens).	
image distance	The image distance d_I is the distance along the principal axis between the image and the center of a lens or mirror.	d_I
image height	The image height h_I is the height, measured in the direction perpendicular to the principal axis, of the image formed by a lens or mirror.	h_I
inverted image	An inverted image is one in which top and bottom have been reversed relative to the orientation of the object.	
lateral magnification	The lateral magnification, or simply magnification, of an image formed by a lens or mirror is the ratio of the image height to the object height. A positive value of magnification indicates an upright image, while a negative value indicates an inverted image.	$m = \dfrac{h_I}{h_O}$ $\quad$ (24-1) $m = -\dfrac{d_I}{d_O}$ $\quad$ (24-10)
law of reflection	A ray of light reflects from a surface so that the angle of incidence θ_i equals the angle of reflection θ_r. This is sometimes referred to as the law of reflection.	$\theta_i = \theta_r$
lens	A lens is an optical device that changes the direction of rays of light as they refract on entering and then exiting it.	
lens equation	The lens equation provides a relationship between the image distance d_I, object distance d_O, and the focal length f of a lens. The lens equation is identical to the mirror equation.	$\dfrac{1}{d_O} + \dfrac{1}{d_I} = \dfrac{1}{f}$ $\quad$ (24-7)

lens maker's equation	For a thin lens made of material of index of refraction n and surrounded by air, the focal length f depends on the radius of curvature of the two surfaces of the lens; the relationship is given by the lens maker's equation.	$\dfrac{1}{f} = (n-1)\left(\dfrac{1}{R_1} - \dfrac{1}{R_2}\right)$ (24-16)
magnification	The magnification, or lateral magnification, of an image formed by a lens or mirror is the ratio of the image height to the object height. A positive value of magnification indicates an upright image, while a negative value indicates an inverted image.	$m = \dfrac{h_I}{h_O}$ (24-1) $m = -\dfrac{d_I}{d_O}$ (24-10)
mirror	A mirror is an optical device that changes the direction of rays of light as they reflect from its surface.	
mirror equation	The mirror equation provides a relationship between the image distance d_I, object distance d_O, and the focal length f of a mirror. The mirror equation is identical to the lens equation.	$\dfrac{1}{d_O} + \dfrac{1}{d_I} = \dfrac{1}{f}$ (24-7)
object distance	The object distance d_O is the distance along the principal axis between the object and the center of a lens or mirror.	d_O
object height	The object height h_O is the height, measured in the direction perpendicular to the principal axis, of an object placed in front of a lens or mirror.	h_O
optical device	An optical device is a piece of either transparent or reflective material shaped so that it changes the direction of rays of light in a regular way.	
plane mirror	A plane mirror is a mirror with a flat reflective surface. The image formed by a plane mirror forms the same distance behind the mirror as the distance the object is in front of the mirror.	
power	The power of an optical device is the inverse of its focal length.	$P = \dfrac{1}{f}$ (24-19)
principal axis	The principal axis runs through the center of a lens or mirror and also through the center of the full sphere which gives the lens or mirror its shape.	
real image	A real image forms where reflected or refracted light rays come together.	
thin lens	Each surface of a thin lens is a small part of a sphere of relatively large radius, so the distance across the lens along the principal axis is negligibly small.	
upright image	An upright image is one in which the orientation of the image is the same as the orientation of the object.	
virtual image	A virtual image is one formed when light rays reflected by a mirror or refracted by a lens appear to meet when the rays are traced back through the lens or mirror. No light rays actually meet where a virtual image forms.	

QUESTIONS AND PROBLEMS

In a few problems, you are given more data than you actually need; in a few other problems, you are required to supply data from your general knowledge, outside sources, or informed estimate.

Interpret as significant all digits in numerical values that have trailing zeros and no decimal points.

For all problems, use $g = 9.8 \text{ m/s}^2$ for the free-fall acceleration due to gravity. Neglect friction and air resistance unless instructed to do otherwise.

• Basic, single-concept problem

•• Intermediate-level problem, may require synthesis of concepts and multiple steps

••• Challenging problem

SSM *Solution is in Student Solutions Manual*

Conceptual Questions

1. •A plane mirror seems to invert your image left and right but not up and down. Why is this? SSM

2. •Explain why reflected light is usually diffuse and scattered.

3. •What is the difference between a real image and a virtual image?

4. •When you view a car's side mirror, you see a smaller image than you would if the mirror were flat. Is the mirror concave or convex? Explain your answer.

5. •Explain the meaning, in terms of physics, of the phrase etched on the right side mirror of most cars: "Objects in mirror are closer than they appear."

6. •What is the radius of curvature of a plane mirror? Explain your answer.

7. •For a certain lens in air, both radii of curvature are positive. Is it a converging lens or a diverging lens, or do you need additional information to tell? Explain your answer. SSM

8. •If we move a glass lens from air into water, what will happen to the focal length of the lens? Explain your answer.

9. •A laptop computer is connected to a video projector which projects an image on a screen. If the lens of the projector is half covered, what happens to the image? Explain your answer.

10. •**Biology** The image focused on your retina is actually inverted (sketch a simple ray trace diagram showing this observation). What does this fact say about our definitions of "right side up" and "upside down"?

11. •**Biology** Experimental subjects who wear inverting lenses (glasses that invert all images) for several days adapt to their new perception of the world so well that they can even ride a bicycle. Several days after the glasses are removed, their perceptions return to normal. Discuss this phenomenon and comment.

12. •**Medical** Explain why converging lenses are used to correct farsightedness while diverging lenses are used for nearsightedness.

13. •**Biology** Discuss why nearsightedness is not found in all people, but virtually everyone eventually becomes farsighted as they age.

14. ••Explain why looking through a small opening often provides visual acuity even to an extremely nearsighted person.

Multiple-Choice Questions

15. •Which is true when an object is moved farther from a plane mirror?
   A. The height of the image decreases and the image distance increases.
   B. The height of the image stays the same and the image distance increases.
   C. The height of the image increases and the image distance increases.
   D. The height of the image stays the same and the image distance decreases.
   E. The height of the image decreases and the image distance decreases. SSM

16. •A real image can form in front of
   A. a plane mirror.
   B. a concave mirror.
   C. a convex mirror.
   D. any type of mirror.
   E. no mirrors.

17. •When an object is placed a little farther from a concave mirror than the focal length, the image is
   A. magnified and real.
   B. magnified and virtual.
   C. smaller and real.
   D. smaller and virtual.
   E. smaller and reversed.

18. •If you want to start a fire using sunlight, which kind of mirror would be most efficient?
   A. a plane mirror
   B. a concave mirror
   C. a convex mirror
   D. any type of plane, concave, or convex mirror
   E. It is not possible to start a fire using sunlight and a mirror; you must use a concave lens.

19. •An object is placed at the center of curvature of a concave mirror. Which of the following is true about the image?
  A. real and upright
  B. real and inverted
  C. virtual and upright
  D. virtual and inverted
  E. real, inverted, and reversed SSM

20. •When an object is placed farther from a convex mirror than the focal length, the image is
  A. larger and real.
  B. larger and virtual.
  C. smaller and real.
  D. smaller and virtual.
  E. smaller and reversed.

21. •Medical When a dentist needs a mirror to see an enlarged, upright image of a patient's tooth, what kind of mirror should he use?
  A. a plane mirror
  B. a concave mirror
  C. a convex mirror
  D. either a plane mirror or a concave mirror
  E. either a plane mirror or a convex mirror

22. •A magnifier allows one to look at a very near object by forming an image of it farther away. The object appears larger. To create a magnifier, one would use a
  A. short focal length (≪1 m) convex lens.
  B. short focal length (≪1 m) concave lens.
  C. long focal length (>1 m) convex lens.
  D. long focal length (>1 m) concave lens.
  E. either a convex or a concave lens.

23. •A compound microscope is a two-lens system used to look at very small objects. Which of the following statements is correct?
  A. The objective lens and the eyepiece both have the same focal length and both serve as simple magnifiers.
  B. The objective lens is a short focal length, convex lens and the eyepiece functions as a simple magnifier.
  C. The objective lens is a long focal length, convex lens and the eyepiece functions as a simple magnifier.
  D. The objective lens is a short focal length, concave lens and the eyepiece functions as a simple magnifier.
  E. The objective lens is a long focal length, concave lens and the eyepiece functions as a simple magnifier. SSM

Estimation/Numerical Analysis

24. •Estimate the focal length (in cm) of the corrective lenses in an average pair of glasses.

25. •Estimate the focal length of a convex blind spot mirror that is often added to a vehicle's outside mirror.

26. •Estimate the radius of curvature of a spherical mirror that is typically found in the corridors of busy hospitals, for example, to help prevent collisions when going around a corner.

27. •Give three examples of spherical concave mirrors in your daily life and estimate their approximate focal length. SSM

28. •Give three examples of spherical convex mirrors in your daily life and estimate their approximate focal length.

29. •Determine the focal length for an unknown lens with the following object and image distances:

Object Distance (cm)	Image Distance (cm)
30	98
35	67
40	53
45	47
50	42
55	38
60	37
65	35
70	34
75	33
80	32

Problems

24-1: Plane Mirrors

30. •The angle of incidence on a flat mirror is 0°; what is the angle of reflection?

31. •Two flat mirrors are perpendicular to each other. An incoming beam of light makes an angle of 30° with the first mirror as shown in Figure 24-59. What angle will the outgoing beam make with respect to the normal of the second mirror? SSM

Figure 24-59 Problem 31

32. •A 1.8-m-tall man stands 2.0 m in front of a vertical plane mirror. How tall will the image of the man be?

33. •What must be the minimum height of a plane mirror in order for a 1.80-m tall person to see a full image of himself?

34. ••A plane mirror is 10 m away from and parallel to a second plane mirror (Figure 24-60). Find the location of the first five images formed by each mirror when an

This page is intentionally left blank.

For complete end of chapter problem sets, please go to
www.whfreeman.com/kestentauck

48. ••An object is positioned 24 cm from a spherical concave mirror of unknown focal length. The image that is formed is 30 cm from the mirror. (a) Calculate the focal length. (b) Is this answer unique? (c) Is the image real or virtual? (d) If the object is 10 cm tall, determine the height of the image(s).

49. ••Construct the ray trace diagrams to locate the images in the following cases. (a) A 10-cm-tall object is located 5 cm in front of a spherical concave mirror with a radius of curvature of 20 cm. (b) A 10-cm-tall object is located 10 cm in front of a spherical concave mirror with a radius of curvature of 20 cm. (c) A 10-cm-tall object is located 20 cm in front of a spherical concave mirror with a radius of curvature of 20 cm. In each case, start by estimating the height of the image from a sketch.

50. ••Derive a relationship between the radius of curvature of a spherical, concave mirror and the object distance that gives an upright image that is four times as tall as the object.

24-4: Spherical Convex Mirrors, a Qualitative Look

51. •Describe the difference between the images seen in a spherical, convex mirror when the object is "up close" (a shorter distance from the mirror than the focal distance of the mirror) compared to "far away" (a longer distance from the mirror than the focal distance of the mirror).

52. •Are there any situations where a real image is formed in a spherical, convex mirror? Why or why not?

53. •How can you remember the difference between the shapes of a spherical *concave* mirror and a spherical *convex* mirror?

54. ••Construct the ray trace diagrams to locate the images and estimate the image height in each of the following cases. (a) A 10-cm-tall object is located 5 cm in front of a spherical convex mirror with a radius of curvature of 20 cm. (b) A 10-cm-tall object is located 10 cm in front of a spherical convex mirror with a radius of curvature of 20 cm. (c) A 10-cm-tall object is located 20 cm in front of a spherical convex mirror with a radius of curvature of 20 cm.

24-5: Spherical Convex Mirrors, a Quantitative Look

55. ••The radius of curvature of a spherical convex mirror is 20 cm. Describe the image formed when a 10-cm-tall object is positioned (a) 20 cm from the mirror, (b) 50 cm from the mirror, and (c) 100 cm from the mirror. For each case, provide the image distance, the image height, the type of image (real or virtual), and the orientation of the image (upright or inverted). SSM

56. ••The radius of curvature of a spherical convex mirror is 15 cm. Describe the image formed when a 20-cm-tall object is positioned (a) 5 cm from the mirror, (b) 20 cm from the mirror, and (c) 100 cm from the mirror. For each case, provide the image distance, the image height, the type of image (real or virtual), and the orientation of the image (upright or inverted).

57. •A car's convex rearview mirror has a radius of curvature equal to 15 m. What are the magnification, type, and location of the image that is formed by an object that is 10 m from the mirror?

58. •An 18-cm-long pencil is placed beside a convex spherical mirror and its image is 10.5 cm in length. If the radius of curvature of the mirror is 88.4 cm, find the image distance, the object distance, and the magnification of the pencil.

59. •A 1-cm-long horse fly hovers 1 cm from a shiny sphere with a radius of 25 cm. Describe the location of the image of the fly, its type (real or virtual), and its length.

60. •A spherical convex mirror is placed at the end of a driveway on a corner with a limited view of oncoming traffic. The mirror has a radius of curvature of 1.85 m. Where will the image of a car that is 12.6 m from the mirror appear?

61. ••Using the mirror equation, prove that all images in spherical convex mirrors are virtual. SSM

62. •A girl sees her image in a shiny glass sphere tree ornament that has a diameter of 10 cm. The image is upright and is located 1.5 cm behind the surface of the ornament. How far from the ornament is the child located?

63. •A shiny sphere, 30 cm in diameter, is placed in a garden for aesthetic purposes. Determine the type, location, and height of the image of a 6.00-cm-tall squirrel that is located 40 cm in front of the sphere.

24-6: Lenses, a Qualitative Look

64. •Under what circumstances will the images formed by converging or diverging lenses be designated as "real"? Indicate the type or types of lenses, and the required position of the object.

65. •A real image that is created due to reflection in a spherical mirror appears in front of the mirrored surface. Is this also the case for a real image that is created due to refraction in a lens? SSM

66. •Where does the bending of light physically take place in a typical biconcave or biconvex lens? Is this how we draw ray trace diagrams? Why or why not?

67. •Which type of lens is used in eyeglasses to correct (a) nearsightedness? (b) farsightedness?

68. •How many rays are required to trace out the image that is formed by a lens in a ray trace diagram?

24-7: Lenses, a Quantitative Look

69. •A 10-cm-tall object is located in front of a converging lens with a power of 5 diopters. Describe the image that is created (type, location, height) and draw the ray trace diagrams for the object located (a) 5 cm from the lens, (b) 10 cm from the lens, (c) 20 cm from the lens, and (d) 50 cm from the lens. **SSM**

70. ••A 10-cm-tall object is located in front of a diverging lens with a power of 5 diopters. Describe the type, location, and height of the image that is created, and draw the ray trace diagrams if the object is located (a) 5 cm from the lens, (b) 10 cm from the lens, (c) 20 cm from the lens, and (d) 50 cm from the lens.

71. ••A 2.00-cm-tall object is 18.0 cm in front of a double convex lens with a focal length of 30.0 cm. (a) Use the lens equation and (b) a ray trace diagram to describe the type, location, and height of the image that is formed.

72. ••A lens is formed from a plastic material that has an index of refraction of 1.55. If the radius of curvature of one surface is 1.25 m and the radius of curvature of the other surface is 1.75 m (**Figure 24-65**), use the lens maker's equation to calculate the focal length and the power of the lens.

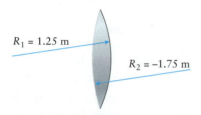

Figure 24-65 Problem 72

73. ••A glass lens ($n = 1.60$) has a focal length of -31.8 cm and a plano-concave shape. (a) Calculate the radius of curvature of the concave surface. (b) If the lens is constructed from the same glass to form a plano-convex shape with the same radius of curvature, what will the focal length be? **SSM**

74. ••A 2.00-cm-tall object is 30.0 cm in front of a double convex lens that has a focal length of 18.0 cm. (a) Use the lens equation and (b) a ray trace diagram to describe the type, location, and height of the image that is formed.

75. ••**Biology** Calculate the overall magnification of a compound microscope that uses an objective lens with a focal length of 0.50 cm, an eyepiece with a focal length of 2.50 cm, and a distance of 18 cm between the two.

General Problems

Note: In these problems "infinity" means a very large distance compared to the focal length of a lens.

76. ••The opposite walls of a barber shop are covered by plane mirrors, so that multiple images arise from multiple reflections, and you see many reflected images of yourself, receding to infinity. The width of the shop is 6.50 m, and you are standing 2.00 m from the north wall. (a) How far apart are the first two images of you behind the north wall? (b) What is the separation of the first two images of you behind the south wall? Explain your answer.

77. ••An object is 40.0 cm from a concave spherical mirror whose radius of curvature is 32.0 cm. Locate and describe the type and magnification of the image formed by the mirror (a) by calculating the image distance and lateral magnification and (b) by drawing a ray trace diagram. On the ray trace diagram draw an eye in a position from which it can view the image.

78. ••Derive the lens maker's equation:

$$\frac{1}{f} = (n - 1)\left(\frac{1}{R_1} - \frac{1}{R_2}\right)$$

Be sure to carefully define the sign conventions for R_1 and R_2 (the respective radius of curvature of each surface).

79. ••**Biology** A typical human lens has an index of refraction of 1.43. The lens has a double convex shape, but its curvature can be varied by the ciliary muscles acting around its rim. At minimum power, the radius of the front of the lens is 10.0 cm while that of the back is 6.00 mm. At maximum power the radii are 6.00 mm and 5.50 mm, respectively. (The numbers can vary somewhat.) If the lens were in air, (a) what would be the ranges of its focal length and its power (in diopters)? (b) At maximum power, where would the lens form an image of an object 25 cm in front of the front surface of the lens? (c) Would the image fall on the retina of a human eye? The retina is located approximately 2.5 cm from the lens.

80. ••**Biology** A typical person's eye is 2.5 cm in diameter and has a near point (the closest an object can be and still be seen in focus) of 25 cm, and a far point (the farthest an object can be and still be in focus) of infinity. (a) What is the range of the effective focal lengths of the focusing mechanism (lens plus cornea) of the typical eye? (b) Is the equivalent focusing mechanism of the eye a diverging or a converging lens? Justify your answer without using any mathematics, and then see if your answer is consistent with your result in part (a).

81. •••A geneticist looks through a microscope to determine the phenotype of a fruit fly. The microscope is set to an overall magnification of 400× with an objective lens that has a focal length of 0.60 cm. The distance between the eyepiece and objective lenses is 16 cm. Find the focal length of the eyepiece lens assuming a near point of 25 cm (the closest an object can be and still be seen in focus). **SSM**

82. ••A thin lens made of glass that has a refractive index equal to 1.60 has surfaces with radii of curvature that have magnitudes equal to 12.0 and 18.0 mm. What are the possible values for its focal length? Sketch a cross-sectional view of the lens for each possible combination, making sure to label the radii of curvature of each surface of the lens and the associated focal length of the entire lens.

83. ••You are designing lenses that consist of small double convex pieces of plastic having surfaces with radii of curvature of magnitudes 3.50 cm on one side and 4.25 cm on the other side. You want the lenses to have a focal length of 1.65 cm in air. What should be the index of refraction of the plastic to achieve the desired focal length?

84. •••A lens of focal length +15.0 cm is 10.0 cm to the left of a second lens of focal length −15.0 cm. (a) Where is the final image of an object that is 30.0 cm to the left of the positive lens? (b) Is the image real or virtual? (c) How do the image's size and orientation compare to those of the original object? Explain your answer. (d) How must you be oriented to see the image?

85. •••A thin, plano-concave lens having a focal length of magnitude 45.0 cm has the same principal axis as a concave mirror with a radius of 60.0 cm. The vertex of the mirror is 20.0 cm from the lens, with the lens in front of the mirror. An object is placed 15.0 cm in front of the lens. (a) Where is the final image due to the lens–mirror combination? (b) Is the final image real or virtual? Upright or inverted? (c) Suppose now that the concave mirror is replaced by a convex mirror of the same radius. Repeat parts (a) and (b) for the new lens–mirror combination. SSM

86. •••When you place a bright light source 36.0 cm to the left of a lens, you obtain an upright image 14.0 cm from the lens and also a faint inverted image 13.8 cm to the left of the lens that is due to reflection from the front surface of the lens. When the lens is turned around, a faint inverted image is 25.7 cm to the left of the lens. What is the index of refraction of the material?

87. •••A thin, double convex lens having a focal length of magnitude 25.0 cm is placed 1.000 m from a plane mirror that is oriented perpendicular to the principal axis of the lens. A flower, 8.40 cm tall, is 1.450 m from the mirror on the principal axis of the lens. (a) Where is the final image of the flower produced by the lens–mirror combination? Is it real or virtual? Upright or inverted? How tall is the image? (b) If the convex lens is replaced by a double concave lens having a focal length of the same magnitude as the original lens, what will be the answers to part (a)?

88. •••Biology A compound microscope has a tube length of 20.0 cm and an objective lens of focal length 8.0 cm. (a) If it is to have a magnifying power of 200×, what should be the focal length of the eyepiece? (b) If the final image is viewed at infinity, how far from the objective should the object be placed?

89. •••Calc Using the lens maker's equation, find the radius of curvature needed for the leading surface (in other words, $R_1 = ?$) to optimize the power of the lens. Assume that the secondary surface has a fixed radius of curvature, R_2. *Hint:* Write the power (P) as a function of R_1, and apply the maximization condition.

90. ••Biology A typical human eye is nearly spherical and usually about 2.5 cm in diameter. Suppose a person first looks at a coin that is 2.3 cm across and located 30.0 cm from her eye and then looks up at her friend who is 1.8 m tall and 3.25 m away. (a) Find the approximate size of each image (coin and friend) on her retina. (*Hint:* Just look at rays from the top and bottom of the object that pass through the center of the lens.) (b) Are the images in part (a) upright or inverted and are they real or virtual?

91. ••Biology You may have noticed that the eyes of cats appear to glow green in low light, as shown at the beginning of chapter 23. This effect is due to the reflection of light by the tapetum lucidum, a highly reflective membrane just behind the retina of the eye. Light that has passed through the retina without hitting photoreceptors is reflected back to the retina, thus enabling the animal to see much better than humans in low light. The eye of a typical cat is about 1.25 cm in diameter. Assume that the light enters the eye traveling parallel to the principal axis of the lens. (a) If some of the light reflected off the tapetum lucidum escapes being absorbed by the retina, where will it be focused? (b) The refractive index of the liquid in the eye is about 1.4. How does this affect the location of the image in part (a)? SSM

92. ••Medical A nearsighted eye is corrected by placing a diverging lens in front of the eye. The lens will create a virtual image of a distant object at the far point (the farthest an object can be and still be in focus) of the myopic viewer where it will be clearly seen. In the traditional treatment of myopia, an object at infinity is focused to the far point of the eye. If an individual has a far point of 70 cm, prescribe the correct power of the lens that is needed.

93. ••Medical A farsighted eye is corrected by placing a converging lens in front of the eye. The lens will create a virtual image that is located at the near point (the closest an object can be and still be in focus) of the viewer when the object is held at a comfortable distance (usually taken to be 25 cm). If a person has a near point of 75 cm, what power reading glasses should be prescribed to treat this hyperopia?

94. •••**Medical** (a) Prove that when two thin lenses are pressed next to one another, the combined focal length ($f_{combined}$) of the two lenses, acting together, is given by

$$\frac{1}{f_{combined}} = \frac{1}{f_1} + \frac{1}{f_2}$$

(b) Describe how this relates to a prescription for a contact lens which is placed directly on the eye? (Assume there is no significant separation between the contact lens and the lens of the eye.) (c) Why would an eyeglass prescription that is identical to a contact prescription give a very subtle difference in the image seen?

95. ••**Medical** An optometrist tests a person and finds that without glasses, he needs to have his eyes 15.0 cm from a book to read comfortably and can focus clearly only on distant objects up to 2.75 m away, but no farther. A typical normal eye should be able to focus on objects that are between 25.0 cm (the near point) and infinity (the far point) from the eye. (a) What type of correcting lenses does the person need: single focal length or bifocals? Why? (b) What should the optometrist specify as the focal length of the correcting contact lens or lenses? (c) What is the power (in diopters) of the correcting lens or lenses? SSM

96. •••**Medical** One of the inevitable consequences of aging is a decrease in the flexibility of the lens. This leads to the farsighted condition called *presbyopia* (elder eye). Almost every aging human will experience it to some extent. However, for the myopic person, at some point, it is possible that far vision will be limited by a subpar far point *and* near vision will be hampered by an expanding near point. One solution is to wear bifocal lenses that are diverging in the upper half to correct the nearsightedness and converging in the lower half to correct the farsightedness. Suppose one such individual asks for your help. The patient complains that he can't see far enough to safely drive (his far point is 112 cm) and he can't read the font of his smart phone without holding it beyond arm's length (his near point is 83 cm)! Prescribe the bifocals that will correct the visual issues for your patient.

97. ••A common zoom lens for a digital camera covers a focal length range of 18 mm to 200 mm. For the purposes of this problem, treat the lens as a thin lens. If the lens is zoomed out to 200 mm and is focused on a petroglyph that is 15.0 m away and 38 cm wide, (a) how far is the lens from the photosensor array of the camera and (b) how wide is the image of the petroglyph on the sensors? (c) If the closest that the lens can get to the sensors at its 18 mm focal length is 5.2 cm, what is the closest object it can focus on at that focal length? SSM

98. ••A macro lens is designed to take very close-range photographs of small objects such as insects and flowers. (See Figure 19-25 for a picture taken with a macro lens.)

At its closest focusing distance, a certain macro lens has a focal length of 35.0 mm and forms an image on the photosensors of the camera that is 1.09 times the size of the object. (a) How close must the object be to the lens to achieve this maximum image size? (b) What is the magnification if the object is twice as far from the lens as in part (a)? For this problem, treat the lens as a thin lens.

99. •**Astronomy** An astronomical telescope consists of an eyepiece lens at one end of a cylindrical tube and an objective lens at the other end. The objective lens gathers light from a distant object (such as a star or a planet) and focuses it at the focal point of the eyepiece lens. The eyepiece basically acts as a simple magnifier to create a virtual image of the objective's image. The overall magnification, M, is found to be

$$M = -\frac{f_o}{f_e}$$

(a) Calculate the magnification of the 36-in Lick Refractor that is housed in the Lick Observatory on Mount Hamilton about 30 min from San Jose, California. The focal length of the objective lens is 17.37 m and the focal length of the eyepiece is 22 mm. (b) What is the significance of the negative sign in the magnification equation? (The telescope was constructed in the 1880s, saw first light in January 1888, and is one of the few user-operated telescopes in use today.)

MCAT® Section
The section that follows includes material from previously administered MCAT® items and is reprinted with permission of the Association of American Medical Colleges (AAMC).

Passage 4 (Questions 19–26)
The Hubble telescope was carried into Earth's orbit and released by the space shuttle *Discovery* while orbiting at 7.8×10^3 m/s. Before releasing the telescope, it was necessary for *Discovery* to achieve an orbit higher than any shuttle had reached in the past. This was because high levels of solar activity had heated Earth's upper atmosphere, causing it to expand outward. In order to avoid atmospheric drag, which could cause the telescope to fall back to Earth, the telescope had to be placed as high as possible above the outer fringes of the atmosphere. The Hubble telescope was placed in orbit at an altitude of 6.11×10^5 m, a distance equivalent to approximately 4.8% of Earth's diameter. This altitude was as high as *Discovery's* chemically powered engines could lift the 1.1×10^4-kg telescope.

The Hubble telescope is the largest telescope ever placed into orbit. Its primary concave mirror has a diameter of 2.4 m and a focal length of approximately 13 m.

In addition to having optical detectors, the telescope is equipped to detect ultraviolet light, which does not easily penetrate Earth's atmosphere. (*Note:* Use $g = 9.8 \text{ m/s}^2$.)

19. When the Hubble telescope is focused on a very distant object, the image from its primary mirror is

A. real and erect.

B. real and inverted.

C. virtual and erect.

D. virtual and inverted.

20. Assuming that the Hubble telescope is traveling in a circular orbit, which of the following expressions is equal to its orbital speed? (*Note: G* is the gravitational constant, M_E is the mass of Earth, M_t is the mass of the telescope, and *r* is the distance of the telescope from the center of Earth.)

A. $\dfrac{GM_E}{r}$

B. $\sqrt{\dfrac{GM_E}{r}}$

C. $\dfrac{GM_t}{r}$

D. $\dfrac{GM_t}{r^2}$

21. If the acceleration due to gravity were nearly constant from Earth's surface to the altitude of the Hubble telescope, what would the telescope's approximate potential energy be relative to Earth's surface?

A. 3.3×10^9 J

B. 6.6×10^9 J

C. 3.3×10^{10} J

D. 6.6×10^{10} J

22. The magnification of a telescope is determined by dividing the focal length of the primary mirror by the focal length of the eyepiece. If an eyepiece with a focal length of 2.5×10^{-2} m could be used with the primary mirror of the Hubble telescope, it would produce an image magnified approximately how many times?

A. 10

B. 96

C. 520

D. 960

23. Which of the following is the best explanation of why ultraviolet light does not penetrate Earth's atmosphere as easily as visible light does?

A. Ultraviolet light has a shorter wavelength and is more readily absorbed by the atmosphere.

B. Ultraviolet light has a lower frequency and is more readily absorbed by the atmosphere.

C. Ultraviolet light contains less energy and cannot travel as far through the atmosphere.

D. Ultraviolet light is reflected away from Earth by the upper atmosphere.

24. As the Hubble telescope was transported into orbit aboard *Discovery*, which of the following best describes the main energy conversion(s) taking place?

A. kinetic to gravitational potential

B. gravitational potential to kinetic

C. gravitational potential to kinetic to chemical

D. chemical to kinetic to gravitational potential

25. What is the approximate kinetic energy of the Hubble telescope while in orbit?

A. 4.3×10^7 J

B. 3.3×10^{11} J

C. 6.6×10^{11} J

D. 4.3×10^{12} J

26. The image of a very distant object that is produced by a mirror such as that used in the Hubble telescope will be at what location in relationship to the mirror and focal point?

A. behind the mirror

B. between the mirror and the focal point

C. at the focal point

D. outside the focal point

Passage IV (Questions 94–99)

Retroreflecting arrays consist of spherical beads. Light is refracted as it enters a bead, then reflected off the back of the bead. Arrays of retroreflectors are attached to flexible sheets used as safety reflectors on clothing and bicycles. Ideal retroreflectors return a beam to its source regardless of the angle of incidence of the beam to it. Each light ray is returned on a path no farther than the diameter of a bead from the source ray. Thus, if a distortion that changes the path of light is placed in front of the retroreflecting array, the incident and reflected rays will pass through the same distortion. When this occurs, the ray perfectly retraces itself, thereby canceling the distortion.

An experiment is conducted with the setup illustrated in **Figure 1.** A light source covered by a screen with a pinhole in it provides a point source of light. A glass pane acts as a beam splitter. Some of the source light is reflected out of the experiment; the remainder of the

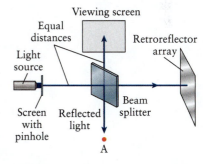

Figure 1

light is incident upon the array. Some of the light returning from the array is reflected, at a right angle to a viewing screen by the beam splitter. The wave front reflected from a retroreflecting array will be irregular. However, a pinhole image will still appear on the screen because the human eye cannot perceive the separation of the wave fronts.

94. The beam splitter in Figure 1 is set at what angle to the incident beam?

    A. 15°
    B. 30°
    C. 45°
    D. 60°

95. Which of the following figures shows a possible path for a ray of light passing through a retroreflecting bead? (Note: Assume that the index of refraction of the bead is greater than that of air.)

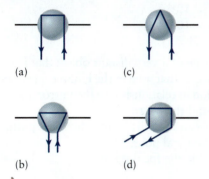

  (a)   (c)

  (b)   (d)

96. The glass that is used as a beam splitter is replaced with glass that is identical except that it has a 10% higher index of refraction. Which of the following changes will occur to the pinhole?

    A. It will move.
    B. It will become larger.

    C. It will become smaller.
    D. It will become clearer.

97. What is the approximate number of wavelengths of light that can travel in one direction within a retroreflecting bead that has a diameter of 5×10^{-5} m? (Note: The speed of light is 3×10^8 m/s and its frequency is approximately 10^{15} Hz.)

    A. 0.6
    B. 1.7×10^2
    C. 1.5×10^4
    D. 3.3×10^6

98. If a single-wavelength laser is used in place of the light source and pinhole, how will the image on the viewing screen change?

    A. It will be larger.
    B. It will be less bright.
    C. It will be shifted.
    D. It will show stronger patterns of interference.

99. With the screen removed, an observer at point A sees a virtual image of the pinhole. Which of the following indicates that the observer is viewing a virtual image?

    A. It moves as the observer changes the angle of view.
    B. It is larger than the real image.
    C. It is dimmer than the real image.
    D. It is fuzzier than the real image.

25 Relativity

(Eric Draper / Getty Images)

Based on our understanding of projectile motion (see Chapter 3), we can easily imagine the parabolic paths that the pins follow as a stationary juggler tosses them hand to hand. This juggler, however is riding a unicycle; describing the motion of the pins becomes more complicated! Relative to the juggler we are at rest, so naturally we consider our observations of the situation to be made from a stationary frame of reference. The juggler observes the motion of the pins from a frame of reference moving relative to us. Yet he and we discover that the laws of physics don't change depending on the frame from which they are observed!

Most of our everyday experiences involve objects that are either not moving with respect to us or are moving at relatively low speed. "Low" here is relative to the speed of light in a vacuum, so even the speed of a supersonic aircraft would be considered low in this context. About 150 years ago, scientists began to realize that although the laws of physics they knew work well at low speeds, the laws of classical physics that we've developed so far in this book don't properly describe light or objects moving at near the speed of light. Albert Einstein first understood the way to extend classical physics into the regime of high speed. In this chapter, we focus on his theory of special relativity. You'll probably find many of the implications of special relativity to be astonishing. For example, a clock moving at high speed relative to us runs slowly, and a ruler moving at high speed relative to us gets shorter along the direction of motion. Prepare for a wild ride!

25-1 Newtonian Relativity

Have you ever experienced this disorienting feeling? You're on a train ready to leave the station, looking out the window at the train right next to yours;

Figure 25-1 This is the view of one train as seen from the window of another. Which train is moving?
(*Eitan Simanor / Alamy*)

one of the trains starts to move, but you can't tell which (**Figure 25-1**). Is your train moving and the other one stationary or vice versa? Perhaps both are moving. How can you tell?

A curiosity that we mostly overlooked in our introduction to kinematics is that Newton's second law ($\Sigma \vec{F}_{\text{dir}} = m\vec{a}_{\text{dir}}$, Equation 4-1) does not distinguish between an object in uniform motion with respect to you and one at rest with respect to you. An object in uniform motion travels at constant speed; zero acceleration means that the object experiences no net force. This must also be true for an object at rest—no acceleration, so no net force. We can turn this realization around and note that an object that experiences no net force could either be at rest or in motion at a constant speed. Newton's laws therefore treat both cases identically; so although we might make a distinction between an object at rest and one in (uniform) motion with respect to us, the laws of physics do not. This statement has a powerful consequence. In the case of the two trains, if acceleration is small enough to be ignored, there is no way to design an experiment or make a measurement that would tell you whether the other train is moving with respect to yours or your train is moving with respect to the other one.

To study this conclusion and the related physics in more detail, we introduce the concept of a **frame of reference**. A frame of reference (or *reference frame* or simply *frame*) is a coordinate system with respect to which we will make observations or measurements. In particular, we will look for frames of reference such that, regardless of the frame from which we make an observation, the laws of physics hold true. For example, imagine yourself zooming along on a bullet train. Someone across the aisle is tossing a ball up and down, so that it goes straight up and then falls straight down into his hand. This is a classic example of the linear motion we explored in Chapter 2. Now imagine that same scene viewed from a vantage point outside the train, say, on the side of the tracks. As the train goes past, the ball does not appear to go straight up and straight down, but instead it traces out a parabolic path, a path that combines linear motion of the ball with the forward motion resulting from the motion of the train. However, while the path of the ball looks different from your seat on the train compared to the view of someone standing on a walkway along the side of the tracks, the physics is the same. The parabolic path seen from the side of the tracks is exactly the one predicted by the physics of projectile motion we explored in Chapter 3. The difference between the observed path of the ball in the two cases results from the frame of reference from which observations are made. If we associate a particular frame of reference with the interior of the train car, in that frame the ball is at rest if not thrown in the air. If we associate another frame of reference with the stationary walkway, in that frame the ball could be undergoing linear motion or, if someone inside the train is tossing the ball up, tracing out a parabolic path.

We'll represent a reference frame by a set of axes. In addition to three orthogonal spatial dimensions (that is, x, y, and z), in any frame we define, we'll also place a clock in the reference frame in order to measure time. We can attach one frame of reference to the train and a second frame to the person standing to the side of the train tracks in our previous example. We have sketched this in **Figure 25-2**, in which we attached a frame labeled S to an observer outside the train and a frame labeled S' ("S prime") to the train. (This is the common convention, but S doesn't stand for anything.) Following the notation of using a prime to indicate a measurement made in the S' frame, a position measured in the S frame is x or y or z, while in S' we measure x', y', and z'. Similarly, time is t in S and t' in S'.

The man to whom we have attached the S frame in Figure 25-2 has visual and other clues that suggest to him that he is at rest and the train is moving. As a passenger on the train you would have clues suggesting the same thing, that the train was moving and the man outside was stationary. Imagine, however, that the man and the train are out in deep space, where there are no clues that suggest motion, such as noises and shaking, or nearby objects moving relative to the background. (Yes, a train whizzing through deep space is a bit far-fetched, but the physics doesn't mind!) Because there is no experiment that can be done to distinguish the S frame from the S' frame, we must conclude that there is no way to tell whether the train is moving relative to the man, or the man is moving relative to the train. We cannot, therefore, pick a single, unique frame to be the frame "at rest" or one to call the "moving frame." At best we can say that the S' frame is moving relative to the S frame, and vice versa. In general, when we consider two frames moving at constant velocity relative to each other, we will use the variable V to denote the relative speed.

Newton's first law tells us that an object at rest tends to stay at rest and an object in uniform motion tends to stay in motion with the same speed and in the same direction unless acted on by a net force. Inertia is the tendency of an object to resist a change in motion, that is, to resist acceleration. For this reason, we refer to a frame of reference attached to a nonaccelerating object as an **inertial frame**. Said another way, an inertial frame is one that moves at constant speed relative to another.

Just as a nonaccelerating frame of reference is an inertial frame, a frame that is accelerating is a *noninertial frame*. If you find yourself in a noninertial frame you

Figure 25-2 An observer on the platform in a train station watches a train move past. We attach a frame S, with coordinate axes to measure x, y, and z and a clock to measure t, to the observer, and a frame S', with coordinate axes to measure x', y', and z' and a clock to measure t', to the train. (*OJO Images / SuperStock*)

might observe objects accelerating even though no force appears to act on them. For example, say that you're riding a slowly rotating merry-go-round at an amusement park. If the ride is inside a featureless room you might not have any sensation that you're rotating, and yet if you place a marble on the platform it rolls away. Even if the rotational speed of the merry-go-round is constant, the direction of your velocity vector changes as you rotate, so the frame attached to you is accelerating and is therefore not an inertial frame. The marble accelerates, even though you observe no force acting on it in your noninertial frame. The surface of Earth is a noninertial frame, although this is hard to detect in our everyday experiences. Earth rotates around its axis and revolves around the Sun, and both of these motions involve acceleration. Because the accelerations associated with the rotational motions of Earth are small compared to g, we often approximate the Earth's surface as an inertial frame.

Why is the distinction between an inertial and a noninertial frame important? First, any frame moving at constant velocity with respect to an inertial frame must itself be an inertial frame. For the man and the train in Figure 25-2, we declared that the train was moving at constant speed so S and S' are both inertial frames. We conclude, therefore, that the laws of motion, that is, Newtonian physics, apply equally well and in the same way in the S' frame as they do in the S frame. The formal statement of this is embodied in the principle of **Newtonian relativity:**

The laws of motion are the same in all inertial frames of reference.

The laws of physics do not distinguish between S and S', so we cannot tell which frame is moving and which is at rest. Indeed, the very idea of absolute motion is meaningless; just as in the train example at the start of this section, we can only describe motion relative to a selected frame of reference. The principle of Newtonian relativity is often expressed in terms of motion:

There is no way to detect absolute motion. Only motion relative to a selected frame of reference can be detected.

As a result of Newtonian relativity, both we and the juggler shown on the opening page of this chapter observe that the juggling clubs follow parabolic paths as prescribed by the laws of motion. The juggler sees the clubs rise and fall along a parabolic path in two dimensions, that is, a single plane; however, because the juggler is moving relative to us, we see the clubs trace out a path in three dimensions. In the language of frames, we place ourselves in a frame S and attach the S' frame to the juggler. To separate the motion of the clubs from the motion of the frame, we need a way to change from the variables measured in the S' frame to those measured in S and vice versa. For example, watching the clubs as the juggler rides by, we make observations of the x, y, and z positions of each club as a function of time. The juggler sees a different motion than we do, so he would observe the pins in different positions. A mathematical relationship between the positions he measures in S' (that is, x', y', and z') and the positions we measure in S is a *transformation*. The **Galilean transformation** of Newtonian relativity relates position and time as measured in the S frame (x, y, z, and t) to position and time as measured in the S' frame (x', y', z', and t') according to these equations:

$$x' = x - Vt \tag{25-1}$$

$$y' = y$$

$$z' = z$$

$$t' = t$$

where V is the relative speed of the two frames. In addition, we have chosen to align the x and x' axes in the two frames, so that the relative velocity vector points in the x and x' direction. Because the frames move in the x (and x') direction relative to each other, the positions and displacements in the other two directions are the same in the two frames.

The Galilean transformation relies on the relationship between the speeds of an object as observed in the two frames. Imagine yourself driving on the highway, either passing or being passed by another car. Its speed as seen by you can be expressed as its speed relative to the road (which is what the driver would read on her speedometer) minus your speed relative to the road (the speed you read on *your* speedometer). Let's say a car is traveling at speed v_x in the positive x direction, relative to coordinate axes attached to a tree on the side of the road. We'll call those axes the S frame. If you observe the car while you are also in the S frame (maybe leaning against the tree) you would of course observe the car moving at speed v_x. But if you make the observation while driving another car in the same direction, relative to you the first car is moving at a speed less than v_x. If your car is moving at V relative to the tree, then in your S' frame, the speed v'_x you observe for the other car is

$$v'_x = v_x - V \qquad . \qquad (25\text{-}2)$$

The speed v'_x is lower than v_x.

Watch Out

There is nothing special about either the *S* or *S'* frame.

In the previous example, and indeed most examples, we have chosen to think about the S frame as at rest and S' as moving. This selection is purely our choice. The physics that applies depends only on the relative motion, so that if we treated the S' frame as being at rest and the S frame as moving, all of the physics we uncover would be the same. So above, where we conclude that the speed v'_x observed in the S' frame of an object traveling at speed v_x in the S frame is lower than v_x, it is implied that we have let S' be in motion relative to S. You might think of this choice as equivalent to letting the S frame be attached to *you*, because usually we think of ourselves as stationary, with objects moving around us!

Equation 25-1 is equivalent to the x component relationship of the Galilean transformation; the time derivative of that equation yields Equation 25-2:

$$\frac{d}{dt}(x' = x - Vt) \rightarrow \frac{d}{dt}x' = \frac{d}{dt}(x - Vt)$$

 See the Math Tutorial for more information on Calculus

which simplifies to

$$\frac{dx'}{dt} = \frac{dx}{dt} - \frac{d(Vt)}{dt}$$

and then finally

$$v'_x = v_x - V \qquad (25\text{-}2)$$

Got the Concept 25-1

Going in the Opposite Direction

An object moves at speed v_x in the x direction in frame S. An observer in the S' frame, which moves in the positive x direction with respect to S at velocity $\vec{V}$, sees the object moving at v'_x. Consider two situations: (a) when the magnitude of $\vec{V}$ is less than the magnitude of $\vec{v}_x$; and (b) when the magnitude of $\vec{V}$ is greater than the magnitude of $\vec{v}_x$. In each case is the direction of $\vec{v}'_x$ the same as or opposite to the direction of $\vec{v}_x$?

As just one example that the laws of motion are the same in all inertial frames of reference, consider the collision of an object that has a mass m_1 and is moving at speed $v_{1,i}$ toward an object that has a mass m_2 and is moving at speed $v_{2,i}$. For simplicity, let's consider the case in which the objects are moving along the same line. Given our new understanding of frames, we declare that the speeds $v_{1,i}$ and $v_{2,i}$ are measured in a frame S. We know that momentum is conserved in S. Is momentum also conserved when the speeds are measured in frame S' moving at V relative to S?

After the collision m_1 moves away at speed $v_{1,f}$ and m_2 moves away at speed $v_{2,f}$, so we express the conservation of momentum, from Equation 7-7, as

$$m_1\vec{v}_{1,i} + m_2\vec{v}_{2,i} = m_1\vec{v}_{1,f} + m_2\vec{v}_{2,f} \tag{25-3}$$

We then ask whether it is also true that

$$m_1\vec{v}'_{1,i} + m_2\vec{v}'_{2,i} = m_1\vec{v}'_{1,f} + m_2\vec{v}'_{2,f} \tag{25-4}$$

Because we are considering motion along a straight line, we can apply Equation 25-2 directly to transform measurements made in the S' frame to the S frame. Written as vectors, Equation 25-2 can be written generally as

$$\vec{v}' = \vec{v} - \vec{V}$$

Transformed to the S frame, Equation 25-4 then becomes

$$m_1(\vec{v}_{1,i} - \vec{V}) + m_2(\vec{v}_{2,i} - \vec{V}) = m_1(\vec{v}_{1,f} - \vec{V}) + m_2(\vec{v}_{2,f} - \vec{V})$$

This equation can be rearranged so that all the terms that contain $\vec{V}$ are grouped together:

$$m_1\vec{v}_{1,i} + m_2\vec{v}_{2,i} - (m_1 + m_2)\vec{V} = m_1\vec{v}_{1,f} + m_2\vec{v}_{2,f} - (m_1 + m_2)\vec{V}$$

The two $(m_1 + m_2)\vec{V}$ terms cancel, leaving

$$m_1\vec{v}_{1,i} + m_2\vec{v}_{2,i} = m_1\vec{v}_{1,f} + m_2\vec{v}_{2,f}$$

This expression is just Equation 25-3, the statement that momentum is conserved when the speeds are measured in the S frame. Because this is true, so then is Equation 25-4; we have shown that momentum is conserved in the S' frame.

Example 25-1 Two Cars

You observe two cars approaching you. A red car is in one lane traveling with a speed of 12 m/s relative to you. In another lane, a blue car is traveling with a speed of 18 m/s with respect to you (**Figure 25-3**). (a) What is the speed of the blue car as measured by the driver of the red car, relative to her frame of reference? (b) What is the speed of the red car as measured by the driver of the blue car, relative to his frame of reference?

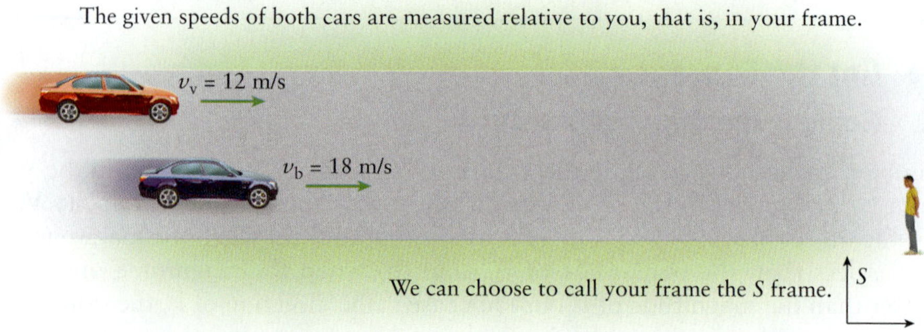

The given speeds of both cars are measured relative to you, that is, in your frame.

$v_v = 12$ m/s

$v_b = 18$ m/s

We can choose to call your frame the S frame.

S

Figure 25-3 A red car and a blue car move relative to one another.

SET UP

The speed of either car, as observed by the driver of the other car, must obey the Galilean transformation, that is,

$$v'_x = v_x - V \qquad (25\text{-}2)$$

We will apply this first to the speed of the blue car as measured by the driver of the red car, and then to the speed of the red car as measured by the driver of the blue car.

SOLVE

(a) We attach the S' frame of reference to the red car, so that the speed of the frame relative to you (in S) is 12 m/s, the speed of the red car relative to you. In the terminology of the Galilean transformation, this is V. The speed we want to transform is v_b (18 m/s), the speed of the blue car as measured in S. Thus, from Equation 25-2,

$$v'_b = v_b - V = 18 \text{ m/s} - 12 \text{ m/s} = 6 \text{ m/s}$$

The driver of the red car sees the blue car moving at 6 m/s relative to her.

(b) We now want to transform the speed of the red car, 12 m/s relative to you, to a reference frame attached to the blue car. With the S' frame moving along with the blue car, the speed of S' relative to you and relative to the S frame, equals 18 m/s. So from Equation 25-2,

$$v'_r = v_r - V = 12 \text{ m/s} - 18 \text{ m/s} = -6 \text{ m/s}$$

To the driver of the blue car, the red car is moving *backward* at 6 m/s.

REFLECT

The driver of the red car sees the blue car moving farther and farther in front of her at a rate of 6 m/s; in other words, relative to her, the blue car is moving at +6 m/s. The driver of the blue car sees the red car falling farther and farther behind him at a rate of 6 m/s; in other words, relative to him, the red car is moving at −6 m/s. Of course, because in both cases we have one car's speed relative to the other, the *magnitude* of the two answers must be same in each case, as it is.

Practice Problem 25-1 A spectator at a race watches as Juan and Bob sprint down the final stretch. He sees Juan running at 7.0 m/s and Bob, trying to catch him, at 8.0 m/s. (a) What is Juan's speed as observed by Bob? (b) What is Bob's speed as observed by Juan?

What's Important 25-1

Newton's laws treat all nonaccelerating objects identically whether they are in constant motion or at rest. The laws of motion are the same in all inertial frames of reference. There is no way to detect absolute motion; only motion relative to a selected frame of reference can be detected.

25-2 The Michelson and Morley Experiment

What's the same? We've asked this question many times as we've studied and worked problems in a broad range of physics topics. A single quantity that doesn't change value over time, or two or more quantities that are equal, are powerful tools for understanding physics. In the last section, we discovered that Newton's laws of motion are the same in all inertial frames, which is an important addition to our array of what's-the-same tools. Given the invariance of Newton's laws under the Galilean transformation that takes us from one inertial frame to another (Equation 25-1), it makes sense to ask whether other or all laws of physics remain the same after a Galilean transformation from one frame to another. Physicists asked this very question during the second half of the nineteenth century, specifically about the laws of electromagnetism. These are the laws embodied in Maxwell's equations.

As we saw in Section 22-2, when applied to a general electromagnetic wave traveling in a vacuum, Maxwell's equations lead to

$$\frac{\partial^2 B}{\partial x^2} = \mu_0 \epsilon_0 \frac{\partial^2 B}{\partial t^2} \tag{22-24}$$

and

$$\frac{\partial^2 E}{\partial x^2} = \mu_0 \epsilon_0 \frac{\partial^2 E}{\partial t^2} \tag{22-25}$$

We found that the product of the permittivity of free space, ϵ_0, and the permeability of free space, μ_0, is related to the speed of the wave:

$$v = \sqrt{\frac{1}{\mu_0 \epsilon_0}} \tag{22-26}$$

Because ϵ_0 and μ_0 are constants, so too is the speed of the wave; specifically, v equals 299,792,458 m/s. In other words, the speed of our generic electromagnetic wave equals c, the speed of light in a vacuum. If we can only measure relative but not absolute motion, however, the speed should be relative to some specific frame.

Relative to what frame is the speed of light equal to c? The most common nineteenth century answer to this question was to imagine a medium filling all space through which light travels, and to measure the speed of light relative to the frame in which that medium is at rest. Although it would need to have some special properties, you can imagine the medium, which scientists of that time called the "luminiferous ether," as something like the stretched string that supports the propagation of a pulse along its length. The frame relative to which the speed of light is c, according to physicists at the time, is the frame in which the luminiferous ether is at rest. In addition, scientists imagined that the speed at which light moved through the ether would depend on its motion relative to the ether. For example, they assumed light would slow down if it were moving in the direction opposite to the motion of the ether, in the same way that you would go more slowly trying to swim against the current in a river. How could there be an ether current? Scientists imagined that as Earth traveled in its orbit around the Sun, the effect of our motion through the ether would be equivalent to an ether wind blowing past us.

In 1887, American physicist Albert Michelson and his colleague, chemist Edward Morley, carried out the first definitive experiment intended to observe effects of the ether. Their goal was to measure the difference in speed between light traveling with the ether wind and light traveling against it. The apparatus Michelson designed for this purpose split a beam of light, sent the two beams along perpendicular paths, and then allowed them to recombine. The interference pattern formed when the beams came back together would depend on the difference in length of the two paths and also on whether the speed at which light traveled along each one was the same or different. Michelson and Morley placed a plate of partially silvered glass at the center of their *interferometer*, shown schematically in **Figure 25-4**, so that some of the light coming from the source was

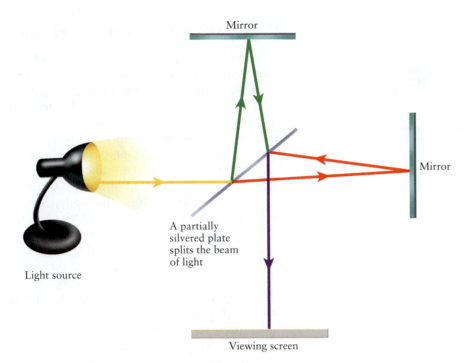

Mirror

Mirror

A partially
silvered plate
splits the beam
of light

Light source

Viewing screen

Figure 25-4 In this simplified version of a Michelson and Morley interferometer, a partially silvered glass plate splits a beam of light so that some light passes through it to a mirror and some is reflected to another mirror. The two beams are then recombined on a viewing screen.

reflected while the rest continued straight through. Mirrors placed along the two paths reflected the light back to the partially silvered glass plate, where again, some was reflected and some transmitted. In this way the original beam of light was split and then recombined at the viewing screen, on which an interference pattern was displayed.

Michelson and Morley expected that the speed with which light traveled along one leg of their interferometer would be different than the speed along the other leg, assuming that an ether wind existed and that one leg was more aligned with the wind than the other. In analogy to the swimmer in the river, light traveling along the more-aligned leg of the interferometer would be moving mostly either upstream or downstream, while along the other leg it would be moving mostly across the current. Michelson and Morley intended to measure this speed difference based on the location of the fringes in the interference pattern formed when light traveling along the two legs recombined.

Michelson and Morley measured no effect due to ether wind. They and other scientists refined and repeated the experiment many times, but the results were always the same: no evidence for luminiferous ether. Although some scientists, including Michelson himself, refused to accept it, the inescapable conclusion of the experiment is that luminiferous ether does not exist. Recall that under Newtonian relativity, to explain a constant value of the speed of light in a vacuum required that Maxwell's equations only applied in a specific reference frame: at rest with respect to a medium through which light could propagate. Without the luminiferous ether, physicists had no choice but to conclude that Newtonian relativity does not apply to light. How then, does light act? Fundamentally, we now understand that although light has wave properties, it requires no medium through which to propagate. For the rest of the answer we turn to Albert Einstein.

What's Important 25-2
Light does not obey Newtonian relativity.

During one tick of this light clock,

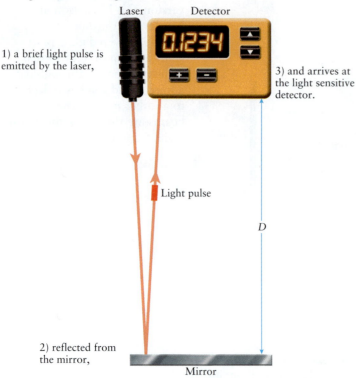

Laser   Detector

1) a brief light pulse is
emitted by the laser,

3) and arrives at
the light sensitive
detector.

Light pulse

D

2) reflected from
the mirror,

Mirror

Figure 25-5 This clock works by sending brief pulses of light to a mirror; the length of each "tick" of the clock is the time between the start of the pulse and the time when the reflected light returns to the detector.

25-3 Special Relativity, Time Dilation

German theoretical physicist Albert Einstein published his theory of special relativity in 1905. Einstein based his theory on two postulates, first, that all laws of physics are the same in all inertial frames, and second, that the speed of light in a vacuum is the same in all frames, independent of both the speed of the source of the light and the speed of the observer. The first postulate extends the principle of relativity beyond Newton's laws of motion to encompass electricity and magnetism, thermodynamics, and electromagnetic waves. (The name "special relativity" also derives from the first postulate; "special" in this case means the theory applies to inertial frames and constant speed motion. Einstein's theory of general relativity encompasses acceleration as well as constant speed.) As we concluded at the end of the last section, Einstein's second postulate, a seemingly straightforward statement, is what clarifies that Newtonian relativity does not apply to light. The implications of the second postulate shook the foundations of early twentieth century scientists' understanding of the natural world.

Perhaps the most astounding consequence of the second postulate of special relativity is that the rate at which time runs depends on the relative speed of two frames from which a measurement of time is made. In the next section, we will treat the issue mathematically, but first we'll do a thought experiment. Einstein was well known for his ability to solve problems by carrying out imaginary experiments, some of which were not technologically possible.

Imagine a special clock that uses short bursts, or pulses, of light to determine each interval of time. A sketch of such a clock is shown in **Figure 25-5**. A laser fires an extremely brief burst of light which is reflected by a mirror before arriving at a light-sensitive detector. The length of time between the start of the pulse and its arrival at the detector is Δt_{proper}, the time interval of one tick of the clock. We use the term *proper* to indicate that the measurement is made in the **rest frame** of the clock, that is, the frame which is at rest with respect to the clock. Physicists say that the rest frame of an object is "attached" to it. The **proper time** is a time measurement made at rest with respect to the clock.

> **! Watch Out**
>
> **Every object has a rest frame, but this does not imply that the object is not moving.**
>
> An object is always at rest with respect to itself, which means that an object is not moving in a reference frame that is attached to it. Notice, however, that the statement is a relative one—the object is not moving *relative* to its rest frame. We can still define any number of other frames with respect to which the object *is* in motion. As a result, we cannot establish the object as either at rest or in absolute motion.

? Got the Concept 25-2

Proper Time

A clock is placed on the Starship *Alpha*, which Albert flies past Earth at half the speed of light. Betty flies Starship *Beta* beside Starship *Alpha*, at the same speed, while George pilots Starship *Gamma* at half the speed of light in the opposite direction. Ellen observes from Earth. Which of the observers measures the proper time of the clock?

In order for our light clock to tick once, a light pulse travels the distance D to the mirror and then the same distance back to the detector, for a total distance of $2D$. We imagine that the clock is placed in a vacuum, so that the speed of the light pulse is c. The time interval of the tick is then

$$\Delta t_{proper} = \frac{2D}{c} \qquad (25\text{-}5)$$

We now let the light clock move at speed V relative to us. In keeping with the way we name frames in discussing Newtonian relativity, we attach a frame S' to the clock and consider ourselves in the S frame. The S' frame therefore moves at speed V relative to us. In the next section, we explore the circumstances that arise when V is a significant fraction of c; however, that detail is not a requirement for this thought experiment.

Figure 25-6a shows the process of a single tick of the clock as observed from the S frame. During the time that it takes for the light pulse to get from the laser to the mirror, the mirror (and indeed the entire clock) has moved. During the time between the reflection from the mirror and the arrival of the light pulse at the detector, the clock has moved again. The distance L that the clock has moved during this time equals the product of the speed V and the time interval of one clock tick. We cannot assume that the length of one tick is Δt_{proper}, however, because Δt_{proper} is measured at rest with respect to the clock. Instead we let the time interval of one tick as observed from the S frame be Δt, so

$$\Delta t = \frac{L}{V} \qquad (25\text{-}6)$$

From our vantage point in the S frame, the light pulse traces out two sides of a triangle during the time Δt, as sketched in **Figure 25-6b**. The time interval is therefore also equal to the sum of these two sides—the distance the light pulse travels—divided by the speed of light. The second postulate of special relativity assures us that the speed of the light pulse is c, so

$$\Delta t = \frac{2\sqrt{D^2 + \left(\frac{L}{2}\right)^2}}{c} \qquad (25\text{-}7)$$

We used the fact that each leg of the path followed by the light pulse is the hypotenuse of a right triangle of sides D and $L/2$.

Let's compare Δt_{proper} and Δt. The distance L does not appear in the expression for Δt_{proper}; to eliminate L from Δt note that, from Equation 25-6,

$$L = V\Delta t$$

So Equation 25-7 becomes

$$\Delta t = \frac{2\sqrt{D^2 + \left(\frac{V\Delta t}{2}\right)^2}}{c}$$

Figure 25-6 (a) When the light clock depicted in Figure 25-5 moves, the distance the light pulse travels is longer than twice the distance from the laser to the mirror. (b) The light ray forms the hypotenuse of two right triangles that can be used to analyze the time it takes the clock to tick.

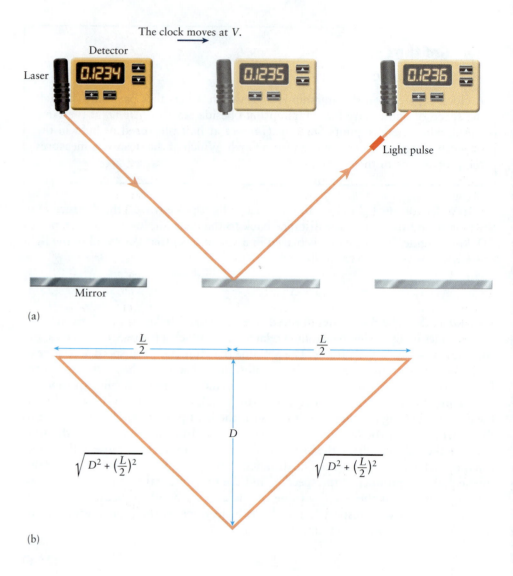

To express Δt in terms of the other quantities, square both sides and rearrange to bring the two Δt terms together:

$$c^2\Delta t^2 = 4D^2 + 4\left(\frac{V\Delta t}{2}\right)^2 = 4D^2 + V^2\Delta t^2$$

or

$$(c^2 - V^2)\Delta t^2 = 4D^2$$

so

$$\Delta t^2 = \frac{4D^2}{c^2 - V^2}$$

We can express Δt in a form similar to Δt_{proper} (Equation 25-5) by taking the square root of both sides and then taking a factor of $\sqrt{c^2}$ out of the denominator:

$$\Delta t = \frac{2D}{\sqrt{c^2 - V^2}} = \frac{2D}{c\sqrt{1 - \dfrac{V^2}{c^2}}}$$

so

$$\Delta t = \frac{1}{\sqrt{1 - \dfrac{V^2}{c^2}}} \Delta t_{\text{proper}}$$ (25-8)

An observer moving relative to the clock measures the duration between successive ticks of the clock to last a time Δt. An observer at rest with respect to the clock measures the time interval between ticks to be Δt_{proper}, regardless of whether the observer and clock are moving with respect to some *other* observer. By using Equation 25-8, Δt and Δt_{proper} are not equal, that is, an observer in the S frame sees time running at a different rate than an observer in the S' frame. This is astounding! This effect of special relativity is known as **time dilation**.

For any nonzero value of V, the speed of one inertial frame relative to another, the denominator in the relationship between Δt and Δt_{proper} is smaller than 1. The factor that relates the two time intervals is therefore always greater than 1, so Δt is always greater than Δt_{proper}. The longer the time interval of one clock tick, the more slowly the clock runs. Because Δt is greater than Δt_{proper}, an observer in the S frame sees the moving clock as running slowly. For any given time interval measured on a clock at rest in S, less time elapses on the clock moving with the S' frame. A moving clock runs slowly.

Watch Out

A moving clock runs slowly.

According to Equation 25-8, Δt is always greater than Δt_{proper} for any nonzero value of the relative speed V. To make clear the effect this has on a time measurement, imagine a pendulum clock, ticking once every time the pendulum makes one full oscillation. If the clock is on your desk you see it swinging back and forth once every second. If the clock is moving, then the swing of the pendulum is slower, so the clock . . . ticks . . . more slowly. Unless V is large, however, you will not be able to perceive this effect.

Watch Out

Time dilation occurs regardless of whether a clock is present.

No clock need be present for the effects of time dilation to occur. Time runs slowly in one frame moving with respect to another frame, even if there is no clock in the moving frame, and even if there is no observer in the other frame. Imagine, say, that you can place a *Caenorhabditis elegans* worm on a spacecraft that will fly past Earth at high speed. *C. elegans* is an organism popular among geneticists because it grows to adulthood in a series of easily identifiable developmental stages that altogether take less than 3 days (Figure 25-7). Each stage lasts 8 to 12 h. Were you to observe a *C. elegans* worm as it moved past you at a relativistic speed, each of the stages might take days or even years, measured on a clock at rest with respect to you.

Figure 25-7 The worm *Caenorhabditis elegans* is an organism popular among geneticists because it grows to adulthood in a series of easily identifiable developmental stages that each lasts a well-defined time. *(Courtesy of Dr. John Sulston)*

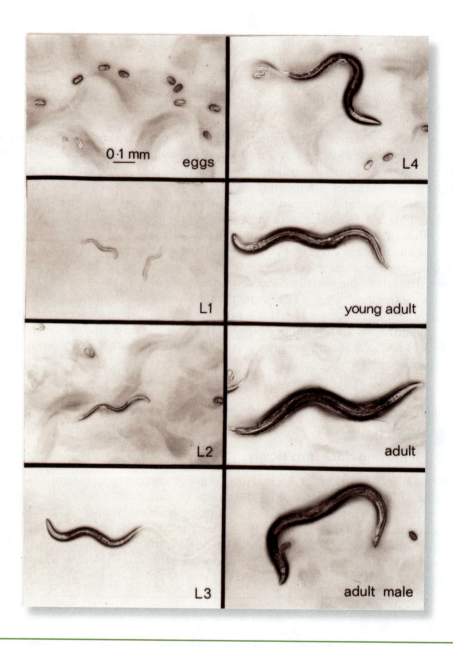

> **❗ Watch Out**
>
> **The effect of time dilation only arises when time in one frame is viewed from another which is in motion relative to the first.**
>
> We imagined, above, a pendulum clock that ticks once every time the pendulum makes one full oscillation. If the clock is on your desk you see it swinging back and forth once every second, and if the clock is moving, the swing of the pendulum is slower. Suppose, however, that you and your desk are moving together relative to some other specific object or frame. Regardless of how fast you and the desk are moving (relative to that other object or frame), you will see no time dilation effect. You will still see, for example, the pendulum swinging back and forth once every second. This is because although you and the clock are moving, neither you nor the clock is moving *relative to each other*.

? Got the Concept 25-3
A Clock Moving Faster

A clock that keeps good time (in other words, if you are at rest with respect to the clock the seconds hand would tick off one second every second) is placed on an imaginary spaceship. As the spacecraft flies past Earth you observe the clock on the bridge through a telescope (this is imaginary, too). You see that the clock on the ship is running slowly. The ship turns around and heads back past Earth again, this time at a higher speed. Do you see the clock running more or less slowly compared to its first pass?

Example 25-2 Muon Decay

Cosmic rays, subatomic particles and radiation, continuously stream down to Earth. One such particle is the muon, created in large quantities in the upper atmosphere due to collisions between incoming cosmic rays and air particles. The muons travel at high speed, around $0.994c$. Muons naturally decay, however; if you create a number of muons in the lab, on average one-half of them will decay after 1.56 μs. This time is known as the half-life of the muon. If 10^6 muons are created at an altitude of 10 km, (a) how many would you expect to strike Earth if relativistic effects were ignored? (b) How many would you expect to strike Earth when the effect of time dilation is considered?

SET UP
We determine the number of muons that strike Earth by determining how long it takes for the muons to get to the surface, in terms of the amount of time it takes for half of them to decay. For example, if we found that three half-lives would elapse, then the number N of muons remaining from the original number of muons N_0 would be

$$N = \frac{1}{2} \times \frac{1}{2} \times \frac{1}{2} N_0 = \frac{1}{2^3} N_0 = \frac{1}{8} N_0 \qquad (25\text{-}9)$$

SOLVE
(a) The time is takes muons to reach Earth's surface is their initial altitude divided by their speed; from an altitude of 10 km at a speed of $0.994c$ it takes muons

$$\Delta t = \frac{10000 \text{ m}}{(0.994)(3.00 \times 10^8 \text{ m/s})} = 3.35 \times 10^{-5} \text{ s}$$

or 33.5 μs. This is 21.5 half-lives. Using Equation 25-9 as an example, the number of muons left after 21.5 half-life intervals is

$$N = \frac{1}{2^{21.5}} 10^6 = 0.337$$

Less than 1 muon, on average, survives the trip.

(b) The time Δt we computed in part (a) is the time we would measure in our frame. The muons, however, experience time in the frame that moves along with them; this is the proper time. According to Equation 25-8, in terms of Δt the proper time interval required for the muons to reach Earth is

$$\Delta t_{\text{proper}} = \sqrt{1 - \frac{v^2}{c^2}} \Delta t = \sqrt{1 - \frac{(0.994c)^2}{c^2}} \frac{10000 \text{ m}}{(0.994)(3.00 \times 10^8 \text{ m/s})}$$

$$= 3.67 \times 10^{-6} \text{ s}$$

or

$$\Delta t_{\text{proper}} = 3.67 \ \mu\text{s}$$

This is only 2.35 half-lives ($3.67 \ \mu\text{s}/1.56 \ \mu\text{s}$), so the fraction of muons left in the time it takes for them to reach the surface (from Equation 25-9) is

$$N = \frac{1}{2^{2.35}} 10^6 = 0.196 \times 10^6$$

Nearly 20%, or about 200,000, of the original 10^6 muons make it to the surface.

REFLECT

One of the earliest confirmations of Einstein's theory of special relativity was a 1941 experiment that compared the number of muons observed in an hour at the top of Mt. Washington in New Hampshire to the number observed at the base of the mountain. Without including the effect of time dilation, it would have been expected that the muon rate at an altitude of about 1100 m at the base of the mountain would only be about 10% of the muon rate at the top (around 2000 m), yet the measured rate at the base was over 80%. As we found in this example, the effect of time dilation causes time to run more slowly for the muon than we measure in our own frame, so fewer half-lives elapse for a given distance traveled. Because fewer half-lives have elapsed, fewer muons have decayed.

Practice Problem 25-2 The charged pion, a subatomic particle, has a half-life of about 1.8×10^{-8} s. One thousand charged pions are sent down a 20-m-long pipe at $0.95c$. (a) How many pions would you expect to reach the end of the pipe if relativistic effects were ignored? (b) How many would you expect to reach the end of the pipe when the effect of time dilation is considered?

Example 25-3 Moving Clock

A clock moves past you at speed V. What must V be so that you see the clock as running at one-half the rate of the clock on your cell phone, which you're holding in your hand?

SET UP

For a moving clock to be observed as running at one-half the rate of a clock at rest with respect to you, the multiplicative factor between Δt and Δt_{proper} (Equation 25-8) must equal 2. Doubling the time interval results in a change by one-half of the rate at which time runs.

SOLVE

Setting the factor in Equation 25-8 equal to 2 yields

$$\frac{1}{\sqrt{1 - \dfrac{V^2}{c^2}}} = 2$$

Squaring both sides of the equation and inverting the terms on each side gives

$$1 - \frac{V^2}{c^2} = \frac{1}{4}$$

which we can rearrange to give

$$\frac{V^2}{c^2} = \frac{3}{4}$$

or

$$V = \sqrt{\frac{3}{4}}c$$

REFLECT

For you to observe the moving clock running a factor of 2 slower than the clock at rest with respect to you, the relative speed between you and the clock would need to be 0.87c. This is about twenty thousand times faster than the top speed of any craft built by humans.

Practice Problem 25-3 The second hand on a clock moving past you at high speed ticks off 1.00 s for every 4.00 s that tick off on your wristwatch. Find the speed of the clock relative to you.

★ What's Important 25-3

Einstein based his theory of special relativity on two postulates: first, that all laws of physics are the same in all inertial frames, and second, that the speed of light in a vacuum is the same in all frames and is independent of both the speed of the source of the light and the speed of the observer. "Special" in this case means the theory applies to inertial frames and constant speed motion. The rate at which time runs depends on the relative speed of the two frames from which a measurement of time is made.

25-4 The Lorentz Transformation, Length Contraction

The speed of light in a vacuum is the same in all frames. Einstein's postulate leads to a remarkable conclusion. Time is not the same from frame to frame. In a real sense, the rate at which a clock runs is different when observed from a frame in relative motion with respect to the frame of the clock. In this section, we'll apply Einstein's postulate to distances as measured in different frames and, more generally, to the transformation that takes us from one frame to another.

The Galilean transformation that describes Newtonian relativity (Equation 25-1) works well enough for objects and frames moving at low speed, say, the speed of a car or even the space shuttle. It does not work well for light, however, or for objects traveling close to the speed of light. Can you guess what the speed might be above which an object is moving too fast to use the Galilean transformation, but below which it would be slow enough that the transformation would work? Careful, there is no correct answer! For any speed you might choose for the motion of one frame relative to another, would you expect the laws of physics to change for a speed just a bit lower, say, an infinitesimal amount lower? Certainly not. Rather, it is more reasonable that the change in the physics from lower to higher speed would be gradual and smooth. Mathematically that means that whatever transformation replaces the Galilean transformation for objects moving at close to the speed of light must also reproduce the Galilean transformation when applied to a low frame-to-frame relative speed.

Consider this possible form of a transformation that could apply when the relative speed between the frames is high:

$$x' = \gamma(x - Vt) \tag{25-10}$$

We have taken the x component of the Galilean transformation from the S frame to the S' frame (Equation 25-1) and added a multiplicative factor γ. For now we won't specify any particular form for γ, but as long as γ approaches 1 as the relative speed between the frames approaches 0, Equation 25-10 correctly reproduces the Galilean transformation in the low-speed, or classical, limit.

We will also need the inverse of Equation 25-10 to explore the nature of γ, that is, a transformation for x in terms of variables in the S' frame. We form the inverse transformation by reversing the terms in the x component equation of Equation 25-1 (to get $x = x' + Vt$), distinguishing between t and t' (to get $x = x' + Vt'$) and then adding the multiplicative factor γ. The inverse of Equation 25-10 is then

$$x = \gamma(x' + Vt') \qquad (25\text{-}11)$$

To uncover a relationship for γ we apply Equations 25-10 and 25-11 to the distance that a ray of light travels in a given time interval as observed in two frames. In doing so we again apply Einstein's second postulate, that the speed of light is independent of the observer's frame.

In one frame, which we label S, the displacement from x_1 to x_2 is related to the time interval $t_2 - t_1$ by the speed of light:

$$x_2 - x_1 = c(t_2 - t_1)$$

According to Einstein's second postulate, the speed of the light ray is also c in any frame S' moving at speed V relative to S. The distance the light ray travels in S' is therefore

$$x_2' - x_1' = c(t_2' - t_1')$$

Because we follow the convention that the x axis in S is aligned with the x' axis in S', there must be some time when the S and S' coordinate axes are aligned; we let that time be zero in both frames. So x_1 equals zero when t_1 equals zero and x_1' equals zero when t_1' equals zero. The previous two equations then simplify to

$$x = ct$$

and

$$x' = ct'$$

To eliminate position from the two equations, we substitute them into Equations 25-10 and 25-11. Equation 25-10 becomes

$$ct' = \gamma(ct - Vt) = \gamma t(c - V) \qquad (25\text{-}12)$$

and Equation 25-11 becomes

$$ct = \gamma(ct' + Vt') = \gamma t'(c + V) \qquad (25\text{-}13)$$

Equations 25-12 and 25-13 contain both t and t', so we can eliminate those variables to find an expression for γ that depends only on V. We first reverse the order of the terms in Equation 25-12, so that t is on the left-hand side of the equation and t' is on the right, matching Equation 25-13:

$$\gamma t(c - V) = ct'$$

Both t and t' cancel when the equation is divided by Equation 25-13:

$$\frac{\gamma t(c - V)}{ct} = \frac{ct'}{\gamma t'(c + V)}$$

or

$$\frac{\gamma(c - V)}{c} = \frac{c}{\gamma(c + V)}$$

We now rearrange to gather the γ terms together:

$$\gamma^2 = \frac{c^2}{(c - V)(c + V)} = \frac{c^2}{c^2 - V^2}$$

or

$$\gamma = \sqrt{\frac{c^2}{c^2 - V^2}}$$

This factor, often called **relativistic gamma**, is more often written as

$$\gamma = \sqrt{\frac{c^2}{c^2 - V^2}} = \sqrt{\frac{1}{\frac{c^2 - V^2}{c^2}}} = \frac{1}{\sqrt{1 - \frac{V^2}{c^2}}} \qquad (25\text{-}14)$$

This is the same factor that appears in Equation 25-8, the relationship that describes the phenomenon of time dilation.

Does this γ approach 1 as the relative speed approaches zero, so that Equations 25-10 and 25-11 reproduce the Galilean transformation in the classical limit? For values of V close to zero, the fraction V^2/c^2 in the denominator of Equation 25-14 approaches zero, so that the denominator approaches one and therefore γ also approaches one. So yes, the transformation expressed in Equations 25-10 and 25-11 correctly reduces to the Galilean transformation for low speeds. More important, these relationships also work for high speeds, that is, speeds close to c. We have uncovered the **Lorentz transformation**, which applies to any two inertial frames regardless of the magnitude of their relative speed. We leave the derivation of the time component of the Lorentz transformation to a problem at the end of this chapter, and summarize the Lorentz transformation as follows:

$$x' = \gamma(x - Vt) \qquad (25\text{-}10)$$
$$y' = y$$
$$z' = z$$
$$t' = \gamma\left(t - \frac{V}{c^2}x\right)$$

The inverse transformation, from the S' frame to the S frame is

$$x = \gamma(x' + Vt') \qquad (25\text{-}15)$$
$$y = y'$$
$$z = z'$$
$$t = \gamma\left(t' + \frac{V}{c^2}x'\right)$$

In other words, the Lorentz transformation provides a means of determining the values of quantities measured in one inertial frame based on the values measured in another inertial frame. The Lorentz transformation depends on relativistic gamma, which as you can verify from Equation 25-14, depends only on the relative speed of the two inertial frames.

? Got the Concept 25-4
y and z

In the Lorentz transformation (Equation 25-10) and its inverse (Equation 25-15), no transformation is carried out between y and y' or z and z'. Why?

> ## ? Got the Concept 25-5
> ### Slow and Fast
> What is the smallest value that relativistic gamma (γ) can have? What is the largest possible value of γ?

Example 25-4 Time Dilation

A time interval is defined as the difference between two times measured at the same position. That is, Δt equals $t_2 - t_1$ at some position x_0 in the S frame and $\Delta t'$ equals $t_2' - t_1'$ at some x_0' in S'. Show that the time dilation relationship (Equation 25-8) we derived for the light clock in Section 25-3 can be obtained from the Lorentz transformation applied to a time interval in S'.

SET UP
The time interval as measured in S is

$$\Delta t = t_2 - t_1$$

We will apply the Lorentz transformation to the two times on the right side of the equation, taking care to make both time measurements at the position x_0.

SOLVE
Substituting the time component of Equation 25-15 in Δt yields

$$\Delta t = \gamma\left[\left(t_2' + \frac{V}{c^2}x_0'\right) - \left(t_1' + \frac{V}{c^2}x_0'\right)\right]$$

Because the transformation was applied at x_0 in both cases, the position terms cancel, leaving

$$\Delta t = \gamma\left[t_2' - t_1'\right]$$

The term in brackets is just the time interval measured in the S' frame, so

$$\Delta t = \gamma\Delta t'$$

$\Delta t'$ is the time interval measured at rest with respect to S', so it is the proper time interval. We have reproduced Equation 25-8.

REFLECT
The Lorentz transformation is a direct mathematical way to determine the value of a quantity measured in one inertial frame based on the value measured in another inertial frame.

We have used the quantity of length frequently in this book, based on a definition that is self-evident. The length of an object is the difference in the position of the two ends. This definition works well regardless from which frame measurements are made as long as the measurements of the two ends are made *at the same instant in time*. This stipulation is required first, so that the length of an object moving relative to you can be measured, and second, because time does not run at the same rate in a frame in relative motion with respect to another.

We now have the tools necessary to find the length of an object in two frames. Let's consider a rod, aligned with the x' direction in frame S', and let S' move at constant speed V in the x'/x direction with respect to S. In S' the length of the rod is

$$L_{\text{proper}} = x_2' - x_1'$$

where x_1' and x_2' are the positions of the two ends of the rod, and where we indicated by the subscript "proper" that the length of the rod is measured in a frame at rest with respect to the rod.

To find the length of the rod as observed from the S frame, we apply the Lorentz transformation in the x direction, the first relationship in Equation 25-10, to the position of each end of the rod, so

$$L_{\text{proper}} = \gamma(x_2 - Vt_2) - \gamma(x_1 - Vt_1)$$

The two position measurements must be made at the same time, however, so t_1 equals t_2 and in this way the two terms that depend on time cancel. Therefore,

$$L_{\text{proper}} = \gamma x_2 - \gamma x_1 = \gamma(x_2 - x_1)$$

or

$$L_{\text{proper}} = \gamma L \qquad \text{(25-16)}$$

where L is the length of the rod as measured in the S frame. Recall that the phenomenon of time dilation is described, from Equation 25-8, by

$$\Delta t = \gamma \Delta t_{\text{proper}}$$

We therefore rewrite Equation 25-16 as

$$L = \frac{1}{\gamma} L_{\text{proper}} \qquad \text{(25-17)}$$

The factor γ depends on the speed of an object or frame (Equation 25-14) and is always equal to or greater than 1, so length L is always equal to or less than the proper length L_{proper}. Said another way, when an object moves relative to an observer, its length in the direction of motion is seen to be contracted compared to the length measured at rest with respect to the object. This phenomenon is referred to as **length contraction**.

Watch Out

An object is shorter when moving than when it is at rest.

To be precise, when an object is in motion relative to an observer, its length along the direction of motion is shorter than its proper length. According to Equation 25-17, L is always less than L_{proper} for any nonzero value of the relative speed V because γ is always greater than 1 in that case.

Watch Out

Length contraction only occurs along the direction of motion.

In developing and applying the Lorentz transformation (Equation 25-10), we chose to let the x axis of the S frame and the x' axis of the S' frame be aligned. This results in a transformation between x and x', but y and y', and z and z', are equal. So the relationship we describe as length contraction (Equation 25-17) applies only in the x direction, that is, in the direction of motion. There is no length contraction in any direction other than the direction of motion, for example, in the directions orthogonal to the motion.

Example 25-5 Meter Stick

A meter stick hurtles through space at a speed of 0.80c relative to you, with its length aligned with the direction of motion. What do you measure as the length of the meter stick?

SET UP

You see the meter stick as contracted because it is in motion relative to you. We attach the S frame to you and the S′ frame to the meter stick, so that V, the relative speed of the frames, is 0.80c. Because the meter stick is at rest when observed from the S′ frame, in S′ it has a measured length of 1.0 m. That is, its proper length L_{proper} equals 1.0 m. The length of the meter stick as observed from the S frame is given by Equation 25-17:

$$L = \frac{1}{\gamma} L_{proper}$$

where γ is given by

$$\gamma = \frac{1}{\sqrt{1 - \dfrac{V^2}{c^2}}} \tag{25-14}$$

SOLVE

For V equal to 0.80c, γ is then

$$\gamma = \frac{1}{\sqrt{1 - \dfrac{(0.80c)^2}{c^2}}} = \frac{1}{\sqrt{1 - 0.64}} = \frac{1}{\sqrt{0.36}} = \frac{1}{0.60}$$

Notice we didn't evaluate the final fractional form; there's no need because the relationship between L_{proper} and L depends on the reciprocal of γ. So from Equation 25-17, using 1.0 m as the proper length,

$$L = \frac{1}{(1/0.60)}(1.0 \text{ m}) = 0.60 \text{ m}$$

REFLECT

A meter stick moving at 0.80c (in the direction of its length) relative to an observer has a measured length equal to 0.60 m. Its length is contracted to 60% of its proper length.

Practice Problem 25-5 You measure the length of a meter stick to be 0.995 m as it hurtles through space with its length aligned with the direction of motion. What is its speed relative to you?

Example 25-6 Meter Stick, Again

A meter stick hurtles through space at a speed of 0.80c relative to you, and at an angle of 30° with respect to the direction of motion in its rest frame (**Figure 25-8a**). What do you measure as the length of the meter stick?

SET UP

As in the previous problem we attach the S frame to you and the S′ frame to the meter stick, so that V, the relative speed of the frames, is 0.80c. Because the meter stick is moving relative to you, you see its length as contracted. Notice, however, that only positions and therefore lengths in *the direction of motion* undergo a transformation. This is evident in the Lorentz transformation (Equation 25-10), in which positions in the directions orthogonal to the direction of motion are the same in *S* and *S'*. In Figure 25-8a we have elected to align the direction of motion with the *x'* direction of the *S'* frame, so it is therefore also aligned with the *x* direction of the *S* frame. With this choice, only the projection of the meter stick in the *x* direction experiences relativistic length contraction.

Because the meter stick is at rest when observed from the *S'* frame, in *S'* it has a measured length of 1.0 m. That is, its proper length L_{proper}, shown in **Figure 25-8b**, equals 1.0 m. Treating the meter stick as a vector of length L_p directed along its length, the *x* and *y* components of L_{proper} are

$$L_{proper,x} = L_{proper} \cos \theta$$

and

$$L_{proper,y} = L_{proper} \sin \theta$$

where θ equals 30° in this case. We apply relativistic length contraction, Equation 25-17, to $L_{proper,x}$ but not $L_{proper,y}$.

SOLVE

So

$$L_x = \frac{1}{\gamma} L_{proper,x} = \frac{1}{\gamma} L_{proper} \cos \theta$$

and

$$L_y = L_{proper,y} = L_{proper} \sin \theta$$

From the previous example, $1/\gamma$ equals 0.60 for *V* equal to 0.80*c*. So with L_{proper} equal to 1.0 and θ equal to 30°:

$$L_x = (0.60)(1.0 \text{ m}) \cos 30° = 0.52 \text{ m}$$

and

$$L_y = (1.0 \text{ m}) \sin 30° = 0.50 \text{ m}$$

The length of the meter stick as measured in the *S* frame is therefore

$$L = \sqrt{L_x^2 + L_y^2} = \sqrt{(0.52 \text{ m})^2 + (0.50 \text{ m})^2} = 0.72 \text{ m}$$

REFLECT

The length of the meter stick is contracted. However, because its length is not aligned with the direction of motion, the amount of contraction, about 72%, is not as severe as in Example 25-5, in which the meter stick was completely aligned with the direction of motion. Also, notice that while the meter stick is oriented at an angle of 30° with respect to the direction of motion as observed from its rest frame, because the *x* and *y* components in *S* are about the same, the meter stick is seen to be oriented about 45° from the direction of motion when observed from the *S* frame.

Practice Problem 25-6 As a meter stick hurtles past you at 0.45*c*, you measure its angle with respect to the direction of motion to be 40°. What angle does the meter stick make with respect to the direction of motion in its rest frame?

A meter stick moves at speed 0.80*c* relative to you at an angle of 30° from its direction of motion. The meter stick is shown in the *S'* frame, in which it is at rest.

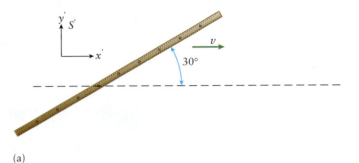

(a)

The length of the meter stick in s' is 1.0 m, its proper length. Treated as a vector of length L_p pointing along its length, the meter stick has components of length x' and y'.

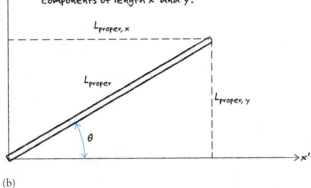

(b)

Figure 25-8 (a) When a meter stick is at an angle relative to its direction of motion, (b) only the projection of its length in the direction of motion experiences a relativistic length contraction.

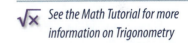 *See the Math Tutorial for more information on Trigonometry*

> ### ? Got the Concept 25-6
> ### No Contraction?
>
> A rod whizzes past you at close to the speed of light. Is it possible that you could measure its length to be equal to its proper length? Explain your answer.

Example 25-7 Muon Decay Revisited

In Example 25-2 we considered cosmic ray muons streaming down to Earth from the upper atmosphere at a speed of $0.994c$. Muons naturally decay; the muon half-life, the average time it takes for one-half of a group of muons to decay, is $1.56\ \mu s$. If 10^6 muons are created at an altitude of 10 km, how many would you expect to strike Earth considering that in the frame of the muons the distance to the surface is length-contracted? Compare your result to the answer to Example 25-2b.

SET UP

We determine the number of muons that strike Earth by determining how long, in terms of the muon's half-life, it takes for the muons to get to the surface. From Equation 25-9, of an initial number of muons N_0, the number N that remain after n half-lives have elapsed is

$$N = \frac{1}{2^n} N_0 \tag{25-18}$$

Our goal, then, is to determine the number of half-lives that elapse during the muons' trip from the upper atmosphere down to the surface.

SOLVE

You are at rest with respect to Earth, so your measurement of the distance from the upper atmosphere to Earth's surface is the proper distance. That is, D_{proper} equals 10 km. The distance is length-contracted in the frame of the muons according to Equation 25-17:

$$D = \frac{1}{\gamma} D_{proper}$$

For muons traveling at $0.994c$ and from Equation 25-14, relativistic gamma is

$$\gamma = \frac{1}{\sqrt{1 - \dfrac{V^2}{c^2}}} = \frac{1}{\sqrt{1 - \dfrac{(0.994c)^2}{c^2}}} = \frac{1}{0.109}$$

so

$$\frac{1}{\gamma} = 0.109$$

In the muons' frame of reference, then, the initial distance to Earth's surface is

$$D = (0.109)(10\ km) = 1.09\ km = 1090\ m$$

Earth's surface approaches at a speed of $0.994c$, so the time it takes the muons to strike the surface is

$$\Delta t = \frac{1090\ m}{(0.994)(3.00 \times 10^8\ m/s)} = 3.67 \times 10^{-6}\ s = 3.67\ \mu s$$

This is only 2.35 half-lives ($3.67\ \mu s/1.56\ \mu s$), so the fraction of muons left in the time it takes for them to reach the surface (from Equation 25-18) is

$$N = \frac{1}{2^{2.35}}10^6 = 0.196 \times 10^6$$

Nearly 20%, or about 200,000, of the original 10^6 muons make it to the surface.

REFLECT

This is the same result we found in Example 25-2, part (b). Our approach in that problem was to transform the measurement of the duration of the flight from the upper atmosphere to the surface. Our approach in this problem is to transform the distance the muons travel. The two approaches are equivalent.

Practice Problem 25-7 A subatomic particle called a *charged pion* has a half-life of about 1.8×10^{-8} s. One thousand charged pions are sent down a 20-m-long pipe at $0.95c$. (a) How long is the pipe in the rest frame of the pions? (b) How many would you expect to reach the end of the pipe considering that in the frame of the charged pions the distance is length contracted?

Example 25-8 A Galactic Competition

A competitor flies a spaceship of proper length 50.0 m toward a large cylinder of proper length 30.0 m in an event at the Galactic Games. Doors at either end of the cylinder open instantly when an object approaches and close instantly when the opening is clear. The goal is to get the ship to fit completely inside the cylinder, so that at least for an instant the ship is inside and both doors are closed. The competitor flies the ship at $0.800c$. How long is the ship as measured by spectators who are in the rest frame of the cylinder? Does the ship fit inside the cylinder, at least momentarily, with both doors closed?

SET UP

The ship is moving relative to the spectators, so they see its length as contracted. Because the spectators are at rest with respect to the cylinder, they measure the length of the cylinder to be its proper length. We therefore first apply relativistic length contraction to the ship, and then compare the contracted length to the proper length of the cylinder to determine if the ship fits inside the cylinder, at least momentarily, with both doors closed.

SOLVE

The ship moves at V equal to $0.800c$ relative to the spectators, so they measure its length to be, from Equation 25-17,

$$L_{ship} = \frac{1}{\gamma}L_{ship,\ proper}$$

where γ is

$$\gamma = \frac{1}{\sqrt{1 - \dfrac{V^2}{c^2}}} \tag{25-14}$$

So the spectators measure the length of the ship to be

$$L_{ship} = \sqrt{1 - \frac{V^2}{c^2}}L_{ship,\ proper} = \sqrt{1 - \frac{(0.800c)^2}{c^2}}(50.0\ \text{m}) = 30.0\ \text{m}$$

REFLECT

The contracted length of the ship is 30.0 m, the same as the proper length of the cylinder. The ship does, for an instant, fit inside the cylinder, so for that instant the ship is inside with both doors closed.

Practice Problem 25-8 A spaceship of proper length 1.00×10^2 m flies through a tunnel drilled through an asteroid, at $0.600c$ relative to the asteroid. What does an observer on the asteroid measure as the length of the spaceship?

Example 25-9 A Galactic Competition, Reconsidered

A competitor flies a spaceship of proper length 50.0 m toward a large cylinder of proper length 30.0 m in an event at the Galactic Games. Doors at either end of the cylinder open instantly when an object approaches and close instantly when the opening is clear. The goal is to get the ship to fit completely inside the cylinder, so that at least for an instant the ship is inside and both doors are closed. The competitor, who flies the ship at $0.800c$, is justified from the standpoint of physics, in treating herself as at rest and the cylinder in motion relative to her. What is the length that she measures for the cylinder?

SET UP

The cylinder is moving relative to the competitor, so she sees its length as contracted. The competitor is at rest with respect to her ship, however, so she measures its length to be its proper length. We therefore apply relativistic length contraction to the cylinder.

SOLVE

The cylinder moves at V equal to $0.800c$ relative to the competitor, so she measures its length to be, from Equation 25-17,

$$L_{cylinder} = \frac{1}{\gamma} L_{cylinder,\ proper}$$

where γ is

$$\gamma = \frac{1}{\sqrt{1 - \dfrac{V^2}{c^2}}} \tag{25-14}$$

The competitor therefore measures the length of the cylinder to be

$$L_{cylinder} = \sqrt{1 - \frac{V^2}{c^2}}\, L_{cylinder,\ proper} = \sqrt{1 - \frac{(0.800c)^2}{c^2}}\,(30.0\ \text{m}) = 18.0\ \text{m}$$

REFLECT

It is not surprising that the competitor sees the length of the cylinder as contracted because it is in motion relative to her. It would appear, however, that her ship, the length of which she measures as 50.0 m, is far too long to fit inside the cylinder.

Practice Problem 25-9 A spaceship of proper length 1.00×10^2 m flies through a tunnel drilled through an asteroid at $0.600c$ relative to the asteroid. In the rest frame of the asteroid, the tunnel is 4.50×10^2 m long. (a) What does the pilot of the spaceship measure as the length of the tunnel? (b) At what speed relative to the asteroid would the spaceship need to travel so that the pilot measures the tunnel to be the same as the length of his ship?

?   **Got the Concept 25-7**
A Relativistic Paradox

The previous two problems appear to present a paradox. A fast-moving ship passes through a cylinder. In the reference frame at rest with respect to the cylinder, the length of the ship is contracted, so it fits entirely in the cylinder at some instant, and at that instant the doors at both ends of the cylinder are closed. In the frame of the ship, however, it is the cylinder that is length contracted, so it is far too short for the ship to fit inside. How can the ship be inside the cylinder with both doors closed in one frame, but not in the other?

✱   **What's Important 25-4**

The Lorentz transformation is a direct mathematical way to determine the value of a quantity measured in one inertial frame based on the value measured in another inertial frame. When an object moves relative to an observer, its length in the direction of motion is seen to be contracted compared to the length measured at rest with respect to the object.

25-5 Lorentz Velocity Transformation

We begin this section with a puzzle. You are traveling north on the local highway at 2.5×10^1 m/s. Your friend is traveling south at 2.0×10^1 m/s. Because there is no absolute motion, nor a frame that might define absolute motion, physics allows you to consider yourself at rest and your friend in motion at 4.5×10^1 m/s relative to you. That is, your relative speed is 4.5×10^1 m/s. What if you were both traveling in futuristic spaceships, you at 2.5×10^8 m/s and your friend at 2.0×10^8 m/s. It isn't possible that you could measure her speed relative to you as 4.5×10^8 m/s —that speed is far greater than c. *No object can move, and no information can be transmitted, at a speed that exceeds the speed of light in a vacuum.* (At the time of this writing, evidence has begun to emerge which suggests that some particles can travel at speeds that slightly exceed the speed of light in a vacuum. Physicists are not yet ready, however, to draw a conclusion from the evidence.) What then is the relative speed of you and your friend in this futuristic scenario, in which the speeds are high rather than low?

We used intuition to determine the relative speed of the two cars in our puzzle. We could have also applied the Galilean transformation by assigning the S frame to you and the S' frame to your friend, and by aligning the motion with the x and x' axes. Then according to the x component of Equation 25-1,

$$x' = x - Vt$$

V is the relative speed of the two frames, and therefore the relative speed of you and your friend. We can relate the relative speed to the individual speeds by taking the time derivative:

$$\frac{d}{dt}(x' = x - Vt)$$

or

$$\frac{dx'}{dt} = \frac{dx}{dt} - V\frac{dt}{dt}$$

 See the Math Tutorial for more information on Calculus

Because t' equals t in Newtonian relativity, the derivative on the left is dx'/dt' or v'_x. The first derivative on the right-hand side equals v_x, and dt/dt equals 1. So

$$v'_x = v_x - V$$

The relative speed is therefore

$$V = v_x + v'_x$$

Our intuition is correct. The relative speed of the two cars in the direction of motion is the sum of their individual speeds.

To determine the relative speed of the two spaceships in our puzzle, we follow the same process as with the two cars above. We again assign the S frame to you and the S' frame to your friend, and align the motion with the x and x' axes. Because the relative speed of the two frames is relativistic, however, we must apply the Lorentz transformation rather than the Galilean transformation. The x component of the Lorentz transformation (Equation 25-10) is

$$x' = \gamma(x - Vt)$$

where γ is relativistic gamma (Equation 25-14) and V is the relative speed of you and your friend. We again take the time derivative in order to relate the relative speed to the individual speeds:

$$\frac{d}{dt}[x' = \gamma(x - Vt)]$$

Note that γ is a constant for a fixed value of V, so this becomes

$$\frac{dx'}{dt} = \gamma\left(\frac{dx}{dt} - V\right) = \gamma(v_x - V) \tag{25-19}$$

Unlike in the case of the Galilean transformation, t' is not equal to t, so we need to be a bit clever in order to make v'_x appear on the left-hand side. However, note that the left-hand side of Equation 25-19 can be written

$$\frac{dx'}{dt} = \frac{dx'}{dt'}\frac{dt'}{dt} = v'_x\frac{dt'}{dt} \tag{25-20}$$

We can use the time component of the Lorentz transformation for the right-most derivative. The fourth (time) relationship in Equation 25-10 is

$$t' = \gamma\left(t - \frac{V}{c^2}x\right)$$

so

$$\frac{dt'}{dt} = \gamma\left(1 - \frac{V}{c^2}\frac{dx}{dt}\right) = \gamma\left(1 - \frac{V}{c^2}v_x\right)$$

From Equation 25-20, dx'/dt is then

$$\frac{dx'}{dt} = v'_x\frac{dt'}{dt} = v'_x\gamma\left(1 - \frac{V}{c^2}v_x\right)$$

Finally, we apply this to Equation 25-19 and find

$$v'_x\gamma\left(1 - \frac{V}{c^2}v_x\right) = \gamma(v_x - V)$$

Although it is not in its simplest form, this relationship contains only three variables, the relative speed V and the speeds v_x and v'_x of the two ships in their individual frames. In other words, we have accomplished the goal of relating the three speeds. One standard rearrangement is

$$v'_x = \frac{v_x - V}{1 - \dfrac{V}{c^2} v_x} \tag{25-21}$$

Equation 25-21 enables us to find the speed v'_x of an object as observed in the S' frame when its speed v_x is known in a frame S moving at relative speed V to S'.

Let's now solve the puzzle that arose when we imagined that you are traveling at 2.5×10^8 m/s and your friend at 2.0×10^8 m/s in the opposite direction. We'll place an observer who isn't moving with either of you in a frame S to watch the action. This observer measures your speed (v_{you}) to be 2.5×10^8 m/s and your friend's speed (v_{friend}) to be -2.0×10^8 m/s. To be clear, both of the speeds are measured in the S frame. The negative value indicates that your friend's *velocity* is in a direction opposite to yours. If we then attach the S' frame to you and transform your friend's speed from S to S', the result is v'_{friend}, her speed in your frame. This is her speed relative to you, the speed we are trying to find. The assignment of the frames and speeds is shown in **Figure 25-9**.

According to Equation 25-21,

$$v'_{friend} = \frac{v_{friend} - V}{1 - \dfrac{V}{c^2} v_{friend}} = \frac{v_{friend} - v_{you}}{1 - \dfrac{v_{you}}{c^2} v_{friend}}$$

where we set the relative speed of the frames equal to your speed because the S' frame moves along with you. Substituting the values of the speeds gives

$$v'_{friend} = \frac{(-2.0 \times 10^8 \text{ m/s}) - 2.5 \times 10^8 \text{ m/s}}{1 - \dfrac{2.5 \times 10^8 \text{ m/s}}{c^2}(-2.0 \times 10^8 \text{ m/s})} = \frac{-4.5 \times 10^8 \text{ m/s}}{1 + \dfrac{5 \times 10^{16} \text{ (m/s)}^2}{(3 \times 10^8 \text{ m/s})^2}}$$

or

$$v'_{friend} = -2.9 \times 10^8 \text{ m/s}$$

Your friend's speed relative to you is quite high. It is not greater than the speed of light in a vacuum, however. The result of this application of the Lorentz transformation is therefore consistent with the stipulation that no object can move at a speed greater than c. This should not be a surprise because we depended on this assumption to obtain the Lorentz transformation.

You (in the red ship) and your friend (in the blue ship) fly space ships toward each other.

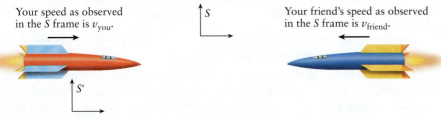

Your speed as observed in the S frame is v_{you}.

Your friend's speed as observed in the S frame is v_{friend}.

The S' frame moves with you, so its speed relative to the S frame is v_{you}. Your friend's speed relative to you is v_{friend} transformed to the S' frame.

Figure 25-9 To determine the speed of one spaceship relative to another, we measure the speeds of both relative to a frame S and attach S' to one of the ships.

Example 25-10 Light Speed

As Mary passes Jack traveling at $0.80c$ relative to him, she fires a laser straight forward, as in **Figure 25-10**. Jack sees the speed of the light as c. By applying Equation 25-21, find an expression for the speed of the laser light as observed by Mary.

Figure 25-10 As Mary's spaceship moves relative to Jack, she fires a laser straight forward. Jack measures the speed of the light to be c. What speed does Mary measure?

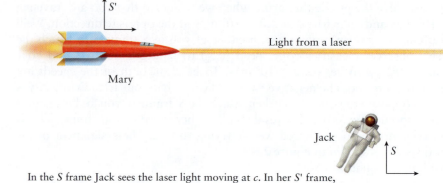

In the S frame Jack sees the laser light moving at c. In her S' frame, what speed does Mary observe for the laser light?

SET UP

To apply Equation 25-21 we need to assign two frames S and S' from which to make observations. We choose to attach S to Jack and S' to Mary. The speed of one frame relative to the other, V, is therefore $0.80c$. In the S frame Jack observes the speed of the light along the direction of motion as c; this is v_x in Equation 25-21. The speed of the light as observed by Mary is v'_x.

SOLVE

Setting V equal to $0.80c$ and v_x equal to c in Equation 25-21 gives

$$v'_x = \frac{c - 0.80c}{1 - \frac{0.80c}{c^2}c}$$

Notice that c^2 cancels in the denominator, and that c is a common factor of the two terms in the numerator, so

$$v'_x = \frac{c(1 - 0.80)}{1 - 0.80} = c$$

REFLECT

Aha! Just as Einstein's second postulate requires, both Mary and Jack observe the speed of the laser beam to be c. The speed of light is independent of the speed of the observer. And notice that we would obtain the same result regardless of the speed of Mary's ship. Perhaps we shouldn't be too surprised, however. After all, it was Einstein's second postulate that initially led us to the Lorentz transformation, and therefore to the velocity transformation of Equation 25-21.

✳ What's Important 25-5

For two objects moving at speeds much less than the speed of light in a vacuum, one's speed relative to the other is additive, as described by the Galilean transformation. At relativistic speeds, we must apply the Lorentz transformation rather than the Galilean transformation to find relative speed.

25-6 Relativistic Momentum and Energy

The laws of physics hold in all frames. We showed in Section 25-1 for example, that the momentum of an isolated system of colliding objects is conserved regardless of the inertial frame from which the collision is observed. We did this, however, using Newtonian relativity and the Galilean transformation, that is, when the frames are moving at low speed relative to each other. Does the law of conservation of momentum hold in a frame when observed from another if the two frames are moving at a high speed relative to one another?

The total momentum of the system *is* conserved, even for objects traveling at relativistic speeds and even for collisions observed from a frame moving at relativistic speed. To ensure this is true, however, it is necessary to generalize our definition of momentum into the realm of relativity. A careful analysis of a collision between objects reveals that to account for relativistic effects, the definition of momentum must be expanded from

$$\vec{p} = m\vec{v}$$

to

$$\vec{p} = \gamma m_0 \vec{v} \tag{25-22}$$

A comparison of the last two relationships for momentum suggests that the γm_0 term is a relativistic description of mass. Indeed, as a general statement we write the **relativistic mass m** as

$$m = \gamma m_0 \tag{25-23}$$

where m_0 is an object's **rest mass**, the mass we measure when at rest with respect to an object. The rest mass of any object is a constant and so is not frame-dependent; to emphasize this we often refer to rest mass as the **invariant mass**. Unless you are at rest with respect to an object you would measure its mass to be m. We haven't had to consider this generalization up until now, however, because at low speeds, for which γ is barely more than 1, relativistic mass m is essentially equal to m_0. In addition, because there is no absolute motion, nor a frame that defines whether an object is in motion or at rest except relative to another frame, a measurement of mass made from one frame moving relative to another will also result in the relativistic mass. The only way that a measurement of mass gives an object's rest mass is to make the measurement at rest with respect to the object.

Watch Out

Momentum and mass are a function of the object's speed relative to an observer.

The observed mass of an object is a product of relativistic gamma and the object's rest mass, as in Equation 25-23. The same is true for an object's momentum (Equation 25-22). Relativistic gamma is, in turn, a function of speed, so both momentum and mass are a function of the object's speed relative to an observer. This is not intuitive because the relativistic effect is only detectable at high speeds, speeds typically outside the realm of our everyday experiences.

As we have seen from Equation 25-14, γ is always either equal to or greater than 1. (When V, the speed of an object or the speed of the frame from which an observation is made, equals zero, γ equals 1. For all other values of V, γ must be greater than 1.) *The relativistic mass m is therefore always equal to or greater than*

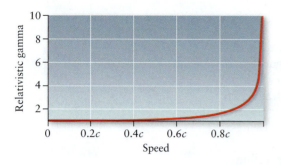

Figure 25-11 Relativistic gamma increases dramatically as speed increases.

the rest mass m_0. Moreover, as speed increases, the relativistic momentum and the relativistic mass increase dramatically, following the rapid rise of γ as a function of V (**Figure 25-11**). Energy must be added to the system (by doing work on it) to increase an object's momentum, so the amount of energy required to increase an object's speed increases for higher and higher speeds. As the speed of an object approaches c, the amount of energy required to increase the speed approaches infinity. We are forced to conclude that it is not possible for an object of nonzero rest mass to accelerate up to the speed of light. Only objects with zero rest mass, such as photons of light, can reach the speed of light in a vacuum!

Just as momentum and mass are a function of the object's speed relative to an observer, so too is the energy of an object. The **relativistic energy E** is proportional to the rest energy E_0:

$$E = \gamma E_0 \qquad (25\text{-}24)$$

An observer at rest with respect to an object measures its **rest energy**.

Wait! This rest energy is something we have not yet encountered. It is not potential energy; potential energy is associated with a force an object experiences. (See, for example, Equation 6-25.) It is not kinetic energy; kinetic energy is associated with motion. Rest energy is neither of those, but rather an outcome of the equivalence of mass and energy. Einstein showed that rest energy is proportional to rest mass, and that the proportionality constant is c^2:

$$E_0 = m_0 c^2 \qquad (25\text{-}25)$$

Combining this with Equation 25-24 yields

$$E = \gamma m_0 c^2$$

or

$$E = mc^2 \qquad (25\text{-}26)$$

You may be familiar with this form of Einstein's famous equation. The physical interpretation of Equation 25-26 is that mass is simply one possible manifestation of energy. Said another way, the mass of an object is a measure of its energy content. **Figure 25-12** provides evidence of the equivalence of mass and energy. This image, recorded using a detector called a *bubble chamber*, shows the conversion of

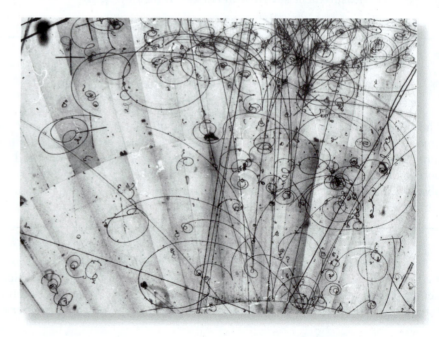

Figure 25-12 This bubble chamber image captures the conversion of the energy of a photon, which has zero rest mass and which leaves no trace in the chamber, into the mass of two particles. (*G.T.Jones, Birmingham University / Fermi National Accelerator Laboratory*)

energy into mass. A photon, which has zero rest mass and is therefore energy, comes up from the bottom center of the image. It leaves no trace in the bubble chamber, so the region looks empty. At some moment the photon's energy is converted into a pair of particles both of which have nonzero rest mass, leaving two traces in the bubble chamber that form what looks like a fancy letter "V." Two particles, both of which have mass, have been created from the energy of the photon!

? Got the Concept 25-8
Energy and Mass

Each colored line emanating from the center of **Figure 25-13** traces the path of a particle created when two highly energetic gold nuclei collide in the Relativistic Heavy Ion Collider in Brookhaven, New York. Unlike the collision of two cars in a crash test (**Figure 25-14**) in which various bits of two cars fly off in every direction, the particles leaving traces in Figure 25-13 are not small pieces of the gold nuclei. Can you explain the source of the particles?

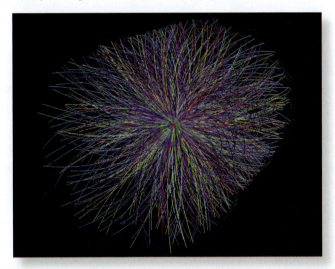

Figure 25-13 Each colored line emanating from the center of the image traces the path of a different particle created when the nuclei of two highly energetic gold atoms collide. *(CERN)*

Figure 25-14 When two cars collide, car bits fly off in every direction. This is *not* what happens when gold nuclei collide (Figure 25-13). *(CTK Photo/Rene Fluger / Alamy)*

Example 25-11 Rest Energy

A typical packet of granulated sugar you might find in a college dining hall contains 4.0 g of the sweetener. The 16 food calories (16 kcal) in the packet represent about 67 kJ of energy. Find the rest energy of 4.0 g of granulated sugar.

SET UP

Equation 25-25 will return energy in joules as long as the mass and the speed of light in a vacuum are also entered in SI units. So we only need to apply

$$E_0 = m_0 c^2$$

setting m_0 equal to the mass of the sugar in kilograms.

SOLVE

Substituting the values given in the problem into Equation 25-25, we find that

$$E_0 = (4.0 \times 10^{-3}\,\text{kg})(3.0 \times 10^8\,\text{m/s})^2$$

or

$$E_0 = 3.6 \times 10^{14}\,\text{J}$$

REFLECT

There's quite a difference between the energy contained in the sugar and the energy equivalent of the mass of the sugar. Obviously, our bodies do not harvest energy from the atoms that make up sugar molecules; we only use some of the energy stored in the chemical bonds between adjacent carbon atoms.

Practice Problem 25-11 A cup of pecans has a mass of 0.108 kg and contains 720 food calories, or 3.0×10^6 J, of energy. (a) What is the rest mass equivalent of the energy content of one cup of pecans? (b) What is the rest energy equivalent of one cup of pecans?

Example 25-12 The Sun

The Sun radiates about 4×10^{26} J of energy every second. (a) At what rate is the Sun's mass decreasing? (b) Show that the units you get in the calculation are equivalent to kilograms per second (kg/s).

SET UP

The mass equivalent of the energy radiated by the Sun can be found by solving Equation 25-25 for rest mass:

$$m_0 = \frac{E_0}{c^2}$$

The change in mass Δm_0 for a given amount of radiated energy ΔE_0 is then

$$\Delta m_0 = \frac{\Delta E_0}{c^2}$$

so the mass loss rate is

$$\text{mass loss rate} = \frac{\Delta m_0}{\text{s}} = \frac{\Delta E_0/\text{s}}{c^2}$$

SOLVE

(a) So

$$\text{mass loss rate} = \frac{4 \times 10^{26}\,\text{J/s}}{(3 \times 10^8\,\text{m/s})^2} = 4 \times 10^9 \frac{\text{J/s}}{(\text{m/s})^2}$$

(b) The units of our answer are

$$[\text{mass loss rate}] = \frac{\text{J/s}}{(\text{m/s})^2} = \frac{\text{J}}{\text{s}}\frac{\text{s}^2}{\text{m}^2} = \text{J}\frac{\text{s}}{\text{m}^2}$$

Using any expression for energy as a guide, for example, $K = \frac{1}{2}mv^2$, we see that units of joules are equivalent to $\text{kg}(\text{m/s})^2$. So

$$[\text{mass loss rate}] = \frac{\text{kg m}^2}{\text{s}^2}\frac{\text{s}}{\text{m}^2} = \frac{\text{kg}}{\text{s}}$$

We should expect to find the mass loss rate given in kg/s because we used SI units for both the rate of energy loss rate (J/s) and c (m/s^2). The Sun is losing about 4×10^9 kg every second due to the energy it radiates.

REFLECT

Should we be worried that the Sun is losing more than 10^9 kg of mass every second? The mass of the Sun is about 2×10^{30} kg. At the mass loss rate we calculated, the Sun would shrink to half its mass in the time it takes to radiate away 1×10^{30} kg. At a rate of 4×10^9 kg/s, the time $t_{1/2}$ it would take the Sun to shed 1×10^{30} kg is

$$t_{1/2} = \frac{1 \times 10^{30}\,\text{kg}}{4 \times 10^9\,\text{kg/s}} = \frac{1}{4} \times 10^{21}\,\text{s}$$

That's about 8×10^{12} y. No need to worry!

Practice Problem 25-12 A 1250-W hair dryer is accidentally left on at full power for one year. Assuming it doesn't overheat and turn off, what is the mass equivalent of the energy output during that time?

✴ What's Important 25-6

Energy, mass, and momentum are a function of an object's speed relative to an observer. The relativistic effect is only detectable at high speeds, speeds typically outside the realm of our everyday experiences. Objects of nonzero rest mass cannot accelerate up to the speed of light.

25-7 General Relativity

Sitting in a chair in the patent office in Bern, Germany, in 1907, a young Albert Einstein had what he would call "the happiest thought of my life." He imagined a man falling freely from the roof of a house and realized that "at least in his immediate surroundings—there exists no gravitational field." If the man released an object, for example, it would accelerate at the same rate as the man accelerated, and that because the man would "not feel his own weight" it would appear to him that neither he nor the object were experiencing a gravitational force. Einstein found this thought startling; it took him down a path that eventually led to his theory of general relativity. Fundamental to this theory is the equivalence of gravity and acceleration, and therefore a context for relativistic physics that includes noninertial (accelerating) frames of reference as well as inertial frames.

According to Newton's second law, acceleration arises from an application of force:

$$\sum \vec{F}_{\text{dir}} = m\vec{a}_{\text{dir}} \qquad (4\text{-}1)$$

For a single object, the acceleration in the direction of the applied force is

$$F = m_{\text{inertial}}a$$

where m_{inertial} is the **inertial mass**. The inertial mass is a measure of an object's inertia, the property of all objects that have mass to resist being accelerated.

We also encountered the quantity mass in the context of gravity. Newton's universal law of gravitation,

$$\vec{F} = -\frac{Gm_1m_2}{r^2}\hat{r} \qquad (10\text{-}1)$$

tells us that the gravitational force that one particle exerts on another is proportional to the product of their masses. For an object of mass m_{grav}, a distance r away from another of mass M, the magnitude of the gravitational force is

$$F = \frac{GMm_{\text{grav}}}{r^2}$$

where we applied a subscript "grav" to the object's mass to indicate that it is a measure of the gravitational effect on it due to the other object. The **gravitational mass** is directly related to an object's weight.

Up until now we have not distinguished between inertial mass and gravitational mass, and so have effectively treated them as one and the same thing. Are they? Physicists as far back as Newton asked this question, and experimental evidence suggested that inertial mass and gravitational mass are numerically equal. It is a remarkable outcome of Einstein's "happiest thought" that the two quantities, which involve rather different concepts, should not only be numerically equal but also be indistinguishable. Einstein postulated that inertial mass and gravitational mass are equivalent; this principle of equivalence, which guides general relativity, is summarized as follows:

A gravitational field is equivalent to an accelerated frame of reference in the absence of gravity.

The principle of equivalence dictates that it is not possible to distinguish experimentally between a system in an accelerating frame and a system under the influence of gravity. In other words, if you were to drop a ball in a windowless elevator car that makes no noise and doesn't shake, you could not tell from the motion of the ball whether the elevator car were sitting stationary on the surface of a planet and experiencing its gravity or accelerating in empty space, far from sources of gravity. For example, if you were standing in the elevator car and experienced an acceleration of 9.8 m/s^2, you wouldn't be able to tell whether the car was stationary on Earth or accelerating at 9.8 m/s^2 in the absence of Earth's gravity.

Let's explore physics in the imaginary elevator car further. In **Figure 25-15a** a ball is thrown horizontally while the elevator car is stationary near Earth's surface. Due to the force of gravity, the ball accelerates downward, following a familiar parabolic arc. The figure shows the positions of the ball at five instants, spanning four equal time intervals. What if the elevator car were far from Earth and from any other massive object that could exert a noticeable gravitational force on the ball? In that case the ball would travel along a straight line, as shown in **Figure 25-15b**. No gravity means no acceleration—or does it? Consider the situation shown in **Figure 25-16**. The elevator car is again far from any object that could exert a

(a) This elevator car is stationary near the surface of the Earth. A ball thrown horizontally follows a parabolic path as it falls.

(b) This elevator car is stationary, far from Earth or any other large massive object. A ball thrown horizontally travels in a straight line.

Figure 25-15 (a) When a ball is thrown horizontally in an elevator car on Earth's surface, it follows a parabolic path. (b) When a ball is thrown horizontally in an elevator car far from any massive object, it follows a straight path.

noticeable gravitational force on the ball, but now the car is accelerating. The direction of the acceleration is "up" (toward the top of the page). Because there is no discernable gravitational force, the ball travels in a straight line as in Figure 25-15b. However, because the elevator car is accelerating, an observer in the car sees the ball trace out a parabolic arc with respect to the walls and floor of the car. Remember that there are no windows in the elevator car, and it makes no noise and does not vibrate as it moves. An observer in the car cannot, therefore, detect its motion. The observer feels the effect of the principle of equivalence: It appears that the ball is falling under the influence of gravity.

The real power of Einstein's postulate about the equivalence of acceleration and gravity is seen when we substitute a beam of light for the ball in our elevator car thought experiment. We expect that a beam of light travels in a straight line, so in analogy to throwing the ball, shining the light horizontally should result in a horizontal beam of light. Certainly that is the case when the elevator is stationary and far from any massive objects. If the elevator were accelerating, however, as in the case of the ball depicted in Figure 25-16, an observer in the elevator car would *not*

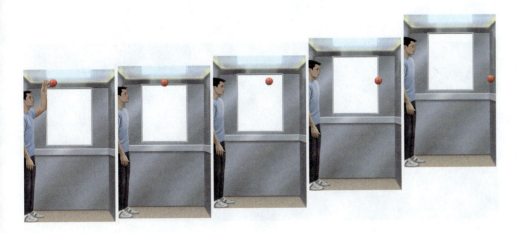

Figure 25-16 When a ball is thrown horizontally in an elevator car which is both far from any massive object and also accelerating, it follows a straight path. An observer in the elevator, however, would see the ball follow a parabolic path with respect to the floor and walls of the elevator car.

see the light move straight toward the far wall. You might need a precise measuring device to detect it, but the light would hit the opposing wall below the height at which it started. Did the light bend or is it that the elevator car has moved during the time that the light travels from one side of the car to the other? The principle of equivalence tells us that the observer has no way of knowing whether the car is accelerating or whether it is at rest near an object that exerts a gravitational force on it. We are therefore justified in declaring that the car is stationary and that the light is deflected! In addition, we must conclude that the beam's parabolic path results from the influence of a gravitational force. Both the beam of light and the ball accelerate due to a gravitational field.

The deflection of light in the presence of massive objects is small, an effect you might imagine would have been impossible to measure in Einstein's time. Einstein realized, however, that because the Sun's mass is large, if the light from a distant star were to pass close to the Sun, the change in its direction would be measurable. Light passing close to the Sun would only be visible during a total eclipse, so in order to test Einstein's theory the English astronomer Sir Arthur Eddington led an expedition to an island off the west coast of Africa in 1919 to photograph a total eclipse of the Sun. Eddington compared the position relative to other stars of a star cluster that was optically close to the Sun during the eclipse to its position at other times. His measurements confirmed Einstein's theory that massive objects deflect light, making both men international celebrities!

A more stunning phenomenon associated with gravitational bending of light occurs when multiple images of a distant star form as light is bent by a closer, massive celestial object, such as the black hole at the center of a galaxy. Such **gravitational lensing** is depicted in **Figure 25-17**, in which a massive object is positioned directly between Earth and a distant star. Light from the star cannot reach Earth directly, but the lensing effect results in light initially not propagating toward Earth to be bent back toward us. In this way light from the star can approach Earth from many directions, for example, the two directions shown in the figure. A good number of examples of this phenomena have been discovered, most often showing four distinct images as in **Figure 25-18a**. When light is bent around the intervening object and reaches Earth from a full circle around it, the star's light is spread out into a ring around the lensing object, as in **Figure 25-18b**.

Earlier in this chapter we saw that Einstein's theory of special relativity addresses time dilation, where a clock moving at constant speed relative to an observer is seen to be running slowly. Imagine a scenario in which *two* observers moving at constant speed relative to each other can each see a clock carried by the other. Each observes a moving clock, so each measures the other's clock to be running slowly *compared to* her own clock. Is this possible? If the two observers stopped their relative motion and brought the clocks together, certainly both clocks could not be slow relative to the other!

Figure 25-17 A massive object positioned between Earth and a distant star bends light from the star so that it can be seen on Earth.

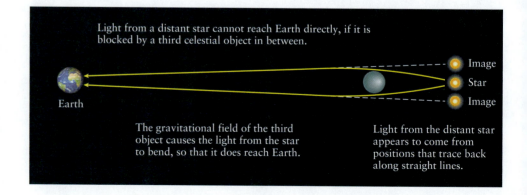

Light from a distant star cannot reach Earth directly, if it is blocked by a third celestial object in between.

Image

Star

Image

Earth

The gravitational field of the third object causes the light from the star to bend, so that it does reach Earth.

Light from the distant star appears to come from positions that trace back along straight lines.

Gravitational Lens G2237+0305

(a)   (b)

Figure 25-18 (a) Four separate images of a single star are formed by a massive gravitational lens. *(NASA, ESA, and STScI)* (b) Light leaving a distant star travels in every direction. When the light is bent toward Earth by a massive gravitational lens, the image formed can be a ring. *(NASA, ESA, A. Bolton (Harvard-Smithsonian CfA) and the SLACS Team)*

Einstein's theory of general relativity is required to resolve this puzzle. When the two observers are in relative motion at constant speed, their frames are symmetric, that is, either observer can consider the other to be in motion. To bring the two clocks together to compare their times, however, one of the observers must decelerate. The two frames are therefore no longer symmetric. One is inertial and the other, the frame that decelerates, is not. In addition, general relativity must be applied to understand the measurement of time from one frame to the other. Time runs more slowly in the frame that experiences the change in speed.

This issue of acceleration and deceleration is central to the so-called "twins paradox," one of the most well-known puzzles in relativity. We'll state it this way: Bertha takes a trip to the star Proxima Centauri on a fast spaceship, while her twin Eartha stays at home. Eartha's clock ticks off 10 years in the time Bertha is gone. Special relativity suggests that were Eartha to observe Bertha's clock, she would see it running slowly compared to her own. After 10 h has elapsed on Eartha's clock, for example, she observes that only 8 h has elapsed on Bertha's. Eartha expects that when Bertha returns, she will have aged 8 years, not 10—Bertha will be younger than Eartha at the end of the trip. But according to the puzzle, Bertha considers herself at rest and sees Eartha moving away and then returning. In the telling of the paradox, the claim is made that the two views are equally valid, so Bertha sees Eartha's clock running slowly and expects that Eartha will be younger than she is at the end of the trip. But the two views are not symmetric! Bertha must accelerate to leave Earth, decelerate when she reaches Proxima Centauri, and then accelerate and decelerate again in order to return home. For this reason her frame can be distinguished from Eartha's. It is simply not correct to declare that each twin can consider the other to be moving in an inertial frame; special relativity does not apply. Although the mathematical details are beyond the scope of this book, general relativity predicts that the clock that accelerates and decelerates (Bertha's clock) will run slowly, so Bertha will be younger than Eartha on her return.

★ **What's Important 25-7**
The principle of equivalence, a central postulate of Einstein's theory of general relativity, declares that a gravitational field is equivalent to an accelerated frame of reference in the absence of gravity. It is not possible to distinguish between a system in an accelerating frame and a system under the influence of gravity. One important prediction of general relativity is that light bends as it passes through a gravitational field.

Answers to Practice Problems

25-1   (a) -1.0 m/s, (b) $+1.0$ m/s

25-2   (a) 67 pions, (b) 430 pions

25-3 $0.968c$

25-5 $0.100c$

25-6   37°

25-7   (a) 6.2 m, (b) 430 pions

25-8 0.800×10^2 m

25-9   (a) 360 m, (b) $0.975c$

25-11   (a) 3.3×10^{-11} kg, (b) 9.7×10^{15} J

25-12 4.38×10^{-7} kg

Answers to Got the Concept Questions

25-1   (a) The speed v'_x as measured by the observer in the S' frame is less than the speed v_x because the observer moves along in the same direction as the object. The object nevertheless moves away from the observer, so $\vec{v}'_x$ is in the same direction as $\vec{v}_x$. (b) Here, the speed of the S' frame causes the observer to overtake the object. The object therefore appears to be moving backward toward the observer as seen from the S' frame, so the direction of $\vec{v}'_x$ is opposite to that of $\vec{v}_x$.

25-2   The proper time is measured at rest with respect to the clock. Albert, aboard the Starship *Alpha*, is at rest with respect to the clock, so he measures the clock's proper time. Although Betty is not on the same craft as the clock, she is also not moving relative to it, so she also measures the clock's proper time. George's speed is the same as the clock's, but not in the same direction. So George is moving with respect to the clock and therefore does not observe proper time intervals. The same is true for Ellen on Earth, who is moving relative to the clock.

25-3   The time interval you observe for one tick of the clock on the ship (Δt) will always be larger than the time interval (Δt_{proper}) you would observe were you at rest with respect to it. The factor that relates the two depends on the relative speed, and from Equation 25-8, the larger V becomes, the greater Δt is compared to Δt_{proper}. So on the second pass, when V is larger, Δt increases compared to Δt_{proper}. The time interval you observe is therefore longer on the second pass, which means the clock appears to run more slowly on the second pass of the ship than on the first.

25-4   We have arbitrarily chosen to make the relative motion between S and S' be along the x/x' direction, so there is no relative motion in the other two directions. Other than this choice, there is nothing special about the y and z (or y' and z') directions.

25-5   The lowest possible speed of one frame relative to another is zero for two frames that are not moving relative to each other. As V approaches this limit the fraction V^2/c^2 in the denominator of Equation 25-14 approaches 0; the denominator in γ and therefore γ itself approaches

1. The smallest possible value of γ is therefore equal to 1. The fastest possible speed of one frame relative to another is c because we will always attach one frame to an object, and no object can exceed the speed of light. In this case, the fraction V^2/c^2 approaches 1, so the denominator of Equation 25-14 approaches zero. Thus as V approaches c, γ becomes infinitely large. We summarize this behavior with the mathematical statement $1 < \gamma < \infty$.

25-6   It *is* possible to measure the length of the rod to be the same as the proper length. If the rod is oriented so that its length is perpendicular to the direction of motion, no contraction occurs along its length, so you would measure its length to be the same as its proper length.

25-7   There is no paradox. In the frame of the cylinder, the time at which the front of the ship is leaving the cylinder is the same as the time at which the rear of the ship is entering the cylinder. A spectator would see the two events as simultaneous. In a real sense, the cylinder serves as a ruler, so a spectator can measure the length of the ship by comparing it to the length of the cylinder. The competitor, in her ship, observes something quite different. To her, the front of her ship leaving the cylinder is not simultaneous with the rear of the ship entering. Instead she sees a sequence of events. At some moment the front of her ship is about to leave the cylinder; the exit door opens. The rear of her ship has not entered the cylinder. Only later, after the front of the ship has exited, does the rear of the ship get inside the cylinder. She cannot use the cylinder as a ruler. Said another way, because time does not run at the same rate in the frame attached to the cylinder as it does in the frame attached to the ship, two events (the front of the ship leaving the cylinder and the rear of the ship entering it) that occur simultaneously in one frame are not simultaneous in the other.

25-8   The particles seen flying off from the collision of the two gold nuclei are the manifestation as mass of the energy the gold nuclei carried. The energy of the colliding nuclei is converted to mass, and because the nuclei carried so much energy before colliding, it is possible to create a large number of less massive or less energetic particles as a result of the collision.

SUMMARY

Topic	Summary	Equation or Symbol
frame of reference	A frame of reference (or "reference frame" or simply "frame") is a coordinate system with respect to which we will make observations or measurements of position and time.	
Galilean transformation	The Galilean transformation relates position and time $(x, y, z,$ and $t)$ as measured in one inertial frame with position and time $(x', y', z',$ and $t')$ as measured in another inertial frame. The Galilean transformation is applicable when V, the speed of one frame relative to the other, is much smaller than the speed of light in a vacuum.	$x' = x - Vt$ (25-1) $y' = y$ $z' = z$ $t' = t$
gravitational lensing	Gravitational lensing occurs when light is bent as it passes near a massive celestial object.	
gravitational mass	The mass of an object is termed "gravitational" when its value is determined by the strength of the gravitational force it feels due to another object.	$F = \dfrac{GMm_{grav}}{r^2}$
inertial frame	An inertial frame is one that moves at constant speed relative to another. A frame that is accelerating is a noninertial frame.	
inertial mass	The inertial mass is a measure of an object's inertia, the property of all objects that have mass to resist being accelerated.	$F = m_{inertial}a$
invariant mass	Invariant mass is another name for rest mass, the mass we measure when at rest with respect to an object.	
length contraction	When an object moves relative to an observer, its length in the direction of motion is seen to be contracted compared to the length measured at rest with respect to the object. The amount of contraction (shortening) depends on the speed of the object; the proportionality constant between the contracted length L of an object and it's proper length L_{proper} is the inverse of relativistic gamma.	$L = \dfrac{1}{\gamma}L_{proper}$ (25-17)
Lorentz transformation	The Lorentz transformation relates position and time $(x, y, z,$ and $t)$ as measured in one inertial frame with position and time $(x', y', z',$ and $t')$ as measured in another inertial frame. The Lorentz transformation is applicable for any value of V, the speed of one frame relative to the other. The Lorentz transformation should be used when V is on the same order as c.	$x' = \gamma(x - Vt)$ (25-10) $y' = y$ $z' = z$ $t' = \gamma\left(t - \dfrac{V}{c^2}x\right)$
Newtonian relativity	Newtonian relativity demands that the laws of physics are the same in all inertial frames when the speed of one frame relative to another is small compared to c.	
proper time	Proper time is a time measurement made at rest with respect to a clock.	Δt_{proper}
relativistic energy	The energy of an object moving with relativistic speed relative to an observer is its relativistic energy. Relativistic energy E is proportional to the object's rest energy E_0.	$E = \gamma E_0$ (25-24)

relativistic gamma	Relativistic gamma (γ) is a proportionality constant that describes the magnitude of relativistic effects like time dilation and length contraction. γ depends only on the speed of an object or frame, and it is always greater than or equal to 1.	$\gamma = \dfrac{1}{\sqrt{1 - \dfrac{V^2}{c^2}}}$	(25-14)
relativistic mass	The mass of an object moving with relativistic speed relative to an observer is its relativistic mass. Relativistic mass m is proportional to the object's rest mass m_0.	$m = \gamma m_0$	(25-23)
rest energy	An observer at rest with respect to an object measures its rest energy. Rest energy E_0 is proportional to rest mass m_0; the proportionality constant is c^2.	$E_0 = m_0 c^2$	(25-25)
rest frame	The rest frame is the frame of reference which is at rest with respect to an object.		
rest mass	An observer at rest with respect to an object measures its rest mass. The rest mass of any object is a constant.	m_0	
time dilation	Time is seen to flow at a slower rate in an inertial frame moving with respect to an observer, compared to the rate at which time flows in the observer's frame. An observer who compares a moving clock and one at rest with respect to himself observes less time to pass on the moving clock, that is, the moving clock runs more slowly. A time interval Δt measured by the observer is longer than the proper time interval Δt_{proper} in the moving frame, so in any given time Δt measured by the observer, less time runs off the moving clock. The amount that Δt and Δt_{proper} differ depends on the speed of the object; the proportionality constant between them is relativistic gamma.	$\Delta t = \dfrac{1}{\sqrt{1 - \dfrac{V^2}{c^2}}}\Delta t_{\text{proper}}$ (25-8) or $\Delta t = \gamma \Delta t_{\text{proper}}$	

QUESTIONS AND PROBLEMS

In a few problems, you are given more data than you actually need; in a few other problems, you are required to supply data from your general knowledge, outside sources, or informed estimate.

Interpret as significant all digits in numerical values that have trailing zeros and no decimal points.

For all problems, use $g = 9.8 \text{ m/s}^2$ for the free-fall acceleration due to gravity. Neglect friction and air resistance unless instructed to do otherwise.

• Basic, single-concept problem
•• Intermediate-level problem, may require synthesis of concepts and multiple steps
••• Challenging problem
SSM *Solution is in Student Solutions Manual*

Conceptual Questions

1. •Why do you think there was so much resistance from the established physics community when Einstein proposed his new theory of relativistic motion in 1905?

2. •How would you change the following Galilean transformation equations in the case where a frame was moving both in the x and the y direction?

$$x' = x - Vt \qquad (25\text{-}1)$$
$$y' = y$$
$$z' = z$$
$$t' = t$$

3. •What is a frame of reference? What is an inertial frame of reference? SSM

4. •What kind of reference frame is Earth's surface? Explain your answer.

5. •What does the phrase "all motion is relative, there is no absolute motion" mean?

6. •Describe the Michelson and Morley experiment. Why do you think that it was repeated so many times? (It may have been the most often performed experiment in the history of physics.)

7. •Give three explanations that justify the "null result" of the Michelson and Morley experiment.

8. •What are the two postulates of the special theory of relativity?

9. •Explain how the measurement of time enters into the determination of the length of an object. SSM

10. •Is it possible for one observer to find that event A happens after event B and another observer to find that event A happens before event B? Explain your answer.

11. •Which is larger, relativistic mass or rest mass? Explain your answer.

12. •For what range of energies can you use $p = E/c$ to a good approximation? Explain your answer.

13. •Some physicists refer to the relativistic increase in momentum as the "mass increase." Describe why this reference is reasonable. Describe why this reference is unreasonable.

14. •Is it possible to accelerate an object to the speed of light in a real situation? Explain your answer.

15. •Gene Roddenberry was the creator of the popular TV and film series *Star Trek*. Once at a public lecture he was asked if he thought we would ever be able to travel faster than light. He responded, "No, and that's a good thing because it means that we can always do it in science fiction." How have science fiction writers gotten around the cosmic speed limit?

16. •What is the fundamental postulate of the general theory of relativity?

17. •Describe one of the predictions of general relativity.

Multiple Choice Questions

18. •Which of the following statements are true?
    A. The laws of motion are the same in all inertial frames of reference.
    B. The laws of motion are the same in all reference frames.
    C. There is no way to detect absolute motion.
    D. All of the above statements are true.
    E. Only two of the above statements are true.

19. •Michelson and Morley concluded from their experimental results that the experiment
    A. was successful in detecting a shift in the interference pattern.
    B. was a failure because they didn't detect a shift in the interference pattern.
    C. was successful because they didn't detect a shift in the interference pattern.
    D. was a failure because they detected a shift in the interference pattern.
    E. lacked the necessary precision to determine a shift in the interference pattern.

20. •Time dilation means that
    A. the slowing of time in a moving frame of reference is only an illusion resulting from motion.
    B. time really does pass more slowly in a frame of reference moving relative to a frame of reference at relative rest.
    C. time really does pass more slowly in a rest frame of reference relative to a frame of reference that is moving.
    D. time is unchanging regardless of the frame of reference.
    E. no two clocks at rest can ever read the same time.

21. •If we use radio waves to communicate with an alien spaceship approaching Earth at 10% of the speed of light, we would receive their signals at a speed of
    A. $0.10c$.
    B. $0.90c$.
    C. $0.99c$.
    D. $1.00c$.
    E. $1.10c$. SSM

22. •If we use radio waves to communicate with an alien spaceship approaching Earth at 10% of the speed of light, the aliens would receive our signals at speed of
    A. $0.10c$.   D. $1.00c$.
    B. $0.90c$.   E. $1.10c$.
    C. $0.99c$.

23. •A meter stick hurtles through space at a speed of $0.95c$ with its length perpendicular to the direction of motion. You measure its length to be equal to
    A. 0 m.
    B. 0.05 m.
    C. 0.95 m.
    D. 1.00 m.
    E. 1.05 m.

24. •A particle has a relativistic momentum of p. If its speed doubles, the relativistic momentum will be
    A. greater than $2p$.
    B. equal to $2p$.
    C. less than $2p$
    D. equal to $2p/bc$.
    E. equal to p.

25. •Consider two atomic clocks, one at the GPS ground control station near Colorado Springs (elevation 1830 m) and the other one in orbit in a GPS satellite (altitude 20,200 km). According to the general theory of relativity, which atomic clock runs slow?
    A. The clock in Colorado runs slow.
    B. The clock in orbit runs slow.
    C. The clocks keep identical time.
    D. The orbiting clock is 95% slower than the clock in Colorado.
    E. The orbiting clock alternately runs slow and then fast depending on where the Sun is. SSM

This page is intentionally left blank.

For complete end of chapter problem sets, please go to
www.whfreeman.com/kestentauck

30 m according to coordinate system S. Compute the corresponding values as measured in the frame S' if the relative velocity between S' and S is 150,000 m/s along the x axis. The reference frames start together, with their origins coincident at $t = 0$.

41. •••An airplane lands in a river during a hurricane. Assume the velocity of the airplane relative to the air is 30 m/s, due east; the velocity of the wind is 20 m/s, due north; and the current in the river is 5 m/s, due south. Calculate the velocity of the airplane relative to the water. SSM

42. •A spaceship moves by Earth at 2.4×10^8 m/s toward the right. A satellite moves by Earth in the opposite direction at 1.6×10^8 m/s. Use the Galilean transformation to calculate the speed of the satellite relative to the spaceship and comment on your answer.

25-2: The Michelson and Morley Experiment

43. •The Michelson–Morley experiment was performed hundreds of times in a futile attempt to find the luminiferous ether. Why was the experiment performed on an enormous slab of marble that was floated in a pool of mercury?

44. ••Consider an airplane traveling at 25 m/s airspeed between two points that are 2000 km apart. What is the round-trip time for the plane (a) if there is no wind? (b) If there is a wind blowing at 10 m/s, perpendicular to the line joining the two points? (c) If there is a wind blowing at 10 m/s along the line joining the two points?

45. ••Suppose that the length L in the Michelson–Morley experiment is 20,000 m. What would be the difference in light travel times on the two legs of the Michelson–Morley experiment *if the ether existed* and if Earth moved relative to it at the following speeds: (a) at its orbital speed around the Sun, (b) at $0.01c$, (c) at $0.1c$, (d) at $0.5c$, and (e) at $0.9c$?

25-3: Special Relativity, Time Dilation

46. ••An observer in reference frame S observes that a lightning bolt A strikes the x axis and 10^{-4} s later a second lightning bolt B strikes the x axis 1.50×10^5 m farther from the origin than A. What is the time separation between the two lightning bolts determined by a second observer in reference frame S' moving at a speed of $0.8c$ along the collinear x–x' axis?

47. •A radioactive particle travels at $0.80c$ relative to the laboratory observers who are performing research. Calculate the half-life of the particle as measured in the laboratory frame compared to the half-life according to the proper frame of the particle.

48. •Muons at rest in the laboratory have an average lifetime of about 2.2 μs. What is the average lifetime, measured in the laboratory, of muons traveling at $0.99c$ with respect to the laboratory?

49. ••When muons traveled through the laboratory at a speed of $0.98c$, scientists obtained an average lifetime value of 11 μs before a muon decayed. If the muons were at rest in the laboratory, what would be the average lifetime of muons? SSM

50. ••The time dilation effect is measured in the lab to have a 0.01% difference between the relative time and the proper time $[\Delta t = (t_r - t_p)/t_p \times 100\% = 0.01\%]$. Calculate the relative speed.

51. ••Astronomy The Andromeda galaxy is a spiral galaxy that is a distance of 2.54 million light-years from Earth. Is there any possible speed that a spaceship can achieve to deliver a human being to this galaxy? Consider the lifetime of a human to be 80 years.

52. ••Astronomy The nearest star to our own Sun is *Proxima Centauri*, at 4.24 light-years. If a spaceship travels at $0.75c$, how much time is required for a one-way trip (a) according to an Earth observer, and (b) according to the captain of the ship?

53. ••A subatomic particle is traveling with a "γ factor," $\gamma = 1/\sqrt{1 - (v^2/c^2)}$, of 20 as observed by a radiation monitor in a nuclear power plant. The particle is measured to "decay" 30 ns after it is observed by the plant. What is the proper lifetime of such a particle?

54. ••A spaceship travels at $0.95c$ toward Alpha Centauri. According to Earthlings, the distance is 4.37 light-years. (a) From the perspective of the space travelers, how long does it take to reach this "next star over" from our own Sun if the ship starts at Earth? (b) How long do Earthlings measure for the trip?

25-4: The Lorentz Transformation, Length Contraction

55. ••A car is moving in the x direction in the reference frame S. The reference frame S' moves at a speed of $0.80c$, along the x axis. The proper length of the car is 3.20 m. Calculate the length of the car according to observers in the S' frame. SSM

56. •The length of a spaceship is measured to have a relative length that is 2/3 of its proper length. What is the speed of the spaceship?

57. •An observer measures a stick to move by at $0.44c$. The stick is oriented parallel to the direction of motion and is found to be 0.88 m long. Determine the proper length of the stick.

58. ••A standard tournament domino is 1.5 in wide and 2.5 in long. Describe how you might orient a domino so that it will measure 1.5 in by 1.5 in as it moves by. What relative speed is required?

59. ••How fast must a pion be moving to travel 100 m (according to the laboratory frame) before it decays? The average lifetime, at rest, of a pion is 2.60×10^{-8} s. Give your answer in units of meters per second (m/s) and as a fraction of the speed of light (in other words, $v = ?$ and $v/c = ?$). SSM

25-5: Lorentz Velocity Transformation

60. ••Spaceship A moves at $0.8c$ toward the right, while spaceship B moves in the opposite direction at $0.7c$ (both speeds are measured relative to Earth). (a) Calculate the velocity of Earth relative to spaceship A. (b) Calculate the velocity of Earth relative to spaceship B. (c) Calculate the velocity of spaceship A relative to spaceship B.

61. ••A car moves by an observer sitting on the side of the road at a velocity of $+0.35c$. A truck moves by the same observer at a velocity of $+0.25c$. Determine the relative velocity between the car and the truck. Give your answer in terms of "the velocity of the truck relative to the car is . . ." and "the velocity of the car relative to the truck is. . . ."

62. ••A spaceship flies by Earth at $0.92c$. It fires a rocket at $0.75c$ in the forward direction, relative to the spaceship. What is the velocity of the rocket relative to Earth?

63. ••Suppose the spaceship in problem 62 continues to fly by Earth at $0.92c$. This time, however, it fires a rocket at $0.75c$ in the backward direction relative to the spaceship. What is the velocity of the rocket relative to Earth? SSM

64. ••Prove that the relative velocity of a laser fired from a spaceship that is moving at $0.92c$ past Earth will be c from the perspective of Earth.

65. ••A proton travels in the accelerator at Fermilab at 99.999954% of the speed of light. An anti-proton travels in the opposite direction at the same speed. What is the relative velocity of the two particles?

25-6: Relativistic Momentum and Energy

66. ••A 2.00-kg object moves at 400,000 m/s. (a) Calculate the classical momentum of the object. (b) Calculate the relativistic momentum of the object. (c) Which of the answers is correct? What is the percent difference?

67. ••An electron travels at $0.444c$. Calculate (a) the relativistic momentum, (b) the relativistic kinetic energy, (c) the rest mass energy, and (d) the total energy of the electron. SSM

68. •A proton ($m = 1.673 \times 10^{-27}$ kg) is traveling at $0.5c$. Calculate the relativistic momentum and the relativistic kinetic energy of the particle.

69. •A particle is traveling with respect to an observer such that its relativistic energy is twice its rest energy. How fast is it moving with respect to the observer?

70. •••Calc A particle is moving along the x axis under the influence of a constant force F_x. Assuming that the initial velocity of the particle is zero ($v_0 = 0$ at $t = 0$), derive a relativistic expression for the acceleration of the particle.

71. •••Calc Using problem 70, derive an expression for the relativistic speed of the particle as a function of time.

72. •A particle has a rest energy of 3.33 MeV and a total energy of 6.00 MeV. Calculate the momentum of the particle.

73. •A proton has a rest energy of 938.3 MeV and a momentum of 200 MeV/c. Calculate its speed. SSM

25-7: General Relativity

74. ••What would an observer measure for the free-fall acceleration in an elevator near the surface of Earth if the elevator accelerates downward at 8 m/s²?

75. ••What would an observer measure for the free-fall acceleration in an elevator near the surface of Earth if the elevator accelerates downward at 18 m/s²? SSM

General Problems

76. ••A super rocket car traverses a straight track 2.40 $\times 10^5$ m long in 10^{-3} s as measured by an observer next to the track. (a) How much time elapses on a clock in the rocket car during the run? (b) What is the distance traveled in traversing the track as determined by the driver of the rocket car?

77. ••Observers in reference frame S see an explosion located at $x_1 = 580$ m. A second explosion occurs 4.5 μs later at $x_2 = 1500$ m. In reference frame S', which is moving along the $+x$ axis at speed v, the explosions occur at the same point in space. What is the separation in time between the two explosions as measured in S'? SSM

78. ••A spaceship departs from Earth for the star Alpha Centauri, which is 4.37 light-years away. The spaceship travels at $0.77c$. (a) What is the time required to get there as measured by a passenger on the spaceship? (b) How long does it take for the spaceship to arrive at Alpha Centauri as measured on Earth?

79. ••A radioactive nucleus traveling at a speed of $0.8c$ in a laboratory decays and emits an electron in the same direction as the nucleus is moving. The electron travels at a speed of $0.6c$ relative to the nucleus. (a) How fast is the electron moving according to an observer in the laboratory? (b) Rocket A travels away from Earth at $0.6c$, and rocket B travels away from Earth in exactly the opposite direction at $0.8c$. What is the speed of rocket B as measured by the pilot of rocket A? (c) Why did you get the same answer that you did for part (a)?

80. ••A beam of pions has a speed of 0.88c. Their mean lifetime, as measured in the reference frame of the laboratory, is 2.6×10^{-8} s. What is the distance traveled by the laboratory, as measured by the pion, during its lifetime?

81. •••Muons have a proper lifetime of 2.20×10^{-6} s. Suppose a muon is formed at an altitude of 3000 m and travels at a speed of 0.950c straight toward Earth. (a) Does the muon reach Earth's surface before it decays? Complete the problem from both perspectives: the muon's point of view and Earth's reference frame. (b) Calculate the minimum speed of the muon so it *just barely* reaches Earth's surface as it decays.

82. ••Two students, Annabelle and Allison, are both the same age when Allison hops aboard a flying saucer and blasts off to achieve a cruising speed of 0.800c for 20 years (according to Allison). Neglecting the acceleration of the ship during blastoff and turnarounds (technically this is impossible), find the difference in age between Annabelle and Allison when they are reunited. Who is younger?

83. ••Suppose a jet plane flies at 300 m/s relative to an observer on the ground. Using only special relativity, determine the distance of the flight, as measured by an observer on the ground, before the clocks aboard the plane are 10 s behind clocks on the ground. Assume the two clocks were originally synchronized to start. SSM

84. ••At the end of the linear accelerator at the Stanford Linear Accelerator Center (SLAC), electrons have a speed of 0.99999999995c. (a) Calculate the value of γ for an electron at SLAC. (b) What time interval would an observer at rest relative to the accelerator measure for a time interval of 1.66 μs measured from the electron's perspective?

85. ••The proper half-life of the K$^+$ meson is about 1×10^{-8} s. How fast is a beam of K$^+$ particles moving if one-half of them decay in 6×10^{-8} s?

86. ••Based on experiments in your lab, you know that a certain radioisotope has a half-life of 2.25 μs. As a high speed spaceship passes your lab, you measure that the same isotope at rest inside the spaceship takes 3.15 s for one-half of it to decay. (a) What is the half-life of the isotope as measured by an astronaut working inside the spaceship? (b) How fast is the spaceship traveling relative to Earth?

87. ••Recent home energy bills indicate that a household used 411 kWh of electrical energy and 201 therms for gas heating and cooking in a period of 1 month. Given that 1.0 therm is equal to 29.3 kWh, how many milligrams of mass would need to be converted directly to energy each month to meet the energy needs for the home? SSM

88. ••Twin astronauts, Jessica and Marsha, are involved in exploration of the planet Jupiter. When Jupiter and Earth are at their closest approach of 6.28×10^{11} m, Jessica flies to Jupiter and back at 0.75c while Marsha remains on Earth. You can neglect the turnaround time and the time for acceleration as well as the motion of the two planets. What is the *difference* in their ages when Jessica returns to Earth? Which twin is now older?

89. ••Astronomy High speed cosmic rays strike atoms in Earth's upper atmosphere and create secondary showers. Suppose a particle in one of the showers is created 25.0 km above the surface traveling downward at 90.0% the speed of light. (a) Consider the following two events: "particle is created in the upper atmosphere" and "particle strikes the ground." We can view the events from two reference frames, one fixed on Earth and one traveling with the created particle. In which of the reference frames are the proper time and the proper length between the two events measured? Explain your reasoning. (b) In the particle's reference frame, how long after its creation does it take it to reach the ground? (c) In Earth's reference frame, how long after creation does it take the particle to reach the ground? (d) Show that the times in parts (b) and (c) are consistent with time dilation.

90. •••A rocket 1.00 km long is traveling parallel to Earth's surface at 0.5c from left to right. At time $t = 0$, a light flashes for an instant at the center of the rocket. Detectors at opposite ends of the rocket record the arrival of the light signal. Call event A the light striking the left detector and event B the light striking the right detector. Observers at rest in the rocket and on Earth record the events. (a) What is the speed of the light signal as measured by the observer (i) at rest in the rocket and (ii) at rest on Earth? (ii) At what time after the flash do events A and B occur as measured by (i) the observer in the rocket and (b) the observer at rest on Earth? Which event occurs first in each case? (c) Show that the results in part (b) are consistent with time dilation.

91. ••Biology Two twin astronauts, Harry and Larry, have identical pulse rates of 70 beats/min on Earth. Harry remains on Earth, but Larry is assigned to a space voyage during which he travels at 0.75c relative to Earth. What will be Larry's pulse rate as measured by (a) Harry on Earth and (b) the doctor in Harry's rocket?

92. ••A rocket is traveling at speed v relative to Earth. Inside a lab in the rocket, two laser beams are turned on, one pointing in the forward direction and the other pointing in the backward direction relative to the rocket's velocity. (a) What is the speed of each laser beam relative to the laboratory in the rocket? (b) Use the relativistic velocity addition formula to find the velocity of each laser beam as measured by an observer at rest on Earth. (c) As observed from Earth, how fast are the two laser beams separating from *each other*?

93. The Newtonian formula for kinetic energy ($K = \frac{1}{2}mv^2$) is not valid for speeds approaching the speed of light. (a) Using the facts that the total energy of an object is $E = mc^2$ and that the total energy is the sum of its rest mass energy and its kinetic energy, show that the kinetic energy of an object can be written as $K = (m - m_0)c^2 = m_0c^2(\gamma - 1)$. (b) A 1000-kg rocket is flying at $0.90c$ relative to your lab. Calculate the kinetic energy of the rocket using the relativistic formula you just derived and the ordinary Newtonian formula. What is the percent error if we use the Newtonian formula? Does the Newtonian formula overestimate or underestimate the kinetic energy?

94. ••We know 1 kg of trinitrotoluene (TNT) yields an energy of 4.2 MJ. The energy released comes from the chemical bonds in the material. How much rest mass would be required to create an explosion equivalent to 1.8×10^9 kg TNT. Assume all the energy comes simply from the rest mass of the material.

95. •••**Astronomy** In December 2009, the discovery was announced of a planet that may contain a large percentage of water in its composition, and hence would be a good candidate for possible life. The planet, GJ 1214b, orbits a small star that is 42 light-years from Earth. In the future, we might decide to send some astronauts to explore the planet. When they arrive there, we want them to be young enough to perform tests. Suppose that the captain is 25 years old at launch time and we want her to be no more than 60 years old when she arrives at the planet. You can ignore acceleration times and any motion of Earth and GJ 1214b. (a) What is the minimum speed the spaceship will need for the captain to be no more than 60 years old at arrival? (b) As soon as the spaceship arrives at the planet, the captain has orders to send a radio signal to Earth to notify Mission Control that the trip was successful. How many years after launch from Earth will it be when the signal arrives at Earth? **SSM**

96. •••A spaceship is traveling at $0.50c$ relative to Earth. Inside the ship, a cylindrical piston that is 50.0 cm long and 4.50 cm in diameter contains 1.25 mol of ideal gas at 25.0 °C under 2.20 atm of pressure. The cylinder is oriented with its axis parallel to the direction in which the spaceship is flying. What is the particle density (in molecules per cubic meter) of the gas in the cylinder as measured by (a) an astronaut in the rocket ship's lab and (b) a scientist in an Earth lab?

97. •••**Calc** In a normal Euclidean geometry, triangles obey the Pythagorean theorem. This concept, extended to three dimensions, would require that the differential element of distance, ds, satisfy the familiar form: $ds^2 = dx^2 + dy^2 + dz^2$. If a curved space is considered, this formula is extended as $ds^2 = \Phi(x,y,z)\, dx^2 + \Gamma(x,y,z)\, dy^2 + H(x,y,z)\, dz^2$, where the functions Φ, Γ, and H are related to the geometry of the situation. This can be simplified with the introduction of the *metric tensor* that is written as $ds^2 = g_{xx}\, dx^2 + g_{xy}\, dx\, dy + g_{xz}\, dx\, dz + \ldots + g_{zy}\, dz\, dy + g_{zz}\, dz^2$. The nine values of $g_{\mu\nu}$ specify the curvature of space. Einstein, in his special theory of relativity, showed that the separation between two points includes time or $ds^2 = c^2\, d\tau^2 = c^2\, dt^2 - dx^2 - dy^2 - dz^2$; he showed that this expression is invariant. In a world in which this expression is not quite true (one that might include the curvature of space–time), there is an extension to this idea: $c^2\, d\tau^2 = g_{\mu\nu}\, dx^\mu\, dx^\nu$. Einstein's general theory of relativity is designed to predict the values of the metric tensor, $g_{\mu\nu}$. Starting with a spherical geometry, in terms of spherical–polar coordinates $c^2\, d\tau^2 = c^2\, dt^2 - dr^2 - r^2\, d\theta^2 - r^2 \sin^2\theta\, d\omega^2$, derive the sixteen values of the metric tensor: $g_{tt}, g_{tr}, g_t\theta, g_t\omega, g_rr, g_r\theta$, etc. Take heart, most of them are zero!

98. •••**Calc** Starting with the definition of work ($W = \int F\, dx$) and the definition of force ($F = dp/dt$), derive the formula for the relativistic kinetic energy (assume the initial velocity is zero).

26 Modern and Atomic Physics

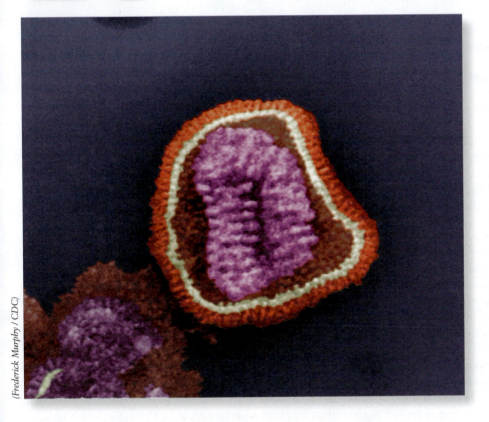

(Frederick Murphy / CDC)

Electrons passing through atoms are diffracted, much like sound waves diffract through, or around the edges of, an open door. When a beam of low energy electrons is shot through a thin slice of a specimen in a transmission electron microscope (TEM), the pattern formed by the diffracted electrons forms an image. A TEM captured the (false-color) image of this influenza virus particle, which is only about 100 nm in diameter. Because the wavelengths of low energy electrons are so much shorter than those of light, a TEM is capable of significantly higher resolution than light microscopes (better than 0.005 nm compared to about 0.2 μm with the most powerful optical microscopes).

We have uncovered hints that light and other electromagnetic waves have a particle-like nature. We know, for example, that light comes in bundles of energy called photons. How far can we extend a treatment of photons as particles? Can they strike a particle, such as an electron, and "bounce" off, or can they scatter the way a cue ball does when it strikes a billiard ball? Do photons carry momentum? If waves have a particle-like nature, do particles ever exhibit properties we associate with waves? In this chapter, we will explore the particle nature of light, the wave nature of particles, and the structure of atoms in more detail.

26-1 Blackbody Radiation

Hot objects glow. As the temperature of a piece of pottery heating in a kiln increases, for example, it begins to glow red, then orange, yellow, and perhaps white. The color of the radiated light is correlated with the temperature. Thomas Wedgwood, an eighteenth-century English inventor and scientist (and son of the founder of Josiah Wedgwood and Sons china company), observed the relationship between temperature and the color of radiated light in the process of making porcelain. He conducted experiments on a wide variety of materials and noted in 1792 that the color of light emitted by a hot object depended only on temperature and not on its size, shape, or the material from

Figure 26-1 The color of the light radiated by a hot object depends on its temperature. *(Courtesy of David Tauck)*

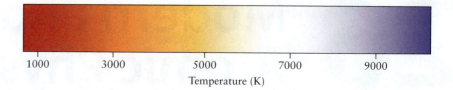

1000   3000   5000   7000   9000

Temperature (K)

which it was made. **Figure 26-1** shows the correlation between color and temperature; using it we can tell, for example, that the temperature of the element of the space heater in **Figure 26-2** is well over 3000 K! Over the course of the next hundred years or so, the energy radiated from hot objects became a topic of some interest to physicists. It is still of interest today, underlying our ability to remotely sense body temperature and to see in the dark.

One way to increase the temperature of an object high enough for it to radiate light energy and appear to glow is for it to absorb energy that strikes it. In general, some of that energy will be reflected and some absorbed; the ideal object with which to study these processes is one that absorbs all of the energy that strikes it. If the object absorbs all incident energy we are guaranteed that any energy it eventually radiates must have first been internal to it. Such a perfect absorber reflects no energy, and in particular no light, so its surface appears black. We term such an object an **ideal blackbody**. In addition, an object or surface that absorbs all incident energy must eventually reach the temperature of the surrounding environment. Recall from Section 14-1 that, by definition, two objects at the same temperature are in thermal equilibrium. Once thermal equilibrium is reached, for every amount of energy absorbed by an object, the same amount must be emitted. So, in addition to being a perfect absorber of energy, an ideal blackbody in thermal equilibrium with its surroundings is also a perfect emitter of energy because it emits as much energy as it absorbs. Neither the absorption nor the emission process of an ideal blackbody depends on properties of the surface, such as how smooth or how clean it is.

As the name suggests, an ideal blackbody is an idealized concept; no object exists that is both a perfect absorber and a perfect emitter. However, some materials, such as graphite and carbon powders, are good approximations, as are the surfaces of stars. Another good approximation to an ideal blackbody would be a closed cavity with a small opening. Imagine a sealed, empty box which is thermally isolated from the outside environment (so that it can neither absorb energy

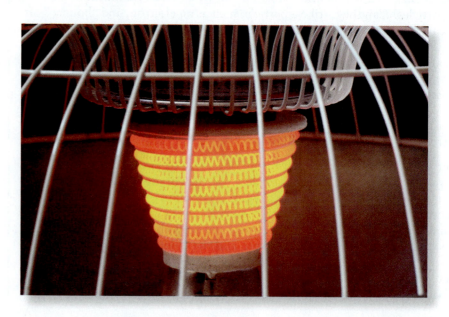

Figure 26-2 There is a direct correlation between the temperature and color of hot object. The color of the glowing heating element of this space heater suggests that its temperature is over 3000 K.

on nor lose energy from the outside surfaces, and with inner walls cool relative to the inside environment. We cut a small hole on one side, through which we can pump energy into the box by shining light (or, more generally, electromagnetic waves) into the opening. Although the light waves may initially be reflected off one of the inner walls of the box, it is highly unlikely that a reflected wave will find its way directly back out through the hole. Instead, a light wave experiences multiple reflections until eventually its energy is absorbed (Figure 26-2). As we pump more and more energy into the box, more and more waves are reflected back and forth, eventually being absorbed and slowly increasing the temperature of the walls of the box. When the walls come to thermal equilibrium with the environment inside the box, however, no more net energy can be absorbed, because for every amount of energy absorbed by the walls, the same amount must be emitted. In addition, because the energy of the walls can no longer increase, even though energy emitted from them will likely bounce around inside for a while, this energy must eventually find its way out of the box through the hole. We have created an ideal blackbody: All of the energy that enters the cavity is first absorbed but eventually re-radiated out through the hole (**Figure 26-3**).

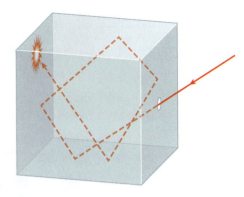

Figure 26-3 A blackbody cavity can be modeled by a box with a small hole in one side. Light shined through the hole experiences many reflections, so its energy is almost always absorbed by one of the inner walls of the box. After the walls reach thermal equilibrium with the environment inside the box, so that no more net energy can be absorbed, the walls begin to radiate. This energy eventually finds its way out of the box through the hole. Because all of the energy that enters the box is first absorbed before it is re-radiated, this is a blackbody cavity. The re-radiation of the absorbed energy is not shown in this figure.

Although the inner walls of the box participate in the absorption and emission processes, we consider the ideal blackbody to be the hole in the box. Energy "strikes" it, is absorbed, and eventually is re-radiated. The most direct way to study the energy it radiates is through the **spectral radiancy** R, defined as the energy emitted per time per area. In the case of our box, this area is the area of the hole.

Like any radiating object, the inside walls of the box, and therefore the hole in the box, emit energy over a range of frequencies. The spectral radiancy R is therefore a function of the frequency f of emitted energy. In particular, at any given temperature T, spectral radiancy R is proportional to the square of f.

The energy waves bouncing back and forth between the inner walls of the box interfere with each other. Like waves on a string (Section 13-5), this interference results in standing waves. Each standing wave pattern (or mode) has a particular energy; R is also proportional to the average energy $\overline{E}$ of all of the standing wave patterns in the box, so

$$R(T,f) \propto f^2\overline{E} \tag{26-1}$$

The spectral radiancy R represents the distribution over all frequencies of the energy emitted from the hole in the cavity, or from any radiating object, per unit time. In Chapter 13, we defined energy per unit time as power; the spectral radiancy is the power spectrum of the radiation from the box. R has units of watts per square meter (W/m^2).

To determine spectral radiancy R, we need to know the average energy of all of the standing waves in the box. We find this by multiplying each possible value of energy by the probability with which that energy appears. This is the same weighted average strategy we applied in Chapter 7, in which we multiplied each possible value in a set by the probability of that value appearing. Generalizing from Equation 7-35 we write

$$\overline{E} = \sum_{i=1}^{N} P(E_i)E_i \tag{26-2}$$

where $P(E_i)$ is the probability that standing waves in the box have energy equal to E_i. (Many values of energy are possible. Here each value of the subscript i identifies one specific energy value. We might have E_1 equal to 0 J, E_2 equal to 5 J, E_3 equal to 10 J, and so on.) Nineteenth-century physicists imagined that the energy bouncing around in a blackbody cavity could be modeled as a gas, so that any value of energy is allowed for each standing wave mode, but with an exponentially falling probability for higher energies. The probability for the energy of gas molecules is also a function of the temperature T of the gas:

$$P(E) = \frac{1}{kT} e^{-E/kT} \tag{26-3}$$

where k is the same Boltzmann constant that appears in the ideal gas law (Equation 14-6). The multiplicative constant $1/kT$ ensures that the total probability, that is, the sum (or integral) of $P(E)$ over all possible values of energy, is 100%. (See Physicist's Toolbox 26-1 for the mathematical basis for this claim.) Also, notice that the value of the exponent of e in Equation 26-3 is closer and closer to zero for smaller and smaller values of energy E, so $P(E)$ is larger for smaller values of energy. In other words, the smaller the energy for a given temperature, the more likely it is to find a standing wave mode of that energy in the box. Saying it the other way around makes it clear that Equation 26-3 is physically reasonable; it is less and less probable to find waves of higher and higher energy in the box.

Physicist's Toolbox 26-1 Forcing the Total Probability to Be 100%

The probability of finding a standing wave of energy E in the blackbody cavity is a function of temperature and is expressed by Equation 26-3. Once there is at least some energy in the cavity, however, the probability of finding *some* wave (or waves), however, must be 100%. Mathematically, this means that we if we add the probabilities of finding waves of each possible value of energy, the total must equal 1, regardless of the temperature.

In the case of the cavity, any energy from 0 to ∞ is allowed, although Equation 26-3 makes it clear that only relatively small energies are probable. The total probability, then is

$$P_{\text{tot}} = \int_0^\infty \frac{1}{kT} e^{-E/kT} dE$$

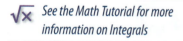

See the Math Tutorial for more information on Integrals

In general $\int e^{ax} dx = (1/a) e^{ax}$, so

$$P_{\text{tot}} = \left(\frac{1}{kT}\right)\left(-\frac{1}{1/kT}\right) e^{-E/kT} \Big|_0^\infty$$

Notice that the factor of $1/kT$ in the original integral cancels with the factor of $1/(1/kT)$ from the integration, so

$$P_{\text{tot}} = -(e^{-\infty} - e^0) = -(0 - 1) = 1$$

Because we included the factor of $1/kT$ in the probability function $P(E)$ (Equation 26-3), the total probability is indeed 100%.

As expressed in Equation 26-2, the weighted average assumes discrete values in the set to be averaged. Equation 26-3, however, assumes a smooth continuum of possible energies, so in order to compute the weighted average of energies, the sum of Equation 26-2 must be converted into an integral:

$$\overline{E} = \int_0^\infty P(E) E dE = \int_0^\infty \frac{1}{kT} e^{-E/kT} E dE \tag{26-4}$$

where we allow a smooth continuum of energy values from zero to ∞. By assuming that the energy distribution of standing waves in a blackbody cavity is well modeled as the distribution of energies of molecules of a gas, physicists of the time were able to determine the average energy of the standing waves in the ideal black-body cavity, and thus its spectral radiancy. The average energy is

$$\overline{E} = \int_0^\infty \frac{1}{kT} e^{-E/kT} E dE = \frac{1}{kT}(-E - kT)e^{-E/kT}\Big|_0^\infty = kT$$

Math Box 26-1 shows how the method of integration by parts is easily applied to solve this integral.

Math Box 26-1   The Power of Integration by Parts

The approach of integration by parts provides a way to change an integral into a form that is (sometimes) easier to integrate. According to the rule of integration by parts, an integral that can be written in the form $\int u dv$ can be transformed to

$$\int u dv = uv - \int v du$$

For the integral in Equation 26-4, let u be E so that dv is $e^{-E/kT}$. Using this choice, du is dE and v is $-kTe^{-E/kT}$. You can verify the result by taking the derivative of v with respect to E. By applying the integration by parts rule the integral becomes

$$\overline{E} = \int_0^\infty \frac{1}{kT} e^{-E/kT} E dE = \frac{1}{kT} E(-kTe^{-E/kT})\Big|_0^\infty - \frac{1}{kT}\int_0^\infty (-kTe^{-E/kT}) dE$$

Look carefully if you need to in order to spot the u, v, and du terms. The integral in the second term is now straightforward:

$$\overline{E} = -Ee^{-E/kT}\Big|_0^\infty - kTe^{-E/kT}\Big|_0^\infty$$

You might be concerned about evaluating the first term at E equal to ∞ because it contains E as a multiplicative factor. However, because $e^{-E/kT}$ falls more rapidly to zero than E rises, as E tends to ∞, the first term gives zero when evaluated at ∞. The first term is also zero when evaluated at zero, because $e^{-E/kT}$ equals 1 when E equals zero. The second term is zero when evaluated at ∞, because $e^{-\infty}$ equals zero, so in all, the average becomes

$$\overline{E} = 0 + 0 - 0 - (-kT) = kT$$

Example 26-1   Weighted Average

A certain parameter x can take any value from zero to one, however, values of x closer to 1 occur more often that values closer to 0. For this example we take $W(x)$, the relative likelihood of a particular value of x to be 3 multiplied by x^2:

$$W(x) = 3x^2$$

$W(x_0)$ is not the probability of finding x to have a certain value x_0, because for some values of x_0, the function $W(x_0)$ is greater than 1. $W(x)$ is instead a probability *density*, so that the probability of finding values of x in the range from a to b is the integral from a to b of $W(x)dx$:

$$P(x \text{ between and b}) = \int_a^b 3x^2 dx$$

$W(x)$ therefore serves as the probability function in finding the weighted average. (a) Show that the total probability of finding *some* value of x is 100%. (b) Determine $\bar{x}$, the average value of x.

SET UP

(a) The total probability is the sum of $P(x)$ over all values of x, that is, the integral of the probability function from zero to one. (b) To find the average value, we convert the sum in Equation 26-2 to an integral, in the way we did in Equation 26-4.

SOLVE

(a) The total probability P is then

$$P = \int_{0}^{1} W(x)\,dx = \int_{0}^{1} 3x^2\,dx = 3\frac{x^3}{3}\Big|_{0}^{1}$$

So, by substituting in the values of the bounds of integration we get

$$P = 3\left(\frac{1^3}{3} - \frac{0^3}{3}\right) = 1$$

The total probability of finding some value is indeed 100%.

(b) The average value of x is found by adding up every possible value of x multiplied by the probability of observing that value. Using the same strategy that we applied in part (a), we compute the weighted average using Equation 26-4:

$$\bar{x} = \int_{0}^{1} W(x)x\,dx = \int_{0}^{1} 3x^2\,x\,dx$$

Evaluating the integral between the bounds gives

$$\bar{x} = \int_{0}^{1} 3x^3\,dx = 3\frac{x^4}{4}\Big|_{0}^{1}$$

or

$$\bar{x} = 3\left(\frac{1^4}{4} - \frac{0^4}{4}\right) = \frac{3}{4} = 0.75$$

The average value of x is 0.75.

REFLECT

Because x is defined only in the range zero to one, had the probability of finding any particular value of x been the same for all values the average value would have been halfway between the extremes or 0.5. The weighting function we were given is not uniform, however; as you can see in **Figure 26-4**, values of x closer to one are more likely than values closer to zero. This causes the weighted average to be pulled closer to one than zero, that is, the weighted average value must be greater than 0.5. Our result of $\bar{x}$ equal to 0.75 is certainly reasonable.

Based on modeling the energy inside the blackbody cavity as if it were molecules of gas bouncing off the inner walls, using Equation 26-1 and using kT as the average energy per standing wave mode, nineteenth-century physicists found the spectral radiancy to be

$$R(T,f) \propto f^2 kT$$

From this equation we see that the theory predicted that at any fixed temperature there would be more and more power emitted per surface area at higher and higher frequencies of radiated light.

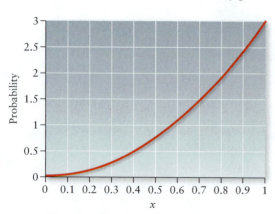

Figure 26-4

Wait—this conclusion simply can't be! If the power emitted were to rise with the frequency of the radiated light, any glowing object would emit an infinite amount of energy when all frequencies were taken together. This result is physically nonsensical and clearly a catastrophic failure of the theory. Because the frequency of ultraviolet light is above the frequencies of visible light, this failure of the prevailing theory was dubbed the "ultraviolet catastrophe.".

Figure 26-5 shows the relationship between radiated power and frequency as predicted by treating the energy in a blackbody cavity as a gas, labeled "Theoretical (gas model).", The figure also shows an example of a spectrum experimentally measured from a glowing object, labeled "Experimental (glowing object)." Both represent the spectrum at the same, fixed temperature. Notice that the two curves lie on top of each other for energy emitted at low frequencies, suggesting that the gas model works well for radiated energy at such frequencies. It would seem that using the gas model to determine the average energy of standing waves in the cavity fails only as the frequency of radiated energy gets larger. For Equation 26-1 to describe the spectrum of our blackbody cavity correctly, the average energy per standing wave mode inside it must fall to zero as frequency increases. This would also cause the spectral radiancy function to fall back down to zero for large frequencies, rather than growing infinitely large in proportion to the square of the frequency. If a theoretical basis could be found that predicted that the average energy per standing wave mode inside a blackbody cavity falls to zero as frequency increases, the ultraviolet catastrophe would be resolved. It was the German physicist Max Planck who, at the turn of the twentieth century, proposed just such a theory.

Planck proposed a new approach to determine the energy of standing waves in a blackbody cavity. The application of his theory predicts that the average energy equals kT for low frequencies of radiated energy and approaches zero for high frequencies of radiated energy. In other words, his method correctly reproduces the experimentally measured spectrum shown in Figure 26-5. What is striking about Planck's approach is that it assumes that the waves inside the cavity can carry only *discrete quantities of energy*. (It was so startling, in fact, that he and others saw it only as mathematically expedient but not physically believable!)

We found the average energy of standing waves in the cavity by adding up each possible energy value weighted by its probability. Modeling the standing waves in the cavity as a gas allows for a smooth continuum of energy values, so Equation 26-4 is an integral. Said another way, the average energy determined by Equation 26-4 is the area under the curve of the probability $P(E)$ multiplied by the energy E. If only specific, discrete values of energy are allowed, however, Equation 26-4 must be converted to a sum:

$$\overline{E} = \sum_{i=0}^{\infty} P(E_i) E_i \, \Delta E \qquad (26\text{-}5)$$

Here, only the specific energy values E_i are present in the cavity, and the probability of a wave of energy E_i is $P(E_i)$. ΔE is the difference between successive allowed values of energy.

Figures 26-6a, 26-6b, and **26-6c** show how the assumption that the energy values in the cavity are discrete rather than continuous addresses the failure of the gas model. In all three, the blue curve is the product of energy values and their associated probability *assuming that all energies are allowed*. The difference between the curves in the three parts of Figure 26-6 is that in Figure 26-6a the allowed energies, marked E_0, E_1, and so on, are separated by a small energy difference,

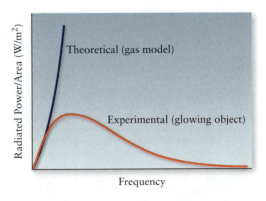

Figure 26-5 Both curves show radiated power versus frequency of a radiating blackbody cavity at the same, fixed temperature. The theoretical curve is that predicted by treating the energy in the cavity as a gas. The experimental curve is the spectrum measured from a glowing object.

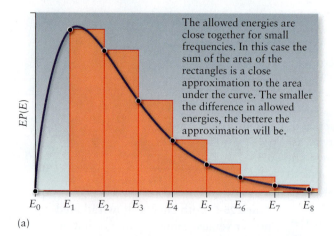

The allowed energies are close together for small frequencies. In this case the sum of the area of the rectangles is a close approximation to the area under the curve. The smaller the difference in allowed energies, the better the approximation will be.

(a)

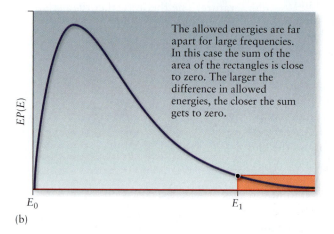

The allowed energies are far apart for large frequencies. In this case the sum of the area of the rectangles is close to zero. The larger the difference in allowed energies, the closer the sum gets to zero.

(b)

Figure 26-6 When energy is quantized, the area under the curve of energy multiplied by probability is determined by summing the areas of rectangles. (a) When the rectangles are narrow, that is, when the allowed energies are close together, the sum does a good job of approximating the area under the curve. (b) When the rectangles are wider, that is, when the allowed energies are farther apart, the sum is smaller. (c) As the difference between the allowed energies gets larger, the sum of the areas of the rectangles approaches zero.

in Figure 26-6b the difference between allowed energies is larger, and the difference between allowed energies is larger still in Figure 26-5c. Because the energy values are discrete, each term in the sum in Equation 26-5 therefore represents the area of a rectangle of height $E_i P(E_i)$ (marked by the dots in the figures) and width ΔE. The rectangles are superimposed on the curves.

When the rectangles are narrow (Figure 26-6a), the sum of their areas is a good approximation to the area under the curve of energy multiplied by the probability of finding that energy in the cavity. Indeed, in the limit of ΔE becoming infinitesimally small, the sum is exactly equal to the area under the curve—this is one definition of an integral. The area under the curve of the function $E_i P(E_i)$ is the average energy as defined by Equation 26-4. In other words, when the separation between allowed energies is small, requiring that only specific energy values are allowed in the cavity, kT is the average energy of the standing waves in the cavity, the same result we found using a gas model.

Notice that the energy associated with the leftmost rectangle that has non-zero height in Figures 26-6a, 26-6b, and 26-6c is E_1, not E_0. That's because in all cases E_0 equals zero, so the height of the rectangle that begins at energy equal to E_0 is zero. So when the rectangles are wide (Figure 26-6b), the sum of their areas is small because even the largest rectangle has a relatively small height. As the width of the rectangles becomes wider, as in Figure 26-6c, the sum of their areas becomes even smaller, and in the limit of ΔE becoming very large, the sum of the areas of the rectangles approaches zero. In that case, then, we find that the average value of the energy states in the cavity equals zero. Again, the average value of energy states in the cavity is the sum of the areas of the rectangles as defined by Equation 26-5.

Planck proposed that only discrete values of energy be allowed in the cavity, and that the energy separation between them be proportional to the frequency. This means that ΔE is small for low frequencies and large for high frequencies. In this way, we find that average energy equals kT for energy radiated at low frequencies and that it approaches zero for energy radiated at high frequencies. With this behavior of the average energy of the standing waves of energy in the blackbody cavity, Equation 26-1 correctly describes the spectrum of a glowing body.

Mathematically, Planck's statement that the separation between successive allowed energy states is proportional to the frequency is

$$\Delta E = hf \tag{26-6}$$

where the proportionality constant h is Planck's constant, approximately 6.63×10^{-34} J · s. We discussed this relationship and Planck's constant in Chapter 22. The energy states in the cavity are therefore

$$E = 0, hf, 2hf, 3hf, \ldots$$

The quantum, or discrete building block unit of energy, is then

$$E = hf \tag{26-7}$$

This seems simple enough. Why did Planck find it to be physically unreasonable? That only discrete, specific amounts of energy are allowed in the cavity implies that the energy in the cavity is in the form of bundles or packets, each with a fixed amount of energy. This result is most certainly not consistent with the understanding that physicists at the turn of the twentieth century had about waves. It was a radical, and to many, an unacceptable conclusion that the waves bouncing around in the blackbody cavity would have discrete energies, and that only certain energies were allowed, rather than a smooth continuum of energies.

We know now that the energy of electromagnetic radiation is *quantized*, both visible light as well as radiation across the entire electromagnetic spectrum. As we discussed in Chapter 22, a photon is a well-defined packet of energy, and the energy of the photon is related to the frequency of the radiation according to Equation 26-7.

Estimate It! 26-1 How Many Photons?

A typical 100 W incandescent lightbulb releases about 2% of its energy in the visible spectrum. Based on an assumption that the average wavelength of emitted visible light is yellow light, estimate the number of visible photons per second radiated from a 100 W lightbulb.

SET UP
Our starting point is that 100 W is equivalent to 100 J/s. If we estimate the energy of one photon in joules, we can ask how many photons are required, all together, to radiate 100 J of energy every second.

SOLVE
The energy of a photon as a function of its wavelength, from Equation 22-1, is

$$E = hf = \frac{hc}{\lambda}$$

For the purposes of our estimate, we pick a wavelength that is yellow light. We like to choose big, round (but reasonable) numbers for estimates; a good choice is λ equal to 600 nm. The energy of one (average) photon is then

$$E = \frac{(6.63 \times 10^{-34} \text{ Js})(3 \times 10^{8} \text{ m/s})}{600 \times 10^{-9} \text{ m}} \approx 3 \times 10^{-19} \text{ J}$$

If every photon radiated were in the visible, the number of photons (N) radiated per second so that their total energy is 100 J would be

$$N/s \approx \frac{100 \text{ J/s}}{3 \times 10^{-19} \text{ J/photon}} \approx 3 \times 10^{20} \text{ photons/s}$$

Only about 2% of the photons are in the visible, however, so a better estimate is

$$N/s = 0.02 \, (3 \times 10^{20} \text{ photons/s}) \approx 6 \times 10^{18} \text{ photons/s}$$

About 6×10^{18} visible photons are radiated every second from a 100 W incandescent bulb.

REFLECT
Each visible photon carries only a relatively small amount of energy, so it takes a lot of them, around 10^{19} photons every second, to deliver the light energy of a 100 W incandescent bulb.

As an aside, the United States, Canada, and a number of other countries have passed legislation that will soon phase out incandescent light bulbs because they are so inefficient. Other technologies, such as compact fluorescent bulbs and tungsten–halogen bulbs convert much more of the energy they use into visible light. A compact fluorescent bulb that draws 25 W, for example, produces as much visible light as the 100 W incandescent bulb we considered in this problem.

> ### ✳ What's Important 26-1
> A perfect absorber reflects no energy, and in particular no light, so its surface appears black. A perfect absorber is also a perfect emitter, because all energy absorbed must be emitted once the object is in thermal equilibrium with its surroundings. We term such an object an ideal blackbody. Electromagnetic radiation is quantized, both visible light as well as radiation across the entire electromagnetic spectrum; the photon is a well-defined packet of electromagnetic radiation.

26-2 Photoelectric Effect

When light of a certain color strikes certain materials, electrons can be ejected from the surface. This observation is the **photoelectric effect,** and its discovery is a remarkable story. In 1886 the German physicist Heinrich Hertz set about to study Maxwell's theory of electromagnetic waves. He designed a device to give off electric sparks that he hoped would, in turn, create electromagnetic waves. He placed his electromagnetic wave generator far from a copper wire bent into a ring with a small gap. A schematic diagram of Hertz's experiment is shown in **Figure 26-7.** According to Maxwell (and Faraday), electromagnetic waves passing through the ring should induce a potential between the two ends, resulting in electrons—a spark—jumping across the gap. Hertz's experiment was a success; a spark was visible in the gap of the copper ring when the electromagnetic wave generator was running. The spark was faint, however, so Hertz enclosed the copper ring inside a dark box. He expected to see the spark in the gap of the ring more easily, but instead it became *fainter!* Although it was not related to his primary work, Hertz spent many months studying this curious phenomenon, eventually concluding that ultraviolet light coming from the sparks created in his electromagnetic wave generator made it easier for sparks to jump across the gap in the copper ring. Ultraviolet light helped eject electrons from the surface—Hertz had discovered the photoelectric effect.

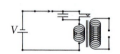

Electromagnetic wave generator   Copper ring

Figure 26-7 In 1886 Heinrich Hertz designed an experiment to study Maxwell's theory of electromagnetic waves. When a high voltage is applied across the gap in the device on the left, sparks jump across the gap, creating electromagnetic waves. When the waves pass through the copper wire bent into a ring with a small gap, on the right, an electric potential is created between the ends of the wire, resulting in a second spark that jumps across the gap in the ring.

The photoelectric effect is also notable as one of the milestones in the development of quantum physics. It also provides yet another way to study organisms and biological tissues. We are able to see the structure of biological tissue by passing light through a thin slice and then magnifying the image using the optics of a microscope. We use changing magnetic fields to induce eddy currents in the body with which we can, for example, monitor activity in different regions of the brain. Devices that rely on the photoelectric effect allow us to examine the molecules that make up the *surface* of a biological specimen.

Studying the photoelectric effect is commonly done by applying a potential across a pair of conducting plates placed inside a tube filled with gas, as shown schematically in **Figure 26-8.** When a positive voltage is applied, negative charges accumulate on one of the plates and are drawn away from the other. As labeled in the sketch in the figure, the cathode is the plate on which negative charges accumulate and the anode is the plate that ends up with a net positive charge. When the potential difference between the plates is high enough, charge can flow from one to the other by ionizing the gas in the tube. In the setup shown in the sketch, the magnitude of the current is measured by an ammeter.

Shining ultraviolet light on the cathode dramatically increases the current in the tube, as shown in **Figure 26-9**, and the brighter the light, the greater the current. In addition and perhaps more startling, electrons can even flow toward the anode when the voltage is reversed so that electrons are repelled from it. Only when the magnitude of such a negative voltage exceeds that of the **stopping potential** V_0 does the current stop.

How is it possible for electrons to flow from cathode to anode when the voltage is reversed? The answer is the key to the photoelectric effect. When an ultraviolet photon interacts with an electron on or near the surface of the cathode, the electron may acquire all of the photon's energy. That additional energy may be enough to liberate the electron from the surface, with any excess increasing the electron's kinetic energy. With enough kinetic energy, an electron ejected from the cathode can overcome a negative potential difference and find its way to the anode. An electron ejected from a surface by the photoelectric effect is referred to as a photoelectron.

The photoelectric effect surprised physicists of the nineteenth century. Here's another surprise. Although increasing the brightness of the light increases the photoelectric current; the stopping potential doesn't change as the brightness increases. An increase in brightness results from a higher number of photons per second falling on the cathode, which in turn means that more energy per second impinges on the cathode. Yet the increase in energy per time does not allow the liberated electrons to overcome a higher stopping potential. Said another way, the increase in the power of the light does not increase the energy of the electrons ejected from the surface.

Einstein explained this phenomenon by realizing that the same quantization approach suggested by Planck in explaining blackbody radiation also applied to the photoelectric effect. Indeed, Einstein proposed that the relationship between the frequency and energy of a photon ($E = hf$, Equation 26-7 applied universally to light of all wavelengths. Einstein was awarded the Nobel Prize in Physics in 1921 for his explanation of the photoelectric effect.

Imagine a particular photon of frequency f that transfers all of its energy to a specific electron. The photon carries energy hf (Equation 22-2), where h is Planck's constant. The kinetic energy K imparted to the electron is the difference between hf and Φ, the energy required to just barely liberate that particular electron from the surface:

$$K = hf - \Phi \tag{26-8}$$

The specific amount of energy required to free the electron from the surface depends on the proximity of that electron to the surface, the cleanliness and

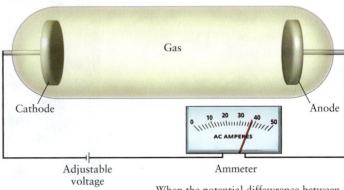

When voltage is applied between the plates in the tube, negative charge accumulates on the cathode and the anode becomes positively charged.

Gas

Cathode   Anode

10 20 30 40
0 \\\\\\\\||||||||||////// 50
AC AMPERES

Adjustable voltage   Ammeter

When the potential diffewrence between the cathode and anode is high enough, charge flows from cathode to anode. The gas glows as it is ionized by the current.

Figure 26-8 When a positive voltage is applied across the plates, negative charges accumulate on one of the plates (the cathode) and are drawn away from the other (the anode). When the potential difference between the plates is high enough, charge can flow from one to the other by ionizing the gas in the tube.

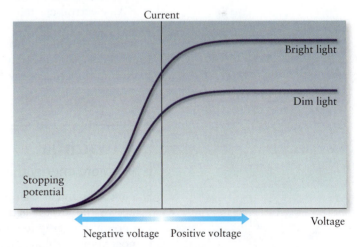

Current

Bright light

Dim light

Stopping potential

Negative voltage   Positive voltage   Voltage

Figure 26-9 Shining ultraviolet light on the cathode in the tube shown in Figure 26-7 dramatically increases the current in the tube, and the brighter the light, the greater the current. Notice also that electrons can flow toward the anode even when the voltage is reversed (negative) so that electrons are repelled from it.

Figure 26-10 The spatial distribution of fibrinogen, a blood protein, is seen in blue on the surface of a special substrate, using photoemission electron microscopy (PEEM). *(Adam Hitchcock, McMaster University)*

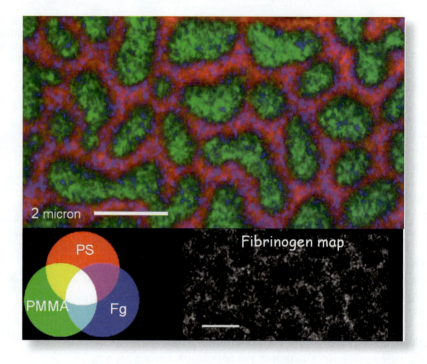

roughness of the surface, as well as other factors. Even under ideal circumstances, however (for example, the electron is at the surface and the surface is clean and smooth), some minimum amount of energy is required to eject a photoelectron. This minimum value is termed the **work function** Φ_0; the work function depends on the material that resides on a surface from which electrons can be ejected. Biological surfaces can be mapped on the basis of this dependence; photoemission electron microscopy (PEEM, also called photoelectron microscopy or PEM) relies on the variation of the work function from place to place on a surface to provide contrast between regions of different composition. **Figure 26-10** shows the spatial distribution of fibrinogen, a blood protein, on the surface of a special substrate recorded using photoemission electron microscopy.

An electron ejected by the photoelectric effect acquires the maximum possible kinetic energy when the energy required to liberate it is as small as allowed by the material. The smallest amount of energy required to liberate an electron is the work function, so setting Φ in Equation 26-8 equal to the work function Φ_0 gives the maximum kinetic energy $K_{\max}$:

$$K_{\max} = hf - \Phi_0 \qquad (26\text{-}9)$$

 Watch Out

More energy than Φ_0 is required to eject some photoelectrons.

Some amount of energy is always required to liberate an electron from a surface by the photoelectric effect. The work function Φ_0 is the smallest possible amount of energy required to eject an electron from a given material under the most favorable conditions. Of course, conditions aren't always favorable, more energy is sometimes required to eject some photoelectrons.

From Equation 17-3 the change in electric potential energy ΔU of an electron that experiences a potential V is

$$\Delta U = e\Delta V$$

where e is the magnitude of the charge of an electron, the fundamental charge. If this change in energy results in the electron acquiring kinetic energy K, then

$$K = e\Delta V$$

So when the voltage between cathode and anode is set to the stopping potential V_0, the kinetic energy is eV_0. It must also be the maximum energy that the electron can acquire, because the largest possible opposition to the electron traveling from cathode to anode can only be overcome by the electron carrying the largest possible kinetic energy. So we can set the energy eV_0 equal to the maximum kinetic energy given by Equation 26-9:

$$eV_0 = hf - \Phi_0 \qquad (26\text{-}10)$$

The potential that will prevent an electron ejected from the cathode from reaching the anode therefore depends on the frequency of the incoming photon. Solving Equation 26-10 for the stopping potential V_0 yields

$$V_0 = \frac{h}{e}f - \frac{\Phi_0}{e} \qquad (26\text{-}11)$$

The relationship between stopping potential V_0 and frequency f is a straight line, as shown in **Figure 26-11**; the slope of the line is h/e. In other words, the slope of the V_0 versus f lines is proportional to Planck's constant h. This provides a mechanism for measuring the value of h that is independent of the blackbody phenomenon that led to the initial formulation of Planck's constant. In addition, notice from the figure that when the stopping potential is set to zero, the photon frequency must be at least f_0, the lowest frequency (and energy) of light that causes a photoelectric effect. So from Equation 26-11,

$$0 = \frac{h}{e}f_0 - \frac{\Phi_0}{e}$$

or

$$h = \frac{\Phi_0}{f_0} \qquad (26\text{-}12)$$

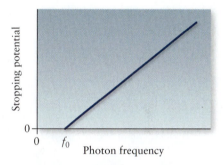

Figure 26-11 Stopping potential versus photon frequency is a straight line. The slope of the line is h/e.

For example, in experiments using a cathode made from tungsten, which has a measured work function Φ_0 equal to 7.24×10^{-19} J, the lowest frequency of light f_0 that causes a photoelectric effect is equal to 1.093×10^{15} Hz. This corresponds to the lowest energy photon that results in a photoelectric effect. Based on these values, from Equation 26-12 Planck's constant is measured to be

$$h = \frac{7.24 \times 10^{-19}\,\text{J}}{1.093 \times 10^{15}\,\text{Hz}} = 6.62 \times 10^{-34}\,\text{Js}$$

This agrees with the accepted value within the precision of the measured values. (Note that a physicist would use the slope of the line, rather than a single data point, to measure h. See Physicist's Toolbox 26-2.)

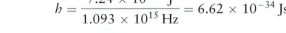

Physicist's Toolbox 26-2 More Information Is Often Better

It is not uncommon to measure the value of one quantity at several values of some other variable when doing a scientific experiment. The result of measuring stopping potential V_0 at several values of frequency f, for example, is shown in Figure 26-11. We determined a value for h using the value of frequency for which the stopping potential equals zero. Doesn't that give the same result as determining the slope of the line, and with less work? For a real experiment, the answer is "no."

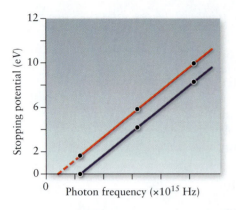

Figure 26-12 Uncertainty in the absolute value of stopping potential can shift the measured line up or down, changing the value of the minimum frequency but not the value of the slope.

It is difficult to calibrate an experimental instrument so that exact values are measured. For example, the exact point at which stopping potential is zero may not be known exactly. In such a case the vertical position of all data points would be shifted up (or down) by the same amount. In **Figure 26-12**, the blue data points and the red data points represent similar measurements. Because it is not possible to calibrate the vertical scale exactly, however, it is possible that the two sets of data points were collected with a different vertical shift. The leftmost data point, for example, was taken at frequency approximately equal to 1.093×10^{15} Hz. When the values plotted as blue points were measured, the stopping potential was recorded as 0 J. When the data graphed as red points were collected, the stopping potential for the value of photon frequency was recorded as 3×10^{-19} J. This was not a mistake; it is simply a result of being unable to set the vertical scale exactly. You can see that if we were to determine h using the point where each line crosses the x axis and therefore the stopping potential V_0 is equal to zero, we would get two different values. The *slope* of both the red and blue lines are the same, however, so measuring h using the slope gives the same result in both cases. What's the fundamental difference? The first approach uses just a single data point from each set of measurements (the frequency for which the stopping potential equals zero) while the second approach relies on at least two points in each set of measurements in order to find the slope of the line. A better result is obtained by using more of the measured data points.

Example 26-2 Photoelectric Effect with Cesium

The work function Φ_0 for a sample of cesium is 3.43×10^{-19} J. (a) What is the minimum frequency of light f_{min} that will result in electrons being ejected from the sample by the photoelectric effect? (b) What is the maximum wavelength of light λ_{max} that will result in electrons being ejected from the sample by the photoelectric effect?

SET UP
The key to solving this problem is recognizing that the minimum energy E_{min} required to eject a photoelectron is equal to the work function, or 3.43×10^{19} J for Cs. We can then use the relationships between the energy of a photon and its frequency and wavelength, from Equations 22-1 and 22-2, to find a solution:

$$E_{min} = hf_{min}$$

and

$$E_{min} = \frac{hc}{\lambda_{max}}$$

SOLVE
(a) Using Equation 22-2, the minimum frequency of a photon that will eject a photoelectron from the Cs surface is

$$f_{min} = \frac{E_{min}}{h} = \frac{\Phi_0}{h} = \frac{3.43 \times 10^{-19}\,\text{J}}{6.63 \times 10^{-34}\,\text{Js}} = 5.17 \times 10^{14}\,\text{Hz}$$

(b) Using Equation 22-1 and recognizing that the work function is the minimum energy required to liberate a photoelectron from the surface, the maximum wavelength of a photon that will eject an electron from the Cs surface is

$$\lambda_{max} = \frac{hc}{E_{min}} = \frac{hc}{\Phi_0} = \frac{(6.63 \times 10^{-34}\,\text{Js})(3.00 \times 10^8\,\text{m/s})}{3.43 \times 10^{-19}\,\text{J}} = 5.80 \times 10^{-7}\,\text{m}$$

REFLECT

The wavelength of light that produces the photoelectric effect in Cs is 5.80×10^{-7} m, or, recalling that 1 nm equals 10^{-9} m,

$$\lambda_{max} = 580 \text{ nm}$$

The range of frequencies of visible light varies between about 380 nm and about 750 nm, so light of a wavelength equal to 580 nm is visible. Light that is 580 nm is a yellow-green color.

★ What's Important 26-2

In the photoelectric effect, an electron in a material can absorb energy of a photon that strikes the surface and as a result be ejected from the surface. The work function Φ_0 is the smallest possible amount of energy required to liberate an electron in this way; the work function depends on the material that resides in a surface. An electron ejected by the photoelectric effect acquires the maximum possible kinetic energy when the energy required to liberate it is minimum.

26-3 Compton Scattering

The phenomena of blackbody radiation and the photoelectric effect suggest that photons can be treated like tiny bundles. We have therefore made the claim that, as quantized units of energy, photons have a particlelike nature. Are there limitations on this claim? For example, an electron absorbs the energy of a photon in the photoelectric effect. But if a photon is like a particle, is it possible that, like in the collisions we studied in Chapter 7, a photon could strike an electron and bounce off? If so, based on our experience with collisions, we ought to expect that linear momentum would be conserved in such an interaction. But photons are massless, and we have defined linear momentum to be the product of the mass and velocity of an object. Do photons have linear momentum? How are we to resolve this puzzle?

First, the answers. Yes, a photon can scatter during the collision with an electron. And yes, photons *do* have linear momentum.

Our discussion of relativistic mass provides a hint about the momentum of a photon. Equation 25-21 defines the mass m of a moving object in terms of its rest mass m_0:

$$m = \gamma m_0$$

Because we are interested in momentum, let's write out relativistic gamma (γ) in terms of the speed v, using Equation 25-12:

$$\gamma = \sqrt{\frac{c^2}{c^2 - v^2}}$$

Substituting this into Equation 25-21 and squaring both sides to eliminate the square root, we have

$$m^2 = \frac{c^2}{c^2 - v^2} m_0^2$$

or

$$m^2 c^2 - m^2 v^2 = m_0^2 c^2$$

Recall that $E = mc^2$ (Equation 25-24), so the first term on the left-hand side is E^2/c^2. That allows us to rewrite the expression in terms of energy, and also momentum! Notice that the second term on the left-hand side is the square of momentum, so

$$E^2 = p^2c^2 + m_0^2c^4 \tag{26-13}$$

This is effectively another form of Einstein's famous statement of the equivalence of mass and energy, but we can apply it to particles, such as photons, that have no rest mass. When m_0 equals zero, Equation 26-13 tells us that

$$E = pc \tag{26-14}$$

The momentum of a photon can therefore be expressed as

$$p = \frac{E}{c} \tag{26-15}$$

The conclusion? A photon's momentum depends on the energy it carries.

In the early 1920s, American physicist Arthur Compton showed conclusively that photons have momentum. During his experiments, an x-ray photon collided with an electron in a carbon atom. In this process, now called **Compton scattering**, he detected both the electron, which is knocked out of the atom, and also the scattered photon. Compton could only account for the directions and energies of the electron and the scattered photon by associating a momentum vector with the incoming and outgoing photon and then requiring that momentum be conserved. In this way, Compton showed that photons do indeed have momentum; for revealing this fundamental aspect of light Compton was awarded the Nobel Prize in Physics in 1927.

Compton's results are most concisely expressed in terms of the change in wavelength of the photon as a result of the collision with the electron. The collision of a photon (γ_i) and a stationary electron (e^-) is depicted in **Figure 26-13**. The electron is scattered at angle φ relative to the initial direction of the photon, and the photon scatters at angle θ relative to its initial direction. Although we have labeled the photon γ_f after the collision, the scattered photon is the same photon that collided with the electron. The subscripts "i" for "initial" and "f" for "final" instead imply that the energy, wavelength, and other quantities associated with the photon have changed as a result of the collision. For example, because energy is conserved during the collision and because some of the photon's energy is almost always transferred to the electron, the outgoing photon carries less energy than it had initially. Recall that the energy and wavelength of a photon are related by Equation 22-1:

$$E = \frac{hc}{\lambda}$$

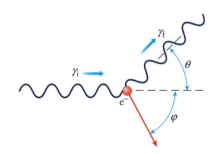

Figure 26-13 After a photon γ_i collides with a stationary electron e^-, the photon is scattered at angle θ relative to its initial direction and the electron is scattered at angle φ relative to the initial direction of the photon. The scattered photon is the same as the incident photon, but we label it γ_f after the collision to indicate that it has lost energy.

Because the photon has less energy after the collision than it had initially, E_f is less than E_i, and the final wavelength λ_f is more than the initial wavelength λ_i. In other words, as the energy of the photon decreases, its wavelength increases. Compton found that the increase in wavelength $\Delta\lambda$ from λ_i to λ_f is a function of the angle θ at which the photon scatters:

$$\Delta\lambda \propto 1 - \cos\theta$$

The proportionality constant depends only on Planck's constant h, the mass of an electron m_e, and the speed of light in a vacuum c:

$$\Delta\lambda = \frac{h}{m_ec}(1 - \cos\theta) \tag{26-16}$$

Because we have defined the scatter angle θ relative to the initial direction of the photon (Figure 26-13), θ ranges from 0° (straight forward) to 180° (straight back). When the photon continues in the same direction after the collision, so that θ equals 0° and $\cos \theta$ equals 1, $\Delta \lambda$ equals zero. In other words, there is no change in the photon's wavelength and no change in the photon's energy.

The proportionality constant in Equation 26-16 is known as the **Compton wavelength** λ_C:

$$\lambda_C = \frac{h}{m_e c} \qquad (26\text{-}17)$$

so we write the Compton scattering relationship as

$$\Delta \lambda = \lambda_C (1 - \cos \theta) \qquad (26\text{-}18)$$

The maximum change in a photon's wavelength and energy when it undergoes Compton scattering occurs when it scatters straight back, in the direction opposite to the one in which it approached the electron. In this case θ equals 180°, so $\cos \theta$ equals -1 and the term in parentheses in Equation 26-18 equals 2. The maximum possible change in the photon's wavelength is therefore $2\lambda_C$. The value of the shift depends on the value of the Compton wavelength. Using the known values of h, m_e, and c we find

$$\lambda_C = \frac{h}{m_e c} = \frac{6.63 \times 10^{-34} \, \text{Js}}{(9.11 \times 10^{-31} \, \text{kg})(3.00 \times 10^8 \, \text{m/s})} = 2.43 \times 10^{-12} \, \text{m}$$

or

$$\lambda_C = 0.00243 \, \text{nm}$$

The value of λ_C plays a significant role in Compton scattering. For example, consider the possibility of a *visible* photon scattering in a collision with an electron in a process described by Equation 26-18. The wavelength of visible light ranges between about 380 nm and 750 nm. The Compton wavelength is clearly considerably smaller. What would it mean, for example, to say that a photon of wavelength 400 nm is scattered and changes by a value on the order of 0.00243 nm? Because its new wavelength (and therefore energy) is effectively the same as its initial wavelength (and energy), we would justifiably say that no Compton scattering has occurred. Photons in the visible range do not interact with electrons through the Compton scattering process. Photons that do Compton scatter fall into the energy regime of x rays. For this reason, Compton scattering plays a significant role in the application of radiation therapy.

Example 26-3 The Units of the Compton Wavelength

Show that when Planck's constant h, the mass of an electron m_e, and the speed of light in a vacuum c are combined to form the Compton wavelength according to Equation 26-17, the result has units of meters (m).

SET UP
The units of h, m_e, and c are $J \cdot s$, kg, and m/s, respectively. Simply combining them does not immediately yield the correct units of m. We need to find a way to express joules in terms of units of fundamental quantities.

The joule is a unit of energy. We can therefore write joules in terms of other units by recalling any equation that involves energy and writing out the quantities, and therefore the units, on which it relies. For example, kinetic energy is given by $K = \frac{1}{2}mv^2$, so 1 J is equivalent to $1 \, \text{kg}(\text{m/s})^2$.

SOLVE

Dimensional analysis of our equation for Compton's wavelength ($\lambda_C = h/m_e c$, Equation 26-17) and substituting the units of kinetic energy for joules gives

$$[\lambda_C] = \frac{[h]}{[m_e][c]} = \frac{\text{J}\,\text{s}}{\text{kg}\,\text{m/s}} = \frac{\text{kg} \cdot (\text{m/s})^2 \cdot \text{s}}{\text{kg} \cdot \text{m/s}}$$

First cancel kilogram (kg) from both the numerator and the denominator, as well as one power of m/s. That gives

$$[\lambda_C] = \frac{\text{m}}{\text{s}}\,\text{s} = \text{m}$$

So all of the units cancel except for one power of m.

REFLECT

The SI unit of the Compton wavelength is the meter. Wavelength is distance, so this is what we should expect. It is always useful to check that the units (or dimensions) of a computed value agree with what you expect.

Practice Problem 26-3 Use Equation 22-1 to express the wavelength of a photon in terms of its energy, and use your result to show that the SI unit of wavelength is the meter (m).

? **Got the Concept 26-1**

Three Photons

Three photons of wavelengths 5×10^{-3} nm, 5 nm, and 5×10^3 nm strike a piece of carbon. Which, if any, could undergo Compton scattering? Which has the highest, and which has the lowest energy?

Example 26-4 How Large an Effect?

A photon that carries 2.00×10^{-14} J of energy Compton scatters in a block of carbon. What is the largest fractional change in energy the photon can undergo as a result?

SET UP

The energy of a photon is given by Equation 22-1:

$$E = \frac{hc}{\lambda}$$

The change in energy is related to the change in wavelength:

$$E_i - E_f = \frac{hc}{\lambda_i} - \frac{hc}{\lambda_f}$$

The fractional change in energy is then

$$\text{fractional } \Delta = \frac{E_i - E_f}{E_i} = \left(\frac{hc}{\lambda_i} - \frac{hc}{\lambda_f}\right) / \left(\frac{hc}{\lambda_i}\right)$$

or

$$\text{fractional } \Delta = \frac{\lambda_i}{\lambda_i} - \frac{\lambda_i}{\lambda_f} = 1 - \frac{\lambda_i}{\lambda_f}$$

Also, because the increase in wavelength $\Delta\lambda$ is the change from λ_i to λ_f, we can express the final wavelength λ_f in terms of the initial wavelength λ_i and the Compton wavelength using Equation 26-18:

$$\Delta\lambda = \lambda_f - \lambda_i = \lambda_C(1 - \cos\theta)$$

The largest change in wavelength is $2\lambda_C$, which occurs when θ equals 180° and cos $\theta = -1$, so

$$\lambda_{f,\text{max}} - \lambda_i = 2\lambda_C$$

or

$$\lambda_{f,\text{max}} = \lambda_i + 2\lambda_C$$

SOLVE

The largest fractional change in energy is then

$$\text{max fractional } \Delta = 1 - \frac{\lambda_i}{\lambda_{f,\text{max}}}$$

so

$$\text{max fractional } \Delta = 1 - \frac{\lambda_i}{\lambda_i + 2\lambda_C}$$

All that's left is to express λ_i in terms of the initial energy using Equation 22-1:

$$\lambda_i = \frac{hc}{E_i}$$

The wavelength of a photon that carries 2.00×10^{-14} J of energy is

$$\lambda_i = \frac{(6.63 \times 10^{-34}\,\text{Js})(3.00 \times 10^8\,\text{m/s})}{2.00 \times 10^{-14}\,\text{J}} = 9.95 \times 10^{-12}\,\text{m}$$

or

$$\lambda_i = 0.00995\,\text{nm}$$

And, finally,

$$\text{max fractional } \Delta = 1 - \frac{\lambda_i}{\lambda_i + 2\lambda_C} = 1 - \frac{0.00995\,\text{nm}}{0.00995\,\text{nm} + 2(0.00243\,\text{nm})}$$

or

$$\text{max fractional } \Delta = 1 - 0.672 = 0.328$$

REFLECT

The photon can lose as much as about 33% of its initial energy when it undergoes Compton scattering.

Practice Problem 26-4 A photon of wavelength 0.0031 nm Compton scatters in a block of carbon, leaving the collision at an angle 50° from its initial direction. Find the wavelength of the scattered photon.

✴ What's Important 26-3

Even though a photon has no mass, it does have linear momentum, which depends on the energy it carries. The phenomenon of Compton scattering, in which a photon collides with and scatters from an electron, conserves both momentum and energy. The Compton scattering relationship describes the increase in the photon's wavelength as a function of the angle at which the photon scatters.

26-4 Wave Nature of Particles

In 1924, French graduate student Louis de Broglie (pronounced "de broy") had an inspiration. Physicists had begun to accept the notion that waves, such as light, had a particle nature. De Broglie reasoned that because the universe consists entirely of radiation and matter, and because "wave–particle dualism for light was an absolutely general phenomenon" (Louis de Broglie, *New Perspectives in Physics*, 1962), this duality should be extended to particles of matter. De Broglie proposed that particles such as electrons should therefore also have wave properties.

Let us extend our understanding of the wave nature of light to particles in the way that de Broglie did. The wavelength λ of a photon is directly related to its energy. From Equation 22-1,

$$\lambda = \frac{hc}{E}$$

Because the momentum p of a photon is also directly related to its energy (Equation 26-14),

$$E = pc$$

the photon's wavelength is also a function of its momentum:

$$\lambda = \frac{hc}{pc}$$

or

$$\lambda = \frac{h}{p} \tag{26-19}$$

De Broglie argued that this relationship, now known as the **de Broglie wavelength**, holds for particles as well as for radiation.

What is the de Broglie wavelength for a typical electron? Equation 26-19 allows us to compute a value. Before we make such a computation, however, we ought to ask a fundamental question: Is this wavelength a physically meaningful quantity? Does an electron really act like a wave, or in practical terms, can we observe electrons exhibiting any of the properties we associate with waves? For example, can we observe the wave phenomenon of diffraction when a beam of electrons passes through a narrow slit, in the way that light and sound waves do (Section 23-6)? Diffraction occurs when a wave passes through a slit of width about the same size as the wavelength of the wave, so this observation raises a second question: If electrons can diffract when passing through a slit, what size slit is required? This brings us back to the original question: What is the de Broglie wavelength for a typical electron?

Consider a relatively slow-moving electron, carrying 1×10^{-18} J of kinetic energy. To find the de Broglie wavelength we must first determine its momentum. For a slow (nonrelativistic) particle, momentum is simply the product of mass and velocity. From the definition of kinetic energy K, the magnitude of velocity v of the electron is

$$v = \sqrt{\frac{2K}{m_e}}$$

where m_e is the electron mass. So momentum p is

$$p = m_e v = m_e \sqrt{\frac{2K}{m_e}} = \sqrt{2m_e K} \tag{26-20}$$

The de Broglie wavelength (Equation 26-19) of the slow electron is then

$$\lambda = \frac{h}{\sqrt{2m_e K}}$$

or numerically,

$$\lambda = \frac{6.63 \times 10^{-34}\,\text{Js}}{\sqrt{2\,(9.11 \times 10^{-31}\,\text{kg})\,(1 \times 10^{-18}\,\text{J})}} \approx 5 \times 10^{-10}\,\text{m}$$

The wavelength of a slow-moving electron is therefore on the order of 10^{-10} m, or 0.1 nm. This is much smaller than openings through which light diffracts, but not so small that nature does not provide slits of such dimensions. Atoms, and the spacing between them, are about the same size. And slow-moving, or low energy, electrons do indeed diffract when they pass through the "slits" formed by the atoms in a material.

As beams of electrons pass through a thin slice of a specimen mounted on a transmission electron microscope (TEM), they diffract and form an image of the material. Because the wavelengths of low energy electrons are so much shorter than those of light, a TEM is capable of significantly higher resolution than light microscopes. Figure 26-14 shows a thin cross section of cilia in a human lung, imaged using a transmission electron microscope. Precisely arranged microtubules that run down the cilia are clearly resolved; the microtubules are on the order of 10 nm in diameter. Transmission electron microscopes are capable of resolutions of better than 0.005 nm.

The transmission electron microscope provides convincing evidence that low energy electrons have wave properties, and that the de Broglie wavelength of a low energy electron is a meaningful quantity. Do all particles have a de Broglie wavelength? Does, for example, a baseball exhibit wave properties? We approach the answer in the same way we did for an electron, by determining the de Broglie wavelength using Equation 26-19 and asking whether it is possible to observe diffraction of a wave of this wavelength.

The official rules of major league baseball require the ball to have a mass of about 0.145 kg. A professional major league pitcher can throw a pitch at perhaps

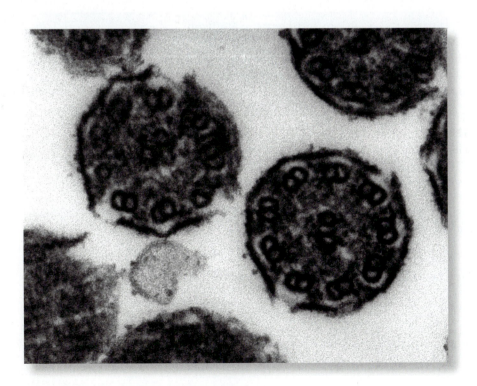

Figure 26-14 An image of a thin cross section of cilia in a human lung, obtained using a transmission electron microscope, shows precisely arranged microtubules running down each cilium. The microtubules, which are on the order of 10 nm in diameter, are clearly resolved. (*Dartmouth Electron Microscope Facility*)

45 m/s (about 100 mph). The momentum of a major league fastball is therefore about

$$p = mv = (0.145 \text{ kg})(45 \text{ m/s}) = 6.5 \text{ kg} \cdot \text{m/s}$$

The de Broglie wavelength, from Equation 26-19, is then

$$\lambda = \frac{6.63 \times 10^{-34} \text{ Js}}{6.5 \text{ kg} \cdot \text{m/s}} \approx 1 \times 10^{-34} \text{ m}$$

Can we construct a slit that has a width about this size? No! In fact, we know of no physical objects that are either separated by distances this small or have such a small size, or even a size 10^{10} times larger than this. And in any event, a baseball certainly would not fit through a slit this size, even if we could find one.

Notice that the higher the velocity and momentum of the baseball, the smaller the de Broglie wavelength. Perhaps we can throw a baseball slowly enough that its wavelength would be on the order of its size, so that we could construct a slit through which it would diffract, say about 0.1 m. For an object of mass 0.145 g to have a de Broglie wavelength of 0.1 m requires the magnitude of its velocity, from Equation 26-19 and using $p = mv$, to be

$$v = \frac{h}{m\lambda} = \frac{6.63 \times 10^{-34} \text{ Js}}{(0.145 \text{ kg})(0.1 \text{ m})} \approx 5 \times 10^{-32} \text{ m/s}$$

At this rate it would take more than 10^{32} s for the baseball to travel from the pitcher's hand to the batter. Our best estimate for the age of our solar system is about 1.4×10^{17} s, so 10^{32} s is a very (very!) long time. It simply doesn't make sense to talk about times this long, or speeds this slow.

The consideration of the wavelength of a high speed baseball and speed required for a baseball to fit through a slit that would result in diffraction lead to the same conclusion: It is not physically reasonable to attribute wave properties to a major league fastball. Particles such as electrons exhibit wave properties. Objects such as baseballs do not.

Example 26-5   Room Temperature Neutron

The thermal kinetic energy of a particle at room temperature is around 4.0×10^{-21} J. (This is the product of the Boltzmann constant k, 1.381×10^{-23} J/K, and room temperature T, 293 K.) Determine the de Broglie wavelength of a thermal neutron, that is, a neutron in thermal equilibrium with its environment.

SET UP

The de Broglie wavelength is given by Equation 26-19:

$$\lambda = \frac{h}{p}$$

Using Equation 26-20, the momentum of a nonrelativistic particle is related to its kinetic energy:

$$p = \sqrt{2m_e K}$$

So a neutron's wavelength in terms of its kinetic energy is

$$\lambda = \frac{h}{\sqrt{2m_n K}}$$

where m_n is the mass of a neutron. The neutron mass is approximately 1.7×10^{-27} kg.

SOLVE

The de Broglie wavelength is then

$$\lambda = \frac{6.63 \times 10^{-34}\,\text{Js}}{\sqrt{2\,(1.7 \times 10^{-27}\,\text{kg})\,(4.0 \times 10^{-21}\,\text{J})}} \approx 1.8 \times 10^{-10}\,\text{m}$$

REFLECT

The de Broglie wavelength of a thermal neutron is about 1.8×10^{-10} m, a distance that is typical of the size of atoms.

Practice Problem 26-5 The mass of a gold atom is about 3.27×10^{-25} kg. Find the de Broglie wavelength of a thermal gold atom.

Example 26-6 A Slow Hummingbird

The ruby-throated hummingbird, the most common hummingbird found in Canada, is typically 0.085 m long and has a mass of 0.0040 kg. At what speed must a hummingbird of such mass and length fly so that its computed de Broglie wavelength would be equal to its length?

SET UP

Using $p = mv$, Equation 26-19 for the de Broglie wavelength becomes

$$\lambda = \frac{h}{m_{hb}v_{hb}}$$

where m_{hb} and v_{hb} are the mass and speed of the hummingbird, respectively. Then

$$v_{hb} = \frac{h}{m_{hb}\lambda}$$

SOLVE

Letting the de Broglie wavelength be equal to the length of the hummingbird L_{hb}, the speed is

$$v_{hb} = \frac{6.63 \times 10^{-34}\,\text{Js}}{(0.0040\,\text{kg})\,(0.085\,\text{m})} = 2.0 \times 10^{-30}\,\text{m/s}$$

REFLECT

This speed is so low as to be nonsensical!

Practice Problem 26-6 At what speed would a student of mass 58 kg need to move through a 0.90-m-wide doorway so that she (her wave equivalent) would be diffracted? Is this a reasonable speed for a student?

★ What's Important 26-4

Particles can exhibit wave properties; for example, electrons can be diffracted when passing through a lattice of atoms. The wave properties of a particle are set by the de Broglie wavelength, which depends on the particle's momentum. Particles such as electrons exhibit wave properties. Objects such as baseballs do not.

26-5 Atoms: Spectra

In Example 1-1 (the very first example in this book!) we considered the rate of human hair growth. The elemental composition of your hair reflects the chemistry of your body as the cells that become hair are formed. Yet because hair is not living tissue, few changes to its chemical makeup occur over time. In a real sense, then, each hair on your head is a record of your biomedical history, starting from the time that the tips of your hair emerged from your head. As we saw in Example 1-1, the rate at which hair grows is about 20 μm/h, so the ability to measure the level to which atoms of a particular element are present along the length of your hair would provide an hour-by-hour record of the chemical changes taking place in your body. A technique to do just this, proton-induced x-ray emission (PIXE), has been studied by physicists since the middle of the last century. PIXE takes advantage of the structure of electrons associated with nuclei to measure the presence of trace elements at the level of parts per million with a spatial resolution of 1 μm or so.

The concept of the atom, a unit of matter so small that it could not be subdivided, comes to us from the early Greeks. (The word is derived from the Greek *átomos*, meaning "indivisible.") By the turn of the twentieth century physicists had some clues about the nature of atoms, including that they are electrically neutral. J.J. Thomson, the discoverer of the electron, proposed that atoms had a positive component in addition to negative electrons. In his model, sometimes called the "plum pudding model" of the atom, electrons were scattered throughout a relatively uniform distribution of positive charge much like raisins in the traditional English dessert.

What experiment might be conducted to test Thomson's model? One of Thomson's former students, Ernest Rutherford, designed one possible experiment to test it. In about 1910, Rutherford and his colleagues fired a subatomic particle, which Rutherford had named the *alpha particle*, into very thin gold foil. If the gold atoms were formed according to Thomson's model, Rutherford expected that the alpha particles would be slightly deflected from their initial direction as a result of passing through the diffuse, positive "pudding" of the atoms. Instead, Rutherford was startled to discover that alpha particles were sometimes scattered at large angles with respect to the initial direction, occasionally leaving the gold foil directly *backward*. (In reflecting on the experiment, Rutherford later said, "It was quite the most incredible event that ever happened to me in my life. It was as incredible as if you fired a 15-in shell [a large projectile fired from a military weapon] at a piece of tissue paper and it came back and hit you. On consideration, I realized that this scattering backwards must be the result of a single collision, and when I made calculations I saw that it was impossible to get anything of that order of magnitude unless you took a system in which the mass of the atom was concentrated in a minute nucleus."[1]) To account for this, Rutherford proposed a model of the atom in which negatively charged electrons orbit a small, positively charged nucleus that contains nearly all of the atom's mass. In Rutherford's model most of the volume of each atom is empty, so most alpha particles fired at the gold foil would experience only slight deflections as they passed through. But once in a while, about one time out of every 10,000, an alpha particle would approach a gold nucleus almost head on and be scattered at a large angle, sometimes directly backward.

Unlike planets orbiting the Sun, electrons orbiting an atomic nucleus fit into a well-defined organizational structure. We'll explore this structure from three perspectives and in historical order: first, the early clues that hinted at the structure, then the development of mathematical models that describe the structure, and finally the theory that explains it.

---

[1] Ernest Rutherford, "The Development of the Theory of Atomic Structure," in J. Needham and W. Pagel (eds.), *Background to Modern Science*, The Macmillan Company, New York, 1938.

In the early part of the nineteenth century, the English scientist William Hyde Wollaston noted the appearance of dark lines in the spectrum of visible light coming from the Sun (**Figure 26-15**). While conducting experiments on glassmaking, a German optician-turned-physicist, Joseph von Fraunhofer, independently discovered these lines, now referred to as Fraunhofer lines. Von Fraunhofer used the reliability of the position of the lines within the spectrum to study the refractive properties of the glass he created.

Why are the lines always in the same locations? An explanation of sorts was provided some years later by Gustav Kirchhoff, the same physicist we encountered in our study of electric circuits. Based on experiments he conducted with Robert Bunsen, a German chemist after whom the laboratory burner he developed is named, Kirchhoff recognized that the black lines in the solar spectrum resulted from specific wavelengths of light being absorbed by the gas through which light passes as it leaves the Sun. Kirchhoff confirmed this conclusion by passing light from a lamp made to simulate sunlight through vapors created by heating sodium, thereby duplicating the closely spaced pair of lines in the yellow-orange region of the Sun's spectrum. (Try to find these lines in Figure 26-15!) The lines result from some property or properties of the elements in the gas through which light passes. This observation, combined with the regularity and repeatability of the lines, was a hint of some underlying structure of those elements.

The spectrum of the visible light given off by the Sun is an **absorption spectrum**. An absorption spectrum appears as a smooth spectrum with narrow, dark lines corresponding to the wavelengths of light that have been absorbed by material through which the light has passed before being observed. The portion of the absorption spectrum of hydrogen that lies in the visible range is shown in **Figure 26-16a**. Physicists of the nineteenth century discovered that light created by heating a substance gives rise to an **emission spectrum**, a spectrum that consists only of specific emitted wavelengths. **Figure 26-16b** shows the visible portion of the emission spectrum created by passing an electric current through hydrogen gas, thus causing it to glow. Clearly, light of only specific wavelengths—colors–is emitted. Notice that the black lines in the absorption spectrum correspond exactly to the hydrogen emission lines, which is another hint of some underlying structure. Although for most of the rest of this section we will confine the discussion to an exploration of the absorption and emission spectra of hydrogen, the underlying physics applies to all elements.

Johann Balmer, a Swiss mathematician, provided a mathematical foundation for the atomic structure hinted at by the regularity of the lines in absorption and emission spectra. Others had precisely measured the wavelengths of the visible lines in the hydrogen spectra; Balmer devised a formula that both reproduced the wavelengths of lines that had been reported and also correctly predicted the wavelengths of spectral lines of which Balmer was unaware. The *Balmer formula* is

$$\lambda = h\frac{m^2}{m^2 - 4} \tag{26-21}$$

Here m is any integer greater than 2. (The variable m does not represent mass in the Balmer formula! Note also that h, which was Balmer's choice, is not Planck's

(a)

(b)

Figure 26-16 (a) When light passes through a gas, light of specific wavelengths is absorbed, forming dark lines. In this absorption spectrum of hydrogen made with visible light, all wavelengths (colors) of light are seen except for dark bands corresponding to the specific wavelengths of light that have been absorbed by the hydrogen atoms. (b) When a gas is made to glow by passing an electric current through it, it emits only specific wavelengths of light. Notice that the lines in this emission spectrum of hydrogen correspond exactly to the hydrogen absorption lines.

constant.) Balmer determined the value of the constant h empirically to be 3.6456×10^{-7} m, in order to reproduce the known, visible hydrogen spectral line wavelengths.

Swedish physicist Johannes Rydberg extended Balmer's work. The **Rydberg formula,**

$$\frac{1}{\lambda} = R_{\mathrm{H}}\left(\frac{1}{n^2} - \frac{1}{m^2}\right) \qquad (26\text{-}22)$$

where all integer values of m and n are allowed as long as m is greater than n, makes it clearer that groups of lines beyond the visible are possible. The Rydberg constant R_{H} is approximately 1.09737×10^7 m^{-1}, or 1.10×10^7 m^{-1} to three significant digits. It is straightforward to show that the Balmer formula is a special case of the Rydberg formula with n set equal to 2.

The Rydberg formula depends on two integers, m and n. We understand today that the integers identify electron energy levels in an atom. Consider the formation of an absorption spectrum as white light passes through hydrogen gas. When a photon of one of the wavelengths described by the Rydberg formula passes through the gas, the energy of that photon is absorbed by an electron in an atom of the gas. As we saw in Chapter 8, for example, in **Watch Out, The Moon is getting farther, not closer, to Earth,** increasing the energy of an orbiting object increases the radius at which it orbits. So the electron moves to an orbit farther from the nucleus when it absorbs the photon's energy. Because most of the photons of the wavelengths described by the Rydberg formula are absorbed as they pass through the gas, light of those wavelengths is not present in the spectrum formed by light exiting the gas. The region of the spectrum at each of those wavelengths is therefore dark. In the visible region of the spectrum this results in black lines against the colors of the light that does pass through the gas, as seen in Figure 26-16a.

The physics that underlies the formation of an emission spectrum is the same as that for an absorption spectrum. In this case, exciting the atoms in hydrogen gas, say, by passing an electric current through it, causes electrons to make transitions to higher energy orbits. Eventually the electrons fall back down to a lower energy orbit, and the energy difference is released in the form of photons of the energies and wavelengths predicted by the Rydberg equation. Because only light of those wavelengths is emitted, an emission spectrum consists only of lines at these wavelengths, as we saw in Figure 26-16b.

Why do the lines in the absorption and emission spectra of hydrogen occur only at specific wavelengths, and always at the same wavelengths? By itself, the absorption and emission phenomenon we just described does not answer these questions. The full explanation requires an additional constraint, that the electrons only orbit the hydrogen nucleus with specific, well-defined energies. As a result, an electron can only absorb or emit a photon when the photon energy corresponds to the energy difference between two electron orbits. As an example, let's say the three lowest energy orbits of electrons in a hydrogen atom have energies 2, 8, 12, and 13 in arbitrary energy units. An electron orbiting with energy equal to 2, say, could absorb a photon of energy equal to 6 because the electron's new energy would be 8 (that is, 2 plus 6), which is an allowed orbital energy. Likewise the electron could absorb a photon of energy 10 or 11, because doing so would also excite the electron into another allowed orbit. The electron could not, however, absorb a photon carrying energy equal to 5, because 7 (that is, 2 plus 5) is not an allowed orbital energy. (Note that, by convention, the energies of atomic electrons are usually given as negative values, with the lowest energy orbit having the most negative value of energy and orbits farther from the nucleus having energies closer and closer to zero. We chose not to follow that convention in this example in order to make it clearer.)

We can now understand the integers m and n in the Rydberg formula. **Figure 26-17** shows schematically the four electron orbits closest to the nucleus of a hydrogen atom. We label these with integers, starting from 1 and starting from the innermost

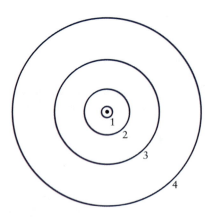

Figure 26-17 Allowed electron orbits are numbered using integers 1, 2, 3, and so on, starting with the orbit closest to the nucleus.

orbit. Associated with each possible orbit is a specific value of energy, which we refer to here as E_1, E_2, E_3, and E_4. An electron can only absorb or emit a photon if its energy would result in the electron making a **transition** to another allowed orbit. For example, an electron in the lowest energy orbit could absorb a photon carrying energy $E_2 - E_1$, which would result in a transition to the second orbit. In the Rydberg formula, the integers m and n describe the two orbits involved in an atomic transition; m corresponds to the higher energy orbit and n to the lower energy orbit. This gives rise to the requirement in the Rydberg formula that m must be greater than n.

An atomic transition can occur between any two orbits n and m, as suggested by **Figure 26-18**. The group of all transitions with a specific value of n are said to be a series, and the wavelengths of the lines in each of the hydrogen series are also grouped together. The series of lines associated with the n equal 1 orbit form the Lyman series, with wavelengths in the ultraviolet. The lines associated with the n equal 2 orbit form the Balmer series, many of which are in the visible, and lines of the Paschen series, associated with the n equal 3 orbit, are in the infrared.

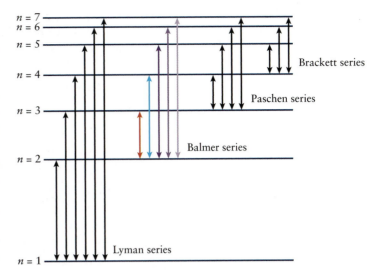

Figure 26-18 The lines in the spectrum of hydrogen fall into a sequence of series, each defined by the lowest energy electron orbit of the atomic transitions in the series. The lines in the Balmer series are mostly in the visible range.

? **Got the Concept 26-2**

The First Hydrogen Lines to Be Discovered

The first lines of the hydrogen spectrum to be discovered are part of the Balmer series. Why do you think the Balmer series was discovered before the others?

Example 26-7 Lyman, Balmer, and Paschen

Give the values of m that correspond to the longest and shortest wavelengths, and determine, to three significant digits, those wavelengths in the (a) Lyman, (b) Balmer, and (c) Paschen series.

SET UP

The wavelengths of the lines in any spectral series in hydrogen are given by the Rydberg formula (Equation 26-22):

$$\frac{1}{\lambda} = R_H \left(\frac{1}{n^2} - \frac{1}{m^2} \right)$$

The value of n determines the series, and the lines in the series correspond to integer values of m greater than n. Note that the shortest wavelength in a series corresponds to the largest possible value of m. The largest possible value of m in all cases is infinity!

SOLVE

(a) For the Lyman series n is equal to 1, so the smallest allowed value of m, corresponding to the longest wavelength in the series, is 2. The wavelength of the photon absorbed or emitted for this atomic transition is

$$\frac{1}{\lambda_{long}} = R_H \left(\frac{1}{1^2} - \frac{1}{2^2} \right) = R_H \left(1 - \frac{1}{4} \right) = \frac{3}{4} R_H$$

The value of the Rydberg constant R_H is $1.09737 \times 10^7 \ \text{m}^{-1}$, so

$$\lambda_{\text{long}} = \frac{4}{3R_H} = \frac{4}{3(1.09737 \times 10^7 \ \text{m}^{-1})} \approx 1.22 \times 10^{-7} \ \text{m} = 122 \ \text{nm}$$

The shortest wavelength in the series corresponds to setting m equal to infinity:

$$\frac{1}{\lambda_{\text{short}}} = R_H \left(\frac{1}{1^2} - \frac{1}{\infty^2} \right) = R_H$$

so

$$\lambda_{\text{short}} = \frac{1}{R_H} = \frac{1}{1.09737 \times 10^7 \ \text{m}^{-1}} \approx 0.911 \times 10^{-7} \ \text{m} = 91.1 \ \text{nm}$$

(b) In a similar way, the longest wavelength in the Balmer series (n equal to 2) corresponds to m equal to 3:

$$\frac{1}{\lambda_{\text{long}}} = R_H \left(\frac{1}{2^2} - \frac{1}{3^2} \right) = R_H \left(\frac{1}{4} - \frac{1}{9} \right) = \frac{5}{36} R_H$$

or

$$\lambda_{\text{long}} = \frac{36}{5R_H} = \frac{36}{5(1.09737 \times 10^7 \ \text{m}^{-1})} \approx 6.56 \times 10^{-7} \ \text{m} = 656 \ \text{nm}$$

To find the shortest wavelength in the series we set m equal to infinity:

$$\frac{1}{\lambda_{\text{short}}} = R_H \left(\frac{1}{2^2} - \frac{1}{\infty^2} \right) = \frac{R_H}{4}$$

or

$$\lambda_{\text{short}} = \frac{4}{R_H} = \frac{4}{1.09737 \times 10^7 \ \text{m}^{-1}} \approx 3.65 \times 10^{-7} \ \text{m} = 365 \ \text{nm}$$

(c) The longest wavelength in the Paschen series (n equal to 3) corresponds to m equal to 3. Thus,

$$\frac{1}{\lambda_{\text{long}}} = R_H \left(\frac{1}{3^2} - \frac{1}{4^2} \right) = R_H \left(\frac{1}{9} - \frac{1}{16} \right) = \frac{7}{144} R_H$$

or

$$\lambda_{\text{long}} = \frac{144}{7R_H} = \frac{144}{7(1.09737 \times 10^7 \ \text{m}^{-1})} \approx 1.87 \times 10^{-6} \ \text{m} = 1870 \ \text{nm}$$

For the shortest wavelength, we again set m equal to infinity:

$$\frac{1}{\lambda_{\text{short}}} = R_H \left(\frac{1}{3^2} - \frac{1}{\infty^2} \right) = \frac{R_H}{9}$$

or

$$\lambda_{\text{short}} = \frac{9}{R_H} = \frac{9}{1.09737 \times 10^7 \ \text{m}^{-1}} \approx 8.20 \times 10^{-7} \ \text{m} = 820 \ \text{nm}$$

REFLECT

The wavelengths of the lines in the Lyman series are in the ultraviolet, and the wavelengths of the lines in the Paschen series are in the infrared. Visible light falls in the range from about 380 nm (violet) to 750 nm (red). So most of the Balmer series transitions either absorb or emit visible light.

Practice Problem 26-7 The Brackett series in the hydrogen spectrum includes transitions to (or from) the n equal 4 orbit. Give the values of m that correspond to the longest and shortest wavelengths in the Brackett series, and determine those wavelengths to three significant digits.

We can now understand how proton-induced x-ray emission that we introduced at the start of this section works. To measure the quantity of trace elements in a sample, such as a strand of hair, scientists focus a beam of energetic protons on a small region of the sample. A beam of protons can be focused down to a spot size on the order of 1 μm in diameter. The energy of the protons causes atomic excitations in the material, which quickly result in the emission of photons, in this case x rays, as the electrons fall back down to lower energy levels. The photons form an emission spectrum in which the wavelength of each line is the result of a transition in an atom of one of the elements present in the sample. The wavelengths emitted by any element are well known, so all elements present can be determined from the spectrum. In addition, the number of photons emitted at the wavelengths characteristic of a particular element is proportional to the relative amount of that element in the sample. The emission spectrum that results from bombarding a sample with energetic protons therefore allows the identification of elements present in a sample as well as the determination of how much of each is present.

We add one final note on emission and absorption spectra. The Rydberg formula for hydrogen implies that each electron orbit carries with it one specific amount of energy. This is not strictly correct, not just for hydrogen but for all elements. Instead, a variety of subtle effects gives rise to a splitting of, or *fine structure* of, atomic energy levels. Perhaps the most well known is the hyperfine splitting of the lowest energy orbit in hydrogen into two energy states that differ by about one part in 10^{-7}. Transitions between the two orbits (energies) result in the emission of photons of wavelength approximately equal to 21 cm. A representation of the "21 centimeter line" was engraved onto the plaque carried by the *Pioneer 10* spacecraft launched by NASA in 1972, in the hope that some other intelligent civilization would intercept it and recognize that we humans know a thing or two.

★ What's Important 26-5

Atoms are composed of electrons orbiting a dense, positive nucleus, following an organized structure. Electrons that make a transition from one allowed orbit to another either radiate or absorb a photon of a well-defined energy, giving rise to the lines in emission and absorption spectra.

26-6 Atoms: The Bohr Model

The force that pulls the Moon toward Earth is directed toward Earth's center, yet the Moon does not fall into Earth. In the same way, the negatively charged electron in an atom experiences a Coulomb force that points directly toward the positively charged protons at the center of the atom. Why does the electron orbit rather than fall into the nucleus? You likely have an intuitive answer, that both the Moon and the electron *orbit* rather than fall in. (Although in a real sense each *is* falling in, but it is always missing the target!) The Danish physicist Niels Bohr based his model of the hydrogen atom, for which received the Nobel Prize in Physics in 1922, on the physics of atomic orbits. The Bohr model provides a theoretical foundation for the physics of atomic spectra that we encountered in the last section.

Bohr made two fundamental assumptions in describing the orbit of an electron around a positive atomic nucleus. First, he modeled the orbit as uniform circular motion, that is, as an electron moving at constant speed in a circular path. This is not exactly correct, but it is more than satisfactory to provide a broad and correct understanding of the atom. Bohr's second assumption was that only specific values are allowed for the angular momentum associated with each orbiting electron. We will return to this second assumption shortly.

> ## ❗ Watch Out
>
> ### Spectral lines are more correctly modeled by treating electron orbits as elliptical.
>
> Bohr's model, in which electrons follow circular orbits of fixed radii around the hydrogen nucleus, provides a good approximation to the hydrogen atom. However, studies of heavier elements, and ultimately careful study of the hydrogen spectrum, showed subtle complexities that could not be predicted by Bohr's model. In particular, individual spectral lines are in fact made up of a number of separate lines that, while closely spaced, are nevertheless distinct. One aspect of corrections to Bohr's model proposed by the German theoretical physicist Arnold Sommerfeld was to consider the electron orbits as ellipses. This introduced a splitting of the single energy levels of Bohr's atom into a group of finely separated energies, an effect known as *fine structure*. The fine structure accounts for the splitting of a given spectral line into a group of closely spaced spectral lines.

Bohr considered a single electron orbiting a nucleus of charge $+Ze$, where Z, the *atomic number of an atom*, is the number of protons in the nucleus. The orbiting electron experiences only one force, the Coulomb attraction between it and the protons in the atomic nucleus. The Coulomb force on an electron orbiting at radius r is, from Equation 16-1,

$$\vec{F} = \frac{k(-e)(Ze)}{r^2}\hat{r} = -\frac{kZe^2}{r^2}\hat{r}$$

Newton's second law (Equation 4-1) requires that the force equal the mass of the electron m_e multiplied by the acceleration it experiences, which for an object in uniform circular motion at speed v is, from Equation 3-34,

$$\vec{a_r} = \frac{v^2}{r}(-\hat{r})$$

so

$$-\frac{kZe^2}{r^2}\hat{r} = -m_e\frac{v^2}{r}\hat{r}$$

By rearranging we find that

$$\frac{kZe^2}{r} = m_e v^2 \tag{26-23}$$

We set this aside for a moment, and examine Bohr's second assumption, a requirement that only specific values are allowed for the angular momentum associated with the electron. Bohr recognized that the dimensions of Planck's constant

h are those of angular momentum, so he constrained the electron's angular momentum to be only multiples of h. Specifically, this requirement is

$$L = n\frac{h}{2\pi} = n\hbar \tag{26-24}$$

where L is the electron's angular momentum and n is any integer starting from 1. The constant $\hbar$ (pronounced "h bar") is defined to be h divided by 2π. Including a factor of 2π offers mathematical advantages when working with physics that involves circles.

To relate angular momentum to Equation 26-23, we express L in terms of the mass of the electron m_e, its speed v, and its distance from the center of the atom r using Equation 8-34. Because the radial vector is perpendicular to the velocity vector, the sine term in the cross product in Equation 8-34 equals one, so

$$L = rm_e v \tag{26-25}$$

Equation 26-23 can be written

$$\frac{kZe^2}{r} = \frac{r^2 m_e^2 v^2}{m_e r^2} = \frac{L^2}{m_e r^2}$$

Bohr's requirement that the electron's angular momentum is an integer multiple of $\hbar$ then gives

$$\frac{kZe^2}{r} = \frac{(n\hbar)^2}{m_e r^2}$$

or

$$r_n = \frac{n^2 \hbar^2}{m_e kZe^2}, \text{n} = 1, 2, 3, \ldots \tag{26-26}$$

We add the subscript "n" to the variable r to indicate that the radius can take on only specific values, and that the allowed values of radius depend on n. Because the values of r_n are proportional to an integer, the orbital radii of electrons in an atom are quantized.

Notice that both n and Z in Equation 26-26 are dimensionless. Because r is a distance, all of the other terms on the right-hand side, taken as they appear in the equation, must have dimensions of distance as well. This distance, usually written as a_0 and called the **Bohr radius**, is

$$a_0 = \frac{\hbar^2}{m_e ke^2} \tag{26-27}$$

Using the best measured values for $\hbar$, m_e, k, and e, the value of the Bohr radius a_0 is approximately

$$a_0 = 0.529 \times 10^{-10} \, \text{m} = 0.0529 \, \text{nm}$$

In terms of a_0, the quantized radii of the electron orbits (Equation 26-26) is

$$r_n = \frac{n^2}{Z} a_0, \text{n} = 1, 2, 3, \ldots \tag{26-28}$$

The integer n identifies the orbit, where the n equal to 1 orbit is the closest to the nucleus. Because every element is distinguished by the number of protons it carries, the atomic number Z specifies a particular element. So, for example, setting Z equal to 1 gives the radii of the electron orbits in a hydrogen atom, and setting Z equal to 1 and n equal to 1 gives the radius of the first electron orbit in hydrogen. This is the normal state, usually referred to as the ground state, of hydrogen. Moreover, notice that the radius of the ground state orbit of hydrogen is the Bohr radius. In other words, the radius of a typical hydrogen atom is about 0.05 nm. We can also conclude from Equation 26-28 together with the value of the Bohr radius that, in general, atoms are on the order of one nanometer in radius. The Bohr model sets the scale for atomic sizes.

What is the energy of an electron orbiting an atomic nucleus according to the Bohr model? Certainly the energy of the electron is the sum of its kinetic energy and its Coulomb potential energy, so

$$E = \frac{1}{2} m_e v^2 + \frac{k(-e)(Ze)}{r_n}$$

where we used Equation 17-10 to arrive at the potential energy term. We will substitute Equation 26-26 for r_n, but first, notice that the kinetic energy term can be written in terms of angular momentum in a way similar to our approach in developing the relationship for r_n. Using Equation 26-25, the kinetic energy term becomes

$$\frac{1}{2} m_e v^2 = \frac{1}{2} \frac{L^2}{m_e r_n^2} = \frac{n^2 \hbar^2}{2 m_e r_n^2}$$

so the total electron energy is

$$E = \frac{n^2 \hbar^2}{2 m_e r_n^2} - \frac{k Z e^2}{r_n}$$

Substituting Equation 26-26 gives the energy in terms of only the physical constants and the counting integer n,

$$E = \frac{n^2 \hbar^2}{2 m_e} \left(\frac{m_e k Z e^2}{n^2 \hbar^2} \right)^2 - k Z e^2 \frac{m_e k Z e^2}{n^2 \hbar^2}$$

Simplifying this is straightforward when you notice that the numerator of both terms is $m_e(kZe^2)^2$, and that both terms have $n^2\hbar^2$ in the denominator, so

$$E = \frac{m_e(kZe^2)^2}{2n^2\hbar^2} - \frac{m_e(kZe^2)^2}{n^2\hbar^2}$$

or

$$E_n = -\frac{m_e(kZe^2)^2}{2n^2\hbar^2}, \, n = 1, 2, 3, \dots \tag{26-29}$$

We add the subscript "n" to the variable E to indicate that the orbital energy of the electron can take on only specific values, and that the allowed values of energy depend on n. Because the values of E_n are inversely proportional to the square of an integer, the orbital energy of electrons in an atom is quantized. This is in agreement with our conclusion in the last section that for atomic spectral lines to occur only at specific, and always the same, wavelengths, electrons must orbit the hydrogen nucleus with specific, well-defined energies.

The value of energy of an electron in orbit around an atomic nucleus is negative, but it gets closer and closer to zero for increasing values of n. In other words, the lowest energy orbit is the one closest to the nucleus, as we would expect. You might imagine that the negative values of energy suggest that the electron is at the bottom of a hole, and its orbital energy is insufficient to allow it to escape. That the electron energy is taken as negative emphasizes that the electron is bound to the nucleus, and that energy must be supplied in order to either move the electron to a higher orbit or to break the electron free from the nucleus altogether.

Because E is an energy, and because both n and Z are dimensionless, all of the other terms on the right-hand side, taken as they appear in the equation, must have dimensions of energy as well. This energy, usually written as E_0 and called the **Rydberg energy**, is

$$E_0 = \frac{m_e(ke^2)^2}{2\hbar^2} \qquad (26\text{-}30)$$

Using the best measured values for $\hbar$, m_e, k, and e, the value of the Rydberg energy E_0 is approximately

$$E_0 = 2.18 \times 10^{-18}\,\text{J}$$

The value of the Rydberg energy is often given in non-SI units; more commonly,

$$E_0 = 13.6\,\text{eV}$$

where eV is the abbreviation for electron volts. The energy that an electron acquires when it experiences a potential of 1 V is 1 eV, or approximately 1.602×10^{-19} J. To give you an idea of the amount of energy 1 eV represents, a single visible light photon carries between about 1.5 and 3 eV.

Equation 26-30 for the quantized electron orbital energy can be written in terms of the Rydberg energy E_0:

$$E_n = -\frac{Z^2}{n^2}E_0, \text{n} = 1, 2, 3, \ldots \qquad (26\text{-}31)$$

Again, the integer n identifies the orbit and the atomic number Z specifies a particular element. Setting Z equal to 1 and n equal to 1 therefore tells us that the energy of the ground state of hydrogen is -13.6 eV. We can also conclude from Equation 26-31 together with the value of the Rydberg energy that, in general, the energy of electrons in orbit around an atomic nucleus is between about 10 eV and, for the largest elements, 10^5 eV. The Bohr model sets the scale for atomic electron energies.

What does the Bohr model predict for the atomic spectral lines of hydrogen? Every line results from a transition between two electron orbits, that is, the energy of the emitted photon is the energy difference ΔE between two allowed orbits. From Equation 22-1, the wavelength of the photon is then

$$\lambda = \frac{hc}{\Delta E}$$

Recalling that the Rydberg formula is written in terms the inverse of the photon wavelength, we write

$$\frac{1}{\lambda} = \frac{\Delta E}{hc} \qquad (26\text{-}32)$$

Let's consider the transition of an electron from a higher orbit m down to a lower orbit n. We can determine the energy difference between the two orbits by applying Equation 26-31. Atomic number Z equals 1 for hydrogen, so

$$\Delta E = -\frac{1}{m^2}E_0 - \left(-\frac{1}{n^2}\right)E_0$$

or

$$\Delta E = \left(\frac{1}{n^2} - \frac{1}{m^2}\right)E_0$$

Substituting this into Equation 26-32 yields

$$\frac{1}{\lambda} = \frac{E_0}{hc}\left(\frac{1}{n^2} - \frac{1}{m^2}\right)$$

Compare this to the Rydberg formula, Equation 26-22—it has exactly the same form!

How does the value of E_0/hc compare to the value of R_H, which equals approximately 1.10×10^7 m^{-1}? We can estimate E_0/hc using known values of E_0, h, and c in SI units:

$$\frac{E_0}{hc} = \frac{2.18 \times 10^{-18} \text{ J}}{(6.63 \times 10^{-34} \text{ Js})(3.00 \times 10^8 \text{ m/s})} = 1.10 \times 10^7 \text{ m}^{-1}$$

The ratio E_0/hc equals R_H. We conclude that the Rydberg formula for the spectral lines of hydrogen is derived directly from the Bohr model.

Example 26-8 Ground State of Lithium

Find the (a) radius, (b) energy, and (c) speed of an electron in the ground state of lithium.

SET UP

Consult the periodic table to find that lithium has atomic number Z equal to 3. The ground state of any atom corresponds to n equal to 1, so for parts (a) and (b) we apply Equations 26-28 and 26-31 with those values of Z and n. To find the speed, notice that our exploration of the Bohr model depends on uniform circular motion, which in turn depends on speed. For example, we set the Coulomb force equal to the product of the electron mass and acceleration, which led to Equation 26-23:

$$\frac{kZe^2}{r} = m_e v^2$$

The speed of the electron, after substituting the quantized Bohr radius from Equation 26-26, is

$$v^2 = \frac{kZe^2}{m_e} \frac{m_e kZe^2}{n^2 \hbar^2} = \frac{(kZe^2)^2}{n^2 \hbar^2}$$

or

$$v = \frac{kZe^2}{n\hbar} \tag{26-33}$$

SOLVE

(a) The radius of the electron orbit is given by Equation 26-28. With Z equal to 3 and n equal to 1, the electron's radius is

$$r_n = \frac{1^2}{3} a_0 = \frac{a_0}{3} = \frac{0.0529 \text{ nm}}{3} = 0.0176 \text{ nm}$$

(b) The energy of the electron orbit is given by Equation 26-31. With Z equal to 3 and n equal to 1, the electron's orbital energy is

$$E_n = -\frac{3^2}{1^2} E_0 = -9E_0 = -9(13.6 \text{ eV}) = -122 \text{ eV}$$

(c) Using Equation 26-33,

$$v = \frac{kZe^2}{n\hbar} = \frac{(8.99 \times 10^9 \text{ N} \cdot \text{m}^2/\text{C}^2)(3)(1.60 \times 10^{-19} \text{ C})^2}{1(6.63 \times 10^{-34} \text{ Js}/2\pi)}$$

or

$$v = 6.54 \times 10^6 \text{ m/s}$$

REFLECT

The radius of an electron in the ground state of lithium is one-third the radius of an electron in the ground state of hydrogen. Because there are three protons in the lithium nucleus, compared to one for hydrogen, the Coulomb force on the electron is greater, so it is reasonable that the orbital radius is smaller. Also, notice that the speed of the electron is about 50 times smaller than the speed of light in a vacuum.

Practice Problem 26-8 Find the (a) radius, (b) energy, and (c) speed of an electron in the n equal to 4 orbit of hydrogen.

Although the Bohr model successfully predicts atomic spectra and other phenomena associated with hydrogen atoms, it nevertheless leaves us with an outstanding question. Why should the electron orbits be quantized in multiples of $\hbar$? For the answer we turn back to Louis de Broglie. Recall that de Broglie postulated that particles, such as electrons, have a wavelike nature. As a result, the electron's wave interferes with itself as the electron makes multiple revolutions around the nucleus. The only allowed orbits are those for which the electron waves constructively interfere, and form standing waves.

The wavelength of a nonrelativistic electron (from Equation 26-19) is

$$\lambda = \frac{h}{m_e v}$$

For an electron orbiting an atomic nucleus we put this in terms of angular momentum by making use of Equation 26-25:

$$\lambda = \frac{hr}{m_e vr} = \frac{hr}{L}$$

To connect this more directly with the circular path of the orbit, we introduce a factor of 2π:

$$\lambda = \frac{hr}{m_e vr} = \frac{2\pi\hbar r}{L}$$

Here's where it gets interesting! Let L be an integer multiple of $\hbar$, as Bohr required; then

$$\lambda = \frac{2\pi\hbar r}{n\hbar}$$

or

$$n\lambda = 2\pi r$$

We recognize $2\pi r$ as the circumference of the orbital path of the electron. An integer number of full electron waves fit into the circumference of the orbit only when Bohr's requirement is met. In a real sense, it is the wavelike nature of particles that results in the quantization of the energy of atomic electrons.

Bohr's model of the atom works relatively well for hydrogen, and can be applied reasonably well to a singly ionized helium atom, that is, an atom with Z equal to 2 but only one electron. For more complicated atoms, it isn't possible to make calculations of energy levels using Bohr's physics. However, the general picture of the atom it provides, with electrons in quantized energy states, applies to all atoms. This has been confirmed experimentally, perhaps most notably by German physicists James Franck and Gustav Hertz in 1914. (Franck and Hertz were awarded the Nobel Prize in Physics in 1925 for their work. Gustav Hertz was a nephew of Heinrich Hertz, whom we encountered earlier.)

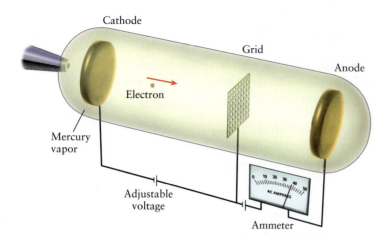

Figure 26-19 As the voltage between the cathode and the mesh grid in the Franck-Hertz experiment increases, the anode current increases. When the kinetic energy of the accelerated electrons equals the energy required to excite an electron in an Hg atom to a higher energy level, accelerated electrons lose energy. Not as many reach the anode, so at the corresponding cathode-grid voltages, the anode current decreases.

In what is now known as the Franck–Hertz experiment, Franck and Hertz used a device similar to the one shown schematically in **Figure 26-19** to measure the effect of bombarding atoms in a gas with electrons. The cathode is heated in order to give off electrons, which are accelerated by a variable voltage toward the mesh grid. Some electrons pass through the grid and arrive at the anode, where the electron current is detected by the ammeter. Notice that a voltage is also applied between the grid and the anode, which acts against the electrons; only electrons that carry sufficient energy as they pass the grid will make it to the anode. These electrical components sit inside a tube filled with low pressure mercury vapor, so collisions between an electron and a mercury atom can occur. Franck and Hertz used mercury vapor because the spectral lines of low pressure mercury gas was well studied; one prominent excitation line has a wavelength of 253.7 nm, corresponding to a photon of energy equal to 4.9 eV. This is the energy emitted when an atomic electron in one of the excited states of a mercury atom to its ground state.

Example 26-9 A Mercury Line

Show that a photon of wavelength equal to 253.7 nm carries kinetic energy equal to 4.9 eV.

SET UP
To find the energy E of a photon of a given wavelength λ, we apply Equation 22-1:

$$E = \frac{hc}{\lambda}$$

SOLVE
To keep the units straight, we use λ in meters, and take Planck's constant as 4.14×10^{-15} eV · s. Then

$$E = \frac{(4.14 \times 10^{-15} \text{ eVs})(3.00 \times 8 \text{ m/s})}{253.7 \times 10^{-9} \text{ m}} = 4.90 \text{ eV}$$

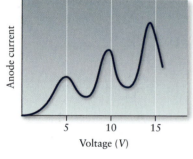

Figure 26-20 As the voltage between the cathode and the mesh grid in the Franck–Hertz experiment increases, the anode current increases. When the kinetic energy of the accelerated electrons equals the energy required to excite an electron in an Hg atom to a higher energy level, accelerated electrons lose energy. Not as many reach the anode at the corresponding cathode-grid voltages, and the anode current decreases.

REFLECT
Yes, the photon corresponding to the 253.7 nm spectral line of Hg is 4.9 eV.

What happens in the collision of an electron and a mercury atom in the Franck–Hertz tube? In general, the more kinetic energy an electron carries as it approaches the grid, the more likely it will make it through. So anode current should grow as the voltage between the cathode and the grid is increased. However, the Bohr model of the atom suggests that because the energy levels of electrons in Hg (and all) atoms are quantized, when the kinetic energy of the incident electron equals the energy difference between two Hg energy levels, the Hg atom can absorb the electron's energy. When this happens it is less likely that the electron will reach the anode, so the anode current should decrease. Consider, then, a typical current versus voltage curve from the Franck–Hertz experiment, shown in **Figure 26-20**. The general trend, as we would expect, is that the anode current increases as the voltage is increased. But the current drops dramatically for voltages immediately greater than certain values of the accelerating voltage, in this case, right above 4.9 V, 9.8 V, and 14.7 V. Notice that 9.8 V and 14.7 V are multiples of 4.9 V, and that 4.9 V corresponds to an electron energy of 4.9 eV—the very same energy of the well-known mercury spectral line!

When the cathode-grid voltage is set to 4.9 eV, the kinetic energy of electrons that leave the cathode reaches 4.9 eV just as they approach the mesh grid. (Recall that the electron volt is the amount of energy gained by an electron when it experiences a potential difference of 1 V.) For this reason, electrons that collide with a mercury atom in close proximity to the grid give up their energy in the process of causing an electron in the atom to jump to a higher energy level. But because the collision occurs close to the grid, there is no opportunity for the electron to undergo another acceleration, in other words, there is no opportunity for the electron to acquire enough energy to reach the anode. For this reason, when the cathode-grid voltage is set to 4.9 eV, so that the collisions between electrons and Hg atoms occur near the grid with electron kinetic energy equal to 4.9 eV, the number of electrons reaching the anode decreases. In addition, when the voltage is set to 9.8 eV, accelerated electrons that collide with an Hg atom halfway between the cathode and the grid lose their energy, and then accelerate up to a kinetic energy of 4.9 eV again by the time they reach the grid. A collision there once again results in the electron transferring its energy to an Hg atom and being unable to reach the anode. A similar phenomenon occurs when the voltage is set to any integer multiple of 4.9 eV. The rise and fall in anode current versus cathode-grid voltage is shown dramatically in Figure 26-20. Underlying is quantization of electron energy levels in the atom, as predicted by Niels Bohr.

As powerful as the Bohr model is at providing an understanding of the atom and atomic spectra, it does not tell the whole story. The Bohr atom treats the electrons that orbit atomic nuclei as particles, yet as we know and as the scientists of early twentieth century were beginning to learn, electrons exhibit the properties of waves. No theoretical description of the atom can be complete unless it accounts for the wave nature of the electrons. The theoretical underpinning of our understanding of atoms that includes their wave properties is found in an equation developed by Austrian physicist Erwin Schrödinger. The **Schrödinger equation** relies on matter waves to describe the state of a system as a function of time, much like Newton's laws do while treating physical systems as particles.

Perhaps the most notable difference between Schrödinger's wave, or quantum mechanical, description and Newton's classical description of physics is seen in the ability to specify the position and velocity of objects. At any instant of time we can identify a specific position in space and a specific velocity vector for any particle in a system, for example, an electron orbiting an atomic nucleus. A wave, however, is not localized in space, even after applying advanced mathematical notions that squeeze it into a relative well-defined region of space. The result is that while Newton's laws predict the position and velocity of an object, the Schrödinger equation predicts the *probability* of finding a certain value of position or velocity. For this reason, electrons in the quantum model of the atom are described not as tiny marbles orbiting the nucleus at fixed radii, but rather as a charge distribution. This distribution, sometimes called a *probability cloud*, gives the probability of finding the electron at any given position; the denser the cloud in some region the more likely it is that the electron will be found there. The probability cloud associated with the ground state of hydrogen is shown in **Figure 26-21a**. The more dense the color in any region in this figure, the more likely it is that the electron will found in that region. The probability distribution is spherically symmetric, that is, it only varies as a function of radius from the center of the nucleus. For that reason we can also express the same information in a curve of probability versus radius, as in **Figure 26-21b**. Notice that the most probable radius of an electron in the ground state of a hydrogen atom is a_0, the Bohr radius.

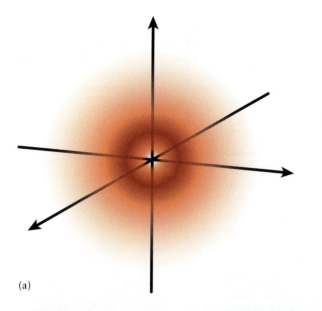

(a)

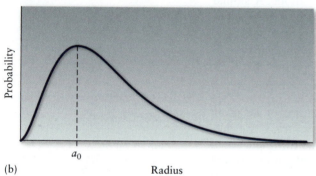

(b)

Probability

a_0

Radius

Figure 26-21 (a) The ground state electron in a hydrogen atom can be found anywhere relative to the nucleus. The darker (denser) the color in a region of the probability cloud, the more likely it is that the electron will be found in that region. (b) For the ground state electron in hydrogen, the probability cloud is spherically symmetric, so we can represent the probability of finding the electron as a function of radius from the center of the nucleus. The probability peaks at a radius equal to a_0, the Bohr radius.

The Bohr atom employs a single integer, or quantum number, to describe electron states. For the Bohr atom this integer is n, which determines the energy level of the electron. In the quantum view of the atom, four **quantum numbers** are required. These are n, the principal quantum number; ℓ, the angular momentum (or orbital) quantum number; m_ℓ, the magnetic quantum number; and m_s, the electron spin quantum number. A set of specific values of each of the four quantum numbers completely describes the state of an electron in an atom.

The principal quantum number plays a role in the quantum atom similar to the one that n plays in the classical Bohr atom; in particular, it specifies the energy level or electron **shell**. The lowest energy state, or ground state, corresponds to n equal to 1.

The angular momentum quantum number arises due to the angular momentum the electron carries. For any value of n, ℓ varies in integer steps from 0 to $n - 1$; each value of ℓ specifies a subshell, or electron **orbital**, within the energy level specified by n. The shape of each orbital is different. By convention we refer to the orbitals with letters rather than integer numbers; the values of ℓ equal to 0, 1, 2, 3, 4, and 5 correspond to the orbitals s, p, d, f, g, and h. It is also standard to refer to an electron subshell by giving both n and the orbital letter code together. For example, an electron in the p subshell of the n equal to 2 energy level is said to be in the 2p subshell. Finally, recall our discussion of fine structure in atomic energy levels at the end of the last section. The slight dependence of the electron's energy on the value of ℓ is in large measure responsible for this splitting.

The magnetic quantum number m_ℓ, which arises due to the interaction of a moving charge—the electron—in the presence of the magnetic field of the nucleus, specifies an orientation of an electron's subshell. The larger the value of ℓ, the more orientations are allowed; m_ℓ can take integer values between $-\ell$ and $+\ell$, including 0.

Figure 26-22 shows the shapes a number of electron orbitals.

Electrons have an intrinsic characteristic called spin, which is akin to angular momentum. Even electrons that do not orbit an atomic nucleus possess spin, which can take on one of two values, often called spin "up" and spin "down." To fully describe an atomic electron, then, we must also specify its spin state. This is described by the electron spin quantum number m_s.

Let's do some counting. For the n equal to 1 shell only one value of ℓ (equal to 0, which is the s orbital), and therefore only one value of m_ℓ, is allowed. Two values of m_s are always possible, so the maximum number of electrons that can occupy the n equal to 1 shell is 1×2, the product of the number of possible m_ℓ values and the number of possible m_s values. That is, two electrons can occupy the 1s state. For n equal to 2, ℓ is allowed to be either 0 or 1. When ℓ equals 0, m_ℓ equals 0, so including the factor of 2 for the electron spin quantum number, one possible value of m_ℓ and two possible values of m_s means two electrons can occupy the 2s state. However, when ℓ equals 1 (the p orbital), m_ℓ can be -1, 0, or $+1$. Including the factor of 2 for the electron spin quantum number, that means that the number of electrons that can occupy the 2p state is 3×2, or 6. Because two electrons can occupy the 2s state

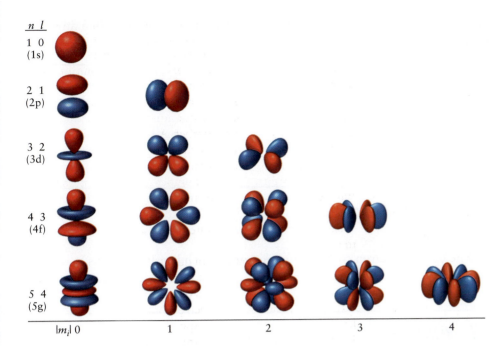

Figure 26-22 The probability cloud for each atomic suborbital, specified by a specific value of the quantum numbers n, ℓ, and m_ℓ, has a different shape.

and six electrons can occupy the 2p state, an atom can have at most eight electrons in the n equal to 2 energy level.

The pattern above repeats for n equal to 3, up through ℓ equals 2, the d orbital. For this orbital five values $(-2, -1, 0, +1, +2)$ are possible for m_ℓ, so the maximum number of electrons in the orbital is 5×2 or 10. So two electrons can occupy 3s, six can occupy 3p, and ten can occupy 3d. The maximum number of electrons in the n equal to 3 energy shell of an atom is therefore 18.

The arrangement of electrons, according to how many occupy which orbitals, is directly correlated to the chemical properties of the elements. Consider the 18 lightest elements and their electron configurations, listed in Table 26-1. (In the table, the number of electrons that occupy a particular orbital is given as a superscript, for example, $2p^4$ indicates that four electrons occupy the 2p orbital.) In hydrogen, lithium, and sodium (as well as potassium, rubidium, cesium, and francium), the outermost, or valence, electron is a single electron in an s orbital. These *alkali metals* share common chemical properties, and occupy a single column in the periodic table. (Take a look!)

Three of the *noble gases* are listed in Table 26-1. These are helium, neon, and argon; in each, the outermost subshell is completely full. As a result all electrons are relatively tightly bound, so these elements do not easily gain, lose, or share electrons. For that reason the noble gases are relatively inert. Helium, neon, and argon, as well as the other noble gases, occupy a single column in the periodic table.

Halogens are elements that are highly reactive, that is, they easily form bonds with certain other elements, especially the alkali metals, to form molecules. Two halogens, fluorine and chlorine, are listed in Table 26-1. The outermost shell in both is one electron short of being full, which means that an atom of one of these elements can readily share an electron with another atom. This is particularly true for an atom of an alkali metal,

Table 26-1   Electron Configurations of Light Elements

Atomic number	Element	Electron configuration
1	Hydrogen (H)	$1s^1$
2	Helium (He)	$1s^2$
3	Lithium (Li)	$1s^2 2s^1$
4	Beryllium (Be)	$1s^2 2s^2$
5	Boron (B)	$1s^2 2s^2 2p^1$
6	Carbon (C)	$1s^2 2s^2 2p^2$
7	Nitrogen (N)	$1s^2 2s^2 2p^3$
8	Oxygen (O)	$1s^2 2s^2 2p^4$
9	Fluorine (F)	$1s^2 2s^2 2p^5$
10	Neon (Ne)	$1s^2 2s^2 2p^6$
11	Sodium (Na)	$1s^2 2s^2 2p^6 3s^1$
12	Magnesium (Mg)	$1s^2 2s^2 2p^6 3s^2$
13	Aluminum (Al)	$1s^2 2s^2 2p^6 3s^2 3p^1$
14	Silicon (Si)	$1s^2 2s^2 2p^6 3s^2 3p^2$
15	Phosphorus (P)	$1s^2 2s^2 2p^6 3s^2 3p^3$
16	Sulfur (S)	$1s^2 2s^2 2p^6 3s^2 3p^4$
17	Chlorine (Cl)	$1s^2 2s^2 2p^6 3s^2 3p^5$
18	Argon (Ar)	$1s^2 2s^2 2p^6 3s^2 3p^6$

which has one valence electron that is easily shared. So bring a sodium atom near a chlorine atom and they will readily bond to form NaCl—table salt!

Got the Concept 26-3
Noble Gases

Identify all of the noble gases.

Got the Concept 26-4
Zinc

Write out the electron configuration for zinc, which has 30 electrons.

! Watch Out

As atomic number increases, electron orbitals do not fill in a continuous fashion.

The configuration of the 18 electrons of Ar, the heaviest element listed in Table 26-1, is $1s^2 2s^2 2p^6 3s^2 3p^6$. The next heaviest element is potassium; the configuration of the 19 electrons in a K atom is $1s^2 2s^2 2p^6 3s^2 3p^6 4s^1$. Notice that the additional electron does not occupy the 3d orbital, although d follows p in our ordering (s, p, d, f, and so on). Orbitals fill according to the increase in energy required and, for that reason, do not fill according to counting up linearly in ℓ (orbital letters) and m_ℓ. Instead, orbitals fill in this order:

1s, 2s, 2p, 3s, 3p, 4s, 3d, 4p, 5s, 4d, 5p, 6s, 4f, 5d, 6p, 7s, 5f, 6d, 7p

Notice that 4s and not 3d follows the 3p orbital in the sequence, which is why the valence electron in K is in a 4s orbital.

✱ What's Important 26-6

Atomic spectral lines arise from transitions between two electron orbits, which according to the Bohr model of the atom are quantized. Because only certain orbital energies are allowed, the energy of an emitted photon, equal to the energy difference between two allowed orbits, is also restricted to certain allowed values.

Answers to Practice Problems

26-3 $\lambda = \dfrac{hc}{E}$, so $[\lambda] = \dfrac{[h][c]}{[E]} = \dfrac{\text{Js} \cdot \text{m/s}}{\text{J}} = \text{m}$

26-4 0.0040 nm

26-5 1.3×10^{-11} m

26-6 1.3×10^{-35} m/s; no!

26-7 $\lambda_{\text{long}} = 4050$ nm for $m = 5$, $\lambda_{\text{short}} = 1460$ nm for $m = \infty$

26-8 0.846 nm, -0.850 eV, 5.45×10^5 m/s

Answers to Got the Concept Questions

26-1 The photon of wavelength 5×10^{-3} nm undergoes Compton scattering. The other two do not. In the second and third cases, the change in the photon's wavelength as described by the Compton scattering relationship (Equation 26-18) would be imperceptible. The photon of the smallest wavelength carries the most energy. In this example, the photons of lower energy do not Compton scatter.

26-2 The Balmer series was discovered before others in the hydrogen spectrum because the wavelengths in the Balmer series lie mostly in the visible part of the electromagnetic spectrum. For this reason little specialized equipment is necessary to observe the lines of the Balmer series. That is not the case for lines in the UV (the Lyman series) or IR (the Paschen series).

26-3 Look down the rightmost column in the periodic table. In addition to helium, neon, and argon (listed above), the noble gases include krypton, xenon, and radon.

26-4 $1s^2 2s^2 2p^6 3s^2 3p^6 3d^{10} 4s^2$

SUMMARY

Topic	Summary	Equation or Symbol	
absorption spectrum	An absorption spectrum is formed when light of a range of wavelengths passes through a gas or diffuse material. Light of energies that correspond to the transitions between allowed atomic orbits is absorbed, resulting in narrow, dark lines in an otherwise smooth spectrum.		
blackbody (or ideal blackbody)	An ideal blackbody absorbs all of the energy that impinges on it and, when in thermal equilibrium with the environment, emits all of the energy it absorbs.		
Bohr radius	The Bohr radius a_0 is the radius of the lowest energy (ground state) electron orbit in hydrogen, as predicted by the Bohr model of the atom.	$a_0 = \dfrac{\hbar^2}{m_e k e^2}$	(26-27)
Compton scattering	A photon, typically an x ray, undergoes Compton scattering in an elastic collision with an electron. Momentum is conserved in the Compton scattering process, an observation that provided the first confirmation that photons carry momentum.		
Compton wavelength	The proportionality constant between the change in a photon's wavelength and the angle at which it scatters in Compton scattering is the Compton wavelength λ_C. Only photons of wavelengths on the order of λ_C undergo Compton scattering.	$\lambda_C = \dfrac{h}{m_e c}$	(26-17)
de Broglie wavelength	Particles such as electrons have wave properties. Their wavelength, known as the de Broglie wavelength, is inversely proportional to momentum p.	$\lambda = \dfrac{h}{p}$	(26-19)
emission spectrum	An emission spectrum is formed when the atoms in a gas or material are excited and glow. Only light of energies that corresponds to the transitions between allowed atomic orbits is emitted, resulting in a spectrum that consists only narrow lines of color on a black background.		

orbital	Electron subshells, or orbitals, arise due to the angular momentum an electron carries. In increasing order, orbitals are classified as s, p, d, f, g, and h.	
photoelectric effect	When light of sufficient energy strikes certain materials, electrons can absorb the energy and be ejected from the surface, a phenomenon known as the photoelectric effect.	
quantum numbers	Four quantum numbers are required to describe the state of an atomic electron. These are n, the principal quantum number; ℓ, the angular momentum (or orbital) quantum number; m_ℓ, the magnetic quantum number; and m_s, the electron spin quantum number. A specific value of each of the four quantum numbers completely describes the state of an electron in an atom.	$n = 1, 2, 3, \ldots$ $\ell = 0, 1, \ldots (n-1)$ $m_\ell = -\ell, -(\ell-1), \ldots, (\ell-1), \ell$ $m_s = \text{up, down}\left(\text{or} +\frac{1}{2}, -\frac{1}{2}\right)$
Rydberg formula	The wavelengths λ of the spectral lines of hydrogen are described by the Rydberg formula. The relationship depends on integers m and n; m corresponds to the higher energy orbit and n to the lower energy orbit in the atomic transition that corresponds to the line. The proportionality constant R_H is known as the Rydberg constant.	$\dfrac{1}{\lambda} = R_H\left(\dfrac{1}{n^2} - \dfrac{1}{m^2}\right)$ (26-22)
Rydberg energy	The Rydberg energy E_0 is ground state energy—the energy of the lowest energy electron orbit—in hydrogen, as predicted by the Bohr model of the atom.	$E_0 = \dfrac{m_e(ke^2)^2}{2\hbar^2}$ (26-30)
Schrödinger equation	The Schrödinger equation describes physical states as waves, and allows the determination of the change of the state or system as a function of time. The Schrödinger equation, and the wave formalism in general, provides information about the probability that a particle will be found in a region of space at a given time, rather than a specific value of its position.	
shell	An electron's shell in an atom is the energy level it occupies.	
spectral radiancy	Spectral radiancy is the energy emitted per time per area of a surface.	R
stopping potential	Electrons with sufficient kinetic energy can flow in a direction opposed by a negative potential. For any given kinetic energy there is some maximum negative potential, the stopping potential, that the electron can overcome.	V_0
transition	An electron can only absorb or emit a photon the energy of which would result in the electron making a transition from one allowed atomic orbit to another.	
work function	The work function, which depends on the material or a molecule that resides in a surface, is the minimum amount of energy required to eject an electron from the surface.	Φ_0

QUESTIONS AND PROBLEMS

In a few problems, you are given more data than you actually need; in a few other problems, you are required to supply data from your general knowledge, outside sources, or informed estimate.

Interpret as significant all digits in numerical values that have trailing zeros and no decimal points.

For all problems, use $g = 9.8 \text{ m/s}^2$ for the free-fall acceleration due to gravity. Neglect friction and air resistance unless instructed to do otherwise.

• Basic, single-concept problem
•• Intermediate-level problem, may require synthesis of concepts and multiple steps
••• Challenging problem
SSM Solution is in Student Solutions Manual

Conceptual Questions

1. •Explain why the difficulty in describing the data from the measurement of blackbody radiation was labeled as the ultraviolet catastrophe.

2. •How does the intensity of light from a blackbody change when its temperature is increased? What changes occur in the body's radiation spectrum?

3. •According to classical electromagnetic theory, an accelerated charge emits electromagnetic radiation. What would this mean for the electron in the Bohr atom? What would happen to its orbit?

4. •Are there quantities in classical physics that are quantized?

5. •What is the shortest wavelength of electromagnetic radiation that can be emitted by a hydrogen atom? SSM

6. •Prior to Einstein's description of the photoelectric effect, light was thought to act like a wave. Explain why the existence of a cutoff frequency favors a description of light as a particle instead.

7. •Describe how the number of photoelectrons emitted from a metal plate in the photoelectric effect would change if (a) the intensity of the incident radiation was increased, (b) the wavelength of the incident radiation was increased, and (c) the work function of the metal was increased.

8. •Is there a relativistic photoelectric effect? That is, is it possible to observe photoelectrons emitted from a metal plate to have relativistic speeds?

9. •Consider the photoelectric emission of electrons induced by monochromatic incident light. The incoming photons all have the same energy, but the emitted electrons have a range of kinetic energies. Why?

10. •A markedly nonclassical feature of the photoelectric effect is the fact that the energy of the emitted electrons doesn't increase as you increase the intensity of the light striking the metal surface. What change does occur as the intensity is increased?

11. •The Compton effect is practically unobservable for visible light. Why?

12. •Which of the two Compton scattering experiments more clearly demonstrates the particle nature of electromagnetic radiation: a collision of the photon with an electron or with a proton? Explain your answer.

13. •An electron and a proton have the same kinetic energy. Which has the longer wavelength? SSM

14. •Is the wavelength of an electron the same as the wavelength of a photon if both particles have the same total energy?

15. •Why do you think Bohr's model was originally designed for the element hydrogen?

16. •Why do you think that Bohr's model of the hydrogen atom is still taught in undergraduate physics classes?

17. •Why do we never observe the wave nature of particles for everyday objects such as birds or bumblebees, for example? SSM

18. •Does the de Broglie wavelength of a particle increase or decrease as its kinetic energy increases?

Multiple-Choice Questions

19. •An ideal blackbody is an object that
    A. absorbs most of the energy that strikes it and emits a little of the energy it generates.
    B. absorbs a little of the energy that strikes it and emits most of the energy it generates.
    C. absorbs half of the energy that strikes it and emits half of the energy it generates.
    D. absorbs all the energy that strikes it and emits all the energy it generates.
    E. neither absorbs nor emits energy except at ultraviolet ("black light") wavelengths.

20. •The color of light emitted by a hot object depends on
    A. the size of the object.
    B. the shape of the object.
    C. the material from which the object is made.
    D. the temperature of the object.
    E. the color of the object.

21. •Which photon has more energy?
    A. a photon of ultraviolet radiation
    B. a photon of green light
    C. a photon of yellow light
    D. a photon of red light
    E. a photon of infrared radiation SSM

22. •Light that has a wavelength of 600 nm strikes a metal surface and a stream of electrons is ejected from the surface. If light of wavelength 500 nm strikes the surface, the maximum kinetic energy of the electrons emitted from the surface will
  A. be greater.
  B. be smaller.
  C. be the same.
  D. be 5/6 smaller.
  E. be unmeasurable.

23. •In the Compton effect experiment, the change in a photon's wavelength depends on
  A. the scattering angle.
  B. the initial wavelength.
  C. the final wavelength.
  D. the density of the scattering material.
  E. the atomic number of the scattering material.

24. •The maximum change in a photon's energy when it undergoes Compton scattering occurs when its scatter angle is at
  A. 0°.
  B. 45°.
  C. 90°.
  D. 135°.
  E. 180°.

25. •As the scattering angle in the Compton effect increases, the energy of the scattered photon
  A. increases.
  B. stays the same.
  C. decreases.
  D. decreases by $\sin\theta$.
  E. increases by $\sin\theta$. SSM

26. •The de Broglie wavelength depends only on
  A. the particle's mass.
  B. the particle's speed.
  C. the particle's energy.
  D. the particle's momentum.
  E. the particle's charge.

27. •An element emits a spectrum that
  A. is the same as all other elements.
  B. is evenly spaced.
  C. is unique to that element.
  D. is evenly spaced and unique to that element.
  E. is indistinguishable from most other elements.

Estimation/Numerical Analysis

28. •What is the approximate size of an atom?

29. •Estimate the amount of blackbody energy radiated from your body. SSM

30. •On average, an electron will exist in any given state in the hydrogen atom for about 10^{-8} s before jumping to a lower level. Estimate the number of revolutions about the nucleus that an electron makes in 10^{-8} s.

31. •What is the typical range of work functions that you will encounter in the majority of situations in this textbook?

32. •Estimate the order of magnitude of atomic ionization energies. Compare this with the order of magnitude for the nuclear binding energy, which is in the megaelectron volt (MeV) range.

33. •Estimate the size of the electrostatic forces that hold the atom together compared to the size of the force due to gravity.

34. •Estimate the ratio of the mass of the proton to the mass of the electron.

35. •Estimate the speed of an electron at which the de Broglie wavelength approximates the size of an atom. SSM

36. •In the early twentieth century, Max Planck used an empirical mathematical equation to describe the data that was collected in measuring blackbody radiation. Here's the formula he conceived:

$$\frac{\Delta I}{\Delta\lambda} = \frac{2\pi hc^2}{\lambda^5}\left(\frac{1}{e^{hc/\lambda k_{\mathrm{B}}T} - 1}\right)$$

where I is the intensity of light emitted, λ is the wavelength of light emitted, h is Planck's constant (6.63×10^{-34} J·s), c is the speed of light (3.00×10^8 m/s), k is the Boltzmann constant (1.38×10^{-23} J/K), and T is the absolute temperature of the radiating cavity in kelvin. Create a spreadsheet that generates at least 15 values (choose a range of 15+ values of wavelength and calculate the value of the above function for $\frac{\Delta I}{\Delta\lambda}$ for some constant temperature). Once you have the set of data, plot it in a graph and apply an appropriate curve fit to depict the unique shape of the curve. Now, change the temperature several times to see how the curve changes.

37. •Using Einstein's explanation of the photoelectric effect ($K = hf - \Phi$), derive a numerical value for Planck's constant (h) by plotting a graph of the following data. Give your answer for h in both joule-seconds and electron volt-seconds.

Wavelength of light (nm)	Stopping potential (V)
400	1.00
450	0.660
500	0.380
550	0.125

Problems

26-1: Blackbody Radiation

38. ••Suppose a blackbody at 400 K radiates just enough heat in 15 min to boil water for a cup of tea. How long will it take to boil the same water if the temperature of the radiator is 500 K?

This page is intentionally left blank.

For complete end of chapter problem sets, please go to
www.whfreeman.com/kestentauck

62. ••A 0.075-nm photon Compton scatters off of a stationary electron. Determine the maximum speed of the scattered electron.

63. ••A photon Compton scatters off of a stationary electron at an angle of 60°. The electron moves away with 80 eV of kinetic energy. Determine the initial wavelength of the photon.

64. ••An x-ray source is incident on a collection of stationary electrons. The electrons are scattered with a speed of 4.50×10^5 m/s, and the photon scatters at an angle of 60° from the incident direction of the photons. Determine the wavelength of the x-ray source.

65. ••Arthur Holly Compton scattered photons that had wavelengths of 0.0711 nm off of a block of carbon during his famous experiment of 1923 at Washington University in St. Louis, Missouri. (a) Calculate the frequency and energy of the photons. (b) What is the wavelength of the photons that are scattered at 90°? (c) What is the energy of the photons that are scattered at 90°? (d) What is the energy of the electrons that recoil from the Compton scattering with $\theta = 90°$?

26-4: Wave Nature of Particles

66. •Calculate the de Broglie wavelength of a 0.150-kg ball moving at 40 m/s. Comment on the significance of the result.

67. •Calculate the de Broglie wavelength of an electron that has a speed of 0.00730c. SSM

68. •What is the de Broglie wavelength of a proton ($m = 1.67 \times 10^{-27}$ kg) moving at 400,000 m/s?

69. ••Restate de Broglie's formula for particle waves in the case that the speeds are relativistic.

70. •A proton ($m = 1.67 \times 10^{-27}$ kg) has a de Broglie wavelength of 1.2×10^{-15} m. Calculate the speed of the proton. Be careful of any relativistic corrections that might apply!

71. ••What is the de Broglie wavelength of an electron that has a kinetic energy of (a) 1 eV, (b) 10 eV, (c) 100 eV, (d) 1 keV, (e) 1 MeV, and (f) 1 GeV? SSM

72. ••Calculate the de Broglie wavelength of an alpha particle ($m_\alpha = 6.64 \times 10^{-27}$ kg $= 3730$ MeV/c^2) that has a kinetic energy of (a) 1 MeV, (b) 5 MeV, and (c) 10 MeV.

73. •Calculate the de Broglie wavelength of a thermal neutron that has a kinetic energy of about 0.04 eV.

74. •A relativistic electron has a de Broglie wavelength of 346 fm (1 fm $= 10^{-15}$ m). Determine its velocity.

75. ••Write an expression that relates the classical kinetic energy ($K = \frac{1}{2}mv^2$) and mass of a nonrelativistic particle to its de Broglie wavelength. [Complete the expression $\lambda(K,m) = ?$]

26-5: Atoms: Spectra

76. ••The Balmer formula can be written as follows:

$$\lambda = (364.56 \text{ nm})\left(\frac{m^2}{m^2 - 4}\right)$$

where m is equal to any integer larger than 2. This represents the wavelengths of visible colors that are emitted from the hydrogen atom. Calculate the first four colors (wavelengths) that are observed in the spectrum of hydrogen.

77. ••**Astronomy** The famous 21-cm line emitted by hydrogen in interstellar gas clouds is the spectral line that is produced when an electron in the ground state of hydrogen switches spin states. Determine the energy difference between the two spin states in the hyperfine transition. SSM

78. ••Prove that the Balmer formula is a special case of the Rydberg formula with n set equal to 2.

Rydberg formula: $\dfrac{1}{\lambda} = R_H\left(\dfrac{1}{n^2} - \dfrac{1}{m^2}\right)$ $R_H = 1.09737 \times 10^7$ m^{-1}

Balmer formula: $\lambda = b\dfrac{m^2}{m^2 - 4}$ $b = 364.56$ nm

79. •••A hypothetical atom has four unequally spaced energy levels in which a single electron can be found. Suppose a collection of the atoms are excited to the highest of the four levels. (a) What is the maximum number of unique spectral lines that could be measured as the atoms relax and return to the lowest, ground state? (b) Suppose the previous hypothetical atom has ten energy levels. Now what is the maximum number of unique spectral lines that could be measured in the emission spectrum of the atom?

80. •Using the Rydberg formula for the Lyman series, calculate the wavelengths of the photons emitted in the transitions from the next five highest energy states, and indicate the initial and final state of the transition corresponding to each wavelength.

81. •Using the Rydberg formula for the Balmer series, calculate the wavelengths of the photons emitted in the transitions from the next four highest energy states, and indicate the initial and final state of the transition corresponding to each wavelength. SSM

82. •Using the Rydberg formula for the Paschen series, calculate the wavelengths of the photons emitted in the transitions from the next three highest energy states, and indicate the initial and final state of the transition corresponding to each wavelength.

83. ••Calculate the shortest wavelength (and the highest energy) associated with emitted photons in the (a) Lyman, (b) Balmer, and (c) Paschen series.

84. ••Reexpress the Balmer formula in terms of the *frequency* of the photons that are emitted (rather than the wavelength). Extend all numerical values out to five significant figures.

85. ••Reexpress the Rydberg formula in terms of the *frequency* of the photons that are emitted (rather than the wavelength). Extend all numerical values out to five significant figures. SSM

26-6: Atoms: The Bohr Model

86. •For an electron in the nth state of the hydrogen atom, write an expression for (a) the angular momentum of the electron, (b) the radius of the electron's orbit, (c) the kinetic energy of the electron, (d) the total energy of the electron, and (e) the speed of the electron.

87. ••Set up a chart for the five quantities listed in problem 86 and calculate the values for $n = 1, 2, 3, 4$, and 5 (4 significant figures, SI units). See the table below.

n	L_n	r_n	K_n	E_n	v_n
1					
2					
3					
4					
5					

88. ••Devise a straightforward method that allows you to calculate (a) the speed and (b) the angular momentum of an electron in the nth Bohr orbit. Your expressions should contain number of the orbit, n, and constants.

89. ••Using the formula that you devised in problem 88, calculate the speed of an electron and the angular momentum of an electron in the tenth Bohr orbit. SSM

90. ••**Astronomy** Radio astronomers use radio frequency waves to identify the elements in distant stars. One of the standard lines that is often studied is designated the 272α line. This spectral line refers to the transition in hydrogen from $n_i = 273$ to $n_f = 272$. Calculate the wavelength and frequency of the electromagnetic radiation that is emitted for the 272α transition.

91. ••**Astronomy** For carbon ($Z = 6$), the frequencies of spectral lines resulting from a single electron transition are increased over those for hydrogen by the ratio of the Rydberg constants:

$$\frac{R_C}{R_H} = \frac{(1 - m_e/m_C)}{(1 - m_e/m_H)}$$

The mass of a hydrogen atom is 1837 times greater than the mass of the electron and the mass of the carbon atom is 12 times greater than the mass of the hydrogen atom. Find the shift in frequency for the carbon 272α

transition compared to the hydrogen 272α transition. Express your answer in megahertz (MHz).

92. ••The Lyman series ($n_f = 1$), the Balmer series ($n_f = 2$), the Paschen series ($n_f = 3$) are commonly studied in basic chemistry and physics classes. The Brackett series ($n_f = 4$) and the Pfund series ($n_f = 5$) are not so well known. (a) Calculate the shortest and longest wavelengths for the spectral lines that are part of the Brackett series. (b) Calculate the shortest and longest wavelengths for the spectral lines that are part of the Pfund series.

93. •A hydrogen atom that has an electron in the $n = 2$ state absorbs a photon. (a) What wavelength must the photon possess to send the electron to the $n = 4$ state? (b) What possible wavelengths would be detected in the spectral lines that result from the deexcitation of the atom as it returns from $n = 4$ to the ground state? SSM

94. •How much energy is needed to ionize a hydrogen atom that starts in the Bohr orbit represented by $n = 3$? If an atom is ionized, its outer electron is no longer bound to the atom.

95. ••Derive the first ten energy levels (sketch and label an energy level diagram) for He$^+$.

General Problems

96. •••**Calc** A certain parameter x is allowed to have any value from zero to one. The probability of observing any particular value is proportional to the cube of the value:

$$P(x) = 4x^3$$

(a) Show that the probability of finding *some* value of x is 100%. (b) Determine the average value of x. (c) Determine the average value of x^2.

97. •••**Calc** Assume the energy probability of gas molecules inside a cavity that is radiating as a blackbody is

$$P(E) = \frac{1}{kT} e^{-E/kT}$$

(a) Prove that the probability of finding *some* energy for the standing waves inside the cavity is 100%. (b) Determine the average value of E. (c) Determine the average value of E^2.

98. ••**Biology** Vitamin D is produced in the skin when 7-dehydrocholesterol reacts with UVB rays (ultraviolet B) having wavelengths between 270 nm and 300 nm. What is the energy range of the UVB photons in joules and electron volts?

99. ••A typical laboratory laser is a He–Ne laser that produces light of wavelength 632.8 nm. The laser beam carries a power of 0.50 mW and strikes a target perpendicular to the beam. (a) How many photons per second strike the target? (b) At what rate does the laser beam deliver linear momentum to the target if the photons are all absorbed by the target?

100. ••**Astronomy** In 2009 astronomers detected gamma ray photons having energy ranging from 700 GeV to around 5 TeV coming from supernovas (exploding giant stars) in the galaxy M82. (a) What is the range of wavelengths of the gamma ray photons detected from M82? (b) Calculate the ratio of the energy of the 5 TeV photons to the energy of visible light having a wavelength of 500 nm.

101. ••Derive the Compton scattering formula for the change in wavelength between the scattered photon and the incident photon when they are exposed to a free electron, initially at rest.

102. ••A photon of frequency 4.81×10^{19} Hz scatters off of a free stationary electron. Careful measurements reveal that the photon goes off at an angle of 125° with respect to its original direction. (a) How much energy does the electron gain during the collision? (b) What percent of its original energy does the photon lose during the collision?

103. ••Derive an expression for the frequency of an electron (f_n) revolving in the nth orbit of a hydrogenlike atom. SSM

104. ••A hydrogen atom makes a transition from the $n = 5$ state to the ground state and emits a single photon of light in the process. The photon then strikes a piece of silicon, which has a photoelectric work function of 4.8 eV. Is it possible that a photoelectron will be emitted from the silicon? If not, why not? If so, find its maximum possible kinetic energy.

105. ••Suppose the electron in the hydrogen atom were bound to the proton by gravitational forces (rather than electrostatic forces). Find (a) the radius and (b) the energy of the first orbit.

106. ••**Biology** The *E. coli* bacterium is about 2.0 μm long. Suppose you want to study it using photons of that wavelength or electrons having that de Broglie wavelength. (a) What is the energy of the photon and the energy of the electron? (b) Which one would be better to use, the photon or the electron? Explain why.

107. ••**Chemistry** A laboratory oven that contains hydrogen molecules H_2 and oxygen molecules O_2 is maintained at a constant temperature. Each oxygen atom is 16 times as massive as a hydrogen atom. Find the ratio of the de Broglie wavelength of the hydrogen molecule to that of the oxygen molecule.

108. •••Use the de Broglie wave concept to fit circular standing waves into the orbits of the Bohr model of the hydrogen atom to prove Bohr's hypothesis of the quantization of angular momentum. Assume that in the first Bohr orbit, exactly one de Broglie wavelength matches up with the circumference, in the second orbit, two waves matches up, in the third orbit, 3 waves match up, and so on (**Figure 26-23**).

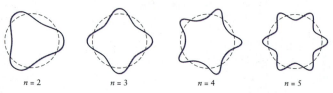

Figure 26-23 Problem 108

109. •**Astronomy** The Gamma-ray Large Area Space Telescope (GLAST) detects high-energy gamma rays. To do this, its detectors convert the gamma rays to an electron and a positron (the electron's identical-mass antiparticle). What is the maximum wavelength of a gamma ray that can produce an electron–positron pair? SSM

27 Nuclear Physics

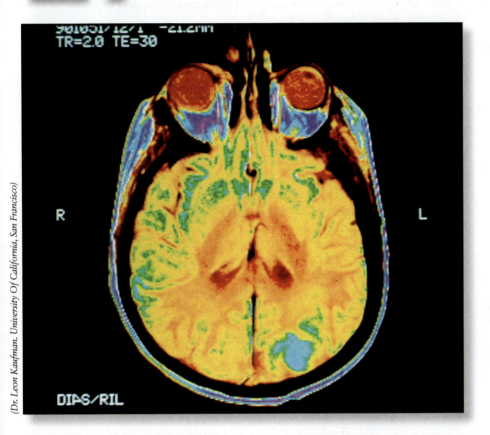

96105171271   -21.2mm
TR=2.0 TE=30

R

L

DIAS/RIL

(Dr. Leon Kaufman, University Of California, San Francisco)

An x-ray image can show hard structures such as bones inside a body, but to study soft tissue, such as internal organs or this human brain, the most common technique makes use of magnetic resonance imaging or MRI. MRI takes advantage of certain properties of the particles that make up atomic nuclei, in particular, that they act like magnetic dipoles. An image such as this one is formed by placing the body in a strong magnetic field to force the nuclear "magnets" to align, and then repeatedly flipping the alignment by pulsing them with electromagnetic energy. The MRI image is formed by detecting the electromagnetic waves that the oscillations create.

Chemical interactions, for example, the bonds that hold molecules together, are governed primarily by the number and arrangement of the electrons in atoms. The energies of the electrons are small, however, compared to the energy associated with the particles that make up atomic nuclei. The nucleus is therefore a significant source of energy. The emission of nuclear radiation when nuclei undergo a transformation from one energy state to another can carry enough energy to harm living things or to be used for therapeutic purposes.

27-1   The Nucleus

You are probably already comfortable with a model of the atom that, as we saw in the last chapter, includes a small, hard, positively charged nucleus at its center. The conclusion that the nucleus must be positively charged arises from two considerations. First, because atoms are neutral, positive charge is required to balance the negative charge of the electrons. In addition, the Coulomb attraction between the nucleus and the electrons provides the force that holds the atom together. The Coulomb *repulsive* force between the protons, however, must be large because they are closely packed. We estimate the magnitude of this repulsive force in Estimate It 27-1. What holds the nucleus together?

In addition to protons, nuclei also have neutrons, the neutrally charged particle we mentioned briefly in Chapter 16. Neutrons and protons share many similar properties, including similar masses: the masses of the neutron and proton are 1.6749×10^{-27} kg and 1.6726×10^{-27} kg, respectively. The **nuclear force**, a force we have not yet encountered, binds **nucleons** (protons and neutrons) together in an atomic nucleus by attracting each nucleon to other nucleons. The presence of neutrons holds the nucleus together.

The nuclear force is characterized by its strength, which over short distances is hundreds of times stronger than the electrostatic force. Because the nuclear force is so much stronger than the electrostatic force that repels protons from each other, and because neutrons as well as protons participate in holding the protons to the nucleus, the nuclear attraction can overcome electrostatic repulsion in nuclei. Nuclear protons are prevented from flying apart due to the nuclear force that attracts them to the neutrons and to other protons in the nucleus. Note that because neutrons are neither positively nor negatively charged, they do not experience electrostatic forces.

If the nuclear force is so strong compared to the electrostatic force, and if protons attract other protons by that force, why are neutrons necessary to help overcome the Coulomb repulsion between protons? The answer lies in the *range* of the nuclear force or the distance beyond which one nucleon no longer experiences a force due to another. The nuclear force diminishes rapidly as the separation between two nucleons increases, while the electrostatic force has a theoretically infinite range. (The electrostatic force between charges separated by a distance r decreases as $1/r^2$, so it is not zero even for large values of r. The nuclear force is approximately proportional to e^{-r}, however, so it decreases rapidly to zero as the separation between nucleons increases.) Because the range of the nuclear force (about 1.5×10^{-15} m or 1.5 fm) is smaller than the diameter of most nuclei (a few to perhaps 15 fm), each proton (and neutron) only exerts an attractive nuclear force on its nearest neighbors. Each proton in a nucleus, however, exerts a repulsive force on *every other* proton. As a result, the nuclear force between neighboring protons cannot overcome the Coulomb repulsion between all of the protons. To prevent a nucleus from spontaneously breaking apart (that is, for it to be *stable*), the nucleus must have about as many neutrons as protons or more (for larger nuclei). An unstable nucleus will eventually undergo a spontaneous transformation, termed a *decay*, in which it either splits apart or gives off energy in some other way. In addition, atoms that have more than about 20 protons require more neutrons than protons to be stable. **Figure 27-1** shows the number of neutrons versus the number of protons in known, stable atomic nuclei. Notice that as the number of protons increases, more additional neutrons are required for stability.

Atoms of each element have a unique number of protons (and electrons): hydrogen has 1, helium 2, lithium 3, and so on. Many of properties of atoms are related to this number, usually designated as the **atomic number** Z. In a similar way, many properties of nuclei

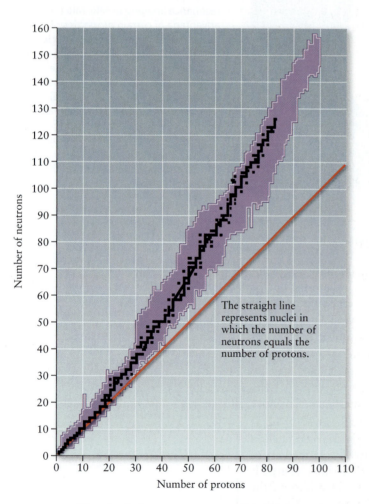

The straight line represents nuclei in which the number of neutrons equals the number of protons.

Figure 27-1 The shaded area represents all known nuclei; the dots indicate nuclei that are stable. For most nuclei, more neutrons than protons must be present to prevent the protons from flying apart due to electrostatic repulsion.

arise from the **neutron number**, that is, the number of neutrons N that the nucleus has. Although the value of Z of an elemental species is fixed (changing the number of protons changes the element), the value of N is not fixed for each element. Each combination of N and Z specifies a *nuclide*. For each element, the one, most common configuration of N and Z corresponds to the most stable nuclide; nuclei of that elemental species with a different number of neutrons are termed *isotopes*. For example, potassium has 19 protons (and 19 electrons). The most stable, and most common, nuclide of potassium has 20 neutrons, $N = 20$. We refer to this nuclide with the symbol ^{39}K. The number of protons is understood from the symbol K for potassium, and the total number of protons and neutrons, termed the **mass number** A, is given in the presuperscript, 39. This potassium nuclide can also be referred to as potassium-39. More than 93% of all potassium atoms are ^{39}K. About 7% of potassium atoms are ^{41}K, however, which contain 19 protons, 19 electrons, and 22 neutrons.

Got the Concept 27-1
N, Z, and A

(a) What is the symbol and name of the element that has (i) 50 protons, (ii) 74 protons, (iii) 82 protons? (b) How many neutrons are found in the nucleus of (i) ^{6}He, (ii) ^{17}C, (iii)^{18}O?

The size of a nucleus, that is, its **nuclear radius**, volume, and mass, is related to the mass number. For example, all nuclei are approximately spherical and have radii proportional to $A^{1/3}$:

$$r = r_0 A^{1/3} \qquad\qquad (27\text{-}1)$$

where the proportionality constant r_0 is 1.2 ± 0.2 fm. Because the volume of a sphere is proportional to r^3 and therefore proportional to $(A^{1/3})^3$, the volume of a nucleus is proportional to mass number A. In other words, the volume of a nucleus is proportional to the number of nucleons that the nucleus has. This is reasonable because the volume of the nucleus is related to how many protons and neutrons comprise the nucleus.

Is there a limit on the size of a nucleus? Said another way, can we simply add more protons, and more and more neutrons, to make larger and larger nuclei? The answer is "no," and for the same reason that neutrons are required for nuclear stability. We have seen that as we work our way up the periodic table to atoms that have more and more protons, more *additional* neutrons are required for stability. Each additional proton exerts a repulsive force on all the others, but the nucleons can only attract their nearest neighbors. The additional neutrons cause the size of the nucleus to grow, so that eventually, too many neutrons are near the surface of the nucleus and therefore not completely surrounded by neighbors. At that point the nuclear forces holding the nucleus together are not large enough to overcome the Coulomb repulsion between the protons and the nucleus is not stable. The largest stable nucleus is bismuth-209, which has 126 neutrons and 83 protons. To be precise, the ^{209}Bi nucleus is not completely stable—it does spontaneously break apart. The time it would take one-half of the ^{209}Bi nuclei in a lump of bismuth to break apart, however, is about one billion times longer than the age of the universe. That's plenty long for us, but because ^{209}Bi eventually decays, some physicists therefore identify lead-208 as the largest stable nucleus. This isotope of lead contains 82 protons and 126 neutrons. Lead-208 does not spontaneously decay.

Estimate It! 27-1 Pushing on a Proton

Estimate the magnitude of the Coulomb force that one proton exerts on the other in the nucleus of a helium atom. Express your answer as a ratio of the force to the force a proton feels due to the influence of gravity on Earth's surface.

SET UP

The magnitude of the Coulomb force between two objects each of charge q separated by a distance r is (Equation 16-1)

$$F = \frac{kq^2}{r^2}$$

where the Coulomb constant k is equal to $8.99 \times 10^{-9}\,\text{N}\cdot\text{m}^2/\text{C}^2$. Because this calculation is an estimate, we want to use big, round (but reasonable) values, so for k we'll use $10^{-8}\,\text{N}\cdot\text{m}^2/\text{C}^2$ ($9 \times 10^{-9}\,\text{N}\cdot\text{m}^2/\text{C}^2$ is approximately $10 \times 10^{-9}\,\text{N}\cdot\text{m}^2/\text{C}^2$ or $10^{-8}\,\text{N}\cdot\text{m}^2/\text{C}^2$). The charge of a proton is the fundamental charge, 1.6×10^{-19} C; so for the estimate we'll use 10^{-19} C as the order of magnitude of the value of the proton charge. We let two protons be separated by the diameter of the helium nucleus. Recall from the discussion above that nuclei are typically a few femtometers in diameter. Helium is at the smaller end of the atomic size scale, so for this estimate it's reasonable to use 1 fm, or 1×10^{-15} m as the order of magnitude of r, the separation of the protons. The mass of the proton is 1.67×10^{-27} kg or about 10^{-27} kg; to compare the force to the proton's weight, we'll let $m_\text{p}g$ be approximately 10^{-26} N.

SOLVE

The magnitude of the Coulomb repulsive force between the protons is then

$$F = \frac{(10^{-8}\,\text{N}\cdot\text{m}^2/\text{C}^2)(10^{-19}\,\text{C})^2}{(10^{-15}\,\text{m})^2} = 10^{-16}\,\text{N}$$

This is rather a small force. However, the ratio of the force to the weight of the proton,

$$\frac{F}{mg} = \frac{10^{-16}\,\text{N}}{10^{-26}\,\text{N}} = 10^{10}$$

is not small. We consider the implication of this ratio below.

REFLECT

We have estimated that the proton in a helium nucleus experiences a Coulomb force that is 10^{10} times bigger than its weight. Imagine if you were to feel a force that much larger (10 billion times larger!) than your weight. For a proton of a helium nucleus, or more generally, for any proton of any nucleus, the repulsive Coulomb force it experiences is enormous. It's also interesting to note that the magnitude of the nuclear (attractive) force between two protons separated by 1 fm is more than 400 times the magnitude of the Coulomb (repulsive) force.

Example 27-1 Nuclear Radii

Estimate the radius of the nucleus of (a) ^{12}C, a relatively small nucleus, (b) ^{59}Co, a larger nucleus, (c) ^{118}Sn, a nucleus of modest relative size, and (d) ^{236}U, a relatively large nucleus. The nuclides ^{12}C, ^{59}Co, and ^{118}Sn are stable nuclear configurations of that element. Uranium-236 is not; all isotopes of uranium are unstable.

SET UP

The radius of any nucleus is given approximately by Equation 27-1:

$$r = r_0 A^{1/3}$$

The value of A for each of the nuclei is given in the presuperscript.

SOLVE

(a) The mass number of ^{12}C is 12 because the nucleus of ^{12}C has 12 nucleons (six protons and six neutrons). We take r_0 equal to 1.2 fm, so

$$r(^{12}\text{C}) = (1.2 \text{ fm})(12)^{1/3} = 2.7 \text{ fm}$$

(b) The mass number of ^{59}Co is 59. (Cobalt has Z equal to 27, so ^{59}Co has 27 protons and 32 neutrons.)

$$r(^{59}\text{Co}) = (1.2 \text{ fm})(59)^{1/3} = 4.7 \text{ fm}$$

(c) The mass number of ^{118}Sn is 118. (The value of Z for tin is 50, so ^{118}Sn has 50 protons and 68 neutrons.)

$$r(^{118}\text{Sn}) = (1.2 \text{ fm})(118)^{1/3} = 5.9 \text{ fm}$$

(d) The mass number of ^{236}U is 236. (Uranium has 92 protons, so ^{236}U also has 144 neutrons.)

$$r(^{236}\text{U}) = (1.2 \text{ fm})(236)^{1/3} = 7.4 \text{ fm}$$

REFLECT

Notice that we chose to examine four nuclei of increasing mass number. The number of neutrons in carbon, the smallest of the four nuclei, equals the number of protons. The stable isotope of cobalt has about 20% more neutrons than protons. In tin, which has twice the mass number of cobalt, the number of neutrons required for nuclear stability is 36% higher than the number of protons. And even with nearly 60% more neutrons than protons, the relatively large ^{236}U nucleus is not stable. So we see direct evidence that more and more additional neutrons, compared to the number of protons, are required for nuclear stability, and that at some size a nucleus is too large for the strong nuclear force to overcome the Coulomb repulsion between the protons.

Also, in this problem we have set a reasonable scale for the size of nuclei. For nuclides that have small mass numbers, their radii will be close to the value of r_0 or about 1.2 fm. For very massive nuclei, such as in the uranium atom, the radii will be around 7.5 fm. As round numbers to put into your back pocket then, we can take the radius of nuclei to be in the range of 1 fm to 10 fm.

Practice Problem 27-1 What is the mass number of a nuclide that is half the diameter of a ^{238}U nucleus?

? Got the Concept 27-2
Nuclei Are Dense

The density of the atomic nuclei of all elements is approximately the same. Give a reason why this is true.

Example 27-2 Nuclear Density

Estimate to two significant digits the mass density of the nucleus of an atom that has mass number A.

SET UP

Mass density ρ is defined (Equation 11-1) as the mass of an object divided by its volume. The volume of the nucleus is

$$V = \frac{4}{3}\pi R^3 = \frac{4}{3}\pi (r_0 A^{1/3})^3 = \frac{4}{3}\pi r_0^3 A$$

where we used Equation 27-1 for the radius R. In addition, we take the average mass of one nucleon as the average of the proton mass and the neutron mass:

$$m_{avg} = \frac{m_p + m_n}{2}$$

so that the mass of the nucleus is equal to Am_{avg}. The mass density is then

$$\rho = \frac{Am_{avg}}{4\pi r_0^3 A/3}$$

Notice that the variable A appears in both the numerator and denominator, so it cancels. The nuclear mass density is then approximately

$$\rho = \frac{m_{avg}}{4\pi r_0^3/3} = \frac{3m_{avg}}{4\pi r_0^3}$$

SOLVE

The average mass of a nucleon is

$$m_{avg} = \frac{1.6726 \times 10^{-27}\,\text{kg} + 1.6749 \times 10^{-27}\,\text{kg}}{2} = 1.6738 \times 10^{-27}\,\text{kg}$$

We'll keep the extra significant digits until the last step of the problem. The mass density of the nucleus is then

$$\rho = \frac{m_{avg}}{4\pi r_0^3/3} = \frac{3(1.6738 \times 10^{-27}\,\text{kg})}{4\pi (1.2 \times 10^{-15}\,\text{m})^3} = 2.3 \times 10^{17}\,\text{kg/m}^3$$

REFLECT

First, notice that because the final expression for nuclear density ρ does not depend on A, the mass density is approximately independent of the mass number of the atom. So the nuclear density of atoms of all elements is about the same, with the value of ρ equal to about $2.3 \times 10^{17}\,\text{kg/m}^3$. Also, recall that the density of water is $1000\,\text{kg/m}^3$; nuclei are on the order of 10^{14} as dense as water. Nuclei are dense! That's not so surprising, however, if you consider that atoms, such as those in water molecules, are mostly empty. Atomic radii are on the order of 1×10^{-10} m to 10×10^{-10} m, while nuclear radii are on the order of 1×10^{-15} m to 10×10^{-15} m. These radii are in a ratio of about a factor of 10^5, so the volumes are in a radio of about $(10^5)^3$, or 10^{15}. It is therefore reasonable that nuclear mass density would be about 10^{15} times greater than atomic mass density. This is consistent with our result in this problem.

Practice Problem 27-2 Estimate to two significant digits the number of nucleons per cubic femtometer (fm^3) in a typical nucleus.

Protons and neutrons, and therefore some atomic nuclei, have angular momentum, a property physicists refer to as *spin*. (This is a misnomer, however, because the angular momentum does not correspond directly to a spinning motion of the protons and neutrons.) The spin of a proton or neutron can take on one of two values, often referred to as "spin up" and "spin down." Recall that angular momentum is a vector quantity (Chapter 8). You can therefore sum spin values by using vector addition, so that only a nucleus with an odd number of nucleons has an overall (noninteger) spin. In particular, the nucleus of hydrogen—a single proton—has overall spin.

Measurements that make use of the spin of hydrogen nuclei enable us to localize hydrogen in an object or body, and also to get information about the material in which the hydrogen atoms are embedded. In a magnetic resonance imaging (MRI) device, spin information is used to form a three-dimensional map of the density of hydrogen atoms in a body. Because living organisms are composed largely of water (for example, somewhere between 60 and 70% of our bodies are made up of water), and because each water molecule contains two hydrogen atoms, living organisms are packed full of protons. As such, MRI is an ideal way to probe the internal structures in the body. **Figure 27-2** shows cracks in the cartilage that supports the knee cap in a patient with knee pain.

Protons and neutrons also have associated magnetic fields. The field is related to the spin, so because spin is either up or down, the field is a dipole in nature. Like a bar magnet, the magnetic field has a direction, and that direction is parallel to the direction of the angular momentum vector. As a result, when a nucleus that has an odd number of nucleons A is placed in a uniform external magnetic field, the magnetic force tends to align the spin direction either generally parallel to or antiparallel to the external field. To be precise, the spin direction aligns so that its component along the direction of the external field has a fixed positive or negative value. As such, the spin direction can rotate, or precess, around an axis defined by the field direction. The precession is shown in **Figure 27-3a** for a nucleus with spin aligned with the external field and in **Figure 27-3b** for a nucleus with spin antialigned with the external field. The energy of the spin-aligned orientation of a nucleus in an external field is higher than the energy of the antialigned orientation.

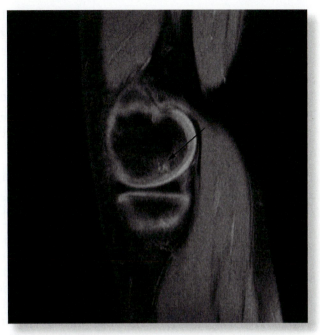

Figure 27-2 When a nucleus of odd A is placed in a uniform external magnetic field it experiences a magnetic force that tends to align the spin direction either generally parallel to or antiparallel to the field. The spin direction aligns so that its component along the direction of the external field has a fixed positive or negative value. The spin direction can rotate, or precess, around an axis defined by the field direction for (a) a nucleus with spin aligned with the external field and (b) a nucleus with spin antialigned with the external field. *(Uniformed Services University. Obtained from MedPix Database)*

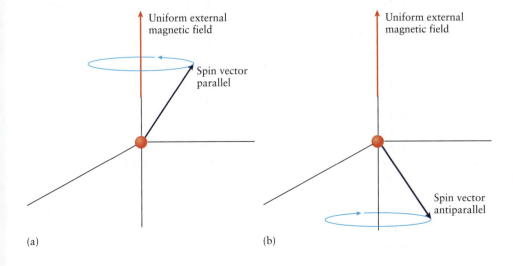

(a)   (b)

Figure 27-3 The arrow in this MRI image indicates cracks in the cartilage that supports the knee cap in a patient with knee pain.

Consider a large number of hydrogen atoms placed in a uniform magnetic field. About one-half of the protons will end up with spin aligned with the field and one-half with spin antialigned. Something curious happens if the atoms are exposed to electromagnetic energy that carries an amount of energy equal to the energy difference between the two states of spin orientation. Protons that are initially in the lower energy state, with their spin antialigned with the field, can absorb that energy and flip their spin to align with the external field. Protons that are initially in the higher energy state, with their spin aligned with the field, can be stimulated to flip their spin to be *antialigned* with the field through the emission of a photon of energy equal to the difference between the states. (This stimulated emission is related to the way a laser works. The name *laser* originally came from the initials of the process by which the device works: light amplification by the stimulated emission of radiation. A laser tube is filled with a gas in which atomic electrons are pumped to excited states. The excited electrons naturally want to fall back down to a lower state, but if left on their own, would do so at random times. However, when a photon of energy equal to the difference between two states, for example, the lowest energy state and the first excited state, are sent into the gas, it can stimulate an electron in the excited state to *drop* to the lower energy state. In the process energy is released, in the form of a second photon of that same energy. Because each of the emitted photons is the same energy as the photon that caused the emission, both photons can stimulate further emission. In this way, the number of photons in the tube increases. In addition, the emitted photons are all in phase with the electrons that stimulated their emission, and the emitted photons travel generally in the same direction as the electrons that stimulated their emission. The result is a powerful, tight beam of light.)

The difference in energy between the two spin states of a hydrogen atom depends on the strength of the external field. With the high field strength required to be able to clearly observe the spin-flipping phenomenon, typically around 3 T, the energy difference is about 5×10^{-7} eV. From Equation 22-2, the frequency of a photon of that energy is

$$f = \frac{E}{h} = \frac{5 \times 10^{-7}\,\text{eV}}{4.15 \times 10^{-15}\,\text{eV} \cdot \text{s}} = 1.2 \times 10^8\,\text{s}^{-1} = 120\,\text{MHz}$$

(We used the value of h in eV·s.) In general, the frequency of the electromagnetic waves used to induce the hydrogen atoms to flip their spins in a strong magnetic field is around 100 MHz. These are radio frequency waves.

How do we "see" the spin flipping that enables an image of the location of hydrogen to be formed? There will always be more atoms initially in the lower energy state than in the higher energy state (lower energy is more likely for a system than higher energy), so when the hydrogen atoms in an external magnetic field are exposed to radio frequency waves, there is a net absorption of the electromagnetic energy. When the radio frequency signal is turned off, the spins begin to return to their initial (equilibrium) state, and in doing so emit as radio frequency waves the energy they had absorbed. The MRI device detects that radio frequency emission and uses it to form a map of the density of hydrogen atoms in the body.

Recall that the difference in energy between the two hydrogen spin states depends on the strength of the external magnetic field. In an MRI device, the magnetic field is made to vary over a body's volume, so that the energy absorbed and then reemitted in the spin-flip process also varies in different parts of the body. The exact frequencies of the radio energy detected by the MRI device therefore provide the information necessary to create images of high spatial resolution. In addition, the time it takes for the spins of the hydrogen nuclei to return to their equilibrium state depends on the particular molecules in the tissue. So, timing information in an MRI device provides the means to differentiate one type of tissue from another.

★ What's Important 27-1

The nuclear force binds nucleons (protons and neutrons) together in the nucleus of an atom. The volume of a nucleus is proportional to the number of nucleons it has. To be stable, the nucleus must have about as many neutrons as protons, and atoms that have more than about 20 protons require more neutrons than protons to be stable. In very large nuclei (mass number greater than 208 or 209), too many of the additional neutrons required for stability are close to the nuclear surface and as a result, the short-ranged nuclear force does not reach enough protons to overcome the Coulomb repulsion. Such large nuclei are therefore always unstable.

27-2 Binding Energy

Release a ball at the top of a hill and it rolls down. Pull an object attached to the free end of a spring away from its equilibrium position and it tends to return to that position. In both cases the systems are finding their way to a more stable configuration. All physical systems do the same; if a more stable configuration exists for a system, it will eventually find itself in that configuration as long as nature provides a mechanism for the transition to take place.

In this context, consider the nucleus of a helium atom, which consists of two protons and two neutrons. The four nucleons remain bound together in a configuration that must be more stable than when they are separate, otherwise the helium nucleus would end up broken apart. This stability results because the attractive nuclear force between the four nucleons overwhelms the electrostatic repulsive force between the two protons. But let's look at stability from another perspective.

The mass M_{tot} of the two protons and two neutrons in the nucleus of ^{4}He is

$$M_{tot} = 2m_p + 2m_n$$

or

$$M_{tot} = 2(1.6726 \times 10^{-27}\,kg) + 2(1.6749 \times 10^{-27}\,kg) = 6.695 \times 10^{-27}\,kg$$

Is this the mass of the helium nucleus? No! The mass of the nucleus of ^{4}He is 6.645×10^{-27} kg. Look closely—this value is *lower* than the mass of the protons and neutrons added together. How can the mass of four protons and neutrons, bound together as a helium nucleus, be less than the mass of the particles taken separately? The answer is the key to nuclear stability; the energy equivalent of the difference in mass is tied up in binding the nucleons together. Recall that energy and mass are equivalent, a conclusion drawn from Einstein's equation $E = mc^2$ (Equation 25-24). We call the energy that holds the nucleus together the **binding energy** E_B. The greater the binding energy per particle in a nucleus, the more tightly the nucleus is bound and therefore the more stable it is.

Figure 27-4 is a graph of the binding energy per nucleon (E_B/A) as a function of mass number in the atoms of all elements The energy values on the vertical axis are given in megaelectron volts (MeV). An MeV is one million electron volts (eV);

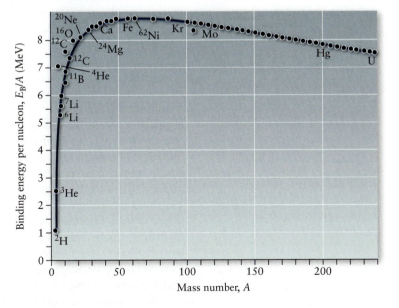

Figure 27-4 The binding energy per nucleon is a measure of the stability of the nucleus of any atom; the more binding energy per nucleon, the more stable the nucleus.

one eV is the amount of energy acquired by an electron when it experiences a potential difference of one volt. We have seen the relationship between energy and potential difference in, for example, Equation 17-3. From this equation the change in the electric potential energy V of a charge q_0 is equal to the electric potential difference ΔU divided by q_0:

$$V = \frac{\Delta U}{q_0}$$

So for an electron of charge e, the change in energy when it experiences an electric potential energy V is

$$\Delta U = eV$$

Electron volts are therefore a natural way to express energy when dealing with particles. One eV is a relatively small amount of energy, equivalent to 1.60×10^{-19} J.

Consider the shape of the curve in Figure 27-4. As A increases E_B/A increases rapidly, peaking at about 8.8 MeV for A in the range of 56 to 62, and then decreases slowly for higher and higher values of A. This shape leads us to an understanding of two general nuclear phenomena, **fission**, the splitting of a nucleus into smaller fragments, and **fusion**, the joining together of two smaller nuclei to form one or more larger ones. The most stable nuclei are ^{56}Fe and ^{58}Fe (two isotopes of iron) and ^{62}Ni; the binding energy per nucleon of these nuclei place them near the peak of the E_B/A curve in Figure 27-4.

? Got the Concept 27-3
Stability

Rank these nuclides in order from most to least stable: (a) ^{11}B, (b) ^{20}Ne, (c) ^{100}Kr, (d) ^{200}Hg.

The processes by which nuclei of relatively large values of mass number A become more stable often result in a decrease in the number of nucleons. In nuclear fission, nuclei split into smaller pieces. For example, atoms of curium-244 can spontaneously fission into xenon-135 and molybdenum-109. Curium has 96 protons, so ^{244}Cm has 148 neutrons. There are 54 protons in xenon and 42 in molybdenum or 96 total. The total number of nucleons in ^{135}Xe and ^{109}Mo, 244, equals the number of nucleons in ^{244}Cm. In other words, ^{135}Xe and ^{109}Mo are the same 96 protons and 148 neutrons in the original ^{244}Cm; the curium atom has split into two fragments.

The processes by which nuclei of relatively small values of A become more stable often involve increasing the number of nucleons, usually as the result of two nuclei fusing together. Fusion processes result in the formation of an atom of a different, more massive element. For example, the 12 protons and 12 neutrons in two carbon-12 nuclei can fuse to form a magnesium-24 nucleus. More than one atom can also be formed by fusion processes. For example, when a helium-3 nucleus fuses with a lithium-6 nucleus, the reaction forms two helium-4 nuclei and one hydrogen-1 nucleus. (Count the nucleons: ^{3}He has two protons and one neutron, and ^{6}Li has three protons and three neutrons, for a total of five protons and four neutrons. The two ^{4}He nuclei have two protons and two neutrons each, leaving one more proton to form a ^{1}H nucleus.)

What is the binding energy per nucleon of the ^{4}He nucleus? We could determine it by taking the difference between the mass of the two protons and two neutrons separately that we found above and the mass of the ^{4}He nucleus, and then converting the mass difference to the equivalent energy. It is instructive, however, to redo the mass determination in units more appropriate for this calculation, units that are derived from the energy units of electron volts. The units eV/c^2 (or MeV/c^2) arise from the equivalence of mass and energy as described by $E = mc^2$.

From $E = mc^2$, mass can be expressed in terms of its energy equivalent:

$$m = \frac{E}{c^2}$$

A natural unit of mass is therefore eV/c^2 and $1\ eV/c^2$ is equivalent to about 1.7827×10^{-27} kg. In these units, the mass of a proton is $938.27\ MeV/c^2$ and the mass of a neutron is $939.57\ MeV/c^2$. In addition, in these units the mass of a helium nucleus is $3727.4\ MeV/c^2$. The difference Δ between the mass of the two protons and two neutrons separately, and the mass of the helium nucleus is

$$\Delta = 2(938.27\ MeV/c^2) + 2(939.57\ MeV/c^2) - 3727.4\ MeV/c^2$$

or

$$\Delta = 28.3\ MeV/c^2$$

The energy equivalent of any mass is obtained by multiplying it by c^2, so the energy equivalent of this difference is 28.3 MeV. (Notice how straightforward it is to find the energy equivalent of mass when we write mass in units of MeV/c^2.) In other words, the binding energy of ^4He is 28.3 MeV. There are four nucleons in the helium nucleus, so E_B/A is about 7.1 MeV. You can verify this result from the curve in Figure 27-4.

The binding energy per nucleon is much higher for ^4He than for other light nuclei. Using Figure 27-4, E_B/A is about 2.5 MeV for ^3He, for example, and about 5.3 MeV for ^6Li. Because of its relatively high binding energy per nucleon, ^4He is far more stable than other light nuclei. For this reason, when large nuclei break apart to transform to a more stable, more energetically favorable state, in many cases they do so by emitting two protons and two neutrons bound together. In such processes the protons and neutrons are referred to as an *alpha particle* (α particle), and we say that the alpha particle is emitted as **nuclear radiation**. Nuclear radiation is the emission by a nucleus of either energy or one of a small number of particles. We discuss nuclear radiation in more detail in Section 27-5.

To find the binding energy of ^4He, we subtracted the mass of the nucleus from the mass of the two protons and two neutrons separately, and then multiplied by c^2 to find the equivalent energy. In general, for a nucleus consisting of N neutrons and Z protons, E_B is

$$E_B = (Nm_n + Zm_p - m_{nucleus})c^2$$

where m_n is the mass of a neutron, m_p is the mass of a proton, and $m_{nucleus}$ is the mass of the nucleus. Because it is common to work with the masses of neutral atoms rather than the masses of their nuclei, we will write this as

$$E_B = (Nm_n + Zm_{^1H} - m_{atom})c^2 \tag{27-2}$$

where $m_{^1H}$ (the mass of a ^1H atom) and m_{atom} (the mass of the neutral atom under consideration) include the mass (and equivalent energy) of the electrons bound to the atom. The masses of neutral atoms are given in Appendix C. Note that the values are given in units of atomic mass unit (u or amu), where

$$1\ u = 931.494\ MeV/c^2$$

Example 27-3 ^4He the Atomic Way

We have previously determined the binding energy per nucleon in the ^4He nucleus using the mass of the nucleus. Use the values from Appendix C and Equation 27-2 to determine the binding energy per nucleon in the ^{4}He nucleus using the atomic mass. Give your answer in MeV/c^2.

SET UP

For ^{4}He, N and Z both equal 2. The mass of a neutron, the ^{1}H atom, and the ^{4}He atom are provided in Appendix C:

$$m_n = 1.008665 \text{ u}$$
$$m_{^1H} = 1.007825 \text{ u}$$
$$m_{^4He} = 4.002602 \text{ u}$$

SOLVE

Using these values, Equation 27-2 yields

$$E_B = [2(1.008665 \text{ u}) + 2(1.007825 \text{ u}) - 4.002602 \text{ u}]c^2 = 0.030378 \text{ u}c^2$$

This value is converted to MeV/c^2 by multiplying by the equivalence of 931.494 MeV/c^2 and 1 u:

$$E_B = 0.030378 \text{ u}c^2 \left(\frac{931.494 \text{ MeV}/c^2}{1 \text{ u}} \right) = 28.3 \text{ MeV}$$

We report the value to three significant figures in order to compare it directly with the value we obtained for E_B of ^{4}He above. The binding energy per nucleon for each of the four protons and neutrons is one-fourth of this value, or approximately 7.1 MeV.

REFLECT

Our answer is the same as the result of the calculation we carried out earlier using the mass of the ^{4}He nucleus, at least to the number of significant figures we used. If you wonder why we would do this kind of calculation using atomic masses rather than nuclear masses, it is simply that, in general, the mass of neutral atoms have been well measured, but we don't have precise measurements of the masses of atomic nuclei.

Practice Problem 27-3 Use the values from Appendix C and Equation 27-2 to determine the binding energy per nucleon in the ^{56}Fe nucleus. Give your answer in MeV/c^2 to three significant digits.

Example 27-4 Breaking Up Is Hard to Do

Imagine that a nucleus of erbium-164 spontaneously breaks into two equal pieces or two nuclei of selenium-82. Use the values from Appendix C and Equation 27-2 to compare the energy tied up in binding the ^{164}Er nucleus together and the total binding energy of the two ^{82}Se nuclei.

SET UP

The atomic number of erbium (Z_{Er}) is 68, and the atomic number of selenium (Z_{Se}) is 34. So each atom of ^{164}Er has 96 neutrons (N_{Er} equals $164 - 68$ or 96). For each Se nucleus, N_{Se} equals $82 - 34$ or 48. We apply Equation 27-2 once for ^{164}Er and twice for ^{82}Se.

SOLVE

The mass of a neutron and the ^{1}H atom, from Appendix C, are 1.008665 u and 1.007825 u, respectively. We can also find the mass of the neutral ^{164}Er and ^{82}Se atoms in Appendix C:

$$m_{Er} = 163.929198 \text{ u}$$
$$m_{Se} = 81.916697 \text{ u}$$

Using Equation 27-2, the binding energy of the ^{164}Er nucleus is

$$E_{B,Er} = (N_{Er}m_n + Z_{Er}m_{^1H} - m_{Er})c^2$$

or

$$E_B = (96(1.008665\ u) + 68(1.007825\ u) - 163.929198\ u)c^2 = 1.434742\ uc^2$$

Similarly, the binding energy of each ^{82}Se nucleus is

$$E_{B,Se} = (N_{Se}m_n + Z_{Se}m_{^1H} - m_{Se})c^2$$

or

$$E_B = [48(1.008665\ u) + 34(1.007825\ u) - 81.916697\ u]c^2 = 0.765273\ uc^2$$

The total energy tied up in binding the two Se nuclei is twice this value or 1.530546 uc^2. This is greater than the binding energy of the Er nucleus; the difference ΔE is 0.095804 uc^2 or

$$\Delta E = 0.095804\ uc^2 \left(\frac{931.494\ MeV/c^2}{1\ u} \right) = 89.2\ MeV$$

REFLECT

Because the binding energy of the two Se nuclei together is greater than that of the original Er nucleus, the configuration of the 96 neutrons and 68 protons (and 68 electrons) is more stable as two ^{82}Se atoms than as one ^{164}Er atom. In addition, were the Er nucleus to split into two as suggested in the problem statement, 89.2 MeV of energy would be released. We note that this process does not, however, occur naturally. It is possible to split an Er nucleus into two equal fragments by smashing it with an energetic particle, however.

★ **What's Important 27-2**

Binding energy holds the nucleus together. The greater the binding energy per particle in a nucleus, the more tightly the nucleus is bound and therefore the more stable it is.

27-3 Fission

As we noted in the last section, nuclei with higher values of binding energy per nucleon (E_B/A) are more stable than those with lower values. Glance back at Figure 27-4, a plot of the binding energy E_B per nucleon in nuclei versus mass number A, to remind yourself that E_B/A tends to get smaller and smaller as A increases beyond 60 or so. In other words, large nuclei are less stable than smaller ones for A greater than about 60. This instability results in spontaneous nuclear processes that result in the fragmentation of the nucleus.

A large nucleus can also be induced to fragment by imparting energy to it through a collision. In either case, the smaller fragments of a large nucleus have a higher value of E_B/A and are therefore more stable.

The collision of a neutron, even one of essentially zero energy, with the nucleus of ^{235}U will cause it to fission:

$$n + {}^{235}U \rightarrow {}^{236}U^* \rightarrow {}^{134}Te + {}^{99}Zr + 3n + E_{released} \qquad (27-3)$$

For a brief time the neutron and ^{235}U nucleus remain stuck together as ^{236}U*, but this clumping of protons and neutrons quickly fissions into fragments. (The asterisk in ^{236}U* indicates that it is an excited and short-lived state of ^{236}U.) In this particular reaction the fragments are an isotope of tellurium, an isotope of zirconium, and three neutrons. Because the fragments are all more stable than the original nucleus, energy is released. The fission of ^{235}U occurs even when the kinetic energy of the colliding neutron is essentially zero, so $E_{released}$ is either mostly or completely due to the change in binding energy between the initial ^{235}U nucleus and the fission products.

Estimate It! 27-2 Energy Released in ^{235}U Fission

Estimate the energy released in the fission process described by Equation 27-3 from the curve in Figure 27-4.

SET UP

The energy released in the fission process is the difference between the energy tied up in the binding of the fragments and the energy tied up in the binding of the ^{235}U nucleus. The binding energies can be estimated by multiplying the number of nucleons in each by the binding energy per nucleon for that isotope, that is, by taking the product of A and E_B/A evaluated at A:

$$E_{released} = 134\, E_B/A\,(A = 134) + 99\, E_B/A\,(A = 99) - 235\, E_B/A\,(A = 235)$$

You can use a ruler to read the values of E_B/A for A equal to 235, 134, and 99 from the curve. There is no binding energy associated with the three neutrons that result from the fission, because each is a separate particle not bound to any other. [As we will discover in the next chapter, the neutron (and the proton) are actually a bound state of three constituent particles. For the purposes of nuclear processes, however, we can ignore such contributions and treat both the neutron and proton as single particles.]

SOLVE

Measuring E_B/A for uranium-235, tellurium-134, and zirconium-99 gives, respectively,

$$E_B/A\,(235) \approx 7.6\,\text{MeV}$$

$$E_B/A\,(134) \approx 8.4\,\text{MeV}$$

$$E_B/A\,(99) \approx 8.7\,\text{MeV}$$

The difference in binding energy ΔE is then

$$\Delta E \approx 134\,(8.4\,\text{MeV}) + 99\,(8.7\,\text{MeV}) - 235\,(7.6\,\text{MeV})$$

or

$$\Delta E \approx 200\,\text{MeV}$$

That is, the energy released is

$$E_{released} \approx 200\,\text{MeV}$$

REFLECT

Our answer is both relatively close to the actual amount of energy released during the process (about 185 MeV), and also typical of the amount of energy released during the fission of ^{235}U. In contrast, a combustion process yields only a few electron volts for every atom of fuel consumed.

There are many different processes by which a heavy nucleus like ^{235}U can undergo fission, for example,

$$\text{n} + {}^{235}\text{U} \rightarrow {}^{236}\text{U}^* \rightarrow {}^{143}\text{Ba} + {}^{90}\text{Kr} + 3\text{n} + E_{\text{released}}$$

and

$$\text{n} + {}^{235}\text{U} \rightarrow {}^{236}\text{U}^* \rightarrow {}^{140}\text{Xe} + {}^{92}\text{Sr} + 2\text{n} + E_{\text{released}}$$

Notice that the total number of protons and the total number of neutrons both remain the same in these processes. The ^{235}U nucleus has 92 protons and therefore 235 minus 92, or 143 neutrons. In the first of the two decay processes, the barium isotope ^{143}Ba has 56 protons and 87 neutrons, and the krypton isotope ^{90}Kr has 36 protons and 54 neutrons. The total number of protons after the fission has occurred is then 56 plus 36, or 92. The total number of neutrons before the fission is 143 plus 1 (accounting for the neutron that starts the process), or 144. After the fission, the number of neutrons is 87 plus 54, plus 3 to account for the three neutrons released in the process; the total is 144, as expected. You can easily verify that the number of protons and the number of neutrons remains the same in the second process above.

$$E_B = (Nm_n + Zm_{^1H} - m_{\text{atom}})c^2 \tag{27-2}$$

❓ Got the Concept 27-4
Fission Processes

Fill in the missing information for the direct decay products in each of these fission processes.

(a) $^{235}\text{U} + \text{n} \rightarrow {}^{144}? + {}^{89}\text{Kr} + ?\,\text{n} + E_{\text{released}}$
(b) $^{239}\text{Pu} + \text{n} \rightarrow {}^{137}\text{Xe} + ? + 3\text{n} + E_{\text{released}}$
(c) $^{239}\text{Pu} + \text{n} \rightarrow {}^{148}? + {}^{?}\text{Kr} + 3\text{n} + E_{\text{released}}$

Example 27-5 Fission to Xenon and Beyond

Determine the energy released when a ^{235}U nucleus undergoes fission to ^{140}Xe, ^{92}Sr, and two neutrons. A table of isotopes lists the binding energy per nucleon in the nuclei of ^{235}U, ^{140}Xe, and ^{92}Sr as 7.59 MeV, 8.29 MeV, and 8.65 MeV, respectively.

SET UP
The energy released during the fission process is the difference between the binding energy of ^{140}Xe and ^{92}Sr nuclei and the binding energy of the ^{235}U nucleus. (There is no binding energy associated with the two neutrons; they are not bound to any other particle.) The total binding energy for each is determined by multiplying the number of nucleons A in each by the value of E_B/A for that value of A. The energy released is then

$$E_{\text{released}} = 140\, E_B/A\,(A = 140) + 92\,\&E_B/A\,(A = 92) - 235\, E_B/A\,(A = 235)$$

SOLVE
We have been given values of E_B/A for the three isotopes involved in the process, so we can determine E_{released}:

$$E_{\text{released}} = 140\,(8.29\text{ MeV}) + 92\,(8.65\text{ MeV}) - 235\,(7.59\text{ MeV})$$

or

$$E_{\text{released}} = 173 \text{ MeV}$$

REFLECT

This result is consistent with our earlier claim that a typical amount of energy released in the fission of ^{235}U is around 200 MeV.

Practice Problem 27-5 Determine the energy released when a ^{235}U nucleus undergoes fission to ^{134}Te, ^{99}Zr, and three neutrons. A table of isotopes lists the binding energy per nucleon in the nuclei of ^{235}U, ^{134}Te, and ^{99}Zr as 7.59 MeV, 8.38 MeV, and 8.54 MeV, respectively.

All of the examples of fission reactions above result in the release of two or three neutrons. Imagine what can happen if a large number of ^{235}U atoms are close to each other. Should one nucleus be struck by a neutron and fission according to Equation 27-3, there would then be three neutrons moving through the uranium. Should each of these strike a ^{235}U nucleus and start a fission process, there would be nine neutrons, so possibly nine more fissions. With a sufficient number of ^{235}U atoms present, this *chain reaction* quickly grows, with an accompanying rapid increase in energy released.

Isotopes that are capable of sustaining a fission chain reaction are used as nuclear fuels. Such isotopes are termed *fissile*; the most common fissile nuclear fuels are ^{233}U, ^{235}U, ^{239}Pu, and ^{241}Pu. The fission reactions they undergo are characterized by the production of, typically, two or three neutrons in addition to larger fragments. In addition, a chain reaction in a fissile material can be induced by a neutron carrying essentially zero kinetic energy. Indeed, in fuels such as ^{235}U, slower, less energetic neutrons are more efficiently absorbed by the fissile nuclei.

The process that produces electricity in most nuclear reactors uses the energy released in a chain, fission reaction to heat water and produce steam to drive generators not unlike the turbines beneath the Grand Coulee Dam (Figure 20-13). There are several challenges to producing energy in this way. First, a minimum amount, or *critical mass*, of fissile material must be present to sustain a chain reaction. As it happens, only a small percentage, about 0.7%, of the naturally occurring uranium in the world is ^{235}U, and the other three commonly used fissile isotopes do not occur naturally. Most naturally occurring uranium is ^{238}U. To use uranium as a nuclear fuel, then, it is necessary to separate the ^{235}U atoms from the ^{238}U, a costly and difficult process known as *enrichment*. In addition, the ^{238}U atoms that inevitably remain tend to absorb free neutrons and thereby inhibit a chain reaction. Once the critical mass of ^{235}U has been assembled, controlling the chain reaction is another challenge. If too many of the neutrons produced in the fissions result in a second fission, the energy released increases so rapidly that the fuel and whatever vessel is used to contain it can be damaged or even melt. To control a chain fission reaction in a nuclear reactor, control rods made of a substance that is a good absorber of neutrons are inserted between pieces of fuel.

Operating a fission nuclear reactor safely is perhaps the most significant challenge posed by nuclear power generation. First, the fission fragments are radioactive. Many of the fragments, or the fragments produced when they decay, are long-lived and tend to produce dangerous radiation for years, centuries, or even longer. In addition, reactors commonly use water as a moderator, a material that tends to slow the free neutrons (in order to make them more easily absorbed by a ^{235}U nucleus), but should the containment vessel rupture, the hot water can be released into the atmosphere in the form of steam carrying radioactive particles. Perhaps the most significant nuclear accident occurred at the Chernobyl Nuclear

Power Plant in Ukraine in 1986, in which an uncontrolled chain reaction caused a catastrophic power increase, leading to a series of explosions and the release of large quantities of radioactive steam, fuel, and smoke into the environment.

★ What's Important 27-3

The binding energy per nucleon of nuclides with more than about 60 neutrons and protons is smaller for increasing values of A. For this reason, the fission process, in which a nucleus breaks up into smaller fragments, leads to more stable configurations of the nucleons in bigger nuclei.

27-4 Fusion

In the last section, we explored nuclear fission, processes by which a large nucleus splits into fragments to move to a more stable configuration of the neutrons and protons it contains. Is smaller always better? The larger the binding energy per nucleon (E_B/A) in a nucleus, the more stable it is, and on the plot of E_B/A versus mass number (Figure 27-4), the very largest nuclei have lower values of E_B/A than smaller ones when A is greater than about 60. This is not the case, however, for small nuclei. With a few exceptions, as we move up toward a mass number A of around 60, the binding energy per nucleon increases. In other words, for small nuclei, moving to a more stable configuration requires processes that make the nucleus larger. These are *fusion* processes.

Consider the process by which two ^{3}He nuclei fuse together to form ^{4}He. First, simply by counting the protons and neutrons, we know that something else must be formed. Each ^{3}He nucleus contain two protons (by definition, helium always has two protons) and one neutron, so between the two ^{3}He nuclei there are four protons and two neutrons. There are two protons and two neutrons in ^{4}He; we need to account for two more protons. Without more neutrons, there is no way for the protons to be bound together in a single nucleus, so the extra protons leave the reaction separately. We label them as ^{1}H nuclei, and then write the process as

$$^3\text{He} + {}^3\text{He} \rightarrow {}^4\text{He} + {}^1\text{H} + {}^1\text{H} + \gamma \qquad (27\text{-}4)$$

Energy is released during the process because the final configuration of the protons and neutrons is more stable than the initial configuration. The γ represents a photon that carries away the energy released.

How much energy is released when two ^{3}He nuclei fuse together? As in fission processes, the energy released in fusion processes is the difference between the energy tied up in the binding of the result of the fusion and the energy tied up in the binding of the original nuclei, in this case, two ^{3}He nuclei. We can read the binding energy per nucleon for ^{3}He directly from Figure 27-4; the value of E_B/A is approximately 2.5 MeV. Each ^{3}He has three nucleons, which results in a total binding energy of 7.5 MeV, so the two ^{3}He nuclei together have binding energy equal to 15.0 MeV. In Section 27-2, we determined the binding energy of a ^{4}He nucleus to be 28.3 MeV, which is consistent the value of about 7.1 MeV per nucleon on the plot in Figure 27-4. The two ^{1}H nuclei are single particles, so they make no contribution to the binding energy. The difference between the energy tied up in binding at the end of the fusion process and initially is therefore the difference between 28.3 MeV and 15.0 MeV. About 13 MeV of energy is released when two ^{3}He fuse to form ^{3}He. The actual value is closer to 12.86 MeV.

The process described in Equation 27-4 is the final step in the **proton–proton cycle**, the fusion process that is the source of the Sun's energy. The cycle begins with the fusing of two protons (the nuclei of ^{1}H) to form ^{2}H. The ^{2}H nucleus fuses with another proton, forming ^{3}He, and finally, two ^{3}He nuclei fuse to form ^{4}He. The three steps are summarized as

$$^1\text{H} + {^1\text{H}} \rightarrow {^2\text{H}} + e^+ + \nu_e \tag{27-5}$$

$$^2\text{H} + {^1\text{H}} \rightarrow {^3\text{He}} + \gamma \tag{27-6}$$

$$^3\text{He} + {^3\text{He}} \rightarrow {^4\text{He}} + {^1\text{H}} + {^1\text{H}} + \gamma \tag{27-7}$$

In the first step, the e^+ particle is a *positron*, a particle similar to an electron but with positive charge, and ν_e is a *neutrino*, a nearly massless, neutral particle. The first step, and the subsequent interaction of the positron with an electron in the Sun, releases 1.44 MeV of energy. The energy released in the second step is 5.49 MeV. Both of the steps must occur twice before the third can occur (because two ^{3}He are required for the third step), so a total of 2(1.44 MeV) plus 2(5.49 MeV) plus 12.86 MeV from the fusion of two ^{3}He, or 26.7 MeV of energy is released.

Fission processes release more energy than the proton–proton fusion cycle, around 200 MeV compared to 26.7 MeV. It might seem therefore, that fission is a more effective way to convert fuel to energy. Consider, however, that while 235 nucleons in ^{235}U are spent in order to release 200 MeV, in the fusion process described above, the 26.7 MeV released come at the expense of only four nucleons. (Six nucleons must be available for the first two steps to occur twice, but two protons are left over at the end of the third step.) Comparing energy per nucleon (think miles per gallon), fission provides less than 1 MeV per nucleon, while proton–proton fusion gives 26.7 MeV divided by 4, or nearly 7 MeV per nucleon. Fusion processes are more efficient at releasing the energy held in binding nuclei together.

Hydrogen makes up about 75% of the Sun's mass of approximately 2×10^{30} kg. That's a lot of hydrogen! If the proton–proton cycle leads to more stability, why doesn't all of that hydrogen quickly fuse to form ^{4}He, leaving the Sun a gigantic (and cool) ball of helium gas? The answer lies in the same forces at play within nuclei: the Coulomb force that repels protons from each other and the strong but short-ranged nuclear force that draws them together. In order for two protons to fuse to form ^{2}H, they must be less than about 10 fm apart so that the nuclear force attraction is able to overcome the Coulomb repulsion. This requires the protons to have considerable energy, some of which comes from being at high temperature. A temperature of more than 4×10^6 K is required for the proton–proton cycle to start. That's hot, but not as hot as the temperature at the core of the Sun, around 15×10^6 K. Even at that temperature, however, the probability that two nearby protons will fuse is small, which means that only a small fraction—about 4×10^{-19}—of the hydrogen in the Sun is undergoing fusion at any one time. The Sun won't burn out for a long time.

In massive stars, more than about 1.2 times the mass of the Sun, core temperatures can be high enough for a second fusion cycle to occur. At temperatures above about 13×10^6 K, a carbon nucleus can fuse with a proton to form an isotope of nitrogen, the first step in a series that ultimately results in the production of another carbon and also a helium nucleus. There are a number of variations of the cycle, but the most common way it evolves is

$$^{12}\text{C} + {^1\text{H}} \rightarrow {^{13}\text{N}} + \gamma$$

$$^{13}\text{N} \rightarrow {^{13}\text{C}} + e^+ + \nu_e$$

$$^{13}\text{C} + {^1\text{H}} \rightarrow {^{14}\text{N}} + \gamma$$

$$^{14}\text{N} + {^1\text{H}} \rightarrow {^{15}\text{O}} + \gamma$$

$$^{15}\text{O} \rightarrow {^{15}\text{N}} + e^+ + \nu_e$$

$$^{15}\text{N} + {^1\text{H}} \rightarrow {^{12}\text{C}} + {^4\text{He}}$$

Because the process has intermediate states of carbon, nitrogen, and oxygen it is referred to as the carbon, nitrogen, and oxygen cycle (or the CNO cycle), or simply the **carbon cycle**. Notice that only four protons are consumed in the process, which releases energy of about 26.7 MeV.

The core temperature of the Sun is higher than the temperature at which the carbon cycle turns on, but not so high that the carbon cycle is more probable than the proton–proton cycle. The carbon cycle produces about 8% of the energy radiated by the Sun.

Estimate It! 27-3 Carbon and Nitrogen

Estimate the amount of energy released in the last step of the carbon cycle.

SET UP
The last step in the carbon cycle is

$$^{15}N + {}^{1}H \rightarrow {}^{12}C + {}^{4}He$$

The energy released is the difference between the binding energy of the ^{15}N nucleus and the energy tied up in the binding of the ^{12}C nucleus and the binding of the ^{4}He nucleus. (There is no binding energy in the ^{1}H nucleus because we treat it is a single proton.) We have previously determined that the binding energy of the ^{4}He nucleus is 28.3 MeV. The binding energy of ^{15}N and ^{12}C is the product of the number of nucleons and the binding energy per nucleon for each. So the energy released is

$$E_{released} = 12\, E_B/A\,(A = 12) + 28.3\,\text{MeV} - 15\, E_B/A\,(A = 15)$$

From Figure 27-4, the binding energy per nucleon is about 7.6 MeV for both ^{12}C and ^{15}N.

SOLVE
With values of E_B/A read from the curve in Figure 27-4, the energy released in the last step of the carbon cycle is

$$E_{released} = 12\,(7.6\,\text{MeV}) + 28.3\,\text{MeV} - 15\,(7.6\,\text{MeV})$$

or

$$E_{released} = 5.5\,\text{MeV}$$

REFLECT
Our answer is close to the accepted value of 4.96 MeV of energy released in the last step of the carbon cycle.

★ What's Important 27-4
The binding energy per nucleon of nuclides that have fewer than about 60 neutrons and protons is smaller for decreasing values of mass number A. The fusion process, in which two nuclei join together to form one larger one, leads to more stable configurations of the nucleons in smaller nuclei.

27-5 Radioactivity

For many people, including you perhaps, phrases such as radioactivity and nuclear radiation have come to be synonymous with "danger." The association is not exactly accidental, given the continued development of nuclear weapons and well-publicized accidents at nuclear power plants. Not all nuclear radiation is dangerous, however. In addition, the level of radiation in our environment is not limited to that released during the explosion of a bomb or after an incident at a power plant. In this section, we explore a number of aspects of nuclear radiation.

All naturally occurring nuclear processes take place because the final state is more stable, more energetically favorable, than the initial state. This is true of fission and fusion processes, and it is true of radiation processes, too. A relatively few nuclides, 266 out of over 3000, are stable. All the rest are radioactive, that is, they decay into another nuclide by radiating away one or more particles. Depending on the process, the radiation can carry away a small or a large amount of energy.

The three most common modes by which radiation occurs are alpha, beta, and gamma radiation. The terms are historical; at the close of the 19th century Ernest Rutherford classified radiation according to the depth that a radiation particle was able to penetrate other objects. Alpha particles penetrated the least, beta particles more, and gamma particles the most. We now know that alpha particles are the nuclei of ^{4}He, beta particles are electrons, and gamma particles are electromagnetic waves (photons). Gamma rays do not fall in the visible spectrum; their energies can be millions of times higher than visible photons.

Nuclear radiation is a statistical process. Consider the emission of a beta particle in the decay of ^{90}Sr, which proceeds as

$$^{90}\text{Sr} \rightarrow ^{90}\text{Y} + \text{e}^- + \bar{\nu}_e \tag{27-8}$$

The beta particle is the electron (e$^-$), and $\bar{\nu}_e$ is an *antineutrino*, related to the light, neutral neutrino particle we discussed in the context of fusion in the last section. Every ^{90}Sr can, and eventually will, decay radioactively to yttrium. In addition, we know that if a sample of ^{90}Sr contains, say, 1000 atoms, after 28.78 years, only about 500 will be left. However, if we select one of the 1000 ^{90}Sr atoms in particular, we have no idea when its nucleus will decay. Perhaps right . . . now! Or perhaps not for 30 or 40 years, or more. We can't make a definitive claim about when any particular atomic nucleus will decay, but we can quantify the rate at which a group of radioactive atoms decays. The standard way to describe that rate is with the time it takes, statistically, for one-half of a large number of the atoms to decay. This time is known as the isotope's *half-life*. The half-life of a radioactive isotope quantifies the number of decays per second, that is, the rate at which radioactive decays occur.

The number of decays per second in a sample depends on how much of the radioactive material is present. This observation underlies the mathematical tools we need to understand radioactive decay. Imagine we place samples of two different radioactive elements, separately, into an experimental apparatus that detects beta particles. After the first sample has been in the device for 1 s we observe 100 decays. In the same amount of time, only one decay is observed in the second sample. Can we conclude that the decay is 100 times higher for the first sample? No, and in fact, it could even be the case that the decay rate for the first sample is lower than that for the second. For example, if there were initially one million radioactive atoms in the first sample and only 100 radioactive atoms in the second, only one out of 10,000 atoms of the first sample decay per second while one out of a 100 decay per second in the other.

Each decay in a sample represents a change in the number N of radioactive atoms present. As we have just seen, the rate of change dN/dt depends on N. We characterize this mathematically as

$$\frac{dN}{dt} \propto -N$$

The rate of change in the number of atoms is negative because as time increases the number of atoms decreases. The proportionality constant in this expression, termed the *decay constant*, gives the probability per unit time that any particular nucleus will decay. The decay constant is usually represented by λ, so

$$\frac{dN}{dt} = -\lambda N$$

We can solve the equation as an expression for the number of atoms present at any time by first gathering the terms associated with amount (N) on one side of the equation and the terms associated with time (t, λ) on the other:

$$\frac{dN}{N} = -\lambda dt$$

and then integrating both sides from some initial time, when t is set equal to zero and the number of atoms is N_0, to a later time t when the number of atoms is $N(t)$, so

$$\int_{N_0}^{N(t)} \frac{dN}{N} = \int_0^t -\lambda dt$$

and therefore

$$\ln N \Big|_{N_0}^{N(t)} = -\lambda t \Big|_0^t$$

or

$$\ln N(t) - \ln N_0 = -\lambda t \tag{27-9}$$

We can simplify the equation by first recognizing that $\ln a - \ln b$ equals $\ln a/b$, and also that $e^{\ln a} = a$. Then Equation 27-9 becomes

$$\ln (N(t)/N_0) = -\lambda t$$

Taking both sides as a power of e removes the natural logarithm:

$$e^{\ln (N(t)/N_0)} = e^{-\lambda t}$$

so

$$N(t)/N_0 = e^{-\lambda t}$$

and finally

$$N(t) = N_0 e^{-\lambda t} \tag{27-10}$$

The number of atoms in a radioactive sample decreases exponentially with time.

From $N(t)$ we can also write an expression for the decay rate, the time derivative of $N(t)$. By convention decay rate R is defined as the negative of the time derivative, that is,

$$R = -\frac{dN(t)}{dt} = -\frac{d}{dt} N_0 e^{-\lambda t} = \lambda N_0 e^{-\lambda t}$$

It will be useful to define the initial decay rate R_0,

$$R_0 = \lambda N_0 \tag{27-11}$$

so

$$R = R_0 e^{-\lambda t} \tag{27-12}$$

We see that the decay rate, as well as the number of atoms in a radioactive sample, decreases exponentially with time. The SI unit of decay rate is the becquerel (Bq), after the 19th century French physicist Antoine Henri Becquerel, who along with Marie and Pierre Curie won the Nobel Prize in Physics for their discovery of radioactivity. The becquerel is equivalent to one radioactive decay per second.

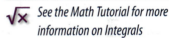

See the Math Tutorial for more information on Integrals

See the Math Tutorial for more information on Exponents and Logarithms

Physicists also commonly use units of curies (Ci) for decay rate; 1 Ci is equivalent to 3.7×10^{10} Bq.

The **half-life** $\tau_{1/2}$ is defined as the time it takes for half of the atoms in a sample to radioactively decay. That is, $N(\tau_{1/2})$ equals $N_0/2$. In addition, from Equation 27-10,

$$N(\tau_{1/2}) = N_0 e^{-\lambda \tau_{1/2}}$$

Putting the two relationships for $N(\tau_{1/2})$ together,

$$N_0 e^{-\lambda \tau_{1/2}} = \frac{N_0}{2}$$

N_0 cancels, leaving

$$e^{-\lambda \tau_{1/2}} = \frac{1}{2}$$

Or, by taking the natural log of both sides,

$$-\lambda \tau_{1/2} = \ln \frac{1}{2}$$

Recall that $-\ln a$ equals $\ln(1/a)$, so

$$\lambda \tau_{1/2} = \ln 2$$

and then

$$\tau_{1/2} = \frac{\ln 2}{\lambda} \tag{27-13}$$

The half-life of radioactive sources varies widely, from far less than one second to billions of years or more. Commonly used radioactive sources include ^{32}P, a beta emitter used in DNA research that has a half-life of 14.3 days, ^{241}Am, an alpha emitter often found in household smoke detectors that has a half-life of 432.2 y, and ^{238}U, an alpha emitter with a half-life of 4.47×10^9 y. Notice that none of these have a half-life on the order of a second or less; radioactive isotopes with half-lives that short aren't terribly useful because they don't stay around long enough.

? Got the Concept 27-5
Half and Half Again

A certain radioactive isotope has a half-life of 100 days. (a) What fraction of the initial number of atoms remains after 200 days? (b) What fraction of the initial number of atoms remains after 500 days? (c) About how long would you need to wait before only one-thousandth of the initial number of atoms remain?

Example 27-6 Technetium

A form of an isotope of technetium, ^{99m}Tc, is a beta emitter with a half-life of 6 h. (Technetium-99m is widely used for medical purposes, partially because it doesn't stay in the body for long.) (a) What is the decay constant λ, the probability that an atom of ^{99m}Tc will decay per second, in a sample? (b) Does λ change as time goes on? (c) What fraction of the initial number of ^{99m}Tc will be left after 1 day? (d) What fraction will be left after 4 days have elapsed?

SET UP

(a) The decay constant λ is proportional to the half-life. From Equation 27-13,

$$\lambda = \frac{\ln 2}{\tau_{1/2}} \tag{27-14}$$

(c) The number of atoms left at time t is given by Equation 27-10,

$$N(t) = N_0 e^{-\lambda t}$$

So the fraction that remain is

$$\frac{N(t)}{N_0} = e^{-\lambda t}$$

To find a numeric value for this fraction at any time we could find λ and then substitute a value for t. It makes the relationship between λ and $\tau_{1/2}$ clearer, however, if we first substitute the relationship for λ into the previous equation for the fraction of atoms left at time t:

$$\frac{N(t)}{N_0} = e^{-\frac{\ln 2}{\tau_{1/2}}t} = \left[e^{\ln 2}\right]^{-\frac{t}{\tau_{1/2}}} = \left[2\right]^{-\frac{t}{\tau_{1/2}}}$$

where we made use of the fact that $e^{\ln 2}$ equals 2. Also, any number taken to a negative power is equivalent to the inverse of that number taken to the absolute value of the power, so

$$\frac{N(t)}{N_0} = \left[\frac{1}{2}\right]^{\frac{t}{\tau_{1/2}}} \tag{27-15}$$

From this we see that the fraction of atoms left after a time t has passed is $\frac{1}{2}$ raised to the power of the number of half-lives that have elapsed, even when that number is not an integer.

SOLVE

(a) To get the decay constant, we need to convert the half-life to seconds before applying Equation 27-14:

$$\lambda = \frac{\ln 2}{(6\text{ h})(3600\text{ s/h})} = 3.21 \times 10^{-5}\text{ s}^{-1}$$

(b) The decay constant λ depends only on the half-life so it does not vary with time.
(c) We can apply Equation 27-15 to determine how many atoms are left after 1 day. Notice that we don't need to convert the time to seconds; because the fraction $t/\tau_{1/2}$ is dimensionless, we'll get the correct answer as long as we use the same units for both. So after 1 day (24 h),

$$\frac{N(t)}{N_0} = \left[\frac{1}{2}\right]^{\frac{24\text{ h}}{6\text{ h}}} = \left[\frac{1}{2}\right]^4 = \frac{1}{16}$$

(d) After 4 days (96 h), the fraction of atoms that remain is

$$\frac{N(t)}{N_0} = \left[\frac{1}{2}\right]^{\frac{96\text{ h}}{6\text{ h}}} = \left[\frac{1}{2}\right]^{16} = \frac{1}{65536}$$

REFLECT

Technetium-99m is widely used for medical purposes partially because it doesn't stay in the body for long. After 4 days, only about one hundred-thousandth (very little) of whatever amount of $^{99\text{m}}$Tc was used in a procedure could remain in the patient's body. Also, more generally, the fraction of radioactive atoms left after a time t has passed is $\frac{1}{2}$ raised to the power of the number of half-lives that have elapsed, even when that number is not an integer.

Practice Problem 27-6 Iodine-131, sometimes referred to as radioiodine, has a half-life of 8.02 days and decays by beta and gamma radiation. (a) What is its decay constant λ, the probability that an atom of ^{131}I will decay per second in a sample? (b) What fraction of the initial number of ^{131}I will be left after 1.00 day?

Alpha Radiation

When a large nucleus breaks into smaller fragments during a nuclear fission process, the new configuration of the protons and neutrons is more stable because the total binding energy increases. There are many ways, of course, for a large nucleus to split into smaller pieces, but the more likely decay products would be those that are more stable, that is, those with larger binding energy per nucleon. It is also probable, however, for smaller fragments to be ejected from a large nucleus. The radioactive emission of a ^{4}He nucleus is therefore the most likely decay process for a large nucleus. ^{4}He carries more binding energy than any other nucleus with a small value of A. (Look again at Figure 27-4.) So the α particle, the nucleus of the ^{4}He atom, is far more stable than other small nuclei and therefore a far more probable decay product of large nuclei.

The alpha radiation process reduces the number of protons (and electrons) Z of the initial, or parent, nucleus by two, and reduces the number of neutrons of the parent by two. The result is a daughter nucleus that has an atomic number of $Z-2$ and a mass number of $A-4$, accompanied by an alpha particle. The *alpha decay process* can be summarized as

$$^A\text{P}^Z \rightarrow {}^{A-4}\text{D}^{Z-2} + \alpha \tag{27-16}$$

where P and D represent the parent and daughter nucleus, respectively. The daughter nucleus and the alpha particle carry away the energy released, although because the daughter is far more massive than the alpha particle, the kinetic energy of the alpha particle is far greater than the kinetic energy of the daughter. If you're surprised that the more massive daughter has far less kinetic energy than the alpha particle, we can find its kinetic energy K_D in terms of the kinetic energy K_α of the alpha particle. First,

$$K_D = \frac{1}{2}m_D v_D^2$$

where m_D is the mass of the daughter nucleus and v_D is its speed after the alpha particle is emitted. Momentum is conserved in the alpha decay process, so

$$m_D v_D = m_\alpha v_\alpha$$

or

$$v_D = \frac{m_\alpha}{m_D}v_\alpha$$

So the kinetic energy of the daughter nucleus can be written

$$K_D = \frac{1}{2}m_D\left(\frac{m_\alpha}{m_D}v_\alpha\right)^2 = \left(\frac{m_\alpha}{m_D}\right)\frac{1}{2}m_\alpha v_\alpha^2$$

In other words, the kinetic energy of the daughter nucleus is equal to the product of the fraction m_α/m_D and the kinetic energy of the alpha particle. The mass of the alpha particle is small compared to the mass of the daughter nucleus, making the fraction m_α/m_D much less than one, so K_D is small compared to K_α.

The element thorium, which contains 90 protons, is an alpha emitter. Radium contains two fewer protons (88), so following Equation 27-16 the α decay of ^{228}Th, for example, is

$$^{228}\text{Th} \rightarrow {}^{224}\text{Ra} + \alpha$$

All heavy nuclei (above Z of 82 or 83) are unstable, so they have some probability of decay by alpha emission. Just as ^{228}Th decays to ^{224}Ra, ^{224}Ra undergoes α decay to ^{220}Rn. The process continues:

$$^{228}\text{Th} \rightarrow {}^{224}\text{Ra} + \alpha$$
$$\hookrightarrow {}^{220}\text{Rn} + \alpha$$
$$\hookrightarrow {}^{216}\text{Po} + \alpha$$
$$\hookrightarrow {}^{212}\text{Pb}$$

Notice that 228 is an integer number of four, and because α decay reduces mass number by four, the mass number of each daughter nucleus in the chain is also an integer multiple of four. For that reason these decays are part of the sequence known as the *4n series*. Three other such series are possible, the $4n+1$, $4n+2$, and $4n+3$ series. Table 27-1 lists the longest-lived parent nucleus in each chain, along with its half-life. In each case, because there is no nucleus in the chain with a longer half-life than that particular nuclear species, there is no way to naturally replenish its supply. Notice that the half-life of neptunium-237, the longest lived parent nucleus in the $4n+1$ series, is short compared to the age of Earth, 4.5×10^9 y. Since Earth formed, about 2000 half-lives of ^{237}Np have elapsed (4.5×10^9 y divided by 2.14×10^6 y), which means that, from Equation 27-15, the ratio of the current amount of ^{237}Np in the world to the amount present when Earth formed, is $\left(\frac{1}{2}\right)^{2000}$. Put that into your calculator—it's so close to zero that there simply is no naturally occurring neptunium-237 on Earth.

Table 27-1	Alpha Decay Chains	
Series	**Longest-lived nucleus**	**Half-life (y)**
$4n$	^{232}Th	1.41×10^{10}
$4n + 1$	^{237}Np	2.14×10^6
$4n + 2$	^{238}U	4.47×10^9
$4n + 3$	^{235}U	7.04×10^8

? Got the Concept 27-6
Alpha Decay Processes

Write out the α decay of the nucleus of (a) ^{222}Rn, (b) ^{210}Po, and (c) ^{252}Cf.

Example 27-7 Alpha Decay of ^{238}U

^{238}U undergoes α decay to ^{234}Th. The binding energy per nucleon in ^{238}U is 7.570 MeV and 7.597 MeV in ^{234}Th. Find the energy released in the process ^{238}U $\rightarrow {}^{234}$Th $+ \alpha$.

SET UP

The energy released is the difference between sum of the binding energy of the alpha particle and the binding energy of ^{234}Th, and the energy tied up in the binding of the parent ^{238}U nucleus:

$$E_{\text{released}} = E_B({}^{234}\text{Th}) + E_B({}^4\text{He}) - E_B({}^{238}\text{U})$$

SOLVE

We have previously determined that the binding energy of the ^{4}He nucleus is 28.3 MeV. (Note that to four significant digits the binding energy of ^{4}He is 28.30 MeV.)

The binding energy of ^{234}Th and ^{238}U is the product of the number of nucleons and the binding energy per nucleon for each. So the energy released is

$$E_{\text{released}} = 234(7.597\,\text{MeV}) + 28.30\,\text{MeV} - 238(7.570\,\text{MeV})$$

or

$$E_{\text{released}} = 4.338 \text{ MeV}$$

REFLECT

The alpha particles emitted by radioactive isotopes that have long half-lives tend to have kinetic energies in the 4 to 5 MeV range. The half-life of ^{238}U is relatively long, about 4.47×10^9 y (see Table 27-1).

Practice Problem 27-7 Thorium-234 undergoes α decay to ^{230}Ra. The binding energy per nucleon in ^{234}Th is 7.597 MeV and 7.622 MeV in ^{230}Ra. Find the energy released in the process ^{234}Th $\rightarrow$ ^{230}Ra $+ \alpha$.

Beta Radiation

In all nuclei other than ^{1}H, the Coulomb repulsion between protons is counteracted, to a greater or lesser extent, by the attraction between both protons and neutrons by the nuclear force. The stability of any particular nuclear isotope is therefore related to the number of nucleons A. However, a number of nuclides exist for any particular value of A; for example, in addition to technetium (see Example 27-6), molybdenum, ruthenium, and rhodium have an isotope that has 99 nucleons. Molybdenum-99 has 42 protons and 57 neutrons, ^{99}Tc has 43 protons and 56 neutrons, ^{99}Ru has 44 protons and 55 neutrons, and ^{99}Rh has 45 protons and 54 neutrons. (It is also possible to create isotopes of other elements that have A equal to 99.) The binding energy per nucleon in each is slightly different, however, so that only one is the most stable. For A equal to 99, the most stable isotope is ^{99}Ru, but this raises an interesting question. If, for example, ^{99}Tc is not the most stable, does a process exist by which ^{99}Tc can become ^{99}Ru, and thereby attain a more stable configuration of its 99 nucleons?

Any process that transforms ^{99}Tc into ^{99}Ru must satisfy these requirements:

- The total number of nucleons remains the same.
- The number of protons increases by one, that is, Z becomes $Z + 1$.
- The net charge remains the same.

You might guess that this process involves the transformation of a neutron into a proton, which would satisfy both the first and second requirements. That's correct, but just that alone would violate the third requirement. As part of this transformation an electron is also produced. Both the initial state (a neutron) and the final state (a positively charged proton and a negatively charged electron) have a net charge of zero; the net charge remains the same in a process in which a neutron transforms into a proton and an electron. To account for other conservation requirements a third particle, the neutral and nearly massless antineutrino ($\bar{\nu}_e$) is also created in the process. The full process, then, is

$$\text{n} \rightarrow \text{p} + \text{e}^- + \bar{\nu}_e \tag{27-17}$$

Because the electron is negative, we refer to this as *beta minus* or β^- *decay*.

The same underlying physics relationships allow a proton in an atomic nucleus to transform into a neutron. The β^+ *decay* process is

$$\text{p} \rightarrow \text{n} + \text{e}^+ + \nu_e \tag{27-18}$$

The e$^+$ particle is a positron, a particle identical to an electron in every way except that it carries positive charge. The ν_e neutrino is, for our purposes, identical to the $\bar{\nu}_e$ antineutrino.

We can now understand the β^- decay of ^{90}Sr (Equation 27-8) discussed earlier. Strontium contains 38 protons, so ^{90}Sr contains 52 neutrons. When ^{90}Sr undergoes β^- decay,

$$^{90}\text{Sr} \rightarrow {}^{90}\text{Y} + e^- + \bar{\nu}_e$$

a neutron in the ^{90}Sr nucleus transforms to a proton according to Equation 27-17. The nucleus then contains 39 protons, which is the value of Z for the element yttrium, but the number of protons plus the number of neutrons remains 90. So the ^{90}Sr nucleus becomes a ^{90}Y nucleus. According to Equation 27-17, an electron and an antineutrino is also produced, which provides the full detail of the decay.

How, then, does ^{99}Tc attain a more stable configuration of its 99 nucleons by transforming to ^{99}Ru? It is through a β^- decay:

$$^{99}\text{Tc} \rightarrow {}^{99}\text{Ru} + e^- + \bar{\nu}_e$$

We now also know how ^{99}Rh, which has 45 protons and 54 neutrons, transforms to the more stable ^{99}Ru:

$$^{99}\text{Rh} \rightarrow {}^{99}\text{Ru} + e^+ + \nu_e$$

One of the protons in the ^{99}Rh nucleus undergoes β^+ decay, resulting in a nucleus of 44 protons and 55 neutrons (ruthenium). According to Equation 27-18, a positron and a neutrino are also produced. So ^{99}Rh decays to ^{99}Ru by β^+ emission.

? Got the Concept 27-7
Beta Decay Processes

Write out (a) the β^- decay of the nucleus of ^{3}H (tritium) and (b) the β^+ decay of the nucleus of ^{59}Ni.

A common application of the occurrence of beta decay is in carbon-14 dating, a technique used to measure the age of objects that are composed, or partially composed, of organic matter. Some small fraction of the carbon in Earth's atmosphere, for example, the carbon in CO_2, is ^{14}C. This radioactive isotope of carbon is a β^- emitter:

$$^{14}\text{C} \rightarrow {}^{14}\text{N} + e^- + \bar{\nu}_e$$

Carbon-14 is constantly produced in the atmosphere by cosmic rays slamming into ^{14}N atoms, and as a result, even though ^{14}C radioactively decays, the ratio of ^{14}C to ^{12}C in the atmosphere has remained relatively constant for at least tens of thousands of years. The $^{14}\text{C}/^{12}\text{C}$ ratio is the same in living organisms, for example, plants that breathe in CO_2, as it is in the atmosphere. However, once an organism dies, it no longer replenishes its supply of carbon, so the $^{14}\text{C}/^{12}\text{C}$ ratio decreases as the ^{14}C decays. The half-life of ^{14}C is 5730 y. A measurement of the $^{14}\text{C}/^{12}\text{C}$ ratio in, say, the smoke stains in the Chauvet-Pont-d'Arc Cave in southern France (Figure 27-5), which contains the earliest known cave paintings, allows a determination of time since the firewood that created the smoke was

Figure 27-5 The age of a cave painting on a wall in the Chauvet-Pont-d'Arc Cave in southern France is determined by carbon-14 dating. *(HTO / Wikimedia)*

part of a living tree. Carbon-14 dating places the time the paintings were made between 30,000 and 33,000 years ago.

Example 27-8 Ötzi the Iceman

In 1991, two German hikers discovered a human corpse in the Ötztal Alps in Italy. The remains were not those of the victim of a climbing accident, rather they were a well-preserved natural mummy of a man who lived during the last Ice Age. The rate of radioactive decay of ^{14}C in the mummy of "Ötzi the Iceman" was measured to be 0.121 Bq/g. In a living organism, the rate of radioactive decay of ^{14}C is 0.231 Bq/g. How long ago did Ötzi the Iceman live?

SET UP

Decay rate R decreases exponentially over time as given by Equation 27-12:

$$R = R_0 e^{-\lambda t}$$

Here R is the current decay rate of a sample, and R_0 is the initial decay rate (that is, the decay rate when the sample was alive). We don't directly know a value for λ, but we can write λ in terms of the half-life through Equation 27-14:

$$\lambda = \frac{\ln 2}{\tau_{1/2}}$$

so

$$R = R_0 e^{-\frac{\ln 2}{\tau_{1/2}}t}$$

This equation can be rearranged as an expression for t.

SOLVE

To rearrange the expression for R to solve for the elapsed time, divide both sides by R_0 and then take the natural logarithm of both sides:

$$\ln \frac{R}{R_0} = \ln e^{-\frac{\ln 2}{\tau_{1/2}}t} = -\frac{\ln 2}{\tau_{1/2}}t$$

so

$$t = -\frac{\tau_{1/2}}{\ln 2}\ln \frac{R}{R_0}$$

The time since Özi the Iceman last walked in the Italian Alps is then

$$t = -\frac{5730 \text{ y}}{\ln 2}\ln\frac{0.121 \text{ Bq}}{0.231 \text{ Bq}} = 5350 \text{ y}$$

REFLECT

Notice that the critical part of the carbon-14 dating calculation involves the ratio of the current decay rate to the initial decay rate. Imagine we measured the decay rate of an artifact to be 0.001 Bq, quite low compared to that of a living organism. The time since the artifact was alive would be

$$t = -\frac{5730 \text{ y}}{\ln 2}\ln\frac{0.001 \text{ Bq}}{0.231 \text{ Bq}} = 45,000 \text{ y}$$

This is about eight half-lives of ^{14}C. Because the decay has dropped so low after eight or ten half-lives, ^{14}C dating is limited to making an accurate age measurement on objects that are less than about 50,000 years old.

Practice Problem 27-8 The Dead Sea Scrolls, a collection of texts written mostly on dried animal skins, were found in caves along the northwest shore of the Dead Sea in the late 1940s and 1950s. Carbon-14 dating indicates that the scrolls were written about 2100 years ago. What is the rate of radioactive decay of ^{14}C in the scrolls?

Gamma Radiation

A nucleus in an excited state radiates energy in a way analogous to the emission of an x ray when an electron in an excited atomic state falls to a state of lower energy. The most common way that a nucleus can become excited is following an α or β decay. Although the α and β decay processes result in a more stable configuration of the nucleons, the nucleons that remain in the daughter nucleus may not be, initially, in the most stable arrangement for that particular nuclide. This excited state decays to a more stable configuration, giving off energy in the form of a gamma (γ) ray.

The energy carried away when a nucleus in an excited state decays is on the order of a megaelectron volt, which makes the radiation fall into the γ radiation regime. From Equation 22-1, the relationship between the energy and wavelength of a photon, and taking the photon energy to be 1 MeV, a typical value of the wavelength is therefore

$$\lambda = \frac{hc}{E} = \frac{(4.15 \times 10^{-15}\,\text{eV} \cdot \text{s})\,(3.00 \times 10^8\,\text{m/s})}{1 \times 10^6\,\text{eV}} = 1.25 \times 10^{-12}\,\text{m}$$

(We used the value of h in eV $\cdot$ s.) In other words, the wavelength of the γ-ray photon emitted as nuclear radiation is around 1×10^{-12} m, or 0.001 nm. Recall that the wavelength of visible light photons is 400 to 700 nm or so; nuclear γ radiation is far from the visible.

Gamma radiation does not change the atomic number Z or the mass number A of a nucleus; that is, the number of protons and the number of neutrons remain the same after a γ ray is emitted. For example, the β^- decay of ^{14}C we discussed in the context of radioactive dating actually proceeds through an excited state of ^{14}N:

$$^{14}\text{C} \rightarrow \,^{14}\text{N}^* + e^- + \bar{\nu}_e$$

The * indicates that ^{14}N* is an excited state of ^{14}N. ^{14}N* decays to ^{14}N by the emission of a gamma photon:

$$^{14}\text{N}^* \rightarrow \,^{14}\text{N} + \gamma$$

The values of Z and A do not change in the second step of the radiation process.

Estimate It! 27-4 Change in Mass after α Radiation

The mass of a nucleus changes when it emits a γ ray. Estimate the fractional change in the nuclear mass as a result of the emission of a γ.

SET UP
The change in mass ΔM of the nucleus is the mass equivalent of the energy E_γ carried away by the γ, or, from $E = mc^2$,

$$\Delta M = \frac{E_\gamma}{c^2}$$

The fractional change in the mass of the nucleus is then

$$f = \frac{\Delta M}{M_{\text{nucl}}}$$

where M_{nucl} is the mass of the nucleus before the γ-ray emission. We need some round but reasonable numbers for both E_γ and M_{nucl}. For the purposes of this estimate, we'll use 1 MeV for E_γ. Nuclear masses vary over a large range; however, we know that the mass of ^{4}He is about 3700 MeV/c^2 (see Section 27-2). So let's take 10^4 MeV/c^2 as the mass of a small nucleus.

SOLVE

The fractional change in the mass of the nucleus is

$$f = \frac{1 \text{ MeV}/c^2}{10^4 \text{ MeV}/c^2} = 0.0001$$

REFLECT

Although γ rays are relatively high energy photons, emission of a γ results in only a small change in the mass of the nucleus.

✴ What's Important 27-5

The three most common modes by which radiation occurs are alpha, beta, and gamma radiation. Alpha particles are the nuclei of ^{4}He, beta particles are electrons, and gamma particles are electromagnetic waves (photons).

Answers to Practice Problems

27-1   30

27-2   0.14 nucleon/fm^3

27-3   8.79 MeV

27-4   185 MeV

27-5   (a) 1.00×10^{-6} s^{-1}, (b) 0.917

27-6   3.662 MeV

27-7   0.179 Bq

Answers to Got the Concept Questions

27-1   (a) (i) Sn, tin; (ii) W, tungsten; (iii) Pb, lead; (b) (i) 4; (ii) 11; (iii) 10

27-2   Density is defined as the mass divided by the volume. The mass of any nucleus is approximately proportional to the number of nucleons A. The volume is proportional to the radius r cubed, and by Equation 27-1, r^3 is proportional to A. Because both mass and volume are (approximately) proportional to A, density is (approximately) independent of A, that is, it is the same for all nuclides.

27-3   (c) ^{100}Kr, (b) ^{20}Ne, (d) ^{200}Hg, (a) ^{11}B

27-4   (a) ^{235}U + n → ^{144}Ba + ^{89}Kr + 3n + $E_{released}$, (b) ^{239}Pu + n → ^{137}Xe + ^{100}Zr + 3n + $E_{released}$, (c) ^{239}Pu + n → ^{148}Ce + ^{89}Kr + 3n + $E_{released}$

27-5   (a) Two half-lives elapse in 200 days, and in each about half the sample decays. So the fraction left after 200 days is $\frac{1}{2}$ multiplied by $\frac{1}{2}$, or $\frac{1}{4}$ of the initial number. (b) In a time period of 500 days, 5 half-lives elapse. The fraction of the initial number or atoms left is then $\frac{1}{2} \cdot \frac{1}{2} \cdot \frac{1}{2} \cdot \frac{1}{2} \cdot \frac{1}{2}$, or $\frac{1}{32}$. (c) One-thousandth, or $\frac{1}{1000}$, is approximately equal to $\left(\frac{1}{2}\right)^{10}$. So the number of atoms falls to one-thousandth of its initial number in about 10 half-lives, or 1000 days.

27-6   (a) ^{222}Rn → ^{218}Po + α, (b) ^{210}Po → ^{206}Pb + α, and (c) ^{248}Cf → ^{224}Cn + α

27-7   (a) ^{3}H → ^{3}He + e$^-$ + $\bar{\nu}_e$, (b) ^{59}Ni → ^{59}Co + e$^+$ + ν_e

SUMMARY

Topic	Summary	Equation or Symbol
alpha radiation	An alpha particle is the nucleus of a ^{4}He atom. When a parent nucleus P emits an alpha particle (α), the resulting daughter nucleus D has two fewer protons and two fewer neutrons than the parent.	$^A\text{P}^Z \rightarrow {}^{A-4}D^{Z-2} + \alpha$ (27-16)
atomic number	The number of protons (and electrons) an atom contains is its atomic number. Each element has a unique atomic number.	Z
beta radiation	In the beta decay process a nucleus emits an electron (β^-) through the transformation of a neutron to a proton, or a positron (β^+) through the transformation of a proton to a neutron.	$\text{n} \rightarrow \text{p} + \text{e}^- + \bar{\nu}_\text{e}$ (27-17) $\text{p} \rightarrow \text{n} + \text{e}^+ + \nu_\text{e}$ (27-18)
binding energy	Nuclear binding energy holds the nucleus together. The greater the binding energy per particle in a nucleus, the more tightly the nucleus is bound and therefore the more stable it is. The binding energy E_B of the nucleus of an atom is the difference between the energy equivalent of the mass of the constituent nucleons and the energy equivalent of the mass of the nucleus, or alternatively, the difference between the energy equivalent of the mass of N neutrons and Z ^{1}H atoms, and the energy equivalent of the mass of the atom.	$E_B = (Nm_\text{n} + Zm_{^1\text{H}} - m_\text{atom})c^2$ (27-2)
carbon cycle	The carbon cycle is a multistep fusion process that occurs in stars with core temperatures hotter than about 13×10^6 K. In the carbon cycle, carbon, nitrogen, and oxygen atoms participate in a series of fusion processes that begins with the fusing of ^{12}C with ^{1}H and ends with production of ^{4}He and ^{12}C.	
fission	The binding energy per nucleon of nuclei with mass number greater than 60 or so decreases and the nuclei are less stable for larger values of A. A process by which such nuclei become more stable is nuclear fission, in which the nucleus splits into smaller, more stable, fragments.	
fusion	The binding energy per nucleon of relatively small nuclei tends to grow as A increases. Fusion is the joining together of two smaller nuclei to form one or more atoms with larger, more stable nuclei.	
gamma radiation	Gamma radiation is the emission of a γ-ray, a high energy photon.	
half-life	The half-life $\tau_{1/2}$ of a radioactive source is the time t that elapses during which the number of atoms $N(t)$ falls to one-half the number initially present (N_0).	$\dfrac{N(t)}{N_0} = \left[\dfrac{1}{2}\right]^{\frac{t}{\tau_{1/2}}}$ (27-15)
mass number	Mass number is the total number of protons and neutrons in an atomic nucleus.	A
nucleon	Nucleon is a general term for protons and neutrons, the two particles of which nuclei are comprised.	

neutron number	The neutron number is the number of neutrons a nucleus contains.	N	
nuclear force	The nuclear force, an attractive force between nucleons, is responsible for holding the protons and neutrons in an atomic nucleus together. The nuclear force is strong, but acts only over distances shorter than the size of nuclei.		
nuclear radiation	Nuclear radiation carries energy away from a nucleus in a process that transforms it to a more stable configuration. The three most common modes of nuclear radiation are alpha radiation, the emission of an α particle; beta radiation, the emission of a β particle; and gamma radiation, the emission of a γ ray. An α is a ^{4}He nucleus, a β is an electron or positron, and a γ is a high energy photon.		
nuclear radius	The nuclear radius r is related to the number of nucleons A, specifically, it is proportional to $A^{1/3}$. The proportionality constant r_0 equals 1.2 ± 0.2 fm.	$r = r_0 A^{1/3}$	(27-1)
proton–proton cycle	Proton–proton fusion is the predominant process by which the Sun releases energy. The cycle begins with the fusing of two protons (the nuclei of ^{1}H) to form ^{2}H. The ^{2}H nucleus fuses with another proton, forming ^{3}He, and finally, two ^{3}He nuclei fuse to form ^{4}He.		

QUESTIONS AND PROBLEMS

In a few problems, you are given more data than you actually need; in a few other problems, you are required to supply data from your general knowledge, outside sources, or informed estimate.

Interpret as significant all digits in numerical values that have trailing zeros and no decimal points.

For all problems, use $g = 9.8$ m/s^2 for the free-fall acceleration due to gravity. Neglect friction and air resistance unless instructed to do otherwise.

• Basic, single-concept problem

•• Intermediate-level problem, may require synthesis of concepts and multiple steps

••• Challenging problem

SSM *Solution is in Student Solutions Manual*

Conceptual Questions

1. •What is an isotope?

2. •What is the difference between atomic number and mass number?

3. •Describe how the approximate radius of the nucleus depends on the mass number raised to the one-third power.

4. •(a) Describe what is meant by the phrase "larger nuclei are neutron rich." (b) Why do most nuclei contain at least as many neutrons as protons?

5. •Describe two characteristics of the binding energy that are comparable to the work function (from the photoelectric effect) and two characteristics that are dissimilar to the concept of the work function.

6. •A simple idea of nuclear physics can be stated as follows: "The whole nucleus weighs less than the sum of its parts." Explain your answer.

7. •Using a graph of binding energy per nucleon versus mass number, determine the most stable isotope in the universe.

8. •Describe the basic characteristics of the nuclear force that exists between nucleons. What other, competing, force is present in the nucleus?

9. •Some historians would claim that without Einstein's special theory of relativity, nuclear physics would never have developed. Explain your answer. SSM

10. •What is the difference between fission and fusion?

11. •(a) Which elements in the periodic table are more likely to undergo nuclear fission? (b) Which are more likely to undergo nuclear fusion?

12. •**Astronomy** (a) Describe the nuclear reactions that are occurring in our Sun. (b) Discuss how the equilibrium state of the Sun is not permanent and discuss the eventual future of our solar system.

13. •Explain how conservation of energy and momentum would be violated if a neutrino were not emitted in beta decay. SSM

14. •To date, the decay constant of a radioactive nucleus is just that, *constant*. It does not depend on the size of the nuclear sample or the temperature or any external fields (such as gravity, electricity, or magnetism). Define the decay constant and comment on how nuclear radioactivity would change if the quantity were dependent on temperature.

17. •At any given instant, a sample of radioactive uranium contains many, many different isotopes (that are *not* uranium!). Explain your answer.

18. •(a) Explain how radioactive ^{14}C is used to determine the age of ancient artifacts. (b) Which types of artifacts can be radiodated and which types cannot?

19. •If atomic masses are used, explain why the mass of a beta particle (an electron!) is *not* accounted for in the basic beta decay. Assume that the mass of the antineutrino ($\bar{\nu}_e$) is very small and can be neglected. SSM

$$n \rightarrow p + e^- + \bar{\nu}_e$$

20. •**Medical** Describe, in broad terms, the health risks associated with the four major forms of radioactivity: alpha, beta, gamma, and neutrons. Focus on the dangers due to inherent health risks and the ability of each to penetrate shielding material.

Multiple-Choice Questions

21. •In an atomic nucleus, the nuclear force binds _____ together.
    A. electrons
    B. neutrons
    C. protons
    D. neutrons and protons
    E. neutrons, protons, and electrons

22. •The mass of a nucleus is _____ the sum of the masses of its nucleons.
    A. always less than
    B. sometimes less than
    C. always more than
    D. always equal to
    E. sometimes equal to

23. •Which of the following statements is true?
    A. Fusion absorbs energy and fission releases energy.
    B. Fusion releases energy and fission absorbs energy.
    C. Both fusion and fission absorb energy.
    D. Both fusion and fission release energy.
    E. Both fusion and fission can release or absorb energy. SSM

24. •In fission processes, which of the following statements is true?
    A. Only the total number of mass nuclei remains the same.
    B. Only the total number of protons remains the same.
    C. Only the total number of neutrons remains the same.
    D. The total number of protons and the total number of mass nuclei both remain the same.
    E. The total number of protons and the total number of neutrons both remain the same.

25. •In a spontaneous fission reaction, the total mass of the products is _____ the mass of the original elements.
    A. greater than
    B. less than
    C. the same as
    D. double
    E. one-half

26. •What is the source of the Sun's energy?
    A. chemical reactions
    B. fission reactions
    C. fusion reactions
    D. gravitational collapse
    E. both fusion reactions and fission reactions

27. •In a spontaneous fusion reaction, the total mass of the products is _____ the mass of the original elements.
    A. greater than
    B. less than
    C. the same as
    D. double
    E. one-half SSM

28. •The decay constant λ depends only on
    A. the number of atoms at the initial time.
    B. the initial decay rate.
    C. the half-life.
    D. the binding energy per nucleon.
    E. whether the decay is alpha, beta, or gamma.

29. •The number of radioactive atoms in a radioactive sample
    A. decreases linearly with time.
    B. increases linearly with time.
    C. decreases exponentially with time.
    D. increases exponentially with time.
    E. remains constant.

30. •The decay rate for any isotope
    A. decreases linearly with time.
    B. increases linearly with time.
    C. decreases exponentially with time.
    D. increases exponentially with time.
    E. remains constant.

This page is intentionally left blank.

For complete end of chapter problem sets, please go to
www.whfreeman.com/kestentauck

45. ••Given that a nucleus is approximately spherical, with a radius $R = R_0 A^{1/3}$, where R_0 is about 1.2 fm, determine its approximate mass density. Express your answer in SI units and convert to tons per cubic inch, units that might be used in a news report. SSM

46. ••Determine the density of the following nuclei (assume the nucleus is a sphere):

a.	^{3}H	e.	^{63}Cu
b.	^{9}Be	f.	^{226}Ra
c.	^{16}O	g.	^{209}Po
d.	^{56}Fe	h.	^{238}U

47. ••If our Sun (mass = 1.99×10^{30} kg, radius = 6.96×10^8 m) were to collapse into a neutron star (an object composed of tightly packed neutrons with roughly the same density as neutrons within a nucleus), what would the new radius of our "neutron-sun" be?

27-2: Binding Energy

48. ••Calculate the atomic mass of each of the isotopes listed below. Give your answer in unified mass units (u) and in grams (g). The values will include the mass of Z electrons.

a.	^{1}H	e.	^{56}Fe
b.	^{4}He	f.	^{90}Sr
c.	^{9}Be	g.	^{131}I
d.	^{12}C	h.	^{238}U

49. •What is the binding energy of carbon-12? Give your answer in MeV. SSM

50. ••What is the binding energy per nucleon for the following isotopes?

a.	^{2}H	e.	^{56}Fe
b.	^{4}He	f.	^{90}Sr
c.	^{6}Li	g.	^{129}I
d.	^{12}C	h.	^{235}U

51. ••Make a rough plot of the binding energy per nucleon (in MeV/nucleon.) vs. mass number for the following nuclei:

a.	^{1}H, ^{2}H	e.	^{32}S
b.	^{4}He	f.	^{56}Fe
c.	^{6}Li	g.	^{197}Au
d.	^{12}C	h.	^{208}Pb

52. ••Using the plot of binding energy per nucleon (from the previous problem), determine the binding energy of boron-10.

53. •What minimum energy is needed to remove a neutron from ^{40}Ca and convert it to ^{39}Ca? The atomic masses of the two isotopes are 39.96259098 u and 38.97071972 u, respectively.

54. •What is the binding energy of the last neutron of carbon-13? The atomic mass of carbon-13 is 13.003355 u.

55. ••**Medical** Iodine-131 is a radioisotope that is used in the treatment of cancer of the thyroid. The natural tendency of the thyroid to take up iodine creates a pathway for which the radioactivity (β^- and γ) that are emitted from this unstable nucleus can be directed onto the cancerous tumor with very little collateral damage to surrounding healthy tissue. Another advantage of the isotope is its relatively short half-life (8 days). Calculate the binding energy of iodine-131 and the binding energy per nucleon. The mass of iodine-131 is 130.906124 u.

27-3: Fission

56. •Calculate the energy released the following nuclear fission reaction:

$$^{239}\text{Pu} + {}^1\text{n} \rightarrow {}^{98}\text{Tc} + {}^{138}\text{Sb} + 4{}^1\text{n}$$

Recall, the atomic masses are $^1\text{n} = 1.008665$ u, $^{239}\text{Pu} = 239.052157$ u, $^{98}\text{Tc} = 97.907215$ u, and $^{138}\text{Sb} = 137.940793$ u.

57. •Complete the following nuclear fission reaction of thorium-232 and calculate the energy released in the reaction:

$$^{232}\text{Th} + {}^1\text{n} \rightarrow {}^{99}\text{Kr} + {}^{124}\text{Xe} + \underline{\quad}?$$

The atomic masses are thorium-232 = 232.038051 u, krypton-99 = 98.957606 u, and xenon-124 = 123.905894 u. SSM

58. •Complete the following fission reactions:
a. $^{235}\text{U} + {}^1\text{n} \rightarrow {}^{128}\text{Sb} + {}^{101}\text{Nb} + \underline{\quad}?\underline{\quad}$
b. $^{235}\text{U} + {}^1\text{n} \rightarrow \underline{\quad}?\underline{\quad} + {}^{116}\text{Pd} + 4{}^1\text{n}$
c. $^{238}\text{U} + {}^1\text{n} \rightarrow {}^{99}\text{Kr} + \underline{\quad}?\underline{\quad} + 11{}^1\text{n}$
d. $\underline{\quad}?\underline{\quad} + {}^1\text{n} \rightarrow {}^{101}\text{Rb} + {}^{130}\text{Cs} + 8{}^1\text{n}$

59. •Complete the following fission reactions:
a. $^{242}\text{Am} + \underline{\quad}?\underline{\quad} \rightarrow {}^{90}\text{Sr} + {}^{149}\text{La} + 4{}^1\text{n}$
b. $^{244}\text{Pa} + {}^1\text{n} \rightarrow \underline{\quad}?\underline{\quad} + {}^{131}\text{Sb} + 12{}^1\text{n}$
c. $\underline{\quad}?\underline{\quad} + {}^1\text{n} \rightarrow {}^{92}\text{Se} + {}^{153}\text{Sm} + 6{}^1\text{n}$
d. $^{262}\text{Fm} + {}^1\text{n} \rightarrow {}^{112}\text{Rh} + \underline{\quad}?\underline{\quad} + 9{}^1\text{n}$

60. •Calculate the energy (in MeV) released in the following nuclear fission reaction:

$$^{242}\text{Am} + \underline{\quad}?\underline{\quad} \rightarrow {}^{90}\text{Sr} + {}^{149}\text{La} + 4{}^1\text{n}$$

Start by completing the reaction and use the following nuclear masses: americium-242 = 242.059549 u, strontium-90 = 89.9077387 u, and lanthanum-149 = 148.934733 u.

61. ••Assuming that in a fission reactor a neutron loses half its energy in each collision with an atom of the moderator, determine how many collisions are required to slow a 200-MeV neutron to an energy of 0.04 eV.

62. ••Knowing that the binding energy per nucleon for uranium-235 is about 7.6 MeV/nucleon and the binding energy per nucleon for typical fission fragments is about 8.5 MeV/nucleon, find an average energy release per fission reaction in MeV.

63. ••How many kilograms of uranium-235 must completely fission to produce 1000 MW of power continuously for one year? SSM

64. ••Repeat problem 61 in the more realistic case where the fission reactions are about 30% efficient in producing 1000 MW of power over 1 year of continuous operation.

65. ••Calculate the number of fission reactions per second that take place in a 1000-MW reactor. Assume that there are 200 MeV/reaction released.

27-4: Fusion

66. •Complete the following fusion reactions:
   a. $^2H + ^3H \rightarrow ^4He + \_\_?\_\_$
   b. $^4He + ^4He \rightarrow ^7Be + \_\_?\_\_$
   c. $^2H + ^2H \rightarrow ^3He + \_\_?\_\_$
   d. $^2H + ^1H \rightarrow ^0\gamma + \_\_?\_\_$
   e. $^2H + ^2H \rightarrow ^3H + \_\_?\_\_$

67. ••Calculate the energy released in each of the fusion reactions in the previous problem. Give your answers in MeV.

68. ••**Astronomy** Consider the proton-proton cycle that occurs in most stars (including our own Sun):

$$\text{Step 1:} \quad ^1H + ^1H \rightarrow ^2H + e^+ + \nu_e$$

$$\text{Step 2:} \quad ^2H + ^1H \rightarrow ^3He + \gamma$$

$$\text{Step 3:} \quad ^3He + ^3He \rightarrow ^4He + 2\,^1H + \gamma$$

Calculate the net energy released from the three steps. Do *not* ignore the mass of the positron in the calculations and take special care in accounting for the particle in the beta decay of Step 1.

69. ••Each D–T fusion reaction releases about 20 MeV. How much tritium (T) is needed to create 10^{14} J of energy, assuming that an endless supply of deuterium (D) is available? SSM

70. ••How many fusion reactions per second must be sustained to operate a D–T fusion power plant that outputs 1000 MW, operating at 33% efficiency?

27-5: Radioactivity

71. •Complete the following conversions (cpm=counts per minute):
   a. $100\ \mu Ci = $ _____ Bq
   b. 1500 cpm = _____ Bq
   c. 16,500 Bq = _____ Ci
   d. 7.55×10^{10} Bq = _____ cpm

72. •The curie unit is defined as 1 Ci = 3.7×10^{10} Bq, which is about the rate at which radiation is emitted by 1.00 g of radium. (a) Calculate the half-life of radium from the definition. (b) What does your calculation tell you about the radiation emission rate of radium?

73. ••**Calc** Starting with the concept that the time rate of decay for a sample of N radioactive nuclei is proportional to the negative value of the number that exist,

$$\frac{dN}{dt} = -\lambda N$$

derive the radioactive decay formula. Recall, λ is the decay constant, $\lambda = (\ln 2)/t_{1/2}$.

74. •The half-life of tritium (^3H) is 12.3 y. Calculate the corresponding decay constant in units of y^{-1} and s^{-1}.

75. •A certain radioisotope has a decay constant equal to 0.00334 s^{-1}. Find the half-life in seconds and days. SSM

76. ••A radioactive sample is monitored with a radiation detector to have 5640 counts per minute. Twelve (12) hours later, the detector reads 1410 counts per minute. Calculate the decay constant and the half-life of the sample.

77. •What fraction of a sample of ^{32}P will be left after 4 months? Its half-life is 14.3 days.

78. •What fraction of a radioactive sample will be left after 6 half-lives? What about 7.5 half-lives?

79. ••**Medical** A patient is injected with 7.88 μCi of radioactive iodine-131 with a half-life of 8.02 days. Assuming that 90% of the iodine ultimately finds its way to the thyroid, what decay rate do you expect to find in the thyroid after 30 days? SSM

80. ••The ratio of carbon-14 to carbon-12 in living wood is 1.3×10^{-12}. How many decays per second are there in of 550 g of wood?

81. ••You take a course in archaeology that includes field work. An ancient wooden totem pole is excavated from your archaeological dig. The beta activity is measured at 150 cpm. If the totem pole contains 225 g of carbon and the ratio of carbon-14 to carbon-12 in living trees is 1.3×10^{-12}, what is the age of the pole?

82. ••Twelve centuries after a tree limb is cut, determine the decay rate for 500 g of carbon from the tree.

83. ••How many nuclei of radon-222 are present in a sample with an activity of 485 cpm? SSM

84. ••Fossils that contain rocks can be radiodated using the isotope ^{87}Rb. This isotope of rubidium undergoes beta decay with a half-life of 4.75×10^{10} y. Ancient samples contain a ratio of ^{87}Sr to ^{87}Rb of 0.0225. Given that ^{87}Sr is a stable product of the beta decay of ^{87}Rb, and there was no ^{87}Sr present in the rocks to begin, calculate the age of the rock sample. Assume that the decay rate is constant over the relativity short lifetime of the rock compared to the half-life of ^{87}Rb.

85. •Complete the following alpha decays:
   a. $^{238}U \rightarrow ^4\alpha + \_\_?\_\_$
   b. $^{234}Th \rightarrow \_\_?\_\_ + ^{-?}Ra$

c. __?__ → $^4\alpha$ + ^{236}U

d. ^{214}Bi → $^4\alpha$ + __?__

86. •Complete the following beta decays:

a. ^{14}C → e$^-$ + $\bar{\nu}$ + __?__

b. ^{239}Np → e$^-$ + $\bar{\nu}$ + __?__

c. __?__ → e$^-$ + $\bar{\nu}$ + ^{60}Ni

d. ^{3}H → e$^-$ + $\bar{\nu}$ + __?__

e. ^{13}N → e$^+$ + __?__ + __?__

87. •Complete the following gamma decays:

a. ^{131}I* → γ + __?__

b. ^{145}Pm* → ^{145}Pm + __?__

c. __?__ → γ + ^{24}Na

88. ••Nickel-64 has an excited state 1.34 MeV above the ground state. The atomic mass of this isotope of nickel is 63.927967 u. (a) What is the mass of the excited state of the nucleus? (b) What is the wavelength of the gamma ray that is emitted when the nucleus decays to the ground state?

General Problems

89. ••(a) What is the approximate radius of the ^{238}U nucleus? (b) What electrical force do two protons on opposite ends of the ^{238}U nucleus exert on each other? (c) If the electrical force in part (b) were the only force acting on the protons, what would be their acceleration just as they left the nucleus? (d) Why do the protons in part (b) not accelerate apart? SSM

90. ••The semiempirical binding energy formula is given as follows:

$$E_B = (15.8\,\text{MeV})A - (17.8\,\text{MeV})A^{2/3}$$
$$- (0.71\,\text{MeV})\frac{Z(Z-1)}{A^{1/3}} - (23.7\,\text{MeV})\frac{(N-Z)^2}{A}$$

where A = mass number, N = number of neutrons, and Z = number of protons. Using the formula, calculate the binding energy per nucleon for fermium-252. Compare your answer with the standard common expression for the binding energy: $E_B = (Nm_n + Zm_p - m_{whole})c^2$.

91. ••The fissionability parameter is defined as the atomic number squared divided by the mass number for any given nucleus (Z^2/A). It can be shown that when this parameter is less than 44, a nucleus will be stable against small deformation; essentially the nucleus will be stable against spontaneous fission. Calculate the value of this parameter for (a) uranium-235, (b) uranium-238, (c) plutonium-239, (d) plutonium-240, (e) californium-246, and (f) californium-254.

92. •The stable isotope of sodium is ^{23}Na. What kind of radioactivity would be expected from (a) ^{22}Na and (b) ^{24}Na?

93. ••In 2010 physicists for the first time created the heavy element number 117 by colliding calcium-48 and berkelium-249 at high energy. The result was two isotopes of the new element, one of which had a half-life of 14 ms and contained 176 neutrons. (a) What is the radius of the nucleus of the new element 117? (b) What percent of the newly created isotope was left 1.0 s after its creation? SSM

94. ••Natural uranium is made up of two isotopes: uranium-235 and uranium-238. The half-life of uranium-235 is 7.04×10^8 y and the half-life of uranium-238 is 4.47×10^9 y. Assuming that all uranium isotopes were created simultaneously and in equal amounts at the same time that Earth was formed, estimate the age of Earth. The current percent abundance of uranium-235 is 0.72% and for uranium-238 it is 99.28%.

95. •••**Astronomy** The atom technetium (Tc) has no stable isotopes, yet its spectral lines have been detected in red giant stars (stars at the end of their lifetimes). Tc can be produced artificially on Earth. Its longest-lived isotope, ^{98}Tc, has a half-life of 4.2 million years. (a) If any ^{98}Tc was present when Earth formed 4.5×10^9 y ago, what percentage of it is still present? (Careful! You cannot do this calculation with your calculator. You must use logarithms to express the answer in scientific notation.) (b) What percent of the original ^{98}Tc would be present in a red giant that is 10 billion years old? (Careful again! You'll need to use logarithms.) (c) Explain why the detection of technetium in old stars is strong evidence that the stars manufacture the atoms in the universe.

96. ••An old wooden bowl in your archeological dig is unearthed and found to have one-fourth of the amount of carbon-14 that a similar sample of fresh wood would. Determine the age of the bowl.

97. ••In an attempt to determine the age of the cave paintings in Chauvet Cave in France, scientists used radiocarbon dating to measure the age of bones of cave bears found in the cave. The bears are depicted in the paintings, so presumably the bones are approximately the same age as the paintings. The results showed that the level of ^{14}C was reduced to 2.35% of its present-day level. How old were the bones (and presumably the paintings)? SSM

98. ••In one common type of household smoke detector, the radioisotope americium-241 decays by alpha emission. The alpha particles produce a small electrical current since they are charged. If smoke enters the detector, it blocks the alpha particles, which reduces the current and causes the alarm to go off. The half-life of ^{241}Am is 433 y and its atomic weight is 241 g/mol. Typical decay rates in smoke detectors are 690 Bq. (a) Write the decay reaction of ^{241}Am and identify the daughter nucleus. (b) By how much does the alpha particle current decrease in 1.0 y due to the decay of the americium? How much in 50 y? (c) How many grams of ^{241}Am are there in a typical smoke detector?

99. ••In March 2011, a giant tsunami struck the Fukushima nuclear reactor in Japan, resulting in very large radiation leaks, including cesium-137. The isotope has a 30-y half-life and is a beta-minus emitter. (a) What daughter nucleus is left after cesium-137 decays? (b) How long after the release will it take for the decay rate of the cesium-137 to be reduced by 99%?

100. ••Three isotopes of aluminum are given in the following table:

Isotope	Atomic mass (u)	BE/nucleon	Decay process
^{26}Al	25.986892		
^{27}Al	26.981538		Stable
^{28}Al	27.981910		

Calculate the binding energy per nucleon for each isotope and make a prediction of the decay processes for the unstable isotopes above and below aluminum-27.

101. ••**Biology** In February 2010, the discovery was announced of leaks due to aging pipes of the carcinogen tritium (^{3}H) at 27 U.S. nuclear reactors. In one well in Vermont, contaminated water registered 70,500 pCi/L, whereas the federal safety limit was 20,000 pCi/L. Tritium is a β^- emitter with a half-life of 12.3 y. (a) How many protons and neutrons does the tritium nucleus contain? (b) Write out the decay equation for tritium and identify the daughter nucleus. (c) If the leak at the Vermont site is stopped, how long will it take for the water in the contaminated well to reach the federal safety level?

102. ••**Medical** Iodine-125 is used to treat, among other things, brain tumors and prostate cancer. It decays by gamma decay with a half-life of 59.4 days. Patients who fly soon after receiving ^{125}I implants are given medical statements from the hospital verifying such treatment because their radiation could set off radiation detectors at airports. If the initial decay rate was 525 μCi, (a) what will the rate be at the end of the first year, and (b) how many months after the treatment will the decay rate be reduced by 90%?

103. ••**Medical** Ruthenium-106 is used to treat melanoma in the eye. This isotope decays by β^- emission with a half-life of 373.59 days. One source of the isotope is reprocessed nuclear reactor fuel. (a) How many protons and neutrons does the ^{106}Ru nucleus contain? (b) Could we expect to find significant amounts of ^{106}Ru in ore mined from the ground? Why or why not? (c) Write the decay equation for ^{106}Ru and identify the daughter nucleus. (d) How many years after ^{106}Ru is implanted in the eye does it take for its decay rate to be reduced by 75%?

104. ••**Medical** You are asked to prepare a sample of ruthenium-106 for a radiation treatment. Its half-life is 373.59 days, it is a beta emitter, its atomic weight is 106 g/mol, and its density at room temperature is 12.45 g/cm^3. (a) How many grams will you need to

prepare a sample having an activity rate of 125 μCi? (b) If the sample in part (a) is a spherical droplet, what will be its radius?

105. ••The electron capture by a proton is *not* allowed in nature. Discuss this fact. Specifically, describe why the following nuclear reaction does *not* occur (and for good reason!): SSM

$$e^- + p \not\rightarrow n + \nu$$

106. ••Taking into account the recoil of the daughter nucleus, calculate the energy of the alpha particle in the following decay:

$$^{235}U \rightarrow {}^4\alpha + {}^{231}Th$$

107. ••There are several radioactive decay series that are observed in nature. Four of the most well-known are as shown. The neptunium decay series actually is extinct. The first three all end in different, stable isotopes of lead.
   a. Thorium Decay Series: $A = 4n$
      Starting isotope: ^{232}Th   Ending isotope: ^{208}Pb
   b. Radium or Uranium Decay Series: $A = 4n + 2$
      Starting isotope: ^{238}U   Ending isotope: ^{206}Pb
   c. Actinium Decay Series: $A = 4n + 3$
      Starting isotope: ^{235}U   Ending isotope: ^{207}Pb
   d. Neptunium Decay Series: $A = 4n + 1$
      Starting isotope: ^{237}Np   Ending isotope: ^{209}Bi

Trace out the pathway (keeping count of the total number of alpha and beta emissions) that terminates with a stable nucleus (somewhere at or below lead-208), for each of the radioactive decay series above.

108. ••Natural uranium contains three isotopes, described in the tables below. (a) If a sample of natural uranium, in its pure metallic form, has an activity of 4.48 μCi, what is the total mass of the sample? (b) What fraction of the total activity is due to each of the three isotopes?

Properties of the Natural Uranium Isotopes

	Uranium-234	Uranium-235	Uranium-238
Half-life	244,500 y	703.8×10^6 y	4.468×10^9 y
Specific activity	231.3 MBq/g	80,011 Bq/g	12,445 Bq/g

Isotopic Composition of Natural Uranium

	Uranium-234	Uranium-235	Uranium-238	Total
Atom %	0.0054%	0.72%	99.275%	100%
Weight %	0.0053%	0.711%	99.284%	100%
Activity %	48.9%	2.2%	48.9%	100%
Activity in 1 g U$_{nat}$	12,356 Bq	568 Bq	12,356 Bq	25,280 Bq

109. ••In a baseless spectacle, a biology professor futilely attempts to explain to his physics professor colleague the fundamentals of nuclear physics. In his lame effort, he declares that the world's energy problems could be solved if only physicists were to pursue the fusion of *heavy* nuclei, rather than the fusion of *light* nuclei. To prove his point, he suggests that the following fusion reaction should be considered:

$$^{157}\text{Nd} + {}^{80}\text{Ge} \rightarrow {}^{235}\text{U} + 2n$$

Using the insights that you have acquired in this chapter, show that his argument is flawed. The mass of neodymium-157 is 156.939032 u and the mass of germanium-80 is 79.925373 u.

110. •••Calc A radioactive nucleus is being produced in a nuclear reactor at a constant rate of R_P (in units of Bq). At the same time, it is decaying at a rate of $R_D = \lambda N$. If $N_0 = 0$ at $t = 0$, integrate to show that

$$N(t) = \frac{R_P}{\lambda}(1 - e^{-\lambda t})$$

111. •••Calc A radioactive (parent) nucleus decays with a decay constant λ_P. The daughter is also radioactive and decays with constant λ_D. Therefore, in a small interval of time (dt), the net gain in daughter nuclei will be equal to the number of daughter nuclei created by a decaying parent minus the number of daughter nuclei decaying:

$$dN_D = \lambda_P N_P \, dt - \lambda_D N_D dt$$

Assuming that the number of parent nuclei at $t = 0$ is N_0, derive a formula for the number of daughter nuclei as a function of time.

112. •••Calc Using the results from the previous problem, show that the number of daughter nuclei reaches a maximum at a time t_{max}. Assume that there are N_0 parent nuclei at $t = 0$ s.

$$t_{max} = \left(\frac{1}{\lambda_P - \lambda_D}\right) \ln\left(\frac{\lambda_P}{\lambda_D}\right)$$

28 Particle Physics

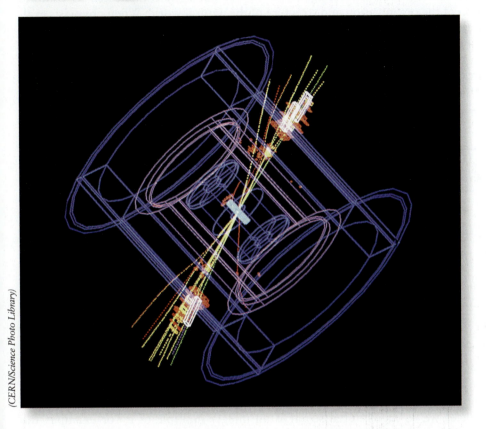

(CERN/Science Photo Library)

When a matter particle and an antimatter particle pair collide, a large number of other subatomic particles can be created from the energy the two carried. Here, the collision of an electron and an antimatter electron (a positron) produces two sprays of particles at the center of an experiment at CERN, the European particle physics laboratory. The blue and purple lines show the outline of the experimental apparatus, and the other lines show the paths of the particles.

Physics is the science of how the universe works; that is the opening sentence of this book. In the last thousand pages, we've considered many phenomena that arise from the interactions between forces, matter, and energy. Then, in the last chapter we discussed the nucleus of the atom. Now we look inside the nucleus and explore its components. We will see examples of Einstein's realization of the equivalence of matter and energy. It may surprise you that this discussion of the tiniest things in the universe leads us to our last and perhaps our biggest topic: the Big Bang and the beginning of the universe.

28-1 The Standard Model: Particles

The English physicist and chemist John Dalton began to use the word *atom* at the start of the 19th century to describe the most fundamental particles of nature. "Atom" comes from the Greek word *atomos*, meaning indivisible; scientists believed that all matter was composed of these building blocks, and that they could not be subdivided into more elementary particles. By the early part of the 20th century, however, physicists had begun to uncover evidence that atoms were indeed composed of smaller particles. First, J.J. Thomson discovered the electron in 1898. Ernest Rutherford found evidence for a small, hard atomic nucleus in 1911, and later, he found evidence for the existence of the proton that makes up the nucleus. In 1932, English physicist

James Chadwick observed the neutron. Throughout the 1940s and 1950s a host of other particles, such as the pion, the kaon, and the delta, were discovered. Each of these particles are different from the others based on their charge, their mass, and the way they interact with other particles, as well as a range of other properties some of which only became known when a particular particle was studied.

All of this was interesting, certainly, but not as uncomplicated as Dalton's vision of the organization and structure of the natural world. When physicists evaluate competing theories or alternative models for a process, they are far more likely to choose the more elegant, concise, and uncomplicated one. (A physicist might tell you he is more likely to select the theory that is more *beautiful*!) Long tables of particle names and properties are neither elegant nor simple. Is there organization that provides a structure to the properties of particles that abound in nature?

The organizing structure of protons, neutrons, pions, and the bulk of the other particles that have been discovered lies in an understanding of the constituents of which all of these are comprised. These constituents are the **quarks**, of which there are six: up, down, charm, strange, top, and bottom. Table 28-1 lists the six quarks, along with some of their properties. The charges in Table 28-1 are given in terms of the fundamental charge e, equal to 1.602×10^{-19} C.

Table 28-1 is presented as three groups, or *generations*, of two quarks. Physicists usually write the quark generations as

$$\begin{pmatrix} u \\ d \end{pmatrix} \begin{pmatrix} c \\ s \end{pmatrix} \begin{pmatrix} t \\ b \end{pmatrix}$$

The quarks along the top row, the u, c, and t quarks, share certain properties, as do the quarks along the bottom row, the d, s, and b quarks. All attempts to find evidence of a fourth generation quark have failed; as best we know, there are only three generations of quarks.

You might notice two peculiar aspects of Table 28-1. One is that quarks have fractional charge, relative to the fundamental charge e. The charge of all of the particles we've encountered so far is an integer multiple of e, for example, the charge of an electron is $-1 \times e$ and the charge of a proton is $+1 \times e$. The charge on a nucleus is therefore also an integral multiple of e, as is the net charge of any atom, although normally atoms are neutral so the integer multiplier is equal to zero. Indeed, of the host of other particles we know, only the quarks have fractional charge. Something else you might find peculiar in Table 28-1 is that we list only an *approximate* rest mass for each quark. We know quite accurate values of the masses of other particles, for example, we've used 1.6749×10^{-27} kg as the mass of the proton. Why wouldn't the mass of the quarks be known as accurately?

The observations that quarks have fractional charge and lack an exact value of mass stem from the fact that quarks are never found in nature as free, solitary particles. Unlike protons and pions, quarks cannot go zipping around the universe as solitary particles. Despite well over a hundred experimental searches in the last few years for free quarks, not a single one has ever been observed. Quarks exist mostly either in pairs or in groups of three, which means that the net charge of any such grouping is the sum of $-\frac{1}{3}e$ and $+\frac{2}{3}e$ charges taken two or three at a time. (As we will see shortly, $+\frac{1}{3}e$ and $-\frac{2}{3}e$ charges are also possible.) All of the *groups* of quarks we observe have charge equal to an integral multiple of e. These groups comprise nearly all of the so-called subatomic particles we have discovered. Protons, neutrons, pions, kaons—they are all composed of quarks. (As of this writing there is some evidence of the existence of particles comprising four quarks, and experiments have been conducted searching for a theoretically predicted five-quark particle.)

Table 28-1	The Six Quarks		
Quark	**Symbol**	**Charge**	**Approximate Rest Mass**
Up	u	$+\frac{2}{3}e$	2 MeV/c^2
Down	d	$-\frac{1}{3}e$	5 MeV/c^2
Charm	c	$+\frac{2}{3}e$	1.3 GeV/c^2
Strange	s	$-\frac{1}{3}e$	0.1 GeV/c^2
Top	t	$+\frac{2}{3}e$	173 GeV/c^2
Bottom	b	$-\frac{1}{3}e$	5 GeV/c^2

A particle composed of three quarks is classified as a **baryon**. The proton is a baryon, composed of two up quarks and one down quark (uud: up, up, down). Check the net charge: each up contributes $+\frac{2}{3}e$ and the down contributes $-\frac{1}{3}e$, so the total charge is $\frac{2}{3}e + \frac{2}{3}e - \frac{1}{3}e$ or $+e$. The quark composition of the neutron is udd; here, the charge of the baryon is $\frac{2}{3}e - \frac{1}{3}e - \frac{1}{3}e$, which equals zero. The neutron carries no net charge because the sum of the charges of its three constituent quarks is zero. Nevertheless, the neutron has properties associated with charge because the quarks that make it up are themselves charged. As we saw in Chapter 27, for example, the neutron has an associated magnetic field.

A particle composed of two quarks is as classified as a **meson**. It is not possible, however, to form a meson from the quarks listed in Table 28-1, because no combination of two quarks in the table has a net charge equal to an integer multiple of e. To understand the mesons, we must introduce the concept of matter and **antimatter**. Every matter particle has an antimatter partner that is identical to it in every way except that it is oppositely charged. We have already encountered the positron (e^+) produced in β^+ decay (Section 27-4) as similar to an electron except that it is positively charged. In fact, the positron and electron are identical except for the sign of their charges; the electron and the positron are a matter–antimatter pair. Each of the six quarks has an antimatter particle associated with it, so there are really 12 quarks available to form mesons (and baryons, too). The antimatter particle associated with each quark is signified by adding "bar" to the name, or by placing a bar over the quark's symbol. So the antimatter particle associated with the up quark is up-bar or $\bar{u}$. The $\bar{u}$ quark carries charge $-\frac{2}{3}e$. The antimatter d-bar quark $\bar{d}$ carries charge $+\frac{1}{3}e$. So, for example, the quark combination $u\bar{d}$ has charge $+\frac{2}{3}e + \frac{1}{3}e$ or $+e$; this is the positively charged pion π^+. The negatively charged pion π^- is $\bar{u}d$, and the neutral pion π^0 exists as both $u\bar{u}$ and $d\bar{d}$. [Pions are the least massive mesons; the rest mass of the π^0 is 135.0 MeV/c^2, and the rest mass of the π^+ (and π^-) is 139.6 MeV/c^2.]

When a particle and its antimatter partner meet their total mass can be converted to its equivalent energy. For example, the collision of an electron and a positron results in one or more photons that carry the total energy of the two quarks, a process known as *annihilation*.

? Got the Concept 28-1

Pions

(a) The quark composition of the π^- is $\bar{u}d$. Show that the charge of the π^- is $-e$.
(b) The neutral pion π^0 exists as both $u\bar{u}$ and $d\bar{d}$. Show that the charge of the π^0 is zero.

? Got the Concept 28-2

Doubly Charged?

(a) Is it possible for a meson that has a charge equal to $+2e$ (or $-2e$) to exist?
(b) Is it possible for a baryon that has a charge equal to $+2e$ (or $-2e$) to exist? Explain.

Let's return to John Dalton's vision that the organizational structure of matter and its properties could be understood from the properties of a small number of fundamental building blocks. Dalton was correct, although it is not atoms but

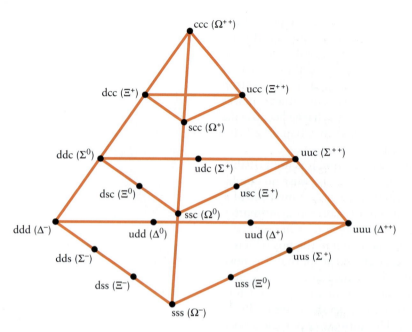

Figure 28-1 The particles formed by the quark combinations share certain fundamental properties derived from the properties of the individual quarks that comprise them. This family of baryons shares a number of properties. The quark combination for each particle is given, with the name of the particle in parentheses. The organizational structure is simple and elegant.

quarks that serve as the fundamental elements of nature. Just as an example, consider the family of baryons shown in **Figure 28-1**. All of the particles share certain properties, and the structure of this geometric pattern correctly describes the way in which other properties vary from particle to particle. For example, the charges of the particles are zero on the left and increase by one moving from left to right, and the property of "strangeness" increases by one, starting from zero, from back to front.

Patterns such as the one shown in Figure 28-1 were codified into the quark model of particle physics, and ultimately into what we now call the *Standard Model*. In this model, nearly all of the particles fit into such geometric patterns; the particles are composed of quarks and draw their properties from the quarks of which they are composed. Physicists will tell you this model is beautiful.

You may have noticed that we have been careful not to imply that not *all* particles are composed of quarks. In addition to quarks there is another category of fundamental particles. These are the **leptons**; as best we know, leptons are not made up of small constituent particles. The electron is a lepton, as is the muon that we encountered in Chapter 25. The neutrino that is created in the beta decay process is also a lepton. There are six leptons, listed in **Table 28-2**, as well as their antimatter partners.

In many ways, the muon and tau are more massive versions of the electron, so they share many properties and interact with other particles in similar ways. For example, scientists have been able to create muonic atoms, in which a μ^- replaces an e^- bound to an atomic nucleus. One way in which electrons, muons, and tau particles *are* significantly different (in addition to the differences in mass) is that while the electron is stable, the other two are not. Electrons do not decay, whereas the muon has a half-life of 1.56 μs (see Example 25-2) and the tau has a half-life of about 2.0×10^{-13} s.

Notice that we have listed the leptons in three groups of two. As is the case for the quarks, leptons form three generations, which we usually write as

$$\begin{pmatrix} e^- \\ \nu_e \end{pmatrix} \quad \begin{pmatrix} \mu^- \\ \nu_\mu \end{pmatrix} \quad \begin{pmatrix} \tau^- \\ \nu_\tau \end{pmatrix}$$

Table 28-2	The Six Leptons		
Lepton	**Symbol**	**Charge**	**Approximate Rest Mass**
Electron	e^-	$-e$	$0.5110 \text{ MeV}/c^2$
Electron neutrino	ν_e	0	$< 2 \text{ eV}/c^2$
Muon	μ^-	$-e$	$105.7 \text{ MeV}/c^2$
Muon neutrino	ν_μ	0	$< 0.18 \text{ MeV}/c^2$
Tau	τ^-	$-e$	$1777 \text{ MeV}/c^2$
Tau neutrino	ν_τ	0	$< 18.2 \text{ MeV}/c^2$

The leptons along the top row in each share certain properties, as do the leptons along the bottom row. Each lepton also has an antimatter particle associated with it. We have already encountered the electron and positron as a matter–antimatter pair. The antimatter muon (μ^+) and antimatter tau (τ^+) are identical to the muon and tau in every way except that μ^+ and τ^+ carry positive rather than negative charge. Sometimes these antimatter leptons are called "antimuons" and "antitaus," respectively.

Over the course of our study of a wide variety of physical phenomena we have uncovered a number of conservation laws, such as the conservation of energy and the conservation of momentum of a system in the absence of external forces. By describing quantities that do not change over time, these rules provide glimpses into the fundamental workings of nature. Not surprisingly, a number of conservation laws describe the physics of particles, too.

Baryon number is conserved. Every matter baryon is assigned *baryon number B* equal to +1, and every antimatter baryon is assigned *B* equal to −1. Particles that are not baryons are assigned *B* equal to zero. In order for baryon number to be conserved in a process, the sum of the values of *B* for all particles that participate in the process must equal the sum of the values of *B* for all particles that are present at the end of the process. Consider, for example, the β^- decay process:

$$n \rightarrow p + e^- + \bar{\nu}_e \qquad (27\text{-}17)$$

The neutron and proton are both matter baryons, so each has *B* equal to +1. The electron and the antineutrino are leptons, so have *B* equal to zero. The total baryon number is therefore equal to 1 before and after the decay; that is, baryon number is conserved.

Lepton number is conserved, too; this conservation law also applies to β^- decay. Every matter electron and electron neutrino is assigned *electron-lepton number* L_e equal to +1, and every antimatter electron and electron neutrino is assigned L_e equal to −1. Muons, muon neutrinos, tau particles, and tau neutrinos each have a similarly defined *muon-lepton number* L_μ or *tau-lepton number* L_τ. A particle that is not a lepton is assigned L_e, L_μ, and L_τ equal to zero. In the β^- decay process, the e^- has L_e equal to +1 and the antimatter neutrino $\bar{\nu}_e$ has L_e equal to −1. Both the neutron and the proton are baryons, so each has L_e equal to zero. The total value of L_e before the decay is therefore zero, and afterward it is zero plus one minus one, or zero. Electron-lepton number is conserved in β^- decay.

? Got the Concept 28-3

Can This Happen?

Charge is conserved in each of these processes, but some never occur. Which are possible, and which are not? Explain your answer. (a) $n + p \rightarrow n + p + p + \bar{p}$ ($\bar{p}$ is an antiproton, or antimatter proton), (b) $p \rightarrow e^+ + \gamma$, (c) $\pi^- \rightarrow \mu^- + \bar{\nu}_\mu$, (d) $\mu^- \rightarrow e^- + \bar{\nu}_e + \nu_\mu$.

✱ What's Important 28-1

The six quarks are the constituents of mesons (two quarks bound together) and baryons (the bound state of three quarks). Leptons, which include electrons and neutrinos, have no constituent particles. In the interactions of subatomic particles, baryon number and lepton number are conserved.

28-2 The Standard Model: Forces

Over the course of your life you have undoubtedly experienced and exerted a force many times. Forces are responsible for all interactions, from the ones we see or experience directly as well as those that underlie every physical phenomenon including the interactions between a proton and a neutron of an atomic nucleus.

Forces are responsible for all interactions, and interactions are responsible for all phenomena.

In this introductory study of physics we have encountered three fundamental kinds of forces: gravity, the electromagnetic force, and now, the strong force. Gravity is an attractive force that draws objects that have mass closer together. Gravitational force attracts you to Earth, and keeps Earth in orbit around the Sun. Electric and magnetic forces, two manifestations of the electromagnetic force, cause charged objects to accelerate. Electrons can be bound to atomic nuclei as a result, and atoms can bond together. The organization of atomic electrons is responsible for many properties of the elements, and how atoms bond in a substance results in the substance's biological, chemical, and physical properties. As we explored in the last chapter, the strong force binds protons and neutrons together to form atomic nuclei, and is at least in part responsible for the radioactive properties of nuclei. To this list, we will add a force that we have not yet named but the effects of which we encountered in Chapter 27; the **weak force** is at the heart of the interaction that governs beta decay.

These four forces, gravity, electromagnetic, strong, and weak, are the only forces known. All interactions and all phenomena result from the four forces. You may see references to the *electroweak force*. The electromagnetic and weak forces are related to each other. The Nobel Prize in Physics in 1979 was awarded to three physicists who developed the theory that underlies both the electromagnetic and weak forces, and the 1984 Nobel Prize in Physics was awarded for the experimental proof that the theory is correct.

Most of the forces we encounter in our daily lives are those that involve one object touching another. You can easily exert enough force on a door, say, to push it open, but in order to do so you must be close enough to the door to touch it. Yet all of the fundamental forces we described above appear to act without any direct contact. We dealt with this phenomenon in a mathematical way for the electric and magnetic forces through the constructs of an electric field and a magnetic field. We said, for example, that every charged object is surrounded by an electric field $\vec{E}$ and that the force another object of charge q experiences in its presence is the interaction of the electric field of the first and the charge of the second ($\vec{F} = q\vec{E}$, Equation 16-2). We did not, however, describe what that electric field is physically. And it has to be *something* physical, because no action can take place at a distance.

All forces result from the exchange of particles. We say that a force is *mediated* by the exchange of a particle, and that the exchanged particle is the *mediator* of the force. (The particles that mediate forces are sometimes called *exchange particles*.) The mediator of the electric force is the photon, for example. In other words, when two charged particles exert an electric force on one another, they do so by exchanging a photon. One of the charged particles emits the photon and the other absorbs it. We can represent the process by a diagram such as the one in **Figure 28-2**. In this slightly simplified version of a *Feynman diagram*, invented by American physicist Richard Feynman both to visualize and to analyze processes that involve subatomic particles, time runs from left to right. The lines associated with each particle do not represent actual paths that the particles take through space; rather, they only indicate which particles interact with which other particles. In Figure 28-2, two electrons exchange a photon and in so doing each exerts a Coulomb force on the other. Physicist's Toolbox 28-1 explains how to read a Feynman diagram. Particles such as electrons are shown as a solid, straight line in Feynman diagrams. Mediator particles are drawn as either a wavy line or a spiral line.

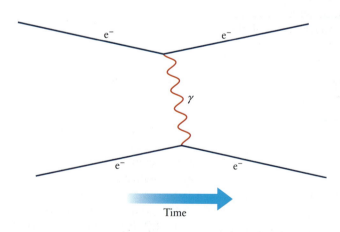

Figure 28-2 This diagram represents the Coulomb force between two electrons (e^-) through the exchange of a photon (γ).

Physicist's Toolbox 28-1   Reading a Feynman Diagram

Feynman diagrams are used to visualize and analyze interactions between sub-atomic particles. We use a slightly simplified version of the diagrams, in which

- Time runs from left to right.
- A straight line represents a particle such a quark or a lepton.
- A wavy line represents a particle that mediates a force, such as a photon.

The lines in a Feynman diagram do not represent the paths that the particles take, but rather the interactions of particles in the diagram.

Notice that in Figure 28-2, at each vertex, that is, a point on the diagram at which more than one particle either enters or leaves, the total charge of all particles entering the vertex equals the total charge of all particles leaving the vertex. Time runs from left to right, so in the upper of the two vertices in Figure 28-2, an electron enters and an electron and a photon leave; charge is conserved. In the lower of the two vertices an electron and a photon enter and an electron leaves; again, charge is conserved. Another vertex in which a photon acts as a mediator of the electromagnetic force is shown in **Figure 28-3**. Here a photon enters the interaction and an electron and a positron leave; we say that the photon *couples* to an electron and a positron. The same force is at play in this process as in Figure 28-2, but because the photon carries no net charge, the net charge of the final state particles in Figure 28-3 must also be zero in order for charge to be conserved. For that reason the two particles that leave the interaction must have opposite charge.

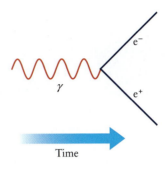

Figure 28-3 The energy of a photon is converted into the mass of an electron and a positron by the electromagnetic interaction.

Got the Concept 28-4
Something from Nothing? No!

The photon that exists at the beginning of the process shown in Figure 28-3 carries no mass. At the end of the interaction we find an electron and a positron, both of which have nonzero mass. How does that happen?

As the mediator of the electric and magnetic forces, the photon provides the mechanism for the interaction of any two charged particles. Quarks all carry charge, so other interactions mediated by the photon include $\gamma \rightarrow u\bar{u}$ and $\gamma \rightarrow d\bar{d}$. Note that in both examples the net charge of the final state particles is zero. These interactions are shown in a more general way in **Figure 28-4a**, in which q represents a quark and $\bar{q}$ represents the accompanying antimatter quark. One of the most basic tools physicists use to study particles is the collision of an electron and a positron to create a photon that then produces a quark–antiquark pair. The process is shown in **Figure 28-4b**. The first step in the interaction, the annihilation of the e^- and e^+, is the reverse of the process shown in Figure 28-2.

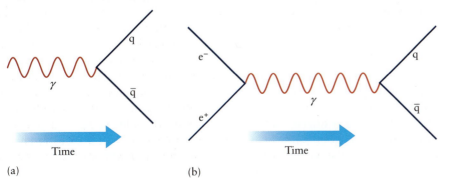

(a)   (b)

Figure 28-4 (a) Because the photon mediates the force between charged particles, it also couples to pairs of charged particles, for example, a quark (q) and its antimatter partner ($\bar{q}$). (b) An electron and a positron annihilate, producing a photon which then couples to a $q\bar{q}$ as shown in part (a).

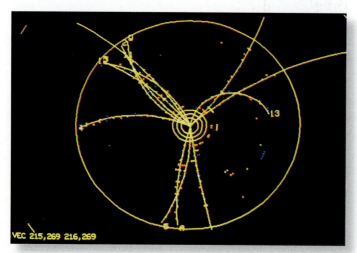

Figure 28-5 The high energy collision between an electron and a positron can create a quark–antiquark pair (see Figure 28-4b), which results in the creation of sprays of particles out of the energy in the collision. *(David Parker/ Science Photo Library)*

Many of the most important discoveries in physics over the past fifty years were made as a result of the interaction shown in Figure 28-4b. Recall that quarks are never found in nature as free, solitary particles. What happens, then, when a q$\bar{\text{q}}$ pair is produced as the result of a collision between a high energy electron and a high energy positron? Through the equivalence of energy and mass, the energy of the electron and positron, and in turn, the energy of the quark and antiquark are converted to a number of subatomic particles, perhaps a large number of subatomic particles. In **Figure 28-5**, for example, the High Resolution Spectrometer detector at the SLAC National Accelerator Laboratory in Menlo Park, California, captured the trails of a dozen particles created as a result of the e$^+$e$^-$ → q$\bar{\text{q}}$ process. Any particle that has a rest mass energy less than the total energy of the colliding particles can be created. So by colliding electrons and positrons that have large energies, physicists can create and study a wide array of subatomic particles.

The weak force is mediated by three particles, the positively and negatively charged W particles (W$^+$ and W$^-$) and the neutral Z^0. (These are really three manifestations of the same particle but have different names for historical reasons.) The Z^0 acts as the mediator of the weak force between an electron and a positron, and between a quark and antiquark, in the same way that a photon does. (Recall that the weak force is related to the electromagnetic force.) Unlike the photon, however, the Z^0 is not massless. The range of a force is directly related to the mass of the mediator particle: the more massive the mediator, the shorter the range of the force. (We discuss the relationship more fully below.) Because the photon has zero rest mass, the range of the force it mediates is infinite. The rest mass of the Z^0 particle is 91.2 GeV/c^2, almost 100 times the mass of a proton. The range of the force it mediates is short, about 0.001 fm.

The photon and the Z^0 are neutral, so they couple to a pair of particles that have a net charge of zero, as in Figures 28-2 and 28-4, for example. The W$^+$ and W$^-$ particles are charged, however, so they couple to a pair of particles that have a net charge that is nonzero. The charged W particles couple to any lepton and the antimatter partner of the other lepton in the same lepton generation (for example, e$^-\bar{\nu}_e$ and $\mu^-\bar{\nu}_\mu$), or to any quark and the antimatter partner of the other quark from the same quark generation (for example, d$\bar{\text{u}}$ and t$\bar{\text{b}}$). In **Figure 28-6a** we show a simplified Feynman diagram for a W$^-$ particle coupling to the e$^-\bar{\nu}_e$ pair, and **Figure 28-6b** shows a W$^-$ particle coupling to a d $\bar{\text{u}}$ pair. Notice that in the first, the sum of the charges of the e$^-$ and $\bar{\nu}_e$ equals -1, and in the second, the sum of the charges of the d and $\bar{\text{u}}$ is -1; in both cases, this number is the charge of the W$^-$. The rest mass of the W particles is 80.4 GeV/c^2; while not quite so large as the mass of their neutral partner, the Z^0, the nonzero mass of the W nevertheless contributes to the short range of the weak force.

The β^- decay process

$$n \rightarrow p + e^- + \bar{\nu}_e \qquad (27\text{-}17)$$

is a weak decay, that is, it is mediated by the weak force, which can be understood by examining the diagram in **Figure 28-7**. A neutron is composed of one u and two d quarks. Following the arrow of time from left to right, one of the d quarks couples to a W$^-$ and a u quark, and the resulting W$^-$ couples to an e$^-$ and $\bar{\nu}_e$. (The W$^-$ can couple to other pairs of particles, which would result in different final states.) Notice that the part of the process in which a d quark couples to a u and a

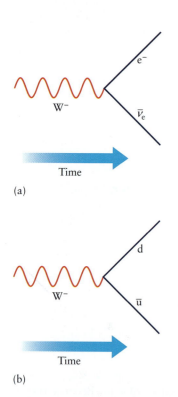

Figure 28-6 (a) The W$^-$ mediates the weak force between, for example, an electron (e$^-$) and an electron neutrino (ν_e), so it couples to an e$^-\bar{\nu}_e$ pair. (b) The W$^-$ couples pairs of quarks from the same quark generation, for example, to a d$\bar{\text{u}}$ pair.

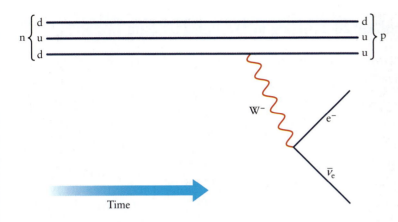

Time

Figure 28-7 A quark diagram shows the process by which a neutron (n) transforms to a proton (p), an electron (e⁻), and an antineutrino ($\bar{\nu}_e$) by β^- decay. The neutron is composed of two down quarks (d) and an up quark (u), and the proton is composed of two ups and a down. In β^- decay, one of the d quarks in the neutron decays to a u quark through the weak interaction, which is mediated by a W⁻ particle. The W⁻ decays to an e⁻ and a $\bar{\nu}_e$.

W⁻ is not quite the process shown in Figure 28-6b. The coupling is the same, but the charges must be rearranged so that charge is conserved from the initial d to the final state u and a W⁻.

? Got the Concept 28-5
Why an Antineutrino?

The W⁻ can couple to an e⁻ and $\bar{\nu}_e$. Explain why the second particle must be an *anti*neutrino, and not a neutrino, even though charge would be conserved in either case.

? Got the Concept 28-6
Charge Is Conserved!

Convince yourself that charge is conserved in the d → W⁻u process that underlies β^- decay.

? Got the Concept 28-7
Three Diagrams

The W⁺ can couple to a c quark and an $\bar{s}$ quark. Draw three diagrams, such as the one in Figure 28-6b, showing the coupling but with each of the three particles serving as the initial state.

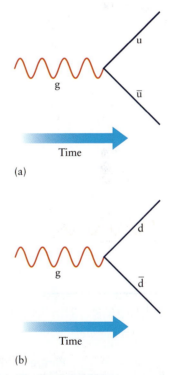

(a)

(b)

Figure 28-8 The gluon (g) mediates the strong force between a quark and its antimatter partner, for example, (a) a u quark and a $\bar{u}$ quark, and (b) a d quark and a $\bar{d}$ quark.

The strong force, through which quarks are attracted to each other, is mediated by the gluon (g). The gluon is neutrally charged, so it must couple to pairs of quarks with a net charge of zero; the gluon couples primarily to any quark and its antimatter partner. **Figure 28-8a** shows, for example, the coupling of a gluon to a u quark and its antimatter partner $\bar{u}$, and **Figure 28-8b** shows the coupling of a gluon to a d quark and its antimatter partner $\bar{d}$.

Physicist's Toolbox 28-2 Gluons in a Feynman Diagram

A gluon, the mediator of the strong force, is represented in a Feynman diagram by spiral line, as in Figure 28-8.

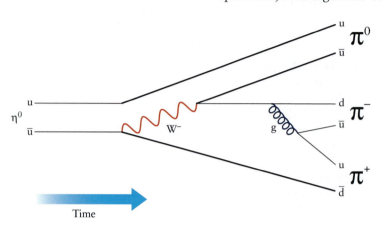

Time

Figure 28-9 A quark diagram shows the process by which a neutral eta (η^0) particle decays to three pions. The η^0 is composed of an up (u) and an antiup ($\bar{u}$) quark. In the decay process the $\bar{u}$ decays to a W$^-$ and a $\bar{d}$ through the weak interaction, and then the W$^-$ produces a $\bar{u}$ and a d. The d quark emits a gluon, which produces a u$\bar{u}$ pair. The six quarks in the final state cannot exist as free, solitary particles; instead they form three mesons, a π^0 meson, a π^- meson, and a π^+ meson.

The strong force interaction can be seen in **Figure 28-9**, a simplified Feynman diagram for the decay of a neutral eta particle to three pions, $\eta^0 \rightarrow \pi^0\pi^-\pi^+$. On the left is the η^0, composed of a u and a $\bar{u}$ quark. As the decay proceeds, the $\bar{u}$ couples to a W$^-$ and a $\bar{d}$ quark; this is a variation on the vertex shown in Figure 28-6b. The W$^-$ then decays to a $\bar{u}$ and a d quark. Follow the d quark; although it appears as if it continues along its way, suddenly a gluon appears as well. This is just the coupling of a gluon to a pair of d quarks. We drew a Feynman diagram for that interaction in Figure 28-8b, although here one of the quarks and not the gluon is the initial state. (In Figure 28-8b, we show the interaction g $\rightarrow$ d$\bar{d}$, while in the η^0 decay shown in Figure 28-9, the process is d $\rightarrow$ dg. When the $\bar{d}$ in Figure 28-8b becomes the initial state of the process, its sign must be changed to conserve charge.) Quarks are never found in nature as free, solitary particles; the six quarks in the final state of this process group together to form three mesons, a π^0 meson, a π^- meson, and a π^+ meson.

? Got the Concept 28-8
Three More Diagrams

The gluon can couple to a b quark and a $\bar{b}$ quark. Draw three diagrams, like the one in Figure 28-6b, showing the coupling but with each of the three particles serving as the initial state.

A particle named the *graviton* mediates the gravitational force. So far, no experimental evidence confirms the existence of the graviton particle, although no experimental evidence excludes it either. The graviton is thought to carry zero charge, and to be massless.

A summary of the forces and the particles that mediate them is presented in **Table 28-3**.

A puzzle has been lurking behind our discussion of forces and their exchange particles. Simply stated, what is the source of the mediator particles? It appears that one particle involved in an interaction emits the mediator and the other particle absorbs it; wouldn't that change the energy of the two particles? The rest mass energy of any particle is constant, and neither its rest energy nor total energy changes as it interacts with other particles. Here's another way to look into the question. In Figure 28-4b,

Table 28-3   The Forces

Force	Range	Propagator(s)	Strength Relative to the Strong Force
Gravity	Infinite	Graviton	10^{-40}
Weak	$\sim10^{-3}$ fm	Z^0, W^+, W^-	10^{-6}
Electromagnetic	Infinite	Photon (γ)	10^{-2}
Strong	~1 fm	Gluon (g)	1

we show an electron colliding with a positron to produce a photon, which then produces a quark–antiquark pair. Suppose an e⁻ and e⁺ collide head-on, with exactly the same magnitude of momentum but in opposite directions. The net momentum of the e⁻ and e⁺ is zero, which would mean that the momentum of the photon is zero. But recall that the momentum of a photon is given by its energy:

$$p = \frac{E}{c} \qquad (26\text{-}15)$$

A photon that carries zero momentum, according to this relationship, has zero energy. A zero energy photon? That doesn't exist! Just as troubling, we considered the collision of two energetic particles (the e⁻ and e⁺) and found that the result (the γ) is a state of zero energy. It would appear that the interaction depicted in Figure 28-4 either does not conserve momentum or does not conserve energy, or both. Throughout our study of a wide array of physical phenomena we have considered the conservation of momentum and the conservation of energy to be inviolate. The way out of this conundrum may surprise you: At the level of sub-atomic particles, nature allows for *temporary* violations of the conservation of momentum and energy. This results from a level of indeterminacy of the state of a particle when it is described as a wave.

In order to deal with the wave properties of particles, we need to find a way to describe particles using waves. Suppose we were to represent a particle as a pure sine wave. Particles are localized, that is, they exist in one, specific location in space. Over the extent of a pure sine wave, however, there is no one region that contains more of the wave than any other region; the wave is not localized in any way and is therefore not a good representation of a particle. We have seen a way to add some localization, however. Recall the phenomenon known as beats, in which waves of slightly different frequency and (or) wavelength interfere to form a pattern such as the one shown in Figure 12-27c. That wave, which is the sum of the two separate waves, is squeezed more into some regions than others, so it is somewhat more localized than either of separate waves individually. A wave that is the sum of more than two separate waves is even more localized; Figure 28-10 shows the result of adding five waves of slightly different wavelengths together. By adding together more and more properly chosen waves, we create a highly localized *wave packet*, which as the number of separate waves increases, specifies more and more precisely where the particle is in space. Such a wave packet provides an excellent way to describe a particle based on waves.

It's not quite that simple, however! Each wave has a different wavelength, so it has a different momentum than each of the others. To achieve greater and greater spatial localization requires adding more and more waves of different momenta, which makes the precision with which we know the momentum of the particle poorer and poorer. This is not a statement about how we as humans can measure the position and momentum of a particle; rather, it is a statement about the inherent properties of a particle that is also a wave. We write this statement mathematically as

$$\Delta x \, \Delta p > \frac{\hbar}{2} \qquad (28\text{-}1)$$

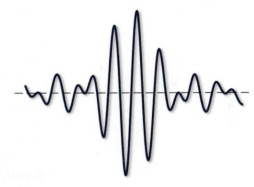

Figure 28-10 The interference of five sine waves of different wavelengths added together results in some degree of localization.

where Δx represents the degree of localization of a particle in space and Δp represents the degree to which the particle's momentum is specified. The left-hand side of Equation 28-1 is the product of Δx and Δp. Also, recall that the constant $\hbar$ is defined to be Planck's constant h divided by 2π. This equation is the **Heisenberg uncertainty principle**, postulated by the 20th-century German theoretical physicist Werner Heisenberg. It is beyond the scope of our book to derive the expression, but its essence is exactly the claim made at the beginning of this paragraph. The more localized in space you make the wave representation of a particle (that is, the smaller you make Δx), the less well you know the particle's momentum (that is, the larger Δp must be to satisfy Equation 28-1).

The relationship between the position and momentum of a particle is similar to the relationship between its energy and time interval during which the particle has that energy. We won't prove this expression, but will simply motivate it by noting that the dimensions of the product of x and p, or Δx and Δp, are the same as those of the product of E and t, or ΔE and Δt:

$$[E][t] = \frac{ML^2}{T^2}T = \frac{ML^2}{T}$$

and

$$[x][p] = LM\frac{L}{T} = \frac{ML^2}{T}$$

With using dimensions as justification, then, we rewrite Equation 28-1 in terms of energy and time:

$$\Delta E\, \Delta t > \frac{\hbar}{2} \tag{28-2}$$

Here ΔE is a measure of the degree to which the energy of a particle is known during the time interval Δt. The left-hand side of Equation 28-2 is the product of ΔE and Δt. The larger the value of ΔE for a particle, the less well known is its energy. To satisfy Equation 28-2, then, the shorter the time interval, the more poorly the energy is specified. The Heisenberg uncertainty principle provides the means for nature to ignore the rules, for example, conservation of energy, for very short periods of time.

Look again at Figure 28-4b. The energy of the final state quark–antiquark pair is the same as the energy of the initial electron and positron. Energy is conserved. In order for the mediator photon to come into existence and then disappear in the creation of the quark and antiquark, however, energy conservation must be temporarily violated. The only way that can happen is if the photon exists only for a short period of time. The time is so short that we can't make a direct observation of it, or interact with it in any way. The photon is said to be a **virtual particle.**

All force mediators are virtual particles. We imagine that every particle is constantly emitting and reabsorbing virtual particles, and that should one be intercepted by another real particle, the two real particles each experience a force as a result of exchanging the virtual particle. How far can a virtual particle stray from the particle that emitted it? Equation 28-2 is the limiting factor. The larger the energy the virtual particle carries, the less time it can exist, so heavier virtual particles therefore can't travel as far as lighter ones. Imagine an electron that is constantly emitting and reabsorbing virtual photons. A photon is massless, so one of the virtual photons could, in theory, travel infinitely far from the electron before being absorbed by another charged particle. In other words, the range of the electromagnetic force is infinitely large. It is more likely, of course, that any particular virtual photon does not travel so far, so more virtual photons stay closer to the electron. For that reason the electromagnetic force is strongest close to the electron, and falls off as distance from the electron increases. The mediators of the weak force and the strong force are not massless, so those forces have a finite range.

Estimate It! 28-1　The Range of the Weak Force

The Z^0 particle is a mediator of the weak force. By taking its energy spread ΔE to be equal to its rest energy, estimate the range of the weak force.

SET UP

The range of the weak force is on the order of the distance a virtual Z^0 can travel in the time Δt allowed by the Heisenberg uncertainty principle. For a particle traveling at the speed of light in a vacuum, the distance R it travels in time Δt is

$$R = c\Delta t$$

We can find Δt from Equation 28-2 in terms of the energy spread:

$$\Delta t > \frac{\hbar}{2\,\Delta E}$$

so

$$R < c\frac{\hbar}{2\,\Delta E}$$

Finally, according to the suggestion in the problem statement, we let ΔE equal $M_{Z^0}c^2$, the rest energy of the Z^0:

$$R < \frac{\hbar c}{2M_{Z^0}c^2}$$

SOLVE

The rest mass (M_{Z^0}) of the Z^0 is 90.2 GeV/c^2, so

$$R < \frac{(4.14 \times 10^{-15}\ \text{eV s}/2\pi)(3.00 \times 10^8\ \text{m/s})}{2(90.2 \times 10^9\ \text{eV})}$$

or

$$R < 1.10 \times 10^{-18}\ \text{m}$$

We estimate that the range of the weak force is on the order of 10^{-18} m or 10^{-3} fm.

REFLECT

Our estimate of the range of the weak force matches the value given in Table 28-3. It should because the range of a force is determined by the mass (energy) of the mediator particle.

★ What's Important 28-2

Particles exert force on each other through the exchange of mediators, virtual particles that are emitted by one particle and absorbed by another. The photon is the mediator of the electromagnetic force, the W^-, W^+, and Z^0 particles mediate the weak force, and the gluon is the mediator of the strong force.

28-3 Matter, Antimatter, Dark Matter

We've uncovered antimatter particles in a variety of physical phenomena, but it might still sound like the stuff of science fiction to you. How much of it is there, really? Is there any antimatter in your environment right now, or could you perhaps run out to a store and buy some? You might guess the answer to the first question, that there isn't very much antimatter in the universe. The answer to the second two might surprise you, however. Yes, there is almost certainly a small amount of antimatter near you right now, and yes, you could go to a store and buy some! (To be more precise, you could go to a store and buy something that is constantly producing antimatter.)

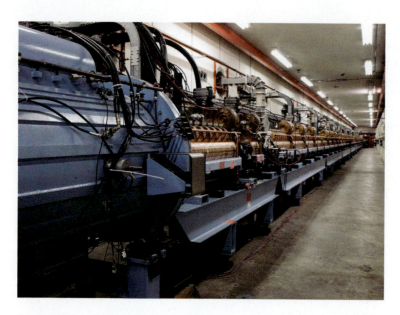

Figure 28-11 *(Reidar Hahn/ Fermilab.)*

Scientists who carry out experiments that require antimatter build large particle accelerators, such as the Tevatron at Fermilab, outside of Chicago, to create large numbers of antimatter particles. Figure 28-11 shows just one small part of the Tevatron accelerator. Although enormous devices are required to produce antimatter on the scale required for physics experiments, many natural sources produce small quantities of antimatter. Many fruits, such as oranges and bananas, for example, are constantly emitting positrons. The fruits contain potassium, and about 0.0118% of all naturally occurring potassium is the radioactive isotope ^{40}K. Its primary decay is by β^- emission, but about one out of every 10^5 decays produces a positron (e^+) by β^+ decay:

$$^{40}K \rightarrow {}^{40}Ar + e^+ + \nu_e$$

The potassium in a typical banana, for example, produces an antimatter positron about once every 2 h. You couldn't do any meaningful experiment with such a low production rate, but if you buy a banana in the store, you are certainly buying a source of antimatter. (Of course, none of the positrons emitted from the ^{40}K in a banana will get very far. As soon as an e^+ encounters an e^- the two will annihilate to create photons.)

About 0.02 g of the approximately 200 g of potassium in your body is ^{40}K. So antimatter, in the form of positrons, is being created inside of you right now, at a rate of about two e^+ particles per minute. There's certainly no cause for concern; the positrons result in no discernible effects.

Estimate It! 28-2 Have an Orange

An average orange contains about a quarter of a gram of potassium, of which about 0.0118% is radioactive ^{40}K and nearly all of the rest is ^{39}K. Potassium-40 has a half-life of 1.28×10^9 years, and about one atom out of every 10^5 decays produces a positron. Estimate the average length of time it takes for a positron to be created in a typical orange.

SET UP

The rate of radioactive decay R is determined by the half-life $\tau_{1/2}$ and the number N of radioactive atoms present. By combining Equations 27-11 and 27-13, we have

$$R = \frac{\ln 2}{\tau_{1/2}} N$$

The units of R are decays per second, of which about 10^{-5} will result in a positron. So the time t_1 required (on average) to produce one positron is then

$$t_1 = \frac{1}{10^{-5}R} = \frac{\tau_{1/2}}{10^{-5} N_{^{40}K} \ln 2} \qquad (28\text{-}3)$$

where $N_{^{40}K}$ is the number of ^{40}K atoms in the orange. To find $N_{^{40}K}$ we will use the mass of potassium in an orange together with an approximate value of the number of grams per mole (molar mass) for potassium.

SOLVE

With 39 protons and neutrons, potassium has a molar mass M_K of about 39 g/mol, so the number of K atoms in an orange is about

$$N_K = \frac{m_K}{M_K} N_A = \frac{0.25 \text{ g}}{39 \text{ g}/1 \text{ mol}} 6.02 \times 10^{23} \frac{\text{atoms}}{\text{mol}} = 3.86 \times 10^{21} \text{ atoms}$$

The number of ^{40}K atoms is then

$$N_{^{40}K} = 0.000118 N_K = (0.000118)(3.86 \times 10^{21} \text{ atoms}) = 4.55 \times 10^{17} \text{ atoms}$$

From Equation 28-3, and converting the half-life to seconds,

$$t_1 = \frac{1}{10^{-5}R} = \frac{1.28 \times 10^9 \text{ y}\left(\dfrac{365 \cdot 24 \cdot 60 \cdot 60 \text{ s}}{\text{y}}\right)}{10^{-5}(4.55 \times 10^{17})0.693} = 12.8 \times 10^3 \text{ s}$$

For our rough estimate, this is about 3.5 h.

REFLECT

If you have an orange sitting on your desk, a positron will be created inside of it, due to the β^+ decay of ^{40}K, about once every 3.5 h. Certainly this quantity isn't a lot of positrons, but it can certainly be said that there is antimatter in your local environment.

Nearly all of the mass in the universe is matter; antimatter is naturally occurring, but only in small quantities. How did the universe come to be composed overwhelmingly of matter? To get a sense of the answer, we need to go back in time to the cosmological beginning of our universe. This is the moment of the *Big Bang*, when the universe was a single hot, dense point that began to expand cataclysmically at the instant we call the beginning of the universe. (It's hard to grasp but important to recognize that the initial state was *not* the compression of everything in the universe into a tiny ball surrounded by emptiness. No, the initial state was simply everything! There is nothing "outside" our universe, so at the moment of the Big Bang, that hot, dense *singularity* was all there was.

Watch Out

The universe is not expanding into somewhere. The universe is of infinite size, and therefore encompasses everything.

The universe is expanding from its initial state, a hot, dense singularity containing everything that is now (still) in it. It is natural to ask what the universe is expanding into, an especially mind-twisting question considering that the universe is of infinite size. A somewhat simplified answer is that the universe was and is of infinite size, but the distance between every point in it is increasing. An analogy offered by 20th-century American mathematician Martin Gardner is to imagine the universe as a lump of dough with various points in it marked by raisins. When the dough is placed in an oven it begins to expand uniformly in every direction. Each raisin remains fixed in the same place within the dough, yet each raisin gets farther and farther from every other raisin. The only drawback of this helpful image is that unlike the dough, which expands to fill a larger volume in the otherwise empty oven, the universe has no equivalent empty space around it.

Immediately after the Big Bang, the universe was so hot and so dense that we have no way to conduct experiments to study the phenomena that occurred. Certainly, however, the universe cooled as it expanded, so that eventually (likely about 10^{-6} s after the Big Bang) quarks that had formed could get close enough to each other for a long enough time so that heavier particles, mesons and baryons, could form. The model of how that happened suggests that there was only the slightest preference for a matter particle forming over an antimatter particle. Theoretically, for every one billion antimatter particles that were created out of the energy of the Big Bang, one billion plus one matter particles were created. We have seen that when a matter particle encounters its antimatter partner they annihilate each other, so when those 10^9 antimatter particles encountered 10^9 matter particles, annihilations resulted in no more particles and lots of energy. And that one extra matter particle? It alone avoided annihilation. In addition, most of the mass in the universe is in the form of protons and neutrons. Baryon number remains constant in every interaction and process, which means that after all matter–antimatter annihilations, the baryons that remained must still remain today. All the matter we see today is a result of the very slight imbalance between matter and antimatter in the early universe.

We close this chapter, and this book, with one last puzzle. Immediately after the Big Bang, the universe began to expand. As physicists began to explore the processes at work in the early universe, it was natural to wonder whether that expansion would continue forever, or whether expansion would slow to the point that, eventually, the universe would begin to collapse. If the mass density of the universe is above some critical value, then the gravitational attraction of every bit of mass on every other bit of mass should eventually pull everything back together. If the mass density of the universe is not greater than that critical value, the thinking was that the rate of expansion of the universe would slow down, but continue forever. However, recent studies of the explosions (supernovae) of stars so far from Earth that it has taken light from the explosions billions of years to reach us now show that the rate at which the universe is expanding is *increasing*. Something is pulling us apart.

Based on the rate at which the universe is expanding, we can estimate the total amount of matter and energy in it. Yet a careful accounting of all the matter and energy we can observe yields only about 4% of the estimated total. In other words, we are unable to see the bulk of what we know is out there in the universe. We don't know the composition of the remaining 96% of the universe. Physicists estimate that most of it is in the form of energy, so-called *dark energy*, and the rest is in the form of *dark matter*. "Dark" is meant to suggest that we can't see it (yet). An understanding of dark energy and dark matter is certainly one of the most exciting pursuits on today's physics frontier, but beyond the scope of this book.

> ✴ **What's Important 28-3**
> Our world is made primarily of matter; only tiny amounts of antimatter occur naturally. The preference for matter over antimatter is a result of a slight imbalance—one part in one billion—between matter and antimatter in the early universe.

Answers to Got the Concept Questions

28-1 (a) The charge of the $\overline{u}$ is $-\frac{2}{3}e$, and the charge of the d is $-\frac{1}{3}e$. The net charge of the π^- is $-\frac{2}{3}e - \frac{1}{3}e$, or $-e$. (b) The combination $u\overline{u}$ has charge $\frac{2}{3}e - \frac{2}{3}e$, or zero. The combination $d\overline{d}$ has charge $-\frac{1}{3}e + \frac{1}{3}e$, which is also zero.

28-2 (a) A meson of charge $+2e$ or $-2e$ cannot exist. Each of the two quarks in a meson has charge equal to either $\pm\frac{2}{3}e$ or $\pm\frac{1}{3}e$, so no combination of those values

results in $+2e$ or $-2e$. (b) It is possible, however, to form a baryon that is doubly charged. For example, the Δ^{++} particle is composed of three u quarks; the total charge of uuu is $\frac{2}{3}e + \frac{2}{3}e + \frac{2}{3}e$, or $2e$.

28-3 (a) Charge and baryon number are conserved (and there are no leptons in the process); it can occur. (b) This process violates baryon number and lepton number conservation; it does not occur. (c) Charge, baryon number, and lepton number are all conserved. This is the most common decay mode of a charged pion. (d) On the left side of the process, the μ^- has L_e equal to zero and L_μ equal to +1. On the right, the ν_μ also has L_e equal to zero and L_μ equal to +1. Also, the electron-lepton numbers of the e^- and $\bar{\nu}_e$ cancel, and both have L_μ equal to zero. So on both the left and right sides of the process, the total electron-lepton number equals zero and the total muon-lepton number equals 1. Charge is also conserved; this process is allowed.

28-4 Mass and energy are equivalent. The mass of the final state e^- and e^+ in the process shown schematically in the Feynman diagram of Figure 28-3 is the mass equivalent of the energy carried by the photon. If the photon carried more energy than the energy equivalent of the rest masses of the e^- and e^+, the excess energy would appear as kinetic energy of the e^- and e^+. We saw evidence of the electromagnetic interaction shown in Figure 28-3 in **Figure 25-12**, in which the conversion of the energy of a photon into an electron and a positron was recorded in a bubble chamber.

28-5 Recall that every matter electron and electron neutrino is assigned electron-lepton number L_e equal to +1, and every positron and antineutrino is assigned L_e equal to -1. The W^- is not a lepton, so it has L_e equal to zero. To conserve lepton number the final state in the process $W^- \rightarrow e^- \bar{\nu}_e$ must therefore also have L_e equal to zero. $W^- \rightarrow e^- \nu_e$ is not allowed because the final state would have L_e equal to +2.

28-6 The charge of the d quark is $-\frac{1}{3}e$, the charge of the u is $\frac{2}{3}e$, and the charge of the W^- is $-1e$. So the sum of the charges of the two particles in the final state is $\frac{2}{3}e$ plus $-1e$, or $-\frac{1}{3}e$. Charge is conserved.

28-7

28-8

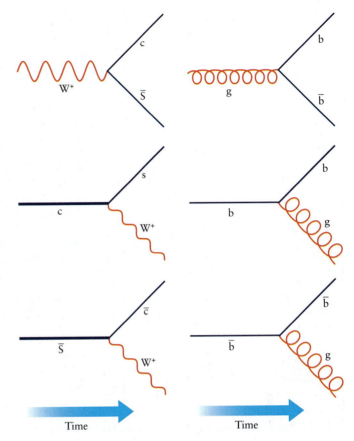

Figure 28-12 Feynman diagrams show the coupling between a W^+ particle and a charm and a strange quark.

Figure 28-13 Feynman diagrams show the coupling between a gluon and a pair of bottom quarks.

SUMMARY

Topic	Summary	Equation or Symbol
antimatter	Every matter particle is coupled with an antimatter particle that is identical to it in every way except that the charge of the antimatter particle is opposite to the charge of the particle.	
baryon	A baryon is a subatomic particle composed of three quarks. Protons and neutrons are baryons.	
Heisenberg uncertainty principle	According to the Heisenberg uncertainty principle, the more well defined the position of a particle in space, the less well defined is the particle's momentum. We express this mathematically by requiring that the product of Δx, the degree of localization of a particle in space, and Δp, the degree to which the particle's momentum is specified, must be greater than a constant. A similar statement can be made relating ΔE, the measure of the degree to which the energy of a particle is known, and Δt, the time interval over which the energy is to be specified.	$\Delta x\, \Delta p > \dfrac{\hbar}{2}$ (28-1) $\Delta E\, \Delta t > \dfrac{\hbar}{2}$ (28-2)

lepton	Leptons are fundamental particles; as best we know, they are not composed of smaller, constituent particles. Electrons, muons, and tau particles (e^-, μ^-, τ^-), their associated neutrinos (ν_e, ν_μ, ν_τ), and the antimatter partners of these particles comprise the three generations of leptons.
meson	A meson is a subatomic particle composed of two quarks. Pions (π^-, π^0, π^+) and kaons (K^-, K^0, K^+) are examples of mesons.
quark	Quarks are the constituents of mesons and baryons. They are fundamental particles in that, as best we know, quarks are not composed of smaller, constituent particles. The six quarks and their antimatter partners form three generations: up and down, charm and strange, and top and bottom. Quarks are fractionally charged, and do not exist in nature as free, solitary particles.
virtual particle	Virtual particles exist only for a very short time, and in so doing, avoid violating conservation of energy and momentum by virtue of the Heisenberg uncertainty principle. All force mediators are virtual particles.
weak force	The weak force is at the heart of the interaction that governs beta decay.

QUESTIONS AND PROBLEMS

In a few problems, you are given more data than you actually need; in a few other problems, you are required to supply data from your general knowledge, outside sources, or informed estimate.

Interpret as significant all digits in numerical values that have trailing zeros and no decimal points.

For all problems, use $g = 9.8 \text{ m/s}^2$ for the free-fall acceleration due to gravity. Neglect friction and air resistance unless instructed to do otherwise.

• Basic, single-concept problem
•• Intermediate-level problem, may require synthesis of concepts and multiple steps
••• Challenging problem
SSM Solution is in Student Solutions Manual

Conceptual Questions

1. •Define these terms: (a) baryon, (b) meson, (c) quark, (d) lepton, and (e) antiparticle.

2. •Neutrinos seem to be everywhere! Regarding the electroweak theory, is it critical that neutrinos have zero mass or a very slight value for mass? Explain your answer.

3. •Discuss the similarities and differences between a photon and a neutrino.

4. •Write the antiparticle of each of the following: (a) e^-, (b) n, (c) p, (d) π^+, and (e) K^0.

5. •How many unique baryons can be produced (neglect the antimatter baryons) with three quarks? If it helps, use the notation of up (u), down (d), and strange (s) quarks to write the different combinations. SSM

6. •A positron is stable, that is, it does not decay. Why, then, does a positron have only a short existence?

7. •When a positron and electron annihilate at rest, why must more than one photon be created?

8. •A meson and a baryon come very close to one another. The particles will interact by which of the fundamental forces (gravity, electroweak, or strong)?

9. •What do exchange particles do?

10. •In your own words, describe the major features of the Standard Model.

Multiple-Choice Questions

11. •A particle composed of two quarks is classified as a
 A. baryon.
 B. meson.
 C. photon.
 D. pion.
 E. particon. SSM

12. •A particle composed of three quarks is classified as a
 A. baryon.
 B. meson.
 C. photon.
 D. pion.
 E. particon.

13. •The quark composition of an antiproton is
 A. uud.
 B. $uu\overline{d}$.

C. $\bar{u}\bar{u}d$.
D. $\bar{u}\bar{u}\bar{d}$.
E. uuu.

14. •The beta decay process is mediated by the
   A. gravity force.
   B. electromagnetic force.
   C. strong force.
   D. weak force.
   E. dark force.

15. •Through which force does the pion interact with other particles?
   A. strong
   B. electromagnetic
   C. weak
   D. gravitational
   E. dark SSM

16. •Through which force does the photon interact with other particles?
   A. strong
   B. electromagnetic
   C. weak
   D. gravitational
   E. dark

17. •Particles that interact using the strong force are
   A. neutrinos.
   B. leptons.
   C. photons.
   D. quarks.
   E. gluons.

18. •Recent observations indicate that the universe is dominated by
   A. matter.
   B. dark matter.
   C. dark energy.
   D. photons.
   E. none of the above

19. •A mysterious energy that seems to cause the expansion of the universe to accelerate is
   A. gravitational potential energy.
   B. electric potential energy.
   C. thermal energy.
   D. dark energy.
   E. kinetic energy.

Estimation/Numerical Analysis

20. •Make an estimate of the rest energy of a gluon that mediates the strong nuclear force between nucleons. *Hint:* The range of a particle inside the nucleus must be less than $R = c\hbar/mc^2$ and the size of the nucleus is about 1.5 fm.

21. •Estimate the mass of each of the six quarks (up, down, charm, strange, top, and bottom). SSM

22. •Given that the weak boson W^+ has an approximate mass of 80 GeV, estimate the range of the electroweak force.

Problems

28-1: The Standard Model: Particles

28-2: The Standard Model: Forces

23. ••Is the reaction $n \rightarrow \pi^+ + \pi^- + \mu^+ + \mu^-$ possible? Explain your answer.

24. ••Is the reaction $e^- + p \rightarrow n + \bar{\nu}_e$ possible? Explain your answer.

25. ••Is the reaction $p \rightarrow e^+ + \gamma$ possible? Explain your answer. SSM

26. ••Determine the unknown particle X in this strong reaction:

$$p + \pi^- \rightarrow K^0 + X.$$

27. ••(a) Determine the particle that is composed of the quark combination *uds*. (b) How would your answer change if the quark combination were instead *uss*? Explain your answer.

28. ••What is the quark structure of a π^- meson?

29. ••What is the quark structure of a K^0 meson? SSM

30. ••Which of the following two possibilities for the weak decay of a sigma particle (a baryon) are possible? Why?
   a. $\Sigma^- \rightarrow \pi^- + p$
   b. $\Sigma^- \rightarrow \pi^- + n$

31. ••Two protons are collided in a particle accelerator to generate new particles from their kinetic energy. Which of the following reactions is possible? Which is not possible, owing to the violation of one or more conservation laws?
   a. $p + p \rightarrow p + p + p + \bar{p}$
   b. $p + p \rightarrow p + p + n + \bar{n}$
   c. $p + p \rightarrow p + K^+$

32. ••Which of the following reactions are possible? If a reaction is not possible, state which conservation law(s) is (are) violated.
   a. $\Lambda^0 \rightarrow p + \pi^-$
   b. $\Delta^+ \rightarrow \Sigma^+ + \pi^0 + \gamma$
   c. $\pi^0 \rightarrow \mu^- + \bar{\nu}_\mu$

33. ••Which of the following reactions is possible? If a reaction is not possible, tell which conservation law(s) is (are) violated.
   a. $n \rightarrow p + e^- + \bar{\nu}_e$
   b. $\mu^- \rightarrow e^- + \bar{\nu}_e + \nu_\mu$
   c. $\pi^- \rightarrow \mu^- + \bar{\nu}_\mu$ SSM

34. ••Using the quark model, explain how the following reaction takes place:

$$\Omega^- \rightarrow \Lambda^0 + K^-$$

This page is intentionally left blank.
For complete end of chapter problem sets, please go to
www.whfreeman.com/kestentauck

59. ••**Chemistry** (a) What is the Bohr radius of muonic hydrogen? Compare this radius with the Bohr radius in ordinary hydrogen. (b) What is the minimum energy needed to ionize muonic hydrogen (hydrogen made with muons instead of electrons) from its ground state? Compare this energy with that needed to ionize ordinary ground-state hydrogen. (c) Explain why there is such a big difference in the ionization energies of muonic hydrogen and ordinary hydrogen. **SSM**

60. •••**Chemistry** (a) What are the three longest wavelengths in the Balmer series for muonic hydrogen (see problem 59)?(b) If you are doing an experiment using muonic hydrogen, how long could you expect to have at least 10% of original muonic hydrogen available to use?

61. •••Suppose than a neutrino and an antineutrino, both of which are just barely moving, encounter each other in space and completely annihilate. In view of the uncertainty about the mass of the neutrino, what is the shortest wavelength of light that could be emitted by the annihilation? Would the light be visible to humans?

62. ••Baryons are said to be "color neutral." This refers to the fact that each of the three quarks that comprise the baryon has a "color" quantum number (which is either red, green, or blue). (There is no actual color to see with quarks; color is only a quantum characteristic that can be likened to the electric charge of an electron, for example). Just as with visible light, when you mix red with green with blue you get white light, so it is with the color nature of quarks. A proton is not simply made up of uud, rather it is $u_R u_G d_B$, for example. All the other quantum numbers that you have learned about (charge, baryon number, strangeness, etc.) still apply. The color quantum numbers "add up" to a neutral white (R + G + B = white). Write down all the possible color quantum states that might exist for an ordinary proton ($u_R u_G d_B$, $u_B u_R d_G$, etc.).

63. ••Like baryons, mesons are also color neutral. (See problem 62 for a discussion of color neutrality.) However, because mesons are made up of quark–antiquark pairs, it is not the combination of colors that cancels, but rather the subtraction of "anticolor" that creates the color neutrality. For example, a π^+ meson is made up of $u_R \bar{d}_R$. The antidown red quark cancels the up red quark much like the negative charge of an electron is canceled by the positive charge of a positron. Write the various possible color combinations for the three pions (π^+, π^0, π^-).

64. ••According to the U.S. Energy Information Administration, the United States used approximately 4.12×10^{12} kWh of electrical energy during 2010. (a) If we could generate all the energy using a matter–antimatter reactor, how many kilograms of fuel would the reactor need, assuming 100% efficiency in the annihilation? (b) Suppose the fuel consisted of iron and anti-iron, each of density 7800 kg/m³. If the iron and anti-iron were each stored as a cubical pile, what would be the dimensions of each cube?

65. •••In July 2011, physicists at the Tevatron at Fermilab near Chicago announced the discovery of a long-predicted but short-lived particle, the *xi-sub-b baryon*, Ξ_b. It is made up of a strange quark, an up quark, and a bottom quark, and is so short-lived that it travels less than a millimeter during its lifetime. (a) What is the charge of the Ξ_b baryon? (b) Estimate its lifetime if it travels 0.5 mm during its lifetime.

66. ••**Astronomy** The universe has evolved through several distinct *epochs* over the course of the last 13.75 billion years. One epoch that occurred very early after the Big Bang is called the *Planck Epoch*. It occurred about 10^{-43} s after the Big Bang. Calculate the *Planck time*, t_P by first using dimensional analysis and finding the proper combination of the fundamental constants: G, c, and $\hbar$. Then use the current numerical values of each to calculate the precise number for t_P. (Recall, G is Newton's gravitational constant and is equal to 6.6738×10^{-11} N·m²/kg², c is the speed of light and is equal to 2.9979×10^8 m/s, and $\hbar$ is Planck's constant divided by 2π or 1.0546×10^{-34} J·s). Start your calculation by asking what power would each of the three constants need to have in order to yield units of seconds:

$$[t_P] = [G]^x [c]^y [\hbar]^z$$

Solve for x, y, and z to derive the formula for t_P.

67. •••**Calc** Two stationary nucleons can be described with a potential function V in a relativistic, quantum mechanical manner with the static case of the Klein–Gordon equation:

$$\nabla^2 V(r) = K^2 V(r)$$

where, $\nabla^2 [V(r)] = \dfrac{1}{r^2} \dfrac{d}{dr}\left[r^2 \dfrac{dV(r)}{dr} \right]$

and $K = \dfrac{mc}{\hbar}$ (m is the mass of the exchanged particle)

(a) By trial and error, show that a correct solution to the differential equation is

$$V(r) = g^2 \frac{e^{-Kr}}{r} \quad (r > 0)$$

where, g gives the strength of the Yukawa potential

$$g^2 = 15\hbar c$$

and $K = \dfrac{1}{R} = \dfrac{1}{1.5 \times 10^{-15}\text{m}}$

(b) Show that the mass of the mediating particle is that of a pion. (c) Using the Heisenberg uncertainty principle ($\Delta E\, \Delta t \geq \hbar/2$), estimate the time during which the pion can exist in the virtual state.

APPENDIX A
SI Units and Conversion Factors

Base Units*

Length	The *meter* (m) is the distance traveled by light in a vacuum in 1/299,792,458 s.
Time	The *second* (s) is the duration of 9,192,631,770 periods of the radiation corresponding to the transition between the two hyperfine levels of the ground state of the ^{133}Cs atom.
Mass	The *kilogram* (kg) is the mass of the international standard body preserved at Sèvres, France.
Mole	The *mole* (mol) is the amount of substance of a system which contains as many elementary entities as there are atoms in 0.012 kg of carbon-12.
Current	The *ampere* (A) is that constant current which, if maintained in two straight parallel conductors of infinite length, of negligible circular cross section, and placed 1 m apart in vacuum, would produce between the conductors a force equal to 2×10^{-7} N/m of length.
Temperature	The *kelvin* (K) is 1/273.16 of the thermodynamic temperature of the triple point of water.
Luminous intensity	The *candela* (cd) is the luminous intensity in a given direction, of a source that emits monochromatic radiation of frequency 540×10^{12} Hz and that has a radiant intensity, in that direction of 1/683 W/steradian.

*These definitions are found on the Internet at http://physics.nist.gov/cuu/Units/current.html.

Derived Units

Force	newton (N)	$1 \text{ N} = 1 \text{ kg} \cdot \text{m/s}^2$
Work, energy	joule (J)	$1 \text{ J} = 1 \text{ N} \cdot \text{m}$
Power	watt (W)	$1 \text{ W} = 1 \text{ J/s}$
Frequency	hertz (Hz)	$1 \text{ Hz} = \text{cy/s}$
Charge	coulomb (C)	$1 \text{ C} = 1 \text{ A} \cdot \text{s}$
Potential	volt (V)	$1 \text{ V} = 1 \text{ J/C}$
Resistance	ohm (Ω)	$1 \Omega = 1 \text{ V/A}$
Capacitance	farad (F)	$1 \text{ F} = 1 \text{ C/V}$
Magnetic field	tesla (T)	$1 \text{ T} = 1 \text{ N/(A} \cdot \text{m)}$
Magnetic flux	weber (Wb)	$1 \text{ Wb} = 1 \text{ T} \cdot \text{m}^2$
Inductance	henry (H)	$1 \text{ H} = 1 \text{ J/A}^2$

Conversion Factors

Conversion factors are written as equations for simplicity;
relations marked with an asterisk are exact.

Length

1 km = 0.6214 mi

1 mi = 1.609 km

1 m = 1.0936 yard = 3.281 ft = 39.37 in.

*1 in. = 2.54 cm

*1 ft = 12 in. = 30.48 cm

*1 yard = 3 ft = 91.44 cm

1 light-year = 1 $c \cdot y$ = 9.461 × 10^{15} m

*1 Å = 0.1 nm

Area

*1 m^2 = 10^4 cm^2

1 km^2 = 0.3861 mi^2 = 247.1 acres

*1 $in.^2$ = 6.4516 cm^2

1 ft^2 = 9.29 × 10^{-2} m^2

1 m^2 = 10.76 ft^2

*1 acre = 43 560 ft^2

1 mi^2 = 640 acres = 2.590 km^2

Volume

*1 m^3 = 10^6 cm^3

*1 L = 1000 cm^3 = 10^{-3} m^3

1 gal = 3.785 L

1 gal = 4 qt = 8 pt = 128 oz = 231 $in.^3$

1 $in.^3$ = 16.39 cm^3

1 ft^3 = 1728 $in.^3$ = 28.32 L
       = 2.832 × 10^4 cm^3

Time

*1 h = 60 min = 3.6 ks

*1 d = 24 h = 1440 min = 86.4 ks

1 y = 365.25 day = 3.156 × 10^7 s

Speed

*1 m/s = 3.6 km/h

1 km/h = 0.2778 m/s = 0.6214 mi/h

1 mi/h = 0.4470 m/s = 1.609 km/h

1 mi/h = 1.467 ft/s

Angle and Angular Speed

*π rad = 180°

1 rad = 57.30°

1° = 1.745 × 10^{-2} rad

1 rev/min = 0.1047 rad/s

1 rad/s = 9.549 rev/min

Mass

*1 kg = 1000 g

*1 tonne = 1000 kg = 1 Mg

1 u = 1.6605 × 10^{-27} kg
      931.49 MeV/c^2

1 kg = 6.022 × 10^{26} u

1 slug = 14.59 kg

1 kg = 6.852 × 10^{-2} slug

Density

*1 g/cm^3 = 1000 kg/m^3 = 1 kg/L

(1 g/cm^3)g = 62.4 lb/ft^3

Force

1 N = 0.2248 lb = 10^5 dyn

*1 lb = 4.448222 N

(1 kg)g = 2.2046 lb

Pressure

*1 Pa = 1 N/m^2

*1 atm = 101.325 kPa = 1.01325 bar

1 atm = 14.7 $lb/in.^2$ = 760 mmHg
        = 29.9 in.Hg = 33.9 ftH_2O

1 $lb/in.^2$ = 6.895 kPa

1 torr = 1 mmHg = 133.32 Pa

1 bar = 100 kPa

Energy

*1 kW · h = 3.6 MJ

*1 cal = 4.186 J

1 ft · lb = 1.356 J = 1.286 × 10^{-3} BTU

*1 L · atm = 101.325 J

1 L · atm = 24.217 cal

1 BTU = 778 ft · lb = 252 cal = 1054.35 J

1 eV = 1.602 × 10^{-19} J

1 u · c^2 = 931.49 MeV

*1 erg = 10^{-7} J

Power

1 horsepower = 550 ft · lb/s = 745.7 W

1 BTU/h = 2.931 × 10^{-4} kW

1 W = 1.341 × 10^{-3} horsepower
     = 0.7376 ft · lb/s

Magnetic Field

*1 T = 10^4 G

Thermal Conductivity

1 W/(m · K) = 6.938 BTU · in./(h · ft^2 · °F)

1 BTU · in./(h · ft^2 · °F) = 0.1441 W/(m · K)

APPENDIX B
Numerical Data

Terrestrial Data

Free-fall acceleration g
    Standard value (at sea level at 45° latitude)*   9.806 65 m/s^2; 32.1740 ft/s^2
    At equator*   9.7804 m/s^2
    At poles*   9.8322 m/s^2

Mass of Earth M_E 5.98×10^{24} kg

Radius of Earth R_E, mean 6.38×10^6 m; 3960 mi

Escape speed 1.12×10^4 m/s; 6.95 mi/s

Solar constant†   1.37 kW/m^2

Standard temperature and pressure (STP):

    Temperature   293.15 K
    Pressure   101.3 kPa (1.00 atm)

Molar mass of air   28.97 g/mol

Density of air (273.15 K, 101.3 kPa), ρ_{air} 1.217 kg/m^3

Speed of sound (273.15 K, 101.3 kPa)   331 m/s

Latent heat of fusion of H_2O (0°C, 1 atm)   334 kJ/kg

Latent heat of vaporization of H_2O (100°C, 1 atm)   2.26 MJ/kg

---

* Measured relative to Earth's surface.
† Average power incident normally on 1 m^2 outside Earth's atmosphere at the mean distance from Earth to the Sun.

Astronomical Data*

Earth
    Distance to the Moon, mean† 3.844×10^8 m; 2.389×10^5 mi
    Distance to the Sun, mean† 1.496×10^{11} m; 9.30×10^7 mi; 1.00 AU
    Orbital speed, mean 2.98×10^4 m/s

Moon
    Mass 7.35×10^{22} kg
    Radius 1.737×10^6 m
    Period   27.32 day
    Acceleration of gravity at surface   1.62 m/s^2

Sun
    Mass 1.99×10^{30} kg
    Radius 6.96×10^8 m

---

* Additional solar system data are available from NASA at http://nssdc.gsfc.nasa.gov/planetary/planetfact.html.
† Center to center.

Physical Constants*

Universal constant of gravitation	G	$6.673\,84(80) \times 10^{-11}$ N·m²/kg²
Speed of light	c	$2.997\,924\,58 \times 10^{8}$ m/s
Fundamental charge	e	$1.602\,176\,565(35) \times 10^{-19}$ C
Avogadro's constant	N_A	$6.022\,141\,29(27) \times 10^{23}$ particles/mol
Gas constant	R	$8.314\,462\,1(75)$ J/(mol·K)
		$1.987\,206\,5(36)$ cal/(mol·K)
		$8.205\,746(15) \times 10^{-2}$ L·atm/(mol·K)
Boltzmann constant	$k = R/N_A$	$1.380\,648\,8(13) \times 10^{-23}$ J/K
		$8.617\,332\,4(78) \times 10^{-5}$ eV/K
Stefan-Boltzmann constant	σ	$5.670\,373(21) \times 10^{-8}$ W/(m²·K⁴)
Atomic mass constant	m_u	$1.660\,538\,921(73) \times 10^{-27}$ kg = 1 u
Permeability of free space	μ_0	$4\pi \times 10^{-7}$ T·m/A
		$1.256\,637\ldots \times 10^{-6}$ T·m/A
Permittivity of free space	ϵ_0	$8.854\,187\,817\ldots \times 10^{-12}$ C²/(N·m²)
Coulomb constant	$k = 1/(4\pi\epsilon_0)$	$8.987\,551\,787\ldots \times 10^{9}$ N·m²/C²
Planck's constant	h	$6.626\,069\,57(29) \times 10^{-34}$ J·s
		$4.135\,667\,516(91) \times 10^{-15}$ eV·s
	$\hbar = h/2\pi$	$1.054\,571\,726(47) \times 10^{-34}$ J·s
		$6.582\,119\,28(15) \times 10^{-16}$ eV·s
Mass of electron	m_e	$9.109\,382\,91(40) \times 10^{-31}$ kg
		$0.510\,998\,928(11)$ MeV/c^2
Mass of proton	m_p	$1.672\,621\,777(74) \times 10^{-27}$ kg
		$938.272\,046(21)$ MeV/c^2
Mass of neutron	m_n	$1.674\,927\,351(74) \times 10^{-27}$ kg
		$939.565\,379(21)$ MeV/c^2
Bohr magneton	$m_B = eh/2m_e$	$9.274\,009\,68(20) \times 10^{-24}$ J/T
		$5.788\,381\,806\,6(38) \times 10^{-5}$ eV/T
Nuclear magneton	$m_n = eh/2m_p$	$5.050\,783\,53(11) \times 10^{-27}$ J/T
		$3.152\,451\,260\,5(22) \times 10^{-8}$ eV/T
Magnetic flux quantum	ϕ_0	$2.067\,833\,758(46) \times 10^{-15}$ T·m²
Quantized Hall resistance	R_K	$2.581\,280\,744\,34(84) \times 10^{4}$ Ω
Rydberg constant	R_H	$1.097\,373\,156\,853\,9(55) \times 10^{7}$ m⁻¹
Josephson frequency–voltage quotient	K_J	$4.835\,978\,70(11) \times 10^{14}$ Hz/V
Compton wavelength	$\lambda_C = h/m_e c$	$2.426\,310\,238\,9(16) \times 10^{-12}$ m

* The values for these and other constants may be found on the Internet at http://physics.nist.gov/cuu/Constants/index.html. The numbers in parentheses represent the uncertainties in the last two digits. (For example, 2.044 43(13) stands for 2.044 43 ± 0.000 13.) Values without uncertainties are exact, including those values with ellipses (such as the value of pi is exactly 3.1415…).

APPENDIX C
Periodic Table of Elements*

1																	18
1 H	2											13	14	15	16	17	2 He
3 Li	4 Be											5 B	6 C	7 N	8 O	9 F	10 Ne
11 Na	12 Mg	3	4	5	6	7	8	9	10	11	12	13 Al	14 Si	15 P	16 S	17 Cl	18 Ar
19 K	20 Ca	21 Sc	22 Ti	23 V	24 Cr	25 Mn	26 Fe	27 Co	28 Ni	29 Cu	30 Zn	31 Ga	32 Ge	33 As	34 Se	35 Br	36 Kr
37 Rb	38 Sr	39 Y	40 Zr	41 Nb	42 Mo	43 Tc	44 Ru	45 Rh	46 Pd	47 Ag	48 Cd	49 In	50 Sn	51 Sb	52 Te	53 I	54 Xe
55 Cs	56 Ba	57–71 Lanthanoids	72 Hf	73 Ta	74 W	75 Re	76 Os	77 Ir	78 Pt	79 Au	80 Hg	81 Tl	82 Pb	83 Bi	84 Po	85 At	86 Rn
87 Fr	88 Ra	89–103 Actinoids	104 Rf	105 Db	106 Sg	107 Bh	108 Hs	109 Mt	110 Ds	111 Rg	112 Cn						

Lanthanoids	57 La	58 Ce	59 Pr	60 Nd	61 Pm	62 Sm	63 Eu	64 Gd	65 Tb	66 Dy	67 Ho	68 Er	69 Tm	70 Yb	71 Lu
Actinoids	89 Ac	90 Th	91 Pa	92 U	93 Np	94 Pu	95 Am	96 Cm	97 Bk	98 Cf	99 Es	100 Fm	101 Md	102 No	103 Lr

*From http://old.iupac.org/reports/periodic_table/IUPAC_Periodic_Table-21Jan11.pdf.

Atomic Numbers and Atomic Weights*

Atomic Number	Name	Symbol	Weight	Atomic Number	Name	Symbol	Weight
1	Hydrogen	H	[1.007 84; 1.008 11]	57	Lanthanum	La	138.90547(7)
2	Helium	He	4.002602(2)	58	Cerium	Ce	140.116(1)
3	Lithium	Li	[6.938; 6.997]	59	Praseodymium	Pr	140.90765(2)
4	Beryllium	Be	9.012182(3)	60	Neodymium	Nd	144.242(3)
5	Boron	B	[10.806; 10.821]	61	Promethium	Pm	
6	Carbon	C	[12.009 6; 12.011 6]	62	Samarium	Sm	150.36(2)
7	Nitrogen	N	[14.006 43; 14.007 28]	63	Europium	Eu	151.964(1)
8	Oxygen	O	[15.999 03; 15.999 77]	64	Gadolinium	Gd	157.25(3)
9	Fluorine	F	18.9984032(5)	65	Terbium	Tb	158.92535(2)
10	Neon	Ne	20.1797(6)	66	Dysprosium	Dy	162.500(1)
11	Sodium	Na	22.98976928(2)	67	Holmium	Ho	164.93032(2)
12	Magnesium	Mg	24.3050(6)	68	Erbium	Er	167.259(3)
13	Aluminum	Al	26.9815386(8)	69	Thulium	Tm	168.93421(2)
14	Silicon	Si	[28.084; 28.086]	70	Ytterbium	Yb	173.054(5)
15	Phosphorus	P	30.973762(2)	71	Lutetium	Lu	174.966 8(1)
16	Sulfur	S	[32.059; 32.076]	72	Hafnium	Hf	178.49(2)
17	Chlorine	Cl	[35.446; 35.457]	73	Tantalum	Ta	180.94788(2)
18	Argon	Ar	39.948(1)	74	Tungsten	W	183.84(1)
19	Potassium	K	39.0983(1)	75	Rhenium	Re	186.207(1)
20	Calcium	Ca	40.078(4)	76	Osmium	Os	190.23(3)
21	Scandium	Sc	44.955912(6)	77	Iridium	Ir	192.217(3)
22	Titanium	Ti	47.867(1)	78	Platinum	Pt	195.084(9)
23	Vanadium	V	50.9415(1)	79	Gold	Au	196.966569(4)
24	Chromium	Cr	51.9961(6)	80	Mercury	Hg	200.59(2)
25	Manganese	Mn	54.938045(5)	81	Thallium	Tl	[204.382; 204.385]
26	Iron	Fe	55.845(2)	82	Lead	Pb	207.2(1)
27	Cobalt	Co	58.933195(5)	83	Bismuth	Bi	208.98040(1)
28	Nickel	Ni	58.6934(2)	84	Polonium	Po	
29	Copper	Cu	63.546(3)	85	Astatine	At	
30	Zinc	Zn	65.38 (2)	86	Radon	Rn	
31	Gallium	Ga	69.723(1)	87	Francium	Fr	
32	Germanium	Ge	72.63 (1)	88	Radium	Ra	
33	Arsenic	As	74.92160(2)	89	Actinium	Ac	
34	Selenium	Se	78.96(3)	90	Thorium	Th	232.03806(2)
35	Bromine	Br	79.904(1)	91	Protactinium	Pa	231.03588(2)
36	Krypton	Kr	83.798(2)	92	Uranium	U	238.02891(3)
37	Rubidium	Rb	85.4678(3)	93	Neptunium	Np	
38	Strontium	Sr	87.62(1)	94	Plutonium	Pu	
39	Yttrium	Y	88.90585(2)	95	Americium	Am	
40	Zirconium	Zr	91.224(2)	96	Curium	Cm	
41	Niobium	Nb	92.90638(2)	97	Berkelium	Bk	
42	Molybdenum	Mo	95.96 (2)	98	Californium	Cf	
43	Technetium	Tc		99	Einsteinium	Es	
44	Ruthenium	Ru	101.07(2)	100	Fermium	Fm	
45	Rhodium	Rh	102.90550(2)	101	Mendelevium	Md	
46	Palladium	Pd	106.42(1)	102	Nobelium	No	
47	Silver	Ag	107.8682(2)	103	Lawrencium	Lr	
48	Cadmium	Cd	112.411(8)	104	Rutherfordium	Rf	
49	Indium	In	114.818(3)	105	Dubnium	Db	
50	Tin	Sn	118.710(7)	106	Seaborgium	Sg	
51	Antimony	Sb	121.760(1)	107	Bohrium	Bh	
52	Tellurium	Te	127.60(3)	108	Hassium	Hs	
53	Iodine	I	126.90447(3)	109	Meitnerium	Mt	
54	Xenon	Xe	131.293(6)	110	Darmstadtium	Ds	
55	Cesium	Cs	132.9054519(2)	111	Roentgenium	Rg	
56	Barium	Ba	137.327(7)	112	Copernicium	Cn	

* Some weights are listed as intervals ([a; b]; $a \le$ atomic weight $\le b$) because these weights are not constant but depend on the physical, chemical, and nuclear histories of the samples used. Atomic weights are not listed for some elements because these elements do not have stable isotopes. Exceptions are thorium, protactinium, and uranium. Elements 113 to 118 are not listed in this table although they have been reported (IUPAC has not named them). From *Atomic Weights of the Elements 2009 (IUPAC Technical Report)*, Pure Appl. Chem., Vol. 83(2), pp. 359–396, 2011.

Table of Atomic Masses

Element	Symbol	Mass number (*indicates radioactive)	Atomic mass	Percent abundance	Half-life and decay mode (if unstable)	
(Neutron)	*n*	1*	1.008665		10.4 m	β^-
Hydrogen	H	1	1.007825	99.985		
Deuterium	D	2	2.014102	0.015		
Tritium	T	3*	3.016049		12.33 y	β^-
Helium	He	3	3.016029	0.00014		
		4	4.002602	99.99986		
		6*	6.018886		0.81 s	β^-
		8*	8.033922		0.12 s	β^-
Lithium	Li	6	6.015121	7.5		
		7	7.016003	92.5		
		8*	8.022486		0.84 s	β^-
		9*	9.026789		0.18 s	β^-
		11*	11.043897		8.7 ms	β^-
Beryllium	Be	7*	7.016928		53.3 d	ec
		9	9.012174	100		
		10*	10.013534		1.5×10^6 y	β^-
		11*	11.021657		13.8 s	β^-
		12*	12.026921		23.6 ms	β^-
		14*	14.042866		4.3 ms	β^-
Boron	B	8*	8.024605		0.77 s	β^+
		10	10.012936	19.9		
		11	11.009305	80.1		
		12*	12.014352		0.0202 s	β^-
		13*	13.017780		17.4 ms	β^-
		14*	14.025404		13.8 ms	β^-
		15*	15.031100		10.3 ms	β^-
Carbon	C	9*	9.031030		0.13 s	β^+
		10*	10.016854		19.3 s	β^+
		11*	11.011433		20.4 m	β^+
		12	12.000000	98.90		
		13	13.003355	1.10		
		14*	14.003242		5730 y	β^-
		15*	15.010599		2.45 s	β^-
		16*	16.014701		0.75 s	β^-
		17*	17.022582		0.20 s	β^-

(Continued)

Element	Symbol	Mass number (*indicates radioactive)	Atomic mass	Percent abundance	Half-life and decay mode (if unstable)	
Nitrogen	N	12*	12.018613		0.0110 s	β^+
		13*	13.005738		9.96 m	β^+
		14	14.003074	99.63		
		15	15.000108	0.37		
		16*	16.006100		7.13 s	β^-
		17*	17.008450		4.17 s	β^-
		18*	18.014082		0.62 s	β^-
		19*	19.017038		0.24 s	β^-
Oxygen	O	13*	13.024813		8.6 ms	β^+
		14*	14.008595		70.6 s	β^+
		15*	15.003065		122 s	β^+
		16	15.994915	99.71		
		17	16.999132	0.039		
		18	17.999160	0.20		
		19*	19.003577		26.9 s	β^-
		20*	20.004076		13.6 s	β^-
		21*	21.008595		3.4 s	β^-
Fluorine	F	17*	17.002094		64.5 s	β^+
		18*	18.000937		109.8 m	β^+
		19	18.998404	100		
		20*	19.999982		11.0 s	β^-
		21*	20.999950		4.2 s	β^-
		22*	22.003036		4.2 s	β^-
		23*	23.003564		2.2 s	β^-
Neon	Ne	18*	18.005710		1.67 s	β^+
		19*	19.001880		17.2 s	β^+
		20	19.992435	90.48		
		21	20.993841	0.27		
		22	21.991383	9.25		
		23*	22.994465		37.2 s	β^-
		24*	23.993999		3.38 m	β^-
		25*	24.997789		0.60 s	β^-
Sodium	Na	21*	20.997650		22.5 s	β^+
		22*	21.994434		2.61 y	β^+
		23	22.989767	100		
		24*	23.990961		14.96 h	β^-
		25*	24.989951		59.1 s	β^-
		26*	25.992588		1.07 s	β^-
Magnesium	Mg	23*	22.994124		11.3 s	β^+
		24	23.985042	78.99		
		25	24.985838	10.00		
		26	25.982594	11.01		
		27*	26.984341		9.46 m	β^-
		28*	27.983876		20.9 h	β^-
		29*	28.375346		1.30 s	β^-

Element	Symbol	Mass number (*indicates radioactive)	Atomic mass	Percent abundance	Half-life and decay mode (if unstable)	
Aluminum	Al	25*	24.990429		7.18 s	β^+
		26*	25.986892		7.4×10^5 y	β^+
		27	26.981538	100		
		28*	27.981910		2.24 m	β^-
		29*	28.980445		6.56 m	β^-
		30*	29.982965		3.60 s	β^-
Silicon	Si	27*	26.986704		4.16 s	β^+
		28	27.976927	92.23		
		29	28.976495	4.67		
		30	28.973770	3.10		
		31*	30.975362		2.62 h	β^-
		32*	31.974148		172 y	β^-
		33*	32.977928		6.13 s	β^-
Phosphorus	P	30*	29.978307		2.50 m	β^+
		31	30.973762	100		
		32*	31.973762		14.26 d	β^-
		33*	32.971725		25.3 d	β^-
		34*	33.973636		12.43 s	β^-
Sulfur	S	31*	30.979554		2.57 s	β^+
		32	31.972071	95.02		
		33	32.971459	0.75		
		34	33.967867	4.21		
		35*	34.969033		87.5 d	β^-
		36	35.967081	0.02		
Chlorine	Cl	34*	33.973763		32.2 m	β^+
		35	34.968853	75.77		
		36*	35.968307		3.0×10^5 y	β^-
		37	36.965903	24.23		
		38*	37.968010		37.3 m	β^-
Argon	Ar	36	35.967547	0.337		
		37*	36.966776		35.04 d	ec
		38	37.962732	0.063		
		39*	38.964314		269 y	β^-
		40	39.962384	99.600		
		42*	41.963049		33 y	β^-
Potassium	K	39	38.963708	93.2581		
		40*	39.964000	0.0117	1.28×10^9 y	β^+, ec, β^-
		41	40.961827	6.7302		
		42*	41.962404		12.4 h	β^-
		43*	42.960716		22.3 h	β^-

(Continued)

Element	Symbol	Mass number (*indicates radioactive)	Atomic mass	Percent abundance	Half-life and decay mode (if unstable)	
Calcium	Ca	40	39.962591	96.941		
		41*	40.962279		1.0×10^5 y	ec
		42	41.958618	0.647		
		43	42.958767	0.135		
		44	43.955481	2.086		
		46	45.953687	0.004		
		48	47.952534	0.187		
Scandium	Sc	41*	40.969250		0.596 s	β^+
		43*	42.961151		3.89 h	β^+
		45	44.955911	100		
		46*	45.955170		83.8 d	β^-
Titanium	Ti	44*	43.959691		49 y	ec
		46	45.952630	8.0		
		47	46.951765	7.3		
		48	47.947947	73.8		
		49	48.947871	5.5		
		50	49.944792	5.4		
Vanadium	V	48*	47.952255			
		50*	49.947161	0.25	15.97 d	β^+
		51	50.943962	99.75	1.5×10^{17} y	β^+
Chromium	Cr	48*	47.954033		21.6 h	ec
		50	49.946047	4.345		
		52	51.940511	83.79		
		53	52.940652	9.50		
		54	53.938883	2.365		
Manganese	Mn	53*	52.941292		3.74×10^6 y	ec
		54*	53.940361		312.1 d	ec
		55	54.938048	100		
		56*	55.938908		2.58 h	β^-
Iron	Fe	54	53.939613	5.9		
		55*	54.938297		2.7 y	ec
		56	55.934940	91.72		
		57	56.935396	2.1		
		58	57.933278	0.28		
		60*	59.934078		1.5×10^6 y	β^-
Cobalt	Co	57*	56.936294		271.8 d	ec
		58*	57.935755		70.9 h	ec, β^+
		59	58.933198	100		
		60*	59.933820		5.27 y	β^-
		61*	60.932478		1.65 h	β^-
Nickel	Ni	58	57.935346	68.077		
		59*	58.934350		7.5×10^4 y	ec, β^+
		60	59.930789	26.223		
		61	60.931058	1.140		
		62	61.928346	3.634		
		63*	62.929670		100 y	β^-
		64	63.927967	0.926		

Element	Symbol	Mass number (*indicates radioactive)	Atomic mass	Percent abundance	Half-life and decay mode (if unstable)
Copper	Cu	63	62.929599	69.17	
		64*	63.929765		12.7 h ec
		65	64.927791	30.83	
		66*	65.928871		5.1 m β−
Zinc	Zn	64	63.929144	48.6	
		66	65.926035	27.9	
		67	66.927129	4.1	
		68	67.924845	18.8	
		70	69.925323	0.6	
Gallium	Ga	69	68.925580	60.108	
		70*	69.926027		21.1 m β−
		71	70.924703	39.892	
		72*	71.926367		14.1 h β−
Germanium	Ge	69*	68.927969		39.1 h ec, β+
		70	69.924250	21.23	
		72	71.922079	27.66	
		73	72.923462	7.73	
		74	73.921177	35.94	
		76	75.921402	7.44	
		77*	76.923547		11.3 h β−
Arsenic	As	73*	72.923827		80.3 d ec
		74*	73.923928		17.8 d ec, β+
		75	74.921594	100	
		76*	75.922393		1.1 d β−
		77*	76.920645		38.8 h β−
Selenium	Se	74	73.922474	0.89	
		76	75.919212	9.36	
		77	76.919913	7.63	
		78	77.917307	23.78	
		79*	78.918497		$\leq 6.5 \times 10^4$ y β−
		80	79.916519	49.61	
		82*	81.916697	8.73	1.4×10^{20} y 2β−
Bromine	Br	79	78.918336	50.69	
		80*	79.918528		17.7 m β+
		81	80.916287	49.31	
		82*	81.916802		35.3 h β−
Krypton	Kr	78	77.920400	0.35	
		80	79.916377	2.25	
		81*	80.916589		2.11×10^5 y ec
		82	81.913481	11.6	
		83	82.914136	11.5	
		84	83.911508	57.0	
		85*	84.912531		10.76 y β−
		86	85.910615	17.3	

(Continued)

Element	Symbol	Mass number (*indicates radioactive)	Atomic mass	Percent abundance	Half-life and decay mode (if unstable)
Rubidium	Rb	85	84.911793	72.17	
		86*	85.911171		18.6 d β^-
		87*	86.909186	27.83	4.75×10^{10} y β^-
		88*	87.911325		17.8 m β^-
Strontium	Sr	84	83.913428	0.56	
		86	85.909266	9.86	
		87	86.908883	7.00	
		88	87.905618	82.58	
		90*	89.907737		29.1 y β^-
Yttrium	Y	88*	87.909507		106.6 d ec, β^+
		89	88.905847	100	
		90*	89.914811		2.67 d β^-
Zirconium	Zr	90	89.904702	51.45	
		91	90.905643	11.22	
		92	91.905038	17.15	
		93*	92.906473		1.5×10^6 y β^-
		94	93.906314	17.38	
		96	95.908274	2.80	
Niobium	Nb	91*	90.906988		6.8×10^2 y ec
		92*	91.907191		3.5×10^7 y ec
		93	92.906376	100	
		94*	93.907280		2×10^4 y β^-
Molybdenum	Mo	92	91.906807	14.84	
		93*	92.906811		3.5×10^3 y ec
		94	93.905085	9.25	
		95	94.905841	15.92	
		96	95.904678	16.68	
		97	96.906020	9.55	
		98	97.905407	24.13	
		100	99.907476	9.63	
Technetium	Tc	97*	96.906363		2.6×10^6 y ec
		98*	97.907215		4.2×10^6 y β^-
		99*	98.906254		2.1×10^5 y β^-
Ruthenium	Ru	96	95.907597	5.54	
		98	97.905287	1.86	
		99	98.905939	12.7	
		100	99.904219	12.6	
		101	100.905558	17.1	
		102	101.904348	31.6	
		104	103.905428	18.6	
Rhodium	Rh	102*	101.906794		207 d ec
		103	102.905502	100	
		104*	103.906654		42 s β^-

Element	Symbol	Mass number (*indicates radioactive)	Atomic mass	Percent abundance	Half-life and decay mode (if unstable)
Palladium	Pd	102	101.905616	1.02	
		104	103.904033	11.14	
		105	104.905082	22.33	
		106	105.903481	27.33	
		107*	106.905126		6.5×10^6 y β^-
		108	107.903893	26.46	
		110	109.905158	11.72	
Silver	Ag	107	106.905091	51.84	
		108*	107.905953		2.39 m ec, β^+, β^-
		109	108.904754	48.16	
		110*	109.906110		24.6 s β^-
Cadmium	Cd	106	105.906457	1.25	
		108	107.904183	0.89	
		109*	108.904984		462 d ec
		110	109.903004	12.49	
		111	110.904182	12.80	
		112	111.902760	24.13	
		113*	112.904401	12.22	9.3×10^{15} y β^-
		114	113.903359	28.73	
		116	115.904755	7.49	
Indium	In	113	112.904060	4.3	
		114*	113.904916		1.2 m β^-
		115*	114.903876	95.7	4.4×10^{14} y β^-
		116*	115.905258		54.4 m β^-
Tin	Sn	112	111.904822	0.97	
		114	113.902780	0.65	
		115	114.903345	0.36	
		116	115.901743	14.53	
		117	116.902953	7.68	
		118	117.901605	24.22	
		119	118.903308	8.58	
		120	119.902197	32.59	
		121*	120.904237		55 y β^-
		122	121.903439	4.63	
		124	123.905274	5.79	
Antimony	Sb	121	120.903820	57.36	
		123	122.904215	42.64	
		125*	124.905251		2.7 y β^-
Tellurium	Te	120	119.904040	0.095	
		122	121.903052	2.59	
		123*	122.904271	0.905	1.3×10^{13} y ec
		124	123.902817	4.79	
		125	124.904429	7.12	
		126	125.903309	18.93	
		128*	127.904463	31.70	$> 8 \times 10^{24}$ y $2\beta^-$
		130*	129.906228	33.87	1.2×10^{21} y $2\beta^-$

(Continued)

Element	Symbol	Mass number (*indicates radioactive)	Atomic mass	Percent abundance	Half-life and decay mode (if unstable)
Iodine	I	126*	125.905619		13 d ec, β^+, β^-
		127	126.904474	100	
		128*	127.905812		25 m β^-, ec, β^+ β^-
		129*	128.904984		1.6×10^7 y
Xenon	Xe	124	123.905894	0.10	
		126	125.904268	0.09	
		128	127.903531	1.91	
		129	128.904779	26.4	
		130	129.903509	4.1	
		131	130.905069	21.2	
		132	131.904141	26.9	
		134	133.905394	10.4	
		136	135.907215	8.9	
Cesium	Cs	133	132.905436	100	
		134*	133.906703		2.1 y β^-
		135*	134.905891		2×10^6 y β^-
		137*	136.907078		30 y β^-
Barium	Ba	130	129.906289	0.106	
		132	131.905048	0.101	
		133*	132.905990		10.5 y ec
		134	133.904492	2.42	
		135	134.905671	6.593	
		136	135.904559	7.85	
		137	136.905816	11.23	
		138	137.905236	71.70	
Lanthanum	La	137*	136.906462		6×10^4 y ec
		138*	137.907105	0.0902	1.05×10^{11} y ec, β^+
		139	138.906346	99.9098	
Cerium	Ce	136	135.907139	0.19	
		138	137.905986	0.25	
		140	139.905434	88.43	
		142	141.909241	11.13	
Praseodymium	Pr	140*	139.909071		3.39 m ec, β^+
		141	140.907647	100	
		142*	141.910040		25.0 m β^-
Neodymium	Nd	142	141.907718	27.13	
		143	142.909809	12.18	
		144*	143.910082	23.80	2.3×10^{15} y α
		145	144.912568	8.30	
		146	145.913113	17.19	
		148	147.916888	5.76	
		150	149.920887	5.64	
Promethium	Pm	143*	142.910928		265 d ec
		145*	144.912745		17.7 y ec
		146*	145.914698		5.5 y ec
		147*	146.915134		2.623 y β^-

Element	Symbol	Mass number (*indicates radioactive)	Atomic mass	Percent abundance	Half-life and decay mode (if unstable)
Samarium	Sm	144	143.911996	3.1	
		146*	145.913043		1.0×10^8 y α
		147*	146.914894	15.0	1.06×10^{11} y α
		148*	147.914819	11.3	7×10^{15} y α
		149	148.917180	13.8	
		150	149.917273	7.4	
		151*	150.919928		90 y β^-
		152	151.919728	26.7	
		154	153.922206	22.7	
Europium	Eu	151	150.919846	47.8	
		152*	151.921740		13.5 y ec, β^+
		153	152.921226	52.2	
		154*	153.922975		8.59 y β^-
		155*	154.922888		4.7 y β^-
Gadolinium	Gd	148*	147.918112		75 y α
		150*	149.918657		1.8×10^6 y α
		152*	151.919787	0.20	1.1×10^{14} y α
		154	153.920862	2.18	
		155	154.922618	14.80	
		156	155.922119	20.47	
		157	156.923957	15.65	
		158	157.924099	24.84	
		160	159.927050	21.86	
Terbium	Tb	158*	157.925411		180 y ec, β^+, β^-
		159	158.925345	100	
		160*	159.927551		72.3 d β^-
Dysprosium	Dy	156	155.924277	0.06	
		158	157.924403	0.10	
		160	159.925193	2.34	
		161	160.926930	18.9	
		162	161.926796	25.5	
		163	162.928729	24.9	
		164	163.929172	28.2	
Holmium	Ho	165	164.930316	100	
		166*	165.932282		1.2×10^3 y β^-
Erbium	Er	162	161.928775	0.14	
		164	163.929198	1.61	
		166	165.930292	33.6	
		167	166.932047	22.95	
		168	167.932369	27.8	
		170	169.935462	14.9	
Thulium	Tm	169	168.934213	100	
		171*	170.936428		1.92 y β^-

(Continued)

Element	Symbol	Mass number (*indicates radioactive)	Atomic mass	Percent abundance	Half-life and decay mode (if unstable)
Ytterbium	Yb	168	167.933897	0.13	
		170	169.934761	3.05	
		171	170.936324	14.3	
		172	171.936380	21.9	
		173	172.938209	16.12	
		174	173.938861	31.8	
		176	175.942564	12.7	
Lutetium	Lu	173*	172.938930		1.37 y ec
		175	174.940772	97.41	
		176*	175.942679	2.59	3.8×10^{10} y β^-
Hafnium	Hf	174*	173.940042	0.162	2.0×10^{15} y α
		176	175.941404	5.206	
		177	176.943218	18.606	
		178	177.943697	27.297	
		179	178.945813	13.629	
		180	179.946547	35.100	
Tantalum	Ta	180	179.947542	0.012	
		181	180.947993	99.988	
Tungsten (Wolfram)	W	180	179.946702	0.12	
		182	181.948202	26.3	
		183	182.950221	14.28	
		184	183.950929	30.7	
		186	185.954358	28.6	
Rhenium	Re	185	184.952951	37.40	
		187*	186.955746	62.60	4.4×10^{10} y β^-
Osmium	Os	184	183.952486	0.02	
		186*	185.953834	1.58	2.0×10^{15} y α
		187	186.955744	1.6	
		188	187.955744	13.3	
		189	188.958139	16.1	
		190	189.958439	26.4	
		192	191.961468	41.0	
		194*	193.965172		6.0 y β^-
Iridium	Ir	191	190.960585	37.3	
		193	192.962916	62.7	
Platinum	Pt	190*	189.959926	0.01	6.5×10^{11} y α
		192	191.961027	0.79	
		194	193.962655	32.9	
		195	194.964765	33.8	
		196	195.964926	25.3	
		198	197.967867	7.2	
Gold	Au	197	196.966543	100	
		198*	197.968217		2.70 d β^-
		199*	198.968740		3.14 d β^-

Element	Symbol	Mass number (*indicates radioactive)	Atomic mass	Percent abundance	Half-life and decay mode (if unstable)
Mercury	Hg	196	195.965806	0.15	
		198	197.966743	9.97	
		199	198.968253	16.87	
		200	199.968299	23.10	
		201	200.970276	13.10	
		202	201.970617	29.86	
		204	203.973466	6.87	
Thallium	Tl	203	202.972320	29.524	
		204*	203.973839		3.78 y β^-
		205	204.974400	70.476	
	(Ra E″)	206*	205.976084		4.2 m β^-
	(Ac C″)	207*	206.977403		4.77 m β^-
	(Th C″)	208*	207.981992		3.053 m β^-
	(Ra C″)	210*	209.990057		1.30 m β^-
Lead	Pb	202*	201.972134		5×10^4 y ec
		204	203.973020	1.4	
		205*	204.974457		1.5×10^7 y ec
		206	205.974440	24.1	
		207	206.975871	22.1	
		208	207.976627	52.4	
	(Ra D)	210*	209.984163		22.3 y β^-
	(Ac B)	211*	210.988734		36.1 m β^-
	(Th B)	212*	211.991872		10.64 h β^-
	(Ra B)	214*	213.999798		26.8 m β^-
Bismuth	Bi	207*	206.978444		32.2 y ec, β^+
		208*	207.979717		3.7×10^5 y ec
		209	208.980374	100	
	(Ra E)	210*	209.984096		5.01 d α, β^-
	(Th C)	211*	210.987254		2.14 m α
	(Ra C)	212*	211.991259		60.6 m α, β^-
		214*	213.998692		19.9 m β^-
		215*	215.001836		7.4 m β^-
Polonium	Po	209*	208.982405		102 y α
	(Ra F)	210*	209.982848		138.38 d α
	(Ac C′)	211*	210.986627		0.52 s α
	(Th C′)	212*	211.988842		0.30 μs α
	(Ra C′)	214*	213.995177		164 μs α
	(Ac A)	215*	214.999418		0.0018 s α
	(Th A)	216*	216.001889		0.145 s α
	(Ra A)	218*	218.008965		3.10 m α
Astatine	At	215*	214.998638		$\approx$100 μs α
		218*	218.008685		1.6 s α
		219*	219.011297		0.9 m α
Radon	Rn				
	(An)	219*	219.009477		3.96 s α
	(Tn)	220*	220.011369		55.6 s α
	(Rn)	222*	222.017571		3.823 d α

(Continued)

Element	Symbol	Mass number (*indicates radioactive)	Atomic mass	Percent abundance	Half-life and decay mode (if unstable)
Francium		221*	221.01425		4.18 m α
	Fr	222*	222.017585		14.2 m β^-
	(Ac K)	223*	223.019733		22 m β^-
Radium	Ra	221*	221.01391		29 s α
	(Ac X)	223*	223.018499		11.43 d α
	(Th X)	224*	224.020187		3.66 d α
		225*			14.9 d β^-
	(Ra)	226*	226.025402		1600 y α
	(MsTh$_1$)	228*	228.031064		5.75 y β^-
Actinium	Ac	225*			10 d α
	(Ms Th$_2$)	227*	227.027749		21.77 y β^-
		228*	228.031015		6.15 h β^-
		229*			1.04 h β^-
Thorium	Th				
	(Rd Ac)	227*	227.027701		18.72 d α
	(Rd Th)	228*	228.028716		1.913 y α
		229*	229.031757		7300 y α
	(Io)	230*	230.033127		75,000 y α, sf
	(UY)	231*	231.036299	100	25.52 h β^-
	(Th)	232*	232.038051		1.40×10^{10} y α
	(UX$_1$)	234*	234.043593		24.1 d β^-
Protactinium	Pa	231*	231.035880		32,760 y α
	(UZ)	234*	234.043300		6.7 h β^-
Uranium	U	231*	231.036264		4.2 d β^+
		232*	232.037131		69 y α
		233*	233.039630		1.59×10^5 y α
	(UII)	234*	234.040946	0.0055	2.45×10^5 y α
	(Ac U)	235*	235.043924	0.720	7.04×10^8 y α
	(UI)	236*	236.045562		2.34×10^7 y α
		238*	238.050784	99.2745	4.47×10^9 y α
		239*	239.054290		23.5 m β^-
Neptunium	Np	235*	235.044057		396 d α
		236*	236.046559		1.54×10^5 y ec
		237*	237.048168		2.14×10^6 y α
Plutonium	Pu	236*	236.046033		2.87 y α, sf
		238*	238.049555		87.7 y α, sf
		239*	239.052157		24,120 y α, sf
		240*	240.053808		6560 y α, sf
		241*	241.056846		14.4 y β^-
		242*	242.058737		3.7×10^5 y α, sf
		244*	244.064200		8.1×10^7 y α, sf
Americium	Am	240*	240.055285		2.12 d ec
		241*	241.056824		432 y α, sf
Curium	Cm	247*	247.070347		1.56×10^7 y α
		248*	248.072344		3.4×10^5 y α, sf

Element	Symbol	Mass number (*indicates radioactive)	Atomic mass	Percent abundance	Half-life and decay mode (if unstable)
Berkelium	Bk	247*	247.070300		1380 y α
		249*	249.074979		327 d β^-
Californium	Cm	250*	250.076400		13.1 y α, sf
		251*	251.079580		898 y α
Einsteinium	Es	252*	252.082974		1.29 y α
		253*	253.084817		2.02 d α, sf
Fermium	Fm	253*	253.085173		3.00 d ec
		254*	254.086849		3.24 h α, sf
Mendelevium	Md	256*	256.093988		75.6 m ec, β^+
		258*	258.098594		55 d α
Nobelium	No	257*	257.096855		25 s α
		259*	259.100932		58 m α, sf
Lawrencium	Lr	259*	259.102888		6.14 s α, sf
		260*	260.105346		3.0 m α, sf
Rutherfordium	Rf	260*	260.160302		24 ms sf
		261*	261.108588		65 s α, sf
Dubnium	Db	261*	261.111830		1.8 s α
		262*	262.113763		35 s α
Seaborgium	Sg	263*	263.118310		0.78 s α, sf
Bohrium	Bh	262*	262.123081		0.10 s α, sf
Hassium	Hs	265*	265.129984		1.8 ms α
		267*	267.131770		60 ms α
Meitnerium	Mt	266*	266.137789		3.4 ms α, sf
		268*	268.138820		70 ms α
Darmstadtium	Ds	269*	269.145140		0.17 ms α
		271*	271.146080		1.1 ms α
		273*	272.153480		8.6 ms α
Roentgenium	Rg	272*	272.153480		1.5 ms α
Copernicium	Cn	277*	?		0.2 ms α
Ununtrium	Unt	284*	?		? α
Ununquadium	Unq	289*	?		? α
Ununpentium	Unp	288*	?		? α
Ununhexium	Unh	292*	?		? α
Ununseptium	Uus				
Ununoctium	Uno	294*	?		? α

Math Tutorial

In this tutorial, we review some of the basic results of algebra, geometry, trigonometry, and calculus. In many cases, we merely state results without proof. Table M-1 lists some mathematical symbols.

M-1   Significant Figures

Many numbers we work with in science are the result of measurement and are therefore known only within a degree of uncertainty. This uncertainty should be reflected in the number of digits used. For example, if you have a 1-meter-long rule with scale spacing of 1 cm, you know that you can measure the height of a box to within a fifth of a centimeter or so. Using this rule, you might find that the box height is 27.0 cm. If there is a scale with a spacing of 1 mm on your rule, you might perhaps measure the box height to be 27.03 cm. However, if there is a scale with a spacing of 1 mm on your rule, you might not be able to measure the height more accurately than 27.03 cm because the height might vary by 0.01 cm or so, depending on which part of the box you measure the height at. When you write down that the height of the box is 27.03 cm, you are stating that your best estimate of the height is 27.03 cm, but you are not claiming that it is exactly 27.030000 . . . cm high. The four digits in 27.03 cm are called **significant figures**. Your measured length, 2.703 m, has four significant digits. Significant figures are also called significant digits.

The number of significant digits in an answer to a calculation will depend on the number of significant digits in the given data. When you work with numbers that have uncertainties, you should be careful not to include more digits than the certainty of measurement warrants. *Approximate* calculations (order-of-magnitude estimates) always result in answers that have only one significant digit or none. When you multiply, divide, add, or subtract numbers, you must consider the accuracy of the results. Listed below are some rules that will help you determine the number of significant digits of your results.

1. When multiplying or dividing quantities, the number of significant digits in the final answer is no greater than that in the quantity with the fewest significant digits.
2. When adding or subtracting quantities, the number of decimal places in the answer should match that of the term with the smallest number of decimal places.
3. Exact values have an unlimited number of significant digits. For example, a value determined by counting, such as 2 tables, has no uncertainty and is an exact value. In addition, the conversion factor 0.0254000 . . . m/in. is an exact value because 1.000 . . . inches is exactly equal to 0.0254000 . . .

Table M-1	Mathematical Symbols
$=$	is equal to
$\neq$	is not equal to
$\approx$	is approximately equal to
$\sim$	is of the order of
$\propto$	is proportional to
$>$	is greater than
$\geq$	is greater than or equal to
$\gg$	is much greater than
$<$	is less than
$\leq$	is less than or equal to
$\ll$	is much less than
Δx	change in x
$\lvert x \rvert$	absolute value of x
$n!$	$n(n-1)(n-2)\ldots 1$
Σ	sum
lim	limit
$\Delta t \rightarrow 0$	Δt approaches zero
dx/dt	derivative of x with respect to t
$\partial x/\partial t$	partial derivative of x with respect to t
$\int$	integral

meters. (The yard is, by definition, equal to exactly 0.9144 m, and 0.9144 divided by 36 is exactly equal to 0.0254.)

4. Sometimes zeros are significant and sometimes they are not. If a zero is before a leading nonzero digit, then the zero is not significant. For example, the number 0.00890 has three significant digits. The first three zeroes are not significant digits but are merely markers to locate the decimal point. Note that the zero after the nine is significant.

5. Zeros that are between nonzero digits are significant. For example, 5603 has four significant digits.

6. The number of significant digits in numbers with trailing zeros and no decimal point is ambiguous. For example 31,000 could have as many as five significant digits or as few as two significant digits. To prevent ambiguity, you should report numbers by using scientific notation or by using a decimal point.

Example M-1 Finding the Average of Three Numbers

Find the average of 19.90, −7.524, and −11.8179.

SET UP
You will be adding 3 numbers and then dividing the result by 3. The first number has three significant digits, the second number has four, and the third number has six.

SOLVE

1. Sum the three numbers.

$$19.90 + (−7.524) + (−11.8179) = 0.55\mathit{81}$$

2. If the problem only asked for the sum of the three numbers, we would round the answer to the least number of decimal places among all the numbers being added. However, we must divide this intermediate result by 3, so we use the intermediate answer with the two extra digits (italicized and red).

$$\frac{0.55\mathit{81}}{3} = 0.18\mathit{60333}\ldots$$

3. Only two of the digits in the intermediate answer, $0.18\mathit{60333}\ldots$, are significant digits, so we must round this number to get our final answer. The number 3 in the denominator is a whole number and has an unlimited number of significant digits. Thus, the final answer has the same number of significant digits as the numerator, which is 2.

The final answer is 0.19.

REFLECT
The sum in step 1 has two significant digits following the decimal point, the same as the number being summed with the least number of significant digits after the decimal point.

Practice Problems

1. $\dfrac{5.3\ \text{mol}}{22.4\ \text{mol/L}}$

2. $57.8\ \text{m/s} − 26.24\ \text{m/s}$

M-2 Equations

An **equation** is a statement written using numbers and symbols to indicate that two quantities, written on either side of an equal sign (=), are equal. The quantity on either side of the equal sign may consist of a single term, or of a sum or difference of two or more **terms**. For example, the equation $x = 1 - (ay + b)/(cx - d)$ contains three terms, x, 1 and $(ay + b)/(cx - d)$.

You can perform the following operations on equations:

1. The same quantity can be added to or subtracted from each side of an equation.
2. Each side of an equation can be multiplied or divided by the same quantity.
3. Each side of an equation can be raised to the same power.

These operations are meant to be applied to each *side* of the equation rather than each term in the equation. (Because multiplication is distributive over addition, operation 2—and only operation 2—of the preceding operations also applies term by term.)

   Caution: Division by zero is forbidden at any stage in solving an equation; results (if any) would be invalid.

Adding or Subtracting Equal Amounts
To find x when $x - 3 = 7$, add 3 to both sides of the equation: $(x - 3) + 3 = 7 + 3$; thus, $x = 10$.

Multiplying or Dividing by Equal Amounts
If $3x = 17$, solve x for by dividing both sides of the equation by 3; thus, $x = \frac{17}{3}$, or 5.7.

Example M-2 Simplifying Reciprocals in an Equation

Solve the following equation for x:

$$\frac{1}{x} + \frac{1}{4} = \frac{1}{3}$$

Equations containing reciprocals of unknowns occur in geometric optics and in electric circuit analysis—for example, in finding the net resistance of parallel resistors.

SET UP
In this equation, the term containing x is on the same side of the equation as a term not containing x. Furthermore, x is found in the denominator of a fraction.

SOLVE

1. Subtract $\frac{1}{4}$ from each side:

$$\frac{1}{x} = \frac{1}{3} - \frac{1}{4}$$

2. Simplify the right side of the equation by using the lowest common denominator:

$$\frac{1}{x} = \frac{1}{3} - \frac{1}{4} = \frac{4}{12} - \frac{3}{12}$$
$$= \frac{4 - 3}{12} = \frac{1}{12} \quad \text{so} \quad \frac{1}{x} = \frac{1}{12}$$

3. Multiply both sides of the equation by $12x$ to determine the value of x:

$$12x\frac{1}{x} = 12x\frac{1}{12}$$

$$\boxed{12} = x$$

REFLECT

Substitute 12 for x in the left side of original equation.

$$\frac{1}{x} + \frac{1}{4} = \frac{1}{12} + \frac{3}{12} = \frac{4}{12} = \frac{1}{3}$$

Practice Problems Solve each of the following for x:

3. $(7.0 \text{ cm}^3)x = 18 \text{ kg} + (4.0 \text{ cm}^3)x$

4. $\dfrac{4}{x} + \dfrac{1}{3} = \dfrac{3}{x}$

M-3 Direct and Inverse Proportions

When we say variable quantities x and y are **directly proportional**, we mean that as x and y change, the ratio x/y is constant. To say that two quantities are proportional is to say that they are directly proportional. When we say variable quantities x and y are **inversely proportional**, we mean that as x and y change, the ratio xy is constant.

Relationships of direct and inverse proportion are common in physics. Objects moving at the same velocity have momenta directly proportional to their masses. The ideal gas law ($PV = nRT$) states that pressure P is directly proportional to (absolute) temperature T, when volume V remains constant, and is inversely proportional to volume, when temperature remains constant. Ohm's law ($V = IR$) states that the voltage V across a resistor is directly proportional to the electric current in the resistor when the resistance remains constant.

Constant of Proportionality

When two quantities are directly proportional, the two quantities are related by a *constant of proportionality*. If you are paid for working at a regular rate R in dollars per day, for example, the money m you earn is directly proportional to the time t you work; the rate R is the constant of proportionality that relates the money earned in dollars to the time worked t in days:

$$\frac{m}{t} = R \quad \text{or} \quad m = Rt$$

If you earn \$400 in 5 days, the value of R is \$400/(5 days) = \$80/day. To find the amount you earn in 8 days, you could perform the calculation

$$m = (\$80/\text{day})(8 \text{ days}) = \$640$$

Sometimes the constant of proportionality can be ignored in proportion problems. Because the amount you earn in 8 days is $\frac{8}{5}$ times what you earn in 5 days, this amount is

$$m = \frac{8}{5}(\$400) = \$640$$

Example M-3 Painting Cubes

You need 15.4 mL of paint to cover one side of a cube. The area of one side of the cube is 426 cm². What is the relation between the volume of paint needed and the area to be covered? How much paint do you need to paint one side of a cube in which the one side has an area of 503 cm²?

SET UP

To determine the amount of paint for the side whose area is 503 cm² you will need to set up a proportion.

SOLVE

1. The volume V of paint needed increases in proportion to the area A to be covered.

    $\boxed{V \text{ and } A \text{ are directly proportional.}}$

    That is, $\dfrac{V}{A} = k$ or $V = kA$

    where k is the proportionality constant

2. Determine the value of the proportionality constant using the given values $V_1 = 15.4$ mL and $A_1 = 426$ cm²:

    $k = \dfrac{V_1}{A_1} = \dfrac{15.4 \text{ mL}}{426 \text{ cm}^2} = 0.0361 \text{ mL/cm}^3$

3. Determine the volume of paint needed to paint a side of a cube whose area is 503 cm² using the proportionality constant in step 1:

    $V_2 = kA_2 = (0.0361 \text{ mL/cm}^2)(503 \text{ cm}^2)$
    $= \boxed{18.2 \text{ mL}}$

REFLECT

Our value for V_2 is greater than the value for V_1, as expected. The amount of paint needed to cover an area equal to 503 cm² to should be greater than the amount of paint needed to cover an area of 426 cm² because 503 cm² is larger than 426 cm².

Practice Problems

5. A cylindrical container holds 0.384 L of water when full. How much water would the container hold if its radius were doubled and its height remained unchanged?

    Hint: The volume of a right circular cylinder is given by $V = \pi r^2 h$, where r is its radius and h is its height. Thus, V is directly proportional to r^2 when h remains constant.

6. For the container in Practice Problem 5, how much water would the container hold if both its height and its radius were doubled?

    Hint: The volume V of a right circular cylinder is given by $V = \pi r^2 h$, where r is its radius and h is its height.

M-4 Linear Equations

A **linear equation** is an equation of the form $x + 2y - 4z = 3$. That is, an equation is linear if each term either is constant or is the product of a constant and a variable raised to the first power. Such equations are said to be linear because the plots of these equations form straight lines or planes. The equations of direct proportion between two variables are linear equations.

Graph of a Straight Line

A linear equation relating y and x can always be put into the standard form

$$y = mx + b \qquad\qquad \textbf{M-1}$$

where m and b are constants that may be either positive or negative. **Figure M-1** shows a graph of the values of x and y that satisfy Equation M-1. The constant b, called the **y intercept**, is the value of y at $x = 0$. The constant m is the **slope** of the

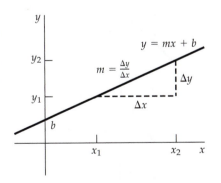

Figure M-1 Graph of the linear equation $y = mx + b$, where b is the y intercept and $m = \Delta y / \Delta x$ is the slope.

line, which equals the ratio of the change in y to the corresponding change in x. In the figure, we have indicated two points on the line, (x_1, y_1) and (x_2, y_2), and the changes $\Delta x = x_2 - x_1$ and $\Delta y = y_2 - y_1$. The slope m is then

$$m = \frac{y_2 - y_1}{x_2 - x_1} = \frac{\Delta y}{\Delta x}$$

If x and y are both unknown in the equation $y = mx + b$, there are no unique values of x and y that are solutions to the equation. Any pair of values (x_1, y_1) on the line in Figure M-1 will satisfy the equation. If we have two equations, each with the same two unknowns x and y, the equations can be solved simultaneously for the unknowns. Example M-4 shows how simultaneous linear equations can be solved.

Example M-4   Using Two Equations to Solve for Two Unknowns

Find any and all values of x and y that simultaneously satisfy

$$3x - 2y = 8 \qquad\qquad \text{M-2}$$

and

$$y - x = 2 \qquad\qquad \text{M-3}$$

SET UP

Figure M-2 shows a graph of the two equations. At the point where the lines intersect, the values of x and y satisfy both equations. We can solve two simultaneous equations by first solving either equation for one variable in terms of the other variable and then substituting the result into the other equation.

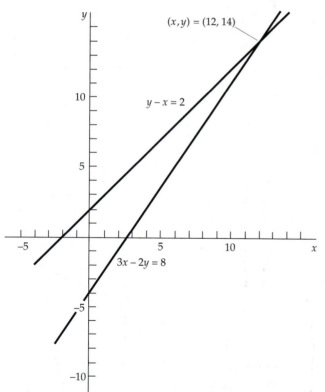

SOLVE

1. Solve Equation M-3 for y: $\quad y = x + 2$

2. Substitute this value for y $\quad 3x - 2(x + 2) = 8$
   into Equation M-2:

3. Simplify the equation and $\quad 3x - 2x - 4 = 8$
   solve for x: $\qquad\qquad\qquad x - 4 = 8$
   $\qquad\qquad\qquad\qquad\qquad x = \boxed{12}$

4. Use your solution for x and $\quad y - x = 2$, where $x = 12$
   one of the given equations $\quad y - 12 = 2$
   to find the value of y: $\qquad y = 2 + 12 = \boxed{14}$

REFLECT

An alternative method is to multiply one equation by a constant such that one of the unknown terms is eliminated when the equations are added or subtracted. We can multiply through Equation M-3 by 2

$$2(y - x) = 2(2)$$

$$2y - 2x = 4$$

and add the result to Equation M-2 and solve for x:

$$\cancel{2y} - 2x = 4$$
$$3x - \cancel{2y} = 8$$
$$\overline{}$$
$$3x - 2x = 12 \Rightarrow x = 12$$

Figure M-2 Graph of Equations M-2 and M-3. At the point where the lines intersect, the values of x and y satisfy both equations.

Substitute into Equation M-3 and solve for y:

$$y - 12 = 2 \Rightarrow y = 14$$

Practice Problems

7. True or false: $xy = 4$ is a linear equation.
8. At time $t = 0.0$ s, the position of a particle moving along the x axis at a constant velocity is $x = 3.0$ m. At $t = 2.0$ s, the position is $x = 12.0$ m. Write a linear equation showing the relation of x to t.
9. Solve the following pair of simultaneous equations for x and y:

$$\frac{5}{4}x + \frac{1}{3}y = 30$$

$$y - 5x = 20$$

M-5 Quadratic Equations and Factoring

A **quadratic equation** is an equation of the form $ax^2 + bxy + cy^2 + ex + fy + g = 0$, where x and y are variables and and a, b, c, e, f, and g are constants. In each term of the equation the powers of the variables are integers that sum to 2, 1, or 0. The designation *quadratic equation* usually applies to an equation of one variable that can be written in the standard form

$$ax^2 + bx + c = 0 \qquad\qquad \text{M-4}$$

where a, b, and c are constants. The quadratic equation has two solutions or **roots**—values of x for which the equation is true.

Factoring

We can solve some quadratic equations by **factoring**. Very often terms of an equation can be grouped or organized into other terms. When we factor terms, we look for multipliers and multiplicands—which we now call **factors**—that will yield two or more new terms as a product. For example, we can find the roots of the quadratic equation $x^2 - 3x + 2 = 0$ by factoring the left side, to get $(x - 2)(x - 1) = 0$. The roots are $x = 2$ and $x = 1$.

Factoring is useful for simplifying equations and for understanding the relationships between quantities. You should be familiar with the multiplication of the factors $(ax + by)(cx + dy) = acx^2 + (ad + bc)xy + bdy^2$.

You should readily recognize some typical factorable combinations:

1. Common factor: $2ax + 3ay = a(2x + 3y)$
2. Perfect square: $x^2 - 2xy + y^2 = (x - y)^2$ (If the expression on the left side of a quadratic equation in standard form is a perfect square, the two roots will be equal.)
3. Difference of squares: $x^2 - y^2 = (x + y)(x - y)$

Also, look for factors that are prime numbers (2, 5, 7, etc.) because these factors can help you factor and simplify terms quickly. For example, the equation $98x^2 - 140 = 0$ can be simplified because 98 and 140 share the common factor 2. That is, $98x^2 - 140 = 0$ becomes $2(49x^2 - 70) = 0$, so we have $49x^2 - 70 = 0$.

This result can be further simplified because 49 and 70 share the common factor 7. Thus, $49x^2 - 70 = 0$ becomes $7(7x^2 - 10) = 0$, so we have $7x^2 - 10 = 0$.

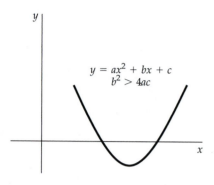

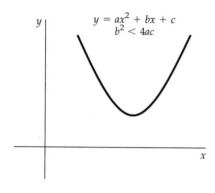

Figure M-3 Graph of y versus x when $y = ax^2 + bx + c$ for the case $b^2 > 4ac$. The two values of x for which $y = 0$ satisfy the quadratic equation (Equation M-4).

Figure M-4 Graph of y versus x when $y = ax^2 + bx + c$ for the case $b^2 < 4ac$. In this case, there are no real values of x for which $y = 0$.

The Quadratic Formula

Not all quadratic equations can be solved by factoring. However, *any* quadratic equation in the standard form $ax^2 + bx + c = 0$ can be solved by the **quadratic formula,**

$$x = \frac{-b \pm \sqrt{b^2 - 4ac}}{2a} = -\frac{b}{2a} \pm \frac{1}{2a}\sqrt{b^2 - 4ac} \qquad \text{M-5}$$

When b^2 is greater than $4ac$, there are two solutions corresponding to the $+$ and $-$ signs. **Figure M-3** shows a graph of y versus x where $y = ax^2 + bx + c$. The curve, a **parabola**, crosses the x axis twice. (The simplest representation of a parabola in (x, y) coordinates is an equation of the form $y = ax^2 + bx + c$.) The two roots of this equation are the values for which $y = 0$; that is, they are the x *intercepts*.

When b^2 is less than $4ac$, the graph of y versus x does not intersect the x axis, as is shown in **Figure M-4**; there are still two roots, but they are not real numbers (see the discussion of complex numbers beginning on page M-21). When $b^2 = 4ac$, the graph of y versus x is tangent to the x axis at the point $x = -b/2a$; the two roots are each equal to $-b/2a$.

Example M-5 Factoring a Second-Degree Polynomial

Factor the expression $6x^2 + 19xy + 10y^2$.

SET UP
We examine the coefficients of the terms to see whether the expression can be factored without resorting to more advanced methods. Remember that the multiplication $(ax + by)(cx + dy) = acx^2 + (ad + bc)xy + bdy^2$.

SOLVE

1. The coefficient of x^2 is 6 which can be factored two ways:

   $ac = 6$
   $3 \cdot 2 = 6$ or $6 \cdot 1 = 6$

2. The coefficient of y^2 is 10 which can also be factored two ways:

   $bd = 10$
   $5 \cdot 2 = 10$ or $10 \cdot 1 = 10$

3. List the possibilities for a, b, c, and d in a table. Include a column for $ad + bc$.
   If $a = 3$, then $c = 2$, and vice versa. In addition, if $a = 6$, then $c = 1$, and vice versa. For each value of a there are four values for b.

a	b	c	d	$ad + bc$
3	5	2	2	16
3	2	2	5	19
3	10	2	1	23
3	1	2	10	32
2	5	3	2	19
2	2	3	5	16
2	10	3	1	32
2	1	3	10	23
6	5	1	2	17
6	2	1	5	32
6	10	1	1	16
6	1	1	10	61
1	5	6	2	32
1	2	6	5	17
1	10	6	1	61
1	1	6	10	16

4. Find a combination such that $ad + bc = 19$. As you can see from the table there are two such combinations, and each gives the same results:

$$ad + bc = 19$$
$$3 \cdot 5 + 2 \cdot 2 = 19 \text{ and}$$
$$2 \cdot 2 + 5 \cdot 3 = 19$$

5. Use the combination in the second row of the table to factor the expression in question:

$$6x^2 + 19xy + 10y^2$$
$$= (3x + 2y)(2x + 5y)$$

REFLECT

As a check, expand $(3x + 2y)(2x + 5y)$.

$$(3x + 2y)(2x + 5y) = 6x^2 + 15xy + 4xy + 10y^2 = 6x^2 + 19xy + 10y^2$$

The combination in the fifth row of the table also gives the step-4 result.

Practice Problems

10. Show that the combination in the fifth row of the table also gives the step-4 result.
11. Factor $2x^2 - 4xy + 2y^2$.
12. Factor $2x^2 + 10x^3 + 12x^2$.

M-6 Exponents and Logarithms

Exponents

The notation x^n stands for the quantity obtained by multiplying x times itself n times. For example, $x^2 = x \cdot x$ and $x^3 = x \cdot x \cdot x$. The quantity n is called the **power**, or the **exponent**, of x (the **base**). Listed below are some rules that will help you simplify terms that have exponents.

1. When two powers of x are multiplied, the exponents are added:

$$(x^m)(x^n) = x^{m+n} \qquad \text{M-6}$$

Example: $x^2 x^3 = x^{2+3} = (x \cdot x)(x \cdot x \cdot x) = x^5$.

2. Any number (except 0) raised to the 0 power is defined to be 1:

$$x^0 = 1 \qquad \text{M-7}$$

3. Based on rule 2,

$$x^n x^{-n} = x^0 = 1$$
$$x^{-n} = \frac{1}{x^n} \qquad \text{M-8}$$

4. When two powers are divided, the exponents are subtracted:

$$\frac{x^n}{x^m} = x^n x^{-m} = x^{n-m} \qquad \text{M-9}$$

5. When a power is raised to another power, the exponents are multiplied:

$$(x^n)^m = x^{nm} \qquad \text{M-10}$$

6. When exponents are written as fractions, they represent the roots of the base. For example,

$$x^{1/2} \cdot x^{1/2} = x$$

so

$$x^{1/2} = \sqrt{x} \quad (x > 0)$$

Example M-6 Simplifying a Quantity That Has Exponents

Simplify $\dfrac{x^4 x^7}{x^8}$.

SET UP

According to rule 1, when two powers of x are multiplied, the exponents are added. Rule 4 states that when two powers are divided, the exponents are subtracted.

SOLVE

1. Simplify the numerator $x^4 x^7$ using rule 1. $x^4 x^7 = x^{4+7} = x^{11}$

2. Simplify $\dfrac{x^{11}}{x^8}$ using rule 4: $\dfrac{x^{11}}{x^8} = x^{11}x^{-8} = x^{11-8} = x^3$

REFLECT

Use the value $x = 2$ to determine if your answer is correct.

$$\frac{2^4 2^7}{2^8} = 2^3 = 8$$

$$\frac{2^4 2^7}{2^8} = \frac{(16)(128)}{256} = \frac{2048}{256} = 8$$

Practice Problems

13. $(x^{1/18})^9 =$

14. $x^6 x^0 =$

Logarithms

Any positive number can be expressed as some power of any other positive number except one. If y is related to x by $y = a^x$, then the number x is said to be the **logarithm** of y to the **base** a, and the relation is written

$$x = \log_a y$$

Thus, logarithms are *exponents*, and the rules for working with logarithms correspond to similar laws for exponents. Listed below are some rules that will help you simplify terms that have logarithms.

1. If $y_1 = a^n$ and $y_2 = a^m$, then

$$y_1 y_2 = a^n a^m = a^{n+m}$$

Correspondingly,

$$\log_a y_1 y_2 = \log_a a^{n+m} = n + m = \log_a a^n + \log_a a^m = \log_a y_1 + \log_a y_2 \quad \text{M-11}$$

It then follows that

$$\log_a y^n = n \log_a y \qquad\qquad \text{M-12}$$

2. Because $a^1 = a$ and $a^0 = 1$,

$$\log_a a = 1 \qquad\qquad \text{M-13}$$

and

$$\log_a 1 = 0 \qquad\qquad \text{M-14}$$

There are two bases in common use: logarithms to base 10 are called **common logarithms**, and logarithms to base e (where $e = 2.718 \ldots$) are called **natural logarithms**.

In this text, the symbol ln is used for natural logarithms and the symbol log, without a subscript, is used for common logarithms. Thus,

$$\log_e x = \ln x \quad \text{and} \quad \log_{10} x = \log x \qquad \text{M-15}$$

and $y = \ln x$ implies

$$x = e^y \qquad \text{M-16}$$

Logarithms can be changed from one base to another. Suppose that

$$z = \log x \qquad \text{M-17}$$

Then

$$10^z = 10^{\log x} = x \qquad \text{M-18}$$

Taking the natural logarithm of both sides of Equation M-18, we obtain

$$z \ln 10 = \ln x$$

Substituting log x for z (see Equation M-17) gives

$$\ln x = (\ln 10)\log x \qquad \text{M-19}$$

Example M-7 Converting between Common Logarithms and Natural Logarithms

The steps leading to Equation M-19 show that, in general, $\log_b x = (\log_b a)\log_a x$, and thus that conversion of logarithms from one base to another requires only multiplication by a constant. Describe the mathematical relation between the constant for converting common logarithms to natural logarithms and the constant for converting natural logarithms to common logarithms.

SET UP
We have a general mathematical formula for converting logarithms from one base to another. We look for the mathematical relation by exchanging a for b and vice versa in the formula.

SOLVE

1. You have a formula for converting logarithms from base a to base b:
$$\log_b x = (\log_b a)\log_a x$$

2. To convert from base b to base a, exchange all a for b and vice versa:
$$\log_a x = (\log_a b)\log_b x$$

3. Divide both sides of the equation in step 1 by $\log_a x$:
$$\frac{\log_b x}{\log_a x} = \log_b a$$

4. Divide both sides of the equation in step 2 by $(\log_a b)\log_a x$:
$$\frac{1}{\log_a b} = \frac{\log_b x}{\log_a x}$$

5. The results show that the conversion factors $\log_b a$ and $\log_a b$ are reciprocals of one another:
$$\frac{1}{\log_a b} = \log_b a$$

REFLECT
For the value of $\log_{10} e$, your calculator will give 0.43429. For ln 10, your calculator will give 2.3026. Multiply 0.43429 by 2.3026; you will get 1.0000.

Practice Problems
15. Evaluate $\log_{10} 1000$.
16. Evaluate $\log_2 5$.

M-7　Geometry

The properties of the most common **geometric figures**—bounded shapes in two or three dimensions whose lengths, areas, or volumes are governed by specific ratios— are a basic analytical tool in physics. For example, the characteristic ratios within triangles give us the laws of *trigonometry* (see Section M-8), which in turn give us the theory of vectors, essential in analyzing motion in two or more dimensions. Circles and spheres are essential for understanding, among other concepts, angular momentum and the probability densities of quantum mechanics.

Basic Formulas in Geometry

Circle The ratio of the circumference of a circle to its diameter is a number π, which has the approximate value

$$\pi = 3.141\ 592$$

The circumference C of a circle is thus related to its diameter d and its radius r by

$$C = \pi d = 2\pi r \quad \text{circumference of circle} \qquad \textbf{M-20}$$

The area of a circle is (**Figure M-5**)

$$A = \pi r^2 \quad \text{area of circle} \qquad \textbf{M-21}$$

Parallelogram The area of a parallelogram is the base b multiplied by the height h (**Figure M-6**):

$$A = bh$$

Triangle The area of a triangle is one-half the base multiplied by the height (**Figure M-7**)

$$A = \frac{1}{2}bh$$

Sphere A sphere of radius r (**Figure M-8**) has a surface area given by

$$A = 4\pi r^2 \quad \text{surface area of sphere} \qquad \textbf{M-22}$$

and a volume given by

$$V = \frac{4}{3}\pi r^3 \quad \text{volume of sphere} \qquad \textbf{M-23}$$

Cylinder A cylinder of radius r and length L (**Figure M-9**) has a surface area (not including the end faces) of

$$A = 2\pi r L \quad \text{surface of cylinder} \qquad \textbf{M-24}$$

and volume of

$$V = \pi r^2 L \quad \text{volume of cylinder} \qquad \textbf{M-25}$$

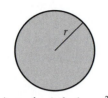

Area of a circle $A = \pi r^2$

Figure M-5 Area of a circle.

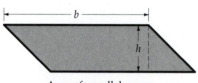

Area of parallelogram
$A = bh$

Figure M-6 Area of a parallelogram.

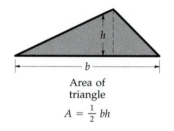

Area of triangle
$A = \frac{1}{2}bh$

Figure M-7 Area of a triangle.

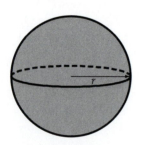

Spherical surface area
$A = 4\pi r^2$
Spherical volume
$V = \frac{4}{3}\pi r^3$

Figure M-8 Surface area and volume of a sphere.

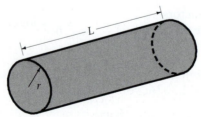

Cylindrical surface area
$A = 2\pi r L$
Cylindrical volume
$V = \pi r^2 L$

Figure M-9 Surface area (not including the end faces) and the volume of a cylinder.

Example M-8 Calculating the Mass of a Spherical Shell

An aluminum spherical shell has an outer diameter of 40.0 cm and an inner diameter of 39.0 cm. Find the volume of the aluminum in this shell.

SET UP

The volume of the aluminum in the spherical shell is the volume that remains when we subtract the volume of the inner sphere having $d_i = 2r_i = 39.0$ cm from the volume of the outer sphere having $d_o = 2r_o = 40.0$ cm.

SOLVE

1. Subtract the volume of the sphere of radius r_i from the volume of the sphere of radius r_o:

$$V = \tfrac{4}{3}\pi r_o^3 - \tfrac{4}{3}\pi r_i^3 = \tfrac{4}{3}\pi (r_o^3 - r_i^3)$$

2. Substitute 20.0 cm for r_o and 19.5 cm for r_i:

$$V = \tfrac{4}{3}\pi [(20.0 \text{ cm})^3 - (19.5 \text{ cm})^3]$$

$$= \boxed{2.45 \times 10^3 \text{ cm}^3}$$

REFLECT

The volume of the shell is expected to be the same order of magnitude as the volume of a hollow cube with an outside edge length of 40.0 cm and an inside edge length of 39.0 cm. The volume of such a hollow cube is $(40.0 \text{ cm})^3 - (39.0 \text{ cm})^3 = 4.68 \times 10^3 \text{ cm}^3$. The result meets the expectation that the volume of the shell is the same order of magnitude as the volume of the hollow cube.

Practice Problems

17. Find the ratio between the volume V and the surface A of a sphere of radius r.
18. What is the area of a cylinder that has a radius that is $1/3$ its length?

M-8 Trigonometry

Trigonometry, which gets its name from Greek roots meaning "triangle" and "measure," is the study of some important mathematical functions, called **trigonometric functions**. These functions are most simply defined as ratios of the sides of right triangles. However, these right-triangle definitions are of limited use because they are valid only for angles between zero and 90°. However, the validity of the right-triangle definitions can be extended by defining the trigonometric functions in terms of the ratio of the coordinates of points on a circle of unit radius drawn centered at the origin of the xy plane.

In physics, we first encounter trigonometric functions when we use vectors to analyze motion in two dimensions. Trigonometric functions are also essential in the analysis of any kind of periodic behavior, such as circular motion, oscillatory motion, and wave mechanics.

Angles and Their Measure: Degrees and Radians

The size of an angle formed by two intersecting straight lines is known as its **measure**. The standard way of finding the measure of an angle is to place the angle so that its **vertex**, or point of intersection of the two lines that form the angle, is at the center of a circle located at the origin of a graph that has Cartesian coordinates and one of the lines extends rightward on the positive x axis. The distance traveled *counterclockwise* on the circumference from the positive x axis to reach the intersection of the circumference with the other line defines the measure of the angle. (Traveling clockwise to the second line would simply give us a negative measure; to

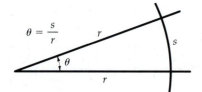

$$\theta = \frac{s}{r}$$

Figure M-10 The angle θ in radians is defined to be the ratio s/r, where s is the arc length intercepted on a circle of radius r.

illustrate basic concepts, we position the angle so that the smaller rotation will be in the counterclockwise direction.)

The most familiar unit for expressing the measure of an angle is the **degree**, which equals $1/360$ of the full distance around the circumference of the circle. For greater precision, or for smaller angles, we either show degrees plus minutes ($'$) and seconds ($''$), with $1' = 1°/60$ and $1'' = 1'/60 = 1°/3600$; or show degrees as an ordinary decimal number.

For scientific work, a more useful measure of an angle is the **radian** (rad). Again, place the angle with its vertex at the center of a circle and measure counterclockwise rotation around the circumference. The measure of the angle in radians is then defined as the length of the circular arc from one line to the other divided by the radius of the circle (**Figure M-10**). If s is the arc length and r is the radius of the circle, the angle θ measured in radians is

$$\theta = \frac{s}{r} \qquad\qquad \text{M-26}$$

Because the angle measured in radians is the ratio of two lengths, it is dimensionless. The relation between radians and degrees is

$$360° = 2\pi \text{ rad}$$

or

$$1 \text{ rad} = \frac{360°}{2\pi} = 57.3°$$

Figure M-11 shows some useful relations for angles.

The Trigonometric Functions

Figure M-12 shows a right triangle formed by drawing the line BC perpendicular to AC. The lengths of the sides are labeled a, b, and c. The right-triangle definitions of the trigonometric functions $\sin \theta$ (the **sine**), $\cos \theta$ (the **cosine**), and $\tan \theta$ (the **tangent**) for an acute angle θ are

$$\sin \theta = \frac{a}{c} = \frac{\text{opposite side}}{\text{hypotenuse}} \qquad\qquad \text{M-27}$$

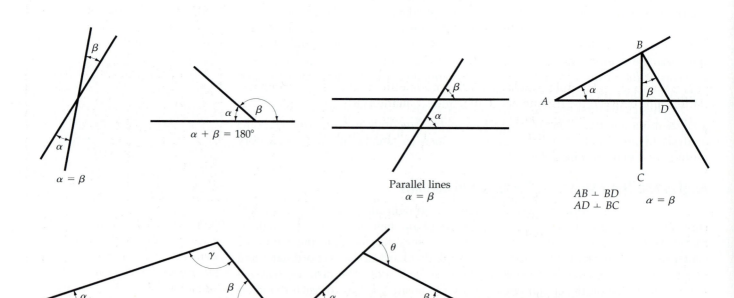

Figure M-11 Some useful relations for angles.

$$\cos \theta = \frac{b}{c} = \frac{\text{adjacent side}}{\text{hypotenuse}} \qquad \text{M-28}$$

$$\tan \theta = \frac{a}{b} = \frac{\text{opposite side}}{\text{adjacent side}} = \frac{\sin \theta}{\cos \theta} \qquad \text{M-29}$$

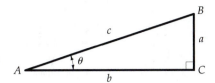

Figure M-12 A right triangle with sides of length a and b and a hypotenuse of length c.

(**Acute angles** are angles whose positive rotation around the circumference of a circle measures less than 90° or $\pi/2$.) Three other trigonometric functions—the **secant** (sec), the **cosecant** (csc), and the **cotangent** (cot), defined as the reciprocals of these functions—are

$$\sec \theta = \frac{c}{b} = \frac{1}{\cos \theta} \qquad \text{M-30}$$

$$\csc \theta = \frac{c}{a} = \frac{1}{\sin \theta} \qquad \text{M-31}$$

$$\cot \theta = \frac{b}{a} = \frac{1}{\tan \theta} = \frac{\cos \theta}{\sin \theta} \qquad \text{M-32}$$

The angle θ, whose sine is x, is called the arcsine of x, and is written $\sin^{-1} x$. That is, if

$$\sin \theta = x$$

then

$$\theta = \arcsin x = \sin^{-1} x \qquad \text{M-33}$$

The arcsine is the inverse of the sine. The inverse of the cosine and tangent are defined similarly. The angle whose cosine is y is the arccosine of y. That is, if

$$\cos \theta = y$$

then

$$\theta = \arccos y = \cos^{-1} y \qquad \text{M-34}$$

The angle whose tangent is z is the arctangent of z. That is, if

$$\tan \theta = z$$

then

$$\theta = \arctan z = \tan^{-1} z \qquad \text{M-35}$$

Trigonometric Identities

We can derive several useful formulas, called **trigonometric identities,** by examining relationships between the trigonometric functions. Equations M-30 through M-32 list three of the most obvious identities, formulas expressing some trigonometric functions as reciprocals of others. Almost as easy to discern are identities derived from the **Pythagorean theorem,**

$$a^2 + b^2 = c^2 \qquad \text{M-36}$$

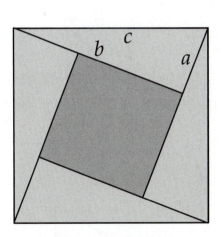

Figure M-13 When this figure was first published, the letters were absent and it was accompanied by the single word "Behold!" Using the drawing, establish the Pythagorean theorem $(a^2 + b^2 = c^2)$.

(**Figure M-13** illustrates a graphic proof of the theorem.) Simple algebraic manipulation of Equation M-36 gives us three more identities. First, if we divide each term in Equation M-36 by c^2, we obtain

$$\frac{a^2}{c^2} + \frac{b^2}{c^2} = 1$$

or, from the definitions of $\sin \theta$ (which is a/c) and $\cos \theta$ (which is b/c)

$$\sin^2 \theta + \cos^2 \theta = 1 \qquad \text{M-37}$$

Table M-2   Trigonometric Identities

$\sin(A \pm B) = \sin A \cos B \pm \cos A \sin B$

$\cos(A \pm B) = \cos A \cos B \mp \sin A \sin B$

$\tan(A \pm B) = \dfrac{\tan A \pm \tan B}{1 \mp \tan A \tan B}$

$\sin A \pm \sin B = 2 \sin\left[\dfrac{1}{2}(A \pm B)\right] \cos\left[\dfrac{1}{2}(A \mp B)\right]$

$\cos A + \cos B = 2 \cos\left[\dfrac{1}{2}(A + B)\right] \cos\left[\dfrac{1}{2}(A - B)\right]$

$\cos A - \cos B = 2 \sin\left[\dfrac{1}{2}(A + B)\right] \sin\left[\dfrac{1}{2}(B - A)\right]$

$\tan A \pm \tan B = \dfrac{\sin(A \pm B)}{\cos A \cos B}$

$\sin^2\theta + \cos^2\theta = 1;\ \sec^2\theta - \tan^2\theta = 1;\ \csc^2\theta - \cot^2\theta = 1$

$\sin 2\theta = 2 \sin\theta \cos\theta$

$\cos 2\theta = \cos^2\theta - \sin^2\theta = 2\cos^2\theta - 1 = 1 - 2\sin^2\theta$

$\tan 2\theta = \dfrac{2\tan\theta}{1 - \tan^2\theta}$

$\sin\dfrac{1}{2}\theta = \pm\sqrt{\dfrac{1 - \cos\theta}{2}};\ \cos\dfrac{1}{2}\theta = \pm\sqrt{\dfrac{1 + \cos\theta}{2}};$

$\tan\dfrac{1}{2}\theta = \pm\sqrt{\dfrac{1 - \cos\theta}{1 + \cos\theta}}$

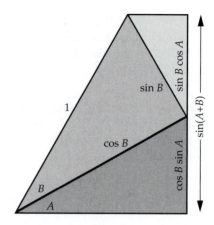

Figure M-14 Using this drawing, establish the identity $\sin(A + B) = \sin A \cos B + \cos A \sin B$. You can also use it to establish the identity $\cos(A + B) = \cos A \cos B - \sin A \sin B$. Try it.

Similarly, we can divide each term in Equation M-36 by a^2 or b^2 and obtain

$$1 + \cot^2\theta = \csc^2\theta \qquad \text{M-38}$$

and

$$1 + \tan^2\theta = \sec^2\theta \qquad \text{M-39}$$

Table M-2 lists these last three and many more trigonometric identities. Notice that they fall into four categories: functions of sums or differences of angles, sums or differences of squared functions, functions of double angles (2θ), and functions of half angles ($\frac{1}{2}\theta$). Notice that some of the formulas contain paired alternatives, expressed with the signs $\pm$ and $\mp$; in such formulas, remember to always apply the formula with either all the upper or all the lower alternatives. Figure M-14 shows a graphic proof of the first two sum-of-angle identities.

Some Important Values of the Functions

Figure M-15 is a diagram of an *isosceles* right triangle (an isosceles triangle is a triangle with two equal sides), from which we can find the sine, cosine, and tangent of 45°. The two acute angles of this triangle are equal. Because the sum of the three angles in a triangle must equal 180° and the right angle is 90°, each acute angle must be 45°. For convenience, let us assume that the equal sides each have a length of 1 unit. The Pythagorean theorem gives us a value for the hypotenuse of

$$c = \sqrt{a^2 + b^2} = \sqrt{1^2 + 1^2} = \sqrt{2} \text{ units}$$

We calculate the values of the functions as follows:

$$\sin 45° = \frac{a}{c} = \frac{1}{\sqrt{2}} = 0.707 \quad \cos 45° = \frac{b}{c} = \frac{1}{\sqrt{2}} = 0.707 \quad \tan 45° = \frac{a}{b} = \frac{1}{1} = 1$$

Another common triangle, a 30°–60° right triangle, is shown in Figure M-16. Because this particular right triangle is in effect half of an *equilateral triangle* (a 60°–60°–60° triangle or a triangle having three equal sides and three equal angles),

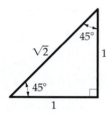

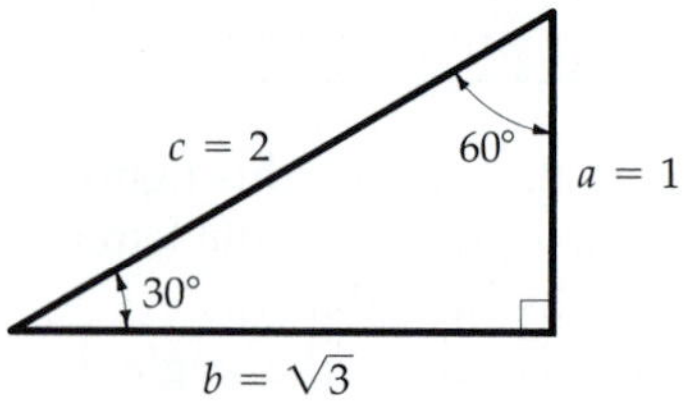

Figure M-15 An isosceles right triangle.

Figure M-16 A 30°–60° right triangle.

we can see that the sine of 30° must be exactly 0.5 (**Figure M-17**). The equilateral triangle must have all sides equal to c, the hypotenuse of the 30°–60° right triangle. Thus, side a is one-half the length of the hypotenuse, and so

$$\sin 30° = \frac{1}{2}$$

To find the other ratios within the 30°–60° right triangle, let us assign a value of 1 to the side opposite the 30° angle. Then

$$c = \frac{1}{0.5} = 2 \qquad\qquad b = \sqrt{c^2 - a^2} = \sqrt{2^2 - 1^2} = \sqrt{3}$$

$$\cos 30° = \frac{b}{c} = \frac{\sqrt{3}}{2} = 0.866 \qquad \tan 30° = \frac{a}{b} = \frac{1}{\sqrt{3}} = 0.577$$

$$\sin 60° = \frac{b}{c} = \cos 30° = 0.866 \qquad \cos 60° = \frac{a}{c} = \sin 30° = \frac{1}{2}$$

$$\tan 60° = \frac{b}{a} = \frac{\sqrt{3}}{1} = 1.732$$

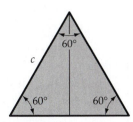

Figure M-17 (a) An equilateral triangle. (b) An equilateral triangle that has been bisected to form two 30°–60° right triangles.

Small-Angle Approximation

For small angles, the length a is nearly equal to the arc length s, as can be seen in **Figure M-18**. The angle $\theta = s/c$ is therefore nearly equal to $\sin \theta = a/c$:

$$\sin \theta \approx \theta \quad \text{for small values of } \theta \qquad\qquad \textbf{M-40}$$

Similarly, the lengths c and b are nearly equal, so $\tan \theta = a/b$ is nearly equal to both θ and $\sin \theta$ for small values of θ:

$$\tan \theta \approx \sin \theta \approx \theta \quad \text{for small values of } \theta \qquad\qquad \textbf{M-41}$$

Equations M-40 and M-41 hold only if θ is measured in radians. Because $\cos \theta = b/c$, and because these lengths are nearly equal for small values of θ, we have

$$\cos \theta \approx 1 \quad \text{for small values of } \theta \qquad\qquad \textbf{M-42}$$

Figure M-19 shows graphs of θ, $\sin \theta$, and $\tan \theta$ versus θ for small values of θ. If accuracy of a few percent is needed, small-angle approximations can be used only for angles of about a quarter of a radian (or about 15°) or less. Below this value, as the angle becomes smaller, the approximation $\theta \approx \sin \theta \approx \tan \theta$ is even more accurate.

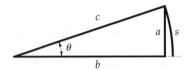

Figure M-18 For small angles, $\sin \theta = a/c$, $\tan \theta = a/b$, and the angle $\theta = s/c$ are all approximately equal.

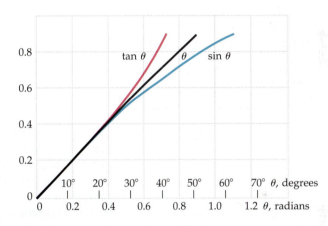

Figure M-19 Graphs of $\tan \theta$, θ, and $\sin \theta$ versus θ for small values of θ.

Trigonometric Functions as Functions of Real Numbers

So far we have illustrated the trigonometric functions as properties of angles. **Figure M-20** shows an *obtuse* angle with its vertex at the origin and one side along the x axis. The trigonometric functions for a "general" angle such as this are defined by

$$\sin \theta = \frac{y}{c} \qquad\qquad \textbf{M-43}$$

$$\cos \theta = \frac{x}{c} \qquad\qquad \textbf{M-44}$$

$$\tan \theta = \frac{y}{x} \qquad\qquad \textbf{M-45}$$

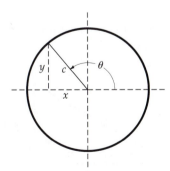

Figure M-20 Diagram for defining the trigonometric functions for an obtuse angle.

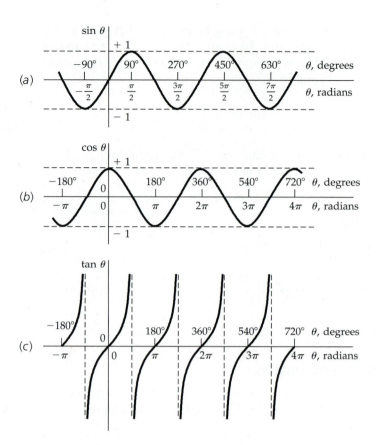

Figure M-21 The trigonometric functions sin θ, cos θ, and tan θ versus θ.

It is important to remember that values of x to the left of the vertical axis and values of y below the horizontal axis are negative; c in the figure is always regarded as positive. **Figure M-21** shows plots of the general sine, cosine, and tangent functions versus θ. The sine function has a period of 2π rad. Thus, for any value of θ, $\sin(\theta + 2\pi) = \sin\theta$, and so forth. That is, when an angle changes by 2π rad, the function returns to its original value. The tangent function has a period of π rad. Thus, $\tan(\theta + \pi) = \tan\theta$, and so forth. Some other useful relations are

$$\sin(\pi - \theta) = \sin\theta \qquad \text{M-46}$$

$$\cos(\pi - \theta) = -\cos\theta \qquad \text{M-47}$$

$$\sin(\tfrac{1}{2}\pi - \theta) = \cos\theta \qquad \text{M-48}$$

$$\cos(\tfrac{1}{2}\pi - \theta) = \sin\theta \qquad \text{M-49}$$

Because the radian is dimensionless, it is not hard to see from the plots in Figure M-21 that the trigonometric functions are functions of all real numbers. The functions can also be expressed as power series in θ. The series for sin θ and cos θ are

$$\sin\theta = \theta - \frac{\theta^3}{3!} + \frac{\theta^5}{5!} - \frac{\theta^7}{7!} + \cdots \qquad \text{M-50}$$

$$\cos\theta = 1 - \frac{\theta^2}{2!} + \frac{\theta^4}{4!} - \frac{\theta^6}{6!} + \cdots \qquad \text{M-51}$$

When θ is small, good approximations are obtained using only the first few terms in the series.

Example M-9 Cosine of a Sum

Using the suitable trigonometric identity from Table M-2, find $\cos(135° + 22°)$. Give your answer in four significant figures.

SET UP

As long as all angles are given in degrees, there is no need to convert to radians, because all operations are numerical values of the functions. Be sure, however, that your calculator is in degree mode. The suitable identity is $\cos(A \pm B) = \cos A \cos B \mp \sin A \sin B$, where the upper signs are appropriate.

SOLVE

1. Write the trigonometric identity for the cosine of a sum, with $A = 135°$ and $B = 22°$:

$$\cos(135° + 22°) = (\cos 135°)(\cos 22°)$$
$$- (\sin 135°)(\sin 22°)$$

2. Using a calculator, find $\cos 135°$, $\sin 135°$, $\cos 22°$, and $\sin 22°$:

$$\cos 135° = -0.7071$$
$$\cos 22° = 0.9272$$
$$\sin 135° = 0.7071$$
$$\sin 22° = 0.3746$$

3. Enter the values in the formula and calculate the answer:

$$\cos(135° + 22°) = (-0.7071)(0.9272)$$
$$- (0.7071)(0.3746)$$
$$= -0.9205$$

REFLECT

The calculator shows that the $\cos(135° + 22°) = \cos(157°) = -0.9205$.

Practice Problems

19. Find $\sin \theta$ and $\cos \theta$ for the right triangle shown in Figure M-12 in which $a = 4$ cm and $b = 7$ cm. What is the value for θ?
20. Find $\sin \theta$ where $\theta = 8.2°$. Is your answer consistent with the small-angle approximation?

M-9 The Binomial Expansion

A **binomial** is an expression consisting of two terms joined by a plus sign or a minus sign. The **binomial theorem** states that a binomial raised to a power can be written, or *expanded*, as a series of terms. If we raise the binomial $(1 + x)$ to a power n, the binomial theorem takes the form

$$(1 + x)^n = 1 + nx + \frac{n(n - 1)}{2!}x^2 + \frac{n(n - 1)(n - 2)}{3!}x^3 + \cdots \quad \text{M-52}$$

The series is valid for any value of n if $|x|$ is less than 1. The binomial expansion is very useful for approximating algebraic expressions, because when $|x| < 1$, the higher-order terms in the sum are small. (The order of a term is the power of x in the term. Thus, the terms explicitly shown in Equation M-52 are of order 0, 1, 2, and 3.) The series is particularly useful in situations where $|x|$ is small compared with 1; then each term is *much* smaller than the previous term and we can drop all but the first two or three terms in the expansion. If $|x|$ is much less than 1, we have

$$(1 + x)^n \approx 1 + nx, \qquad |x| \ll 1 \quad \text{M-53}$$

The binomial expansion is used in deriving many formulas of calculus that are important in physics. A well-known use in physics of the approximation in Equation M-53 is the proof that relativistic kinetic energy reduces to the classic formula when the velocity of a particle is very small compared with the velocity of light c.

Example M-10 Using the Binomial Expansion to Find a Power of a Number

Use Equation M-53 to find an approximate value for the square root of 101.

SET UP

The number 101 readily suggests a binomial, namely, $(100 + 1)$. To approximate the answer using the binomial expansion, we must manipulate the expression to get a binomial consisting of 1 and a term less than 1.

SOLVE

1. Write $(101)^{1/2}$ to give an expression $(1 + x)^n$ in which x is much less than 1:

$$(101)^{1/2} = (100 + 1)^{1/2}$$
$$= (100)^{1/2}(1 + 0.01)^{1/2}$$
$$= 10(1 + 0.01)^{1/2}$$

2. Use Equation M-53 with $n = \frac{1}{2}$ and $x = 0.01$ to expand $(1 + 0.01)^{1/2}$:

$$(1 + 0.01)^{1/2} = 1 + \tfrac{1}{2}(0.01)$$
$$+ \frac{\tfrac{1}{2}(-\tfrac{1}{2})}{2}(0.01)^2 + \cdots$$

3. Because $|x| \ll 1$, we expect the magnitude of terms of order 2 and higher to be significantly smaller than the magnitude of the first-order term. Approximate the binomial (1) by keeping only the zeroth and first-order terms, and (2) by keeping only the first 3 terms:

Keeping only the zeroth and first-order terms gives

$$(1 + 0.01)^{1/2} \approx 1 + \tfrac{1}{2}(0.01)$$
$$= 1 + 0.005\ 000\ 0$$
$$= 1.005\ 000\ 0$$

Keeping only the zeroth, first-, and second-order terms gives

$$(1 + 0.01)^{1/2} \approx 1 + \tfrac{1}{2}(0.01) + \frac{\tfrac{1}{2}(-\tfrac{1}{2})}{2}(0.01)^2$$
$$\approx 1 + 0.005\ 000\ 0 - 0.000\ 012\ 5$$
$$= 1.004\ 987\ 5$$

4. Substitute these results into the equation in step 1:

Keeping only the zeroth and first-order terms gives

$$(101)^{1/2} = 10(1 + 0.01)^{1/2} \approx \boxed{10.050\ 000}$$

Keeping only the zeroth, first-, and second-order terms gives

$$(101)^{1/2} = 10(1 + 0.01)^{1/2} \approx \boxed{10.049\ 875}$$

REFLECT

We therefore expect our answer to be correct to within about 0.001%. The value of $(101)^{1/2}$, to eight figures, is 10.049 876. This differs from 10.050 000 by 0.000 124, or about one part in 10^5, and differs from 10.049 875 by about one part in 10^7.

Practice Problems For the following, calculate the answer keeping the zeroth and first-order terms in the binomial series (Equation M-53), find the answer using your calculator, and show the percentage discrepancy between the two values:

21. $(1 + 0.001)^{-4}$
22. $(1 - 0.001)^{40}$

M-10 Complex Numbers

Real numbers are all numbers, from $-\infty$ to $+\infty$, that can be *ordered*. We know that, given two real numbers, one is always equal to, greater than, or less than the other. For example, $3 > 2$, $1.4 < \sqrt{2} < 1.5$, and $3.14 < \pi < 3.15$. A number that *cannot* be ordered is $\sqrt{-1}$; we cannot measure the size of this number, and so it makes no sense to say, for example, that $3 \times \sqrt{-1}$ is greater than or less than $2 \times \sqrt{-1}$. The earliest mathematicians who dealt with numbers containing $\sqrt{-1}$ referred to these numbers as *imaginary* numbers because they could not be used to measure or count something. In mathematics the symbol i is used to represent $\sqrt{-1}$.

Equation M-5, the quadratic formula, applies to equations of the form

$$ax^2 + bx + c = 0$$

The formula shows that there are no real roots when $b^2 < 4ac$. There are, however, still two roots. Each root is a number containing two terms: a real number and a multiple of $i = \sqrt{-1}$. The multiple of i is called an **imaginary number**, and i is called the **unit imaginary**.

A general **complex number** z can be written

$$z = a + bi \qquad \text{M-54}$$

where a and b are real numbers. The quantity a is called the real part of z or Re(z), and the quantity b is called the imaginary part of z or Im(z). We can represent a complex number z as a point in a plane, called the complex plane, as shown in **Figure M-22**, where the x axis is the **real axis** and the y axis is the **imaginary axis**. We can also use the relations $a = r \cos\theta$ and $b = r \sin\theta$ from Figure M-22 to write the complex number z in **polar coordinates** (a system in which a point is designated by the counterclockwise angle of rotation θ and the distance r in the direction of θ):

$$z = r \cos\theta + ir \sin\theta \qquad \text{M-55}$$

where $r = \sqrt{a^2 + b^2}$ is called the **magnitude** of z.

When complex numbers are added or subtracted, the real and imaginary parts are added or subtracted separately:

$$z_1 + z_2 = (a_1 + ib_1) + (a_2 + ib_2) = (a_1 + a_2) + i(b_1 + b_2) \qquad \text{M-56}$$

However, when two complex numbers are multiplied, each part of one number is multiplied by each part of the other number:

$$z_1 z_2 = (a_1 + ib_1)(a_2 + ib_2) = a_1 a_2 + i^2 b_1 b_2 + i(a_1 b_2 + a_2 b_1)$$
$$= a_1 a_2 - b_1 b_2 + i(a_1 b_2 + a_2 b_1) \qquad \text{M-57}$$

where we have used $i^2 = -1$.

The **complex conjugate** z^* of the complex number z is that number obtained by replacing i with $-i$ when writing z. If $z = a + ib$, then

$$z^* = (a + ib)^* = a - ib \qquad \text{M-58}$$

(When a quadratic equation has complex roots, the roots are **conjugate complex numbers**, in the form $a \pm bi$.) The product of a complex number and its complex conjugate equals the square of the magnitude of the number:

$$zz^* = (a + ib)(a - ib) = a^2 + b^2 = r^2 \qquad \text{M-59}$$

A particularly useful function of a complex number is the exponential $e^{i\theta}$. Using an expansion for e^x, we have

$$e^{i\theta} = 1 + i\theta + \frac{(i\theta)^2}{2!} + \frac{(i\theta)^3}{3!} + \frac{(i\theta)^4}{4!} + \cdots$$

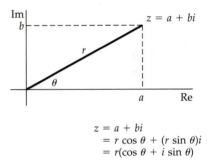

$$z = a + bi$$
$$= r \cos\theta + (r \sin\theta)i$$
$$= r(\cos\theta + i \sin\theta)$$

Figure M-22 Representation of a complex number in a plane. The real part of the complex number is plotted along the horizontal axis, and the imaginary part is plotted along the vertical axis.

Using $i^2 = -1$, $i^3 = -i$, $i^4 = +1$, and so forth, and separating the real parts from the imaginary parts, this expansion can be written

$$e^{i\theta} = \left(1 - \frac{\theta^2}{2!} + \frac{\theta^4}{4!} - \cdots\right) + i\left(\theta - \frac{\theta^3}{3!} + \cdots\right)$$

Comparing this result with Equations M-50 and M-51, we can see that

$$e^{i\theta} = \cos\theta + i\sin\theta \qquad\qquad \textbf{M-60}$$

Using this result, we can express a general complex number as an exponential:

$$z = a + ib = r\cos\theta + ir\sin\theta = re^{i\theta} \qquad\qquad \textbf{M-61}$$

If $z = x + iy$, where x and y are real variables, then z is called a **complex variable**.

Complex Variables in Physics

Complex variables are often used in formulas describing AC circuits: the impedance of a capacitor or an inductor includes a real part (the resistance) and an imaginary part (the reactance). (There are alternative ways, however, of analyzing AC circuits—such as rotating vectors called *phasors*—that do not require assigning imaginary values.) Complex variables are also important in the study of harmonic waves through Fourier analysis and synthesis. The time-dependent Schrödinger equation contains a complex-valued function of position and time.

Example M-11   Finding a Power of a Complex Number

Calculate $(1 + 3i)^4$ by using the binomial expansion.

SET UP
The expression is of the form $(1 + x)^n$. Because n is a positive integer, the expansion is valid for any value of x, and all terms, other than those of order n or lower must equal zero.

SOLVE

1. Write out the expansion of $(1 + 3i)^4$ to show the terms up through the fourth-order term:

$$1 + 4\cdot 3i + \frac{4(3)}{2!}(3i)^2 + \frac{4(3)(2)}{3!}(3i)^3$$
$$+ \frac{4(3)(2)(1)}{4!}(3i)^4$$

2. Evaluate each term, remembering that $i^2 = -1$, $i^3 = -i$, and $i^4 = +1$:

$$1 + 12i - 54 - 108i + 81$$

3. Show the result in the form $a + bi$:

$$(1 + 3i)^4 = \boxed{28 - 96i}$$

REFLECT
We can solve the problem algebraically to show that the answer is correct. We first square $(1 + 3i)$ and then square the result, to get $(1 + 3i)^4$:

$$(1 + 3i)^2 = 1\cdot 1 + 2\cdot 1\cdot 3i + (3i)^2 = 1 + 6i - 9 = -8 + 6i$$

$$(-8 + 6i)^2 = (-8)(-8) + 2(-8)(6i) + (6i^2)$$

$$= 64 - 96i - 36 = 28 - 96i$$

Practice Problems Express in the form $a + bi$:
23. $e^{i\pi}$
24. $e^{i\pi/2}$

M-11 Differential Calculus

Calculus is the branch of mathematics that allows us to deal with instantaneous rates of change of functions and variables. From the equation of a function—say, x as a function of t—we can always find x for a particular t, but with the methods of calculus you can go much further. You can know where x will have certain properties, such as a maximum or a minimum value, without having to try endless values of t. With calculus, if given the proper data, you can find, for example, the location of maximum stress on a beam, or the velocity or position of a falling object at a time t, or the energy a falling object has acquired at the time of impact. The principles of calculus are derived from examining functions at the infinitesimal level—analyzing how, say, x will change when the change in t becomes vanishingly small. We start with **differential calculus**, in which we determine the *limit* of the rate of change of x with respect to t as the change in t becomes closer and closer to zero.

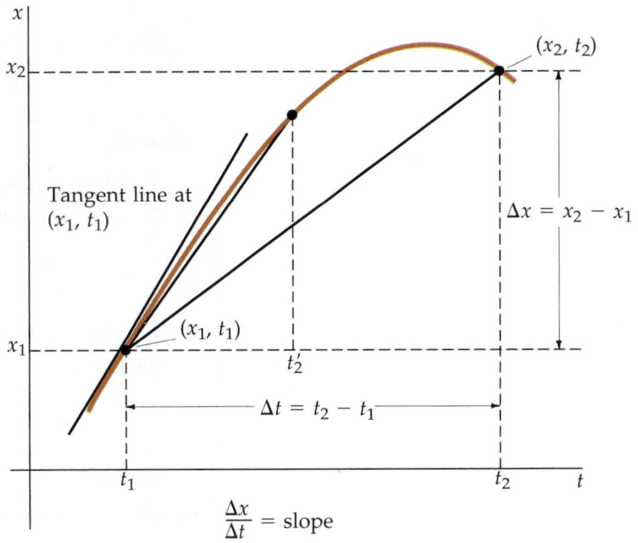

Figure M-23 is a graph of x versus t for a typical function $x(t)$. At a particular value $t = t_1$, x has the value of x_1, as indicated. At another value t_2, x has the value x_2. The change in t, $t_2 - t_1$, is written $\Delta t = t_2 - t_1$; and the corresponding change in x is written $\Delta x = x_2 - x_1$. The ratio $\Delta x / \Delta t$ is the slope of the straight line connecting (x_1, t_1) and (x_2, t_2). If we take the limit as t_2 approaches t_1 (as Δt approaches zero) the slope of the line connecting (x_1, t_1) and (x_2, t_2) approaches the slope of the line that is tangent to the curve at the point (x_1, t_1). The slope of this tangent line is equal to the **derivative** of x with respect to t and is written dx/dt:

$$\frac{dx}{dt} = \lim_{\Delta t \to 0} \frac{\Delta x}{\Delta t} \qquad \text{M-62}$$

(When we find the derivative of a function, we say that we are **differentiating** the function; and the very small dx and dt elements are called **differentials** of x and t, respectively.) The derivative of a function of t is another function of t. If x is a constant and does not change, the graph of x versus t is a horizontal line with zero slope. The derivative of a constant is thus zero. In **Figure M-24**, x is not constant but is proportional to t:

$$x = Ct$$

This function has a constant slope equal to C. Thus the derivative of Ct is C. **Table M-3** lists some properties of derivatives and the derivatives of some particular functions that occur often in physics. It is followed by comments aimed at making these properties and rules clearer. More detailed discussion can be found in most calculus textbooks.

Figure M-23 Graph of a typical function $x(t)$. The points (x_1, t_1) and (x_2, t_2) are connected by a straight line. The slope of this line is $\Delta x / \Delta t$. As the time interval beginning at t_1 is decreased, the slope for that interval approaches the slope of the line tangent to the curve at time t_1, which is the derivative of x with respect to t.

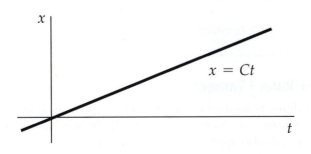

Figure M-24 Graph of the linear function $x = Ct$. This function has a constant slope C.

Table M-3   Properties of Derivatives and Derivatives of Particular Functions

Linearity

1. The derivative of a constant C multiplied by a function $f(t)$ equals the constant multiplied by the derivative of the function:

$$\frac{d}{dt}[Cf(t)] = C\frac{df(t)}{dt}$$

2. The derivative of a sum of functions equals the sum of the derivatives of the functions:

$$\frac{d}{dt}[f(t) + g(t)] = \frac{df(t)}{dt} + \frac{dg(t)}{dt}$$

Chain rule

3. If f is a function of x and x is in turn a function of t, the derivative of f with respect to t equals the product of the derivative of f with respect to x and the derivative of x with respect to t:

$$\frac{d}{dt}f(x(t)) = \frac{df}{dx}\frac{dx}{dt}$$

Derivative of a product

4. The derivative of a product of functions $f(t)g(t)$ equals the first function multiplied by the derivative of the second plus the second function multiplied by the derivative of the first:

$$\frac{d}{dt}[f(t)g(t)] = f(t)\frac{dg(t)}{dt} + g(t)\frac{df(t)}{dt}$$

Reciprocal derivative

5. The derivative of t with respect to x is the reciprocal of the derivative of x with respect to t, assuming that neither derivative is zero:

$$\frac{dt}{dx} = \left(\frac{dx}{dt}\right)^{-1} \quad \text{if } \frac{dt}{dx} \neq 0 \quad \text{and} \quad \frac{dx}{dt} \neq 0$$

Derivatives of particular functions

6. If C is a constant, then $dC/dt = 0$.

7. $\dfrac{d(t^n)}{dt} = nt^{n-1}$ If n is constant.

8. $\dfrac{d}{dt}\sin \omega t = \omega \cos \omega t$ If ω is constant.

9. $\dfrac{d}{dt}\cos \omega t = -\omega \sin \omega t$ If ω is constant.

10. $\dfrac{d}{dt}\tan \omega t = \omega \sin^2 \omega t$ If ω is constant.

11. $\dfrac{d}{dt}e^{bt} = be^{bt}$ If b is constant.

12. $\dfrac{d}{dt}\ln bt = \dfrac{1}{t}$ If b is constant.

Comments On Rules 1 Through 5

Rules 1 and 2 follow from the fact that the limiting process is linear. We can understand rule 3, the chain rule, by multiplying $\Delta f/\Delta t$ by $\Delta x/\Delta x$ and noting that as Δt approaches zero, Δx also approaches zero. That is,

$$\lim_{\Delta t \to 0} \frac{\Delta f}{\Delta t} = \lim_{\Delta t \to 0}\left(\frac{\Delta f}{\Delta t}\frac{\Delta x}{\Delta x}\right) = \lim_{\Delta t \to 0}\left(\frac{\Delta f}{\Delta x}\frac{\Delta x}{\Delta t}\right) = \left(\lim_{\Delta t \to 0}\frac{\Delta f}{\Delta x}\right)\left(\lim_{\Delta t \to 0}\frac{\Delta x}{\Delta t}\right) = \frac{df}{dx}\frac{dx}{dt}$$

where we have used that the limit of the product is equal to product of the limits.

Rule 4 is not immediately apparent. The derivative of a product of functions is the limit of the ratio

$$\frac{f(t + \Delta t)g(t + \Delta t) - f(t)g(t)}{\Delta t}$$

If we add and subtract the quantity $f(t + \Delta t)g(t)$ in the numerator, we can write this ratio as

$$\frac{f(t + \Delta t)g(t + \Delta t) - f(t + \Delta t)g(t) + f(t + \Delta t)g(t) - f(t)g(t)}{\Delta t}$$

$$= f(t + \Delta t)\left[\frac{g(t + \Delta t) - g(t)}{\Delta t}\right] + g(t)\left[\frac{f(t + \Delta t) - f(t)}{\Delta t}\right]$$

As Δt approaches zero, the terms in square brackets become $dg(t)/dt$ and $df(t)/dt$, respectively, and the limit of the expression is

$$f(t)\frac{dg(t)}{dt} + g(t)\frac{df(t)}{dt}$$

Rule 5 follows directly from the definition:

$$\frac{dx}{dt} = \lim_{\Delta t \to 0}\frac{\Delta x}{\Delta t} = \lim_{\Delta x \to 0}\left(\frac{\Delta t}{\Delta x}\right)^{-1} = \left(\frac{dt}{dx}\right)^{-1}$$

Comments on Rule 7

We can obtain this important result using the binomial expansion. We have

$$f(t) = t^n$$

$$f(t + \Delta t) = (t + \Delta t)^n = t^n\left(1 + \frac{\Delta t}{t}\right)^n$$

$$= t^n\left[1 + n\frac{\Delta t}{t} + \frac{n(n-1)}{2!}\left(\frac{\Delta t}{t}\right)^2 + \frac{n(n-1)(n-2)}{3!}\left(\frac{\Delta t}{t}\right)^3 + \cdots\right]$$

Then

$$f(t + \Delta t) - f(t) = t^n\left[n\frac{\Delta t}{t} + \frac{n(n-1)}{2!}\left(\frac{\Delta t}{t}\right)^2 + \cdots\right]$$

and

$$\frac{f(t + \Delta t) - f(t)}{\Delta t} = nt^{n-1} + \frac{n(n-1)}{2!}t^{n-2}\Delta t + \cdots$$

The next term omitted from the last sum is proportional to $(\Delta t)^2$, the following to $(\Delta t)^3$, and so on. Each term except the first approaches zero as Δt approaches zero. Thus

$$\frac{df}{dt} = \lim_{\Delta x \to 0}\frac{f(t + \Delta t) - f(t)}{\Delta t} = nt^{n-1}$$

Comments on Rules 8 to 10

We first write $\sin \omega t = \sin \theta$ with $\theta = \omega t$ and use the chain rule,

$$\frac{d\sin\theta}{dt} = \frac{d\sin\theta}{d\theta}\frac{d\theta}{dt} = \omega\frac{d\sin\theta}{d\theta}$$

We then use the trigonometric formula for the sine of the sum of two angles θ and $\Delta\theta$:

$$\sin(\theta + \Delta\theta) = \sin\Delta\theta\cos\theta + \cos\Delta\theta\sin\theta$$

Because $\Delta\theta$ is to approach zero, we can use the small-angle approximations

$$\sin\Delta\theta \approx \Delta\theta \quad \text{and} \quad \cos\Delta\theta \approx 1$$

Then

$$\sin(\theta + \Delta\theta) \approx \Delta\theta\cos\theta + \sin\theta$$

and

$$\frac{\sin(\theta + \Delta\theta) - \sin\theta}{\Delta\theta} \approx \cos\theta$$

Similar reasoning can be applied to the cosine function to obtain rule 9.

Rule 10 is obtained by writing $\tan\theta = \sin\theta/\cos\theta$ and applying rule 4 along with rules 8 and 9:

$$\frac{d}{dt}(\tan\theta) = \frac{d}{dt}(\sin\theta)(\cos\theta)^{-1} = \sin\theta\frac{d}{dt}(\cos\theta)^{-1} + \frac{d(\sin\theta)}{dt}(\cos\theta)^{-1}$$

$$= \sin\theta(-1)(\cos\theta)^{-2}(-\sin\theta) + (\cos\theta)(\cos\theta)^{-1}$$

$$= \frac{\sin^2\theta}{\cos^2\theta} + 1 = \tan^2\theta + 1 = \sec^2\theta$$

To obtain rule 10, let $\theta = \omega t$ and use the chain rule.

Comments on Rule 11

Again we use the chain rule

$$\frac{de^\theta}{dt} = \frac{b\,de^\theta}{b\,dt} = b\frac{de^\theta}{d(bt)} = b\frac{de^\theta}{d\theta} \quad \text{with} \quad \theta = bt$$

and the series expansion for the exponential function:

$$e^{\theta+\Delta\theta} = e^\theta e^{\Delta\theta} = e^\theta\left[1 + \Delta\theta + \frac{(\Delta\theta)^2}{2!} + \frac{(\Delta\theta)^3}{3!} + \cdots\right]$$

Then

$$\frac{e^{\theta+\Delta\theta} - e^\theta}{\Delta\theta} = e^\theta + e^\theta\frac{\Delta\theta}{2!} + e^\theta\frac{(\Delta\theta)^2}{3!} + \cdots$$

As $\Delta\theta$ approaches zero, the right side of this equation approaches e^θ.

Comments on Rule 12

Let

$$y = \ln bt$$

Then

$$e^y = bt \Rightarrow t = \frac{1}{b}e^y$$

Then, using rule 11, we obtain

$$\frac{dt}{dy} = \frac{1}{b}e^y \therefore \frac{dt}{dy} = t$$

Then, using rule 5, we obtain

$$\frac{dy}{dt} = \left(\frac{dt}{dy}\right)^{-1} = \frac{1}{t}$$

Second- and Higher-Order Derivatives; Dimensional Analysis

Once we have differentiated a function, we can differentiate the resulting derivative as long as terms remain to differentiate. A function such as $x = e^{bt}$ can be differentiated indefinitely: $dx/dt = be^{bt}$ (this function differentiates to give $b^2 e^{bt}$, and so on).

Consider velocity and acceleration. We can define velocity as the rate of change of position of a particle or dx/dt, and acceleration as the rate of change of velocity, or the *second* derivative of x with respect to t, written dx^2/dt^2. If a particle moves at a constant velocity, then dx/dt will equal a constant. The acceleration, however, will be zero: Having constant velocity is the same as having no acceleration, and the derivative of a constant is zero. Now consider a falling object, subject to the constant acceleration of gravity: The velocity itself will be time-dependent, so the *second* derivative, dx^2/dt^2 will be a constant.

The *physical dimensions* of a derivative with respect to a variable are those that would result if the original function of the variable were divided by a value of the variable. For example, the dimension of an equation in which one term is x (for position) is that of length (L); the dimensions of the derivative of x with respect to time t are those of velocity (L/T), and the dimensions of dx^2/dt^2 are those of acceleration (L/T^2).

Example M-12 Position, Velocity, and Acceleration

Find the first and the second derivatives of $x = \frac{1}{2}at^2 + bt + c$ where a, b, and c are constants. The function gives the position (in m) of a particle in one dimension, where t is the time (in s), a is acceleration (in m/s^2), b is velocity (in m/s) at a time $t = 0$, and c is the position (in m) of the particle at $t = 0$.

SET UP
Both the first and the second derivatives are sums of terms; for each differentiation we take the derivative of each term separately and add the results.

SOLVE
1. To find the first derivative, first compute the derivative of the first term:
$$\frac{d\left(\frac{1}{2}at^2\right)}{dt} = \left(\frac{1}{2}a\right)2t^1 = at$$

2. Compute the first derivative of the second and third terms:
$$\frac{d(bt)}{dt} = b, \quad \frac{d(c)}{dt} = 0$$

3. Add these results:
$$\frac{dx}{dt} = at + b$$

4. To compute the second derivative, repeat the process for the result in step 3:
$$\frac{d^2x}{dt^2} = at + 0 = a$$

REFLECT
The physical dimensions show that the answer is plausible. The original function is an equation for position; all terms are in meters—the units of t^2 and t cancel the units of s^2 and s in the constants a and b, respectively. In the function for dx/dt, all terms are similarly in m/s: The constant c has differentiated to zero, and the unit for t cancels one of the units for s in the constant a. In the function for dx^2/dt^2, only the acceleration constant remains; as expected, its dimensions are L/T^2.

Practice Problems
25. Find dy/dx for $y = \frac{5}{8}x^3 - 24x - \frac{5}{8}$.
26. Find dy/dt for $y = ate^{bt}$, where a and b are constants.

Solving Differential Equations Using Complex Numbers

A **differential equation** is an equation in which the derivatives of a function appear as variables. It is an equation in which the variables are related to each other through their derivatives. Consider an equation of the form

$$a\frac{d^2x}{dt^2} + b\frac{dx}{dt} + cx = A\cos\omega t \qquad \text{M-63}$$

that represents a physical process, such as a damped harmonic oscillator driven by a sinusoidal force, or a series RLC combination being driven by a sinusoidal potential drop. Although each of the parameters in Equation M-63 is a real number, the time-dependent cosine term suggests that we might find the steady-state solution to this equation by introducing complex numbers. We first construct the "parallel" equation

$$a\frac{d^2y}{dt^2} + b\frac{dy}{dt} + cy = A\sin\omega t \qquad \text{M-64}$$

Equation M-64 has no physical meaning of its own, and we have no interest in solving it. However, it is of use in solving Equation M-63. After multiplying through Equation M-64 by the unit imaginary i, we add Equation M-64 and Equation M-63 to obtain

$$\left(a\frac{d^2x}{dt^2} + ai\frac{d^2y}{dt^2}\right) + \left(b\frac{dx}{dt} + bi\frac{dy}{dt}\right) + (cx + ciy) = A\cos\omega t + Ai\sin\omega t$$

We next combine terms to get

$$a\frac{d^2(x+iy)}{dt^2} + b\frac{d(x+iy)}{dt} + c(x+iy) = A(\cos\omega t + i\sin\omega t) \qquad \text{M-65}$$

which is valid because the derivative of a sum is equal to the sum of the derivatives. We simplify our result by defining $z = x + iy$ and by using the identity $e^{i\omega t} = \cos\omega t + i\sin\omega t$. Substituting these into Equation M-65, we obtain

$$a\frac{d^2z}{dt^2} + b\frac{dz}{dt} + cz = Ae^{i\omega t} \qquad \text{M-66}$$

which we now solve for z. Once z is obtained, we can solve for x using $x = \text{Re}(z)$.

Because we are looking only for the steady-state solution for Equation M-65, we can assume its solution is of the form $x = x_0\cos(\omega t - \varphi)$, where φ is a constant. This is equivalent to assuming that the solution to Equation M-66 is of the form $z = \eta e^{i\omega t}$, where η, pronounced eta (like beta without the b), is a constant complex number. Then $dz/dt = i\omega z$, $d^2z/dt^2 = -\omega^2 z$, and $e^{i\omega t} = z/\eta$. Substituting these into Equation M-65 gives

$$-a\omega^2 z + i\omega bz + cz = A\frac{z}{\eta}$$

Dividing both sides of this equation by z and solving for η gives

$$\eta = \frac{A}{-a\omega^2 + i\omega b + c}$$

Expressing the denominator in polar form gives

$$(-a\omega^2 + c) + i\omega b = \sqrt{(-a\omega^2 + c)^2 + \omega^2 b^2}\, e^{i\varphi}$$

where $\tan\varphi = \omega^2 b^2/(-a\omega^2 + c)$. Thus,

$$\eta = \frac{A}{\sqrt{(-a\omega^2 + c)^2 + \omega^2 b^2}} e^{-i\varphi}$$

so

$$z = \eta e^{i\omega t} = \frac{A}{\sqrt{(-a\omega^2 + c)^2 + \omega^2 b^2}} e^{i(\omega t - \varphi)}$$

$$= \frac{A}{\sqrt{(-a\omega^2 + c)^2 + \omega^2 b^2}} [\cos(\omega t - \varphi) + i\sin(\omega t - \varphi)] \qquad \text{M-67}$$

It follows that

$$x = \text{Re}(z) = \frac{A}{\sqrt{(-a\omega^2 + c)^2 + \omega^2 b^2}} \cos(\omega t - \varphi) \qquad \text{M-68}$$

The Exponential Function

An **exponential function** is a function of the form a^{bx}, where $a > 0$ and b are constants. The function is usually written as e^{cx}, where c is constant.

When the rate of change of a quantity is proportional to the quantity itself, the quantity increases or decreases exponentially, depending on the sign of the proportionality constant. An example of an *exponentially* decreasing function is nuclear decay. If N is the number of radioactive nuclei at some time, then the change dN in some very small time interval dt will be proportional to N and to dt:

$$dN = -\lambda N\, dt$$

where λ is the *decay constant* (not to be confused with the decay rate dN/dt, which decreases exponentially). The function N satisfying this equation is

$$N = N_0 e^{-\lambda t} \qquad \text{M-69}$$

where N_0 is the value of N at time $t = 0$. **Figure M-25** shows N versus t. A characteristic of exponential decay is that N decreases by a constant factor in a given time interval. The time interval for N to decrease to half its original value is its *half-life* $t_{1/2}$. The half-life is obtained from Equation M-69 by setting $N = \frac{1}{2}N_0$ and solving for the time. This gives

$$t_{1/2} = \frac{\ln 2}{\lambda} = \frac{0.693}{\lambda} \qquad \text{M-70}$$

An example of *exponential increase* is population growth. If the number of organisms is N, the change in N after a very small time interval dt is given by

$$dN = +\lambda N\, dt$$

where λ is now the *growth constant*. The function N satisfying this equation is

$$N = N_0 e^{\lambda t} \qquad \text{M-71}$$

(Note the change of sign in the exponent.) A graph of this function is shown in **Figure M-26**. An exponential increase can be characterized by a doubling time T_2, which is related to λ by

$$T_2 = \frac{\ln 2}{\lambda} = \frac{0.693}{\lambda} \qquad \text{M-72}$$

Very often, we know population growth as an annual percentage increase and wish to calculate the doubling time. In this case, we find T_2 (in years) from the equation

$$T_2 = \frac{69.3}{r} \qquad \text{M-73}$$

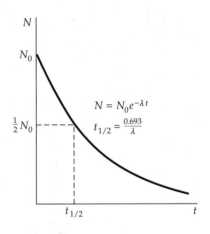

Figure M-25 Graph of N versus t when N decreases exponentially. The time $t_{1/2}$ is the time it takes for N to decrease by one-half.

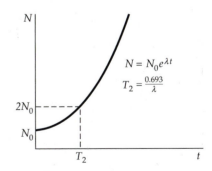

Figure M-26 Graph of N versus t when N increases exponentially. The time T_2 is the time it takes for N to double.

Table M-4 Exponential and Logarithmic Functions

$e = 2.718\,28\ldots$

$e^0 = 1$

If $y = e^x$, then $x = \ln y$

$e^{\ln x} = x$

$e^x e^y = e^{(x+y)}$

$(e^x)^y = e^{xy} = (e^y)^x$

$\ln e = 1; \ln 1 = 0$

$\ln xy = \ln x + \ln y$

$\ln \dfrac{x}{y} = \ln x - \ln y$

$\ln e^x = x; \ln a^x = x \ln a$

$\ln x = (\ln 10) \log x$

$\quad = 2.30\,26 \log x$

$\log x = (\log e) \ln x = 0.434\,29 \ln x$

$e^x = 1 + x + \dfrac{x^2}{2!} + \dfrac{x^3}{3!} + \cdots$

$\ln(1 + x)$

$\quad = x - \dfrac{x^2}{2} + \dfrac{x^3}{3} - \dfrac{x^4}{4} + \cdots$

where r is the percent per year. For example, if the population increases by 2 percent per year, the population will double every $69.3/2 \approx 35$ years. **Table M-4** lists some useful relations for exponential and logarithmic functions.

Example M-13 Radioactive Decay of Cobalt-60

The half-life of cobalt-60 (^{60}Co) is 5.27 y. At $t = 0$, you have a sample of ^{60}Co that has a mass equal to 1.20 mg. At what time t (in years) will 0.400 mg of the sample of ^{60}Co have decayed?

SET UP

When we derived the half-life in exponential decay, we set $N/N_0 = 1/2$. In this example, we are to find the time at which two-thirds of a sample remains, and so the ratio N/N_0 will be 0.667.

SOLVE

1. Express the ratio N/N_0 as an exponential function:

$$\frac{N}{N_0} = 0.667 = e^{-\lambda t}$$

2. Take the reciprocal of both sides:

$$\frac{N_0}{N} = 1.50 = e^{\lambda t}$$

3. Solve for t:

$$t = \frac{\ln 1.50}{\lambda} = \frac{0.405}{\lambda}$$

4. The decay constant is related to the half-life by $\lambda = (\ln 2)/t_{1/2}$ (Equation M-70). Substitute $(\ln 2)/t_{1/2}$ for λ and evaluate the time:

$$t = \frac{\ln 1.5}{\ln 2} t_{1/2} = \frac{\ln 1.5}{\ln 2} \times 5.27 \text{ y} = 3.08 \text{ y}$$

REFLECT

It takes 5.27 y for the mass of a sample of ^{60}Co to decrease to 50% of its initial mass. Thus, we expect it to take less than 5.27 y for the sample to lose 33.3% of its mass. Our step-4 result of 3.08 y is less than 5.27 y, as expected.

Practice Problems

27. The discharge time constant τ of a capacitor in an RC circuit is the time in which the capacitor discharges to e^{-1} (or 0.368) times its charge at $t = 0$. If $\tau = 1$ s for a capacitor, at what time t (in seconds) will it have discharged to 50.0% of its initial charge?

28. If the coyote population in your state is increasing at a rate of 8.0% a decade and continues increasing at the same rate indefinitely, in how many years will it reach 1.5 times its current level?

M-12 Integral Calculus

Integration can be considered the inverse of differentiation. If a function $f(t)$ is *integrated*, a function $F(t)$ is found for which $f(t)$ is the derivative of $F(t)$ with respect to t.

The Integral as an Area Under a Curve; Dimensional Analysis

The process of finding the area under a curve on a graph illustrates integration. **Figure M-27** shows a function $f(t)$. The area of the shaded element is approxi-

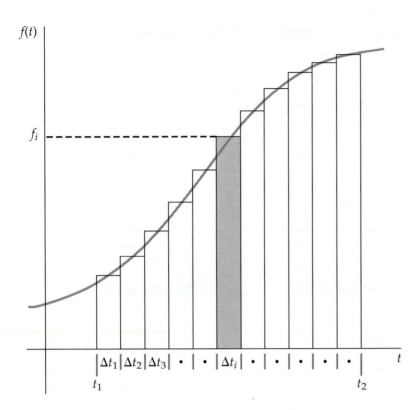

Figure M-27 A general function $f(t)$. The area of the shaded element is approximately $f_i \Delta t_i$, where f_i is evaluated anywhere in the interval.

mately $f_i \Delta t_i$ where f_i is evaluated anywhere in the interval Δt_i. This approximation is highly accurate if Δt_i is very small. The total area under some stretch of the curve is found by summing all the area elements it covers and taking the limit as each Δt_i approaches zero. This limit is called the **integral** of f over t and is written

$$\int f \, dt = \text{area}_i = \lim_{\Delta t \to 0} \sum_i f_i \, \Delta t_i \qquad \textbf{M-74}$$

The *physical dimensions* of an integral of a function $f(t)$ are found by multiplying the dimensions of the *integrand* (the function being integrated) and the dimensions of the integration variable t. For example, if the integrand is a velocity function $v(t)$ (dimensions L/T) and the integration variable is time t, the dimension of the integral is L = (L/T) × T. That is, the dimensions of the integral are those of velocity times time.

Let

$$y = \int_{t_1}^{t} f \, dt \qquad \textbf{M-75}$$

The function y is the area under the f versus t curve from t_1 to a general value t. For a small interval Δt, the change in the area Δy is approximately $f \, \Delta t$:

$$\Delta y \approx f \, \Delta t$$

$$f \approx \frac{\Delta y}{\Delta t}$$

If we take the limit as Δt approaches 0, we can see that f is the derivative of y:

$$f = \frac{dy}{dt} \qquad \textbf{M-76}$$

Table M-5
Integration Formulas†

1. $\int A\,dt = At$

2. $\int At\,dt = \dfrac{1}{2}At^2$

3. $\int At^n\,dt = A\dfrac{t^{n+1}}{n+1},\ n \neq -1$

4. $\int At^{-1}\,dt = A\ln|t|$

5. $\int e^{bt}\,dt = \dfrac{1}{b}e^{bt}$

6. $\int \cos \omega t\,dt = \dfrac{1}{\omega}\sin \omega t$

7. $\int \sin \omega t\,dt = -\dfrac{1}{\omega}\cos \omega t$

8. $\displaystyle\int_0^\infty e^{-ax}\,dx = \dfrac{1}{a}$

9. $\displaystyle\int_0^\infty e^{-ax^2}\,dx = \dfrac{1}{2}\sqrt{\dfrac{\pi}{a}}$

10. $\displaystyle\int_0^\infty xe^{-ax^2}\,dx = \dfrac{2}{a}$

11. $\displaystyle\int_0^\infty x^2 e^{-ax^2}\,dx = \dfrac{1}{4}\sqrt{\dfrac{\pi}{a^3}}$

12. $\displaystyle\int_0^\infty x^3 e^{-ax^2}\,dx = \dfrac{4}{a^2}$

13. $\displaystyle\int_0^\infty x^4 e^{-ax^2}\,dx = \dfrac{3}{8}\sqrt{\dfrac{\pi}{a^5}}$

†In these formulas, A, b, and ω are constants. In formulas 1 through 7, an arbitrary constant C can be added to the right side of each equation. The constant a is greater than zero.

Indefinite Integrals and Definite Integrals

When we write

$$y = \int f\,dt \qquad \text{M-77}$$

we are showing y as an **indefinite integral** of f over t. To evaluate an indefinite integral, we find the function y whose derivative is f. Because that function could contain a constant term that differentiated to zero, we include as our final term a **constant of integration** C. If we are integrating the function over a known segment—such as t_1 to t_2 in Figure M-27—we can find a **definite integral**, eliminating the unknown constant C:

$$\int_{t_1}^{t_2} f\,dt = y(t_2) - y(t_1) \qquad \text{M-78}$$

Table M-5 lists some important integration formulas. More extensive lists of integration formulas can be found in any calculus textbook or by searching for "table of integrals" on the Internet.

Example M-14   Integrating Equations of Motion

A particle is moving at a constant acceleration a. Write a formula for position x at time t given that the position and velocity are x_0 and v_0 at time $t = 0$.

SET UP

Velocity v is the derivative of x with respect to time t, and acceleration is the derivative of v with respect to t. We should be able to write a function $x(t)$ by performing two integrations.

SOLVE

1. Integrate a with respect to t to find v as a function of t. The a can be factored from the integrand because a is constant:

$$v = \int a\,dt = a \int dt$$
$$v = at + C_1$$

where C_1 represents a multiplied by the constant of integration.

2. The velocity $v = v_0$ when $t = 0$:

$$v_0 = 0 + C_1 \Rightarrow C_1 = v_0$$
$$\text{so } v = v_0 + at$$

3. Integrate v with respect to t to find x as a function of t:

$$x = \int v\,dt = \int (v_0 + at)\,dt = \int v_0\,dt + \int at\,dt$$
$$x = v_0 \int dt + a \int t\,dt = v_0 t + \tfrac{1}{2}at^2 + C_2$$

where C_2 represents the combined constants of integration.

4. The position $x = x_0$ when $t = 0$ is:

$$x_0 = 0 + 0 + C_2$$
$$\text{so } x = x_0 + v_0 t + \tfrac{1}{2}at^2$$

REFLECT

Differentiate the step-4 result twice to get the acceleration

$$v = \frac{dx}{dt} = \frac{d}{dt}(x_0 + v_0 t + \tfrac{1}{2}at^2) = 0 + v_0 + at$$

$$a = \frac{dv}{dt} = \frac{d}{dt}(v_0 + at) = a$$

Practice Problems

29. $\displaystyle\int_3^6 3\,dx =$

30. $V = \displaystyle\int_5^8 \pi r^2\,dL =$

Answers to Practice Problems

1. 0.24 L

2. 31.6 m/s

3. $6.0 \, \text{kg/cm}^3$

4. -3

5. 1.54 L

6. 3.07 L

7. False

8. $x = (4.5 \, \text{m/s})t + 3.0 \, \text{m}$

9. $x = 8, y = 60$

11. $2(x - y)^2$

12. $x^2(2x + 4)(x + 3)$

13. $x^{1/2}$

14. x^6

15. 3

16. ~ 2.322

17. $V/A = \frac{1}{3}r$

18. $A = \dfrac{2}{3}\pi L^2$

19. $\sin\theta = 0.496, \cos\theta = 0.868, \theta = 29.7°$

20. $\sin 8.2° = 0.1426, 8.2° = 0.1431 \, \text{rad}$

21. $0.996, 0.996\,00$, close to 0%

22. $0.96, 0.960\,77, \ll 1\%$

23. $-1 + 0i = -1$

24. $0 + i = i$

25. $dy/dx = \frac{15}{8}x^2 - 24$

26. $dy/dt = ae^{bt}(bt + 1)$

27. 0.693 s

28. 51 y

29. 9

30. $3\pi r^2$

This page is intentionally left blank.
In the first edition this page will provide
answers to odd-numbered problems.

This page is intentionally left blank.
In the first edition this page will provide
answers to odd-numbered problems.

This page is intentionally left blank.
In the first edition this page will provide
answers to odd-numbered problems.

This page is intentionally left blank.
In the first edition this page will provide
answers to odd-numbered problems.

This page is intentionally left blank.
In the first edition this page will provide
answers to odd-numbered problems.

This page is intentionally left blank.
In the first edition this page will provide
answers to odd-numbered problems.

This page is intentionally left blank.
In the first edition this page will provide
answers to odd-numbered problems.

This page is intentionally left blank.
In the first edition this page will provide
answers to odd-numbered problems.

This page is intentionally left blank.
In the first edition this page will provide
answers to odd-numbered problems.

This page is intentionally left blank.
In the first edition this page will provide
answers to odd-numbered problems.

This page is intentionally left blank.
In the first edition this page will provide
answers to odd-numbered problems.

This page is intentionally left blank.
In the first edition this page will provide
answers to odd-numbered problems.

This page is intentionally left blank.
In the first edition this page will provide
answers to odd-numbered problems.

This page is intentionally left blank.
In the first edition this page will provide
answers to odd-numbered problems.

This page is intentionally left blank.
In the first edition this page will provide
answers to odd-numbered problems.

This page is intentionally left blank.
In the first edition this page will provide
answers to odd-numbered problems.

This page is intentionally left blank.
In the first edition this page will provide
answers to odd-numbered problems.

This page is intentionally left blank.
In the first edition this page will provide
answers to odd-numbered problems.

This page is intentionally left blank.
In the first edition this page will provide
answers to odd-numbered problems.

This page is intentionally left blank.
In the first edition this page will provide
answers to odd-numbered problems.

This page is intentionally left blank.
In the first edition this page will provide
answers to odd-numbered problems.

This page is intentionally left blank.
In the first edition this page will provide
answers to odd-numbered problems.

INDEX